Introduction to Microfabrication

Introduction to Microfabrication
Second Edition

Sami Franssila

*Professor of Materials Science at Aalto University
and Adjunct Professor of Micro- and Nanotechnology
at University of Helsinki, Finland*

A John Wiley and Sons, Ltd., Publication

This edition first published 2010
© 2010, John Wiley & Sons, Ltd

First Edition published in 2004

Registered office
John Wiley & Sons Ltd, The Atrium, Southern Gate, Chichester, West Sussex, PO19 8SQ, United Kingdom

For details of our global editorial offices, for customer services and for information about how to apply for permission to reuse the copyright material in this book please see our website at www.wiley.com.

The right of the author to be identified as the author of this work has been asserted in accordance with the Copyright, Designs and Patents Act 1988.

All rights reserved. No part of this publication may be reproduced, stored in a retrieval system, or transmitted, in any form or by any means, electronic, mechanical, photocopying, recording or otherwise, except as permitted by the UK Copyright, Designs and Patents Act 1988, without the prior permission of the publisher.

Wiley also publishes its books in a variety of electronic formats. Some content that appears in print may not be available in electronic books.

Designations used by companies to distinguish their products are often claimed as trademarks. All brand names and product names used in this book are trade names, service marks, trademarks or registered trademarks of their respective owners. The publisher is not associated with any product or vendor mentioned in this book. This publication is designed to provide accurate and authoritative information in regard to the subject matter covered. It is sold on the understanding that the publisher is not engaged in rendering professional services. If professional advice or other expert assistance is required, the services of a competent professional should be sought.

Library of Congress Cataloguing-in-Publication Data

Franssila, Sami.
 Introduction to microfabrication / Sami Franssila. – 2nd ed.
     p. cm.
   Includes index.
   ISBN 978-0-470-74983-8 (cloth)
 1. Microelectromechanical systems. 2. Integrated circuits. 3. Semiconductor processing. 4. Nanotechnology.
 5. Microfabrication. I. Title.
 TK7875.F73 2010
 621.381 – dc22

2010010076

A catalogue record for this book is available from the British Library.

ISBN: 978-0-470-74983-8

Set in 9/11pt Times by Laserwords Private Limited, Chennai, India
Printed and Bound in Singapore by Markono Print Media Pte Ltd

Contents

Preface to the First Edition ix
Preface to the Second Edition xiii
Acknowledgements xv

1 Introduction 1
Substrates, Thin Films, Processes, Dimensions, Devices, MOS Transistor, Cleanliness and Yield, Industries, Exercises
References and Related Reading

2 Micrometrology and Materials Characterization 15
Microscopy and Visualization, Lateral and Vertical Dimensions, Optical Techniques, Electrical Measurements, Physical and Chemical Analyses, Practical Issues with Micrometrology, Measurements Everywhere, Exercises
References and Related Reading

3 Simulation of Microfabrication Processes 29
Simulator Types, Levels of Simulation, The 1D Simulators, The 2D Simulators, The 3D Simulators, Other Simulation Needs in Microfabrication, Exercises
References and Related Reading

4 Silicon 35
Silicon Material Properties, Silicon Crystal Growth, Silicon Crystal Structure, Silicon Wafering Process, Defects and Non-Idealities in Silicon Crystals, Advanced Wafers, Exercises
References and Related Reading

5 Thin-Film Materials and Processes 47
Thin Films vs. Bulk Materials, Physical Vapor Deposition, Chemical Vapor Deposition, PECVD: Plasma-Enhanced CVD, ALD: Atomic Layer Deposition, Electrochemical Deposition (ECD), Other Methods, Thin Films Over Topography: Step Coverage, Stresses, Metallic Thin Films, Polysilicon, Oxide and Nitride Thin Films, Polymer Films, Advanced Thin Films, Exercises
References and Related Reading

6 Epitaxy 69
Heteroepitaxy, Epitaxial Deposition, CVD Homoepitaxy of Silicon, Doping of Epilayers, Measurement of Epitaxial Deposition, Simulation of Epitaxy, Advanced Epitaxy, Exercises
References and Related Reading

7 Advanced Thin Films 77
General Features of Thin-Film Processes, Film Growth and Structure, Thin-Film Structure Characterization, Surfaces and Interfaces, Adhesion, Two-Layer Films, Alloys and Doped Films, Multilayer Films, Selective Deposition, Reacted Films, Simulation of Deposition, Thickness Limits of Thin Films, Exercises
References and Related Reading

8 Pattern Generation 93
Pattern Generators, Electron Beam Lithography, Laser Pattern Generators, Photomask Fabrication, Photomask Inspection, Defects and Repair, Photomasks as Tools, Other Pattern Generation Methods, Exercises
References and Related Reading

9 Optical Lithography 103
Lithography Process Flow, Resist Chemistry, Resist Application, Alignment and Overlay, Exposure, Resist Profile, Resolution, Process Latitude, Basic Pattern Shapes, Lithography Practice, Photoresist Stripping, Exercises
References and Related Reading

10 Advanced Lithography 115
Projection Optical Systems, Resolution of Projection Optical Systems, Resists, Thin-Film Optics in Resists, Lithography Over Steps, Optical Extensions

of Optical Lithography, Non-Optical Extension of Optical Lithography, Lithography Simulation, Lithography Triangles, Exercises
References and Related Reading

11 Etching 127
Etch Mechanisms, Etching Profiles, Anisotropic Wet Etching, Wet Etching, Plasma Etching (RIE), Isotropic Dry Etching, Etch Masks, Non-Masked Etching, Multistep and Multilayer Etching, Etch Processes for Common Materials, Ion Beam Etching, Etch Process Characteristics, Selecting Etch Processes, Exercises
References and Related Reading

12 Wafer Cleaning and Surface Preparation 143
Classes of Contamination, Chemical Wet Cleaning, Physical Wet Cleaning, Rinsing and Drying, Dry Cleaning, Particle Removal, Organics Removal, Metal Removal, Contact Angle, Surface Preparation, Exercises
References and Related Reading

13 Thermal Oxidation 153
Thermal Oxidation Process, Deal–Grove Oxidation Model, Oxidation of Polysilicon, Oxide Structure, Local Oxidation of Silicon, Stress and Pattern Effects in Oxidation, Simulation of Oxidation, Thermal Oxides vs. other Oxides, Exercises
References and Related Reading

14 Diffusion 165
Diffusion Process, Diffusion Mechanisms, Doping of Polysilicon, Doping Profiles in Diffusion, Diffusion Applications, Simulation of Diffusion, Diffusion at Large, Exercises
References and Related Reading

15 Ion Implantation 173
The Implantation Process, Implant Applications, Implant Damage and Damage Annealing, Tools for Ion Implantation, Ion Implantation Simulation, Implantation Further, Exercises
References and Related Reading

16 CMP: Chemical–Mechanical Polishing 181
CMP Process and Tool, Mechanics of CMP, Chemistry of CMP, Non-Idealities in CMP, Monitoring CMP Processes, Applications of CMP, CMP as a Whole, Exercises
References and Related Reading

17 Bonding 191
Bonding Basics, Fusion Bonding Blanket Silicon Wafers, Anodic Bonding, Metallic Bonding, Adhesive Bonding, Layer Transfer and Temporary Bonding, Bonding of Structured Wafers, Bond Quality Measurements, Bonding for Packaging, Bonding at Large, Exercises
References and Related Reading

18 Polymer Microprocessing 203
Polymer Materials, Polymer Thermal Properties, Thick-Resist Lithography, Molding Techniques, Hot Embossing, Nanoimprint Lithography, Masters for Replication, Processing on Polymers, Polymer Bonding, Polymer Devices, Polymer Overview, Exercises
References and Related Reading

19 Glass Microprocessing 225
Structure and Properties of Glasses, Glass Substrates, General Processing Issues with Glasses, Glass Etching, Glass Bonding, Glass Devices, Specialty Glasses, Exercises
References and Further Reading

20 Anisotropic Wet Etching 237
Basic Structures on <100> Silicon, Etchants, Etch Masks and Protective Coatings, Etch Rate and Etch Stop, Front-Side Processed Structures, Convex Corner Etching, Membrane Fabrication, Through-Wafer Structures, <110> Etching, <111> Silicon Etching, Comparison of <100>, <110> and <111> Etching, Exercises
References and Related Reading

21 Deep Reactive Ion Etching 255
RIE Process Capabilities, RIE Process Physics and Chemistry, Deep Etching, Combining Anisotropic and Isotropic DRIE, Microneedles and Nozzles, Sidewall Quality, Pattern Size and Pattern Density Effects, Etch Residues and Damage, DRIE vs. Wet Etching, Exercises
References and Related Reading

22 Wafer Engineering 271
Silicon Crystals, Gettering, Wafer Mechanical Specifications, Epitaxial Wafers, SOI Wafers, Bonding Mechanics, Advanced Wafers, Variety of Wafers, Exercises
References and Further Reading

23 Special Processes and Materials 283
Substrates other than Silicon, Pattern Generation, Patterning, Powder Blasting, Deposition, Porous Silicon, Molding with Lost Mold, Exercises
References and Related Reading

24 Serial Microprocessing 299
Focused Ion Beam (FIB) Processing, Focused Electron Beam (FEB) Processing, Laser Direct Writing, AFM Patterning, Ink Jetting, Mechanical Structuring, Chemical and Chemomechanical Machining Scaled Down, Conclusions, Exercises
References and Further Reading

25 Process Integration 313
The Two Sides of the Wafer, Device Example 1: Solar Cell, Device Example 2: Microfluidic Sieves, Wafer Selection, Masks and Lithography, Design Rules, Resistors, Device Example 3: PCR Reactor, Device Example 4: Integrated Passive Chip, Contamination Budget, Thermal Processes, Metallization, Passivation and Packaging, Exercises
References and Related Reading

26 MOS Transistor Fabrication 329
Polysilicon Gate CMOS, Polysilicon Gate CMOS: 10 μm to 1 μm Generations, MOS Transistor Scaling, CMOS from 0.8 μm to 65 nm, Gate Module, SOI MOSFETs, Thin-Film Transistors, Integrated Circuits, Exercises
References and Related Reading

27 Bipolar Transistors 347
Fabrication Process of SBC Bipolar Transistor, Advanced Bipolar Structures, Lateral Isolation, BiCMOS Technology, Cost of Integration, Exercises
References and Related Reading

28 Multilevel Metallization 357
Two-Level Metallization, Planarized Multilevel Metallization, Copper Metallization, Dual Damascene Metallization, Low-k Dielectrics, Metallization Scaling, Exercises
References and Related Reading

29 Surface Micromachining 369
Single Structural Layer Devices, Materials for Surface Micromachining, Mechanics of Free-Standing Films, Cantilever Structures, Membranes and Bridges, Stiction, Multiple Layer Structures, Rotating Structures, Hinged Structures, CMOS Wafers as Substrates, Exercises
References and Related Reading

30 MEMS Process Integration 387
Silicon Microbridges, Double-Sided Processing, Membrane Structures, Piezoresistive Pressure Sensor, Tilting and Bending Through-Wafer Etched Structures, Needles and Tips, Channels and Nozzles, Bonded Structures, Surface Micromachining Combined with Bulk Micromachining, MEMS Packaging, Microsystems, Exercises
References and Related Reading

31 Process Equipment 409
Batch Processing vs. Single Wafer Processing, Process Regimes: Temperature and Pressure, Cluster Tools and Integrated Processing, Measuring Fabrication Processes, Equipment Figures of Merit, Simulation of Process Equipment, Tool Lifecycles, Cost of Ownership, Exercises
References and Related Reading

32 Equipment for Hot Processes 419
High-Temperature Equipment: Hot Wall vs. Cold Wall, Furnace Processes, Rapid Thermal Processing/Rapid Thermal Annealing, Furnaces vs. RTP Systems, Exercises
References and Related Reading

33 Vacuum and Plasmas 425
Vacuum Physics and Kinetic Theory of Gases, Vacuum Production, Plasma Etching, Sputtering, Residual Gas Incorporation into Deposited Film, PECVD, Residence Time, Exercises
References and Related Reading

34 CVD and Epitaxy Equipment 433
Deposition Rate, CVD Rate Modeling, CVD Reactors, CVD with Liquid Sources, Silicon CVD Epitaxy, Epitaxial Reactors, Control of CVD Reactions, Exercises
References and Related Reading

35 Cleanrooms 441
Cleanroom Construction, Cleanroom Standards, Cleanroom Subsystems, Environment, Safety and Health (ESH), Cleanroom Operating Procedures, Mini-Environments, Exercises
References and Related Reading

36 Yield and Reliability 449
Yield Definitions and Formulas, Yield Models, Yield Ramping, Package Reliability, Metallization Reliability, Dielectric Defects and Quality, Stress Migration, Die Yield Loss, Exercises
References and Related Reading

37 Economics of Microfabrication 457
Silicon, IC Costs and Prices, IC Industry, IC Wafer Fabs, MEMS Industry, Flat-Panel Display Industry, Solar Cells, Magnetic Data Storage, Short Term and Long Term, Exercises
References and Related Reading

38 Moore's Law and Scaling Trends 469
From Transistor to Integrated Circuit, Historical Development of IC Manufacturing, MOS Scaling, Departure from Planar Bulk Technology, Memories, Lithography Future, Moore's Law, Materials Challenges, Statistics and Yield, Limits of Scaling, Exercises
References and Related Reading

39 Microfabrication at Large 485
New Devices, Proliferation of MEMS, Microfluidics, BioMEMS, Bonding and 3D Integration, IC–MEMS Integration, Microfabricated Devices for Microfabrication, Exercises
References and Related Reading

Appendix A   Properties of Silicon 499

Appendix B   Constants and Conversion Factors 501

Appendix C   Oxide and Nitride Thickness by Color 503

Index 505

Preface to the First Edition

Microfabrication is generic: its applications include integrated circuits, MEMS, microfluidics, micro-optics, nanotechnology and countless others. Microfabrication is encountered in slightly different guises in all of these applications: electroplating is essential for deep sub-micron IC metallization and for LIGA-microstructures; deep-RIE is a key technology in trench DRAMs and in MEMS; imprint lithography is utilized in microfluidics where typical dimensions are 100 µm, as well as in nanotechnology, where feature sizes are down to 10 nm. This book is unique because it treats microfabrication in its own right, independent of applications, and therefore it can be used in electrical engineering, materials science, physics and chemistry classes alike.

Instead of looking at devices, I have chosen to concentrate on microstructures on the wafer: lines and trenches, membranes and cantilevers, cavities and nozzles, diffusions and epilayers. Lines are sometimes isolated and sometimes in dense arrays, irrespective of linewidths; membranes can be made by timed etching or by etch stop; source/drain diffusions can be aligned to the gate in a mask aligner or made in a self-aligned fashion; oxidation on a planar surface is easy, but the oxidation of topographic features is tricky. The microstructure-view of microfabrication is a solution against outdating: alignment must be considered for both 100 µm fluidic channels and 100 nm CMOS gates, etch undercutting target may be 10 nm or 10 µm, but it is there; dopants will diffuse during high temperature anneals, but the junction depth target may be tens of nanometres or tens of micrometres.

A common feature of older textbooks is concentration on physics and chemistry: plasma potentials, boundary layers, diffusion mechanisms, Rayleigh resolution, thermodynamic stability and the like. This is certainly a guarantee against outdating in rapidly evolving technologies, but microfabrication is an engineering discipline, not physics and chemistry. CMOS scaling trends have in fact been more reliable than basic physics and chemistry in the past 40 years: optical lithography was predicted to be unable to print submicron lines and gate oxides today are thinner than the ultimate limits conceived in the 1970s. And it is pedagogically better to show applications of CVD films before plunging into pressure dependence of deposition rate, and to discuss metal film functionalities before embracing sputtering yield models.

In this book, another major emphasis is on materials. Materials are universal, and not outdated rapidly. New materials are, of course, being introduced all the time, but the basic materials properties like resistivity, dielectric constant, coefficient of thermal expansion and Young's modulus must always be considered for low-k and high-k dielectrics, SnO_2 sensor films, diamond coatings and 100 µm-thick photoresists alike. Silicon, silicon dioxide, silicon nitride, aluminium, tungsten, copper and photoresist will be met again in various applications: nitride is used not only in LOCOS isolation, but also in MEMS thermal isolation; aluminium not only serves as a conductor in ICs but also as a mirror in MOEMS; copper is used for IC metallization and also as a sacrificial layer under nickel in metal MEMS; photoresist acts not only as a photoactive material but also as an adhesive in wafer bonding.

Devices are, of course, discussed but from the fabrication viewpoint, without thorough device physics. The unifying idea is to discuss the commonalities and generic features of the fabrication processes. Resistors and capacitors serve to exemplify concepts like alignment sequence and design rules, or interface stability. After basic processes and concepts have been introduced, process integration examples show a wide spectrum of full process flows: for example, solar cell, piezoresistive pressure sensor, CMOS, AFM cantilever tip, microfluidic out-of-plane needle and super-self-aligned bipolar transistor. Small process-sequence examples include, similarly, a variety of structures: replacement gate, cavity sealing, self-aligned rotors and dual damascene-low-k options are among the others.

Older textbooks present microfabrication as a toolbox of MEMS or as the technology for CMOS manufacturing. Both approaches lead to unsatisfactory views on microfabrication. Ten years ago, chemical–mechanical polishing was not detailed in textbooks, and five years ago discussion on CMP was included in multilevel metallization chapter. Today, CMP is a generic technology that has applications in CMOS front-end device isolation and surface micromechanics, and is used to fabricate photonic crystals and superconducting devices. It therefore deserves a chapter of its own, independent of actual or potential applications. Similarly, wafer cleaning used to be presented as a preparatory step for oxidation, but it is also essential for epitaxy, wafer bonding and CMP. Device-view, be it CMOS or some other, limits processes and materials to a few known practices, and excludes many important aspects that are fruitful in other applications.

The aim of the book is for the student to feel comfortable both in a megafab and in a student lab. This means that both research-oriented and manufacturing-driven aspects of microfabrication must be covered. In order to keep the amount of material manageable, many things have had to be left out: high density plasmas are mentioned, but the emphasis is on plasma processing in general; KOH and TMAH etching are both described, but commonalities rather than differences are shown; imprint lithography and hot embossing are discussed but polymer rheology is neglected; alternatives to optical lithography are mentioned, but discussed only briefly. Emphasis is on common and conceptual principles, and not on the latest technologies, which hopefully extends the usable life of the book.

Structure of the Book

The structure of this book differs from the traditional structure in many ways. Instead of discussing individual process steps at length first and putting full processes together in the last chapter, applications are presented throughout the book. The chapters on equipment are separated from the chapters on processes in order to keep the basic concepts and current practical implementations apart.

The introduction covers materials, processes, devices and industries. Measurements are presented next, and more examples of measurement needs in microfabrication are presented in almost every chapter. A general discussion of simulation follows, and more specific simulation cases are presented in the chapters that follow.

Materials of microfabrication are presented next: silicon and thin films. Silicon crystal growth is shortly covered but from the very beginning, the discussion centres on wafers and structures on wafers: therefore, silicon wafering process, and resulting wafer properties are emphasized. Epitaxy, CVD, PVD, spin coating and electroplating are discussed, with resulting materials properties and microstructures on the centre stage, rather than equipment themselves. Lithography and etching then follow. This order of presentation enables more realistic examples to be discussed early on.

The basic steps in silicon technology, such as oxidation, diffusion and ion implantation are discussed next, followed by CMP and bonding. Moulding and stamping techniques have also been included. In contrast to older books, and to books with CMOS device emphasis, this book is strong in back-end steps, thin films, etching, planarization and novel materials. This reflects the growing importance of multilevel metallization in ICs as well as the generic nature of etch and deposition processes, and their wide applicability in almost all microfabrication fields. Packaging is not dealt with, again in line with wafer-level view of microfabrication. This also excludes stereomicrolithography and many miniaturized traditional techniques like microelectrodischarge machining.

Microfabrication is an engineering discipline, and volume manufacturing of microdevices must be discussed. Discussions on process equipment have often been bogged by the sheer number of different designs: should the students be shown both 13.56 MHz diode etcher, triode, microwave, ECR, ICP and helicon plasmas, and should APCVD, LPCVD, SA-CVD, UHV-CVD and PECVD reactors all be presented? In this book, the process equipment discussion is again tied to structures that result on wafers, rather than in the equipment *per se*: base vacuum interaction with thin-film purity is discussed; the role of RTP temperature uniformity on wafer stresses is considered; and surface reaction versus transport controlled growth in different CVD reactors is analysed. Cleanroom technology, wafer fab operations, yield and cost are also covered. Moore's law and other trends expose students to some current and future issues in microfabrication processes, materials and applications.

In many cases, treatment has been divided into two chapters: for example, Chapter 5 treats thin film basics, and Chapter 7 deals with more advanced topics. Lithography and etching have been divided similarly. This enables short or long course versions to be designed around the book. The figures from the book are available to teachers via the Internet. Please register at Wiley for access www.wileyeurope.com/go/microfabrication.

Advice to Students

This book is an introductory text. Basic university physics and chemistry suffices for background. Materials science and electronics courses will of course make many aspects easier to understand, but the structure of the book does not necessitate them. The book contains 250 homework problems, and in line with the idea of microfabrication as an independent discipline, they are about fabrication processes and microstructures; not about devices. Problems fall mainly in three categories: process design/analysis, simulations and back-of-the-envelope calculations. The problems that are designed to be solved with a simulator are marked by "S". A simple one-dimensional simulator will do. The "ordinary" problems are designed to develop a feeling for orders of magnitude in the microworld: linewidths, resistances, film thicknesses, deposition rates, stresses etc. It is often enough to understand if a process can be done in seconds, minutes or hours; or whether resistance range is milliohms, ohms or kiloohms. You must learn to make simplifying assumptions, and to live with uncertain data. Searching the Internet for answers is no substitute to simple calculations that can be done in minutes because the simple estimates are often as accurate (or inaccurate) as answers culled from Internet. It should be borne in mind that even constants are often not well known: for instance, recent measurements of silicon melting point have resulted in values 1408 °C by one group, 1410 °C by one, 1412 °C by seven groups, 1413 °C by eight groups and 1416 °C by three groups, and if older works are encountered, values range from 1396 °C to 1444 °C. With thin film materials properties are very much deposition process dependent, and different workers have measured widely different values for such basic properties as resistivity or thermal conductivity. Even larger differences will pop up, if, for instance, the phase of metal film changes from body-centered cubic to β-phase: temperature coefficient of resistivity can then be off by a factor of ten. Polymeric materials, too, exhibit large variation in properties and processing. There are also calculations of economic aspects of microfabrication: wafer cost, chip size and yield. A bit of memory costs next to nothing, but the fabs (fab is short for fabrication facility) that churn out these chip are enormously expensive.

Comments and hints to selected homework problems are given in Appendix A. In Appendix B you can find useful physical constants, silicon material properties and unit conversion factors.

Preface to the Second Edition

If you search on "microfabrication" in Google Scholar, there will be over 100 000 hits; if you type in "MEMS" in Science Direct, you will get in excess of 300 000 hits; "transistor" in IEEE Xplore produces 60 000 articles; and "thin film" in Scirus, over a million articles. There is obviously no problem finding the scientific literature.

This book is a primer that prepares you to access the primary literature. It is a textbook, not a reference work or an encyclopedia, and even less a review article. This means that the topics and examples are chosen on pedagogic grounds, so many remarkable seminal works and recent breakthroughs are not addressed. It also means that most fabrication processes are not shown in full detail, rather fundamental ideas are described, and nuances and special features are left out, in order to highlight the main ideas. The articles and books in the "References and Further Reading" sections have been selected to be accessible to students, and serve as supplementary reading for assignments and exercises.

This is a fabrication text, and device physics is discussed only briefly. It would be impossible to include device physics because microfabrication can be applied to hundreds of different microdevices. In this book the examples are drawn mostly from CMOS, MEMS, microfluidics and solar cells, but there are examples from hard disk drives, flat panel displays, optics and optoelectronics, DNA chips, bipolar and power semiconductor devices, and nanotechnology.

What is New in this Second Edition?

There is about 25% more material, and the text and figures have been revised throughout. The basic structure of the book remains unchanged, and the order of chapters is mostly as in the first edition.

Silicon wafer basics are discussed in Chapter 4. More advanced wafers, and wafer behavior during processing, are discussed elsewhere, in Chapters 6 (epitaxy), 17 (bonding) and 22 (wafer engineering). Chapters 5 and 7 on thin films have been reorganized into basic and advanced chapters, with step coverage and stresses now included in the basic chapter.

Pattern generation (Chapter 8) now includes additional material on electron beam lithography. Chapter 9 introduces photoresists and optical lithography, but it is limited to 1× contact/proximity lithography only. Chapter 10 deals with advanced IC fabrication lithography with reduction steppers and scanners.

Chapter 12, on wafer cleaning and surface preparation, has been expanded to include polymer surface treatments and aspects of fluid behavior on surfaces that are important in fluidics.

Chapter 17 on bonding has been completely rewritten. It now covers fusion, anodic, metallic and adhesive bonding, including process physics and chemistry as well as technical implementations. Much of the more specialized material on silicon fusion bonding and SOI wafers has been moved to Chapter 22 on wafer engineering.

Chapter 18 has been completely rewritten and is now called "Polymer Microprocessing." In addition to the old material on embossing, imprinting and replica molding, it now contains more on polymer materials properties, thick-resist lithography, polymer bonding and a great number of device examples, especially in fluidics.

Chapter 19, on glass microprocessing, is new, but it incorporates some elements of a dismantled chapter "Processing on Non-silicon Substrates". The new chapter, however, concentrates on processing glass itself, to make microdevices of glass, in addition to fabricating devices on glass.

Chapter 23, on special processes, is a collection of techniques which are not in the mainstream of microfabrication: namely, niche techniques, non-standard approaches, special materials, and the like. For instance, porous silicon by electrochemical etching has been moved from the etching chapter to this chapter, in order to streamline the basic etching chapter.

Another new chapter, Chapter 24, deals with serial microprocessing. This includes many direct writing and machining techniques, like focused ion beam processing, laser microfabrication and micromilling and machining.

The core of the book, Chapter 25, on process integration, is now hopefully more general because CMOS- and MEMS-specific issues have been moved to their respective chapters, Chapters 26 and 30. Examples include solar cell devices, fluidic filters, resistors and integrated passive devices and PCR chips for DNA amplification. The MOS chapter now covers thin film transistors (TFTs) as well. Advanced CMOS is treated in Chapter 38 in order to keep the introduction to CMOS simple enough.

The old Chapter 22, on sacrificial structures, has been moved, greatly expanded and renamed. It is now Chapter 29, entitled "Surface Micromachining." Chapter 30 concerns MEMS process integration. It has been completely rewritten and contains almost 100% more material with many device examples.

Chapter 31 on process equipment has been rewritten with elements from a dismantled chapter on integrated processing, and with parts from an old Chapter 37, Wafer fab.

Chapter 36 is now called "Yield and Reliability." It draws on the old chapters on yield, process integration and wafer fab. The discussion has been expanded, especially regarding MEMS reliability.

Each of Chapters 37, 38 and 39 can serve as a concluding chapter, with a slightly different emphasis: Chapter 37, on the economics of microfabrication, centers on costs and markets for CMOS, MEMS, solar and flat-panel displays, and magnetic data storage; Chapter 38, on Moore's law, deals with scaling trends in ICs; and Chapter 39, on microfabrication at large, concerns the integration of different technologies, materials and functionalities, scaling with "More than Moore."

Teachers adopting the text will have access to all the figures and tables in the book as PDF slides. Additionally, a solutions manual will be available at www.wiley.com/go/Franssila_Micro2e.

Acknowledgements

A number of people have contributed to this 2nd edition in various capacities. Sardar Bilal Alam, Susanna Aura, Kestas Grigoras, Eero Haimi, Klas Hjort, Ville Joksa Jokinen, Jari Koskinen, Heikki Kuisma, Tomi Laurila, Marianne Leinikka, Antti Niskanen, Victor Ovtchinnikov, Ville Saarela, Lauri Sainiemi, Ali Shah, Gianmario Scotti, Pia Suvanto, Markku Tilli and Santeri Tuomikoski each read part of the manuscript and gave input that led to rewriting and reorganizing of the text.

Numerous colleagues and friends have provided assistance in finding material, editing figures, helping with software, looking for articles, and contributing their SEMs and other primary material. These people include Veli-Matti Airaksinen, Tapani Alasaarela, Florence Amez-Droz, Nikolai Chekurov, Nico de Rooij, Kai-Erik Elers, Jean-Christophe Eloy, Martin Gijs, Leif Grönberg, Ulrika Gyllenberg, Atte Haapalinna, Kalle Hanhijärvi, Ole Hansen, Paula Heikkilä, Ari Hokkanen, Scotten W. Jones, Tord Karlin, Ivan Kassamakov, Hannu Kattelus, Marianna Kemell, Kimmo Kokkonen, Kai Kolari, Jorma Koskinen, Anders Kristensen, Anu Kärkkäinen, Boris Lamontagne, Volker Lerche, Lauri Lipiäinen, Laura Luosujärvi, Merja Markkanen, Jyrki Molarius, Juha Muhonen, Joachim Oberhammer, Peter Ochojski, Antti Peltonen, Tuomas Pensala, Risto J. Puhakka, Mikko Ritala, Tapani Ryhänen, Henrik Rödjegård, Tomi Salo, Anke Sanz-Velasco, Jens Schmid, Pekka Seppälä, Andreas Stamm, Andrey Timofeev, Daniel Tracy, Esa Tuovinen, Albert van den Berg, Brandon Van Leer and Matthias Worgull.

Simone Taylor, Nicky Skinner, Laura Bell and Clarissa Lim at Wiley have been instrumental in pulling it all together.

Last but not least, thanks are due to Anna, Aku, Atte, Kiira and Oliver.

1

Introduction

Integrated circuits, microsensors, microfluidics, solar cells, flat-panel displays and optoelectronics rely on microfabrication technologies. Typical dimensions are around 1 micrometer in the plane of the wafer (the range is rather wide, from 0.02 to 100 μm). Vertical dimensions range from atomic layer thickness (0.1 nm) to hundreds of micrometers, but thicknesses from 0.01 to 10 μm are most typical. Microfabrication is the collection of techniques used to fabricate devices in the micrometer range.

The historical developments of microfabrication-related disciplines are shown below. The invention of the transistor in 1947 sparked a revolution. The transistor was born out of the fusion of radar technology (fast crystal detectors for electromagnetic radiation) and solid state physics. Developments of microfabrication methods enabled the fabrication of many transistors on a single piece of semiconductor and, a few years later, the fabrication of integrated circuits; that is, transistors were connected to each other on the wafer, rather than separated from each other and reconnected on the circuit board.

Microelectronics makes use of the semiconductor properties of silicon, but it is also important that silicon dioxide is such a useful material, for passivating silicon surfaces and protecting silicon during wafer processing. Silicon dioxide is readily formed on silicon, and it is high-quality electrical insulator. In addition to silicon transistors, integrated circuits require multiple levels of metal wiring, to route signals. Silicon microelectronic devices today are characterized by their immense complexity and miniaturization: a billion transistors fit on a chip the size of a fingernail.

Micromechanics makes use of the mechanical properties of silicon. Silicon is extremely strong, and flexible beams, cantilevers and membranes can be made from it. Pressure sensors, resonators, gyroscopes, switches and other mechanical and electromechanical devices utilize the excellent mechanical properties of silicon. Microelectromechanical systems (MEMS) or

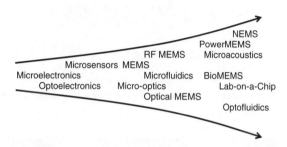

Figure 1.1 Evolution of microtechnology subfields from the 1960s onwards

microsystems, as they are also called, have expanded in every possible direction: microfluidics, microacoustics, biomedical microdevices, DNA microarrays, microreactors and microrockets to name a few. New subfields have emerged: BioMEMS, PowerMEMS, RF MEMS, as shown in Figure 1.1.

Silicon optoelectronic devices can be used as light detectors like diodes and solar cells, but light emitters like lasers and LEDs are made of gallium arsenide and indium phosphide semiconductors. Micro-optics makes use of silicon in another way: silicon, silicon dioxide and silicon nitride are used as waveguides and mirrors. MOEMS, or optical MEMS, utilize silicon in yet another way: silicon can be machined to make tilting mirrors, adjustable gratings and adaptive optical elements. The micromirror of Figure 1.2 takes advantage of silicon's smoothness and flatness for optics and its mechanical strength for tilting.

Microtechnology has evolved into nanotechnology in many respects. Some of the tools are common, like electron beam lithography machines, which were used to draw nanometer-sized structures long before the term nanotechnology was coined. Electron beam and ion beam defined nanostructures are shown in Figure 1.3. Thin films down to atomic layer thicknesses have been grown

Introduction to Microfabrication, Second Edition Sami Franssila
© 2010 John Wiley & Sons, Ltd

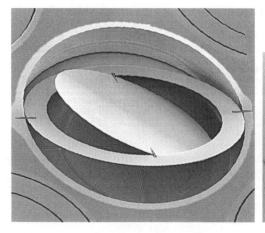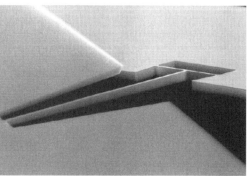

Figure 1.2 Micromirror made of silicon, 1 mm in diameter, is supported by torsion bars 1.2 μm wide and 4 μm thick (detail figure). Reproduced from Greywall *et al.* (2003), Copyright 2003, by permission of IEEE

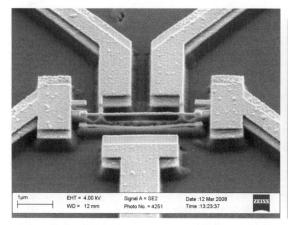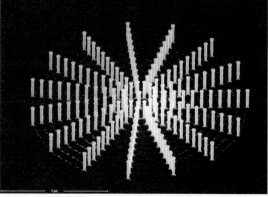

Figure 1.3 Electron microscope image of an electron beam defined gold–palladium horizontal nanobridge (courtesy Juha Muhonen, Aalto University) and vertical ion beam patterned nanopillars (courtesy Nikolai Chekurov, Aalto University); 100 nm minimum dimension in both

and deposited in the microfabrication communities for decades. Novel ways of creating nanostructures by self-assembly (self-organization) are being continuously adopted by the microfabrication community as tools to extend the capabilities of microfabrication. The tools of nanotechnology, such as the atomic force microscope (AFM), have been adopted in microfabrication as a way to characterize microstructures.

Solar cells and flat-panel displays can be large in area, but the crucial microstructures are similar to those in microdevices. Hard disks, and hard disk read/write heads especially, are microfabricated devices, with some of the most demanding feature size and film thickness issues anywhere in microfabrication.

Listed in the references and further reading section at the end of this chapter are a number of books and review articles on microfabrication in diverse disciplines.

1.1 Substrates

Silicon is the workhorse of microfabrication. Silicon is a semiconductor, and the power of microelectronics arises strongly from the fact that silicon is available in both p-type (holes as charge carriers) and n-type (electrons as

Figure 1.4 Silicon wafer, 100 mm in diameter, with about 200 device chips and a dozen test chips on it. Courtesy VTT

charge carriers), and its resistivity can be tailored over a wide range, from 0.001 to 20 000 ohm-cm. Silicon wafers are available in 100, 125, 150, 200 and 300 mm diameters and various thicknesses. Silicon is available in different crystal orientations, and the control of its crystal quality is very advanced.

Bulk silicon wafers (Figure 1.4) are single crystal pieces cut from larger single crystal ingots and polished. Silicon is extremely strong, on a par with steel, and it also retains its elasticity to much higher temperatures than metals. However, single crystalline (also known as monocrystalline) silicon wafers are fragile: once a fracture starts, it immediately develops across the wafer because covalent bonds do not allow dislocation movements.

Many microfabrication disciplines use silicon for convenience: it is available in a wide variety of sizes and resistivities; it is smooth, flat, mechanically strong and fairly cheap. Most of the machinery for microfabrication was originally developed for silicon ICs and newer technologies ride on those developments.

Single crystalline substrates include silicon, quartz (crystalline SiO_2), gallium arsenide (GaAs), silicon carbide (SiC), lithium niobate ($LiNbO_3$) and sapphire (Al_2O_3). Polycrystalline silicon is widely used in solar cell production. Amorphous substrates are also common: glass (which is SiO_2 mixed with metal oxides like Na_2O), fused silica (pure SiO_2; chemically it is identical to quartz) and alumina (Al_2O_3) are used in microfluidics, optics and microwave circuits, respectively. Sheets of polyimides, acrylates and many other polymers are also used as substrates. Substrates must be evaluated for available sizes, purities, smoothness, thermal stability,
mechanical strength, etc. Round substrates are compatible with silicon, but square and rectangular ones need special processing because tools for microfabrication are geared for round silicon wafers.

1.2 Thin Films

More functionality is built on the substrates by deposition (and further processing) of thin films: various conducting, semiconducting, insulating, transparent, superconducting, catalytic, piezoelectric and other layers are deposited on the substrates. Thin films for microfabrication include a wide variety of elements: metals of common usage include aluminum, copper, tungsten, titanium, nickel, gold and platinum. Metallic alloys and compounds commonly encountered include Al–0.5% Cu, TiW, titanium silicide ($TiSi_2$), tungsten silicide (WSi_2) and titanium nitride (TiN). The compound is stoichiometric if its composition matches the chemical formula; for example, there is one nitrogen atom for each titanium atom in TiN. In practice, however, titanium nitride is more accurately described as TiN_x, with the exact value of x determined by the details of the deposition process. The most common dielectric thin films are silicon dioxide (SiO_2) and silicon nitride (Si_3N_4). Other dielectrics include aluminum oxide (Al_2O_3), hafnium dioxide (HfO_2), diamond, aluminum nitride (AlN) and many polymers.

A special case of thin-film deposition is epitaxy: the deposited film registers the crystalline structure of the underlying substrate, and, for example, more single crystal silicon can be deposited on a silicon wafer but with different dopant atoms and different dopant concentration.

The general material structure of a microfabricated devices is shown in Figure 1.5. Interfaces between the thin film and bulk, and between films, are important for the stability of structures. Wafers experience a number

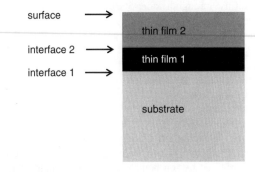

Figure 1.5 Materials and interfaces in a schematic microstructure

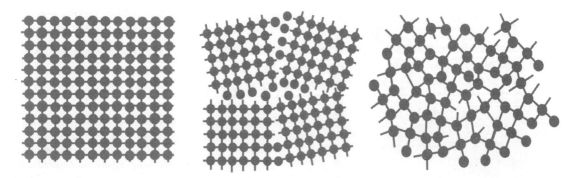

Figure 1.6 Single crystalline, polycrystalline and amorphous materials

of thermal treatments during their fabrication, and various chemical and physical processes are operative at interfaces, for example chemical reactions and diffusion. Sometimes reactions between films are desired, but most often they should be prevented. This can be achieved by adding extra films, known as barriers, in between films.

For example, thin film 1 might present an aluminum conductor and thin film 2 the passivation layer of silicon nitride; or films 1 and 2 are antireflective and scratch-resistant coatings in optics; or film 1 is thin tunnel oxide and film 2 a charge storage layer (as in memory cards).

Surface physical properties like roughness and reflectivity are material and fabrication process dependent. The chemical nature of the surface is important: some surfaces are reactive, others passive. Many surfaces will be covered by native oxide films if left unattended for some time; for example, silicon, aluminum and titanium form surface oxides over a time scale of hours. Water vapor adsorbed on surfaces must be eliminated before the wafers are processed further.

Thick substrates are not immune to thin films: a thin film 0.1 µm thick may have such a high stress that a silicon wafer 500 µm thick will be curved by tens of micrometers; or minute iron contamination on the surface will diffuse through a wafer 500 µm thick during a fairly moderate thermal treatment.

Just like substrate wafers, the grown and deposited thin films can be

- single crystalline
- polycrystalline
- amorphous

as shown in Figure 1.6. During wafer processing, single crystal films usually stay single crystalline, but they can be amorphized by ion bombardment for example. Polycrystalline films experience grain growth, for instance during heat treatments. Foreign atoms, dopants and

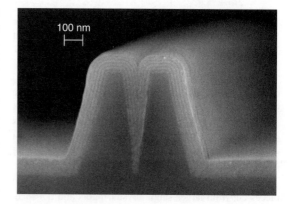

Figure 1.7 Atomic layer deposited aluminum oxide and titanium oxide thin films over silicon waveguide ridges. Courtesy Tapani Alasaarela, Aalto University

alloying atoms do not distribute themselves uniformly in polycrystalline material but aggregate at grain boundaries, which can lead to both beneficial and detrimental effects, depending on the particular materials and process conditions. Amorphous films can stay amorphous or they can crystallize during high-temperature steps, usually into the polycrystalline state.

Sometimes it is enough to deposit films on flat, planar wafers, but most often the films have to extend over steps and into trenches (Figure 1.7). These severe topographies introduce further deposition process-dependent subtleties.

Note on notations

<Si>	single crystal material
c-Si	single crystal material
α-Si/a-Si	amorphous material
a-Si:H	amorphous material with imbedded hydrogen (atomic % sometimes given)

nc-Si	nanocrystalline material (grain size a few nanometers)
μc-Si	microcrystalline material (grain size in the range of tens of nanometers)
mc-Si	multicrystalline (large-grained polycrystalline, grain size ≫ film thickness)
Al–0.5% Cu	aluminum alloy with 0.5% copper
W_2N, Si_3N_4	stoichiometric compounds
SiN_x, $x \approx 0.8$	non-stoichiometric compound
W:N	stuffed material, nitrogen at grain boundaries (non-stoichiometric)
WF_6 (g)	material in gas phase (for WF_6, boiling point 17 °C)
WF_6 (l)	material in liquid phase
W (s)	material in solid phase
H_2SiOF_6 (aq)	aqueous solution
SiH_2 (ad)	material adsorbed on a surface
$Si/SiO_2/Si_3N_4$	film stacks are marked with substrate or bottom film on the left

1.3 Processes

Microfabrication processes consist of four basic operations:

1. High-temperature processes to modify the substrate.
2. Thin-film deposition on the substrate.
3. Patterning of thin films and the substrate.
4. Bonding and layer transfer.

Under each basic operation there are many specific technologies, which are suitable for certain devices, substrates, linewidths or cost levels. Some techniques work well in research, producing a few devices with elaborate features, but completely different methods may be required if those devices are to be mass produced.

Surface preparation and wafer cleaning could be termed the fifth basic operation but, unlike the other four, no permanent structures are made. Surfaces are modified by etching away a few atomic layers, or by depositing one molecular layer. Surface preparation requirements are widely different in different process steps: in wafer bonding it is paramount to eliminate particles that would create voids if left between the wafers, while in oxidation it is important to eliminate metallic contamination and in epitaxy to ensure that native oxides are removed.

High-temperature steps are used to oxidize silicon and to dope silicon by diffusion, and they are crucial for making transistor, diodes and other electronic devices.

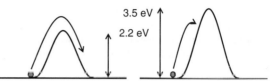

Figure 1.8 Diffusion process: the 2.2 eV barrier can be crossed at ease at 900 °C but the frequency of crossing the 3.5 eV barrier is low. A higher temperature, for example 1050 °C, would be needed for the 3.5 eV barrier to be crossed at ease

Devices like piezoresistive pressure sensors also rely on high-temperature steps, with epitaxy and resistor diffusion as the key processes. High-temperature steps can be simulated extensively, by solving diffusion equations on a computer. The high-temperature regime in microfabrication is ca. 900 °C to 1200 °C, temperatures where dopants readily diffuse and the silicon oxidation rate is technically relevant.

Many chemical and physical processes are exponentially temperature dependent. The Arrhenius equation (Equation 1.1) is a very general and very useful description of the rates of thermally activated processes. Activation energy can be illustrated as a jumping process over a barrier (Figure 1.8). According to the Boltzmann distribution, an atom at temperature T has an excess of energy E_a with a probability $\exp(-E_a/kT)$. Higher temperature leads higher barrier crossing probability:

$$\text{rate} = z(T) \, e^{(-E_a/kT)}$$

$$k = 1.38 \times 10^{-23} \, \text{J/K} = 8.62 \times 10^{-5} \, \text{eV/K} \quad (1.1)$$

The magnitude of the pre-exponential factor $z(T)$ and the activation energy E_a vary a lot.

In etching reactions, the activation energy is below 1 eV, in polysilicon chemical vapor deposition E_a is 1.7 eV, in substitutional dopant diffusion it is 3.5–4 eV and in silicon self-diffusion 5 eV. For a silicon etching process with 0.7 eV activation energy, raising the temperature from 20 to 40 °C results in a rate six times higher.

A great many microfabrication processes show Arrhenius-type dependence: etching, resist development, oxidation, epitaxy, chemical vapor deposition (which are chemical processes) are all governed by exponential temperature dependencies, as are diffusion, electromigration and grain growth (which are physical processes).

Low-temperature processes leave metal-to-silicon interfaces stable, and generally 450 °C is regarded as the upper limit for low temperatures. Between 450 and 900 °C there is a middle range which must be discussed with specific materials and interfaces in mind.

The high-temperature regime is also known as the front-end of the line (FEOL) in the silicon IC business and the low-temperature regime as the back-end of the line (BEOL). But these terms have other meanings as well: for many people in the electronics industry outside silicon wafer fabrication, front-end includes all processing on wafers, and back-end refers to dicing, testing, encapsulation and assembly. We will use the first definition.

Many thin-film steps can be carried out identically on silicon wafers and other substrates; by definition they are layers deposited on top of a substrate. Thin-film steps do not affect the dopant distribution inside silicon; that is, diodes and transistors are unaffected by them.

Processes act on whole wafers – this is the basic premise. The whole wafer is subject to, for instance, diffusion from the gas phase, and metal is evaporated everywhere. Either selected areas must be protected by masks before the process, or else the material must be removed from selected areas afterward, by etching or polishing.

Patterning processes define structures usually in two steps: polymer processing to form an intermediate pattern which then acts as a mask for etching, deposition, ion implantation or other modification of the underlying material; and after the pattern has been transferred to solid material, the intermittent polymer mask is removed.

The main patterning technique in microfabrication is optical lithography, also known as photolithography. In Figure 1.9 photolithography is shown side by side with the thermal imprint/embossing process. In both processes a polymer film is modified locally to create patterns. In lithography, photosensitive polymer film is exposed to UV light, which hardens the polymer by crosslinking (so-called negative resists). In imprinting, a thermoplastic polymer softens upon heating, and a master stamp is pressed against it. The system is allowed to cool down before the stamp is released, and then the polymer retains its imprinted shape.

Many old methods have been successfully scaled down to micrometer and nanometer scales. Etching was once used by knights to engrave their armor with their coats-of-arms, and metal etching with similar acidic solutions can make aluminum patterns in the micrometer range. Once an original microstamp or nanostamp has been made, its replication into polymers is fairly easy (it is actually the detachment that is difficult). Electroplating is likewise easily applicable to nanometer structures. Casting polymers into micromolds is also popular in microfabrication: the elastomeric (rubber-like) material PDMS (poly(dimethyl)siloxane) is a favorite material for simple microfluidic devices.

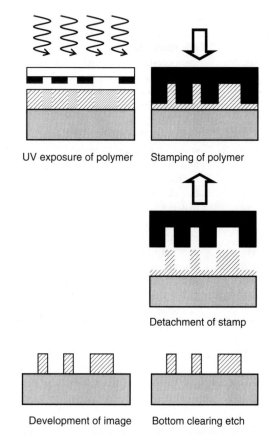

Figure 1.9 Optical lithography (left) and thermal imprint (right): UV light crosslinks photosensitive polymer, and unexposed parts are developed away (in so-called negative resists). In imprinting, softened polymer is forced to shape, and after cooling the shape is retained even though the master is removed. In imprinting, some material remains at the bottom and must be cleared by etching.

Wafer bonding and layer transfer enable more complex structures to be made. Bonding a wafer on top of a trench turns it into a channel, useful for microfluidics. Bonding more wafers can lead to elaborate fluidic channel patterns, as in the burner of a flame ionization detector, Figure 1.10. Bonding two wafers with electrodes creates a capacitor, for instance for pressure sensing. Bonding two different wafers can also be used simply as a method to create a new kind of a starting wafer, with the best properties of the two wafers combined.

These elementary operations of patterning, modification, deposition and bonding are combined many times over to create devices. Process complexity is often discussed in terms of the number of lithography steps (the

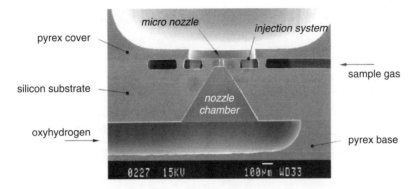

Figure 1.10 Oxyhydrogen burner of a flame ionization detector by Pyrex–glass/silicon/Pyrex–glass bonding. Reproduced from Zimmermann *et al.* (2002), Copyright © 2002 by permission of Elsevier Science Ltd

term mask levels is also used): five lithography steps are enough for a simple PMOS transistor (late 1960s' technology, and still used as a student lab process in many universities), and many MEMS, solar cell and flat-panel display devices can be made with two to six photolithography steps, but 32 nm linewidth microprocessors and logic circuits require over 30 patterning steps.

1.4 Dimensions

Microfabricated systems have minimum dimensions from 20 nm to 50 μm, depending on the device types. Advanced microprocessors and memories and the read/write heads of hard disk drives must have features <100 nm to be competitive. In Figure 1.11 the transmission electron micrograph shows the cross-section of a 65 nm MOS gate. Many other electronic devices like RF and power transistors make do with 100 nm to 1 μm dimensions. MEMS devices typically have 1–10 μm minimum lines and microfluidic devices might have 50 μm as the smallest feature.

Microfabricated device sizes are compared to physical, chemical and biological small objects in Figure 1.12, with microscopy methods capable of observing them.

Narrow individual lines can be made by a variety of methods; what really counts is resolution, the power to resolve two neighboring structures. It determines the device packing density. Resolution usually gets most attention when microscopic dimensions are discussed, but alignment between structures in different lithography steps is equally important. Alignment is, as a rule of thumb, one-third of minimum linewidth. High resolution but poor alignment can result in inferior device packing density compared to poorer resolution but tighter alignment.

As another rule of thumb, vertical and lateral dimensions of microdevices are similar. If the height-to-width

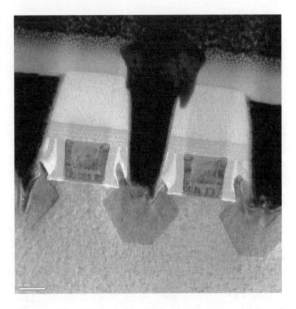

Figure 1.11 Transmission electron microscope image of 65 nm MOS transistor gates. Courtesy Young-Chung Wang, FEI Company

or aspect ratio is more than 2:1, special processing is needed, and new phenomena need to be addressed in such three-dimensional devices. Highly 3D structures are used extensively in both deep submicron ICs and in MEMS, for example in the microneedle of Figure 1.13.

Oxide thicknesses below 5 nm are used in CMOS manufacturing as gate oxides and as flash-memory tunnel oxides. Epitaxial layer thicknesses go down to the atomic layer and up to 100 μm in the thick end. There are also self-limiting deposition processes which enable extremely thin films to be made, often at the expense of deposition

8 Introduction to Microfabrication

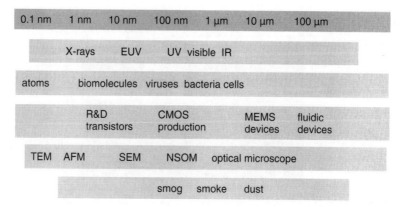

Figure 1.12 Dimensions in the microworld: electromagnetic radiation, natural objects, humanmade devices, microscopy methods and dirt

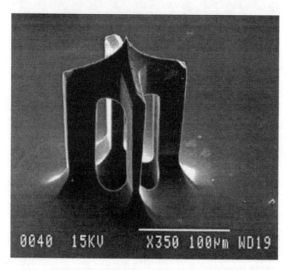

Figure 1.13 Silicon microneedle, about 100 μm. Reproduced from Griss and Stemme (2003), Copyright 2003, by permission of IEEE

rate. Chemical vapor deposition (CVD) can be used for anything from a few nanometers to a few micrometers. Sputtering also produces films from 0.5 nm to 5 μm. Spin coating is able to produce films as thin as 100 nm, or as thick as 100 μm. Typical applications include polymer spinning. Electroplating (galvanic deposition) can produce metal layers of almost any thickness, from a few nanometers up to hundreds of micrometers.

But almost every device includes structures with dimensions of about 100 μm. These are needed to interface the microdevices to the outside world: most devices need electrical connections (by a wire-bonding or bumping process); microfluidic devices must be connected to capillaries or liquid reservoirs; solar cells and power semiconductors must have thick and large metal areas to bring in and take out the high currents involved; and connections to and from optical fibers require structures about the size of fibers, which is also on the order of 100 μm.

1.5 Devices

Microfabricated device can be classified in many ways:

- material: silicon, III–V, wide band gap (SiC, diamond), polymer, glass
- integration: monolithic integration, hybrid integration, discrete devices
- active vs. passive: transistor vs. resistor, valve vs. sieve
- interfacing: externally (e.g., sensor) vs. internally (e.g., processor).

The above classifications are based on device material or functionality. In this book we are concentrating on fabrication technologies, so the following classification is more useful:

- volume (bulk) devices
- surface devices
- thin-film devices
- stacked devices.

Power transistors, thyristors, radiation detectors and solar cells (Figure 1.14) are volume devices: currents are generated and transported (vertically) through the wafer, or, alternatively, device structures extend through the wafer, as in many bulk micromechanical devices. The starting

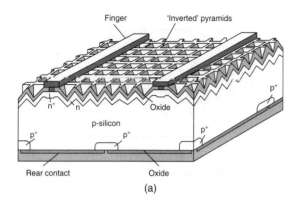

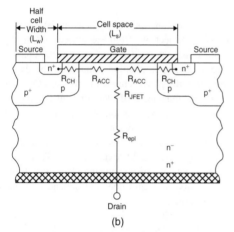

Figure 1.14 Volume devices: (a) passivated emitter, rear locally diffused solar cell, reproduced from Green (1995) by permission of University of New South Wales; (b) n-channel power MOSFET cross-section, reproduced from Yilmaz *et al.* (1991) by permission of IEEE

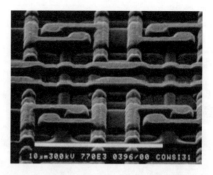

Figure 1.15 Surface devices: 0.5 μm minimum linewidth CMOS in a scanning electron microscope (SEM) view

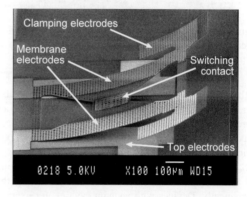

Figure 1.16 Curl switch. Reproduced from Oberhammer and Stemme (2004), Copyright 2004, by permission of IEEE

wafers for volume devices need to be uniform throughout. Patterns are often made on both sides of the wafer and it is important to note that some processes affect both sides of the wafer and some are one sided.

Surface devices make use of the material properties of the substrate but generally only a fraction of wafer thickness is utilized in making the devices. However, device structure or operation is connected with the properties of the substrate. Most ICs fall under this category: namely, MOS and bipolar transistors, photodiodes, CCD image sensors as well as III–V optoelectronic devices.

In silicon CMOS, only the top 5 μm layer of the wafer is used in making the active devices, the remaining 500 μm of wafer thickness being for support: that is, mechanical strength and impurity control. Shown in Figure 1.15 are CMOS polysilicon gates of 0.5 μm width and 0.25 μm height. Surface devices can have very elaborate 3D structures, like multilevel metallization in logic circuits, which can be 10 μm thick, but this is still only a fraction of wafer thickness; therefore the term surface device applies.

Devices can be built by depositing and patterning thin films on the wafers, where the wafer has no role in device operation. Wafer properties like thermal conductivity or optical transparency may be part of device operation, but another substrate could be used instead. Thin-film transistors (TFTs) are most often fabricated on non-semiconductor substrates of glass, plastic or steel. Devices like RF switches and relays, optical modulators or DNA arrays are often fabricated on silicon wafers for convenience, but they could be fabricated on glass or polymer substrates as well. Figure 1.16 shows a RF switch: the silicon nitride/gold thin film flap curls up because of film stresses, but can be forced flat by electrostatic actuation.

In MEMS devices with free-standing elements, membranes and cantilevers pose certain processing limitations

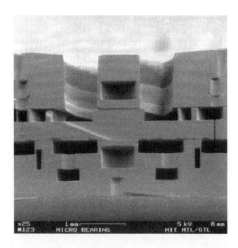

Figure 1.17 A microturbine by five-wafer silicon-to-silicon bonding. Reproduced from Lin *et al.* (1999) by permission of IEEE

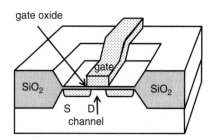

Figure 1.18 Schematic of a MOS transistor: gate, source (S) and drain (D) in an active area defined by thick isolation oxide

of their own. Many processes cannot be done on movable, bending structures because they are not stable enough, and therefore the release step is often the very last process step. Similarly, devices with through-wafer holes pose serious limitations in many process steps.

Stacked devices are made by layer transfer and bonding techniques. Two or more wafers are joined together permanently. Devices with vacuum cavities, for example absolute pressure sensors, accelerometers and gyroscopes, are stacked devices made of bonded silicon/glass wafer pairs. Micropumps and valves are typically stacks of many wafers. Figure 1.17 shows a microturbine. It is made by bonding together five wafers. More and more layer transfer and wafer bonding techniques are being developed, and stacked devices of various sorts are expected to be appear, for example GaAs optical devices bonded to Si-based electronics, or MEMS devices bonded to ICs.

1.6 MOS Transistor

The MOS transistor is a capacitor with a silicon substrate as the bottom electrode, the gate oxide as the capacitor dielectric and the gate metal as the top electrode (Figure 1.18). The MOS transistor has been the driving force of the microfabrication industries. It is the top device by all measures: number of devices sold, the narrowest linewidths and the thinnest oxides in mass production, as well as dollar value of production. Most equipment for microfabrication was originally designed for MOS IC fabrication, and later adapted to other applications.

Despite the name MOS, the gate electrode is usually made of phosphorus-doped polycrystalline silicon, not of metal. The basic function of a MOS transistor is to control the flow of electrons from the source to drain by the gate voltage and the field it generates in the channel. In a NMOS transistor, a positive voltage on the gate pulls electrons from the p-type channel to the Si/SiO$_2$ interface where an overabundance of electrons inverts the region under the gate to n-type, enabling electrons to flow from the n+ source to the n+ drain.

Transistors are isolated electrically from neighboring transistors by SiO$_2$ field oxide areas. This isolation takes up a lot of area, and therefore the transistor packing density on a chip does not depend on transistor dimensions alone.

Scaling down MOS transistor channel length makes the transistors faster. The other main aspect is area scaling: a factor N linear dimension scaling reduces the area to A/N^2. Gate width, gate oxide thickness and source/drain diffusion depths are closely related, and the ratios are more or less unchanged when the transistors are scaled down. As a rough guide, for a gate width of L, the oxide thickness is $L/50$, and the source/drain junction depth is $L/5$.

1.7 Cleanliness and Yield

Microfabrication takes place under carefully controlled conditions of constantly circulating purified airflow, with temperature, humidity and vibrations also under strict control because micrometer-scale structures would otherwise be destroyed by particles or else the lithography process would be ruined by vibrations or temperature and humidity fluctuations. Personnel in cleanrooms wear gowns to prevent particle emissions (Figure 1.19), and work procedures have been developed to minimize all disturbances.

Wafers are cleaned actively during processing: thousands of liters of ultrapure water (UPW, as known as de-ionized water, DIW) is used for each wafer during

Figure 1.19 Working in a cleanroom, courtesy VTT

device fabrication. This is the dynamic part of particle cleanliness: the passive part comes from careful selection of materials for cleanroom walls, floors and ceilings, including sealants and paints, as well as selection of materials for design and for process equipment, wafer storage boxes and all associated tools, fixtures and jigs.

Even though extreme care is taken to ensure cleanliness in microprocessing, some devices will always be defective. As the number of process steps increases, yield Y goes down according to

$$Y = Y_0^n \qquad (1.2)$$

where Y_0 is the yield of a single process step and n is the number of steps. With 100 process steps and 99% yield in each individual step, this results in a 37% yield (representative of a 64 k DRAM chip), but a 99% yield for a 500-step process (representative of a 16 Mbit DRAM) results in a yield <1%. Clearly a 99% yield is not enough for modern memory fabrication. Chip design also affects yield through area:

$$Y = e^{-DA} \qquad (1.3)$$

where A is the chip area and D is the defect density: making small chips is much easier than making big chips.

Yield has two major components: stochastic and systematic. Stochastic (random) defects are unpredictable occurrences of pinholes in protective films, particle adhesion on the wafer, corrosion of metal lines, etc. Systematic defects come from equipment performance limitations, impurities in starting materials and design errors: two features may be placed so close to each other that they inadvertently touch, or impurities in chemicals do not allow low enough leakage currents.

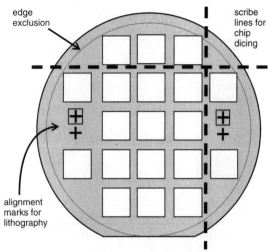

Figure 1.20 Silicon wafer with chips, alignment marks and edge exclusion. The scribe line area is reserved for dicing the chips separately

IC wafers contain typically a hundred or hundreds of chips (also called die). This is depicted in Figure 1.20. It also shows the other elements: alignment marks for registering structures on different layers to each other; and scribe lines, the space reserved for dicing the wafer into separate chips after completing the processing. The number of chips on a wafer has remained more or less unchanged for decades because chip size and wafer size have grown in parallel: $0.2\,\text{cm}^2$ chips were made on 100 mm wafers while $2\,\text{cm}^2$ chips are usual on 300 mm wafers. In extreme cases only one chip fits the wafer, for example a solar cell, a thyristor or a position-sensitive radiation detector. Microfluidic separation devices with channels 5 cm long and optical waveguide devices with large radii of curvature can have a handful of devices per wafer. With standard logic chips or micromechanical pressure sensors and accelerometers, thousands can be crammed together to fit a wafer.

1.8 Industries

Worldwide, about $6 billion is spent on silicon wafers annually. These are used to make $250 billion worth of semiconductor devices, which fuel the $1200 billion electronics industry. In 2009, about 10^{19} transistors were shipped, approximately 1 billion devices for each and every person on Earth. As recently as 1968 it was one transistor per year per person. Price of course explains a

Table 1.1 Wafer size and linewidth distribution in 2009

Wafer size	% of area	% of wafers
300 mm	38	15
200 mm	38	35
150 mm	17	30
<150 mm	7	20

Linewidth distribution in production:

≥400 nm	10%
160–400 nm	15%
80–160 nm	20%
65–80 nm	20%
<65 nm	35%

Data from SEMI and SICAS.

lot: in 1968 transistors cost about 1$ a piece; in 2009 the cost was less than one-millionth of a cent. Device density on chips is quadrupling every three years, a trend known as Moore's law.

Scaling has continued relentlessly for the past 50 years. Linewidths were in the 30 µm range in the early 1960s, and they are 30 nm in the year 2010. Lithographic scaling has thus improved packing density by a factor of a million (linear dimension scaling by a factor of 1000 equals area density scaling by a factor of 1 000 000). The number of transistors on a chip has increased from one to one billion, however. The other two main factors have been an increase in chip size and in circuit cleverness: new designs, new fabrication processes and novel materials use less area for the same functionality.

IC technology generations are classified by their linewidths and each new generation has dimensions roughly 30% smaller than the previous one. In 2010 the minimum linewidth in production is about 30 nm, but this represents just a small fraction of all ICs manufactured. The 200 million wafers (approximately 6 square kilometers) are distributed as shown in Table 1.1. Small wafers are used to fabricate specialty ICs, MEMS and power devices. Gigabit memories and latest generation processors are made on 300 mm wafers. When counted as silicon area, the smaller linewidths gain in importance because linewidth scaling has been accompanied by an increase in wafer size, which means that 65 nm devices are fabricated on 300 mm wafers but 1 µm devices on 100 mm wafers. Only 35% of all devices are fabricated with the latest 45 nm and 32 nm generation technologies, but the fabrication facilities for the 65 nm technology generation (commonly called

"wafer fabs" or simply "fabs") are very new: they were mostly built in 2005–2007.

The microsystems/MEMS industry as such does not exist: microsystems are more a technology than an industry, therefore statistics are erratic. Some estimates put microsystem sales at $8 billion. The flat-panel display industry has sales of close to $100 billion. The solar cell industry is growing rapidly, with sales of $30 billion. The magnetic data storage industry is similar in size but consists of a limited number of established players, while the solar industry is mostly populated by start-up companies.

Note on drawings

The z-dimension is enlarged relative to the x- and y-directions to make drawings easier to read. For example, in bulk micromechanics the diaphragm of a piezoresistive sensor is 20 µm, or 5% of wafer thickness, and the piezoresistor diffusion depth is 5% of diaphragm thickness, that is 1 µm.

1.9 Exercises

1. Silicon atom density is 5×10^{22} cm^{-3}. If the boron dopant concentration is 10^{15} cm^{-3}, how far are the boron atoms from each other?
2. IC chips are getting larger even though the linewidths are scaled down because more functions are integrated on a chip. Calculate the signal line resistance for:
   (a) aluminum conductors 1 µm thick, 3 µm wide and 500 µm long (resistivity 3 µohm-cm);
   (b) copper conductors 0.3 µm wide, 0.5 µm thick and 1 mm long (resistivity 2 µohm-cm).
3. Silicon dioxide can sustain a 10 MV/cm electric field. Calculate oxide thickness regimes for:
   (a) CMOS ICs where the operating voltages are 1–5 V;
   (b) capillary electrophoresis (CE) microfluidic chips where 500–5000 V is used.
4. DRAM memory is a capacitor. How many electrons are stored in a DRAM capacitor if it has an area of 1 µm^2 and a silicon dioxide dielectric 5 nm thick?
5. A micromechanical pressure sensor consists of a 1 mm^2 movable silicon electrode and 1 µm air gap as the dielectric. What is the capacitance? If femtofarad capacitance change can be measured, what is the corresponding displacement of the movable capacitor electrode?
6. Aluminum wires do not tolerate current densities higher than 1 MA/cm^2. What are the maximum currents that can run in micrometer aluminum wiring?

7. If a silicon etching reaction has an activation energy of 0.7 eV, and the etch rate at 80 °C is 1.3 µm/min, how much is it at 100 °C?
8. What defect densities are typical of modern IC production?
9. CMOS linewidths have been scaled down steadily by 30% every three years. In the year 2010 linewidths were in the range of 32 nm. When will the linewidths equal atomic dimensions?

References and Related Reading

Elwenspoek, M. (2001) **Mechanical microsensors**, Springer.
Geschke, O., H. Klank and P. Telleman (2004) **Microsystem Engineering of Lab-on-a-chip Devices**, 2nd edn, Wiley-VCH Verlag GmbH.
Green, A.M. (1995) **Silicon Solar Cells**, University of New South Wales Press.
Greywall, D.S. et al. (2003) Crystalline silicon tilting mirrors for optical cross-connect switches, *J. Microelectromech. Syst.*, **12**, 708–712.
Griss, P. and G. Stemme (2003) Side-opened out-of-plane microneedles for microfluidic transdermal liquid transfer, *J. Microelectromech. Syst.*, **12**, 296.
Hierold, C. (2004) From micro- to nanosystems: mechanical sensors go nano, *J. Micromech. Microeng.*, **14**, S1–S11.
Lin, C.-C. et al. (1999) Fabrication and characterization of a micro turbine/bearing rig, Proceedings of MEMS'99, p. 529.
Maluf, N. and K. Williams (2004) **An Introduction to Microelectromechanical Systems Engineering**, 2nd ed, Artech House.
Motamedi, M.E. (2005) **MOEMS: Micro-Opto-Electro-Mechanical Systems**, SPIE.
Oberhammer, J. and G. Stemme (2004) Design and fabrication aspects of an S-shaped film actuator based DC to RF MEMS switch, *J. Microelectromech. Syst.*, **13**, 421–428.
Plummer, J.D., M.D. Deal and P.B. Griffin (2000) **Silicon VLSI Technology**, Prentice Hall.
Poortmans, J. and V. Arkhipov (eds.) (2006) **Thin Film Solar Cells**, Wiley.
Rebeiz, G. (2002) **RF MEMS: Theory, Design and Technology**, John Wiley & Sons, Inc.
Sikanen, T. et al. (2010) Microchip technology in mass spectrometry, *Mass Spectrom. Rev.*, **29**, 351–391.
Solgaard, O. (2008) **Photonic Microsystems: Micro and Nanotechnology Applied to Optical Devices and Systems**, Springer.
Tanaka, Y. et al. (2007) Biological cells on microchips: new technologies and applications, *Biosens. Bioelectron.*, **23**, 449–458.
Tsuchizawa, T. et al. (2005) Microphotonics devices based on silicon microfabrication technology, *IEEE J. Sel. Top. Quantum Electron.*, 11, 232–240.
Weibel, D.B., W.R. DiLuzio and G.M. Whitesides (2007) Microfabrication meets microbiology, *Nature Rev. Microbiol.*, 5, 209–218.
Yilmaz, H. et al. (1991) 2.5 million cell/in^2, low voltage DMOS FET technology, Proceedings of IEEE APEC 1991, p. 513.
Ziaie, B. et al. (2004) Hard and soft micromachining for BioMEMS: review of techniques and examples of applications in microfluidics and drug delivery, *Adv. Drug Delivery Rev.*, **56**, 145–172.
Zimmermann, S., A. Vogel and J. Müller (2002) Miniaturized flame ionization detector for gas chromatography, *Sens. Actuators*, **B83**, 285–289.

2

Micrometrology and Materials Characterization

When micrometer lines are patterned, and nanometer films grown, measurement tools have to be available to characterize those processes. The measurement spot must sometimes be as small as possible, to obtain information on details of microstructures, but sometimes we can use blanket test wafers for ease of measurement. Often we want to scan and map large areas for uniformity and to get statistical data, and then speed of measurement is paramount. Data is needed on electrical properties like resistivity, optical properties like refractive index, chemical properties like bond identification, physical properties like crystal structure, and mechanical properties like size and shape. Amazingly accurate and precise measurements can be done, but it should be borne in mind that routine monitoring is needed together with complex, highly detailed measurements.

2.1 Microscopy and Visualization

Optical microscopy resolution is roughly a micrometer, i.e. similar to wavelength. This is useful for practically all MEMS and solar cell and display devices, while for modern ICs and nanotechnology it is hopelessly inadequate. It is a rough and ready method for checking, for example, thin-film delamination, color changes due to corrosion, scratches from mechanical handling and similar major faults, and these can be observed even if the details of the structures are smaller than the resolution. Defects can also be made larger by etching or deposition, and in this way very small pinhole defects in thin films, for instance, can be visually observed.

Optical microscope performance has been extended by many simple and ingenious ways over the years. In the near-field scanning optical microscopes (NSOM) light is collected through a small optical aperture, for example 50 nm, and therefore resolution is determined by this size, not by wavelength. In confocal microscopes light is collected only from the focal depth, and vertical information can be obtained. In dark-field microscopes illumination is from the side, and detection is from above as usual. This gives enhanced information about edges and steps on the wafer. Fluorescence microscopes measure fluorescence, and in the best case the signal is generated by a single molecule, which is definitely smaller than the optical resolution of the microscope. Fluorescence is therefore able to detect very small amounts of material, for example photoresist residues.

Electron microscopes can resolve much smaller features: the scanning electron microscope (SEM) can see 5 nm objects. In top view imaging the SEM is very much like the optical microscope, but with two advantages: higher magnification and better depth of field (Figure 2.1). Its real power, however, comes into play in tilted and cross-sectional views (Figure 2.2). Cross-sectional images reveal information like the sidewall angle of microstructures, and sidewall surface quality, or film thickness variation over a step. SEM resolution is sufficient for film thickness determination in the hundreds of nanometers range, but below that its accuracy is inferior to other techniques.

The transmission electron microscope (TEM) provides the ultimate image resolution, down to atomic imaging. The high-resolution TEM (HRTEM) has a special advantage in calibration: the lattice spacing of atoms can be used as accurate internal calibration standards (Figure 2.3). The drawbacks of the TEM are manifold: sample preparation is difficult and time consuming, and only a very small area is imaged. In the TEM each picture takes

---

Introduction to Microfabrication, Second Edition   Sami Franssila
© 2010 John Wiley & Sons, Ltd

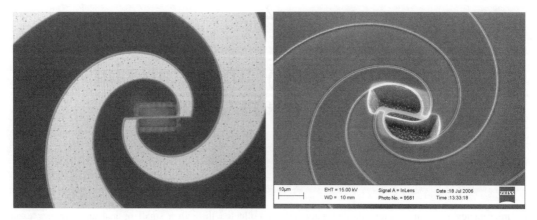

Figure 2.1 Optical microscope and scanning electron microscope (SEM) views of a spiral antenna bolometer. Courtesy Leif Grönberg, VTT

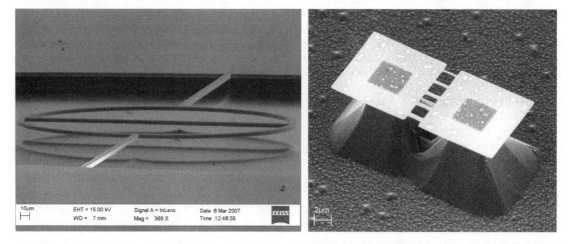

Figure 2.2 SEM: left, a 5 µm minimum linewidth stress test structure, courtesy Lauri Sainiemi Aalto University; right, 100 nm nanobridges, courtesy Nikolai Chekurov Aalto University

hours, but minutes with the SEM and seconds with an optical microscope.

Scanning probe microscopy (SPM) is a collection of techniques which use a sharp nanoneedle tip on a soft cantilever to scan the sample (Figure 2.4), while the interaction of the needle tip with the surface is measured. This interaction can be physical distance, which is sensed as a tunneling current (between the metal surface and metal tip in the scanning tunneling microscope, STM), or magnetic, capacitive, chemical or many other interactions. The STM can have atomic resolution. It is a research tool for surface science, but its relative, the atomic force microscope (AFM), which has nanometer resolution, is a favorite metrology tool in microfabrication. In the AFM the force in question is the repulsive force between atoms at very short distances. Because this repulsive force is universal, many materials can be used for tips, and silicon and silicon nitride are standard choices, with silicon for cantilevers. Either constant force mode or constant height mode can be used. The former is slow and suitable for hard materials, the latter better for rapid scanning. Cantilever deflection is measured (usually optically) to 0.01 nm, and the piezoelectric stage covers a scan area of 5×5 µm typically. The AFM is popular as a surface characterization tool but it is also very good for measuring

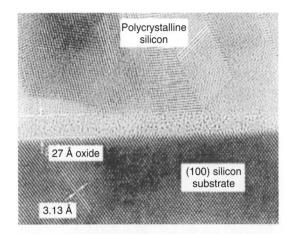

Figure 2.3 High-resolution transmission electron micrograph of single crystal silicon/amorphous silicon dioxide/polycrystalline silicon structure. Reproduced from Buchanan (1999) by permission of IBM

lateral and vertical dimensions in the micrometer and nanometer range.

2.2 Lateral and Vertical Dimensions

For device lateral dimensions, 10% deviation is usually accepted as a fabrication tolerance. Measurement precision should be 10% of that variation, that is 10 nm for 1 μm structures. For 100 nm structures this translates to 1 nm, which is very difficult indeed to obtain.

Linewidth is often known as a critical dimension (CD). All major CD measurements rely on scanning, where an optical slit or aperture, a laser or electron beam spot or a mechanical stylus is scanned over the line. Linewidth measurement depends on edge detection in all these methods. This has both inherent and microstructure-related limitations. The signal from the edge is not a sharp delta function, even in the case of a perfectly vertical sidewall. Beam spot and mechanical stylus alike have finite dimensions which blur line edge detection. Both electromechanical stylus systems and AFMs can be used. The former have a tip radius of curvature of 1–10 μm, the latter 1–10 nm. Stylus instruments are often called profilometers, but, as shown in Figure 2.5, they do not necessarily provide any information about profiles!

Film thicknesses range from one atomic layer to hundreds of micrometers, and no single method can cover such a thickness range. Conductive and dielectric films must often be measured by different techniques even though scanning probe methods are quite universal: a step is formed by etching and a probe tip scans the step. Z-scale precision can be 1 nm or even down to 1 Å (0.1 nm), but in most practical cases surface roughness sets the lower limit for step height/film thickness measurement.

Commonly used optical thickness measuring methods are ellipsometry and reflectometry. In ellipsometry the change in polarization is measured and the amplitude ratio of two different polarizations is computed. Film thickness can be obtained when the optical constants of the film are known. Ellipsometry can be used to measure those constants too, but then additional measurements are needed: for example, multiple angles or multiple wavelengths must be used. Ellipsometry works best for film thicknesses below the measurement wavelength (633 nm, most commonly) because periodicity makes interpretation

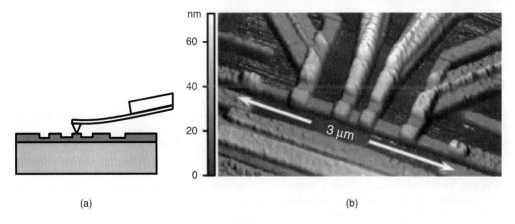

Figure 2.4 (a) AFM cantilever tip scanning the topography; (b) AFM micrograph of a single electron transistor, reproduced from Timofeev *et al.* (2009) by permission of A Timofeev. Copyright 2009 by The American Physical Society

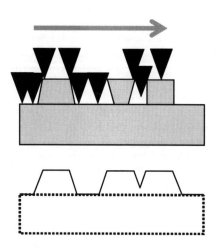

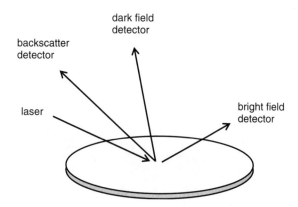

Figure 2.5 Scanning linewidth measurement, with scan profile shown below. Feature dimensions can be measured but no information is obtained on sidewall angle because the measurement instrument tip/spot itself has non-zero size

Figure 2.6 Scatterometry measurement with three different detectors

of thick-film thicknesses difficult. The low end is just a few nanometers, but for very thin films uncertainty is introduced because optical constants are not really constants, but depend on film thickness.

In reflectometry a wavelength scan is made (e.g., 300–800 nm) and this is fitted to a reflection model. Reflectometry can measure films from a few nanometers up to 50 μm thick.

X-ray reflection (XRR) can be used to measure very thin films. Unlike optical methods, XRR is insensitive to changes in film refractive index. Measurement time, however, is minutes, or even hours, compared to seconds for optical tools. XRR is amenable only to surfaces with a roughness no more than a nanometer and thickness less than a hundred nanometers. There are many such applications in microtechnology: for example, CMOS gate oxides on single crystal silicon are thin and extremely smooth.

2.3 Optical Techniques

In addition to microscopy, optics has many attractive qualities for monitoring microfabrication: optical measurements are mostly quick, non-contact and accurate. Lasers can be focused down to micrometer spots, and rapidly scanned over the full wafer, proving both local and global information. Surface roughness, particles and defects can be measured by scatterometry (Figure 2.6). Laser light is scattered from the wafer surface, and multiple detectors measure the reflected and scattered light. Directly reflected (specular) light gives information on surface steps and mounds, backscattered light contains information on subsurface defects and diffuse scattering (by the dark-field detector) is indicative of surface particles and microroughness. A general way to obtain more information in optical measurements is to use two different wavelengths (in laser systems) or to scan wavelengths, and to use multiple incidence and detection angles. In scatterometry varying the incidence and detection angles reveals differences between pits and particles for example.

Optical measurements provide information on layer thicknesses (ellipsometry, reflectometry, interferometry) and on optical properties of thin films and surface layers (refractive index, absorption coefficient). They can measure step heights (optical profilometry), surface planarity and wafer curvature (interferometry), and membrane vibrations (interferometry, vibrometry, holography). Interferometry is useful for both static measurements of membrane planarity and cantilever curvature, as well as dynamic properties. Optical probing can be used to measure vibration fields under real operating conditions without disturbing the device. In this way, the mechanical vibration field can be measured, revealing, for example, unwanted vibration modes or energy leakage from a resonator. Optical measurements can thus provide valuable feedback to refine simulations and the design of vibrating MEMS structures (Figure 2.7).

Optical measurements can provide a wealth of information on materials properties. For example, amorphous and crystalline phases have different optical properties. Porous material optical properties can be gauged by modeling the material as a composite of solid and air (or liquid, if the pores are filled with toluene, for example). Multilayer thin films can be measured, but increased reliance on modeling makes the results somewhat inaccurate.

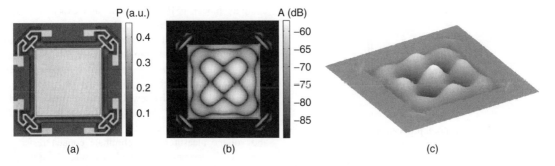

Figure 2.7 Scanning laser interferometer measurement of a vibration mode in a MEMS resonator: (a) light power image of the resonator; (b) relative vibration amplitude field at 68 MHz; (c) instantaneous surface deflection. Courtesy Kimmo Kokkonen and Lauri Lipiäinen, Aalto University

Optical excitation is used to generate charge carriers in semiconductors, and measuring these provides information about electrical properties. Laser pulse generated local heating causes acoustic waves in solids which will be reflected from material interfaces, which is utilized in thickness measurements. Similarly, thermal pulses provide information on changes in material properties, for example amorphization caused by ion implantation.

2.4 Electrical Measurements

A number of electrical measurements can be used to characterize substrates and deposited thin films: namely, resistivity, conductivity type, carrier density and lifetime, mobility, contact resistance or barrier height. Resistivity (ρ) depends on charge, carrier mobility (μ) and carrier number (n), as follows:

$$\rho = \frac{1}{nq\mu} \quad (2.1)$$

This equation has a catch, however, because mobility (μ) decreases with increasing carrier concentration (n) and experimental data is needed to correct for this. Carrier concentration is measured either by mercury probe capacitance–voltage (Hg-CV) measurement or by surface photovoltage (SPV). In the Hg-CV method the capacitance of the depletion region in a mercury semiconductor Schottky diode is measured. This method is accurate but slow and there is the danger of mercury contamination, although no mechanical damage.

Resistivity is an important property of conducting layers but resistance is the property that can be measured easily. For a rectangular piece of conducting material, resistance is given by

$$R = \frac{\rho L}{WT} \quad (2.2)$$

where ρ is the resistivity, L the length, T the thickness and W the width. If we consider a square piece of conductor, $L = W$, we can then define sheet resistance, R_s, as

$$R_s \equiv \rho/T \quad (2.3)$$

R_s is in units of ohms, but it is usually denoted by ohms/square to emphasize the concept of sheet resistance. The resistance of a conductor line can now be easily calculated by breaking down the conductor into n squares: $R = nR_s$ (Figure 2.8). Sheet resistance is independent of square size. Sheet resistances of doped semiconductor layers will be discussed in Chapter 14.

Measurement of R_s can be done in several ways: direct measurement necessitates the fabrication of a metal line (lithography and etching steps), but the result follows easily:

$$R_s = R/n = V/nI \quad (2.4)$$

The four-point probe (4PP) method uses two outer probe needles to feed current through the sample and two inner needles to measure voltage, as shown in Figure 2.9.

In the case of a large and thick substrate, resistivity is given by

$$\rho = 2\pi s \frac{V}{I} \quad (2.5)$$

Figure 2.8 Conceptualizing metal line resistance: four squares with sheet resistance R_s in series gives resistance as $R = 4R_s$

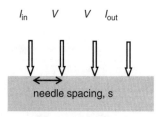

Figure 2.9 The four-point probe measurement set-up with identically spaced needles

In the case of a thin film of thickness T on an insulating substrate (e.g., Al film on SiO_2), the resistivity is given by

$$\rho = \frac{\pi}{\ln 2}\frac{V}{I}T = 4.53\frac{V}{I}T \quad \text{or} \quad R_s = 4.53\frac{V}{I} \quad (2.6)$$

When the sample size is 15 times larger than the probe spacing, resistivity is correct to within 1%. For smaller samples, geometric correction factors need to be applied.

Thickness has to be measured independently, for example by profilometer. Alternatively, sheet resistance can be used to calculate thickness once thin-film resistivity is known (bulk values cannot be used for thin-film resistivities, as will be discussed in Chapter 5).

If depth information of carrier concentration is needed, spreading resistance profiling (SRP) can be used (Figure 2.10). Angle lapping is used to produce a bevel, and resistance is probed along the bevel down to the substrate. An accurate bevel angle is needed to convert data into a thickness profile, and the layer-by-layer resistance data must be converted to resistivity by rather complex models. SRP is difficult and expensive but it provides data that is not available from other methods.

Many electrical test structures have been devised for conductive films and doping structures. These are fast measurements, ideally suited for wafer mapping: sheet resistance measurement requires four pads for probe needles, and electrical linewidth measurement similarly. Contact chains make do with two pads but generally four-pad measurements, with separate feeds for current and voltage measurements, eliminate contact resistance parasitics. A combined six-pad structure (Figure 2.11) can be used to measure both sheet resistance R_s and electrical linewidth.

In the six-terminal structure, current I_c is driven through terminals 2 and 3 and voltage drop V_c is measured across terminals 5 and 6. Sheet resistance R_s is given by

$$R_s = \frac{\pi}{\ln 2}\frac{V_c}{I_c} \quad (2.7)$$

Bridge resistance R_b is the voltage drop between terminals 4 and 5, V_{45}, divided by current I_{13} driven through terminals 1 and 3. Linewidth W is then given

$$W = \frac{R_s \cdot L}{R_b} \quad (2.8)$$

Assuming a rectangular cross-sectional profile usually holds fairly well for plasma-etched lines. Line length L is fixed on the photomask, and if $L \gg W$, minor inaccuracies in lithography (e.g., corner rounding) can be

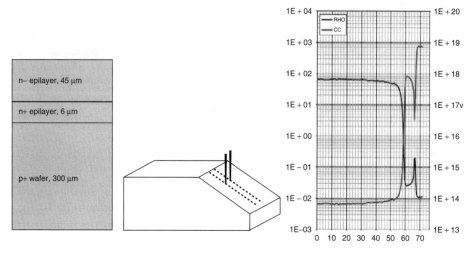

Figure 2.10 Spreading resistance profiling (SRP): two-point electrical measurements on a beveled sample. Resistance and carrier concentration profiles of n^-/n^+ epi on p^+ substrate. Courtesy Atte Haapalinna, Okemtic

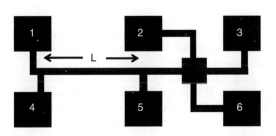

Figure 2.11 Electrical six-terminal test structure for sheet resistance and linewidth

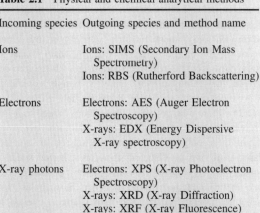

ignored. Diffusion can be measured similarly, but the assumption of the profile needs to be accounted for.

Electrical test structures are implemented in test chips on the wafer, or, alternatively, they can be imbedded in the scribe lines between chips. Test structures for wafer fabrication measurements can thus be discarded after the fabrication is completed. This saves area because the dicing saw requires a margin of about a hundred micrometers between the chips anyway, as shown in Figure 1.20.

2.5 Physical and Chemical Analyses

One special aspect of materials in microfabrication is their extreme purity: impurities are specified even in parts per trillion (ppt, 10^{-12} relative abundance). This is a relief in some cases because background signals are very low, but if the impurities themselves need to be measured, then we face some tough challenges.

Elemental concentrations are often needed: nitrogen in TiN thin films (50% for stoichiometric film), copper in aluminum (0.5–4% Cu), phosphorus in oxide (5% by weight), boron in silicon wafers (1×10^{16} cm^{-3}), oxygen in silicon (10–20 ppma, parts per million atoms), sodium impurities in tungsten sputtering target (ppb, parts per billion), or iron in silicon (ppt). These different concentration levels mean that different analytical methods must be employed.

Elemental detection can be accomplished with many methods quite readily, but quantification is often difficult. Comparative results are often presented: treatments A, B, C vs. reference sample. Treatments might represent new plasma CVD oxide processes and thermal oxide used as a reference, or different thermal treatments are compared to the as-deposited sample as a reference.

Measurement and characterization of microstructures differ from macroscopic structures and bulk materials in many respects. Small analysis areas and volumes limit the available methods and sensitivities. The signal-to-noise ratio (S/N) is proportional to the square root of the number of atoms probed:

$$\text{S/N} \propto \sqrt{\text{number of atoms probed}} \propto R \cdot \sqrt{z} \quad (2.9)$$

where R is the probing radius and z is the depth of analysis (cylinder volume $\propto R^2 z$). Equation 2.9 explains why no single method can fulfill all microcharacterization needs.

A large number of techniques are available for chemical and physical analyses. We can excite the sample with electrons, photons or ions, and measure electrons, photons or ions ejected from the sample. Table 2.1 collects some of the main methods, classified according to excitation/emission species. The list is by no means exhaustive; you can find many more techniques in a dedicated book.

Unfortunately, most methods are limited to certain elements only. The only exception is SIMS, which can detect every element from hydrogen to uranium. Auger spectroscopy cannot detect hydrogen, helium or lithium due to a fundamental limitation of the three electron Auger processes, but all other elements are detectable. X-ray methods are insensitive to light elements: depending on X-ray window design, boron ($m = 11$) can be detected, but sometimes fluorine ($m = 19$) and sodium ($m = 23$) are the lightest detectable elements.

2.5.1 Analysis area and depth

Optical methods cover a wide range of analysis areas: sometimes micrometer laser spots or large areas from lamp illumination, or scanning a wafer by laser beams.

X-rays cannot be focused, and X-ray methods require typically rather large areas, in the millimeter range. Ion beams can be focused to submicron spots in focused ion beam (FIB) equipment, but most applications use broad beams, in the millimeter range. Electron beam excitation is the most accurate, and when combined with electron detection, the smallest possible analysis volume is obtained.

Analysis depths vary a lot. Some methods are sensitive to a few atomic layers only, while others extend to a micrometer or so. Electrons have very small mean free paths in solids, so only electrons from the top few nanometers can escape and be detected, therefore electron spectroscopies are very surface sensitive. X-rays penetrate deep into matter, and the volume probed is much larger. Ion penetration depth is easily varied by changing energy. In all cases the incident angle can be used to modify the penetration depth: glancing angle incidence leads to surface sensitive analysis when total reflection takes place.

Diffusion depths and film thicknesses are often on the order of a micrometer. Analysis techniques that extend this deep would be very useful, but only a few exist. Rutherford backscattering spectrometry (RBS) has a typical analysis depth of around a micron (for helium ion energy of 2 MeV). Electron beams also penetrate by about a micrometer into solids, and generate a signal from that depth. A combination of surface erosion and surface sensitive analysis is commonly adopted for top micrometer analysis: the ion beam removes material and the newly formed surface is probed by AES or SIMS.

Analysis must be done not only on microfabricated structures themselves but also on defects and non-idealities which are smaller than the device dimensions. If the chemical composition or structure of defects has to be identified, this is even more demanding than analysis

of regular microstructures. Contaminants often come in quantities too small for even the best analytical methods. Vacancies and other point defects are smaller than the resolution of even the best microscopic methods. Indirect methods must be used, like carrier lifetime measurements (defects act as traps for charge carriers), positron annihilation spectroscopy (PAS) (positron lifetime is longer in a material with voids) or photoluminescence (identification of defects by their recombination radiation) or Raman spectroscopy (structural defects, implant damage, local stresses shifting photon energy).

2.5.2 Secondary Ion Mass Spectrometry (SIMS)

In SIMS the surface to be analyzed is bombarded by ions which knock atoms from the surface. Some of these are ionized, and subsequently mass analyzed, giving their identity. SIMS is thus a surface sensitive technique, but another important SIMS application is depth profiling: an ion beam erodes the surface and layers beneath the surface become available to analysis. When the erosion rate is known, SIMS data provides information on atomic concentrations as a function of depth (Figure 2.12).

SIMS measurement is slow and expensive, but it is the accepted standard for dopant depth distribution measurement (even though we are most often interested in electrically active dopants, whereas SIMS only counts atoms). SIMS offers nanometer depth resolution and 10^6 dynamic range.

2.5.3 Rutherford Backscattering (RBS)

RBS is based on elastic recoil collisions. Helium ions (alpha particles) penetrate matter and slow down, but one ion

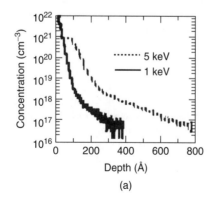
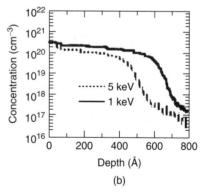

Figure 2.12 SIMS data of arsenic ion depth distribution inside silicon. Two different ion implantation energies have been used: (a) arsenic depth distribution immediately after implantation; (b) after 1050 °C, 10 s heat treatment. Reproduced from Plummer and Griffin (2001) by permission of IEEE

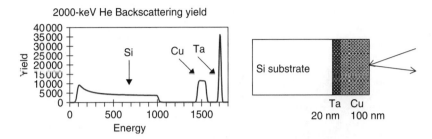

Figure 2.13 RBS spectrum of Si/Ta/Cu (20 nm/100 nm) sample. Even though the tantalum is beneath the copper, its signal is at a higher energy because tantalum is so much heavier. Courtesy Jaakko Saarilahti, VTT

in a million experiences 180° elastic recoil and bounces back toward the surface, slows down on the way, and finally emerges from the solid to reach the detector. All these steps can be handled by calculations and RBS is a quantitative method. Elastic recoil from heavy atoms is more pronounced, and RBS is ideally suited for heavy atoms like arsenic, tantalum, copper and tungsten.

Signal energy is sometimes confusing because it depends not only on the depth from where it originates, but also on the mass of the atom that caused backscattering. In Figure 2.13 a tantalum layer beneath copper has been measured by RBS. The silicon signal is weak because silicon is a light atom and it is deep beneath the copper and tantalum. Copper is the topmost layer, but because it is lighter than tantalum, its peak is lower in energy.

RBS detectivity depends on the matrix: elements lighter than the matrix are not readily detectable. Oxygen and nitrogen analyses on top of silicon wafers are therefore difficult for RBS. Mass separation between neighboring elements is poor in RBS, therefore silicon, aluminum and phosphorus cannot readily be resolved. RBS detection limits are around 10^{20} cm^{-3} (0.5%), but with heavy elements RBS can detect concentrations even down to 10^{17} cm^{-3} (0.001%).

2.5.4 Auger Electron Spectroscopy (AES)

In Auger measurement an electron beam (3–5 keV) hits the surface and an inner core electron is ejected. The vacancy left behind is filled by an electron from an outer shell, which gives off excess energy during transition. Another outer shell electron receives this energy, and escapes. The energy of the escaping electron is determined by the atomic energy levels and can be uniquely identified. This third electron is called the Auger electron. The escape depth of low-energy Auger electrons is on the order of a nanometer only, which makes AES a truly surface sensitive technique. In Figure 2.14 residues from TiW etching are analyzed by AES: etching is incomplete

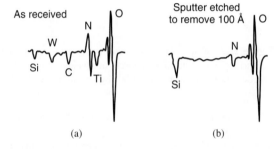

Figure 2.14 Auger analysis of silicon dioxide surface: (a) evidence of titanium and tungsten residues; (b) after sputter etching has removed 100 Å (10 nm) surface layer, the sample has been reanalyzed and found to be free of Ti and W. Reproduced from Schaffner (2000) by permission of IEEE

because both tungsten and titanium are detected on the surface (and carbon, but carbon is a very common contaminant). Etching 100 Å (10 nm) or more removes all traces of tungsten and titanium.

With the aid of sample erosion techniques (similar to SIMS), AES can be transformed into a depth profiling technique: after surface analysis, sputtering removes some material, and Auger measurement of the newly formed surface is carried out. This is continued until the desired sample depth is probed. An example of an Auger depth profile is shown in Figure 7.11.

2.5.5 Energy Dispersive X-ray spectroscopy (EDX)

When the electron energy is high enough to ionize an atom, X-rays can be generated in the recombination process. Electron beams can be focused down to 5 nm spots and devices can be probed for localized analysis. The electron beam diverges as it interacts with the matter. The scattering of electrons spreads the beam over a volume much larger than the beam spot on the surface, as

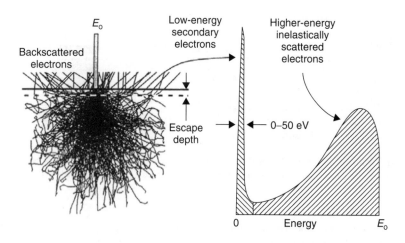

Figure 2.15 Finely focused electron beam hits the sample surface, and low-energy secondary electrons escape from the surface only, but backscattered and inelastically scattered electrons contribute to the signal deep inside the sample. Reproduced from Schaffner (2000) by permission of IEEE

shown in the Figure 2.15. Auger electrons which originate at the very surface are unaffected by this spreading, but X-rays and backscattered electrons generated deep inside the sample can escape and reach the detector.

The radius of the X-ray signal is given by

$$R_x(\mu m) = \frac{0.04 \cdot V^{1.75}}{\rho} \quad (2.10)$$

where acceleration voltage V is in kilovolts and density ρ in g/cm^3. Analysis radius R is given by

$$R = \sqrt{R_x^2 + d^2} \quad (2.11)$$

where d is beam spot diameter.

This analytical radius of EDX (also known as electron microprobe analysis, EMPA) can be orders of magnitude bigger than the electron beam spot size.

EDX can detect elemental concentrations at the 1% level. Examples of suitable analytical tasks include phosphorus determination in CVD oxide (5% wt typical) or copper concentration in aluminum film (0.5–4% Cu typical). EDX is most often connected to a SEM which is used to image the area of interest first and then subjected to elemental analysis by EDX. If the sample is made thin, on the order of 100 nm, electron scattering effects can be eliminated. This is utilized in the TEM and electron energy loss spectroscopy (EELS).

2.5.6 X-ray Photoelectron Spectroscopy (XPS)

XPS is closely related to AES in two ways: low-energy electrons are analyzed, and because their escape depth is so small, the method is surface sensitive, but XPS excitation is by X-rays. This has important ramifications for the analysis area: X-ray spots are fairly large, in the hundred micrometer range, and large areas are needed for analysis.

Primary X-rays (a few kilovolts) eject electrons from the sample. The energy of the ejected electrons is related to their binding energy, and this enables not only elemental identification but also chemical bond identification. The electron energy is slightly different depending on bonding, so for example C–O, C–F and C–C bonds can be distinguished. The other name for XPS, ESCA (Electron Spectroscopy for Chemical Analysis) emphasizes this important feature of XPS. Figure 2.16 shows a XPS analysis of platinum etching in Cl$_2$/CO. Initially the surface is of course covered by platinum, with a trace amount of carbon (from atmospheric contamination). After partial etching some chlorine is detected, indicating that chlorine is active in the surface reaction. After overetching, all traces of platinum and chlorine disappear and only silicon and oxygen from the underlying SiO$_2$ film are detected.

2.5.7 X-ray Diffraction (XRD)

Structural information such as crystal orientation, texture and grain size is important in a number of cases. The resistivity of metal film can increase by an order of magnitude upon phase change, and the initial state, amorphous and polycrystalline, affects thin-film polycrystalline silicon final grain size after annealing. Piezoelectric properties are very sensitive to crystal structure, and

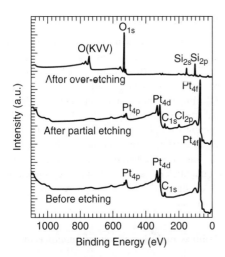

Figure 2.16 XPS analysis of platinum etching. Reproduced from Kim and Woo (1998), Copyright 1998, American Chemical Society

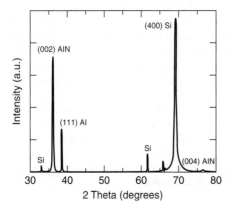

Figure 2.17 Aluminum nitride thin film on aluminum thin film on silicon wafer – crystallinity checked by XRD. Reproduced from Naik *et al*. (1999) by permission of ECS, The Electrochemical Society

XRD is used to measure crystallinity (Figure 2.17). XRD provides structural information of aluminum nitride crystal. Auger could indicate that aluminum and nitrogen are present in 1:1 ratio, but it does not indicate whether AlN crystals are present: amorphous AlN also has 1:1 elemental ratio. TEM also provides similar crystallinity information, but TEM analysis area is only tens on nanometers, whereas XRD gives an average over hundreds of micrometers. TEM requires tedious sample preparation while XRD needs none.

2.5.8 X-ray Fluorescence (XRF)

Each element gives off characteristics X-rays, as discussed above for the electron beam-excited case. But X-ray excitation can be used, too. This is a very simple yet powerful analytical technique throughout science and engineering. It is a bulk analytical method, with limited use in microfabrication. Its variant, total reflection XRF (TXRF), however, is very important. X-rays hit the wafer at a grazing angle (almost parallel to the wafer surface) and penetrate only a few atomic layers. Therefore the fluorescence signal they excite concerns surface atoms only. TXRF can measure surface impurities at the $10^{10}\,\text{cm}^{-2}$ level.

2.5.9 Other methods

Thermal desorption spectroscopy (TDS; also known as Temperature Programmed Desorption) analyses desorption products upon sample heating using a mass spectrometer. TDS can identify and quantify the desorbed specie. If the material from wafer surface can be dissolved in acid, atomic adsorption spectroscopy (AAS), and many other standard methods of chemical analysis become available.

2.6 Practical Issues with Micrometrology

Many analytical methods can produce accurate results only at the expense of great time and effort: for instance, TEM can image individual atoms but the analysis time is days (it consists mostly of tedious sample preparation but also complicated analysis). TEM analysis costs about $1000–2000 per sample if bought as a service.

Monitoring must be preferably so fast that whole wafer mapping can be performed to check for uniformity. Mapping measurement also requires that the analytical equipment can handle whole wafers. Many optical and electrical measurements are suitable for mapping, but most physical and chemical methods require the wafer to be broken into ca. square centimeter pieces.

Uniformity can be defined across the wafer (within-wafer non-uniformity, WIWNU), wafer-to-wafer (WTWNU) and lot-to-lot. The definitions for uniformity are given by

$$U = \frac{\text{max} - \text{min}}{2 \times \text{average}} \quad \text{and} \quad U = \frac{\text{max} - \text{min}}{\text{max} + \text{min}} \quad (2.12)$$

The former is applied when five measurements are taken, one at the wafer centre and four at 90° from each other at half-radius; the latter when the four points are at wafer edges.

A uniformity of 5% with 6 mm edge exclusion was long accepted as a typical process performance nowadays 3% with 3 mm exclusion might be required. Some processes are inherently better, for example thermal oxidation, ALD and photoresist spinning routinely produce better than 1% uniformity. On the other hand, CMP is notoriously non-uniform, with 10% considered as good uniformity.

2.6.1 Contact vs. non-contact measurements

Measurements can be divided into two categories: contact and non-contact (non-invasive). Both laser-induced thermal pulses and a four-point probe can be used to monitor ion implant dose, but 4PP makes physical contact with the wafer by metal (tungsten) needles, and the wafer is deemed contaminated – it is not allowed to continue to high-temperature steps. Linewidth measurement by SEM is non-contact, as opposed to the stylus profiler or AFM, which make contact with the wafer. Because full wafers are analyzed in a linewidth SEM, only top view pictures are possible, and no cross-sectional information can be obtained.

2.6.2 Blanket vs. patterned wafer analysis

Both in R&D and in production, analytical methods are bound by a number of practical constraints related to the number of data points, measurement spot size and speed of measurement. Blanket wafer measurements are simple to perform and many basic studies in film deposition, diffusion, ion implantation, polishing or bonding can be done on blanket wafers, but in many cases structured wafers are indispensable. Linewidths and spacings need to be identical to product wafers, but they must be more accessible to probing, by optical or electron beams, or by mechanical probes. Test structure size needs to be matched to design complexity: if the product chip has 1 million contact holes, how should this be extrapolated from a 1000-hole test structure?

2.6.3 Destructive vs. non-destructive analysis

Equipment and labor costs of measurements range from a few cents to a few dollars per wafer, but if the measurement is wafer destructive, its cost is at least the wafer cost, or $10–100 per sample. Many physical analysis systems require samples of about 1 cm^2, and the wafer has to be broken, for example for RBS, AES or XPS analysis. If cross-sectional information is needed, wafer breakage is also then necessary.

2.6.4 Standards and reference materials

Calibration standards (with traceability to NIST, the National Institute of Standards and Technology) and reference materials (which are supplier certified) are available for all major wafer-level measurements: namely, film thickness and step height, dimensions, electrical resistivity and particles. Reference materials are sufficient for daily work but they must be calibrated regularly against traceable standards. The standards and references are silicon wafers with dedicated test patterns for the quantities in question. A single wafer can provide a series of standards, like different resistivity windows or step heights.

2.6.5 Failure analysis and reverse engineering

Analytical methods are needed not only during fabrication, but also after wafer processing has been completed. When circuits are found to malfunction, either in testing or after field return, the causes must be identified. Hard errors, that is consistent failures, are much easier to locate and understand than soft errors, that is intermittent failures, which may take place only under certain operating conditions (e.g., above a certain temperature or frequency). As in wafer-level analysis, non-destructive methods are tried first and destructive ones only afterward.

In reverse engineering, a chip is "disassembled" step by step, and the structures, materials and functions are recorded (see Figure 28.6 for IC metallization stripped of all dielectric films). This is practiced for example for competitive intelligence or patent infringement examination. Methods like electron beam-induced current (EBIC) and voltage contrast SEM can be used to probe the electrical functions of a circuit.

2.7 Measurements Everywhere

No microfabrication process works without measurements, either in the research phase, or in manufacturing. Measurement needs change when processes evolve from novelty to routine, but there are some constants, such as the need to be able to measure ever finer details: for example, not only linewidth, but also line edge roughness; and not only film thickness, but also interface quality; and not only membrane deflection statically but at megahertz frequency.

It is not unusual to find that no analytical method is able to do the job: either the quantity involved or the analysis area is too small. Sometimes it is possible to use devices themselves as measuring instruments. Device performance degradation is attributed to minute effects which are not amenable to direct physical measurements. MOS transistors are sensitive to metal contamination at levels below analytical detection limits (in the 10^9 cm^{-3} range). Microscopic vacuum cavities are created by wafer bonding or deposition, and no pressure gauge is small enough to probe these cavities, but the mechanical quality factor Q of microfabricated mechanical resonators in the cavities is indicative of cavity pressure.

2.8 Exercises

1. The sheet resistance of typical aluminum metallization is 0.03 ohms/square. What is the aluminum thickness?
2. The resistance of copper lines 200 μm long was measured as 40 ohms. From the copper deposition process we know that the thickness is 300 nm. What is the linewidth?
3. AFM scan area is 1×1 μm, which corresponds to 512×512 pixels. What must the AFM tip radius be so that resolution is tip limited?
4. Estimate the analytical radius of an electron microprobe (EMPA).
5. Can RBS be used to measure dopant profiles?
6. If an electron beam is focused to a 15 nm spot, and at least 100 Auger events (electrons) must be collected to get a signal, what is the detection limit of an Auger microprobe?
7. SIMS raw data consists of ion counts vs. erosion time. How can these be converted to concentration vs. depth data?
8. What is the acceleration voltage of an atomic resolution TEM?

References and Related Reading

Bubert, H. and H. Jenett (eds) (2002) **Surface and Thin Film Analysis**, Wiley-VCH Verlag GmbH.

Buchanan, M. (1999) Scaling the gate dielectric: materials, integration and reliability, *IBM J. Res. Dev.*, **43**, 245.

Bunday, B.D. *et al.* (2007) Value-added metrology, *IEEE Trans. Semicond. Manuf.*, **20**, 266–277.

Diebold, A.C. (1994) Materials and failure analysis methods and systems used in the development of and manufacture of silicon integrated circuits, *J. Vac. Sci. Technol.*, **B12**, 2768.

Holmgren, O. *et al.* (2009) Analysis of vibration modes in a micromechanical square-plate resonator, *J. Micromech. Microeng.*, **19**, 015028.

Kim, J.H. and S.I. Woo (1998) Chemical dry etching of platinum using Cl_2/CO gas mixture, *Chem. Mater.*, **10**, 3576–3582.

Naik, R.S. *et al.* (1999) Low-temperature deposition of highly textured aluminum nitride by direct current magnetron sputtering for applications in thin-film resonators, *J. Electrochem. Soc.*, **146**, 691–696.

Plummer, J.D. and P.B. Griffin (2001) Material and process limits in silicon VLSI technology, *Proc. IEEE*, **89**, 240.

Runyan, W.R. and T.J. Schaffner (1998) **Semiconductor Measurements and Instrumentation**, McGraw-Hill.

Schaffner, T.J. (2000) Semiconductor characterization and analytical technology, *Proc. IEEE*, **88**, 416.

Schroder, D.K. (1998) **Semiconductor Material and Device Characterization**, 2nd edn, John Wiley & Sons, Inc.

Tiggelaar, R.M. *et al.* (2009) Stability of thin platinum films implemented in high-temperature microdevices, *Sens. Actuators*, **A152**, 39–47.

Timofeev, A.V. *et al.* (2009) Electronic refrigeration at the quantum limit, *Phys. Rev. Lett.*, **102**, 200801.

Yu, G.-Q., S.-H. Lee and J.-J. Lee (2002) Effects of thermal annealing on amorphous carbon nitride films by r.f. PECVD, *Diamond Relat. Mater.*, **11**, 1633–1637.

3

Simulation of Microfabrication Processes

Microfabrication processes consist of tens or hundreds of steps which take weeks or months to complete, and development cycles easily become very long. Simulation is one way to shorten this time. Simulation accuracy is strongly dependent on the details of the process to be simulated, and even a simple simulator can be extremely valuable if it saves enough experimentation time and effort. Simulators can provide meaningful trend data and comparisons between different process options, even though accuracy might be less than perfect. Simulators can be used to explore possibilities and narrow down options before experimental work is begun. Simulation can provide information which is not experimentally available or difficult to measure. For instance, there is no dopant profiling method with sub-10 nm resolution in both vertical and lateral directions, and therefore simulation is the de facto method for 2D dopant distribution analysis.

3.1 Simulator Types

There are two breeds of process simulators: integrated packages that can be used to simulate the whole fabrication process with many different steps in sequence; and dedicated simulators for specific process steps. Dedicated simulators are available for almost all processes, ranging from ion implantation damage production to lithography defect modeling, to crystal structure prediction of deposited films. A silicon anisotropic wet etching simulator can use crystal lattice information (bond energies and densities) to predict which atoms will be removed by the etchant. Dedicated simulators are more detailed, more accurate and more computationally intensive. A basic principles diffusion simulation would start from lattice parameters, interatomic potentials, vacancy production and annihilation rates and atom–defect interactions, and provide diffusion data as an output. Integrated simulation packages use simpler models, but offer seamless stitching of different process steps into whole processes. For instance, macroscopic phenomenological diffusion models based on Fick's equations are used. Bulk silicon process steps, that is high-temperature steps that affect the dopant distribution inside silicon, epitaxy, diffusion, implantation and oxidation, can be analyzed by solving the relevant diffusion equations.

Etching and deposition produce the topography on a wafer. This build-up of topography is difficult to simulate because it involves physics (e.g., plasma generation) and chemistry (e.g., surface reactions) at many time scales and physical sizes. Film deposition simulators depend on atom arrival angles which are not physical constants like diffusivities, but parameters sensitive to experimental conditions. Etching reactions are complex interactions between the chemical contributions (spontaneous etching, free energy considerations) and physical processes (e.g., ion bombardment-enhanced desorption). Topography process simulators are usually semiempirical: some important model parameters are extracted from experiments without fundamental physical validation.

Even though simulation is fast, building a simulator is slow and tedious. It is not possible to build simulators for all possible new materials, processes and devices because calibration data needs to be available, and it is readily available only for those materials, processes and devices that are widely studied and used. In this sense the predictive power of process simulation remains poor.

3.2 Levels of Simulation

Process simulation, device simulation and circuit simulation together are termed TCAD, for technology CAD, in contrast to the more established ECAD, or electronic simulations, which involves logic and system simulations. Process simulation deals with physical structures, namely atoms and their distributions; device simulation deals with

Process simulation
-structures
-dopant profiles
-layer thicknesses
==> *input to device simulation*

Device simulation
-electrical, mechanical, thermal, optical behavior
-current-voltage, force-displacement, potential-flow
==> *input to circuit simulation*

Circuit simulation
-output signal and noise
-rise time, speed, delays

Figure 3.1 Levels of simulators

currents and potentials in devices; and circuit simulation is used to study larger circuit blocks (Figure 3.1). The dopant concentrations produced by a process simulator are used as input for the device simulator, and device simulator results form the starting material for circuit simulation.

Circuit simulation is the most advanced of the three, and process simulation is the least developed. Device simulators for CMOS today are predictive because CMOS device physics is well understood. Of course, continuous scaling to smaller linewidths means that new phenomena must be implemented regularly into process and device simulators.

MEMS simulators must deal with a lot more varied phenomena (e.g., electrical, mechanical, thermal, magnetic, optical, etc., phenomena). MEMS fabrication simulators include silicon etching simulators, which take mask design as input and produce silicon 3D shape as output. Most device simulators, however, are FEM simulators which use drawn geometry as a starting point, instead of process simulator output.

3.3 The 1D Simulators

A 1D simulator treats matter as layers, and the simulation outputs are layer thicknesses and dopant distributions in the vertical direction (Figure 3.2). Such 1D simulation has been used since the 1970s when SUPREM from Stanford University emerged. Diffusion, ion implantation, oxidation and epitaxy are treated. Two additional, non-physical process steps are included, namely film deposition and etching, but these are just geometrical steps, like "add 500 nm of undoped oxide on silicon" or "remove top 50 nm of silicon by etching." These steps are needed for more realistic models of surfaces and interfaces, but they do not reveal anything about the deposition or etching processes.

Over the years more layers and more realistic models have been added to 1D simulators. For instance, some simulators can handle the oxidation and doping of polycrystalline silicon. Polycrystalline materials require more inputs than single crystals, for example grain size and texture, and assumptions of grain boundary diffusion vs. bulk diffusion among others. ICECREM (from Fraunhofer Institute IISB, http://www.iisb.fraunhofer.de) is an advanced 1D simulator. It can simulate the following processes:

- epitaxy
- oxidation
- diffusion
- ion implantation.

Additionally, the following steps can be used to generate geometry:

1. Deposition of undoped oxide films (protective capping layers).
2. Deposition of doped oxide films (diffusion sources).
3. Etching (removal of oxide and silicon).

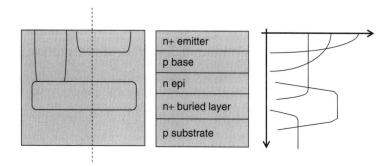

Figure 3.2 Cross-section of a npn bipolar transistor and its 1D simulation model of dopant concentrations along the cut line

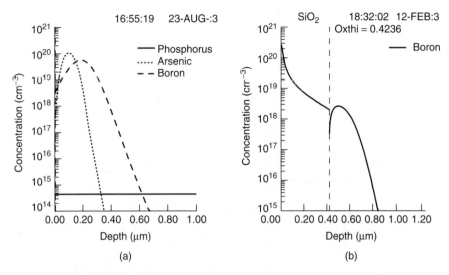

Figure 3.3 A 1D simulation (ICECREM) of (a) arsenic (50 keV energy) and boron (50 keV) ion implantation into silicon, dose 10^{15} ions/cm^2, and (b) dry oxidation of BF$_2^+$ implanted silicon (20 keV, 10^{15} ions/cm^2)

ICECREM models can account for a number of important real-life effects like high phosphorus concentration in diffusion, implantation through the oxide and oxidation-enhanced diffusion (OED). These features will be discussed in the relevant chapters on basic processes. ICECREM output consists of diffusion profiles, oxide thicknesses, sheet resistances and junction depths. Sensitivity analysis can be carried out to study both process parameter and model parameter changes.

A typical simulator input file begins with the substrate definition (crystal orientation <100> or <111>, doping type and level/resistivity). The grid is defined next: that is, the depth into silicon that needs to be simulated (e.g., down to 5 μm), division of matter into layers (e.g., 10 nm thick) and concentrations that need to be calculated (concentrations vary between 10^{21} cm^{-3}, limited by dopant solubility, and 10^{15} cm^{-3}, which is a typical substrate doping level). Process steps are then defined in sequence, followed by output commands. Model parameters can be modified by the user, but default parameters are good for initial simulations and novice users. Simulation examples in the chapters on epitaxy, oxidation, diffusion and implantation are discussed using ICECREM.

A 1D simulator output can visualize dopant depth distributions and film thicknesses, as shown in Figure 3.3. There are two important points in concentration curves: the maximum concentration and its depth, and the junction depth where substrate dopant level and diffused dopant levels match. Junction depths range from tens of nanometers to many micrometers.

3.4 The 2D Simulators

A 2D simulation is indispensable because a 1D simulation of more slices cannot predict 2D profiles. This is illustrated in Figure 3.4 for a simple 5 μm linewidth MOS transistor: the 1D simulation produces accurate doping profiles and oxide thicknesses along lines A, B and D, but it cannot produce any meaningful results for C (where implanted dopant spreads laterally under the mask) or E (where oxidation occurs under a protective nitride layer). The 1D results for A, B and D are valid for 5 μm transistors, but as the device is scaled to smaller linewidths, more and more 2D effects arise, and a 2D simulator will be needed for profiles along B and D as well.

The 2D diffusion simulators take into account oxide and polysilicon structures on top of silicon, and produce dopant profiles that extend under the gate and masking

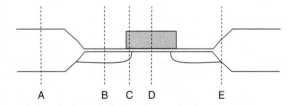

Figure 3.4 Vertical profiles of a MOS transistor: film thicknesses and dopant distributions along lines A, B and D can be simulated with a 1D simulator, but profiles along C and E require 2D simulation

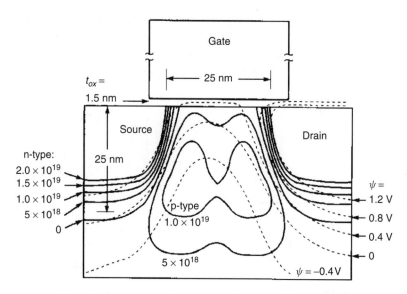

Figure 3.5 A 2D simulation: dopant concentration profiles of a CMOS transistor with a gate length of 25 nm. Reproduced from Taur *et al.* (1998) by permission of IEEE

layer, as shown in Figure 3.5, for example. The structures above the silicon surface are usually not simulated but simply drawn geometries. They are tools to add realism, not unlike the deposition and etching steps in 1D simulators.

The 2D simulators are about cross-sections of structures where 1D ones were only about layers. A 2D simulation enables topography simulation: in a 1D simulation it is not possible to study the deposition of films over patterns, neither are cross-sections relevant. Continuum simulator SAMPLE and atomistic simulator SIMBAD are compared in Figure 3.6 for metal sputtering. In both cases metal is deposited over a trench, and thickness of metal on sidewalls and trench bottom is predicted. A continuum simulator which predicts thicknesses from atom arrival angle distribution and surface mobility considerations is useful but the result is geometric only. An atomistic simulator can show crystal structure, and reveal differences in film quality on horizontal and vertical walls.

2D simulation is computationally intensive, and 2D simulators usually have a 1D simulation tool imbedded in them, for quick and easy initial 1D tests. Savings in computational time can be orders of magnitude. A grid, or simulation mesh, in a 1D simulator is regular and easy to generate, but in 2D simulators mesh generation is much more difficult. In order to reduce computational time, a dense grid is used where abrupt changes are expected, and a sparse grid where gradients are not steep. Instead of a rectangular grid, triangular grids are often employed.

3.5 The 3D Simulators

When scaling to smaller and smaller dimensions continues, 3D simulation becomes mandatory. Narrow but long transistors can be simulated by a 2D simulator, but a narrow and short transistor with similar dimensions in both the x- and y-directions really needs 3D treatment. Again, complexity and time of simulation increase drastically in the 2D case. If a layer 1 µm deep is simulated in a 1D simulator with 10 nm grid spacing, 100 layers need to be calculated. A similar grid size in 2D simulation requires 100×100 squares (10^4), and in 3D simulation it equals 10^6 cubes. Roughly speaking, if 1D simulation takes seconds, 2D takes minutes and 3D hours.

However, a 10 nm grid is no good for 3D simulation because this simulation is used especially for 100 nm devices and the like, and perhaps a 1 nm grid is used. But the question is not just computational: additional physical models need to be developed, because more and more atomistic models must be used, and continuum approximation fails due to the atomic nature of matter. In order to take advantage of 3D process simulation, 3D device simulators must be used, just like 2D process simulators feed into 2D device simulators. Advanced device simulators must similarly account for the fact that electric current is not a continuous variable but a stream of charged packets of 1.6×10^{-19} C.

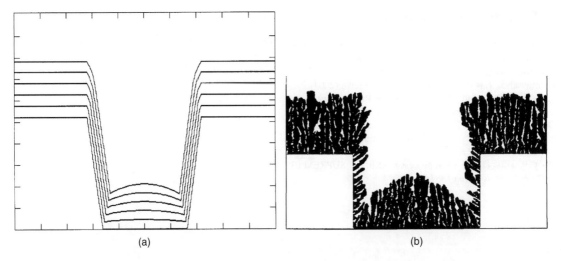

Figure 3.6 Continuum and atomistic metal step coverage simulation: (a) SAMPLE 2D simulation of metal deposition 0.5 μm thick into a trench 1 μm wide and 1 μm deep; only film thickness is simulated; (b) SIMBAD: sputtered tungsten into a trench with prediction of columnar grain structure. Reproduced from Dew *et al.* (1991) by permission of AIP

3.6 Other Simulation Needs in Microfabrication

Simulation needs extend from the atomic scale to reactor scale. At a meter scale, simulation is needed to predict gas flows and temperature distributions inside a reactor; at the centimeter level it can be used to study uniformity across the wafer; at the micrometer scale, simulation is needed to predict doping profiles and step coverage; and atomic-level simulation is needed to understand the details of film growth and diffusion. For thin-film deposition such a simulator would produce a relation between process parameters and film properties. At the present time, such multiscale simulation remains a faraway goal.

Optical lithography simulation is a self-contained regime in process simulation. Its main output is a resist profile with optics, resist photochemistry and development as the main modules. It will be discussed in Chapter 10. Tool simulation will be briefly discussed in Chapter 31 and deposition simulation in Chapter 7.

3.7 Exercises

These exercises require a simulator, for example ICECREM or SUPREM, but any 1D simulator will do. ICECREM can be obtained free of charge from: icecrem@iisb.fraunhofer.de.

1. How much difference is there between the oxidation rates of boron-, phosphorus- and arsenic-doped wafers when all have identical doping levels?
2. Use a 1D simulator to find the step height resulting from oxidation of phosphorus-doped silicon depicted below:

   Phosphorus implant   Thermal oxidation

3. How does thermal oxide thickness on a phosphorus-doped wafer change with dopant concentration?
4. What energy must phosphorus ions have to pass through 200 nm oxide?
5. Compare your simulator to other simulators for ion implantation of arsenic into silicon:

E (keV)	Dose (cm^{-2})	Simulator	Range (Å)	Peak concentration (cm^{-3})
40	1.4×10^{13}	TRIM	332	6.0×10^{17}
40	1.4×10^{13}	PREDICT	268	3.8×10^{18}
40	1.4×10^{13}	CUSTOM	270	4.6×10^{18}
90	7.2×10^{14}	TRIM	636	8.6×10^{18}
90	7.2×10^{14}	PREDICT	603	9.9×10^{19}
90	7.2×10^{14}	CUSTOM	530	1.2×10^{20}

6. Calculate the oxide thickness for 10, 100, 1000 and 10 000 minutes of oxidation at 1100 °C.

References and Related Reading

Bechtold, T., E.B. Rudnyi and J.G. Korvink (2005) Dynamic electro-thermal simulation of microsystems – a review, *J. Micromech. Microeng.*, **15**, R17–R31.

Dabrowski, J., E.R. Weber and J. Dabrowski (eds) (2004) **Predictive Simulation of Semiconductor Processing: Status and Challenges**, Springer.

Dew, S.K. *et al.* (1991) Modeling bias sputter planarization of metal films using ballistic deposition simulation, *J. Vac. Sci. Technol.*, **A9**, 519–523.

Ho, C.P. *et al.* (1983) VLSI process modeling – SUPREM III, *IEEE Trans. Electron Devices*, **30**, 1438.

Law, M. (2002) Process modeling for future technologies, *IBM J. Res. Dev.*, **46**, 339–346.

Lorentz, J. *et al.* (1996) Three-dimensional process simulation, *Microelectron. Eng.*, **34**, 85.

Taur, Y. *et al.* (1998) 25 nm CMOS design considerations, International Electron Devices Meeting, p. 789.

4

Silicon

The first transistor was made of polycrystalline germanium in 1947. Electron mobility in germanium is higher than in silicon, and germanium was readily available. However, silicon, with its larger band gap, was favoured because of smaller leakage currents. Initially there was no consensus whether single crystalline or polycrystalline material was better, but the rapid development of single crystal silicon growth in the 1950's soon dominated the market. The real breakthrough came when the beneficial role of silicon dioxide was recognized: it provided passivation of semiconductor surfaces, and it resulted in improved transistor reliability. When it was further noticed that the SiO_2 layer could act as a diffusion mask and as isolation for integrated metallization, the way was open for the invention of the integrated circuit.

Steady increases in wafer size from half an inch have continued up to this day, with wafers of 300 mm diameter now in production, and the first samples of 450 mm diameter being readied in 2009. For other substrates smaller sizes are still widely used and when new materials like silicon carbide (SiC) or gallium nitride (GaN) are introduced, crystal growth and wafer yield are so low that only small ingots and small wafers make sense.

Silicon is the basis of microelectronics, with over 90% of the semiconductor market. MEMS are largely based on silicon wafers, and silicon is used everywhere in microtechnology because of its availability and excellent properties, even though sometimes only a fraction of those properties are utilized. Over 80% of solar cells are made of crystalline silicon today, but thin-film cells are rapidly gaining popularity. Flat-panel displays are made on glass panels but the transistors controlling the pixels in active matrix LCDs are made of polycrystalline silicon.

Note on Terminology

Single crystal silicon (SCS for short) is known by many names. Some people prefer the term monocrystalline silicon, though the abbreviation mc-Si refers to multicrystalline silicon! As short forms, c-Si is used for crystalline silicon and a-Si for amorphous silicon, while polycrystalline silicon is known simply as poly. In the solar cell industry crystalline silicon is sometimes called X-Si.

4.1 Silicon Material Properties

Silicon material properties are an excellent compromise between performance and stability. An energy gap of 1.12 eV makes silicon devices less prone to thermal noise than germanium devices with a 0.67 eV gap. Silicon is transparent in the infrared (above 1.1 μm wavelength) which means that it can used as an optical material at 1.55 μm telecom wavelength applications.

Silicon source gases can be purified to extremely high degrees of purity. Silicon resistivity ranges from 10 000 Ohm-cm (dopant concentration 10^{12} cm^{-3}) to 0.0001 Ohm-cm (dopant concentrations and 10^{21} cm^{-3}), Figure 4.1. There are slight differences between different dopants but the limits are the same: solid solubility sets the upper dopant concentration limit and materials purity (of silicon itself and of materials used in the fabrication process) sets the lower limit.

Silicon is strong: its Young's modulus (elastic modulus) can be as high as 190 GPa for the <111> orientation and less for other orientations, but it is at least 132 GPa. Young's modulus (E) is the ratio of stress (σ) to strain (ε, elongation):

$$E = \sigma/\varepsilon \qquad (4.1)$$

It is important in assessing for example thermal stresses that the wafers experience during processing.

The excellent mechanical properties of silicon have been utilized since the 1960s in micromechanical pressure and force sensors which rely on bending beams and diaphragms. Piezoresistive detection depends on doped

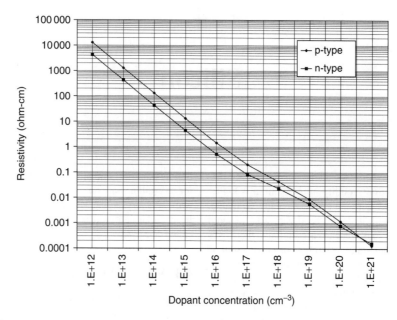

Figure 4.1 Silicon resistivity can be varied over eight orders of magnitude by doping

resistors, and capacitive detection relies on the ability to micromachine shallow air gaps on the order of 1 µm. Both are standard processes in silicon microfabrication.

Silicon is as strong as steel but this fact is disguised by two factors: most of us do not have experience of steel plates 0.5 mm thick; and, since silicon is brittle, the breakage pattern is therefore different from the ductile fracture of multicrystalline steel. Silicon is almost ideally elastic (obeying Hooke's law) up to the yield point, and after that a catastrophic failure takes place. Most metals and oxides obey Hooke's law initially, but then deform plastically before fracture. The yield strength of silicon is 7 GPa at room temperature, different steel varieties have yield strengths of 2–4 GPa, while the yield strength of aluminum is only 0.17 GPa. Fracture strain (elongation before breakage) for single crystal silicon is 4%, an exceptionally large value. More silicon properties are listed in Table 4.1.

4.2 Silicon Crystal Growth

4.2.1 Purification of silicon

Silicon wafer manufacture is a multistep process which begins with sand purification and ends up with final polishing and defect inspection. Silica sand, SiO_2, is reduced by carbon, yielding 98% pure silicon according to the reaction

$$SiO_2 + 2\,C \Rightarrow Si + 2\,CO\,(g) \qquad (4.2)$$

This material is known as metallurgical grade silicon (MGS). MGS is converted to gaseous trichlorosilane $SiHCl_3$ (boiling point 31.8 °C) according to the reaction

$$Si + 3\,HCl \Rightarrow SiHCl_3 + H_2\,(g) \qquad (4.3)$$

The main impurities in MGS (Fe, B, P) react to form $FeCl_3$, BCl_3 and PCl_3/PCl_5. Trichlorosilane gas is purified by distillation during which $FeCl_3$ and PCl_3/PCl_5 are removed as high-boiling-point contaminations and BCl_3 as low-boiling-point contamination. Trichlorosilane is converted back to solid silicon by the decomposition of $SiHCl_3$ on hot silicon rods by the reaction

$$2\,SiHCl_3 + 2\,H_2\,(g) \Rightarrow 2\,Si\,(s) + 6\,HCl\,(g) \qquad (4.4)$$

This material is of extremely high purity and is known as electronic grade silicon (EGS). It is a polycrystalline material which will be used as a source material in single crystal growth.

4.2.2 Czochralski (CZ) crystal growth

In CZ growth a silica crucible (SiO_2) is filled with undoped electronic grade polysilicon. Dopant is introduced by adding pieces of doped silicon (for low doping concentration) or elemental dopants P, B, Sb or As (for high doping concentration). The crucible is heated in a vacuum

Table 4.1 Properties of silicon at 300 K

Structural and mechanical	
Atomic weight	28.09
Atoms, total (cm^{-3})	4.995×10^{22}
Crystal structure	diamond (FCC)
Lattice constant (Å)	5.43
Density (g/cm^3)	2.33
Density of surface atoms (cm^{-2})	(100) 6.78×10^{14}
	(110) 9.59×10^{14}
	(111) 7.83×10^{14}
Young's modulus (GPa)	190 (111) crystal orientation
Yield strength (GPa)	7
Fracture strain	4%
Poisson ratio, ν	0.27
Knoop hardness (kg/mm^2)	850
Electrical	
Energy gap (eV)	1.12
Intrinsic carrier concentration (cm^{-3})	1.38×10^{10}
Intrinsic resistivity (ohm-cm)	2.3×10^5
Dielectric constant	11.8
Intrinsic Debye length (nm)	24
Mobility (drift) (cm^2/V-s)	1500 (electrons)
	475 (holes)
Temperature coeff. of resistivity (K^{-1})	0.0017
Thermal	
Coefficient of thermal expansion (°C^{-1})	2.6×10^{-6}
Melting point (°C)	1421
Specific heat (J/kg-K)	700
Thermal conductivity (W/m-K)	150
Thermal diffusivity	0.8 cm^2/s
Optical		
Index of refraction	3.42	$\lambda = 632$ nm
	3.48	$\lambda = 1550$ nm
Energy gap wavelength	1.1 µm	(transparent at larger wavelengths)
Absorption	$>10^6$ cm^{-1}	$\lambda = 200$–360 nm
	10^5 cm^{-1}	$\lambda = 420$ nm
	10^4 cm^{-1}	$\lambda = 550$ nm
	10^3 cm^{-1}	$\lambda = 800$ nm
	<0.01 cm^{-1}	$\lambda = 1550$ nm

to about 1420 °C to melt the silicon. A single crystalline seed of known crystal orientation is dipped into the silicon melt. Silicon solidifies into a crystal structure determined by the seed crystal. Figure 4.2 describes the crystal puller. A thin neck is quickly drawn to suppress defects which develop because of the high-temperature difference between the seed and the melt, and then the pulling rate is lowered. Both the ingot and the crucible are rotated (in opposite directions), ingot rotation at about 20 rpm, crucible rotation at about 10 rpm. Molten silicon is very fluid, and the rotations induce flows in the crucible; temperature differences are also a source of convection currents. Much engineering has gone into understanding the behavior of molten silicon.

Ingot diameter is determined by ingot pull rate. The pulling rate is limited by heat conduction away from the crystallization interface, and therefore large-diameter ingots have lower pulling rates. While an ingot of 100 mm diameter can be pulled at 1.4 mm/min, the 200 mm ingot pull rate is 0.8 mm/min. Typical pulling time is 30 hours

38 Introduction to Microfabrication

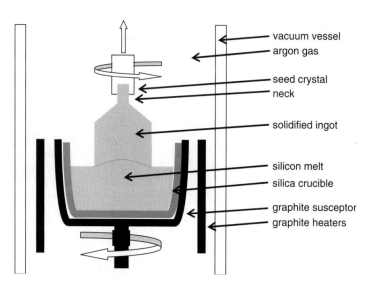

Figure 4.2 Czochralski crystal pulling: silicon (melting point 1421 °C) solidifies as it is pulled up. Pulling speed (mm/min), ingot rotation speed (20 rpm) and crucible counter-rotation speed (10 rpm) together determine ingot diameter

(corresponding to ingot lengths of 2 m), not including heating and cooling, which add another 30 hours to the process, for 200 mm ingots. A finished ingot is seen in Figure 4.3.

Ingot length is determined by the yield strength of the silicon neck and crucible size. The thin neck is not perfect material because it has defects arising from thermal shock, and torsional forces act on it during rotation (Figure 4.3). Silicon yield strength is significantly lower at high temperatures, but 300 mm ingots can weigh up to 300 kg.

Figure 4.3 Ingot: narrow neck, shoulder and body

Not all polysilicon can be utilized: about 10% of original polysilicon remains in the crucible. Crucibles cannot be reused and are extremely expensive disposable objects.

EGS is extremely pure, for instance boron, phosphorus and iron levels can be as low as 0.01–0.02 ppb. However, the crucible is a source of impurities, and for boron, sodium and aluminum it is the crucible and not EGS that determines ingot purity. If synthetic silica is used for crucibles, much higher purity CZ ingots can be pulled.

There is inevitable contamination of the growing silicon crystal from the materials that are essential to the growth set-up: the silica crucible is slightly dissolved during the crystal growth process, and therefore oxygen is always present in CZ silicon. Oxygen concentrations are 5–20 ppma. Carbon from graphite is present at concentrations below 1 ppm.

4.2.3 Dopant incorporation

Impurities are incorporated from the melt into the ingot, but different dopants have widely different segregation coefficients. The segregation coefficient is defined as

$$k_0 = \frac{\text{concentration}_{\text{solid}}}{\text{concentration}_{\text{liquid}}} \quad (4.5)$$

All dopants and metallic impurities are enriched in the melt, with oxygen as perhaps the only material which is incorporated preferentially into the silicon solid phase.

Because dopant segregation coefficients are less than unity (Table 4.2), excess dopant is needed in the melt,

Table 4.2 Segregation of dopants and impurities at silicon melt/solid interface

Dopants		Others	
Boron	$k_0 = 0.8$	Iron	$k_0 = 6.4 \times 10^{-6}$
Phosphorus	$k_0 = 0.35$	Copper	$k_0 = 8 \times 10^{-4}$
Arsenic	$k_0 = 0.3$	Nickel	$k_0 = 1.3 \times 10^{-4}$
Antimony	$k_0 = 0.023$	Gold	$k_0 = 2.25 \times 10^{-5}$
Gallium	$k_0 = 0.0072$	Oxygen	$k_0 = 1.25$
Aluminum	$k_0 = 0.002$	Carbon	$k_0 = 0.07$

compared to the final ingot. This can be calculated from k_0 values easily using Equation 4.6. As pulling advances, the melt volume decreases, dopant concentration in the melt increases and therefore dopant concentration C_s in the ingot increases along its length:

$$C_s = k_0 C_0 (1 - X)^{k_0 - 1} \qquad (4.6)$$

where C_0 is the initial dopant concentration in the melt, X is the fraction solidified and k_0 is the segregation coefficient.

The concentration of oxygen in the ingot decreases as pulling advances because its segregation coefficient is above unity. There is another reason, too: as the melt is consumed, there is less contact between the silica crucible and melt and less dissolution of oxygen. Ingot resistivity and oxygen concentration will thus vary, and if tight specifications are set, not all of the ingot will be within the specifications.

Because the crystal is rotated during growth, there are also radial non-uniformities in the ingots. One source for these is simply cooling rate: the edges cool faster than the centre (because of radiation at 1400 °C) and therefore any process that depends on cooling rate will be affected. For instance, vacancies and many other defects show a radial distribution.

Because molten silicon is electrically conductive, magnetic fields can be used to control melt behavior. Magnetic fields reduce local temperature fluctuations and thermal convection currents, which leads to more stable melt and consequently to more uniform growth. Magnetic Czochralski (MCZ) growth enables better control of oxygen levels in the crystal, probably because it suppresses flows that distribute oxygen from the crucible. At least a more uniform melt enables other process parameters, like argon gas flow, to be varied over a larger range.

4.2.4 Float zone crystal growth

If high-resistivity silicon is needed, float zone (FZ) crystal growth is used. In the FZ method a polysilicon ingot is placed on top of a single crystal seed. The polycrystalline ingot is melted by an external RF coil, and the solidifying silicon copies the crystal structure of the seed. The coil is moved up, the melted zone advances upward, and the single crystal solidifies along the ingot.

The highest FZ silicon resistivities are on the order of 20 000 ohm-cm, compared to 1000 ohm-cm for CZ. Because there is no silica crucible, there is much less oxygen (10^{16} cm^{-3}, compared to 10^{18} cm^{-3} for CZ) and metal contamination from the crucible is also eliminated. FZ wafers, however, are mechanically weaker than CZ wafers because oxygen mechanically strengthens silicon. FZ wafers are available only in smaller diameters, 150 mm maximum, with 200 mm FZ demonstrated but not used in device manufacturing. When doped FZ silicon is made, dopants are introduced by flushing the melt zone with gaseous dopants like phosphine (PH$_3$) or diborane (B$_2$H$_6$). High-resistivity FZ is often doped via neutron transmutation doping (NTD) according to the reaction

$$n + {}^{28}\text{Si} \Rightarrow {}^{29}\text{Si} \Rightarrow {}^{29}\text{P} + e^- \qquad (4.7)$$

A silicon nucleus captures a neutron, and the newly formed nucleus decays by β-decay. This doping method explains why high-resistivity silicon (5–20 kohm-cm) is available only in n-type.

4.3 Silicon Crystal Structure

Silicon has a cubic diamond lattice structure. The unit cell can be thought of as two interleaved face-centered cubic (FCC) lattices with their origins at (0,0,0) and (1/4, 1/4, 1/4). The distance between two atoms is $(\sqrt{3/4})a$, and radius $(\sqrt{3/8})a$, where a is the unit cell edge length, 5.430 95 Å. As shown in Figure 4.4 there are 18 atoms to

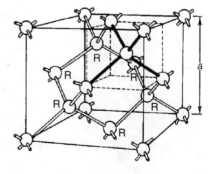

Figure 4.4 Silicon lattice: the unit cell consists of eight atoms. Reproduced from Jenkins (1995) by permission of Prentice Hall

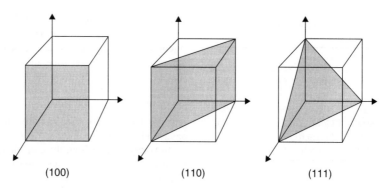

Figure 4.5 Some important silicon crystal planes with their Miller indices

be considered: eight at vertices (they are shared between eight unit cells and therefore contribute one atom to the unit cell); six face atoms are shared between two neighboring unit cells and contribute three atoms; and there are four atoms fully inside the unit cell. The volume fraction of the space filled by silicon atoms is 34%, very low compared to hexagonal close packing, which fills 74% of the space. This open structure leads to fast diffusion of dopants (and impurities) in silicon.

Miller indices define the planes of a crystal. The plane that defines the faces of the cube (see (Figure 4.5)) intersects axes 1, 2, 3 at $(1, \infty, \infty)$, respectively. The Miller index of a plane is given by the reciprocal of these intersects, namely (1,0,0). The planes tying edges are designated (1,1,0) and diagonal planes (1,1,1). The set of six equivalent planes (the six faces of the cube) together are designated {100}. There are twelve (1,1,0) and eight (1,1,1) planes. The unit vectors can also be negative: they are designated with a bar, e.g. $(1,\bar{1},1)$.

Fourfold symmetry of (100) and sixfold symmetry of (110) and (111) (Figure 4.6) will become apparent in anisotropic wet etching of silicon (to be discussed in Chapter 20).

The angles between the planes can be calculated from the scalar product of the normal vectors:

$$\mathbf{a} \cdot \mathbf{b} = |\mathbf{a}| \, |\mathbf{b}| \cos(\mathbf{a}, \mathbf{b}) \quad (4.8)$$

Visual examination shows that (100) and (110) planes meet at 45° and all the other angles can be calculated easily when negative unit vectors are accounted for: $\bar{1}10$ is $(-1,1,0)$. The angle between (111) and (100) planes is calculated from $1 = \sqrt{3} \cos \alpha$, giving $\alpha = 54.7°$.

In order to become familiar with silicon crystal structure, the paper fold model (Figure 4.7) can be handy. Copying the model on an overhead transparency and gluing it together will result in a 26-gon which visualizes the crystal planes nicely. It will be indispensable when

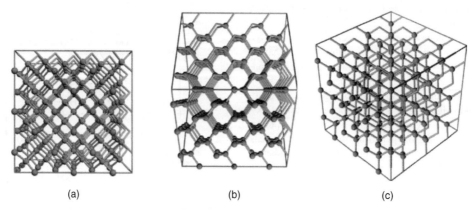

Figure 4.6 Silicon crystal viewed from different angles: (a) face view (100); (b) edge view (110); (c) vertex view (111). Courtesy Ville Voipio, Aalto University

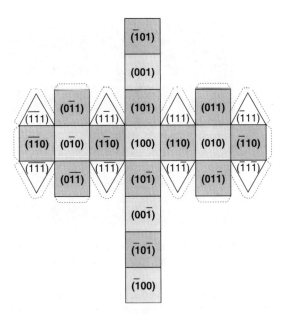 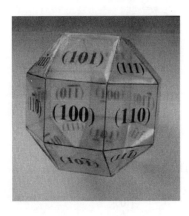

Figure 4.7 Fold-up paper model of silicon crystal planes. Courtesy Hiroshi Toshiyoshi

crystal plane-dependent etching of silicon is discussed in Chapter 20.

Two crystal orientations are widely used in microfabrication: <100> and <111>. The former is the main material for CMOS, MEMS and solar cells; the latter for bipolar transistors, power semiconductor devices and radiation detectors which rely on epitaxial deposition. Wafers for special applications can also be cut to other index planes, like (311) and (511).

4.4 Silicon Wafering Process

Silicon ingots are transformed into wafers by a multistep process which includes mechanical, thermal and chemical treatments, and many cleaning and inspections steps. Some 200 million silicon wafers are produced annually. The main steps in the wafering process are listed in Table 4.3.

The silicon crystal orientation is determined by the seed crystal. After the ingot has cooled down, it is ground to exact diameter and cut into 50 cm stocks, which are measured for crystal orientation by X-ray diffraction. A flat (Figure 1.20) or a notch is then ground into the ingot to establish orientation. The flat or notch of a <100> wafer is oriented along the [110] direction.

The ingot is then sawn into slices. These slices are for example 600 μm thick when the final target thickness of the wafers is 400 μm. The wire saw is made of kilometers-long continuous wire (diameter 80–250 μm) which uses mechanical abrasion and a chemically active slurry to cut through the ingot. In practice multiwire saws are used to increase throughput (Figure 4.8). Thousands of wafers can be cut in a single process 12 hours long.

The surface of a <100> wafer is a (100) plane with [100] surface normal vector, usually cut as precisely as practical (Figure 4.9). Sometimes wafers are intentionally miscut, for example by 3°. The silicon surface resulting from such a miscut is shown in Figure 4.10 (top). In a perfectly cut wafer an atomically flat surface would result, but in a miscut wafer atomic terraces are formed. The terrace length depends on the miscut angle. If the wafer is additionally off-oriented (the flat is not exactly along [110]), kinks will be created, as shown in Figure 4.10

Table 4.3 Silicon wafering process

- Ingot diameter grinding
- Ingot crystal orientation by XRD
- Flat grinding
- Sawing ingot into wafers
- Lapping
- Edge contouring
- Laser marking
- Etching
- Donor annealing
- Final polishing
- Inspections

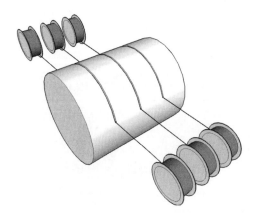

Figure 4.8 Multiwire saw. Courtesy Jorma Koskinen

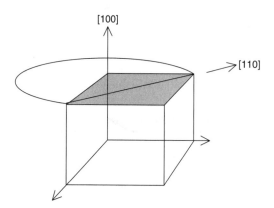

Figure 4.9 A <100> silicon wafer is cut so that one of the (100) planes defines the wafer surface; the vector normal to the surface is in the direction [100] and the flat is along the direction [110]

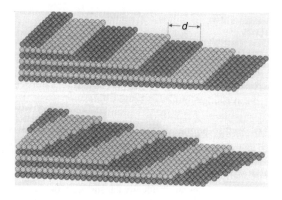

Figure 4.10 Miscut wafer surface in atomistic view (top); miscut and off-orientation wafer with kinks along terrace edges (bottom). Reproduced from Butt *et al.* (2006), by permission of Wiley-VCH GmbH

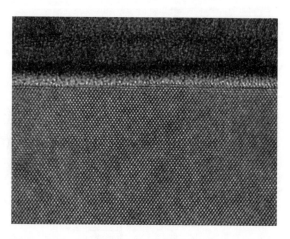

Figure 4.11 HRTEM micrograph of silicon lattice. Courtesy Young-Chung Wang, FEI Company

(bottom). These will act as preferential sites for deposited atoms to attach to, as will be discussed in Chapters 6 and 7. A high-resolution TEM micrograph (Figure 4.11) of a silicon lattice reveals the atomic planes and terraces, and nicely shows the difference between single crystalline silicon and the amorphous oxide on top of it.

Flats and notches are used by automatic wafer handlers to orient wafers inside equipment. In lithography flats enable patterns to be aligned to crystal planes. This latter aspect is especially important in micromechanics where crystal plane-dependent anisotropic etching is a major technique (Chapter 20).

The next step is lapping: waviness and taper from sawing are removed by lapping. In this step the wafers are rotating between two massive steel plates in an alumina slurry. Lapping ensures parallelism of wafer surfaces. The surface roughness is about 0.1–0.3 μm after lapping. The edges of the wafers are then rounded (Figure 4.12a) in order to prevent chipping of silicon during wafer handling and to eliminate water marks during drying steps. Wafer breakage often starts from a crack at the edge and, because silicon is brittle, the crack propagates through the whole wafer. Sometimes the edges are polished in order to improve mechanical strength.

Wafers are marked by laser scribing. Subsequent cleaning steps remove the silicon dust generated by marking. Alphanumeric or bar code marking enable wafer identity tracking during processing, and back tracing to the wafering process and to the ingot, ensuring total traceability.

Etching is then used to remove lapping damage: both alkaline (KOH) and acidic (HF–HNO$_3$) etches can be used. Roughness is reduced somewhat by acid etching, but not by alkaline etching. The SEM micrograph in Figure 4.12b shows the back of a wafer after alkaline etching.

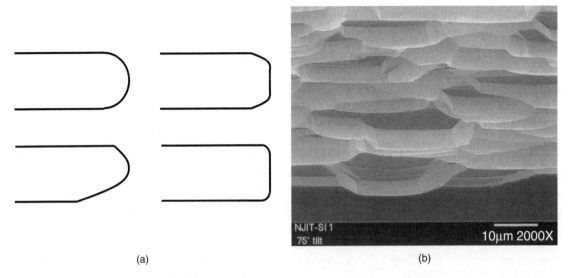

(a) (b)

Figure 4.12 (a) Different edge roundings (b) Back of an alkaline etched (non-polished) silicon wafer in SEM. Reproduced from Abedrabbo *et al.* (1998), copyright © 1999 by permission of Elsevier Science Ltd

Thermal donors are charged interstitial oxygen complexes which behave like n-type dopants. An annealing step at 600–800 °C destroys thermal donors and stabilizes wafer resistivity.

Final polishing with 10 nm silica slurry in alkaline solution removes about 20 μm of silicon and results in a root mean square (RMS) surface roughness of 0.1–0.2 nm. The formulas for calculating roughness are given in Equations 4.9a–c below. Surface height z is measured over a sampling distance L, and RMS roughness is calculated in Equation 4.9c using mean height as a reference plane.

$$\sigma^2 = \frac{1}{L}\int_0^L (z-m)^2\, dx \quad \text{standard deviation} \quad (4.9a)$$

$$m = \frac{1}{L}\int_0^L z\, dx \quad \text{mean height} \quad (4.9b)$$

$$\sigma = \sqrt{\frac{1}{L}\int_0^L z^2\, dx} \quad \text{RMS roughness} \quad (4.9c)$$

Inspection and cleaning steps constitute a major fraction of all wafering steps. Wafers are measured for mechanical and electrical properties. Contactless measurements, for example capacitance, optical and eddy current methods, are preferred because contact methods introduce contamination and damage. Wafers are specified for particle cleanliness. Laser light scattering can be used to measure particle size distributions down to 60 nm, but even the unaided eye can detect particles larger than about 0.3 μm because of their scattering under intense light (e.g., from a slide projector).

Wafers are specified for a number of electrical, mechanical, contamination and other properties as agreed between the wafer supplier and customer. The specifications in Table 4.4 are an example of fairly standard 100 mm wafers. Wafer specifications for ICs and MEMS are discussed in more detail in Chapters 26 and 30, respectively.

Thickness refers to wafer centre point thickness only, and other numbers are needed to account for thickness

Table 4.4 Basic specifications for 100 mm wafers for integrated circuits and microsystems

	IC	MEMS
Growth method	CZ	CZ
Type/dopant	P/boron	P/boron
Orientation	100	100
Off-orientation	0.0 ± 1.0°	0.0 ± 0.2°
Resistivity	16–24 ohm-cm	1–10 ohm-cm
Diameter	100.0 ± 0.5 mm	100.0 ± 0.5 mm
Thickness	525 ± 25 μm	380 ± 10 μm
Front side	Polished	Polished
Back side	Etched	Polished
Flat orientation	<110> ± 1°	<110> ± 0.2°
Oxygen level	13–16 ppma	6–10 ppma
Particles	<20 at 0.3 μm	20 at 0.3 μm

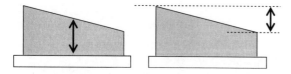

Figure 4.13 Thickness and total thickness variation (TTV). Wafer flattened to chuck; that is, back-side reference

Table 4.5 Resistivity vs. dopant concentration

Dopant level	Short	Concentration (cm^{-3})	Resistivity (ohm-cm)
Lightly doped	n−, p−	10^{13}–10^{15}	>10
Moderately doped	n, p	10^{15}–10^{18}	10–0.1
Highly doped	n+, p+	10^{18}–10^{20}	0.1–0.001
Very highly doped	n++, p++	>10^{20}	0.001

variation and geometric distortions (Figure 4.13). Total thickness variation, TTV, is defined as the difference between the maximum and minimum values of thickness encountered in the wafer. Wafers come in standard sizes and thicknesses, for example 100 mm and 525 μm, or 200 mm and 725 μm. In IC fabrication or many thin-film devices, wafer thickness is not an issue, but in bulk MEMS applications through-wafer etching is standard, and it depends critically on wafer thickness control.

The resistivity of silicon can be varied over many orders of magnitude, but typical wafers come with dopant concentrations of 10^{15} to 10^{18} cm^{-3}, corresponding to resistivities of 0.1–10 ohm-cm. Resistivity ranges and shorthand designations for them are listed in Table 4.5.

EGS purification, CZ crystal growth and the multistep wafering process produce high-quality silicon wafers, but the process is energy intensive and expensive. Silicon for solar cells needs to be considerably cheaper. That topic will be discussed in Chapter 37.

4.5 Defects and Non-Idealities in Silicon Crystals

Even though silicon wafer fabrication results in wafers with extremely well-defined properties, some defects are bound to be found. These defects can be classified according to their origin into three categories:

1. Grown-in defects from crystal pulling.
2. Defects resulting from wafering.
3. Process-induced defects.

The first two classes are the responsibility of the wafer manufacturer and the third class requires wafer and device manufacturer co-operation because process-induced defects depend on the starting material. For instance, heating and cooling rates, and maximum temperatures experienced by the wafers, determine how the defects in the wafers will behave, or how new defects are generated.

Defects are further classified into point defects (e.g., vacancies, empty lattice sites), line defects (e.g., stacking faults, extra rows of atoms in an otherwise perfect crystal), area defects (e.g., dislocation loops) and bulk defects (e.g., precipitates). The main types of defects are shown in Figure 4.14.

There are many sources for vacancies: some result from imperfections of crystal growth, and some are created during wafer processing; for example, ion implantation doping of silicon involves high-energy ion collisions in silicon, and ions displace silicon atoms from their lattice sites, creating vacancies. Some vacancies are always present because of statistics: above absolute zero, lattice vibrations displace atoms and create vacancies and interstitials. This becomes significant for silicon around 900 °C, which is the temperature where diffusion becomes fast enough to be technologically relevant.

Haze is defined as light scattering from surface defects, for example scratches, surface roughness or crystal

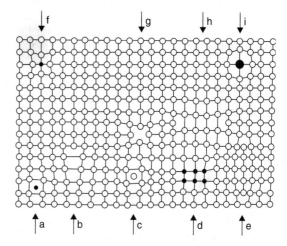

Figure 4.14 Defects in single crystalline material: a, foreign interstitial; b, dislocation; c, self-interstitial; d, precipitate; e, stacking fault (external); f, foreign substitutional; g, vacancy; h, stacking fault (internal); i, foreign substitutional. Reproduced from Green (1995) by permission of University of New South Wales

Table 4.6 Sources of non-idealities in silicon wafers

EGS polysilicon	Dopants (B, P) and other impurities (C, metals)
Czochralski growth	Impurities from quartz
	Oxygen from quartz
	Carbon from graphite and SiC
	Vacancies and interstitials
	Precipitates
	Dislocations
Wafering process	Contamination from tools
	Mechanical distortions
Wafer processing	Contamination
	Crystallinity defects
	Precipitation
	Mechanical distortions
	Dislocations

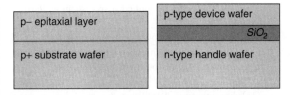

Figure 4.15 Epitaxial wafer with p-epitaxial layer on p+ substrate; and silicon-on-insulator (SOI) wafer with a p-type device layer and n-type handle wafer, separated by an oxide layer

defects. Haze measurement is by done by scatterometry, where the whole wafer is scanned, in contrast to roughness measurement, which is local area only, for instance a $5 \times 5\,\mu m$ area by a AFM.

Table 4.6 lists some non-idealities found in silicon wafers and their most common sources. Some of these can be avoided or limited to acceptable levels by careful choice of processing conditions, including wafer cleaning, temperature ramp rates and materials that make contact with the wafers, for instance wafer boats in furnaces. Defects will be discussed in more detail in Chapter 22.

4.6 Advanced Wafers

ICs are made in the thin surface layer of a silicon wafer, while many MEMS devices extend throughout the wafer. Lithography is carried out also on the back side of the MEMS wafer and double side polished (DSP) wafers are then needed. While single side polished (SSP) wafers exhibit for example $0.5\,\mu m$ RMS roughness on the back side, that of the DSP wafer is nearly as perfect as the front side, but not completely, because it is attached to a holder during front-side polishing and some damage from the holder remains.

Bulk micromechanics relies on crystal plane-dependent etching: silicon (100) planes etch 100 times faster than (111) planes, offering intriguing structuring possibilities. These will be discussed in Chapter 20. Most surface micromachining (Chapter 29) is done on SSP silicon wafers, even though device functionality is determined by thin films deposited on top of silicon wafers, and other substrates could be used. Because most of the processing equipment in microtechnology has been developed for silicon processing, compatibility favors silicon use even when silicon properties per se are not required.

Epitaxy is a process for growing more crystalline silicon on top of a silicon wafer, with doping level and/or dopant type different from the substrate wafer (Figure 4.15). Multilayer structures with many layers of different doping levels and types are equally easy to make. This will be discussed in more detail in Chapters 6 and 22.

Silicon-on-insulator (SOI) wafers can be made for example by bonding a silicon wafer to an oxidized wafer. SOI wafers offer the possibility to optimize processes on top and bottom wafers separately, as shown in Figure 4.15. It is also possible to reduce process steps and minimize leakage currents because pn junctions can be eliminated. Advanced wafers will be discussed in Chapter 22.

The more special wafers naturally cost more. Double side polishing adds 50% to the price, epitaxy can double the wafer price and SOI can cost anything from three to ten times the price of a basic bulk wafer. The price also goes up with wafer size: 100 mm wafers cost $10 and 200 mm wafers (with four times the silicon area compared to 100 mm) cost $40; 300 mm wafers are more expensive per cm^2, and 450 mm wafers are only available in test quantities, so prices will not stabilize before volume production starts.

4.7 Exercises

1. Estimate the silicon lattice constant from atomic mass and density.
2. Consider an Olympic swimming pool filled with golf balls and one squash ball. If golf balls represent silicon atoms, and the squash ball represents a phosphorus atom, what would the resistivity of a silicon piece be with a similar doping level?
3. Electronic grade polysilicon is available with 0.01 ppb phosphorus concentration. What is the highest ingot

4. If 50 kg of ultrapure polysilicon is loaded into a CZ crystal puller, how much boron should be added if the target doping level of the ingot is 10 ohm-cm?
5. Convert CZ and FZ silicon maximum resistivities 1000 ohm-cm and 20 000 ohm-cm into dopant concentrations.
6. If wafer resistivity specifications are 5–10 ohm-cm (phosphorus), calculate the fraction of the ingot that yields wafers within this specification.
7. Calculate the RMS roughness for the three surfaces shown below. Compare RMS to peak-to-valley roughness.

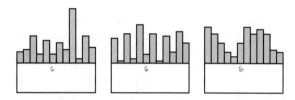

References and Related Reading

Abedrabbo, S. *et al.* (1998) Perspectives on emissivity measurements and modeling in silicon, *Mater. Sci. Semicond. Process.*, **1**, 187–193.
Bullis, W.M. (2000) Current trends in silicon defect technology, *Mater. Sci. Eng.*, **B72**, 93–98.
Butt, H.-J., K. Graf and M. Kappl (2006) **Physics and Chemistry of Interfaces**, Wiley-VCH Verlag GmbH.
Green, M.A. (1995) **Silicon Solar Cells**, Centre for Photovoltaic Devices and Systems, Sydney.
Hahn, P.O. (2001) The 300 mm silicon wafer – a cost and technology challenge, *Microelectron. Eng.*, **56**, 3–13.
Hull, R. (1999) **Properties of Crystalline Silicon**, IEE Publishing.
Irwin, J. (1962) Resistivity of bulk silicon, *Bell. Syst. Tech. J.*, **41**, 387.
Jenkins, T. (1995) **Semiconductor Science**, Prentice Hall.
Lindroos, V. *et al.* (2010) **Handbook of Silicon Based MEMS Materials and Technologies**, Elsevier.
Markov, I.V. (2003) **Crystal Growth for Beginners: Fundamentals of Nucleation, Crystal Growth, and Epitaxy**, World Scientific.
Petersen, K. (1982) Silicon as a mechanical material, *Proc. IEEE*, **70**, 420. Reprinted in W. Trimmer (ed.), **Micromechanics and MEMS: Classic and Seminal Papers to 1990**, IEEE Press, 1997.
Porrini, M. *et al.* (2005) Measurement of boron and phosphorus concentration in silicon by low-temperature FTIR spectroscopy, *Appl. Phys.*, **A81**, 1187–1190.
Shimura, F. (1997) **Semiconductor Silicon Crystal Technology**, Academic Press.

5

Thin-Film Materials and Processes

Thin films are needed to make metal wires and to insulate those wires, to make capacitors, resistors, inductors, membranes, channels, nozzles, mirrors, beams and plates, and to protect those structures against mechanical and chemical damage. Thin films have roles as permanent parts of finished devices, but they are also used intermittently during wafer processing, as protective films, as sacrificial layers, and as etch and diffusion masks.

A great many solid materials are available as thin films: aluminum, gold, copper, tungsten and nickel are routinely used in microfabrication. Oxides of silicon, aluminum, hafnium and tantalum are used, as are nitrides of silicon and titanium. Diamond-like and Teflon-like films offer special properties, as do various alloys like PtMn, TiW, SiGe and CoFe.

The same thin film can serve in many different functions: for example, silicon nitride is the traditional passivation film in integrated circuits, providing both chemical endurance and mechanical scratch protection. But silicon nitride is also used as a mask for oxidation (because oxygen diffusion through nitride is practically nil), as a capacitor dielectric (nitride dielectric constant is higher than that of silicon dioxide), as a suspended membrane (because nitride stress can be made small and tensile), and as optical material (with refractive index matching nicely between silicon and air). Similarly, silicon is used not only for its electronic properties, but also for its mechanical strength (micromechanics), optical absorption at visible wavelengths (solar cells, photodetectors), low absorption in the infrared (waveguides for 1.55 μm optical telecom applications), high Seebeck coefficient (thermoelectric devices) and high thermal conductivity (uniform temperature distribution in micro-hotplate sensors).

This chapter deals with the most common deposition processes for thin films, with the basic characteristics which make thin films different from the bulk, as well as some important applications.

5.1 Thin Films vs. Bulk Materials

In thin films at least one dimension of the material is small, the thickness. For narrow lines, two dimensions are small, and for dots all three dimensions are small. This gives rise to the prominence of surface effects like surface scattering of electrons, leading to size-dependent resistivity or, at very small dimensions, to quantum effects. The size scale for quantum effects is estimated by Debye lengths, which are of the order of 10–100 nm at room temperature.

The density of thin films is often very low compared to bulk materials. Sputtered tungsten films can have a density as low as 12 g/cm^3 compared to a bulk value of 19.5 g/cm^3 and evaporated gold can have a density as low as 3 g/cm^3 but then it was optimized for infrared absorption. Usually porosity should be avoided as it leads to long-term instability: water vapor can adsorb in the pores, and high surface area makes the films reactive: such films oxidize and corrode readily.

Many thin-film properties, resistivity, dielectric constant, coefficient of thermal expansion and refractive index are thickness dependent. Additionally, the film properties depend on deposition processes in profound ways. The example in Table 5.1 gives resistivities of sputtered molybdenum films prepared under slightly different process conditions in different sputtering systems. In addition to thickness dependence, three other facts emerge from

Introduction to Microfabrication, Second Edition Sami Franssila
© 2010 John Wiley & Sons, Ltd

Table 5.1 Resistivity of sputtered molybdenum

Thickness	Underlayer	Conditions	Resistivity (μohm-cm)
Bulk	–	–	5.6
Thin film, 50 nm	SiO$_2$	System 1, 20 °C	17
Thin film, 300 nm	SiO$_2$	System 1, 20 °C	12
Thin film, 300 nm	TiW	System 1, 20 °C	9
Thin film, 300 nm	SiO$_2$	System 2, 20 °C	15
Thin film, 300 nm	SiO$_2$	System 3, 150 °C	9
Thin film, 300 nm	SiO$_2$	System 3, 450 °C	8

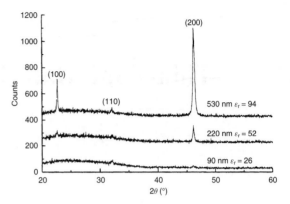

Figure 5.2 Thickness dependence of SrTiO$_3$ films by XRD. The 90 nm thick film is amorphous but thicker films are polycrystalline. Dielectric constant depends on structure, which depends on thickness. Reproduced from Vehkamäki et al. (2001) by permission of Wiley-VCH

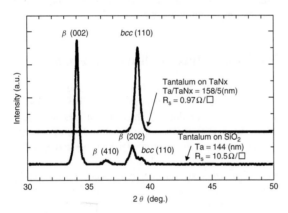

Figure 5.1 X-ray diffraction (XRD) of tantalum thin films: crystal structure and resistivity depend on the underlying material. Reproduced from Ohmi (2001) by permission of IEEE

this table: thin-film resistivity is always more than that of the bulk material; different deposition systems result in very different thin-film properties; and the underlying material has a major effect on film properties.

In Figure 5.1 the tantalum thin-film structure (as measured by X-ray diffraction) and resistivity are similarly seen to depend on the underlying layer: tantalum film on tantalum nitride is low resistivity, while tantalum on oxide has a resistivity an order of magnitude higher. The difference is due to the different crystalline phase, which is due to the different underlying layer.

Structure depends on film thickness, and it may be that very thin films (e.g., 50 nm) are amorphous, yet thicker films (e.g., 200 nm) are polycrystalline. This is shown in Figure 5.2 for SrTiO$_3$ film. XRD peaks indicative of crystallinity only appear for thicker films. The dielectric constant ε_r is also a strong function of thickness.

Thin films can be amorphous, polycrystalline or single crystalline (epitaxial) as deposited. Epitaxial films remain single crystalline during annealing; polycrystalline films experience grain growth and sometimes phase transitions into other crystal structures; amorphous films either stay amorphous or crystallize (into a polycrystalline material, and under very special conditions to single crystalline material). Silicon dioxide, silicon nitride and aluminum oxide are exceptional amorphous films because they remain amorphous in all typical microfabrication processes. Pictured in Figure 5.3 are SEM cross-sections of Al$_2$O$_3$ and SrTiO$_3$ films: aluminum oxide is amorphous and strontium titanate is polycrystalline.

Films prepared by different sputtering systems are different, and films prepared by two completely different deposition processes will differ even more. Copper films made by sputtering, evaporation, electroplating or CVD can have a factor of two differences in resistivity or grain size.

5.2 Physical Vapor Deposition

The general idea of physical vapor deposition (PVD) is material ejection from a solid target material, transported in a vacuum to the substrate surface where film deposition takes place. Atoms can be ejected from the target by various means: resistive heating, electron beam heating, ion bombardment or laser beam bombardment (known as laser ablation). All aluminum films in microfabrication

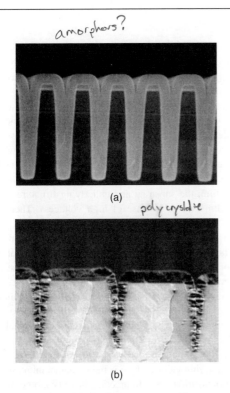

Figure 5.3 SEM micrographs of thin-film structure: (a) amorphous aluminum oxide, reproduced from Ritala *et al.* (1999) by permission of Wiley-VCH; (b) polycrystalline strontium titanate, reproduced from Vehkamäki *et al.* (2001) by permission of Wiley-VCH

are deposited by PVD, and PVD is used for copper, refractory metals and for metal alloys and compounds like TiW, WN, TiN, $MoSi_2$, ZnO and AlN.

5.2.1 Evaporation

Evaporation of elemental metals is fairly straightforward: hot metals have high vapor pressures and in a high vacuum the evaporated atoms will be transported to the substrate.

Typical deposition rates in evaporation are 0.1–1 nm/s, which is very slow.

Evaporation systems are either high vacuum (HV) or ultrahigh vacuum (UHV) systems, with the best UHV deposition systems with 10^{-11} torr base pressures, and 10^{-12} torr oxygen partial pressures. In (ultra)high vacuum the atoms do not experience collisions, and therefore they take a line-of-sight route from source to substrate. The mean free path (MFP) is the measure of collisionless transport, and below about 10^{-4} torr the MFP is larger than the size of a typical deposition chamber (for more discussion on vacuum science and technology, refer to Chapter 33). Low deposition temperature combined with line-of-sight transport means that evaporated films will not coat sidewalls of holes and ridges well, even though film quality on planar surfaces is good.

There are very few parameters in evaporation that can be used to tailor film properties. Atoms arrive at thermal speeds, which results in basically room temperature deposition. There is no bombardment in addition to the thermalized atoms themselves, which bring very little energy to the surface. Substrate heating can be done to improve film quality. This works because impurities are desorbed and adsorbed atoms can diffuse and find energetically favorable lattice sites.

Low-melting-point metals, such as gold and aluminum, can easily be evaporated, but refractory metals require more sophisticated heating methods. Localized heating by an electron beam (Figure 5.4) can vaporize even tungsten (m.p. 3650 K), but deposition rates are, however, very low. Additionally, X-rays will be generated, which can damage sensitive devices.

Evaporation of alloys and compounds is tricky: the component with higher vapor pressure will evaporate more readily, and it can happen that the minority atoms in the starting material end up as the majority atoms in the thin film. Most compounds decompose when heated, therefore evaporation of compounds is limited to a few special cases, like silicon dioxide.

It is possible that the molten metal reacts with the crucible because temperatures are very high, even though this

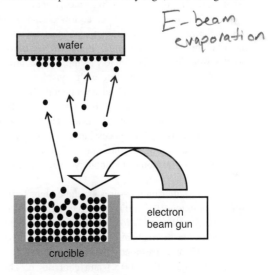

Figure 5.4 Electron beam evaporation: heated metal vaporizes and the evaporated atoms are transported in high vacuum to the substrate wafer

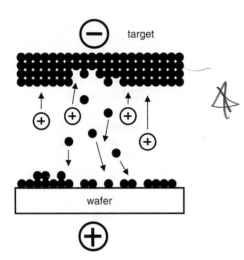

Figure 5.5 Sputtering: argon ions knock atoms out of a target, and the ejected atoms travel in a vacuum and deposit on the wafer

is being minimized by the use of refractory materials for crucibles: namely, Mo, Ta, W, graphite, BN, SiO_2 and ZrO_2. Some crucible material can be incorporated into film also in the case of electron beam misalignment: if a misaligned e-beam hits the crucible, crucible material will be evaporated and incorporated in the deposited film.

5.2.2 Sputtering

Sputtering (Figure 5.5) is the most important PVD method. Argon ions (Ar^+) from a glow discharge plasma hit the negatively biased target and eject typically one target atom. The ejected target atoms will be transported to the substrate wafers in a vacuum. These atoms are energetic and hit the substrate with considerable energy, which has both beneficial and detrimental effects on the growing film. Typical sputtering rates are 1–10 nm/s, significantly higher than in evaporation. Sputtering of nonconductive films necessitates use of RF fields to prevent charging of the target. Further discussion on sputtering systems technology can be found in Chapter 33.

Because sputtering pressures are quite high, 1–10 mtorr (cf. evaporation 10^{-6} torr), sputtered atoms will experience many collisions before reaching the substrate. In a process called thermalization, the high-energy sputtered particles (5 eV corresponds to about 60 000 K!) collide with argon gas ($T = 300$ K) and cool down. Thermalization occurs also in other species present in the plasma, namely the reflected neutrals (some argon ions are neutralized upon target collision). These neutrals provide energy to the substrate. Thermalization reduces the energy of particles reaching the substrate and it reduces the flux of particles to the substrate. Lower flux means lower deposition rate.

In contrast to evaporation, the energy flux to the substrate wafer can be substantial. This has both beneficial and detrimental effects: loosely bound atoms (both film forming atoms as well as unwanted impurities) will be knocked out, improving adhesion and making the film denser. But energies that are too high can cause damage to the film, the substrate and underlying structures (thin-oxide breakdown because of high voltages). There will always be some argon trapped in the film, but its effect can usually be neglected because argon is a noble gas and therefore non-reactive. Incorporation of residual oxygen or nitrogen is much more pronounced because they are reactive and form oxides and nitrides.

Sputtering yield is the number of target atoms ejected per incident ion. Sputtering yields of metals range from about 0.5 (for carbon, silicon and refractory metals Ti, Nb, Ta, W) to 1–2 for aluminum and copper, to 4 for silver at 1000 eV argon ion energy. Refractory metals have low sputtering yields, which is the fundamental reason for lower deposition rates. In practice, there is another reason which further lowers the deposition rate: refractory metals tend to have higher resistivity and thus lower thermal conductivity, which means that high sputtering powers cannot be applied to refractory sputtering targets. For heavy metals like tungsten and tantalum, sputtering yields are higher with xenon and krypton: these heavy gases transfer energy more efficiently to similar mass target atoms. However, argon is almost exclusively used.

If oxygen is added to the sputtering atmosphere intentionally (usually together with argon), oxide films will result. The method is called reactive sputtering. In similar vein, nitrogen additions lead to nitrides. This is the way for example that Ta_2O_5 and TiN are made by sputtering.

5.3 Chemical Vapor Deposition

In chemical vapor deposition (CVD) the source materials are brought into the reactor in the gas phase, they are activated in the plasma, diffuse to the wafer surface, and react there to deposit film. Byproducts are desorbed and pumped away as shown in Figure 5.6. Deposition rates are temperature dependent according to the Arrhenius equation (Equation 1.1), but they are on the order of 0.1–10 nm/s.

Common CVD processes include

$$SiH_4 \text{ (g)} \Longrightarrow Si \text{ (s)} + 2 H_2 \text{ (g)} \quad \sim 600\,^\circ C \quad (5.1)$$

$$\text{SiCl}_4 \text{ (g)} + 2\,\text{H}_2 \text{ (g)}$$
$$\Longrightarrow \text{Si (s)} + 4\,\text{HCl (g)} \qquad \sim 1200\,°\text{C} \quad (5.2)$$

$$\text{SiCl}_4 \text{ (g)} + 2\,\text{H}_2 \text{ (g)} + \text{O}_2 \text{ (g)}$$
$$\Longrightarrow \text{SiO}_2 \text{ (s)} + 4\,\text{HCl (g)} \qquad \sim 900\,°\text{C} \quad (5.3)$$

$$3\,\text{SiH}_2\text{Cl}_2 \text{ (g)} + 4\,\text{NH}_3 \text{ (g)}$$
$$\Longrightarrow \text{Si}_3\text{N}_4 \text{ (s)} + 6\,\text{H}_2 \text{ (g)}$$
$$+ 6\,\text{HCl (g)} \qquad \sim 800\,°\text{C} \quad (5.4)$$

CVD processes depend on both chemical reactions and flow dynamics. There are two main cases: high flow rate supplies enough reactants and film deposition is limited by slow surface chemical reactions (termed "surface reaction limited"); or fast surface reaction consumes source gas rapidly and the deposition rate is limited by gas supply. This is termed "mass transport limited" or "diffusion limited". These two cases will be discussed in more detail in Chapter 34.

Silicon deposition (Equation 5.1 or 5.2) on a single crystalline silicon wafer can result in a single crystalline thin film. This is termed epitaxy and it is an important special case of thin-film deposition. The next chapter is devoted to epitaxial deposition. Most deposition processes lead to amorphous or polycrystalline films.

Silicon dioxide can be deposited by many reactions, Equations 5.3, 5.5, 5.6 and 5.7, for example:

$$\text{SiH}_4 \text{ (g)} + 2\,\text{N}_2\text{O (g)}$$
$$\longrightarrow \text{SiO}_2 \text{ (s)} + 2\,\text{H}_2 \text{ (g)} + 2\,\text{N}_2 \text{ (g)} \quad (5.5)$$

$$\text{SiH}_4 \text{ (g)} + \text{O}_2 \text{ (g)} \longrightarrow \text{SiO}_2 \text{ (s)} + 2\,\text{H}_2 \text{ (g)} \quad (5.6)$$

$$\text{Si}(\text{OC}_2\text{H}_5)_4 \longrightarrow \text{SiO}_2 \text{ (s)} + \text{gaseous byproducts} \quad (5.7)$$

The simple reaction in Equation 5.6 is, however, problematic. Silane and oxygen can already react in the gas phase, which means that solid oxide particles are formed in the gas stream. These will then float around the reactor and sporadically deposit on the wafers. In the nitrous oxide process, oxide is formed by a surface reaction, therefore particle contamination is reduced (but in both cases oxide is formed on the reactor walls, and these films will be a source of flakes and particles if the reactor is not cleaned regularly).

The names for CVD oxides are unfortunately many. LTO, for low-temperature oxide, refers to oxide deposited by the reaction in Equation 5.6. The low deposition temperature of 425 °C is desirable in many cases. HTO obviously stands for high-temperature oxide (Equation 5.3), but the difference is deeper: different source gases are used, and the resulting film quality is much better at high temperatures. TEOS (Equation 5.7) is the name of the precursor molecule tetraethoxysilane Si(OC$_2$H$_5$)$_4$, but it is used as the name for the resulting oxide too (deposited at 700 °C, resulting in high-quality oxide). Sometimes the name USG is used: it stands for undoped silica glass. However, there are no metals in USG, so it is not glass in the traditional sense (see Chapter 19 for glass microprocessing).

The addition of POCl$_3$ gas to the source gas flow leads to phosphorus-doped oxide deposition. The resulting film is called PSG, for phosphorus-doped silica glass. A few

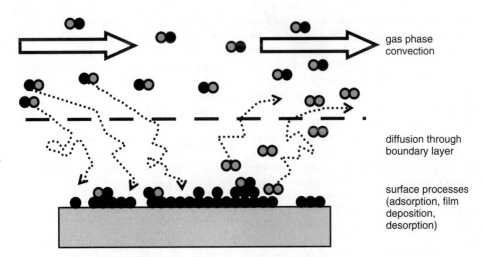

Figure 5.6 CVD: source gas molecules adsorb and react on surface to form a film, and the reaction products are desorbed, diffused and pumped away

percent of phosphorus (5 atomic % maximum) modifies the oxide in many ways. Phosphorous getters sodium ions which are detrimental to MOS transistors, and therefore PSG is used as a passivation layer in integrated circuits. In MEMS PSG is used as a sacrificial layer because its etch rate in hydrofluoric acid is much faster than that of undoped CVD oxide. Phosphorus also lowers the glass transition temperature of PSG, making it possible to flow PSG at about 1000 °C. If both boron and phosphorus are added, we get BPSG. This oxide film flows at about 950 °C, resulting in smoothly sloping walls.

CVD tungsten is deposited in two steps. The silane reduction step (Equation 5.8) deposits a thin nucleation layer over every surface in the system and high-rate blanket deposition with hydrogen reduction (Equation 5.9) is used to achieve the desired total thickness:

$$WF_6 \text{ (g)} + SiH_4 \text{ (g)} \Longrightarrow W \text{ (s)} + 2 \text{ HF} + H_2 \text{ (g)}$$
$$+ SiF_4 \text{ (g)} \quad (5.8)$$

$$WF_6 \text{ (g)} + 3 H_2 \text{ (g)} \Longrightarrow W \text{ (s)} + 6 \text{ HF (g)} \quad (5.9)$$

This process is able to fill holes and trenches (Figure 5.19) and is very important in multilevel metallization (to be discussed in Chapter 28).

5.4 PECVD: Plasma-Enhanced CVD

Because high temperatures cannot be used in many cases, for example when oxide needs to be deposited on aluminum (m.p. 650 °C), one has to find new solutions. New source gas chemistries which enable lower deposition temperatures is one way to go. Another solution is to enhance source gas decomposition and reactions by plasmas. This results in deposition rates similar to thermal CVD, 0.1–10 nm/s, at much lower temperatures, typically around 300 °C, enabling deposition on most metals, for instance. Unfortunately lower deposition temperature results in less dense films.

A simple parallel-plate diode reactor for PECVD is shown in Figure 5.7. Wafers are placed on a heated bottom electrode, the source gases are introduced from the top, and pumped away around the bottom electrode. The operating frequency is often 400 kHz, which is slow enough for ions to follow the field, which means that heavy ion bombardment is present. At 13.56 MHz only the electrons can follow the field, and the ion bombardment effect is reduced.

In thermal CVD, pressure, temperature, flow rate and flow rate ratio are the main variables. In PECVD there is additionally the RF power that can be varied. In advanced

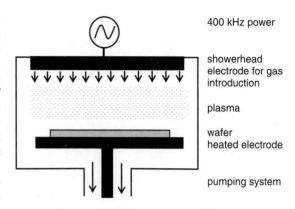

Figure 5.7 Schematic PECVD system

PECVD reactors, RF power can be applied to both electrodes, and the two power sources can supply different frequencies, duty cycles and power levels. The ratio of 13.56 MHz power to kilohertz power is important for film stress tailoring. PECVD shares many beneficial features of both thermal CVD and sputtering.

Whereas thermal oxide or LPCVD nitride is stoichiometric SiO_2 and Si_3N_4, with ratios 1:2 and 3:4 of atoms, many other (PE)CVD films are non-stoichiometric: for example, plasma nitride is best described as SiN_x ($x \approx 0.8$). Amorphous silicon, a-Si, or more specifically designated as a-Si:H, is made by PECVD, the overall reaction being the same as that of LPCVD silicon (Equation 5.1). Hydrogen is incorporated into deposited films up to 30 at. %. Hydrogen release during annealing has to be considered: it has both beneficial and detrimental effects.

PECVD can be used to deposit mixed oxides, nitrides and carbides, as well as doped oxides just like thermal CVD. A mixture of silane, nitrous oxide and ammonia will result in oxynitride, SiO_xN_y, with varying ratios of nitrogen and oxygen, covering the whole range of compositions (and material properties) between oxide and nitride.

Silicon carbide is deposited via the reaction

$$SiH_4 \text{ (g)} + CH_4 \text{ (g)} \longrightarrow SiC \text{ (s)} + 4 H_2 \text{ (g)} \quad (5.10)$$

Carbon is deposited by the reaction (resembling silicon deposition, Equation 5.1)

$$CH_4 \text{ (g)} \longrightarrow C \text{ (s)} + 2 H_2 \text{ (g)} \quad (5.11)$$

Depending on the exact process conditions, many allotropes of carbon can be made. Non-conducting hydrogenated carbon films resemble diamond in some, but not all, respects, and they are known as diamond-like carbon, DLC. Films with less hydrogen have sp^3 bonds similar to diamond, and they are referred to as ta-C, tetrahedral amorphous carbon. If intense plasma or a hot filament is used, highly reactive atomic hydrogen is produced. In this case it is possible to grow polycrystalline diamond films. Under different CVD conditions carbon nanotubes (CNTs) are made. The important factor for CNT deposition is the presence of metallic catalyst particles, for example iron or nickel.

5.5 ALD: Atomic Layer Deposition

In ALD, film is deposited one atomic layer at a time, offering ultimate thickness control. ALD works in pulsed mode: chemical bonds are formed between precursor gas molecules and the surface atoms. Once all possible reaction sites are occupied, no more reactions can take place (Figure 5.8). A purging nitrogen pulse then removes all unreacted precursor molecules. A pulse of second precursor is then introduced. It reacts with the first reacted layer, the surface saturates similarly, and unreacted precursor gases are purged away. Repetition of successive reactant and purge pulses leads to film deposition in a layer-by-layer fashion. The ability of ALD to coat over steps is excellent because all surfaces are coated alike. Figure 1.7 shows ALD alumina and titanium multilayer films deposited over steps. This ability to coat steep topographical features is increasingly in demand as both ICs and MEMS are made more 3D.

As an example of the ALD process, hafnium dioxide deposition is discussed. HfO_2 is a material with a high dielectric constant and is being used as the gate oxide in advanced CMOS. Hafnium chloride reacts with surface hydroxyl groups to form Hf–O bonds. The second precursor is water, and the oxygen in water reacts with the hafnium to form Hf–O bonds again, with hydrogen chloride formed as a byproduct. The overall reaction for hafnium dioxide deposition is given by Equation 5.12. The notation (ad) emphasizes that the reactions take place between adsorbed molecules on the surface, not in the gas phase:

$$HfCl_4 \text{ (ad)} + 2 \, H_2O \text{ (ad)} \Longrightarrow HfO_2 \text{ (s)} + 4 \, HCl \text{ (g)} \tag{5.12}$$

ALD is free of one of the main mechanisms of irreproducibility in CVD: homogeneous gas phase reactions. Because only one gas is introduced at a time, there cannot be gas phase reactions between precursors.

The layer thickness is given by the number of pulses times the monolayer thickness. In theory one monolayer per pulse is deposited, but in many cases sub-monolayer growth is seen. One explanation is steric hindrance: large precursor molecules take up space, so it is simply impossible for another precursor molecule to come close enough, and some surface atoms will not react with precursor molecules. This is depicted in Figure 5.9. It can also be noted that not all surface sites are reactive enough for the ALD reaction to take place.

Both monolayer and sub-monolayer deposition are self-limiting. Practical growth rates range are around 1 Å/cycle (0.1 nm/cycle): for Al_2O_3 deposition they are 1.1 Å/cycle and for TiN, 0.2 Å/cycle. When thickness/cycle numbers are translated into deposition rates, one has to take into account the flushing cycles between the pulses. Overall rates of a few nanometers per minute are typical for ALD. This is slow: for example, the LPCVD rate of polysilicon is typically 10 nm/min. But there are many applications where films of a few nanometers are needed, for example CMOS gate oxides and diffusion barriers in copper metallization.

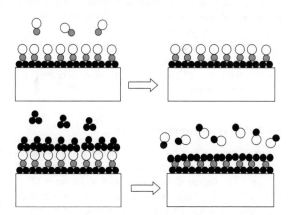

Figure 5.8 ALD: first pulse of precursors saturate wafer surface, and extra precursors are purged away by a nitrogen pulse; second precursor gases react with first layer, and reaction products are purged away

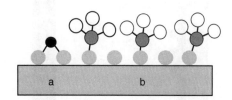

Figure 5.9 Sub-monolayer deposition in ALD: (a) non-reactive surface site; (b) steric hindrance by a large precursor molecule prevents another precursor molecule from approaching the reactive site

5.6 Electrochemical Deposition (ECD)

5.6.1 Electroplating/galvanic deposition

Electroplating takes place on a wafer that is connected as a cathode in metal ion-containing electrolyte solution. The counter-electrode is either passive, like platinum, or made of the metal to be deposited (Figure 5.10).

Electroplating can be very simple: copper is deposited on the cathode according to the reduction reaction, Equation 5.13, while at the anode copper is dissolved into the electrolyte solution:

$$\text{At cathode} \quad Cu^{2+} + 2\,e^- \longrightarrow Cu(s)$$
$$\text{electrolyte solution: } CuSO_4 \quad (5.13)$$
$$\text{At anode} \quad Cu \longrightarrow Cu^{2+} + 2\,e^-$$

Gold is plated in a two-step process (Equation 5.14) with the second, charge transfer reaction, as the rate limiting step:

$$Au(CN)_2^- \rightleftharpoons AuCN + CN^-$$
$$AuCN + e^- \Longrightarrow Au(s) + CN^- \quad (5.14)$$

Electroplating rates vary a lot but are generally in the range of 0.1–10 μm/min. Deposited mass is calculated from

$$\text{mass} = \frac{\alpha I t M}{nF} \quad (5.15)$$

where I is the current, t the time, M the molar mass, n the species charge state, α the deposition efficiency and F the Faraday constant, 96 500 coulombs.

Noble metals can be deposited at 100% efficiency ($\alpha = 1.00$). In less noble metal deposition hydrogen evolution makes α smaller, and for some non-metals like phosphorus co-deposition with cobalt (Co:P 12%, a soft magnetic material), α can be as low as 0.2. Other typical electroplated metals include nickel and iron–nickel (81% Ni, 19% Fe, Permalloy). Many metals have no plating processes available: aluminum, titanium, tungsten, tantalum and niobium cannot be plated.

Three transport processes are active during ECD: diffusion at the electrodes due to local depletion of the reactant via deposition; migration in the electrolyte; and convective transport in the plating bath. The last is connected to electrochemical cell design, and it is affected by factors such as stirring, heating, recirculation and hydrogen evolution.

Macroscopic current distribution is determined by the plating bath electrode arrangement and wafer and bath conductivity. Electrical contact to the wafer also needs careful consideration. Microscopic (local) current distribution depends on pattern density and pattern shapes. The third scale in ECD is the feature scale: potential gradients inside structures are important, especially when deep and narrow grooves are filled.

In practice the plating solutions are complex mixtures of electrolytes, salts (for conductivity control), modifiers for film uniformity and morphology improvement, as well as surfactants. Accelerators (brighteners) are additives that modify the number of growth sites. Suppressors are additives for surface diffusion control. Taken together, these additives increase the number of nucleation sites and keep the size of each nucleation site small, which drives smooth growth. Pulsed plating can also be used in balancing nucleation and grain growth: high overpotential and low surface diffusion favor nucleation, and the opposite conditions favor grain growth. Many plating solutions are proprietary. Plating baths are rather aggressive solutions, and photoresist leaching into the plating bath or adhesion loss are real concerns for reproducible plating.

5.6.2 Plating on structured wafer

Electroplating onto a photoresist pattern easily produces elaborate microstructures, like the gears shown in Figure 5.11. The process is described in Figure 5.12. A conductive seed layer is sputtered on the wafer. This seed layer, also known as the plating base or field metal, can be very thin, tens of nanometers. Photoresist is exposed and developed, and metal plating then follows. Photoresist is then removed and the seed metal is etched away, resulting in metallic microstructures.

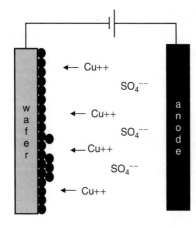

Figure 5.10 Electroplating: $CuSO_4$ electrolyte ionizes to produce Cu^{++} and SO_4^{2+} ions, copper film deposits at the cathode

Figure 5.11 Nickel gear structures (50 μm high) made by electroplating. Reproduced from Guckel (1998) by permission of IEEE

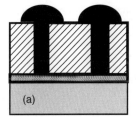 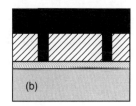

Figure 5.13 (a) Overplating; (b) backplating

plated metals is about 5–10%, so that 50 nm seed layer thickness is less than thickness fluctuations of plated metal 1 μm thick. Electroplating is suitable for extremely small structures, too: modern IC metallization is done by electroplating copper into trenches narrower than 100 nm wide and 200 nm high. Electroplating can fill trenches 500 μm deep and 5 μm wide (aspect ratio 100:1).

Usually plating is allowed to proceed till resist top surface level but not above. It is, however, possible to overplate, and to form mushroom-shaped structures (Figure 5.13). After resist stripping, such a mushroom can be annealed (reflown) to form a ball-like bump. Bumps of Sn–Pb and In are used for flip-chip packaging. Alternatively, plating can be continued until metal fronts touch. Removal of resist underneath results in freestanding metal bridges, or in fluidic channels, depending on design details. The applications can be in RF circuits as air bridges or as cooling channels for high-power electronics.

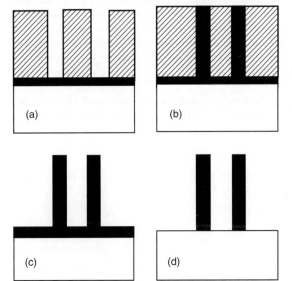

Figure 5.12 Resist masked plating (LIGA, for Lithography and Galvanic plating): (a) seed layer deposition and lithography; (b) plating; (c) resist stripping; (d) seed layer removal

5.6.3 Electroless deposition

Electroless deposition depends on a reduction reaction in an aqueous solution which contains metal salts and a reducing agent. Metal deposition takes place as a result of metal ion reduction. The surface needs to be suitable for electroless deposition and this is achieved by exposing it to a catalyst, such as $PdCl_2$. This reducing agent starts the reduction reaction which then continues locally. Selective deposition is thus possible. Gold, nickel and copper are the usual metals to be deposited by the electroless method. The major advantage of electroless deposition compared to electroplating is elimination of the need to make electrical contacts to the wafer.

The seed layer needs to be removed after plating, otherwise it would electrically short all metallized structures. Often the deposited metal itself can act as an etch mask for seed layer removal because the seed layer is always very thin compared to the plated metal; in many cases the seed layer thickness is less than the plating thickness variation. Thickness uniformity of

Copper electroless deposition chemistries traditionally use sodium hydroxide in the plating bath, but sodium is a contaminant in transistors. Alternative pH adjustment can be done with TMAH (tetramethyl ammonium hydroxide). Copper sulfate $CuSO_4$ in formaldehyde (HCHO) and EDTA (ethylene diamine tetraacetic acid) complexing agent are the basic constituents of the bath. Surfactants (polyethylene glycol) and stabilizers (2,2′-dipyridyl) can be added. The reaction is described by

$$CuEDTA^{2-} + 2\ HCHO + 4\ OH^-$$
$$\Longrightarrow Cu + H_2 + 2\ H_2O + 2\ HCOO^-$$
$$+ EDTA^{4-} \qquad (5.16)$$

The deposition rate is on the order of 100 nm/min. The electroless deposition set-up is extremely simple. Selectivity, however, is difficult to maintain. Hydrogen evolution and incorporation into the film are a problem because hydrogen is mobile; carbon incorporation is another problem.

Gold can be deposited from KOH, KCN, KBH$_4$, KAu(CN)$_2$ mixture at rates exceeding 5 µm/min, even though much lower rates are usually used. Temperatures for electrochemical deposition processes range from room temperature to 100 °C.

5.7 Other Methods

5.7.1 Spin coating

Spin coating is a very widely used method for resist spinning and increasingly for other materials as well, for example spin-on-glasses (SOGs) and polymers (known together as spin-on-dielectrics, SODs) are usually spin coated. Spin coating will be discussed in Chapter 9 in more detail, but briefly the material is dissolved in a suitable solvent, dispensed on a wafer and spun at high speed (e.g., 1000–5000 rpm), see Figure 9.3. Polymeric films can replace inorganic films, especially when thick films are needed. Thicknesses up to 1000 µm can be made by spin coating; inorganic films made either by CVD or by PVD cannot usually be thicker than a few micrometers. Spin-coated films fill cavities and recesses because they are liquids during spin coating. This is advantageous for filling gaps and smoothing, but if a uniform thickness over the topography is desired, spinning is not ideal. Room temperature spinning is always accompanied by baking (in the range 100–250 °C).

5.7.2 Self-limiting methods

After a fashion resembling ALD, monolayer thick polymer films are made by covalently bonding the molecule to a surface. Self-assembled monolayers (SAMs) are made this way. The molecules have a reactive group at one end and a non-reactive group at the other end. In Figure 5.14 the reactive group is the trichloro group (SiCl$_3$) and the polymer chain consists of a chain of 18 carbon atoms (octadecane, or C-18) with methyl group (CH$_3$) at the other end. Silicon reacts with hydroxyl groups on the silicon surface, forming strong Si–O bonds, and HCl is released. When all hydroxyl groups on the surface have reacted, no more reactions can take place: there are no more reactive sites available once the surface is covered by one molecular layer.

Another SAM is shown by Equation 5.17, where fluorinated SAM reacts to form a hydrophobic (Teflon-like) layer on the surface:

$$F_3C-(CF_2)_n-Si-OH + HO-Si\ (surf)$$
$$\longrightarrow F_2C-(CF_2)_n-Si-O-Si\ (surf) + H_2O \qquad (5.17)$$

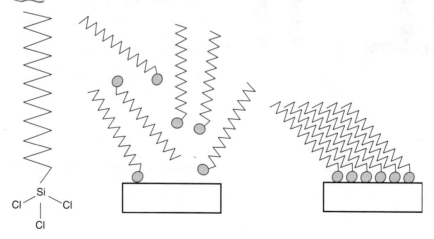

Figure 5.14 Self-assembled monolayer of octadecane trichlorosilane (OTS) on a silicon surface

Fluorinated SAMs have applications in microfluidics, and they are used as antisticking layers in imprinting (Figure 1.9): the non-stickiness of the fluorinated surface makes detachment of the molded piece from the mold master easier.

5.8 Thin Films Over Topography: Step Coverage

Deposition on a patterned substrate introduces new considerations as film must go over steps. These new considerations include the following: atoms arriving in a line-of-sight manner (as in evaporation) and diffusively (as in CVD) will penetrate into grooves and cavities very differently; if there is energetic bombardment present in the process (as in sputtering and PECVD), it will affect horizontal and vertical surfaces differently. And then there is simple geometry: on a horizontal free surface the angle of arrival of atoms is 180°, in convex corners it is 270°, and in the bottom concave corners it is only 90°, as depicted in Figure 5.15. This leads to pronounced deposition at step corners, termed cusping, and reduced deposition in concave (bottom) corners.

If the arriving atoms have high surface diffusivity (e.g., because of high deposition temperature as in thermal CVD) they can move around and find energetically favorable positions and film thickness will be uniform over steps. If the deposition reaction is surface controlled as in ALD, it will naturally follow the existing geometry.

Step coverage is defined as the ratio of film thickness on the sidewall (A) to film thickness on the top horizontal surface (H), Equation 5.18a. Bottom coverage can be defined similarly, Equation 5.18b. These are depicted in Figure 5.15. That is,

$$\text{step coverage} = A/H \quad (5.18a)$$
$$\text{bottom coverage} = B/H \quad (5.18b)$$

Thermal CVD processes (including TEOS, HTO, nitride, LPCVD poly and CVD-W) have very good step coverage, as shown in Figure 5.16a. This kind of 100% step coverage is termed conformal. ALD has also 100% step coverage, even in sharp corners, as shown for TiN film in Figure 5.6b. PECVD step coverage is highly process dependent: in Figure 5.16c oxide step coverage is ca. 50% but in 5.16d only 25%. In Figure 5.16e PECVD oxide step coverage of an overhang structure is shown: ca. 30% has been achieved even in the shadow areas. Overhang coverage is possible in flow systems, but

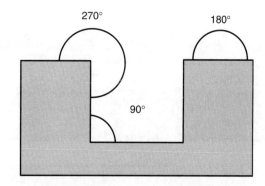

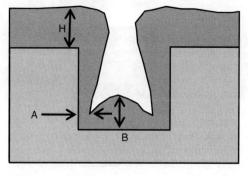

Figure 5.15 Top, arrival angles of depositing species at different positions; bottom, step coverage, A/H, and bottom coverage, B/H

beam-like deposition systems sputtering and evaporation cannot cover overhangs. Sputtering step coverage is highly variable, but generally it is in the 20–50% range. Simulated examples are shown in Figures 3.6 and 7.17. Step coverage in evaporation is very poor.

Conformal deposition is no guarantee that film quality on the sidewalls is equal to that of planar areas. In sputtering and PECVD particle bombardment hits the horizontal surfaces, densifying the film, and the sidewalls are much less dense material, and the film on the sidewall will etch much faster than the film on horizontal surfaces.

Step coverage is usually not a major problem for low-aspect-ratio structures, for, say, a hole 1 μm wide, 0.5 μm deep hole, but at 1:1 and higher aspect ratios, step coverage deteriorates rapidly. It is important to remember that on real microdevices there are always structures of various shapes and spacings, and film deposition over all these spaces needs to be considered. It is far too simple to consider one size or depth only. It has to be remembered that aspect ratio is a dynamic variable: a contact hole that

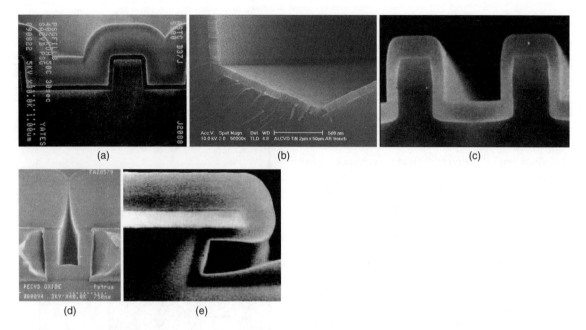

Figure 5.16 Step coverage: (a) conformal coverage in CVD; (b) confrmal coverage in ALD; (c) PECVD TEOS oxide with 50% step coverage; (d) PECVD with 25% coverage; (e) PECVD oxide over an overhang structure. Figures a, c and d from ref. Cote; (b) courtesy Kai-Erik Elers; (e) courtesy Oxford Plasma Technology

has 1:1 aspect ratio initially turns into a 2:1 aspect ratio hole as deposition proceeds, and just before closure the aspect ratio approaches infinity.

Good step coverage in metallization in essential for reliability. Even though metal film will be continuous even with, say, 10% step coverage, current density will increase dramatically at the thinnest point, causing a major reliability problem.

5.9 Stresses

Thin films are under either compressive or tensile stresses when deposited on the wafers. Stresses consist of extrinsic stresses, caused by thermal expansion mismatch between the film and the substrate, and of intrinsic stresses which depend on film microstructure and the deposition process. Stress at room temperature is the sum of intrinsic and extrinsic stresses.

Stresses in thin films cause the wafer to curve, as shown in Figure 5.17. Thin film can be imagined as a spring, where tensile stress equals spring elongation. If this elongated spring is attached to a much more massive wafer in its elongated, tensile state, it will still try to return to its original size, and in doing so it will bend the wafer. A film under tensile stress will result in a concave shape for a wafer + film combination. A compressively stressed film will try to return to its shape, too, and the film + wafer will end up with a convex shape. If tensile stress is too high, the film will crack. And too high compressive stresses lead to buckling.

Figure 5.17 gives a macroscopic depiction of stresses, but the same reasoning works on the atomic level as well: the germanium lattice constant is 4.2% larger than that of silicon, and if germanium film is deposited on silicon, it has to be compressed to fit. Conversely, if silicon film is deposited on germanium, it will be under tensile stress.

Extrinsic stresses can be estimated from thermal expansion coefficient differences by

$$\sigma = \frac{E_f}{(1-\nu)}(\alpha_f - \alpha_s) \cdot \Delta T \qquad (5.19)$$

where the indices f and s refer to film and substrate, respectively, and E is Young's modulus of the film, ν the Poisson ratio of the film, α the coefficient of thermal expansion (CTE) and ΔT the temperature difference.

By convention, negative stresses are compressive. In a first approximation the temperature difference is the difference between deposition and measurement

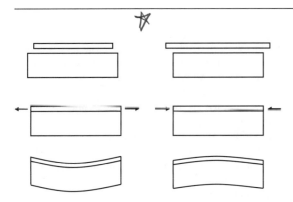

Figure 5.17 Tensile stresses (left) and compressive stresses (right) in thin films. Imagine a free film stretched or compressed to wafer size and attached to a massive wafer. Next, imagine that the films try to return to their original shape. The wafer will curve as a consequence

temperatures, but the situation is really much more complex because stress relaxation can occur during high-temperature deposition.

The coefficient of thermal expansion of silicon is $2.6 \times 10^{-6}/°C$ (at room temperature). The only other materials used in microfabrication that have smaller coefficients are silicon dioxide, silicon nitride and diamond, which have CTEs of 0.5×10^{-6}, 2.4×10^{-6} and $1.1 \times 10^{-6}/°C$, respectively. Oxide, nitride and diamond are therefore the only materials that can develop compressive extrinsic stresses over silicon substrates. The CTE of aluminum is $23 \times 10^{-6}/°C$, which is fairly high, that of tungsten $4 \times 10^{-6}/°C$ and polymers have CTE values in the range of $30–100 \times 10^{-6}/°C$.

The bimetal thermometer is a classic example of thermal expansion coefficient mismatch. Bimorph structures can be used as sensors and actuators in microsystems, but the initial shape has to be known. As shown in Figure 5.18, SiO$_2$/Al and SiO$_2$/Ti cantilevers are bent because of stresses in the structures, without external sensing or actuation force. In a single material cantilever (e.g., CVD polysilicon) stress gradients can lead to similar bending.

Intrinsic stresses are caused by many mechanisms which are not fully understood. Deposited polycrystalline films are not at their energy minimum. Especially, low deposition temperature means that the arriving atoms do not have enough energy to find energetically favorable positions, and the film builds up without relaxation. Voids and incorporated foreign atoms contribute to intrinsic stresses. Bombardment during deposition has a pronounced effect on many film properties, including stresses, because the bombardment pinches off loosely bound atoms, resulting in a more uniform, less stressed film. Too high a bombardment, on the other hand, implants atoms into the film in a non-equilibrium way, and compressive stresses build up. Crystallization and phase transitions, and other processes which lead to volume changes, such as outgassing, lead to stress changes.

Evaporated metal films are usually under tensile stresses (100 MPa to 1 GPa). Sputtered films can be under tensile or compressive stresses. Sputtering, with ion bombardment during deposition, is a much more complex process than evaporation, and stress tailoring can be achieved by:

- bias power
- argon pressure
- sputtering gas mass
- temperature
- deposition rate.

Sputtered film stress can be tailored by deposition pressure: films are usually under compressive stress if

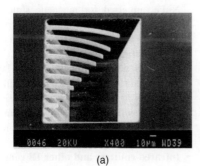
(a)

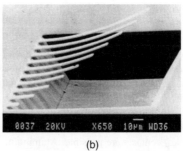
(b)

Figure 5.18 (a) compressive stress in SiO$_2$/Al cantilevers causes downward bending; (b) tensile stress in SiO$_2$/Ti cantilevers leads to upward bending. Reproduced from Fang and Lo (2000) by permission of Elsevier

deposited at low pressure (about 1 mtorr in a magnetron sputtering system) but turn to tensile stress as deposition pressure is raised (to some 10 mtorr). This crossover pressure increases with atomic mass. However, this is not a universal solution, because pressure affects not only film stress but many other properties, like deposition rate and film density.

5.9.1 Stress measurement

Thin film stresses are usually measured by methods based on wafer curvature. Optical techniques or stylus profilometers are used for curvature measurement. Because silicon wafers have curvature (e.g., 30 μm), it needs to be eliminated from the results. Therefore, either of two approaches are used: measure wafer curvature, deposit film, measure composite curvature, and deduct silicon wafer curvature; or else measure composite curvature, etch thin film away and measure wafer curvature.

Film stress is given by the Stoney formula:

$$\sigma = \frac{E_s}{6(1-\nu)} \frac{t_s^2}{t_f} \left(\frac{1}{R} - \frac{1}{R_0} \right) \quad (5.20)$$

where t_s is the substrate thickness, ν the Poisson ratio of the substrate (0.27 for silicon), t_f the film thickness, R the radius of curvature for the substrate+film system (negative when convex) and R_0 the radius of curvature for the substrate without film (infinite for a flat wafer).

Wafers are about 1000 times thicker than films, and because all solids have similar elastic constants, wafer stresses and strains are about 1000 times less than those of thin films. Thin-film stresses are on the order of 10–1000 MPa (1000 MPa = 10^{10} dyn/cm^2).

5.10 Metallic Thin Films

Metallic thin films have various application in microfabricated devices.

- **Conductors:** Resistivity is the main consideration: aluminum and copper are main choices for most applications, and gold is often used in RF devices, like inductor coils, to minimize resistive losses. Doped silicon and polycrystalline silicon can be used as conductors, but their resistivity is very high compared with metals.
- **Contacts to semiconductors:** Ohmic (metal-like) and Schottky (diode-like) contacts are possible. Aluminum, itself p-type dopant in silicon, makes good ohmic contact to p-type silicon. Platinum silicide is one candidate for silicon Schottky contacts
- **Capacitor electrodes:** Capacitor electrodes need not be highly conductive. The most important capacitor electrode, the MOSFET gate, is chosen to be polycrystalline silicon because its interface with silicon dioxide is stable, and its lithography and etching properties are good.
- **Plug fills:** When vertical holes need to be filled with a conducting material, CVD tungsten and electrodeposition of copper are employed. Because distances are short, it is rather step coverage than resistivity which determines the choices.
- **Resistors:** Doped semiconductors, metals, metal compounds and alloys can be used as resistors. Heating resistors can be made of almost any material, but precision resistors are difficult to make.
- **Adhesion layers:** Noble metals like gold and platinum do not adhere well to substrates, and therefore thin (10–20 nm thick) "glue" layers of titanium or chromium are needed.
- **Barriers:** Barriers are needed to prevent unwanted reactions between thin films or diffusion of unwanted atoms. Amorphous metal alloys and compounds like tungsten nitride W:N, titanium–tungsten TiW, TiN and TaN are used as barriers between metals and silicon.
- **Mechanical materials:** Aluminum, nickel and TiAl alloys are materials for micromechanical free-standing beams and cantilevers, in e.g. micromirrors and resonators. Films like TiN can be used as mechanical stiffening layers to prevent mechanical changes in the underlying, softer films, like aluminum.
- **Optical materials:** In image sensors metals act as light shields, and chromium is used in photomasks to block light. TiN is often deposited on top of aluminum to reduce reflectivity, because lithography is difficult of highly reflecting surface. Transparent conductors like indium doped tin oxide (ITO; $In_xSn_yO_2$) are needed in displays and light emitting devices.
- **Magnetic materials:** Nickel and nickel alloys, Ni:Fe, are used for magnetic structures in microactuators. Cores of microtransformers are also made of these materials, which are usually deposited by electroplating.
- **Catalysts and chemically active layers:** Chemical sensors, microreactors and fuel cells use films like palladium and platinum as catalysts.
- **Electron emitters:** Vacuum microemitter tips are often made of molybdenum because of its high melting point and low work function.
- **Infrared emitters and other IR components:** Heated wires emit infrared, and porous metallic films, like aluminum black, act as IR absorbers. Metallic meshes act as IR filters and aluminum is used as an IR mirror.

Table 5.2 Properties of metals

Metal	Resistivity bulk (ohm-cm)	Resistivity thin film (ohm-cm)	CTE (ppm)	Thermal conductivity (W/cm-K)	Melting point (°C)
Al	3	3–4	23	2.4	650
Cu	1.7	2–4	16	4	1083
Mo	5.6	10–20	5	1.4	2610
W	5.6	10–100[a]	4.5	1.7	3387
Ta	12	20–200[a]	6.5	0.6	3000
Ti	48	100–200	8.6	0.2	1660
Co	6.2	10–20	12.5	0.7	1500
Ni	6.8	10–20	13	0.9	1455
Cr	13	20–40	6	0.7	1875
Pt	10	20–100	9	0.7	1769
Au	1.7	2–3	14	3	1064

[a] Tungsten and tantalum can exist in two different phases which have different resistivities; a minor change in sputtering conditions can result in either phase.

- **Sacrificial layers:** Many devices require free-standing structures. These must be fabricated on solid films, which will subsequently be etched away. Copper is often used as a sacrificial material under nickel or gold.
- **Protective coatings:** Sometimes the role of the topmost layer is simply to protect the underlying layers from the ambient: from etching agents or environmental stressors. Nickel and chromium are used as masks for etching.
- **X-ray components:** Masks for X-ray lithography require high atomic mass materials which effectively block X-rays. Tungsten, gold and lead are prime candidates. X-ray mirrors are made by alternating heavy (tungsten, molybdenum) and light materials (carbon or silicon) with layer thicknesses in the nanometer range.
- **Bonding layers:** Gold and tin layers are used in eutectic bonding. Tight, hermetic bonds can be obtained at fairly low temperatures when eutectic alloys are formed.
- **Bonding pads:** Wires have to be attached to chips, and this is best achieved with soft metals like aluminum and gold, while hard metals like tungsten or chromium are unsuitable for wire bonding, and also difficult to probe by probe needles.

Deposition process greatly influences the choice of metals. Not all materials are amenable to all deposition methods, and the resulting film properties (resistivity, phase, texture, adhesion, stress, surface morphology) are closely connected with the details of the deposition process, and may well be idiosyncratic with the equipment. Reproducing results that have been obtained with another piece of equipment can be a nightmare.

5.10.1 Properties of metallic thin films

Low resistivity is required of metals in thin-film form. Thin-film resistivity is usually much higher than bulk resistivity. Aluminum, copper and gold thin-film resistivities are close to bulk values; for most others thin films, resistivities are factor of two higher. Important microfabrication metals are listed in Table 5.2. Resistivities are strongly deposition process dependent, as was shown in Table 5.1, and the tabulated values should be used as guidelines with every deposition process being characterized individually. It should also be borne in mind that thermal conductivity similarly depends on the details of deposition process and film thickness.

Alloys and compounds TiW, TiN_x and TaN_x have resistivities that are even more strongly deposition process dependent than simple metals, and the exact composition will also have a profound effect. Resistivities of TiW, TiN_x and TaN_x are usually in the range of 100–500 µohm-cm.

Young's moduli are the same order of magnitude for all metals, from 100 GPa for soft metals to 600 GPa for refractory metals. Many metal properties are related to the melting point. High melting point equals high bond strength, and a stable atomic arrangement in solids. This translates to a high current density tolerance.

5.11 Polysilicon

Polysilicon (polycrystalline silicon) is deposited in a low-pressure CVD process (LPCVD, at around 1 torr pressure) by the silane decomposition reaction according to Equation 5.1. A TEM micrograph of polycrystalline silicon on oxide is shown in Figure 2.3. Deposition at 630 °C leads to polycrystalline films with a grain size of about 100 nm. Between 580 and 600 °C grain size decreases and deposition at about 570 °C results in an amorphous film (a-Si). Because deposition rate is governed by the Arrhenius law, the rate drops dramatically, to a few nanometers per minute.

In a CVD process we can add dopant gases to the gas feed, so dopants are incorporated in the deposited film. Diborane, B_2H_6, dopes the film p-type and $POCl_3$ results in n-type polysilicon. In situ boron doping increases deposition rate while phosphorus doping reduces it. Both dopants make the process more difficult to control, in terms of uniformity and reproducibility. Quite often poly is deposited undoped, and doped afterward. The doping methods include thermal diffusion (Chapter 14) and ion implantation (Chapter 15), just like the doping of single crystalline silicon. Polysilicon can be oxidized at high temperatures (more on this in Chapter 13). After all, it is silicon.

High doping levels of 10^{21} cm^{-3} result in a polysilicon resistivity of about 500 µohm-cm, or 50 ohm/square for film thicknesses of 100 nm. Polysilicon can be used to make conductive wiring when temperatures are too high for metal wires. Obviously the resistance of polysilicon is much higher than that of metallic conductors, but for short wires it is acceptable. Electron mobility in polysilicon is an order of magnitude less than in single crystalline materials, 10–50 cm^2/V-s. The polysilicon interface with thermal oxide is well characterized and polysilicon is the "Metal" in MOS transistors. The MOS transistor is a capacitor, and the rather high resistivity of polysilicon is not a major disadvantage.

Poly (625 °C film) and a-Si (570 °C film) differ in their surface smoothness, and many differences remain even when the films are processed further. Amorphous silicon will crystallize when heated above about 600 °C, but then grain size will be different from poly that was polycrystalline to begin with. The resistivity of the originally amorphous film will be lower than that of poly, and stresses will be different (see Figure 25.9).

Polysilicon can be used as a mechanical material just like single crystal silicon. Its mechanical constants are similar to single crystalline material: Young's modulus is about 160 GPa for both while the yield strength of poly is 2–3 GPa vs. about 7 GPa for single crystalline. The thermal conductivity of polysilicon is 0.2–0.3 W/cm-K vs. 1.57 W/cm-K for single crystal material, and the coefficients of thermal expansions are identical. But CVD deposition offers possibilities for realizing multilayer structures which cannot be made in single crystal material.

5.11.1 Amorphous silicon

Amorphous silicon can mean different things to different people. Taken literally, it is just the structure that matters. Amorphous silicon can be made for instance by sputtering. LPCVD amorphous silicon deposited at 570 °C contains minor amounts of hydrogen, but the amount is insignificant for most applications.

In solar cell and flat-panel display manufacturing, amorphous silicon is done by PECVD at low deposition temperature. Lots of hydrogen is incorporated in the films, and thus the proper name is hydrogenated amorphous silicon, a-Si:H. In a-Si:H films the hydrogen content can be up to 30 at.% (and much less in wt%), and they contain considerable amounts of oxygen and carbon impurities (10^{19} cm^{-3}). PECVD a-Si can be doped in situ like LPCVD silicon, but resistivities are higher than for poly of identical dopant concentration.

5.12 Oxide and Nitride Thin Films

Dielectric films have, just like metallic films, a plethora of applications in microdevices. Table 5.3 classifies dielectric film applications into three categories: as structural parts in finished devices, as intermittent layers during wafer processing, and as protective coatings for finished devices. Surprisingly, many films can serve all these roles.

5.12.1 Properties of dielectric films

Higher deposition temperature usually leads to denser films which have slower etch and polishing rates, and are less susceptible to moisture absorption. In CVD, and in PECVD in particular, films can have HF etch rates varying enormously depending on particular types of equipment and process conditions (power, flow rate and ratios, temperature). As a rule of thumb, if the thermal SiO_2 etch rate is 100 nm/min, 300–1000 nm/min is expected for (PE)CVD oxides. Densification annealing of CVD films at high temperature can lower this by a factor of two.

Silicon dioxide and nitride are the most widely used dielectrics in microfabrication, but sometimes polymer films offer exceptional properties. One polymer, BCB, benzo cyclo butadiene, is contrasted with oxide and nitride in Table 5.4.

Films should be free of pinholes, small point-like defects, otherwise they are useless as protective coatings. For PECVD, values less than 0.1 pinholes/cm^2 are good. If the film is less dense than the bulk, it can be either because of porosity or because of pinholes.

Table 5.3 Uses of dielectric thin films in microtechnology

Function	Examples
Structural parts of finished devices	
Intermetal insulation	SiO_2, polymers
Gate oxides in MOS transistors	SiO_2, HfO_2
Capacitor dielectrics	SiO_2, Si_3N_4, Ta_2O_5, $BaSrTiO_3$
Tunnel oxide in EPROMs	SiO_2
Spacers in MOS and bipolar transistors	CVD oxide, CVD nitride
Ion barriers	Al_2O_3, Si_3N_4
Gap fill materials	Oxides, spin-on films
Tunnel oxides in Josephson junction devices	AlO_x, NbO_x
Dielectric mirrors	CVD oxide, nitride, polysilicon
Micromechanical beams and plates	LPCVD nitride
Antireflective coatings	PECVD SiN_x, SiO_2
Heat sink for lasers and power devices	Diamond
Hydrophobic surfaces	PTFE, PDMS, other polymers
Microfluidic structures	Polymers, oxide, nitride, diamond
Microlenses	Polymers, spin-on glasses
Protective coatings against ambient in final devices	
Passivation layer and metal ion barrier	SiO_x, SiO_xN_y
Humidity and scratch protecting barriers	PECVD SiN_x, polyimide
Tribological coating (wear, friction)	Diamond, SiC
Corrosion-resistant coatings in harsh environments	Ta_2O_5, SiC
Sacrificial and intermittent layers during wafer processing	
Mask for thermal oxidation	Si_3N_4
Diffusion and ion implantation masks	SiO_2, Si_3N_4
Dopant evaporation barrier	CVD oxide, SiN_x
Etch stop layer	SiN_x
Etch masks in bulk micromechanics	Oxide, Si_3N_4
Dopant sources	PSG, BSG
Sacrificial layers in surface micromechanics	PSG, resist
Release layers	Teflon, SAMs

In order to minimize capacitances ($C = \varepsilon A/L$, $\varepsilon = \varepsilon_r \varepsilon_0$) between metal layers, it is preferable to use films with a low dielectric constant (known as low-k or low-ε materials). Polymeric materials (e.g., parylene), modified CVD oxides (SiO_xF_y, SiO_xC_y) and porous spin-on oxides are low-k materials with ε_r between 2 and 4. The topic of dielectric constant will be discussed in connection with multilevel metallization for ICs in Chapter 28.

Films with a high dielectric constant are required in applications where high capacitance is needed. MOS transistors and DRAMs are capacitors, and in order to make the capacitors smaller, area has been scaled down. To keep capacitance constant, the capacitor dielectric thickness has been scaled down. This approach cannot be continued indefinitely because of tunneling currents through thin oxides. High-k dielectrics for CMOS will be discussed in Chapter 26. Thin-film dielectrics have a breakdown field in the range of 10^5–10^7 V/cm (10–1000 V/μm). This issue is especially important for MOS transistors and flash memories which today have film thicknesses in the sub-10 nm range.

5.12.2 Measurements for dielectric films

Thickness and refractive index are basic measurements for lossless dielectric films. Optical methods are accurate, quick, non-contact and suitable for both research and manufacturing control applications. Accuracy of measurement is a fraction of a nanometer for both ellipsometry and reflectometry.

Reflectometry assumes a known index of refraction, but measures real thickness by fitting reflections over a wide wavelength range to the $d - n_f$ model. Thicknesses

Table 5.4 Properties of silicon dioxide, silicon nitride and BCB (benzo cyclo butadiene)

	SiO$_2$	Si$_3$N$_4$	BCB
Resistivity (Ω-cm), 25 °C	10^{16}	10^{16}	10^{19}
Density (g/cm^3)	2.2	2.9-3.1	1.05
Dielectric constant	3.8–3.9	6–7	2.65
Dielectric strength (V/cm)	12×10^6	10×10^6	3×10^6
Thermal expansion coefficient (ppm/°C)	0.5	1.6	52
Melting point (°C)	1700	1800	350[a]
Refractive index	1.46	2.00	1.54
Specific heat (J/kg-°C)	1000	700	2180
Young's modulus (GPa)	87	300	2–3
Yield strength (GPa)	8.4	14	
Stress in film on Si (MPa)	200–400 C[b]	1000 T[b]	30 T[b]
Thermal conductivity (W/K-m)	1.4	19	0.29
Etch rate in buffered HF (nm/min)	100	1	0

[a] Glass transition point.
[b] T = tensile, C = compressive.

from 10 nm to 50 μm can be measured, depending on the equipment and algorithm.

Ellipsometry measures thickness and refractive index in a single measurement because both the amplitude and phase of reflected polarized light are measured. In the thin limit ellipsometry is not accurate because optical constants of very thin films, below 10 nm, are not really constants, though precision is very good. For thicker films multiple reflections and interference mean that the solution is periodic and additional information is needed as to which period is in question. See Appendix C for an oxide and nitride color chart which can be used as a quick thickness monitoring method.

Ellipsometry needs a fairly large area for the measurement (e.g., 100×100 μm) while reflectometer spots can be as small as a few micrometers, which enables measurement from the structures themselves, without a dedicated test site. The easiest and quickest way to gauge thickness is from interference colors. The accuracy of this approach is about 10 nm, but the colors repeat at regular intervals, and absolute thickness determination requires additional information.

5.13 Polymer Films

Polymer films can be deposited by a number of methods:

- spin coating
- dip coating
- sputtering
- evaporation
- CVD and PECVD
- self-limiting vapor phase and liquid phase reactions.

We will discuss polymer thin films briefly here; more information can be found in Chapter 18. Polymer films can offer exceptional properties, like softness, which may be required for low-pressure sensors or sensitive cantilever sensors. Many sensors use polymers as active parts of the devices: for example, capacitive humidity sensors work on the principle that capacitance changes when the polymeric capacitor dielectric absorbs water. Thin-film polymer is paramount for device operation so that humidity rapidly penetrates the whole film. Polymers are also used as structural materials in microsystems. Those structures can be thin or thick, up to millimeters.

Widely used polymer materials in microfabrication include thermally stable aromatic polymers (BCB), epoxies (SU-8) and polyimides (PI). All of these are available as photoresists, too, acting in the negative mode. Non-photoactive polyimides are also widely used. Thermoplast polymers PMMA, PC and COC are used in embossing/imprinting applications. Parylene is used as a structural material, protective coating and thermal insulator. Various fluoropolymers are used to make hydrophobic surfaces. Perfluorinated films like Teflon have other uses, too, because of their exceptional properties, like low water absorption, low friction, extreme chemical tolerance and very good electrical properties at high frequencies.

Polymers are inferior to inorganic films in terms of mechanical strength. Tensile strengths of polymers are in the range of 100–400 MPa, and Young's moduli on the order of 1–10 GPa, compared to 50–500 GPa for inorganic solids and elemental metals. Stresses in polymers are inherently low (<100 MPa) whereas stress minimization in oxides and nitrides is quite a challenge. In addition to normal process-related variation, polymer properties vary from manufacturer to manufacturer, and the listed properties are indicative of some typical values only.

Polymers have thermal limitations: maximum usable temperatures are often in the range of 100–200 °C, but some exceptional polymers tolerate 400 °C. The coefficients of thermal expansion are in the range of 30–100 ppm/°C vs. 1–20 ppm/°C for elemental metal films and simple inorganic compounds, which is a considerable mismatch.

Many of the thin-film deposition methods described in this chapter can be applied to polymer thin films. Evaporation is used to deposit small organic molecules like pentacene ($C_{14}H_{22}$) which will deposit as a conductive thin film and can be used as a channel material in organic electronics (Figure 26.21). The same process conditions and process performance apply for the evaporation as of organics, as for any other material, for example 5×10^{-7} mbar pressure and 0.1 nm/s deposition rate. Polymers cannot be evaporated: they will decompose rather than evaporate.

Parylene is deposited by thermal CVD. Layer thicknesses are similar to other CVD processes in the thin end, but because of polymer softness, stress build-up is less, and layers tens of micrometers thick can be made. These were the traditional parylene applications in microtechnology: thick protecting layers for finished devices. Today, the conformal deposition of thin films, low deposition temperature, basically room temperature, enable novel applications.

Teflon, an insulator, must be sputtered in a RF system (this applies to inorganic insulators as well). Deposition rate tends to be low (e.g., 0.05 nm/s), but high-density, pinhole-free film can then be obtained. Polyimides, polypropylene and polyethylene have also been sputter deposited.

Teflon-like perfluoropolymer films $(CF_2)n$ can also be plasma deposited. Fluorine-containing source gases CHF_3 or C_4F_8 (which are readily available because they are used to plasma etch silicon dioxide) are broken down in plasma, and fluoropolymer is deposited at wafer, at room temperature. FTIR and XPS analyses will reveal the C–F bonds, and also C–H bonds which indicate incomplete fluorination.

5.14 Advanced Thin Films

Thin-film deposition is seldom the last process step. The films will be modified intentionally or unintentionally in subsequent process steps. For example, all elevated temperature steps will modify thin films.

So far we have been dealing mostly with single layer films. But processes and structures can be made much more functional and reliable by adopting multilayer films. In IC metallization multiple layers of metal are used for various reasons: titanium improves adhesion, TiN acts as a barrier between materials and prevents reactions, CVD-W is used because it can fill contact holes, etc. Dielectrics are similarly used in double layer structures: passivation is provided by PSG/nitride: phosphorus-doped oxide is a good barrier for sodium ion diffusion, and nitride is an excellent mechanical scratch-protective coating. These are depicted schematically in Figure 5.19 and in the SEM micrograph of Figure 5.20.

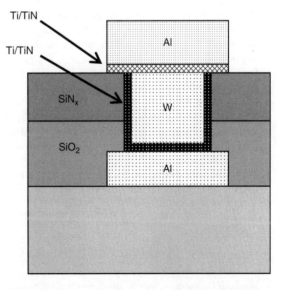

Figure 5.19 Cross-section of multilevel metallization: double layer dielectric (SiO_2/SiN_x), triple layer plug fill metallization (Ti/TiN/W) and triple layer top metallization (Ti/TiN/Al)

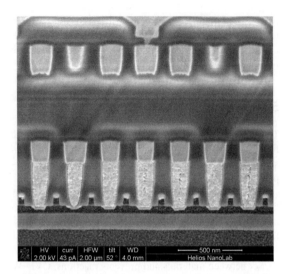

Figure 5.20 Contact plug filled by Ti/TiN/CVD-W. Courtesy Brandon Van Leer, FEI Company

- PECVD of oxide
- electroplated copper
- ALD of aluminum oxide
- reactive sputtering of TiN.

5.15 Exercises

1. What are the resistivities of the tantalum films in Figure 5.1?
2. If silane (SiH_4) flow in a single wafer (150 mm) PECVD reactor is 5 sccm (cm^3/min), what is the theoretical maximum deposition rate of amorphous silicon?
3. If the electroplating current density is 100 mA/cm^2 in nickel deposition, what will be the rate?
4. Calculate the wafer bow that a thin film 100 nm thick with 100 MPa stress induces on a silicon wafer 675 μm thick and 150 mm in diameter. How about a 100 nm thick film of 500 MPa on a wafer 380 μm thick and 100 mm diameter wafer?
5. Draw schematic step coverages for (a) evaporated films and (b) ALD films over the following steps:

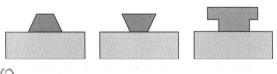

6. Find out (from the scientific literature) typical deposition rates and film thicknesses for the following processes:
   - evaporation of aluminum
   - sputtering of tungsten
   - CVD of tungsten
   - PECVD a-Si:H

References and Related Reading

Briand, D. et al. (1999) In situ doping of silicon deposited by LPCVD, *Semicond. Sci. Technol.*, **14**, 173–180.

Cote, D.R. et al. (1999) Plasma-assisted chemical vapor deposition of dielectric thin films for ULSI semiconductor circuits, *IBM J. Res. Dev.*, **43**, (1–2), 5.

Datta, M. and D. Landolt (2000) Fundamental aspects and applications of electrochemical microfabrication, *Electrochim. Acta*, **45**, 2535–2558.

Doms, M. et al. (2008) Hydrophobic coatings for MEMS applications, *J. Micromech. Microeng.*, 055030.

Ehrfeld, W. (2003) Electrochemistry and microsystems, *Electrochim. Acta*, **48**, 2857–2868.

Fang, W. and C.-Y. Lo (2000) On the thermal expansion coefficients of thin films, *Sens. Actuators*, **84**, 310.

French, P.J. (2002) Polysilicon: a versatile material for microsystems, *Sens. Actuators*, **A99**, 3–12.

Grigoras, K., V.-M. Airaksinen and S. Franssila (2009) Coating of nanoporous membranes: atomic layer deposition versus sputtering, *J. Nanosci. Nanotechnol.*, **9**, 3763–3770.

Guckel, H. (1998) High aspect ratio micromachining via deep X-ray lithography, *Proc. IEEE*, **86**, 1586.

Kamins, T. (1980) Structure and properties of LPCVD polysilicon films, *J. Electrochem. Soc.*, **127**, 686–690.

Kamins, T. (1998) **Polycrystalline Silicon for Integrated Circuits and Displays**, 2nd edn, Springer.

Kim, M.-C. et al. (2004) Characterization of polymer-like thin films deposited on silicon and glass substrates using PECVD method, *Thin Solid Films*, **447–448**, 592–598.

Knez, M., K. Nielsch and L. Niinistö (2007) Synthesis and surface engineering of complex nanostructures by atomic layer deposition, *Adv. Mater.*, **19**, 3425–3438.

Leskelä, M. and M. Ritala (2002) Atomic Layer Deposition (ALD): from precursors to thin film structures, *Thin Solid Films*, **409**, 138.

Maier-Schneider, D. et al. (1996) Elastic properties and microstructure of LPCVD polysilicon films, *J. Micromech. Microeng.*, **6**, 436–446.

Murarka, S.P. (1993) **Metallization: Theory and Practice for VLSI and ULSI**, Butterworth–Heinemann.

Ohmi, T. (2001) A new paradigm of silicon technology, *Proc. IEEE*, **89**, 394–412.

Puurunen, R. (2005) Surface chemistry of atomic layer deposition: a case study for the trimethylaluminum/water process, *J. Appl. Phys.*, **97**, 121301.

Ritala, M. et al. (1999) Perfectly conformal TiN and Al_2O_3 film deposited by atomic layer deposition, *Chem. Vapor Depos.*, **5**, 7.

Ruythooren, W. et al. (2001) Electrodeposition for the synthesis of microsystems, *J. Micromech. Microeng.*, **10**, 101.

Shacham-Diamand, Y. and V.M. Dubin (1997) Copper electroless deposition technology for ultra-large-scale-integration (ULSI) metallization, *Microelectron. Eng.*, **33**, 47.

Smith, D.L. (1995) **Thin-film Deposition: Principles and Practice**, McGraw-Hill.

Vehkamäki, M. *et al.* (2001) Atomic layer deposition of $SrTiO_3$, *Chem. Vapor Depos.*, **7**, 75.

Xia, L.-Q. *et al.* (1999) High temperature subatmospheric chemical vapor deposited undoped silicate glass, *J. Electrochem. Soc.*, **146**, 1181–1185.

6

Epitaxy

Epitaxy is a very special case of thin-film deposition: the deposited film will be single crystalline. This can only take place when special conditions are met. The deposited layer registers the crystalline information from the substrate. In order to do so properly, the crystal lattices of the film and the substrate must be identical or closely matching. The simplest case is homoepitaxy: film and substrate are the same material, for example silicon deposition on silicon. Because crystal information is "transmitted" across the substrate/film interface, the surface quality of the starting wafer is of paramount importance. A residual film a few atomic layers thick can prevent epitaxy. Epitaxy reactors are therefore designed with extreme cleanliness in mind, use the highest purity chemicals and are very delicate and expensive pieces of equipment.

Epitaxy is a demanding process and high-quality epitaxial films are difficult to make. Epitaxial deposition can fail partially and result in defective single crystalline material, or it can fail completely and result in polycrystalline or even amorphous film. Whether the defective material is usable for devices depends on the density and location of those defects: if defects are confined to substrate/film interface, and the deposited layer is mostly defect-free, the material may be usable, but this depends on the device operating principle and engineering judgment is needed to decide on acceptable defect levels.

6.1 Heteroepitaxy

Epitaxy on dissimilar materials is termed heteroepitaxy, with examples such as AlAs on GaAs, GaN on SiC and SiGe on Si. The lattice constants of various semiconductors are shown in Figure 6.1. The $Al_xGa_{1-x}As$ system is favorable because lattice constants of all GaAs and AlAs alloys differ by less than 0.2%, and multiple layers of AlAs/GaAs/AlAs type can be grown easily, with periods down to atomic layer thickness, equipment limitations allowing. Semiconductor lasers and solar cells can have tens of layers grown epitaxially (Figure 6.2).

Heteroepitaxy of silicon–germanium is an important application. Germanium is a group IV element like silicon, and they have identical lattice structures, so one basic requirement of epitaxy is fulfilled. Their lattice constants are, however, different: that is, silicon 0.543 nm, germanium 0.566 nm. Deposition of silicon with a small concentration of germanium will result in a single crystalline film because the silicon lattice will hold a small number of slightly larger germanium atoms in place (Figure 6.3). Deposition of pure germanium will result in a polycrystalline film because the lattice mismatch is too large to be accommodated.

There exists a critical thickness t_c (which depends on the Si_xGe_{1-x} lattice constant and therefore germanium fraction) below which mismatch can be accommodated by elastic deformation, as shown in Figure 6.4. Below the critical thickness the thin epilayer is strained to fit the silicon lattice, and above the critical thickness the lattice relaxes via misfit dislocations, and the film quality may become useless for device applications. In the strained metastable region the epilayer is thermodynamically unstable but kinetically prevented from finding a relaxed state.

It is possible to increase the germanium content gradually, and finally Si_xGe_{1-x} films with 50% germanium can be deposited epitaxially. SiGe films are under compressive stress, and if a silicon layer is deposited on SiGe, it will be under tensile stress.

There are other applications of heteroepitaxy on silicon: SiC on Si is intensely studied because SiC substrates are expensive, and silicon substrates would be readily available, also in large diameters. Shown in Figure 6.5 is a TEM micrograph of epitaxial oxide $<Y_2O_3>$ on silicon. As further proof of epitaxial film quality, epitaxial silicon is grown on $<Y_2O_3>$. Obviously this has to be the

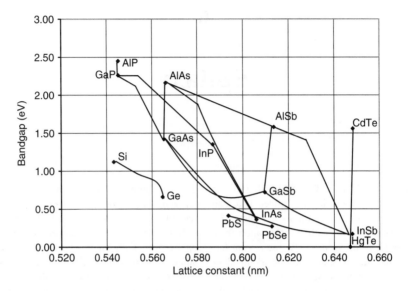

Figure 6.1 Lattice constants and band gaps of various semiconductors

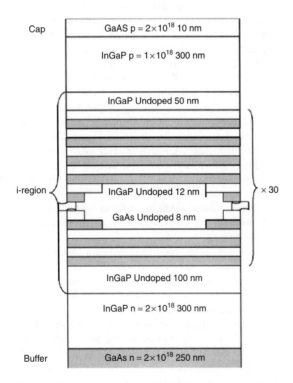

Figure 6.2 Superlattice structure of a quantum well solar cell with 30 periods of GaAs/InGaP. Reproduced from Magnanini *et al.* (2008) by permission of Elsevier

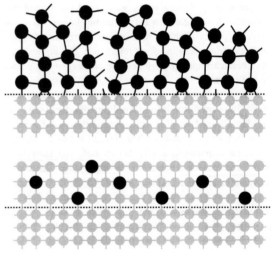

Figure 6.3 Germanium epitaxy on silicon is impossible because the lattice constants are too different. However, an alloy Si_xGe_{1-x} can be deposited because the silicon lattice can accommodate some germanium atoms

case: if the two materials have matching lattices and clean surfaces, epitaxy works both ways. Another example of heteroepitaxy is the growth of high-temperature cuprate superconductors on silicon: first single crystalline YSZ, yttria-stabilized zirconia, is grown on single crystalline

6.2 Epitaxial Deposition

Silicon epitaxy on silicon enables freedom in doping level and doping-type tailoring. A lightly doped epitaxial p-type layer ($\rho \sim 10$ ohm-cm) can be grown on a heavily p-doped substrate wafer ($\rho \sim 0.2$ ohm-cm). These types of wafers are used for microprocessors and other high-performance logic circuits. An n-silicon epitaxial layer on a p-substrate is used in many micromechanical devices because of electrochemical etch stop (Chapter 20).

In a CZ ingot resistivity depends on the position in the ingot as shown by Equation 4.6. Deposition of the epilayer equalizes wafers in this respect. Epilayer uniformity (both thickness uniformity and doping uniformity) is good, and if very tight resistivity specification is needed, epitaxial wafers override bulk silicon wafers. Another benefit of epitaxy is the absence of oxygen and carbon, which are always present in CZ silicon. About 20% of all starting wafers sold are epiwafers. But epitaxy is also as a part of device processing and this is extensively used in making bipolar transistors, see Chapter 27.

Boron dopant atoms are smaller than silicon, and the resulting boron-doped epilayer will be under tensile stress. Arsenic atoms are larger than silicon, and an arsenic-doped epilayer is under compressive stress. But dopant atom concentrations are fairly small (10^{15}–10^{17} cm^{-3} vs. 5×10^{22} cm^{-3} silicon atom density) and the effect is minor.

Epitaxy depends on crystal information and the energy of arriving atoms: they must have enough energy (surface mobility) to find energetically favorable sites on the surface. All epitaxy processes use elevated temperature, to give atoms surface mobility, and deposition rate also goes up with temperature, but too high a deposition rate is no good: there should be enough time for atoms to find their place on the crystal before the next layer is deposited.

CVD epitaxy of silicon with SiH$_{4-x}$Cl$_x$ ($x = 0$–4) source gases has been established since the 1950s. The basic chemical reactions are identical to polysilicon deposition (Equations 5.1 and 5.2). In the compound semiconductor field MOCVD (Metal Organic CVD, also known as MOVPE for Vapor Phase Epitaxy) is common. GaAs is deposited using precursors like GaCl$_5$ and AsH$_3$.

Molecular beam epitaxy (MBE) is a variant of evaporation. Instead of an open crucible, the source material is heated in a Knudsen cell. This cell consists of a crucible that is closed except for a small orifice, and atoms can escape from the cell only through the small orifice. The atom beam (in the molecular flow regime, hence the name MBE) emanating from the orifice is much more

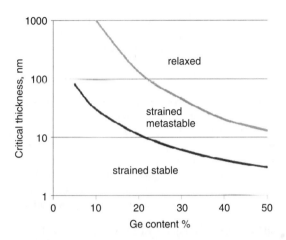

Figure 6.4 Si$_x$Ge$_{1-x}$ epitaxy on silicon: thicknesses and germanium concentrations refer to 600 °C growth. At higher temperatures critical thickness is smaller

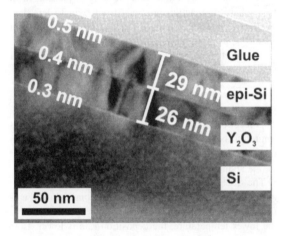

Figure 6.5 Epitaxial silicon on epitaxial yttrium oxide on silicon, TEM cross-section. Reproduced from Borschel et al. (2009), Copyright 2009, American Institute of Physics

silicon, followed by CeO$_2$ and finally by GdBaCuO superconductor film.

In silicon-on-sapphire (SOS), single crystal silicon is deposited on single crystal sapphire. The lattices of sapphire and silicon are different but if the sapphire crystal is properly oriented the apparent lattice constant is close enough to 0.543 nm of silicon. SOS was the first SOI technology, but it has largely been replaced by other SOI technologies.

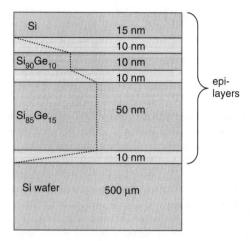

Figure 6.6 Thin heteroepitaxial Si$_{1-x}$Ge$_x$ layers for high-speed bipolar transistors. The hatched layers are graded epilayers with constantly changing germanium content

stable than a beam evaporating from an open crucible. This stability is tantamount to growing films with atomic layer thickness control.

In solid phase epitaxy (SPE) the solid film registers the crystalline structure from the underlying single crystalline substrate. Amorphous films on single crystalline substrates can thus be converted to single crystalline by annealing. Nucleation is important: a single crystal should be grown from one direction only. If there are many nuclei, the resulting film will be polycrystalline. Of course all the limitations of clean interfaces, matching lattice, etc., still apply. Epitaxy from the liquid phase (LPE) is also possible: both saturated solutions and melts can be used as sources for epitaxial growth. LPE was the dominant technology in the early days of III–V semiconductor laser and LED fabrication, but it has largely been superseded by gas phase and vacuum systems.

The number and thickness of epilayers are practically unlimited: in IGBT power transistors a moderately doped n-layer is grown first, followed by a thicker lightly doped layer. Layer thicknesses in power semiconductors are for example 50 μm (see Figure 2.10). In high-speed bipolar transistors (heterojunction bipolar transistors, HBTs) very thin epitaxial layers are used (in the tens of nanometers range), to engineer the band gap and therefore emitter performance (Figure 6.6).

6.3 CVD Homoepitaxy of Silicon

As an example of homoepitaxy, CVD silicon epitaxy is described (epitaxial reactors and growth process details are presented in Chapter 34). A reactor is heated to about 1200 °C under hydrogen flow, which reduces native oxide, so a clean silicon surface is obtained:

$$SiO_2\ (s) + H_2\ (g) \Leftrightarrow SiO\ (g) + H_2O\ (g) \quad (6.1)$$

Growth commences when silane gases of type SiH$_x$Cl$_{4-x}$ ($x = 0$–4) are introduced into the reactor: Equation 6.2 for dichlorosilane (DCS) and Equation 6.3 for silicon tetrachloride. Silane, SiH$_4$, can also be used. Two main differences to poly deposition are temperature and substrate conditioning: higher temperature equals higher deposition rate, higher surface mobility and therefore the possibility to find energy minimum sites on the surface:

$$SiH_2Cl_2\ (g) \Rightarrow Si\ (s) + 2\ HCl\ (g) \quad T = 1150\,°C \quad (6.2)$$

$$SiCl_4\ (g) + 2\ H_2\ (g) \Leftrightarrow Si\ (s) + 4\ HCl\ (g)\ T = 1250\,°C \quad (6.3)$$

Reaction 6.3 is reversible, and cleaning is possible with HCl when the reaction proceeds from right to left (i.e., hydrogen chloride etching of silicon). Excessive etching should be avoided because surface roughness tends to increase in etching. Silicon tetrachloride can also be used as silicon etchant:

$$SiCl_4\ (g) + Si\ (s) \Rightarrow 2\ SiCl_2\ (g) \quad (6.4)$$

This reaction can be prevented when the SiCl$_4$ fraction is limited to below 27% (Figure 6.7), and in practice much more dilute gases are used: a typical incoming gas consists of 1% silane and 99% hydrogen.

The SiCl$_4$ process temperature is, however, very high and undesirable dopant diffusion takes place during epitaxy. The SiH$_4$ reaction is better in this respect, but due to the lower temperature, the deposition rate is lower. Low temperature, and therefore minimal diffusion, are important considerations when sharp interfaces must be made. Trichlorosilane SiHCl$_3$ (TCS) and dichlorosilane SiH$_2$Cl$_2$ (DCS) are good compromises.

Typical epitaxial deposition rates are 1–5 μm/min. They depend on the silane gas chosen, on temperature and on flows. The deposition rate can be increased by operating at higher temperature, but at very high temperatures polycrystalline, rather than epitaxial, film results. At high temperature the growth rate is so high that there is no time for atoms to arrange themselves into a single crystalline lattice. At low temperatures the film atoms do not have enough energy to diffuse and find suitable lattice sites, and polycrystalline films result.

Perfectly flat surfaces offer no preferred sites for atoms to position themselves, and epi growth is therefore difficult. It can be aided by miscut wafers: instead of slicing

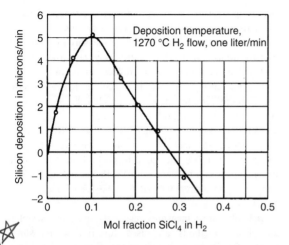

Figure 6.7 Silicon deposition rate as a function of SiCl$_4$/H$_2$ flow ratio. Above about 2–3 µm/min the resulting film is polycrystalline, not epitaxial. For SiCl$_4$/H$_2$ (1%/99%) the typical growth condition is 1 µm/min. Reproduced from Theurer (1961) by permission of Electrochemical Society Inc.

the ingot perfectly, a 3° miscut for example is used (typical for <111> material specifications), see Figure 4.10. Atomic steps so created act as nucleation sites for epitaxy. Arriving atoms can form more bonds at kinks, and therefore they are more stably bound than atoms on flat areas. Newly arrived atoms therefore arrange themselves regularly according to terraces determined by the crystal structure, while on planar areas various island structures can be formed randomly (Figure 6.8).

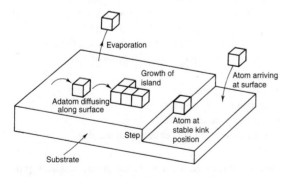

Figure 6.8 Terrace step kink (TSK) growth model of epitaxy: growth proceeds at kinks, and adatoms on flat surface diffuse to energetically favorable positions at kinks. Wafer miscut creates terraced structure. Reproduced from Jenkins (1995)

The term epi-poly is sometimes used in micromechanics. It is an oxymoron: epitaxial films are single crystalline, and poly means polycrystalline. What is meant is that a CVD epireactor has been used to deposit a thick layer of silicon, using epi growth conditions which enable deposition rates 1–5 µm/min versus 10 nm/min for LPCVD poly at 625 °C. But the underlying film is amorphous SiO$_2$, resulting in polycrystalline film. Typical epi-poly thicknesses are 10–50 µm, compared to 0.1–2 µm typical of LPCVD polysilicon, which is used as a CMOS gate and as a structural layer in surface MEMS. Thick poly is a popular material in micromechanics, combining some of the best sides of both polysilicon thin films and thick silicon layers.

6.4 Doping of Epilayers

Epilayer doping level and dopant type can be chosen independently of the substrate. Gaseous dopants, PH$_3$, B$_2$H$_6$, AsH$_3$, are added to source gas flow, enabling doping during epitaxy. Dopant concentration in the epitaxial film can be varied over seven orders of magnitude (10^{13}–10^{20} cm^{-3}). In many applications several epilayers with different doping levels and/or types are grown sequentially, or graded structures where the composition or doping level changes in minor steps, for example from Si to Si$_{0.7}$Ge$_{0.3}$ in tens of increments of germanium concentration.

Epitaxial deposition need not be the first process step: doped silicon is also single crystalline silicon and epitaxy on it works just as well. In bipolar transistor fabrication, buried layer formation by diffusion is the first step (see Figure 3.2), followed by epitaxial deposition of a lightly doped epilayer on top of a heavily doped buried layer. Base and emitter diffusions will then be done in this lightly doped epitaxial layer. Further discussion of epitaxy on structured wafers can be found in Chapter 27 on bipolar technology.

Because of the high temperatures involved, dopant diffusion will inevitably take place during epitaxy. If the epilayer doping level is lower than that of the substrate, the epilayer will be doped from the substrate through two different mechanisms: (1) solid state diffusion across the substrate/epilayer interface; and 2) dopant atom outdiffusion from the substrate into the gas stream and subsequent vapor phase doping, known as autodoping (Figure 6.9). Autodoping depends on the volatility of dopants, with antimony (Sb) giving the least (the lowest vapor pressure, with arsenic and boron somewhat higher, and phosphorus the highest). Autodoping comes both from substrate itself and from any doped regions that have been made in steps preceding epitaxy.

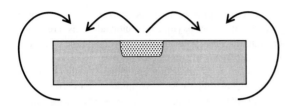

Figure 6.9 Autodoping: in substrate autodoping, dopants evaporated from heavily doped substrate add to intentionally added dopant; in lateral autodoping, dopants from heavily doped regions lead to local doping variation

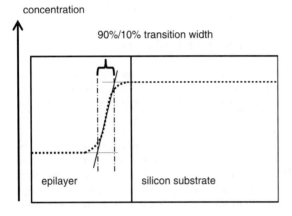

Figure 6.10 Transition width at substrate/epilayer interface; lightly doped epitaxial layer on heavily doped substrate

The abruptness of the doping at the substrate/epilayer interface depends on both deposition temperature and time, as well as dopant type. The transition width can be taken as the distance where doping concentration drops from a 90% value to a 10% value (Figure 6.10). Depending on whether the substrate or the epilayer is more strongly doped, the transition can take place either inside the epilayer or inside the substrate.

6.5 Measurement of Epitaxial Deposition

At least three measurements must be carried out on epitaxial wafers: thickness, resistivity and surface quality. Surface quality is assessed first and foremost by optical inspection: pyramids, mounds and hillocks scatter light which can be detected by optical methods. A Nomarski interference contrast microscope can detect surface height differences and IR depolarization reveals stresses. Laser scattering measures particles and microroughness. Optical methods are fast, and 100% of wafers are inspected.

The thickness of epilayers can be measured by Fourier transform infrared (FTIR) spectroscopy: constructive and destructive interference from reflections at the surface and at the substrate/epilayer interface are detected. FTIR spectroscopy requires, however, a highly doped substrate (resistivity below 0.025 ohm-cm). On resistive substrates, spreading resistance profiling (SRP) is used. SRP requires sample beveling: that is, it is sample destructive – 1 wafer in 25 or 1 in 100 is measured by SRP. Transition width measurement can be done with SIMS but this is a time-consuming measurement and is done for example once for 1000 wafers.

SRP of course measures resistivity also, but simpler and faster methods are used for routine measurements. Resistivity is measured by the mercury probe capacitance–voltage method (Hg-CV method) for p/p and n/n structures and by the four-point probe method for n/p and p/n structures. In both methods a metal contact is made on silicon, even though liquid mercury drop contact is much more benign than tungsten needle contact of 4PP. Wafers are not usable after metal probes. Non-contact measurements would be much better, but most are rather cumbersome and require special conditions to be fulfilled.

6.6 Simulation of Epitaxy

The epitaxy simulators currently used in process integration studies are not physically based. A true physical simulator would use temperature, flow rate and surface reaction rate constants as inputs, and it would reproduce growth rate and dopant distribution as the outputs. Instead, epitaxy simulators are really hybrids between film deposition and diffusion simulators: deposition rate and temperature are given, and dopant profile is calculated from diffusion constants at the relevant temperature.

The input for epitaxy simulator requires:

- dopant type of wafer
- deposition rate and time
- deposition temperature
- dopant type and concentration in the flow.

Such a semiempirical simulator can predict a dopant profile across the substrate/epilayer interface, taking into account both outdiffusion from the substrate and diffusion from the epilayer into the substrate.

Some estimates of gas phase dopant concentration and the resulting epilayer doping are given below:

Dopant in gas phase	Dopant in epitaxial film
10^{-10} bar	10^{15} cm^{-3}
10^{-8} bar	10^{17} cm^{-3}
10^{-6} bar	10^{19} cm^{-3}

Dopant partial pressures are very small indeed because epitaxy is carried out at atmospheric pressure. Extreme dilution in hydrogen is used to incorporate such dilute dopant flows. Note that phosphorus and boron incorporation into growing silicon is very strong: its concentration in film is much higher than its gas phase concentration. Arsenic incorporated into epitaxial film is even more pronounced.

A 1D simulation of epitaxy is shown in Figure 6.11. Epitaxial films nominally 4 µm thick are deposited. A heavily doped epilayer dopes the substrate, and the electrical junction is deeper than the nominal thickness of the epilayer. When the epilayer is lightly doped and the substrate heavily doped, the opposite is true. Note that the same deposition rate, 0.2 µm/min, has been used for all temperatures. This is a limitation in epitaxy simulation: rates are really temperature dependent, but they have to be given manually; they do not follow from first principles.

6.7 Advanced Epitaxy

If there are both oxide and single crystal silicon areas on the wafer, deposition will be epitaxial on silicon and polycrystalline on oxide. In between there is a region with ill-defined material: both its crystallinity and thickness are difficult to predict. In selective epitaxial growth (SEG) film is deposited only in those areas where single crystal silicon is present (Figure 6.12). SEG is a complex process and not fully understood. One process uses SiH$_2$Cl$_2$, HCl and H$_2$ as source gases. Chlorine etches nuclei on oxide faster than on silicon, and deposition is preferentially on silicon. Other factors like temperature, pressure,

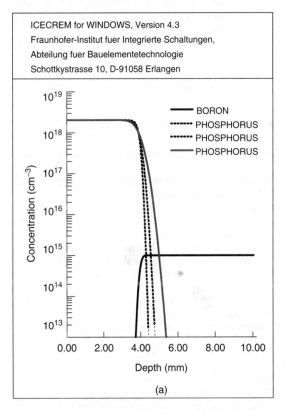

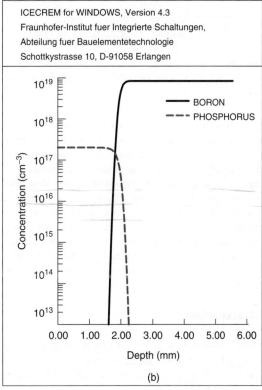

Figure 6.11 ICECREM simulation of epitaxial interface sharpness: three different growth temperatures (1050, 1100, 1150 °C) have been used to grow an epitaxial layer nominally 4 µm thick: (a) heavily doped epilayer on lightly doped substrate; (b) lightly doped epilayer on heavily doped substrate

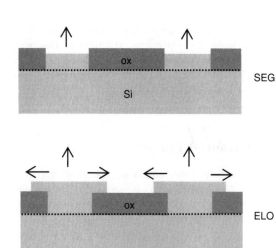

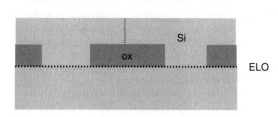

Figure 6.12 SEG, selective epitaxial growth, no deposition on oxide; ELO, epitaxial lateral overgrowth; merging of epitaxial film fronts over oxide

flow rates, partial pressures and the chemical nature of the dielectric (oxide, nitride) play their role in SEG. Selective epitaxy can be done many times over, as long as there is a high-quality single crystalline seed area available. Epitaxy requires crystal orientation information from the substrate, but once this information is registered, epitaxial growth can continue over amorphous or polycrystalline material. The epitaxial lateral overgrowth (ELO) technique incorporates patterned seed areas, oxide isolation and lateral overgrowth. One of the main problems in ELO is the point where the two growth fronts merge: defect density can be very high.

6.8 Exercises

1. Can a laboratory scale with 0.1 mg resolution be used for epilayer thickness measurements?
2. What are the resistivities of the substrates and epilayers in Figure 6.12?
3. Growth rates as a function of temperature are given below for SiH_4 epitaxy. Plot rate as a function of inverse temperature and check if there are two growth regimes!

700	750	800	850	900
0.04	0.09	0.2	0.4	0.5
950	1000	1050	1100 °C	
0.6	0.7	0.75	0.8 µm/min	

4. In Figure 6.7 the deposition rate is negative at high $SiCl_4$ fractions. What does this mean?
5. How abrupt (in nanometers) are transition widths in thick homoepitaxial silicon cases? What about SiGe epitaxy?
6. If a 200 mm wafer bows 20 µm due to an epitaxial layer 5 µm thick, what is the stress in the epilayer?

Simulator exercises:

7. For an n+/n− structure (substrate 10^{18} cm^{-3}, epilayer 10^{15} cm^{-3}) calculate the transition width as a function of epitaxy temperature for an epilayer 4 µm thick.
8. A starting wafer doping level is 10^{15} cm^{-3} phosphorus. The epilayer is dopes with boron doped at 10^{17} cm^{-3} concentration. Calculate the junction depth as a function of growth temperature.
9. If pnp-bipolar transistors are made, the buried layer has to be p-type. Calculate the boron updiffusion for different epitaxy conditions when the buried layer doping is 10^{18} cm^{-3} and the epilayer doping is 10^{15} cm^{-3}.

References and Related Reading

Baliga, J.B. (1986) **Epitaxial Silicon Technology**, Academic Press.
Beaucarne, G. *et al.* (2006) Epitaxial thin-film Si solar cells, *Thin Solid Films*, **511–512**, 533–542.
Borschel, C. *et al.* (2009) Structure and defects of epitaxial Si (111) layers on Y_2O_3 (111)/Si(111) support systems, *J. Vac. Sci. Technol.*, **B27**, 305–309.
Crippa, D., D.R. Rode and M. Masi (2001) Silicon epitaxy, in **Semiconductors and semimetals**, Vol. 72, Academic Press.
Herman, M.A., W. Ricter and H. Sitter (2004) **Epitaxy**, Springer.
Jenkins, T. (1995) **Semiconductor Science**, Prentice Hall.
Magnanini, R. *et al.* (2008) Investigation of GaAs/InGaP superlattices for quantum well solar cells, *Thin Solid Films*, **516**, 6734–6738.
Meyerson, B.S. (1992) UHV/CVD growth of Si and Si:Ge alloys: chemistry, physics, and device applications, *Proc. IEEE*, **80**, 1592.
Theurer, H. (1961) Epitaxial silicon films by the hydrogen reduction of $SiCl_4$, *J. Electrochem. Soc.*, **108**, 649.
Wu, Y.H. *et al.* (1999) The effect of native oxide on epitaxial SiGe from deposited amorphous Ge on Si, *Appl. Phys. Lett.*, **74**, 528.

7

Advanced Thin Films

In this chapter we will discuss thin-film deposition processes in more detail, and explore the relationships between process parameters and the resulting film properties. Process pressure and temperature, and ion bombardment during deposition, play their roles in resulting film qualities: grain size and orientation, or amorphousness; interface sharpness; volume homogeneity and surface smoothness/roughness. We will encounter a wide variety of cases involving double layer, multilayer, alloyed, reacted and other thin films. Some of these are stable, some are reacted intentionally, and others exhibit unwanted interfacial processes. These complex thin-film stacks have applications in acoustics, magnetics, optics, electronics and practically every field of microtechnology.

7.1 General Features of Thin-Film Processes

Thin-film deposition involves thermal physics, fluid dynamics, plasma physics, gas phase chemistry, surface chemistry, solid state physics and materials science. We must deal with source materials (sputtering targets, precursor gases, electrolytes), we must address the transport of source material to the substrate (in high vacuum, low vacuum, at atmospheric pressure or in liquid), and we have to understand surface processes (adsorption, reaction, desorption, ion bombardment-induced effects). Characterization of films entails dozens of techniques ranging from optical to nuclear, electrical to mechanical. This multidisciplinarity leads to a great number of phenomena and models which must taken into account, in both experimental work and simulation.

There are a few basic methods of source excitation and tens of different configurations of these. Thermal activation can be resistive, and electron, ion and laser beams can be used, too. Plasma sources range from simple DC diodes to microwave, helical and inductive configurations. In the liquid phase the choices are less numerous, and electrochemical and chemical potential differences are the main driving forces.

Film deposition on the substrate surface is a sum of many factors. In the first approximation, deposition is independent of substrate. This distinguishes deposition from growth processes, like thermal oxidation, which is intimately coupled with the substrate. But surfaces do interact with deposition processes via available chemical bonds, contamination and crystallography. An important parameter is the sticking coefficient, or the probability that an impinging particle will remain on the surface. A high sticking coefficient means that the particle will come to rest at the point of impingement, and a low sticking coefficient means that only energetically favorable attached species will stick, and others will desorb. Sticking coefficients range from 10^{-5}, typical of CVD, to approaching unity, in evaporation.

Even if no annealing is done immediately after film deposition, films will experience thermal treatments during subsequent processing. Thermal loads from these treatments can be considerable, and they affect many film properties, like grain size, resistivity and stress. Film surfaces and interfaces will be modified during these annealing steps by diffusion, dissolution or chemical reactions.

7.2 Film Growth and Structure

Atoms impinging on a surface attach to the surface either with chemical bonds (≈ 1 eV, chemisorption) or by short range van der Waals forces (≈ 0.3–0.4 eV, physisorption). Chemically bonded atoms tend to stay fixed on the surface, with only a small chance of desorption or diffusion. Physisorbed atoms are only loosely bound. This is essential in ALD: a monolayer is chemisorbed

---

Introduction to Microfabrication, Second Edition   Sami Franssila
© 2010 John Wiley & Sons, Ltd

but the temperature is selected to be so high that any additional physisorbed layer would not adhere.

Thin-film deposition is about phase transition from the vapor phase to solid phase. Atoms condense on a substrate. These adatoms (adsorbed atoms) are subject to desorption and surface diffusion. Some adatoms bond to each other, reducing the desorption probability. More atoms aggregate and some of the bigger clusters avoid desorption. These processes are depicted in Figure 7.1. These small nuclei are still mobile, and they grow by merging with other nuclei, but they can also incorporate atoms from the vapor phase. Nuclei grow in size to become islands, but remain separate, and more nuclei can form between the islands. Coalescence is driven by surface energy and surface area minimization. Island merging creates a continuous film. For sputtered and evaporated metal films this happens at about 10–20 nm film thickness. Films thinner than this are optically transparent but they can form an electrically conductive path (percolation). Such films have applications as permeable electrodes in gas sensors and as top metals in optical devices.

There are many modes of film growth. Island growth is also known as three-dimensional growth. It is common in metal deposition on insulators, where bonds between film atoms are stronger than bonds between film atoms and the substrate. Two-dimensional growth, also called layer-by-layer growth, takes place when the arriving film atoms bond more strongly with the substrate than with each other. It is possible to deposit a single atomic layer of, for example, nickel on mica. A third mode, called the Stranski–Krastanov mode, is a mixture of 2D and 3D modes. A fourth mode is columnar growth. These are pictured in Figure 7.2. When adatom mobility is close to zero, as in the room temperature deposition of tungsten, atoms stick where they land and grains grow upward. Columnar growth looks similar to island growth initially, but the difference is in coalescence: in the island mode there is enough energy for the film atoms to find energetically favorable sites (due to the high temperature in

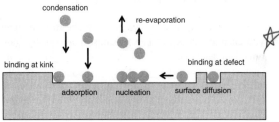

Figure 7.1 Surface processes in thin-film growth

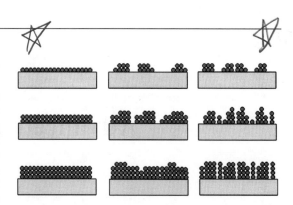

Figure 7.2 Thin-film growth modes: left, 2D (layer-by-layer) growth; middle, 3D (island) growth; right, columnar growth with voids between the grains

epitaxy and CVD, or due to ion bombardment in plasma deposition processes) and the islands will eventually coalesce, resulting in a uniform film. In columnar growth grains continue to grow upward, and voids between the grains remain. The surface will be rough, because grain height is determined by random nucleation.

If we measure the early stages of thin-film growth by surface-sensitive techniques, for example Auger electron spectroscopy or XPS (which probe the topmost nanometer only), we can distinguish 2D and 3D growth: in 2D growth mode the signal from the substrate quickly dies out because the whole surface becomes covered by the deposited layer. In 3D mode the substrate signal slowly decreases as the proportion of open substrate area is diminished.

Zone models explain the basic features of PVD film structure. The first question is: will the film be amorphous or polycrystalline? Silicon and other covalently bonding materials often end up amorphous. Many compounds (like TiN and Al_2O_3) and metal alloys (TiW, SiCr) with dissimilar-sized atoms similarly result in amorphous films. Elemental metal deposition usually results in polycrystalline films, including all the standard thin films in microtechnology (Al, Au, Cu, W, Ti, Cr, Pt).

The crystallinity of sputtered films is determined by a complex interaction between the substrate (its surface chemistry, surface structure and temperature) and the deposited film. In the zone model pressure and temperature are the main variables to explain film microstructure (temperatures are normalized to melting point temperatures, T/T_m, in K), see Figure 7.3. Zone 1 is small grained and porous, zone 2 has larger columnar grains and zone 3 exhibits still larger grains. The intermediate region is termed zone T (for transition).

Zone 1 is the region where the low momentum of impinging species is combined with slow chemical processes

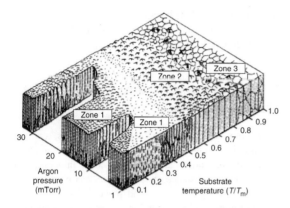

Figure 7.3 Zone model of sputtered thin-film microstructure. Reproduced from Thornton (1986) by permission of American Institute of Physics

due to low temperature: film atoms come to rest almost immediately, leading to columnar grains. Such a structure is under moderate tensile stress. The voids between the grains are nanometer sized, which leads to measurable density reduction and poor stability because atmospheric water and oxygen will be absorbed in the voids. Oxygen impurities in the deposition chamber can change the intrinsic stress from tensile to compressive and complicate the simple model described above.

At lower pressure ion bombardment becomes more important (fewer ion collisions and therefore more energetic bombardment). Ions densify the film by knocking out loosely bound atoms. A further increase in ion bombardment (at lower pressure or higher sputtering power) leads to the disappearance of voids and conversion to compressive stress. Higher temperature leads to enhanced surface diffusion, which will result in denser films as atoms can migrate and find energetically favorable sites to attach to.

Zone 2 occurs at $T/T_m > 0.3$, so surface diffusion is significant. Grains grow larger, and defects are eliminated. Zone 3 occurs at $T/T_m > 0.5$, where diffusion processes are very fast. Elimination of voids enhances diffusion. Films are annealed during deposition. Grains are more isotropic and films "lose memory" of the deposition process details.

Sputtered aluminum grain size is about 0.5 μm, similar to the typical film thickness. In 3 μm lines there are always many grains across the line, but in 0.5 μm lines the situation changes dramatically: there are practically no three-grain boundaries and the grains are end to end, known as the bamboo structure. All processes which depend on grain boundaries, like diffusion and electromigration, are strongly affected.

Film structure can change not only continuously as described above, but also abruptly. Tantalum films sputtered under different conditions can end up in either the body centered cubic (BCC) structure or as tetragonal β-Ta. The resistivity of BCC-Ta is about 20 μohm-cm with a temperature coefficient of resistivity (TCR) of 3800 ppm/°C. Values for β-Ta are about 160 μohm-cm and 178 ppm/°C, respectively (see Figure 5.1 for another tantalum deposition experiment). Phase transitions will also take place during annealing, for example $TiSi_2$ has several phases, see Figure 7.16.

Grains in polycrystalline films can have any crystal orientation, but in practice films are often strongly textured: the distribution of grain orientations is along one or two main crystal planes. For example, aluminum films have usually (111) texture, that is (111) planes parallel to the wafer surface. For undoped LPCVD polysilicon, (110) orientation crystals dominate, but for in situ phosphorus-doped poly, (311) is the dominant orientation. Grain structure of LPCVD polycrystalline silicon is shown in Figure 7.4. Because grains of different orientation grow at slightly different rates, the surfaces of polycrystalline films are generally rough.

Texture is established during deposition, and it is not greatly affected by subsequent annealing steps below $(2/3)T_m$, even though grain size increases. Texture inheritance is common: subsequent films easily acquire the same texture as the underlying film. Thin seed layers can therefore be used to modify thick layers. In hard disks

Figure 7.4 SEM of polycrystalline silicon film 6 μm thick deposited by LPCVD on a 1 μm oxide layer. Courtesy Lauri Sainiemi, Aalto University

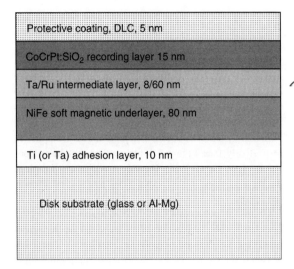

Figure 7.5 Hard disk structure: the Ta/Ru intermediate layer is used to induce crystal orientation in the CoCrPt:SiO$_2$ recording layer. Adapted from Piramanayagan (2007)

the Ta/Ru intermediate layer is deposited to induce [0002] orientation growth of the magnetic CoCrPt:SiO$_2$ recording layer, Figure 7.5.

7.3 Thin-Film Structure Characterization

PVD films, especially sputter-deposited films, can be modified by a number of deposition process parameters. System configuration and geometry come into play via target–substrate distance, base pressure, gas phase impurities, bias voltage. Pressure and power affect the momentum of impinging atoms and ions, and substrate temperature is important for desorption, diffusion and reactions.

Collimated sputtering is a technique where a mechanical grid is placed between the anode and cathode, and off-angle atoms do not contribute to the flux arriving at the wafer, but are deposited on the collimator walls. Collimated sputtering is good for bottom coverage (Figure 5.15). In Table 7.1 a collimated system is

Table 7.1 Sputtered titanium nitride (TiN) film characterization: collimated vs. standard

Film property	Analytical technique	Collimated TiN	Standard TiN
Thickness	RBS (density = 4.94 g/cm^{-3})	81 nm	161 nm
	TEM cross-section	82 nm	178 nm
Sheet resistance	Four-point probe	13.7 ohm/sq	7.4 ohm/sq
R_s uniformity	Four-point probe	3.3%	5%
Resistivity	R_s by four-point probe Thickness by TEM	112 µohm-cm	132 µohm-cm
Density	Thickness by TEM and RBS	4.88 g/cm^{-3}	4.47 g/cm^{-3}
	Density by RBS	93% of bulk	86% of bulk
Stoichiometry (Ti/N)	RBS	1.31	1.00
Phase	Glancing angle XRD	TiN (38–1420)	TiN (38–1420)
(JCPDS card #)	Electron diffraction	TiN (38–1420)	TiN (38–1420)
Preferred orientation	θ–2θ XRD	(220)	(220)
	Electron diffraction		
Net stress	Wafer curvature	2.7 GPa (tensile)	3.1 GPa (tensile)
Grain structure	Cross-section TEM	Columnar	Columnar
	Plan view TEM	2D equiaxial	2D equiaxial
Average grain size	TEM	19.2 nm	18.3 nm
Average roughness	AFM	0.43 nm	1.23 nm
Min/max roughness		8 nm	18.7 nm
Specular reflection (% of Si reference)	Scanning UV	248 nm: 142% 365 nm: 55% 440 nm: 57%	145% 95% 123%
Impurities (at. %)	AES	O < 1% C < 0.5%	O < 1% C < 0.5%

Source: Wang, S.-Q. and J. Schlueter (1996).

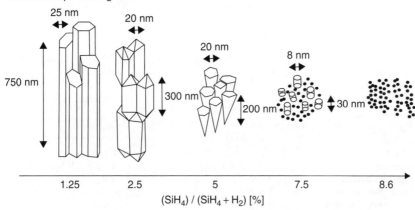

Figure 7.6 Microstructure evolution of silicon films deposited by PECVD. Grain size measurement by TEM; surface roughness by AFM. Reproduced from Vallat-Sauvain et al. (2000) by permission of AIP

compared to a conventional system, and analyzed for a range of film parameters. Such extensive film characterization is done when a new process is being developed and, if adopted in production, a simple monitoring measurement like sheet resistance can be used.

The electrical characterization described in Chapter 2 and in Table 7.1 was at DC, but circuits that operate at gigahertz frequencies must be measured at proper frequencies. The same applies to dielectric films, too.

The main parameters in CVD processes are the reactant gases, flow rates, flow rate ratios of the reactants, temperature and pressure. In PECVD, RF power and RF frequency play important roles. In Figure 7.6 on PECVD silicon deposition, the effects of $SiH_4/(SiH_4 + H_2)$ flow ratio on crystal size were studied. High-frequency (70 MHz) PECVD was employed, and glass wafers were used as substrates at 225 °C. Keeping all other deposition parameters constant, changes in gas ratio have resulted in enormous variations in grain size and surface roughness. In LPCVD polysilicon deposition using SiH_4 as a source gas, a similar grain size variation can be seen as a function of temperature: at 630 °C large grains (on the order of 100 nm) are formed, below 600 °C the grain size is reduced and at 570° the film is amorphous. The effect of temperature on MgF_2 grain structure can be seen in Figure 7.7: the lower the temperature, the smaller the grains.

Figure 7.7 Grain structure of MgF_2 by SEM as a function of deposition temperature: (a) 250 °C; (b) 300 °C; (c) 350 °C; (d) 400 °C. Reproduced from Pilvi et al. (2008), Copyright 2008, American Chemical Society

Dielectric films are also measured for a number of properties (Table 7.2). Again, this kind of extensive characterization relates to the research phase of new materials. Boron nitride is a new material that has been studied because of its potential as an insulator in multilevel metallization: it has a lower dielectric constant than nitride

Table 7.2 Film characterization: PECVD BN (boron nitride)

PECVD conditions:	Process A	Process B
gases	B_2H_6 (1%)/NH_3	$B_3N_3H_6/N_2$
flow rates	1800 sccm/120 sccm	100 sccm/200 sccm
RF power	500 W	200 W
pressure	660 Pa (= 5 torr)	400 Pa (= 3 torr)
temperature	400 °C susceptor	300 °C susceptor
Deposition rate	300 nm/min	370 nm/min
Uniformity	<5% (3σ)	3% (3σ)
Refractive index	1.746	1.732
Stress	−400 MPa (compressive)	−150 MPa (compressive)
Etch rate in RIE	62 nm/min	28 nm/min
Etch rate H_3PO_4 167 °C	1–11 nm/min	–
Etch rate BHF	0.5 nm/min	<1 nm/min
B/N ratio	1.02	1.02
Hydrogen content	<8 at. %	<8 at. %
Density	1.89 g/cm^3	1.904 g/cm^3
Structure	Amorphous	Amorphous
Step coverage	60% (1 × 1 μm)	80% (0.5 × 0.5 μm)
Optical band gap	4.7 eV	4.9 eV
Dielectric constant	3.8–5.7	3.8–5.7
Breakdown potential	6–7 MV/m	6–8 MV/m

Source: Cote, D.R. *et al.* (1995).

(3.8 vs. 6) and low etch and polish rates. It is not used in volume manufacturing. One special feature is the use of etch rate as a quality criterion. With dielectrics, thermal SiO_2 acts as a reference film which can always be used to eliminate etchant concentration or temperature effects.

Many of the measurements listed in Table 7.2 are often laborious and in production control ellipsometric or reflectometric thickness and refractive index measurements would probably be used.

IR spectroscopy measures molecular vibrations around 10 μm wavelength. This is indicative of chemical bonds, because IR vibrations are typically bond stretching and bending vibrations. IR spectroscopy is most often practiced using an interferometric measurement set-up known as FTIR, for Fourier transform IR. In silicon nitride thin films it is used to see the presence and qualitative abundance of Si–N, Si–H, N–H bonds (Figure 7.8).

7.4 Surfaces and Interfaces

The surface roughness of thin films varies considerably. In general, high-temperature deposition results in smoother films. Epitaxial films are of course very smooth, but many amorphous films can also be extremely smooth. There is a strong correlation between surface smoothness and volume homogeneity: thermal oxide, amorphous silicon and TEOS oxide are both smooth and homogeneous, whereas doped polysilicon and columnar tungsten are rough and

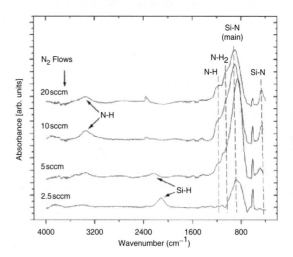

Figure 7.8 FTIR analysis of silicon nitride films deposited by PECVD: changing nitrogen flow results in different proportion of Si–H, Si–N and N–H bonds. Reproduced from Biasotto *et al.* (2008) by permission of Elsevier

inhomogeneous. Volume inhomogeneity makes measurement of thin-film properties difficult. It is usual then to treat the film as if it were a stack of many layers, each with slightly different properties, for example interfacial mixed layer, bulk of film and surface layers modeled as three materials each with materials constants of their own.

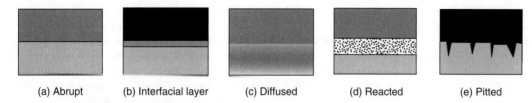

(a) Abrupt (b) Interfacial layer (c) Diffused (d) Reacted (e) Pitted

Figure 7.9 Possible interface structures

Thermodynamics gives hints of interface stability. A change in Gibbs free energy is given by

$$\Delta G = G_{products} - G_{reactants} \quad (7.1)$$

ΔG is positive for a stable pair of materials. For the reaction of titanium with silicon dioxide

$$Ti + SiO_2 \Rightarrow TiO_2 + Si \quad (7.2)$$

the change in Gibbs free energy is $\Delta G = G_{TiO_2} - G_{SiO_2} = (160\text{–}165)\,\text{kcal} = -5\,\text{kcal}$, indicative that the reaction can take place. Thermodynamics, however, is about initial and final states, not about rates: some thermodynamically favorable processes are so slow that practical devices can be made. But if thermodynamics forbids a reaction, it cannot proceed: the change in Gibbs free energy for the cobalt/silicon dioxide reaction is positive, and cobalt does not reduce oxide. This means that cobalt silicide formation, Equation 7.3, is very sensitive to the presence of native oxide, while the titanium–silicon reaction, Equation 7.4, can proceed in spite of a thin oxide layer:

$$Co + 2\,Si \rightarrow CoSi_2 \quad (7.3)$$

$$Ti + 2\,Si \rightarrow TiSi_2 \quad (7.4)$$

Interface types also vary significantly. Abrupt interfaces (Figure 7.9a) are not only idealizations but encountered in epitaxy, and other methods, ALD, CVD, PVD, electrochemical, also produce almost ideally sharp interfaces. However, native oxides are very often encountered on interfaces (Figure 7.9b), but in many cases those about 1 nm films are broken or dissolved in subsequent annealing steps.

The case of silicon dioxide/copper (Figure 7.9c) shows copper diffusion into the oxide. A silicon/titanium pair will react and form silicide (Figure 7.9d). Many metals do form silicides: copper silicides form at very low temperature, 200–300 °C, nickel, cobalt and titanium at 400, 500 and 600 °C respectively. Tungsten, molybdenum and tantalum will also form silicides, not all of them simple $MeSi_x$ compounds but complex mixtures of various silicides, for example Me_2Si_5, Me_2Si_3, $MeSi_2$, $MeSi$.

Aluminum reacts with tungsten to form $Al_{12}W$ and titanium forms Al_3Ti.

Aluminum does not form a silicide. Annealing at 425 °C will dissolve native oxide, ensuring good electrical contact. However, too much annealing will lead to pitting: silicon is soluble in aluminum (as shown in the Al–Si phase diagram, Figure 7.10) and an open volume is left behind as silicon atoms migrate into the aluminum. Aluminum, on the other hand, will diffuse to fill the space left by silicon dissolution. This leads to the case depicted in Figure 7.9e. These aluminum

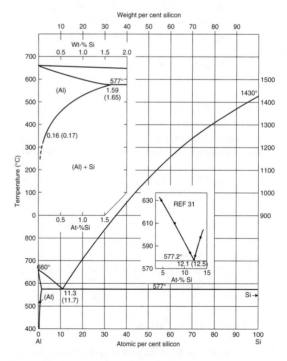

Figure 7.10 Aluminum/silicon phase diagram. Reproduced from Hansen and Anderko (1958) by permission of McGraw-Hill

spikes can be micrometers deep and extend beyond the pn junction. To prevent junction spiking, aluminum can be alloyed with silicon: a silicon concentration of 0.5% (wt %) will saturate aluminum at 425 °C, and 1% Si will prevent silicon dissolution at 500 °C. The other, more general solution is to implement a diffusion barrier, to be discussed shortly.

7.5 Adhesion

Adhesion is a major issue in thin-film technology. As a rule of thumb, poor adhesion is the norm and only special attention will lead to good adhesion. Some materials have poor adhesion due to their chemical nature: noble metals are noble because they do not react, therefore they do not form bonds across interfaces. Adhesion is also related to surface cleanliness: residues or dirt from previous steps will almost inevitably lead to poor adhesion. Deposition process variables do play a role: in sputtering energetic ions and atoms will knock out contamination and loosely bound film atoms, but in evaporation the arriving atoms do not have enough energy to displace weakly bonded atoms.

Adhesion layers are additional films with the role of improving adhesion and, to a first approximation, have no effect on device structure or operation. Adhesion layer films are selected on the basis of their bond forming abilities: titanium and chromium are the two most widely used materials, and other oxide forming metals also show good adhesion, like aluminum. The thickness of the adhesion layer is in the range of 10 nm because it has the role of a surface only. Adhesion layer and structural film are deposited immediately after each other in the same vacuum chamber: the freshly formed adhesion layer surface ensures cleanliness and thus eliminates one main factor of poor adhesion. Typical pairs of adhesion layer/noble metal include Ti/Pt and Cr/Au (and vice versa). Adhesion layers are also useful for near-noble refractory metals like tungsten.

The first adhesion test is the tape pull test: adhesive tape (standard office tape is commonly used) is attached to the thin film and then pulled off. If the film peels off with the tape, it has failed the adhesion test. More advanced tests use quantifiable pull force.

7.6 Two-Layer Films

Two layers of thin films can offer performance benefits compared to single film, as discussed in connection with Figure 5.19 and in connection with adhesion. There are many other reasons for adopting two-layer films. If very thin films are made, depositing two layers reduces the possibility of defects: it would be improbable for two defects to coincide. This is utilized in making masks for deep etching of glass. Similarly, metal wires consisting of two different metals can survive even if one film is destroyed, by for example electromigration or corrosion (see Chapter 36). Stress compensation is often done by inserting a thin silicon dioxide film underneath silicon nitride. The tensile stress in the nitride is compensated by the compressive stress in the oxide. In optical MEMS the mirror membrane material is chosen on mechanical grounds, and the choice is often silicon or nitride. But in order to have high reflectance, the mirror can be coated by, for instance, gold or aluminum.

Bimorph cantilevers (of the type shown in Figure 5.18) can be used as temperature sensors: unequal CTE leads to cantilever bending. In IR sensors (bolometers and thermopiles) there is usually a sensor layer and the absorber layer is separate (with sometimes an extra aluminum layer underneath to reflect non-absorbed light). In electrowetting microfluidics the electrodes are covered by a hydrophobic thin film.

Barriers are additional layers between two materials. Their role is to separate adjacent layers. Many aspects of barriers are similar to adhesion layers: barriers are not needed for device operation as such, but their presence makes either the device fabrication process more robust or the resulting device more stable. Barriers are thin, like adhesion layers, with 10–100 nm as the typical barrier thickness.

Total barriers must prevent all fluxes through them, both atom diffusion and charge carrier transport. In flat-panel displays ion barriers are needed: sodium diffusion from glass sheet to transistors has to be prevented. When solar cells are processed on steel sheets, iron diffusion must be prevented. In flexible electronics, gas barriers are needed: oxygen diffusion through a polymer sheet into active devices must be prevented. The common solution is to deposit an inorganic film, typically oxide or nitride, by PECVD, ALD or spin coating. Barriers are needed on both sides of the polymer because its permeability easily allows oxygen diffusion through polymer sheets hundreds of micrometers thick.

In the case of metallization, current has to flow through the barrier, but metal mixing by diffusion must be prevented. Metallic barriers have relatively loose requirements for resistivity (the distance is less than 100 nm). Most barrier materials have resistivities around 100–500 μohm-cm, one to two orders of magnitude higher than conductors. While resistivity is not a problem for barriers, contact resistivity must be low.

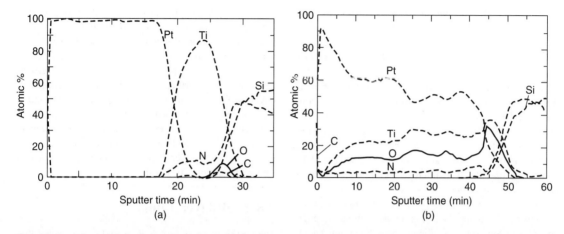

Figure 7.11 Auger depth profile of Pt/Ti/SiN$_x$/Si structure: (a) as deposited; (b) oxygen annealed at 600 °C: interdiffusion of films is almost complete. Oxygen and carbon accumulations on the surface in the as-deposited sample indicate cleaning problems. Reproduced from Kang *et al.* (1999) by permission of Institute of Pure and Applied Physics

The first diffusion barriers to be implemented were TiW films 100 nm thick between aluminum and silicon to prevent Al–Si junction spiking. Early experiments in a poor vacuum led to the incorporation of oxygen and nitrogen, which passivated grain boundaries without the researchers noticing the real effect behind the barrier performance. When the mechanism was elucidated, reactive sputtering of TiW in an argon/nitrogen atmosphere was adopted. It led to 10 nm grains and nitrogen incorporation at grain boundaries, both of which led to improved barrier performance. Amorphous films are preferable as barriers and a-WN is one candidate.

Copper metallization needs barriers between the copper and silicon to prevent a silication reaction (formation of for example Cu$_3$Si). Tantalum nitride, TaN, is one possible choice. But a barrier is needed also between the copper and oxide because copper diffuses into the oxide. Silicon nitride can be used.

Adhesion layer and diffusion barrier stability can be checked by electrical and physical measurements. Sheet resistance increase is a quick and simple measurement. Copper resistivity is very low, ca. 2 μohm-cm, and when the barrier fails, copper reacts to form silicides which are more resistive. They can be identified by X-ray diffraction but the resistance increase is indicative of silicide formation. Diode leakage at the pn junction is another quick electrical measurement.

Auger depth profiling can be used to see barrier reactions. Auger measurement is slow and destroys samples, but it can be done without any sample preparation. Usually an as-deposited sample is compared to an annealed sample(s), and barrier failure is evidenced by intermixing of metal and silicon across the barrier. The accumulation of material at the interfaces, and atom distributions across the film, are helpful in understanding the reactions behind adhesion failure.

Note that the Auger analysis shown in Figure 7.11 does not indicate TiO$_2$ formation even though coexistence of the titanium and oxygen might suggest it: Auger is about atoms and not about compounds. XRD could show TiO$_2$ formation by the appearance of diffraction peaks identified as arising from TiO$_2$.

7.7 Alloys and Doped Films

In addition to elemental metals, many alloys are routinely used in microtechnology: amorphous TiW (about 30 at. % titanium), SiCr and NiCr for resistors. Early integrated circuits used aluminum for metallization. Aluminum–silicon (1% Si) was adopted, and later Al–Si was replaced by Al–Si–Cu for improved electromigration resistance. Similarly, dielectrics can be doped and alloyed. Fluorine-doped oxide, SiO$_x$F$_y$, can be deposited by PECVD, but film instability limits the usable fluorine concentration to about 5% (by weight), because film stability deteriorates and the material becomes hygroscopic. Other materials deposited by PECVD include carbon-doped oxide SiO$_x$C$_y$ and SiC$_x$N$_y$, which are used as intermetal dielectric layers in multilevel metallizations.

CVD oxide can be doped by adding phosphine (PH$_3$) gas to the source gas flow. Phosphorus-doped CVD oxide, PSG (Phosphorus-doped Silica Glass), is a widely used doped film. Phosphorus oxide is formed by CVD and intermixed with silicon dioxide:

$$4\ PH_3\ (g) + 5\ O_2\ (g) \Rightarrow 2\ P_2O_5\ (s) + 6\ H_2\ (g) \quad (7.5)$$

Doped oxide films typically have about 5% dopant by weight. Higher doping levels lead to porous, hygroscopic material. The toxicity of PH_3 (and B_2H_6 for BSG) needs to be accounted for, but CVD reactors use silane, which is a flammable gas, so that the basic designs of CVD reactors are suitable for dangerous gases. TMP (trimethyl phosphite) and TMB (trimethyl borate) are less toxic alternatives to hydrides.

Polycrystalline SiGe (polySiGe) can be used instead of polysilicon. Germane, GeH_4, is mixed with silane, SiH_4, and almost any ratio of silicon and germanium can be made. Most MEMS applications use $Si_{50}Ge_{50}$ or similar. The main benefit of polySiGe is its lower stress annealing temperature, which makes it easier to implement MEMS structures together with ICs. SiGe can also be etched selectively against silicon, enabling free-standing silicon structures to be made.

Sputtering is suitable for alloy deposition. In alloy sputtering the flux is enriched in the component with higher sputtering yield (yields from alloys are even less accurately known than yields from elemental solids; elemental solid yields are used as approximations). The composition of the sputtered flux is given by

$$C_{\text{film}} = \frac{Y_a}{Y_b} \frac{X_a}{X_b} \quad (7.6)$$

where the X_i are the concentration proportions in the target (they add to unity: $X_a + X_b = 1$). Because matter is conserved, the target is enriched in the other component

$$C_{\text{target}} = \frac{Y_b}{Y_a} \frac{X_a}{X_b} \quad (7.7)$$

A steady-state situation develops and composition remains unchanged. Sputtering in a reactive atmosphere, in argon/nitrogen or argon/oxygen mixtures, results in nitride or oxide films, or stuffed films with small amounts of reactive impurities at grain boundaries. Molybdenum sputtering in a high-nitrogen-content atmosphere results in Mo_2N compound formation, but a low-nitrogen sputtering atmosphere results in film designated as Mo:N, nitrogen-stuffed molybdenum. It is very close to molybdenum in chemical composition, but the minuscule nitrogen has an important role: it prevents crystallization, resulting in amorphous films. Typical applications of reactive sputtering are TiN, Ta_2O_5, ZnO, AlN, TiW:N and WO_3. Often reactively sputtered films

are non-stoichiometric, and an annealing step (e.g., in oxygen) is needed to improve film quality.

The introduction of small amounts of nitrogen or oxygen into an argon plasma does not appreciably change the properties of the plasma discharge or of the growing film, but after a critical partial pressure is reached, the target surface transforms into nitride or oxide, and the plasma discharge is established at another equilibrium. If reactive gas flow is then reduced, the target remains nitrided/oxidized, and return to initial conditions takes places at much lower partial pressures, that is reactive sputtering exhibits hysteresis.

7.8 Multilayer Films

The performance of simple elemental or compound films, with or without barriers or adhesion layers, is often not enough, and multilayer films are introduced to offer improvement.

For many generations (0.8–0.5–0.35–0.25 μm) IC metallization was done with a Ti/TiN/Al/TiN film stack. Titanium acts as an adhesion layer, TiN as a diffusion barrier, Al as a current-carrying film and the top TiN has the dual role of mechanically stiffening the structure and reducing reflectivity (important for lithography). Metallization reliability has been greatly improved by adopting such multilayer metallization schemes, but a price has been paid elsewhere: the etching of such multilayer structures is difficult.

Dielectric mirrors with λ/4 layer thicknesses for high-reflectance surfaces are one example of multilayer films. Undoped polysilicon, oxide and nitride are the usual films. For visible wavelengths, layer thicknesses around 100 nm are typical. Similar λ/4 structures are used in thin-film bulk acoustic resonators (TFBARs): multilayers of W:SiO_2, with thicknesses about 1.5 μm, act as acoustic mirrors, see Figure 7.12.

In PECVD, oxynitride films of composition SiO_xN_y can be easily made. By tailoring the composition, the refractive index can be varied from 1.46 to 2, covering the full range of SiO_2 and Si_3N_4 refractive indices. By sandwiching the SiO_xN_y film between two films of lower refractive index, it acts as a waveguide (Figure 7.13). Doping of oxide by for example phosphorus (PSG) or germanium can also be used to tailor the refractive index, but only over a limited range before other film properties change too much.

7.9 Selective Deposition

Both CVD and electrochemical processes can be used for selective deposition, with electroless copper and CVD

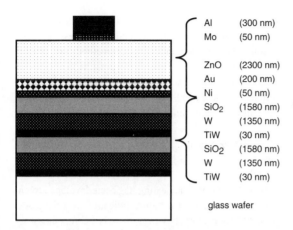

Figure 7.12 Bulk film acoustic resonator (FBAR) structure on a glass wafer: a piezoelectric ZnO resonator is sandwiched between gold and aluminum electrodes. TiW, Ni and Mo are thin adhesion promotion layers. W and SiO_2 form a $\lambda/4$ acoustic wavelength filter. Adapted from VTT Microelectronics Annual Research Review (2001)

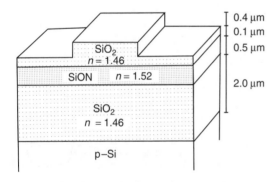

Figure 7.13 Refractive index $SiO_2/SiO_xN_y/SiO_2$ waveguide: n_f 1.46/1.52/1.46. Reproduced from Hilleringmann and Goser (1995) by permission of IEEE

tungsten being the most studied ones. The silicon surface reduction process allows selective CVD tungsten in contact holes:

$$2\ WF_6\ (g) + 3\ Si\ (s) \Rightarrow 2\ W\ (s) + 3\ SiF_4\ (g) \quad (7.8)$$

This reaction is selective because SiO_2 does not reduce WF_6. However, about 20 nm of silicon is consumed, and the reaction is self-limiting: WF_6 cannot diffuse through the growing tungsten layer. Tungsten deposition

Figure 7.14 Selective deposition: problems with unequal hole depths and spurious nucleation

is continued by silane reduction of tungsten hexafluoride on tungsten according to

$$WF_6\ (g) + 2\ SiH_4\ (g) \Rightarrow W\ (s)$$
$$+ 3\ H_2\ (g) + 2\ SiHF_3\ (g) \quad (7.9)$$

This reaction, however, is mass transport limited, and difficult to control. Additionally, it faces problems when contact holes of different depths have to be filled: some are underfilled, some are overfilled (Figure 7.14).

Plug fill can be achieved by continuing deposition in hydrogen reduction mode, Equation 5.7. There is always the problem of selectivity loss. It is usually connected with residues from the preceding process steps, for instance incomplete resist removal. Selective deposition processes are rare in volume manufacturing even though they sometimes offer enormous simplifications in process integration.

7.10 Reacted Films

A rather interesting class of conducting thin films is the silicides, compounds of silicon and metals, for example $TiSi_2$, $CoSi_2$, NiSi, WSi_2 and PtSi. Silicides combine the good properties of silicon and metals, such as high-temperature stability and low resistivity, with the lowest values of about 15 µohm-cm. Silicide CTEs are typically $15 \times 10^{-6}/°C$. Young's moduli for silicides are on the order of 100 GPa.

Silicides are formed by many methods, for instance by CVD or alloy sputtering. Tungsten silicide WSi_2 is made by CVD. Most often silicides are made by solid state reaction of metal film and silicon, for example $Pt + Si \rightarrow PtSi$. The interesting feature of reacted silicides is that they can be made without silicide etching, in a self-aligned

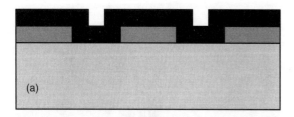
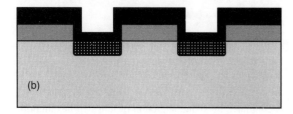
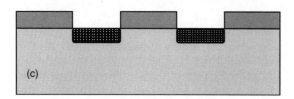

Figure 7.15 Silicide formation by metal/silicon reaction: (a) metal sputtering on oxide pattern; (b) reaction at metal/silicon interface, no reaction on oxide; (c) selective etching of unreacted metal leaves silicide

formed under the original surface. This volume expansion means that reacted silicides are under compressive stress. Various silicides and their formation methods are listed in Table 7.3.

Titanium silicide is formed by annealing in argon or nitrogen. Typical conditions are about 750 °C and 30 s in rapid thermal annealing (RTA) equipment (see Chapter 32 for more on RTA). A simple one-step anneal in argon, which would produce a predictable thickness of titanium silicide, is not usable because silicide grows over oxide laterally.

Two-step nitrogen annealing has been developed to ensure silicide formation without silicidation over the oxide. In nitrogen annealing, however, there is a competing reaction taking place at the surface: titanium nitridation. First, annealing has to be optimized so that the silicon/titanium reaction (TiSi formation) at the interface is faster than the gas phase nitridation of titanium into TiN. TiN film is beneficial because it suppresses the lateral growth of silicide over the oxide. The first anneal results in C49 phase $TiSi_2$ which has fairly high resistivity, see Figure 7.16.

mode, Figure 7.15. This offers possibilities in advanced MOS transistor metallization (to be discussed in Chapter 26). The desired pattern is defined in the oxide, and metal is deposited. Metal/silicon reaction takes place during annealing in those areas where metal and silicon are in contact, but the metal does not react with the oxide. The unreacted metal can be etched away to leave the silicide and oxide.

The surface of the resulting silicide is approximately at the level of the original silicon surface, and the silicide is

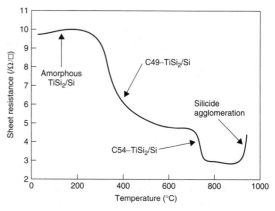

Figure 7.16 $TiSi_2$ C49 to C54 phase transitions followed by sheet resistance measurements. Reproduced from Mann *et al.* (1995) by permission of IBM

Table 7.3 Silicide properties

Silicide	Resistivity (μohm-cm)	Formation	Selective etch
$TiSi_2$	15–20	Ti/Si reaction at about 750 °C	$NH_4OH:H_2O_2$
$TiSi_2$	15–20	CVD $TiCl_4/SiH_2Cl_2/H_2$	–
$CoSi_2$	15–20	Co/Si reaction at 500 °C	$HCl:H_2O_2$ 3:1
NiSi	15–20	Ni/Si reaction at 400 °C	HNO_3
WSi_2	30	CVD WF_6/SiH_2Cl_2 at 400 °C	–
PtSi	30	Pt/Si reaction	$HCl:HNO_3$ 3:1

In the case of nitrogen annealing, we have to remove not only unreacted metallic titanium but also TiN, so we need to know the selectivity for both Ti:TiSi$_2$ and TiN:TiSi$_2$ pairs. The thickness of titanium cannot be simply calculated from the titanium, silicon and TiSi$_2$ densities because some titanium is consumed by the TiN formation reaction. TiSi$_2$ thickness is also reduced by the fact that selective etches are not infinitely selective: some TiSi$_2$ is lost during titanium etching. If titanium thickness is scaled down and the rest of the process is unchanged, TiSi$_2$ thickness will decrease more than predicted by the simple metal-to-silicide relation because surface-nitrided thickness is independent of titanium thickness.

The second anneal transforms silicide C49 into C54 phase which has a resistivity of about 15 µohm-cm. This resistivity of C54 titanium silicide is lower that that of many thin-film metals. Anneal temperature must be at least 700 °C to effectuate the phase transformation but in practice 850 °C for 30 s is usually used, see Figure 7.16. Too high an annealing temperature leads to silicide agglomeration: the silicide balls up and the film breaks down.

7.11 Simulation of Deposition

Topography simulation (for deposition, etching and polishing) works on fluxes and surface processes: at each grid point the incoming flux (from the fluid phase) and surface reaction probability are evaluated (with return flux of reaction products in the case of etching/polishing, or non-sticking species in the case of deposition) to calculate the new surface height. In principle the generation of incoming species could be simulated (for instance, ion and radical production in plasma) but this is usually not integrated into a topography simulator; rather, it is part of a reactor simulator. New surface points are calculated and those points are connected to represent the surface. Accuracy is increased by calculating new points between existing points when the points are far apart and, similarly, by eliminating points that become close to each other.

Deposition models define atom arrival angles, and various models are available in most simulators: fully directional, hemispherical, conical, etc. Etch models include isotropic and anisotropic models and user-definable mixtures of the two. Model selection is very much an empirical question, and the predictive power of topography simulation is diminished by this semiempirical tailoring of model parameters.

The input for a typical topography simulation includes:

- surface topography already made
- material to be deposited
- deposition model (angular distribution of depositing species)
- thickness/rate and time.

Adjustable parameters include surface diffusivity, which determines how much lateral movement an impinging species is allowed before it is "frozen" in the growing film.

The topography simulator SAMPLE 2D, developed at the University of California, Berkeley, has been used to obtain the profiles shown in Figure 7.17. The hemispherical deposition model is an approximation of sputter deposition. Trench dimensions have been varied to see the effect of aspect ratio on step coverage. In a 1:1 aspect ratio trench the step coverage is about 15%, but in a 2:1 aspect ratio trench only a meager 5%. A slightly sloped profile in the 2:1 trench leads to about 10% step coverage.

Note that step coverage over isolated lines is always the same irrespective of line aspect ratio: step coverage

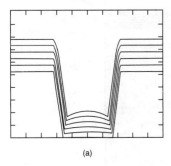
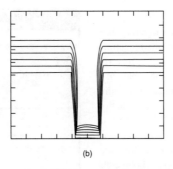
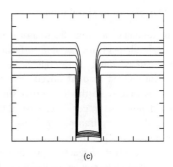

(a)   (b)   (c)

Figure 7.17 Simulation of deposition step coverage with SAMPLE 2D. Hemispherical deposition model corresponds to sputtering. Trench widths are 1 µm and 0.5 µm, depth is 1 µm. Wall angle either 90° or 81°. Film thickness is 0.5 µm on planar top surface.

Figure 7.18 A 3D Monte Carlo simulation of aluminum deposition into a contact hole: (a) high rate deposition; (b) low rate deposition. Both depositions are at the same temperature. The simulation is 3D, but only a cut through the contact hole centerline is shown. Reproduced from Baumann and Gilmer (1995) by permission of IEEE

depends on atom arrival angles and, by definition, isolated lines have large unobstructed spaces next to them, therefore identical step coverage will result.

Monte Carlo (MC) and molecular dynamics (MD) simulations offer more realism, for example the prediction of step coverage based on relaxation (Figure 7.18). Calculations can be speeded up by treating matter as 100 Å cluster spheres instead of individual atoms. Clusters, and the thus the atoms, come to rest at stable positions, for example when touching three other spheres. The arrival of new material and rearrangements of already deposited films can be simulated simultaneously. Temperature and sticking coefficient are used as parameters for surface mobility.

A 2D simulation can overestimate bottom coverage by 40%, compared to a 3D one. This is intuitively easy to understand because 2D simulation treats recesses as infinitely long trenches, with very large acceptance angles along the trenches, whereas 3D simulation takes into account the real acceptance angle.

7.11.1 Scales in simulation

The fundamental simplification of many topography/thin-film simulators is the fact that surface-controlled reactions are assumed. On a microscopic scale this is true: material is being added to or removed from a surface, but on a macroscopic scale this is a gross simplification. Etching and deposition processes can be either surface reaction limited or mass transport process limited. Transport of reactants from gas flow to the surface (as in a CVD reactor) or removal of reaction products by convection (like removal of hydrogen bubbles that result from silicon etching) can be more critical to etching or deposition than the surface processes. Whether surface reaction or mass transport is the mechanism which determines the reaction rate has to be studied for each process. If the reaction is transport limited, then the simulation should be able to model fluid dynamics at the reactor scale, in addition to surface processes at the micrometer scale.

7.12 Thickness Limits of Thin Films

The thinnest films are obviously one atomic layer thick, as evidenced by ALD and MBE deposited films. Very thin films are often discontinuous and the thickness required for continuous films is process and material dependent. One criterion is transparency, which translates to about 10–20 nm.

PVD and CVD techniques are suitable for films in the thickness range of 10–1000 nm. This is partly a practical limitation due to deposition rates, which are generally 1–100 nm/min. In many cases thicker films are desired, and PVD or CVD methods are too slow. In CVD silicon epitaxy, 100 μm layer thickness is feasible because epitaxial growth rates can be as high as 5 μm/min and the limit is economical, physical. For most polycrystalline and amorphous CVD and PVD films, however, stresses build up to unacceptable levels for thicker films, limiting thicknesses to a few micrometers. Electroplated films can be very thick, and through-wafer plating is fairly standard. Deposition rates can be micrometers/minute, even though they are usually much smaller for thinner films.

The read/write heads of hard disk drives involve multilayer structures. Shown in Figure 7.19 is the film structure of a giant magnetoresistive (GMR) read/write head. It is only 39 nm thick, but there are nine layers, the thinnest

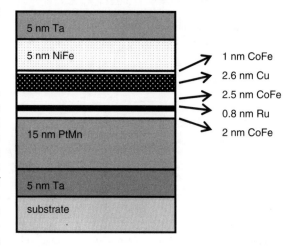

Figure 7.19 Thin-film stack of a giant magnetoresistive (GMR) read head. Redrawn after Kools *et al.* (2000)

of them just 0.8 nm thick. This is no laboratory curiosity, but a volume-manufactured piece.

The minimum thickness/minimum period of multilayer structures depends on growth process characteristics, surface smoothness and the sharpness of interfaces. Interface abruptness depends on the reactor operating principle: if growth is dependent on gas flow in the reactor, the minimum thickness is determined by the gas residence time in the reactor (to be discussed in Chapter 33), which can be fractions of seconds or tens of seconds. Flow systems, like CVD, are thus not suitable for very thin layers. Beam systems, evaporation, sputtering and MBE, with shutters, are able to turn deposition off and on over a time scale of less than a second.

Periodic multilayers have been fabricated for X-ray and extreme ultraviolet (EUV) optics: Si/Mo and W/C and similar light element/heavy element mirrors can have reflectances of 70% at 13 nm wavelength which is being considered for future EUV lithography. Periodicities are on the order of nanometers, identical to X-ray wavelengths. Surface roughness should be limited to a fraction of film thickness, which translates to 0.1–0.2 nm. The number of layers can be 50 for example. Surface smoothness must not be compromised as more and more layers are added. A layer-by-layer growth mode is preferred, and amorphous films are preferred because of their smoothness. Thickness control is important, from layer to layer and day to day.

Additionally, reactions between the layers have to be prevented. Molybdenum and silicon form silicide, $MoSi_2$, and its formation must be prevented. Sputtering is a room temperature process, and silicide formation thermodynamically should not take place. However, sputtering is a high-energy process, and on an atomic level the local temperature may rise, and energetic bombardment will knock atoms forward, enhancing the mixing of layers. If silicide is formed, its volume is about 10% less than that of molybdenum and silicon combined, and this needs to be accounted for. Alternatively, silicide formation must be prevented. Boron carbide has been tested as a barrier. The structure is shown in Figure 7.20. Layer thicknesses were extrapolated from bulk sputtering rates, which introduces some inaccuracy. A reflection of 70% was achieved at 13.5 nm (wavelength chosen for future EUV lithography systems).

ZrO_2/HfO_2 multilayers have been used in order to improve leakage currents in deposited capacitor dielectrics. These polycrystalline multilayers have been termed nanolaminates to separate them from epitaxial superlattices. The hardness of the TiN/AlN multilayer, 1800 on the Knoop hardness scale, is practically the average of TiN and AlN hardnesses for thicker

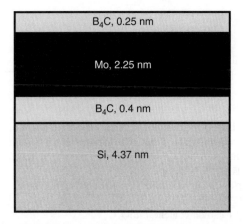

Figure 7.20 A single four-layer element of a periodic EUV multilayer mirror for 13.5 nm lithography. The four-layer structure is repeated 50 times, to a total thickness of about 350 nm. Adapted from Bajt (2002)

multilayers, but when periodicity is reduced below 10 nm, the hardness increases to 4000.

7.13 Exercises

1. Which method would you use to measure the following properties of PECVD silicon nitride film: (a) deposition rate; (b) uniformity; (c) stress; (d) stoichiometry; (e) structure; (f) step coverage?
2. What are the resistivities of the titanium silicide C49 and C54 phases (Figure 7.15)?
3. The speed of sound in ZnO is 5700 m/s. What is the intended operating frequency for the TFBAR shown in Figure 7.11?
4. A periodic lattice of W and C is used as a $\lambda/4$ X-ray mirror. What layer thicknesses should be used for 100 eV X-rays?
5. If 20 nm of nickel reacts with an overabundance of silicon, how thick a layer of NiSi will be formed? Densities: Si, 2.3 g/cm^3; Ni, 8.9 g/cm^3; NiSi, 7.2 g/cm^3.
6. $CoSi_2$ is formed by a cobalt thin-film reaction with silicon. What is the position of the $CoSi_2$ surface relative to the original silicon surface? Densities: Co, 8.9 g/cm^3; $CoSi_2$, 5.3 g/cm^3.
7. Oxygen is soluble in titanium up to 34 at. %. How thick must a silicon dioxide film be in order to be dissolved by a titanium film 50 nm thick? The density of titanium is 4.5 g/cm^3 and that of silicon dioxide 2.3 g/cm^3.
8. Draw a deposited film profile over given topography for:
   (a) sputtered aluminum, 300 nm thick
   (b) CVD TEOS, 0.25 μm thick

(c) PECVD oxide, 0.2 μm thick
(d) LPCVD polysilicon, 100 nm thick.

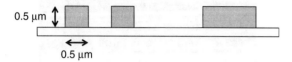

9. TiAl$_3$ is formed in a reaction between aluminum and titanium films. What will happen to the volume of the metal line? Densities: Al, 2.7 g/cm^3; Ti, 4.5 g/cm^3; TiAl$_3$, 3.35 g/cm^3.
10. How would you measure the following film properties:
   (a) crystal size in poly-SiGe
   (b) carbon content in SiO$_x$C$_y$
   (c) stoichiometry of TiSi$_2$
   (d) surface roughness of polysilicon
   (e) resistivity of electroplated copper?

References and Related Reading

Bajt, S. (2002) Improved reflectance and stability of Mo-Si multilayers, *Opt. Eng.*, **41**, 1797–1804.

Baumann, H.F. and G.H. Gilmer (1995) 3D modeling of sputter and reflow processes for interconnect metals, International Electron Devices Meeting, p. 89.

Biasotto, C. *et al.* (2008) Silicon nitride thin films deposited by electron cyclotron resonance plasma enhanced chemical vapor deposition for micromechanical applications, *Thin Solid Films*, **516**, 7777–7782.

Chou, B.C.S. *et al.* (1997) Fabrication of low-stress dielectric thin-film for microsensor applications, *IEEE Electron Device Lett.*, **18**, 599.

Cote, D.R. *et al.* (1995) Low-temperature CVD processes and dielectrics, *IBM J. Res. Dev.*, **39**, 437.

Fang, W. and C.-Y. Lo (2000) On the thermal expansion coefficients of thin films, *Sens. Actuators*, **84**, 310.

Hansen, M. and K. Anderko (1958) **Constitution of binary alloys**, 2nd edn, McGraw-Hill.

Hilleringmann, U. and K. Goser (1995) Optoelectronic system integration on silicon: waveguides, photodetectors, and VLSI CMOS circuits on one chip, *IEEE Trans. Electron Devices*, **42**, 841.

Kang, U. *et al.* (1999) Pt/Ti thin film adhesion on SiN$_x$/Si substrates, *Jpn. J. Appl. Phys.*, **38**, 4147.

Kools, J.C.S. *et al.* (2000) Deposition technology for thin film magnetic recording heads reader fabrication, *Thin Solid Films*, **377–378**, 705–711.

Mann, R.W. *et al.* (1995) Silicides and local interconnections for high-performance VLSI applications, *IBM J. Res. Dev.*, **39**, 403.

Martinu, L. and D. Poitras (2000) Plasma deposition of optical films and coatings: a review, *J. Vac. Sci. Technol.*, **A18**, 2619–2645.

Murarka, S.P. (1993) **Metallization, Theory and Practice for VLSI and ULSI**, Butterworth–Heinemann.

Pilvi, T. *et al.* (2008) Atomic layer deposition of MgF$_2$ thin films using TaF$_5$ as a novel fluorine source, *Chem. Mater.*, **20**, 5023–5028.

Piramanayagam, S.N. (2007) Perpendicular recording media for hard disk drives, *J. Appl. Phys.*, **102**, 01301.

Raaijmakers, I.J. *et al.* (1990) Microstructure and barrier properties of reactively sputtered Ti-W nitride, *J. Electron. Mater.*, **19**, 1221.

Rossnagel, S.M. *et al.* (1996) Thin, high atomic weight refractory film deposition for diffusion barrier, adhesion layer and seed layer applications, *J. Vac. Sci. Technol.*, **B14**, 1819.

Thornton, J.A. (1986) The microstructure of sputter-deposited coatings, *J. Vac. Sci. Technol.*, **A4**, (6), 3059.

Vallat-Sauvain, E *et al.* (2000) Evolution of microstructure in microcrystalline silicon prepared by very high frequency glow-discharge using hydrogen dilution, *J. Appl. Phys.*, **87**, 3137...

Wang, S.-Q. and J. Schlueter (1996) Film property comparison of Ti/TiN deposited by collimated and uncollimated physical vapor deposition techniques, *J. Vac. Sci. Technol.*, **B14**, (3), 1837.

Wang, S.-Q. *et al.* (1996) Step coverage comparison of Ti/TiN deposited by collimated and uncollimated physical vapor deposition techniques, *J. Vac. Sci. Technol.*, **B14**, (3), 1846.

Wang, Y.Y. *et al.* (1998) Synthesis and characterization of highly textured polycrystalline AlN/TiN superlattice coatings, *J. Vac. Sci. Technol.*, **A16**, 3341.

Xu, Y.P. *et al.* (1992) A study of sputter deposited silicon films, *J. Electron. Mater.*, **21**, 373.

8

Pattern Generation

A pattern generation tool transcribes the chip design data into a physical structure. It must be able to expose single pixels to create all possible designs, and expose them fairly fast since designs can consist of millions of pixels. Sometimes these patterns are written directly on the wafer in question, and the method is called direct write, but more often these patterns are written on a glass plate to produce photomasks for optical lithography. UV illumination through a photomask will expose all the patterns simultaneously, enabling very fast patterning, once the mask has been written.

Direct write is akin to writing with a pen: it is easy to write short notes, and you can change your mind on the fly to create new motifs. However, if you want to write a longer story, it can certainly be done, but it will take time. And if you want to make copies of your work, each copy takes the same time to write as the first one. With photomasks the time to make the first pattern is the same as in direct writing, but the subsequent copies can be made very quickly. However, if you want to change something, you need to create another original pattern. So optical lithography is akin to a printing press, with the same strengths and weaknesses as printing, compared to handwriting.

8.1 Pattern Generators

The first pattern generator consisted of a mechanical stage, aperture blades and a UV lamp (Figure 8.1). The wafer is covered with photoresist, a layer of photosensitive polymer. Resists come in two flavors, like photographic films: positive and negative. With positive resists the exposed areas will become soluble in developer, while with negative resists the exposed areas are crosslinked and become insoluble. To create a pixel, aperture blades are sized and positioned, followed by the exposing flash. After mechanical movement of the stage, the aperture sizing

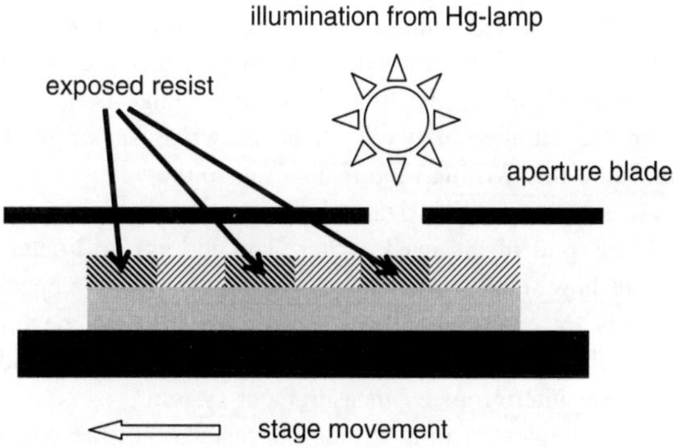

Figure 8.1 Optomechanical pattern generator: stage movement, aperture blade and flash bulb expose the pattern pixel by pixel

operation and flashing are repeated, with an operating frequency of about 1 Hz. This method was employed in the early era of microfabrication when linewidths were above 10 µm.

Today, if simple and fairly large patterns are needed, laser printers can be used to write the patterns. They are, after all, designed to write arbitrary shapes. While office laser printers can produce lines in the 100 µm range, more advanced laser writers are used in the printed circuit board industry, enabling linewidths down to a few tens of micrometers.

Dedicated laser mask enable submicron linewidths to be written. Even smaller features can be exposed by focused electron or ion beams. The beam can be scanned very quickly, up to 500 MHz, but modern integrated circuits can consist of billions of pixels, and writing times can be days for advanced chip designs.

Electron and laser beam systems are the standard tools for pattern generation. We will first discuss issues relevant to both, such as how to break up a complex chip design

Introduction to Microfabrication, Second Edition Sami Franssila
© 2010 John Wiley & Sons, Ltd

into a form that can be exposed pixelwise, and how to stitch together large patterns from very small pixels.

First, the design data has to be fractured to suit the beam writer. All patterns must be broken down to squares or rectangles. This generates enormous data files: even a fairly simple chip design can be gigabytes after fracturing. Integrated circuits are amenable to a Manhattan street plan of x and y coordinates, while for instance CDs require a spiraling "street plan." Pattern generators exist specifically designed for both basic types. A fractured pattern using an xy-oriented machine is shown in Figure 8.2. Note that circular and wedge shapes are approximated. If the beam writer has a small address grid, and a small spot size, this approximation can be made increasingly accurate but at the expense of writing time. If the pixel edge is halved, then the number of pixels to be drawn is quadrupled, meaning that writing time is also fourfold.

In addition to spot size, it is important to consider the address grid of the mask writer. The address grid determines how small increments can be drawn. While a beam spot is for example 100 nm, an address grid can be for example 25 nm. It is determined by stage mechanics and the laser interferometer measurement system.

The simplest writing strategy is raster scan: the beam scans all over the plate and at each spot an exposure/no-exposure decision is made. This works fine for dense patterns, but if the design consists of a few patterns far apart, the raster scan wastes time. On the other hand, writing time is independent of design.

Vector scanning enables skipping of empty (non-exposed) spaces, thus faster writing, at the expense of system complexity: managing gigabytes of data at 100 MHz data rate is a formidable task. Position accuracy is also more difficult to realize than in a raster scan: rapid sweeping from field to field is not as accurate as step by step from pixel to neighboring pixel. Raster scan and vector scan are compared in Figure 8.3.

Basically a single shot can expose a single pixel, but in practice the minimum pixel is exposed by for example

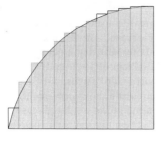

Figure 8.2 Circular shape reconstructed as rectangles ready for pixelwise writing

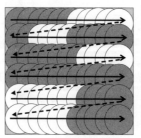 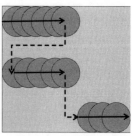

Figure 8.3 Raster scan and vector scan: in raster scan every pixel is scanned while in vector scan empty spaces are skipped

5×5 shots. There are several reasons for this. First of all, edge acuity can be much better if multiple, overlapping shots are used. In the case of a single shot per pixel, there is the danger of statistical errors: a line would be discontinuous if a single pixel were exposed incorrectly.

The ultimate resolution of electron or laser beam writing is only applicable to a small area, for example a field of $250 \times 250\, \mu m^2$. This area can be scanned electromagnetically (e-beam) or acousto-optically (laser beam). If a larger area needs to be drawn, additional movements must be introduced. The stage scan is a mechanical movement, controlled by an interferometer. Patterns at adjacent fields must be accurately stitched together at intersections. For instance, the stitching error of subfields can be made as small as 6 nm in the x-direction, but not in the y-direction, because the former depends on beam scanning but the latter on mechanical stage movement

Writing time can be limited by several factors which depend on pixel size (d), total area (A), required exposure dose (resist sensitivity, S), beam current (I), electronic scan frequency (f), mechanical scan speed (v), electronic scan length (l) and mechanical scan length (L). First of all, the area that needs to be exposed by the current in use gives time τ_1:

$$\tau_1 = AS/I \qquad (8.1)$$

Exposed pixel size d affects writing time via τ_2: when the same area is broken up into smaller pixels, more shots are needed. If the beam writer's electronic scan frequency can be increased, τ_2 can be decreased:

$$\tau_2 = \frac{A}{fd^2} \qquad (8.2)$$

The wafer stage mechanical movement time must be considered for a complete system, which gives τ_3:

$$\tau_3 = A/Lv \qquad (8.3)$$

Assuming the following parameters

$A = 10\,\text{cm}^2$
$f = 100\,\text{MHz}$
$d = 100\,\text{nm}$
$L = 10\,\text{cm}$
$l = 250\,\mu\text{m}$
$v = 10\,\text{cm/s}$

and calculating for three different resist sensitivities (1, 10 and 100 µC/cm²), we get τ_1 exposure times of 100, 1000 and 10 000 s, respectively. Mechanical movements are not limiting (400 s), but a 100 nm pixel size at 100 MHz results in 10 000 s for τ_2. Therefore, resist choice is not very critical because τ_2 is the limiting time. A rough estimate for mask writing time in this case is 3 hours.

8.2 Electron Beam Lithography

In this section we will study issues specific to electron beam writing. Electron beam spots in the 5 nm range are available. This is not limited by the wavelength of electrons ($\lambda = 8\,\text{pm}$ for 25 kV) but rather by electron source size and electron optics aberrations and diffraction for highly collimated beams. Beam current cannot be increased indefinitely because of Coulomb repulsion. Microampere currents are already quite intense, and 2 µA beam current will lead to spreading of 100 nm because of interaction aberration.

One of the distinguishing features of electrons is that electrons are light mass objects and when they hit photoresist with high energy (typically 10–50 kV) they scatter. Even though the beam spot is very small, scattering broadens the beam inside the resist and the exposed area is much larger than the beam spot size (Figure 8.4).

As can be seen, thinner resist would be beneficial because the beam will traverse it with less scattering. However, thin resist is problematic in further processing because it does not block ion implants or withstand harsh etching conditions. Thin resists are also more likely to be defective, with pinholes, lowering yield. Higher beam energy will help: the traversed electrons scatter less if they are energetic. However, these high-energy electrons do not increase resist exposure directly in proportion to their energy, therefore high energy is not a straightforward way to improve resolution.

Electrons scatter both forward and backward. These high-energy electrons are, however, not the major component of resist exposure: most of the resist exposure comes from low-energy secondary electrons which have been created when the beam slows down. These electrons have energies less than 400 eV and mostly less than 200 eV.

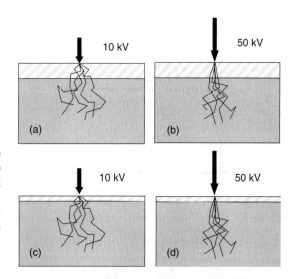

Figure 8.4 The effects of resist thickness and electron beam acceleration voltage on beam scattering and resolution. Note that almost all electrons traverse the resist and penetrate the wafer

The corresponding travel distances of such electrons in the resist are 10 nm and 5 nm, respectively. Therefore the minimum exposed size is limited to about 10 nm. But there are many other factors contributing to final feature size, for example the development of exposed patterns (the analogy to photography is obvious: it is not only the quality of the negative that affects print quality, but also all the other steps, e.g., development time, concentration and temperature, which make a contribution).

An approximation to effective beam diameter in resist is given by

$$d_{\text{eff}}\,(\text{nm}) = 0.9 \times \left(\frac{t}{V}\right)^{1.5} \qquad (8.4)$$

where resist thickness t is in nm and voltage in kV. Ideal photoresist should remain unaffected by the beam until a threshold dose is delivered, and then be fully exposed. Such an idealized resist response is shown as the box-like curve in Figure 8.5 (negative resist is assumed, i.e., exposed resist determines the line). Below threshold dose no pattern is formed, and above it a line of constant width is produced. Real resists and lithographic processes have finite contrast. Because of scattering and natural variability, some parts of the resists are exposed and a narrow line is formed at very low dose, and subsequently wider lines as the dose increases.

Some electrons experience backscattering (large-angle scattering) with about micrometer range. Exposure dose

96 Introduction to Microfabrication

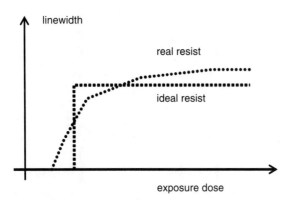

Figure 8.5 Dose–linewidth relationship in e-beam exposure

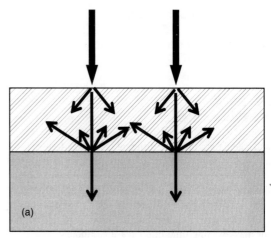

Figure 8.6 Proximity effect: electron backscattering at resist/substrate interface leads to increased exposure in the space between the lines. Dose thus depends on the neighboring structures

thus depends on the neighboring structures. This is known as the proximity effect, see Figure 8.6.

The proximity effect can be combated by biasing structures smaller or larger so that the final pattern is of the desired size and shape. This is, however, a formidable task because there are a myriad of possible combinations of shapes in microstructures. It also slows down writing considerably, because proximity correction means more complex shapes, as will be discussed in Chapter 11, in the context of optical proximity correction.

To determine how many electrons or photons are needed to expose a pixel, we need to understand the sensitivity of photoresists (yes, they are called photoresists also in the case of electron beam lithography). There are dozens of different formulations for e-beam resists, some of which are listed in Table 8.1 below. There is an unfortunate inverse relationship between the resist sensitivity (measured as charge per area needed to expose it) and the minimum linewidth.

This sensitivity–linewidth tradeoff can be gauged as follows. Let us assume 100 nm by 100 nm pixels and resist sensitivity of $0.3\,\mu C/cm^2$. The number of electrons needed to expose the pixels is then 186. This is precious few electrons, and statistical variations are great. In order to write 10×10 nm pixels, two electrons would suffice, but in practice less sensitive resist is used to average out random variation.

While electron beam lithography is the workhorse of nano- and microfabrication, it has many difficulties. The slow writing speed for large areas is a major drawback, but because there really are no competing technologies in the 100 nm feature size range, electron beam writing must be employed. There are also challenges in meeting all the criteria simultaneously: for example, pattern density affects local heating by electron-deposited energy, and the conductivity of the substrate affects charging, which

can lead to the deflection of electrons. Despite all these problems, e-beam lithography is widely practiced as a research tool for writing test structures and devices, and in the production of photomasks for optical lithography. There are also some device manufacturing applications, in devices where very small series are produced. In those cases the cost of a mask is not spread over many devices, and direct writing does make economical sense.

Table 8.1 E-beam resists: sensitivity vs. feature size

Resist	Sensitivity ($\mu C/cm^2$)	Beam energy (kV)	Minimum feature size (nm)
PMMA	300	50	10
EBR-9	25	50	200
AZ5206	6	10	250
COP	0.3	10	1000

8.3 Laser Pattern Generators

Laser beam pattern generators work on similar principles as e-beam systems, but there are a few important differences which make laser beam writing faster and cheaper. First of all, laser writing is done in room ambient, while e-beam systems operate in vacuum. The costs of the vacuum vessel and pumps are eliminated, and temperature control is easier in room ambient than in vacuum. Delicate mechanical movements of the stage are difficult to implement in vacuum, when it is necessary to minimize particles simultaneously. The mechanics of a laser pattern generator are shown in Figure 8.7. Mechanical movements are measured by interferometric techniques, and stage positioning uses piezoelectric translators for achieving nanometer position resolution.

Because laser beams use visible and UV wavelengths, a wide variety of photoresists developed for optical lithography are available to laser beam writing. Laser beam systems can write for example 0.6 μm minimum lines at 1 cm^2/min and 6 μm lines at 100 cm^2/min. Laser pattern generators are used whenever their linewidth resolution is adequate, and only when lines smaller than laser resolution are needed are the slower, more expensive e-beam systems used. In a CMOS process some mask levels may be written by e-beam and some by laser systems. Laser writing completely dominates mask making for flat-panel displays and other large-area applications, which, usually, work with larger linewidths.

8.4 Photomask Fabrication

Instead of directly writing millions of pixels on a wafer, beam writers can be used to write photomasks for optical lithography. Photomasks are glass plates with chromium

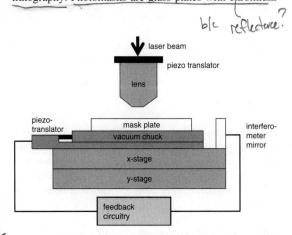

Figure 8.7 Mechanics of the laser mask writer stage

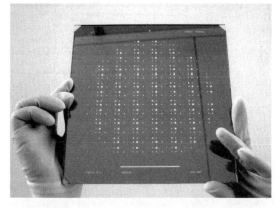

Figure 8.8 A 7 inch photomask: a chromium layer 100 nm thick on a glass plate holds the desired patterns

layers (about 100 nm thick) on them (Figure 8.8). The beam writer exposes the resist on the mask plate, in identical fashion to writing patterns on a wafer (just replace the wafer in Figure 8.1 by a mask plate).

Optical lithography with photomasks is the dominant patterning technology because optical exposure is fast: illumination through a photomask exposes up to 10^{10} pixels in less than 1 second. The enormous throughput advantage of optical exposure warrants the making of mask plates, which can be costly: a set of 15 plates (needed for 1 μm CMOS processes) costs $15 000; and an advanced 45 nm process requires $1 million mask sets ($30 000 dollars each for 35 plates).

Optical lithography can be done with reduction optical systems (to be discussed in Chapter 10), which means that the patterns on the mask are larger than final structures on the wafer. This is a great relief for mask makers: the 100 nm final size on a wafer corresponds to 400 nm on the mask when 4× reduction optics is used. This is a major benefit of optical lithography compared to X-ray lithography and nanoimprint (Figure 1.9) which require the mask pattern size to be the same as the final pattern size on the wafer.

Soda lime glass plates are used for larger linewidths (>3 μm) and quartz (fused silica) is the material of choice for micron and submicron work. This is for both thermal and optical reasons: making smaller patterns calls for shorter wavelength exposure and quartz transmission in ultraviolet wavelengths is superior to soda lime (Figure 8.9). The mask plate heats up during exposure, and thermal expansion needs to be addressed. Quartz has a much smaller coefficient of thermal expansion

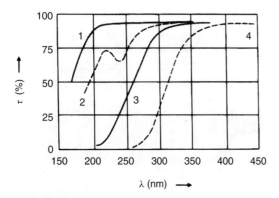

Figure 8.9 Optical transmission of various glasses 1 mm thick: 1, ultrapure quartz; 2, common quartz; 3, very pure soda lime glass; 4, common soda lime. Reproduced from Scholze (1977), with kind permission of Springer

4. Metrology
   CD (critical dimension) control
5. Inspection for pattern integrity
   defects (in chrome)
   pattern fidelity (shape and position)
6. Cleaning
   particle removal
7. Repair
   focused ion beam etching and/or deposition
8. Final defect inspection
   Adapted from Skinner et al. (1997)

than soda lime, therefore it maintains its dimensions more accurately.

Photomasks with chrome-on-glass also go by the name of binary masks, because there is either transmission or blockage of light, but nothing else. In a phase shift mask (PSM), the phase of the light is manipulated while traversing the mask. PSMs will be discussed in Chapter 10. If the mask is mostly covered by chrome, with only a small percentage of transparent area, it is said to be a dark-field (DF) mask (as in Figure 8.8); if it is mostly transparent, with only a small percentage of chrome, it is designated a light-field (LF) mask, also known as a bright field (BF) mask.

The processes needed to fabricate masks is shown in the process flow below. In addition to pattern writing, there are many other process steps, especially etching, which will affect the final pattern quality. There are also many inspection, measurement and repair steps, which, in fact, make up a major part of mask price.

Process flow for mask fabrication

1. Mask blank preparation
   deposition of chrome on quartz
   resist application
2. Pattern writing
   e-beam or laser
3. Pattern processing
   resist development
   chrome etching
   resist stripping

8.5 Photomask Inspection, Defects and Repair

Photomask fabrication requires, in addition to scanning beam equipment, a repertoire of inspection and repair equipment. Three basic control measurements for masks are linewidth, position and defects. Linewidth is a local measurement over a test structure pattern. With linewidths in the micrometer range, measurement should be able to discern about 10 nm. Pattern position is a global measurement and it is usually fixed to a mask writing tool, controlled by a stage interferometer, and measured to about 10 nm accuracy over a mask plate size of 10 cm.

Two basic inspection strategies are used: optical inspection combined with comparison to a known perfect mask plate (known as die-to-die) or comparison between design data and the finished mask plate (die-to-data). Defects on the mask are fatal because they will be reproduced on all the wafers. Defects can be classified into two broad categories of soft defects and hard defects. Soft defects are mainly particles or resist residues that can be cleaned away. Hard defects are permanent spots or scratches in chrome or in quartz.

Defects come in many guises, but from a repair point of view there are two grand classes of defects:

- missing chrome
- extra chrome.

Missing chrome calls for a deposition process. Usually a metallic layer is deposited (e.g., tungsten). Extra chrome calls for an etching process. These repair processes will be discussed in Chapter 24 on serial microprocessing.

The geometric/topological classification of defects (see Figure 8.10 below) is as follows:

- protrusion (extra chrome attached to a feature)
- intrusion (partial loss of chrome in a feature)

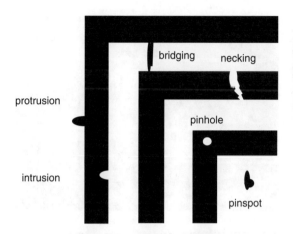
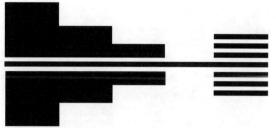

Figure 8.10 Mask defects: extra chrome (protrusion, bridging, pinspot) and missing chrome (necking, intrusion and pinhole). If defect is very small, it may cosmetic only because it does not print in lithography. Redrawn after ref. Skinner

- bridge (chrome connecting two features)
- necking (discontinuity in a line)
- pinhole (hole in a chrome)
- pinspot (extra chrome on a light-field area).

From yield and reliability points of view not all defects are equal. Defect must be understood as a very broad term: anything that prints on the wafer or changes critical dimension by more than 10% counts as a defect. This can be a light transmission error (e.g., roughness in quartz), a pattern error, a stochastic scratch or an undulating line edge.

Defect size is important: not all defects are able to destroy the functionality of the chip. As a rule of thumb, defects greater than one-third of the minimum linewidth are prospective "killer defects." Smaller ones will not print on the wafer in the lithography process.

Optical defects not related to written patterns include:

- transmission variability in glass (LF areas)
- transmission variability in chrome (DF areas).

Transmission defects are subtle and, even if detected, their repair is not straightforward. Phase shift mask making is very expensive partly because of difficulties in the inspection and repair or transmission defects.

8.6 Photomasks as Tools

Photomasks are used to make devices, but they also serve as tools for process and device engineers. As shown in

Figure 8.11 Test structure for lithography and etching: the central line is surrounded by dark-field and light-field areas, and it is found as an isolated line as well as an array line. In an ideal case the linewidth should be independent of its neighborhood

Figure 1.4, there are not only device chips on the wafer, but test chips as well. Process engineers want to see the resolution of the optical lithography process, and this is checked by the linewidth test structures. Process robustness is tested by implementing it on photomask structures which span a range around the baseline fabrication process. For example, if the design linewidth is $3\,\mu m$, test structures may span the range $1-10\,\mu m$. The same applies for spaces between the lines. Linewidth is dependent on the neighboring structures, therefore test structures should include lines of different kinds: isolated, nested, dense and sparse, as shown in Figure 8.11.

Device engineers design different geometries of devices, for example rectangular and octagonal inductor coils, or straight and meandering resistors. For transistor parameter extraction a set of test transistors with dimensions of, say, 2, 3, 5, 10, 20 and $50\,\mu m$ is used.

8.6.1 Photomask requirements

Because of high cost and long delivery times, it is important to specify masks correctly, and not to overpay for them. Not all lithography steps are equal: some are critical for device performance and the highest possible mask quality must be specified. In CMOS this applies to the gate and contact level masks. The metallization levels are usually less demanding, and the masks can be made to looser specifications. And there are levels with even looser requirements: bonding pads for wire bonding or fluidic ports are tens or hundreds of micrometers in size. They can be realized by even cheaper masks. Many packaging applications require compatibility with wafer processing, but make do with cheaper masks. As stated before, the various inspection and measurement steps take up a major part of photomask price. Mask buyers can

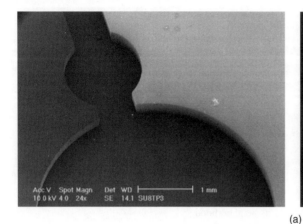
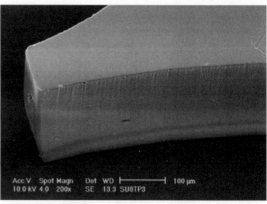

(a)

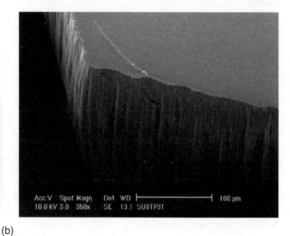

(b)

Figure 8.12 Thick photoresist patterning: with chrome mask (a) and with plastic mask (b). Reproduced from Lee *et al.* (2001), with kind permission of Springer Science and Business Media

specify defects, and accept imperfect plates with some defects which have been classified as non-fatal.

In a university lab, mask costs can be brought down by adopting completely different technologies for different mask levels. The cost of laser-printed overhead transparencies is next to nothing, but they are suitable for structures in the size range of hundreds of micrometers. Polymer-based masks suffer from wear and tear and from dimensional instability. It also takes some effort to insert plastic masks into a mask aligner designed for chrome-on-glass plates.

In addition to linewidth, other issues need to be addressed in selecting mask technology. Sometimes large-linewidth structures require very smooth edges, for example some optical devices. And if rounded structures are needed, the number of polygons used to approximate circular shapes becomes important. These issues are depicted in Figure 8.12, showing the difference between chrome masks and plastic masks. Additionally, plastic masks may look opaque to the eye but still leak light in the UV, rendering them unusable, especially for thick resists which require long exposure times.

8.7 Other Pattern Generation Methods

There are needs for masks at different linewidths and cost levels. There is the consideration between mask making and direct writing when production runs are small. Alternative pattern generation methods have been devised, but so far they have not become competitive with e-beam and laser beam methods for mainstream applications.

One alternative technology is based on SLM (Spatial Light Modulators) technology, namely arrays of micromirrors made by microfabrication. They can be used as "programmable masks" or "virtual masks" which are projected on a wafer. Applications of this approach are shown Figures 23.2 and 39.21, and the actual micromirrors are detailed out in Figures 29.25 and 39.21.

All the above methods are general purpose pattern generation methods: they can produce any shape the designer has imagined. There are also many methods to produce certain shapes very effectively. For instance, interference lithography (holographic lithography) is based on intersecting laser beams, and it is a very convenient way to produce regular arrays of lines and spaces, for example optical gratings. Di-block copolymers are systems that will phase segregate and end up, for example, with dot and hole patterns. Other shapes can be made, but each pattern needs a different block copolymer. Microbeads can form regular hexagonal arrays during liquid drying and have been used to make photonic crystals. Electrochemical etching of aluminum can result in a regular hexagonal lattice of holes. Some of these special techniques will be discussed in Chapter 23.

8.8 Exercises

1. How far will (a) a 10 keV e-beam penetrate into silicon and (b) a 50 keV beam into quartz?
2. What is the difference between writing masks and writing directly on silicon from the viewpoint of an electron beam writer?
3. What is the smallest possible feature size that can be written with a 50 keV electron beam?
4. What is the photomask writing time for a gigabit circuit with 1 000 000 000 contact holes of 90 nm diameter, when the incrementing rate is 500 MHz and the mask plate area 10 cm by 10 cm? The photomask is four times the final size.
5. If a 7 inch mask plate for 1× lithography with 1 μm linewidth is written by e-beam, what is the limiting process and what would be an estimate of the writing time?
6. How is electron beam system throughput affected if 5× masks are drawn, instead of 1× masks?
7. Calculate the number of quanta needed to expose resists: electron beam resist sensitivity is 10 μC/cm^2, and optical resist sensitivity 10 mJ/cm^2.
8. Use a laser printer to make simple line/space test structures with different resolutions (say 600 dpi and 1200 dpi) and check by microscope for linewidths, line edge roughness and reproducibility.

References and Related Reading

Allen, P.C. (2002) Laser scanning for semiconductor mask pattern generation, *Proc. IEEE*, **90**, 1653.

Eynon, B. and B. Wu (2005) **Photomask fabrication technology**, McGraw-Hill.

Hahmann, P. *et al.* (2007) High resolution variable-shaped beam direct write, *Microelectron. Eng.*, **84**, 774–778.

Kawano, H. *et al.* (2003) Development of an electron-beam lithography system for high accuracy masks, *J. Vac. Sci. Technol.*, **B21**, 823–827.

Lee, L.J. *et al.* (2001) Design and fabrication of CD-like microfluidic platforms for diagnostics: polymer-based microfabrication, *Biomed. Microdevices*, **3**, 339–351.

McCord, M.A. and M.J. Rooks (1997) Electron beam lithography, in P. Rai-Choudhury (ed.) **Handbook of Microlithography, Micromachining and Microfabrication**, SPIE.

Meyer, A.R., A.M. Clark and C.T. Culbertson (2006) The effect of photomask resolution on separation efficiency on microfabricated devices, *Lab Chip*, **6**, 1355–1361.

Pease, F.R. (2005) Maskless lithography, *Microelectron. Eng.*, **78–79**, 381–392.

Pugh, G. *et al.* (1998) Impact of high resolution lithography on IC mask design, Custom Integrated Circuits Conference, IEEE, p. 149.

Rizvi, S. (ed.) (2005) **Handbook of Photomask Manufacturing Technology**, CRC Press.

Scholze, H. (1977) **Glas-Natur, Struktur und Eigenschaften**, 2nd edn, Springer.

Skinner, J.G. *et al.* (1997) Photomask fabrication procedures and limitations, in P. Rai-Choudhury (ed.) **Handbook of Microlithography, Micromachining and Microfabrication**, SPIE.

Wu, B. (2006) Photomask plasma etching: a review, *J. Vac. Sci. Technol.*, **B24**, 1–15.

9

Optical Lithography

Optical lithography, also known as UV lithography or photolithography, uses UV lamps or UV lasers to expose photosensitive film through photomasks. Chromium patterns on a photomask block light at selected areas, forming patterns in the photosensitive film (which is called photoresist for historical reasons). Once a photomask with millions of pixels has been written, photoresist can be exposed very quickly: typical exposure times are on the order of 1 second.

After a photoresist pattern is thus formed, various possibilities are open: for instance, the open areas of the underlying material can be etched away, in the case of silicon dioxide, by hydrofluoric acid (Figure 9.1). Photoresist is removed after etching. Wafer processing then continues with thin-film deposition, doping or plating steps, and new lithographic steps. The successive layers have to be aligned to each other, so that desired contacts are formed. Overlay of successive layers is a critical factor in lithography.

There are three rather different elements in the optical lithography process:

1. Optics: radiation generation, propagation, diffraction, interference.
2. Chemistry: photochemical reactions in the resist, development.
3. Mechanics: mask to wafer alignment and parallelism, focusing.

We will first go through the lithography process step by step as the wafer sees it. A more detailed discussion of the various steps then follows. All discussion in this chapter refers to 1× lithography: the patterns on the mask are transferred to the wafer in the same size as they are on the mask. This mode is called contact/proximity lithography. Projection lithography with reduction optics will be dealt with in the next chapter.

9.1 Lithography Process Flow

9.1.1 Bake and prime

Practically all microfabrication processes start with some sort of surface preparation step (these will be discussed in detail in Chapter 12). Surface preparation "initializes" the wafer, and the starting conditions will then be independent of preceding process steps or wait time. In lithography, the first preparation step is a baking process. Adsorbed water is removed by baking the wafer at 100–200 °C.

The next step, wafer priming, is also known as adhesion promotion. Hexamethyl disilazane vapor (HMDS, $(H_3C)_3$–Si–NH–Si–$(CH_3)_3$) is applied at reduced pressure. As shown in Figure 9.2, the oxygen in the hydroxyl group at the wafer surface will form a bond with silicon atom in HMDS, and the amine group reacts with hydrogen, releasing ammonia.

The monomolecular silane layer makes the wafer slightly hydrophobic (contact angles 60°–70° for example), which prevents water readsorption. It also ensures good wetting by photoresist. This is especially important for materials like metals, polysilicon and PSG, because resist adhesion to these materials is poor. Adhesion promotion is also a guarantee against cleanroom humidity variations and a precaution against processing delays, as HMDS primed water retains its water-free condition even for a couple of days.

9.1.2 Spin coating

Spin coating is the standard resist application method. A few milliliters of resist is applied on a static or slowly rotating wafer. Acceleration to about 5000 rpm spreads the resist over the wafer, leaving a very uniform layer (Figure 9.3). Standard resist thicknesses are about 1 μm when contact/proximity lithography is used.

Introduction to Microfabrication, Second Edition Sami Franssila
© 2010 John Wiley & Sons, Ltd

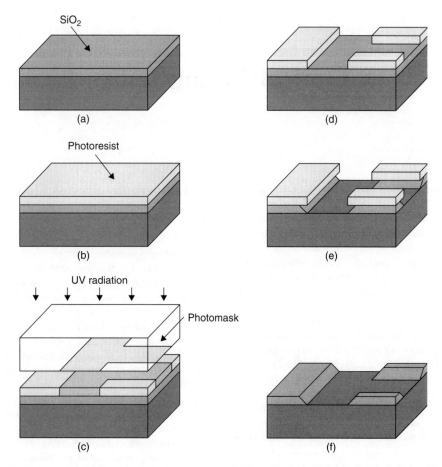

Figure 9.1 Making oxide patterns: (a) oxide film deposition; (b) photoresist application (negative resist used); (c) UV exposure through a photomask; (d) development of resist image; (e) etching of oxide; (f) photoresist removal. Courtesy Esa Tuovinen, University of Helsinki

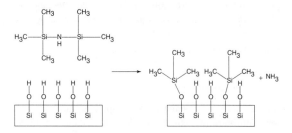

Figure 9.2 HMDS adhesion promotion (priming)

9.1.3 Pre-exposure bake

The spin-coated resist still contains solvent, and a bake step is needed to drive out most of the solvent. This bake is, for example, 90 °C for 30 min in an oven or 90 °C for 60 s on a hotplate. Care is needed to optimize the bake temperature, because too high a temperature will decompose the photoactive compound, leading to lower sensitivity to exposing radiation. Pre-exposure bake is also known as soft bake or post-apply bake.

9.1.4 Alignment

Next, the wafer and the photomask are inserted into a mask aligner. In alignment, the patterns already on the wafer are aligned to the new mask patterns (Figure 9.4).

In the case of the first mask level, no alignment is needed. However, it is useful to print zero-level alignment marks on the wafer using a specific mask, and etch those marks in silicon. This is especially necessary when the first process step does not leave anything visible on the

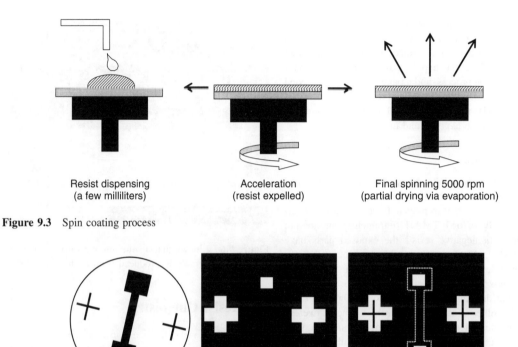

Figure 9.3 Spin coating process

Figure 9.4 Alignment operation: left, wafer with a resistor and alignment marks; middle, photomask with contact holes and alignment marks; right, after linear translation and rotation of the wafer, the alignment marks on the wafer and mask coincide, and consequently the resistor and contact holes too

wafer surface; for example, doping by ion implantation will only insert some dopant atoms inside silicon.

9.1.5 Exposure

The mask aligner is, in spite of its name, also an exposure tool. The simplest lithographic technique is contact lithography: the photomask and the resist-covered wafer are brought into intimate contact, and exposed. A mask aligner is shown in Figure 9.5: mercury lamp UV radiation is filtered and uniformly distributed over the wafer. Mechanical parts ensure wafer and mask distance and parallelism. The lithography section of a cleanroom is similar to a dark room in a photographic lab, except that yellow light is used instead of red light, because photoresists are sensitive below 450 nm only.

Resolution is determined mainly by pattern dimensions on the mask. Extremely small patterns can be made in theory by pressing the mask into intimate contact with the resist (called vacuum contact). This is, however, a source of defects when resist debris adheres to the mask.

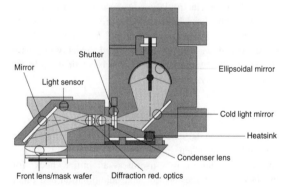

Figure 9.5 Exposure/mask aligner tool: lamp–mirror–shutter–filter–lens system, and mask and wafer. Courtesy Süss Microtech

9.1.6 Post-exposure bake

Some resists require post-exposure bake (PEB). It leads to the diffusion of photogenerated molecules. These molecules are responsible for changing the resist solubility in the developer and this diffusion smoothes out optical interference effects. In so-called chemically amplified resists (CARs) the UV exposure only initiates the photopolymerization process, and a bake is absolutely needed to complete it. During the bake, the catalyst molecules generated by UV exposure react and produce the solubility difference between exposed and unexposed resist.

9.1.7 Development

In development the soluble parts of the resist are etched away. In positive resists the exposed parts turn to carboxylic acid, which is removed by an alkaline developer, typically 0.26 M TMAH (tetramethyl ammonium hydroxide). In negative resists the exposed parts have been crosslinked and rendered stable, and the unexposed parts are removed by a solvent developer. Just as in photography, it is the combined effects of exposure and development that determine the final outcome. One can compensate the other to a certain extent by tuning it: overexposure and underdevelopment, and vice versa, may save the day (but certainly not lead to optimal results). Rinsing and drying complete the development, again in analogy to photography.

9.1.8 Hard bake

Hard bake hardens the resist, which is useful in the subsequent plasma etching and ion implantation steps, which are energetic processes. It is also useful in wet processes like etching or electroplating because it improves resist adhesion. The hard bake temperature is limited by the resist glass transition temperature T_g: above T_g the resist will flow, and shallow sloped walls will result (Figure 18.29), while vertical or nearly vertical walls are practically always desired.

9.1.9 Inspection and metrology

After hard bake the results of the lithography process are checked. Even when linewidths are below optical microscopy resolution, it is useful as an initial check: for instance, resist adhesion loss and delamination and other gross errors can be seen. Linewidth specification is usually set as ±10% of design value. Linewidth measurements by mechanical stylus, AFM or SEM form the basis of lithography process control. SEM micrographs of photoresist patterns can be seen in Figure 9.6. If resist patterns are found to be faulty, the photoresist is removed and the wafers resist coated and exposed again. The rework rate is a few percent in manufacturing. It is one of the few process steps where rework is possible.

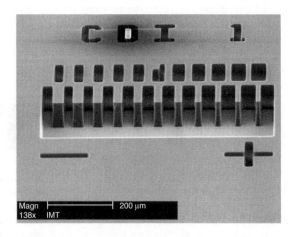

Figure 9.6 Linewidth test structure of positive photoresist in SEM micrograph. Reproduced from Roth *et al.* (1999)

9.2 Resist Chemistry

In this section we will discuss in more detail the working mechanisms of photoresists. Resists have three main components:

1. Base resin, which determines the mechanical and thermal properties.
2. Photoactive compound (PAC), which determines sensitivity to radiation.
3. Solvent, which controls viscosity.

The most common base resin for positive resists is phenolic novolak, which is soluble in alkaline developers. Diazonapthoquinine (DNQ) photoactive compound acts as an inhibitor, and unexposed resist is therefore non-soluble in developer. Upon exposure, the DNQ reacts to form an acidic compound which makes the exposed resist soluble. As can be seen in Figure 9.7, water is needed for the reaction to take place. This is one reason for cleanroom humidity control: cleanroom air is the source of this water vapor. Constant humidity around 40% RH is typical of cleanrooms. Another concern is gas evolution: nitrogen gas can lead to bubble formation inside the resist, especially when working with thick resists.

Negative resists can become insoluble because of an increase in molecular weight, that is polymerization. The resist becomes crosslinked either via free radical or acid-catalyzed polymerization. Negative resists are usually developed by solvents. This is a bit of a problem because solvents can swell the crosslinked resist, reducing pattern fidelity.

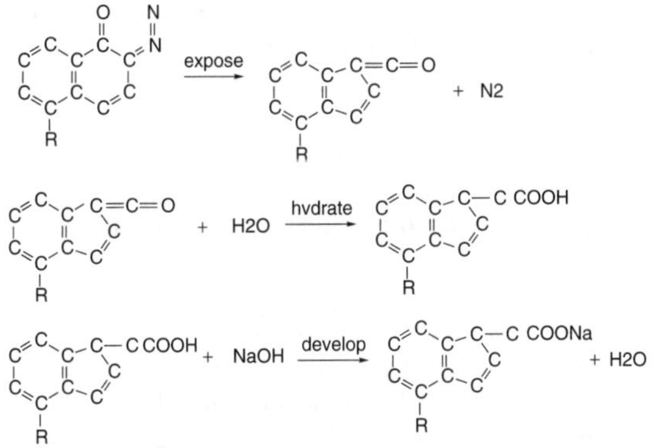

Figure 9.7 Diazonapthoquinine (DNQ) novolak resist reaction upon UV exposure. Photoactive compound reacts to form carboxylic acid, which is soluble in the developer

The crosslinking feature which makes negative resists stable also makes photoresist removal difficult, an obvious dilemma. Many negative resists are stable enough to be used as permanent structural materials in microsystems, especially microfluidics. These will be discussed in Chapter 18.

9.3 Resist Application

Spin coating is a very widely used method for photoresist spinning and polymer deposition in general. Spinning is a simple process for viscous material deposition. Spinners, with typical speeds up to 10 000 rpm, are found in every microfabrication laboratory. The main parameters for film thickness control are viscosity, solvent evaporation rate and spin speed. Spin-coated film thicknesses range from 0.1 μm up to 500 μm, with standard photoresists usually around 1 μm. Special features of thick resists will discussed in Chapter 18 on polymer microprocessing.

Depending on the wafer size and desired film thickness, a drop of 1–10 ml (= cm^3) is dispensed at the wafer center. Rapid acceleration to for example 1000 or 5000 rpm spreads the liquid toward the edges. Half of the solvent can evaporate during the first few seconds, so rapid acceleration is essential because viscosity increases upon drying, and radially non-uniform thickness will result from viscosity differences. Spin speed can be controlled to about ±1 rpm; this is important because ±50 rpm will result in 10% thickness differences.

Film thickness depends mainly on resist viscosity (η) and spin speed (ω), according to

$$t \propto \sqrt{\frac{\eta}{\omega}} \quad (9.1)$$

Spin speed can be used to tailor resist thickness over one decade, for example 0.5–5 μm, but beyond that a new resist formulation with a different viscosity must be used. Viscosity is dependent on resist solid content (which can vary from 20 to 80%) and temperature. The solvent evaporation rate depends on the ambient environment, and a closed spinner bowl with saturated solvent vapor and an adjustable exhaust from the spinner bowl can both be used to control evaporation.

Spin coating is very good in terms of film uniformity: ±5 nm can be achieved for 1 μm films, but for thicker films it is not so good, but for example 100 μm ±5 μm. Because spin coating works on liquid materials, smoothness is excellent: RMS roughness of spin-coated films is comparable to the initial smoothness of a silicon wafer, for example 0.2 nm.

Turbulence (both from the spin process itself and from cleanroom airflows) and ambient humidity (which is affected by the exhaust from the spinner bowl and cleanroom environmental controls) affect evaporation rate and consequently film thickness. Turbulence sets in earlier in larger systems, and spin speeds for 300 mm wafers are limited to about 2000 rpm to avoid turbulence. Pinhole defects in spin-coated films are thickness dependent: thinner films are more defective. Pinholes can be caused by particles on wafers, but also by particles in the dispensed fluid, even though all chemicals in microfabrication have been filtered with submicron filters. Air bubbles formed during dispensing (caused by an unclean dispense tip, for example) can cause either pinholes or large buddles, in the millimeter range.

Resist is expelled over the edge of the wafer during spinning. At the wafer edge, however, a balance is formed where drying increases viscosity and reduces flow, and finally surface tension becomes the dominant force, leading to the formation of a bulge. This bulge is known as edge bead. Edge bead height sets a minimum mask-to-wafer distance and often is so high that an optimum distance (gap) cannot be set. Edge bead removal (EBR) is a process where a directed solvent jet etches the resist away from the wafer edges.

9.3.1 Resist profile over topography

Spin processing over severe topography is difficult: liquid-like film will fill grooves and crevasses, and highly non-uniform thickness results. On the other hand, this planarizing effect is sometimes used to advantage: spin coating fills the gaps and smoothes out topography. Spin-on-glass planarization works this way (see Figure 16.7).

There are other resist coating techniques: namely, spray coating, electrochemical coating and casting.

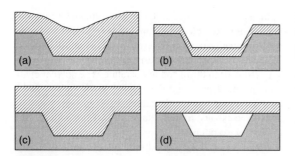

Figure 9.8 Resist over topography: (a) spin coated; (b) spray coated; (c) cast; (d) laminated dry film

Electrochemical coating requires special resist formulations, and is applicable to conductive surfaces only. Spray is applicable to thin resists. Casting is suitable for thick resists only. These techniques are especially suited to applications where resist coverage is needed over severe topography. And if there are holes in the wafer, as in MEMS devices after through-wafer etching, spin coating is unsuitable for many reasons: the resist would spill into the holes, and the vacuum chuck would not be able to hold the wafer during spinning (unless a special design were used). Still another way is to use dry film resists: negative resists are available as sheets that can be laminated on a wafer. These different resist coating techniques over an etched cavity are shown in Figure 9.8.

9.4 Alignment and Overlay

Because microdevices are built up layer by layer, overlay of successive layers on top of previous layers is a paramount performance criterion of optical lithography align/exposure tools. Overlay refers to general pattern placement, and alignment refers to specific spots on the wafer, namely the alignment marks (or alignment keys or targets), which are used for the alignment procedure. Because alignment is limited to specific structures (usually on the wafer or chip edge), it is not a full guarantee of overlay elsewhere. Overlay is affected by lens aberrations, wafer chuck irregularities (equipment-related problems), mask pattern misplacement (mask fabrication problems), or distortions of the wafer itself, such as warping or site flatness. We will, however, use the term alignment because it is an easy operational concept. The term mask aligner nicely underlines the importance of alignment. A contact/proximity aligner that can print 3 μm minimum lines is typically capable of 1 μm alignment accuracy between levels.

Alignment needs to be evaluated over a long time: device fabrication processes take weeks or even months. For example, temperature differences between different exposures will affect alignment because of thermal expansion of the wafer, the wafer stage and the photomask. The lenses in the optical path of the exposure tool are subject to constant UV flood, and they too need to be thermally stabilized.

Alignment needs to be discussed from two rather different points of view:

1. **Equipment view:** This is an optomechanical problem of finding alignment marks on the mask and on the wafer, and manipulating them to coincide.
2. **Device design view:** This is a design issue and it depends on how structures need to overlap or relate to each other, for instance metallization has to overlap contacts and a guard ring has to surround the active device.

Alignment could be done using the devices themselves, but this is impractical because of the micrometer dimensions and multiple identical structures. Therefore separate alignment marks are used. Alignment marks are much larger than device features because they exist only for alignment and have nothing to do with resolution. Alignment is usually done on a wafer level, with two alignment marks as far from each other as possible, to increase theta (rotational) resolution.

The alignment sequence determines which layers are aligned to each other. Layers are not usually aligned sequentially to the previous layer, but to some important previous layer. The capacitor top electrode is aligned to the bottom electrode, even though other steps might have been done in between.

9.5 Exposure

A mercury lamp provides strong spectral peaks at wavelengths of 436, 405 and 365 nm. Typical exposure energies for standard positive novolak resists is in the range of 100 mJ/cm^2, which translates to exposure times of a second or a few seconds. Thicker resists require larger exposure doses, and the times can be even minutes.

Proximity lithography is a modification of contact lithography: a small gap, for example 3–50 μm is left between the mask and the wafer (Figure 9.9). Resolution is now not simply dependent on feature size on mask, but diffraction effects have to accounted for. Both contact and proximity lithography are done in one and same machine: the gap between the mask and the wafer is an adjustable parameter, with values from zero upward. Contact/proximity lithography systems are 1×: the image is the same size as the original.

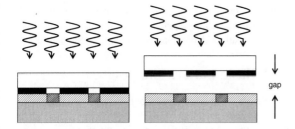

Figure 9.9 Contact lithography: mask and wafer are in intimate contact. Proximity lithography: a gap of typically 5–50 μm is instituted between resist-coated wafer and the photomask

9.6 Resist Profile

Perfectly vertical resist walls (90°) are difficult to make. Positive resists usually have a slightly positive slope, 85°–89°, negative resists have typically similar negative profile. This is a natural consequence of exposure light intensity through the mask (Figure 9.10). Due to diffraction, some light is exposing resist underneath the opaque parts of the mask. With positive resists this exposed part becomes soluble, and a positive slope results. With negative resists the diffracted light contributes to crosslinking reactions, making the exposed parts less soluble in developer, resulting in a negative (also called retrograde) profile.

9.7 Resolution

Making individual narrow lines is not a major problem in microlithography, but rather making closely spaced narrow lines. Line plus space, which is microlithographic resolution, is called pitch. An individual narrow line can be made even accidentally by for example overexposure (but the line shape will be far from ideal). Resolution, or the ability to separate two patterns, is the criterion for lithography capability. Proximity lithography minimum resolvable linewidth (half-pitch) is calculated from Fresnel diffraction and approximated by

$$\text{linewidth} \approx \sqrt{\lambda \times \left(g + \frac{d}{2}\right)} \quad (9.2)$$

where g is the gap and d the resist thickness. Typical values for wavelength λ of exposing radiation are 436, 405 or 365 nm when a mercury lamp is used as the light source. Gaps range from zero (called vacuum contact) to a few tens of micrometers. With a 405 nm wavelength, gap of

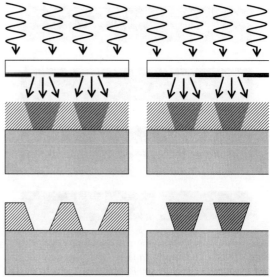

Figure 9.10 Diffraction at chrome pattern edges affects photoresist profile: positive resist (left) with exposed parts developed away leads to positive slope; negative resist (right) with exposed parts hardened leads to negative profile

5 μm and 1 μm resist thickness, minimum lines of 1.5 μm (and 1.5 μm spaces between them) can be obtained.

With thick MEMS resists minimum linewidths are clearly much larger. This is shown in Figure 9.11 where patterns of various sizes are printed in resist 7 μm thick. Four different exposure modes are used and the best results are achieved for vacuum contact between the mask and the resist-coated wafer. Hard contact results are also good, but for soft contact and 20 μm proximity the smallest 2.5 μm feature is not exposed, and the profile of the resist gets fairly shallow.

X-ray lithography is 1× proximity lithography, too. Wavelength is just much smaller, for example 4 nm, enabling much smaller minimum lines. However, making masks with very small dimensions is problematic.

9.8 Process Latitude

Many lithography parameters change and drift over time, including resist thickness (which is affected by for example viscosity change due to temperature change), exposure energy (which changes as the lamp ages), developer concentration (which may increase due to water evaporation), and many more. The process window for lithography can be defined in many ways, but exposure dose–development time is useful (Figure 9.12). The process window is the set of

Figure 9.11 The effect of mask-wafer gap: vacuum contact (topmost) results in good resist profiles, hard contact (2nd from top) in acceptable profile while soft contact and 20 μm proximity gap (bottom) result in sloped walls for 5 μm and 10 μm features and incomplete exposure for 2.5 μm feature. Positive resist 7 μm thick. Courtesy Süss Microtech

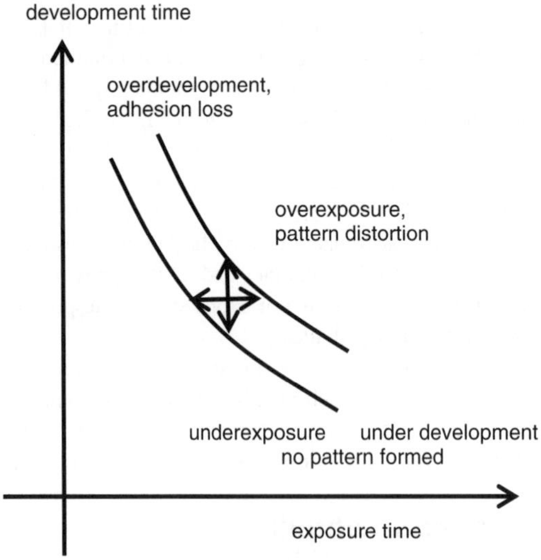

resist in open spaces, but it also clears resist where it should not, leading to patterns with widths different from design widths. Also, the resist sidewall profile is affected.

9.9 Basic Pattern Shapes

There are four basic shapes that have to be patterned: line, trench, hole and dot. More complex shapes can all be patched together from these elementary shapes, as pattern generators do. An opaque chromium line on a mask will end up as a line on the wafer if positive resist is used, but as a trench in the case of negative resist. A transparent opening in chromium will result in a trench with positive mask, and a line with negative resist.

Patterns come in two basic varieties: isolated and array (Figure 9.13). Lithography for these is different, and the ultimate lithographic resolution is also shape dependent. For example, stray light is a major issue for a light-field structures, whereas in dark-field patterns it is not so much of an issue. Dark-field patterns are also less affected by particles: most of them fall on chrome and will not affect patterns.

9.10 Lithography Practice

After lithography, various processes are possible, and all of them exhibit different requirements for resists in terms of optimum thickness and sidewall profile, chemical stability, thermal and mechanical specifications, etc. Resists

Figure 9.12 Process window for lithography process: exposure time vs. development time

exposure dose and development time combinations that results in linewidths which are within specification, typically ±10%.

Underexposure leads to no pattern formation, and it is therefore a fatal processing error. Overexposure clears all

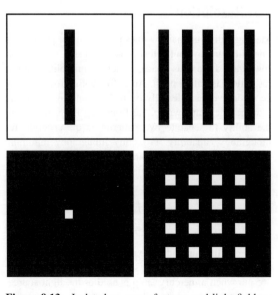

Figure 9.13 Isolated vs. array features and light field vs. dark field

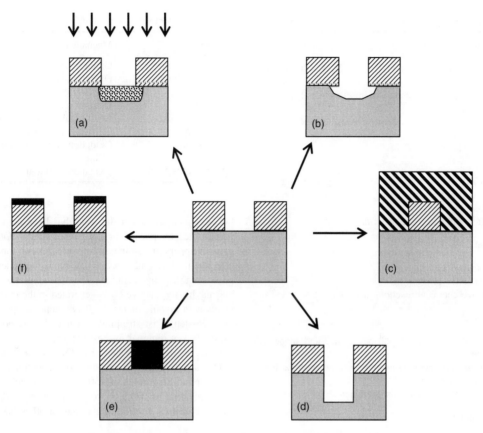

Figure 9.14 Processing after lithography: (a) ion implantation (Chapter 15); (b) wet etching (Chapter 11); (c) molding (Chapter 18); (d) plasma etching (Chapter 11); (e) electroplating (Chapters 5 & 29); (f) lift-off (Chapter 23)

face a serious scaling trade-off: thickness has to be scaled down for better resolution, but etch resistance and implant blocking capability cannot be sacrificed. Also, making resist thinner by a factor of two increases the number of pinhole defects by a factor of 10 or even 100.

Other criteria for resist selection include the following:

- sensitivity (photospeed)
- contrast
- exposure latitude
- shelf life.

In a university setting photospeed is not important, but in manufacturing setting it certainly is. Contrast (to be discussed in the next chapter) is important when vertical walls are needed. Exposure latitude is a measure of linewidth constancy in spite of variations in exposure energy. Shelf life refers to the fact that decomposition of the photoactive compound during storage eventually makes resist useless.

Figure 9.14 shows six different applications for resist patterns. In wet processes (wet etching, electroplating) adhesion is important, while in vacuum processes (ion implantation, plasma etching, lift-off) vacuum compatibility is essential. In molding, the molded piece must be detached from the mold, and surface energy, surface smoothness and sidewall angle are important considerations. The resist requirements for the different applications are listed in Table 9.1.

And of course there is the possibility that the resist structure is the final structure. Such permanent structures have to be evaluated for mechanical and thermal stability, surface properties and chemical tolerance. Chapter 18 will deal with permanent polymer microstructures.

9.11 Photoresist Stripping

After photoresist has served its role as a protective layer, it must be removed. There are a number of methods to

Table 9.1 Resist requirements for different applications

Ion implantation:

- resist thickness of 1 μm will stop B, P, As, Sb ions with <200 keV energy
- beam current heats resist: cooling or current limitation are needed
- resist carbonizes under heavy doses ($>10^{15}$ cm^{-2}), difficult to remove

Wet etching:

- resist adhesion is important, resist may peel off
- resist will not tolerate strong acids or alkaline etch solutions
- hot etch baths degrade resist fast

Molding:

- smooth surface
- non-negative profile
- minimize chemical reactions with polymers

Plasma etching:

- resist will be etched in plasma: its size and shape will change
- resist will be damaged by plasma (both bombardment and thermal effects)
- removal of damaged resist is difficult

Electroplating:

- plating solutions are often chemically aggressive
- adhesion is important

Lift-off:

- thickness of the film needs to be less than resist thickness
- resist sidewall profile preferably retrograde
- deposition process $T < 120\,°C$ because of resist thermal limitation

Table 9.2 Photoresist stripping

Techniques	Mechanism
Oxygen plasma	Oxidation in vacuum
Ozone discharge	Oxidation under atmospheric pressure
Acetone	Dissolution in liquid
Ozonized water	Bond breaking and dissolution
Sulfuric acid	Oxidation in liquid
Organic amines	Oxidation and dissolution in liquid
Hydrogen peroxide	Oxidation in liquid

accomplish this, listed in Table 9.2. The choice depends on the particular process step, the materials present on the wafer, resist nature and established laboratory practice (which may be determined by historical precedence, environmental concerns or other idiosyncratic factors).

Sulfuric acid is a strong oxidant, and therefore an effective remover of organic materials. The photoresist surface is carbonized during ion implantation but hot sulfuric acid/peroxide mixture can remove it. However, sulfuric acid cannot be used if the wafer is metallized because the acid will etch metals too. Acetone is a fairly mild remover, but it cannot be used if the resist has been damaged or transformed by plasma or ion bombardment. Other organic strippers are amine and phenolic based, and more effective than acetone. Oxygen plasma is the general purpose resist stripping technique. Ozone is similar to oxygen plasma: highly reactive O_3 is generated in the reactor and reacts with organic materials. It is a common practice to use two-step resist stripping: plasma (dry) removal followed by wet removal for best results.

Ultrapure ozonized water UPW-O_3 (10–100 ppm ozone in DI water) is potentially a major cost reduction invention in stripping. Strip rates of 150 nm/min can be achieved, and utilization of ozone is very efficient even though a simple chemical reaction might suggest otherwise:

$$CH_2 + 3\ O_3 \Rightarrow CO_2 + H_2O + 3\ O_2 \qquad (9.3)$$

CH_2 can be used as a model molecule for photoresist. This calculation shows that 10.3 g of ozone is needed to remove 1 g of resist; that is, a batch of 25 wafers (200 mm diameter wafers) would need about 10–100 kg of ozonized water depending on pattern density. Fortunately much less is needed: ozone breaks up longer molecules, and the smaller molecules are water soluble.

9.12 Exercises

1. What fraction of resist ends up on the wafer in spin coating?
2. What is the ultimate resolution in optical contact lithography?
3. What is the best possible resolution of X-ray contact/proximity lithography?
4. A silicon wafer of 100 mm diameter has 1 μm lines fabricated on it. The photomask is made of soda lime glass with a coefficient of thermal expansion (CTE) of 10 ppm ($10 \times 10^{-6}/°C$). How accurately must the temperature in the patterning process be controlled in order to keep distortions from thermal expansion over

the 100 mm wafer below 0.3 μm? Silicon CTE is $2.5 \times 10^{-6}/°C$.

5. Plot minimum feature size in proximity lithography as a function of resist thickness from 1 μm to 1000 μm thicknesses.
6. If 5000 rpm spinning produces a resist 1 μm thick, plot resist thickness at other spin speeds!
7. How do light-field and dark-field photomasks differ with respect to particle contamination?
8. If a wafer with a resist 350 μm resist is baked on a hotplate which is 0.1° off-horizontal, what will be the resist non-uniformity due to gravitational flow?
9. What alignments are needed in fabricating the device of Figure 5.19?

References and Related Reading

Chuang, Y.-J., F.-G. Tseng and W.-K. Lin (2002) Reduction of diffraction effect of UV exposure on SU-8 negative thick photoresist by air gap elimination, *Microsyst. Technol.*, **8**, 308–313.

Cui, Z. (2005) **Micro-Nanofabrication**, Springer.

Franssila, S. and S. Tuomikoski (2010) MEMS lithography, in V. Lindroos *et al.*, **Handbook of Silicon Based MEMS Materials and Technologies**, Elsevier.

Mack, C. (2007) **Fundamental Principles of Optical Lithography**, John Wiley & Sons, Ltd.

Moreau, W. (1988) **Semiconductor Microlithography**, Plenum Press.

Pham, N.P. *et al.* (2008) Photoresist coating and patterning for through-silicon via technology, *J. Micromech. Microeng.*, **18**, 125008.

Rai-Choudhury, P. (ed.) (1997) **Handbook of Microlithography, Micromachining and Microfabrication**, SPIE.

Roth, S. *et al.* (1999) High aspect ratio UV photolithography for electroplated structures, *J. Micromech. Microeng.*, **9**, 105–108.

Saha, S.C. *et al.* (2008) Tuning of resist slope with hard-baking parameters and release methods of extra hard photoresist for RF MEMS switches, *Sens. Actuators*, **A143**, 452–461.

Shaw, J.M. *et al.* (1997) Negative photoresists for optical lithography, *IBM J. Res. Dev.*, **41**, 81.

10

Advanced Lithography

This chapter concentrates on the making of narrow lines for modern integrated circuits. Linewidth scaling has been driven by CMOS scaling for four decades, and the trend continues. While in the 1980s it was thought that 1 μm would be the limit of optical lithography, in 2010 lines of 50 nm are manufactured in high volumes by UV exposure (microprocessor gates are 20 nm, but these are not done directly by optics, as will be explained later in this chapter). This tremendous progress is based on a number of innovations in optics, optomechanics, resist chemistry and related processing advances.

From a manufacturing point of view, lithography performance can be pinpointed to a few key elements:

- **CD control:** How accurately is the critical dimension (linewidths) controlled?
- **Overlay/alignment:** This concerns the placement of patterns relative to previous layers.
- **Defectivity:** This applies to pattern fidelity itself, as well to added particles: there should be none.
- **Metrology:** Masks have to be verified for correctness and the resist patterns inspected after lithography; both tasks are becoming formidable because of simultaneous linewidth reduction and chip size increase.
- **Cost:** Because lithography is done many times (25 photomasks in DRAM, 35 in modern microprocessors) it is imperative to keep throughput and yield high.

10.1 Projection Optical Systems

While contact and proximity lithography discussed in the previous chapter relied on 1× reproduction of the mask pattern on the wafer, today 4× reduction optics is the norm. This is a great relief for mask makers as 50 nm features are "only" 200 nm on the mask. The main elements of a reduction optical system are depicted in Figure 10.1.

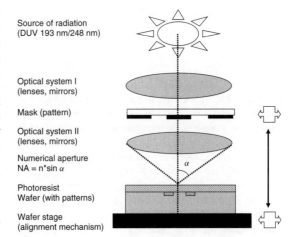

Figure 10.1 Reduction projection optical lithography system

In projection optical systems the optical system II of Figure 10.1 is the key element: it provides an image of the mask on the wafer. To emphasize the difference from 1× masks, reduction masks are often called reticles.

Projection optics today is used for chipwise exposure: one chip is exposed, the wafer is moved to a new position, and another chip is exposed. This approach is termed step-and-repeat, and the systems are known as steppers (Figure 10.2). In spite of the fact that it is easier to expose a chip-sized area than full wafers, stepper lenses are very large and expensive. The lens of a stepper with for example an 8 cm² field (on a wafer) can weigh 500 kg and cost $10 million (Step-and-repeat was an existing technique in the photomask industry: the original chip pattern was written once on a mask blank and the final 1× full wafer mask with hundreds of identical chips was made by copying the original pattern many times over to

Introduction to Microfabrication, Second Edition Sami Franssila
© 2010 John Wiley & Sons, Ltd

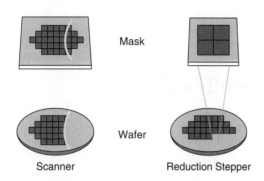

Figure 10.2 Scanning projection and reduction step-and-repeat. Reproduced from Mack (2007) by permission of John Wiley & Sons, Ltd

another mask blank.) Step-and-repeat lithography was the mainstay of IC fabrication in the 1980s and 1990s.

The alternative method is step-and-scan (with tools known as scanners), where the mask and wafer are moved in opposite directions and the exposure is done through a slit. Scanners expose smaller areas, in a sense, because only the narrow slit is exposed, making it easier and cheaper to fabricate optics for scanners. The field size can be made larger because of cheaper optics. And the field size is limited by optics only in the slit height direction; in the scan direction it is limited by stage mechanics. The benefits of cheaper optics are partially lost because the mechanics of wafer stage and mask holder movements must be delicately coordinated. IC fabrication relies mostly on scanners today.

It is certainly slower to expose chips than to expose full wafers. Steppers in the early 1980s exposed 25 wafers per hour (WPH) while the then dominant 1× projection aligners had throughputs of 100 WPH. Today steppers and scanners achieve up to 200 WPH. Assuming a 40 × 20 mm exposure field and 300 mm wafer size, this translates to a fraction of a second per chip. Exposure field size is important for DRAMs and microprocessors, which are large chips, and an integral number have to fit an exposure field. With smaller chips it is possible to reduce reticle costs by including many different designs on the same reticle.

Many other operations in addition to exposure can be done individually for each chip, enabling tighter process specifications. For example, focusing and alignment can be done for each chip, and wafer warping can be compensated for. Additionally, chipwise exposure enables the optimization of lithography in an economical fashion. In test runs all chips can be exposed differently, in order to find the optimum exposure dose and focus conditions, and to check process robustness. The focus–exposure

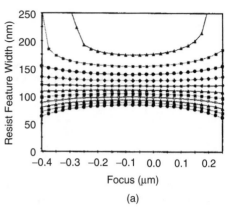

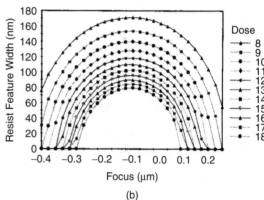

Figure 10.3 Focus–exposure matrix: linewidth plots with different focus depth and exposure dose (mJ/cm^2) combinations: (a) array lines; (b) isolated lines. Reproduced from Mack (2007), figure 10.14, by permission of John Wiley & Sons, Ltd

(FE) matrix results are shown in Figure 10.3. It shows linewidths for different focus settings with exposure dose as a parameter. As can be seen, it is difficult to optimize for both isolated and array lines simultaneously.

Alternatively, the FE matrix can be displayed as constant linewidths in focus–exposure dose coordinates. The process window can then be expressed as the maximum rectangle that fits within the linewidth specification (usually 10%), see Figure 10.4. For the isolated lines the process window is much smaller than for dense array lines.

It is possible to change the reticle between exposures and to have many different chips on one wafer in any proportion. The inclusion of different designs for rapid prototyping or test chips is thus flexible. Because the reticle never makes physical contact with the wafers, its lifetime is infinite in principle. In practice it is determined

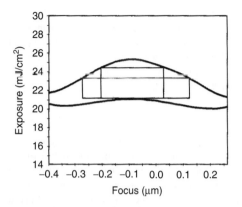

Figure 10.4 Process window in focus–exposure coordinates: all combinations inside the boxes result in acceptable linewidths from Mack (2007) by permission of John Wiley & Sons, Ltd

Table 10.1 Linewidth scaling of CMOS

Wavelength (nm)	Aperture	k factor	Linewidth (μm)	DOF (μm)
$\lambda = 436$	NA = 0.38	$k_1 = 0.8$	1	±1.5
$\lambda = 365$	NA = 0.48	$k_1 = 0.6$	0.5	±0.8
$\lambda = 248$	NA = 0.60	$k_1 = 0.6$	0.25	±0.35
$\lambda = 248$	NA = 0.65	$k_1 = 0.5$	0.18	±0.30

by product lifetime: many chips have such short lifecycles that a reticle set is often used only once, to make a lot of 25 wafers of 1000–10 000 chips depending on size.

Projection optics allows an extra trick to preserve reticle quality. A transparent film, called a pellicle, is framed in front of the reticle. Airborne particles will settle on the pellicle film, which is about 100 μm above the chrome pattern. Particles on the pellicle will be out of focus, and do not print on the wafer.

10.2 Resolution of Projection Optical Systems

The resolution of a projection optical system is approximated by the Rayleigh relation, Equation 10.1, and the depth of focus (DOF) is given by Equation 10.2:

$$\text{resolution} \approx \frac{k_1 \lambda}{\text{NA}} \quad (10.1)$$

$$\text{DOF} \approx \frac{k_2 \times \lambda}{\text{NA}^2} = \pm \frac{\lambda}{2 \times \text{NA}^2} \quad (10.2)$$

where NA is the numerical aperture. From geometrical optics we get k_1 as 0.5, which means that minimum lines will be λ/2NA. One approach to better resolution (smaller linewidths) is by wavelength reduction. This strategy has been steadily used: from 436 nm (g-line from an Hg lamp), to 365 nm (i-line from an Hg lamp) to 248 nm (KrF laser) to 193 nm (ArF laser). All else being equal, this alone would result in a factor of two improvement in resolution and a factor of four improvement in device areal density. NA enhancement is another clear route which has been used. In 20 years NA has increased from about 0.2 to 0.9, a factor of five improvement through better optical design and manufacturing.

Table 10.1 lists CMOS resolution trends assuming $k_2 = 1$ but letting k_1 evolve.

Resolution enhancement by NA increase has been paid for dearly on the focus side: DOF is becoming very small indeed. DOF (Equation 10.2) is an optical concept but resist chemistry and resist profile specifications (which depend on subsequent process steps) must also be considered. Besides optical DOF, other factors must be accounted for: the wafer is not flat, neither is the wafer chuck, and stepper focus mechanisms are not perfect. All these contribute, say, 0.1–0.2 μm errors in focus. Previous etching and deposition steps can easily create topographic variations on the order of half a micrometer, so planarization is critical for lithography (see Chapter 16 on chemical mechanical polishing).

A 4× reduction makes mask making much easier. Errors in both resist image on the mask and the etched chrome image on the mask are reduced, leading to tighter linewidth tolerances on the wafer. Mask writer placement error is also reduced, improving the overlay between two layers. The more complicated optics of reduction systems (in contact printing there is no imaging optics) introduced some distortion but this is a minor price to be paid.

10.3 Resists

In Chapter 9 we treated resists as if they were digital on/off materials which either react to exposure or not. Now we are dealing with more realistic cases: resists have exposure threshold energy, finite contrast and finite selectivity in developers. Resists are also optical materials and part of an optical system with reflections, interference and absorption. All these aspects become more pronounced when resists go over topography; patterning on a planar surface is fairly straightforward.

The calculation of exposure uses normalized remaining inhibitor concentration $M(x, t)$: it describes the fraction of inhibitor left after exposure for a certain time in a certain position inside the resist. Optical absorption α in photoresist is given by

$$\alpha = AM(x,t) + B \qquad (10.3)$$

where A is exposure-dependent and B exposure-independent absorption. A and B are known as Dill parameters, and their values for novolak resists are in the range of 0.4–1/μm for A and 0.01–0.1/μm for B. A decrease of inhibitor concentration depends not only on light intensity $I(x,t)$, but also on sensitivity to exposing radiation C and of course to inhibitor concentration M. Time-dependent inhibitor concentration is given by

$$\frac{\partial M}{\partial t} = -I(x,t)M(x,t)C \qquad (10.4)$$

The sensitivity parameter C is also known as Dill C and its value for novolak resists is on the order of 0.01 cm^2/mJ. A, B and C are all of course wavelength dependent. Analytical solutions to resist exposure are very difficult and simulation is used extensively.

Resist sensitivity can be tailored for different wavelengths (or for electrons, ions or X-rays). Sensitivity is important for productivity. With typical exposure energies on the order of 100–500 mJ/cm^2 for DNQ positive resists, exposure times for standard resists 1 μm thick are on the order of 1 s with 500 W lamps.

The first DUV lasers had intensities that were too low for practical throughputs and this problem led to the development of high-sensitivity chemically amplified resists (CARs) in the 1980s. CAR works in two steps: photoacid generator (PAG) molecules decompose upon photon impact and these decomposition products catalyze more PAG decomposition so that a single photon can lead to 100 decomposition reactions. In the second step, the post-exposure bake, the photoreaction products diffuse (a few nanometers or a few tens of nanometers) and react, and the reaction products are responsible for the solubility difference between exposed and unexposed resist. The coefficient Dill A is zero for CARs, because the exposure itself does not change the resist greatly.

Because the reaction is catalytic, the exposure dose is very small and system throughput is high (the exposure time is just one factor in throughput: the stepper has to move between exposures, align and focus, and all these add up in throughput calculations). CARs need only 10–50 mJ/cm^2 exposure doses, one-tenth of novolak resists. However, the very fact that the reaction is catalytic poses a danger: if the reaction is quenched, and multiplication stops, the resist is not exposed. This can happen due to airborne contaminants which react with the resist. Ammonia is one prime culprit, and it cannot be completely eliminated from cleanroom air because it is such an essential cleaning chemical (see Chapter 12) and it is released by the HMDS priming process (Equation 9.2). The two-step nature makes CAR processing time sensitive: no delays are allowed between exposure and bake.

10.3.1 Contrast

Photoresist contrast is important for both resolution and profile. A sigmoid (nonlinear) response function is essential for patternability. Optical wavefronts after the mask are not ideal square waves but rather attenuated sine waves, and a linear response as a function of exposure dose is rather useless because photoresist patterns would be smoothly curving bumps, not clearly defined rectangular shapes.

Contrast (γ) is calculated for positive and negative resists as the slope of dose required to clear the resist completely, Equation 10.5, which is graphically shown in Figure 10.5:

$$\gamma_p = \frac{1}{\log\left(\frac{d_c}{d_0}\right)} \qquad \gamma_n = \frac{1}{\log\left(\frac{d_0}{d_i}\right)} \qquad (10.5)$$

where d_c is dose to clear all resist and d_0 is extrapolated dose at the kink in the contrast curve. For negative resist d_0 and d_i are defined analogously. Typical contrasts are 2–5 for novolak-based positive resist and 5–10 for deep-UV CARs. High contrast is beneficial for the resist profile: on the one hand it is easy to achieve vertical walls, while on the other hand high contrast resist is easily overexposed, or saturated.

Resist thickness is limited by its mechanical stability: resist patterns that are too tall will collapse during drying due to capillary forces. This same phenomenon is seen in surface MEMS: short stubby ones survive better than long slender ones (Equation 29.8). Resist thickness is constantly reduced, and resist aspect ratios are 2–3:1, which translates to resists 300 nm thick for 90 nm technology and 130 nm thick for 45 nm node.

10.4 Thin-Film Optics in Resists

Photoresist is a part of an optical system involving the illumination light source, the lenses and the photomask, and we have to include the substrate also, because some light reaching the substrate through the resist will be reflected back, contributing to pattern formation.

Photoresist thickness determines the optical path length for incoming and outgoing rays. Constructive and destructive interference inside the photoresist lead to intensity

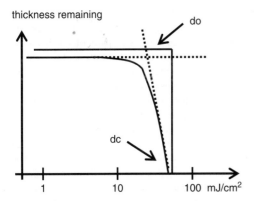

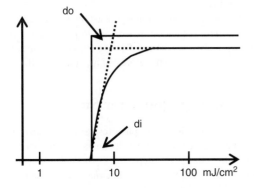

Figure 10.5 Resist contrast plots on a thickness–exposure dose axes for infinite contrast resist and real resists: left, positive resist; right, negative resist

variations in the vertical direction through the resist. This is seen as standing wave patterns in the developed resist (Figure 10.6). In the extreme case, the parts that received least light (in positive resist) will not be developed by a developer that has high selectivity between exposed and unexposed parts (high contrast developer).

Thin-film interference in the resist leads to thickness-dependent exposure doses. Depending on resist thickness, the total dose needed to expose the resist changes. If destructive interference takes place in the resist top surface, almost all of the illumination energy is absorbed in the resist, whereas in the case of constructive interference at the top surface, only half of the energy stays inside the resist. Maxima and minima alternate at $\lambda/(4n)$ intervals; for example, for $\lambda = 365$ nm exposure and 1.6 index of refraction for typical photoresist, this interval is 56 nm.

On a planar surface this problem can easily be solved by better control of photoresist thickness (spinning process), but on a structured surface there is no general solution to the variable resist thickness problem.

Figure 10.6 Reflections at air/resist and resist/substrate interfaces result in interference pattern of standing waves. Reproduced from Peterson *et al.* (1996) by permission of Henley Publishing

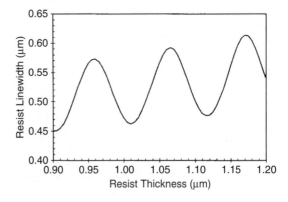

Figure 10.7 The effect of resist thickness variation on linewidth. Reproduced from Mack (2007) by permission of John Wiley & Sons, Ltd

The effect of resist thickness variation of linewidth is shown in Figure 10.7. Minor variation of 50 nm (5% of resist thickness) can lead to 20% linewidth variation.

Post-exposure bake (PEB), which enhances the diffusion of photoproducts, will make standing wave effects smaller. A diffusion distance of about half the period will smooth out standing waves. However, when lines are very narrow, PEB cannot be used: the diffusion distance must be considerably smaller than the linewidth, otherwise it will affect the line shape too much.

Swing ratio is a measure of variation introduced by thin-film optical effects. It is determined as exposure–dose variation (max − min) divided by mean value. It can be defined similarly for the linewidth. It is analogous to a lossy Fabry-Pérot interferometer and can be modeled by the following equation:

$$S = 4 \, e^{(-\alpha D)} \sqrt{R_1 R_2} \qquad (10.6)$$

where R_1 is reflectivity at the air/resist interface, R_2 is reflectivity at the resist/substrate interface, α is the resist absorption coefficient and D is resist thickness.

Obviously there are four ways to minimize the swing ratio. The first strategy is to minimize R_1, which translates to a top antireflective coating (TARC). Light traversing the TARC twice will interfere destructively and minimize reflections if TARC thickness matches the $\lambda/(4n)$ condition. The TARC refractive index should be given by

$$n_{\text{TARC}} = \sqrt{n_{\text{resist}} n_{\text{air}}} \qquad (10.7)$$

With the resist n typically around 1.65, the TARC refractive index should be about 1.3. TARC thickness would then be about 70 nm.

Photoresist-like spinning is a favorite method for coating TARC, and the material is very much photoresist-like (non-absorbing, however); it will be removed by the developer. Added process complexity is small. TARC is insensitive to substrate material, therefore it is a fairly general method to reduce reflections and swing. If, however, TARC is deposited over steps in a way similar to resist, TARC thickness will be variable, and its effectiveness reduced. There are not many materials with $n = 1.3$, so choices are not abundant.

The reduction of R_2 involves bottom antireflective coatings (BARCs). The role of BARCs on absorbing metal surfaces is to reduce reflectivity, and on transparent surfaces (like oxides) to reduce thickness effects. BARCs work by index matching, just like TARCs, but also by absorption: absorbed light will not re-enter the resist. BARC thicknesses are similar to TARC's, in the range of 20–70 nm, but the materials and processes are different. BARCs must tolerate developers because, if they did not, they would undercut the resist patterns. BARCs are therefore patterned by dry etching. Spin-on polymer-based BARCs do exist, but inorganic BARCs that will be left as permanent parts of the finished devices are also used. Titanium nitride TiN is a BARC for aluminum lithography, but it is deposited in the same process as the aluminum, not in conjunction with resist processing. Oxides and nitrides can also be used as BARCs. It is difficult to remove them selectively, and most often they remain as parts of finished devices. Inorganic BARCs can act as hard masks for etching: the resist is used as a mask for BARC etching, and BARC is then used as a mask for film etching.

The absorption strategy involves resist tailoring. The standard α are around 0.2–1/μm. Adding dyes to increase absorptivity α to for example 2/μm means that all radiation will be absorbed in the top resist layer, and the bottom part will not be exposed. So there is an optimum between swing ratio reduction and resist profile. Top surface imaging (TSI), overcomes the absorption dilemma by using very thin and very absorbing resists, which are not sensitive to profile variation like standard resists.

The fourth possible strategy, resist thickness increase, is at odds with resolution: if we wish to print narrow lines, thinner resists are better. Scaling to smaller linewidths with this strategy is therefore not an option at all.

10.5 Lithography Over Steps

The viscous flow of photoresist over steps leads inevitably to uneven resist thickness (Figure 10.8) and linewidth change at step edges. Because spin coating results in variable resist thickness over steps, the linewidth will be dependent on underlying steps via resist thickness changes.

On non-planar surfaces the effect of structures from previous steps causes some problems. Reflections from underlying metal lines can cause resist exposure in unwanted places. This is called reflective notching and is depicted in Figure 10.9.

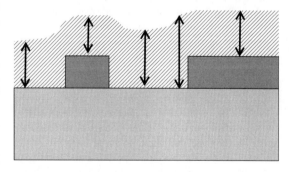

Figure 10.8 Thickness variation of spin-coated resist over topographic features

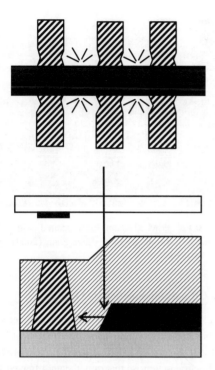

Figure 10.9 Reflective notching: left, top view of distorted resist lines; right, cross-sectional view showing how the underlying metal line reflects incoming light into the resist sidewall

10.6 Optical Extensions of Optical Lithography

Over the years many alternatives have been proposed as replacements for optical lithography, among them X-ray lithography, e-beam projection, ion beam projection and e-beam direct write. Optical lithography has been able to prosper because a constant flux of improvements has been forthcoming. For instance, there is immersion, an old technique that has been known to microscopists since the nineteenth century. NA (Equations 10.1 and 10.2) can be increased by replacing air (refractive index 1.00) with a higher refractive index fluid, usually water ($n = 1.35$). Immersion has been used in the production of 65 nm and more advanced chips.

Optical waves have four basic properties: amplitude, phase, direction and polarization. Engineering these has resulted in many new ways to improve optical lithography performance. Phase shift masks (PSMs), off-axis illumination (OAI), optical proximity correction (OPC) and subresolution assist features (SRAFs) are collectively known as resolution enhancement technologies (RETs).

10.6.1 Phase shift masks (PSM)

Normal masks are called binary masks because there are two possibilities for light: transmission or blockage. In phase shift masks (PSMs) there is a third possibility: transmission with phase shift. Shifters are semitransparent structures that produce a 180° phase shift in the transmitted light. Light along the shifted path will be out of phase with the light going through the unshifted part, and the amplitude will go through zero. Intensity, which is amplitude squared, will be much steeper compared to a binary mask, which improves both resolution and edge contrast.

The phase shift for light traveling in air for a distance L is $\Phi = 2\pi L/\lambda$, and for light traveling in the phase shifter material with index of refraction n, $\Phi = 2\pi n L/\lambda$. For a 180° phase shift, $\Delta\Phi = 180°$, the condition for shifter thickness is given by

$$L = \frac{\lambda}{2(n-1)} \qquad (10.8)$$

In alternating PSM (Figure 10.10) a shifter is either etched or deposited for every second feature, which limits PSM applications to regular arrays. The rim shifter (see Figure 10.11) utilizes undercut and can be applied to any pattern shape and size.

Rim PSM fabrication makes use of ingenious self-alignment with back-side illumination (Figure 10.11): an ordinary binary mask is fabricated first, with chrome patterns on a quartz plate. Shifter material is then deposited all over the plate, and photoresist is spun. The structure is then exposed from the opposite side of the

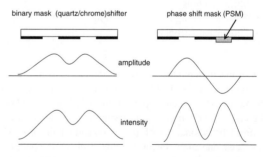

Figure 10.10 Binary mask (left) and alternating phase shift mask (right) compared: amplitude goes through zero for PSM, and intensity (= amplitude squared) is steep

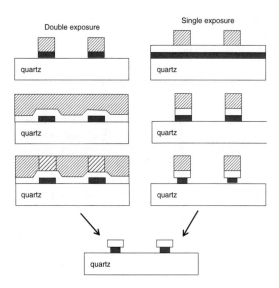

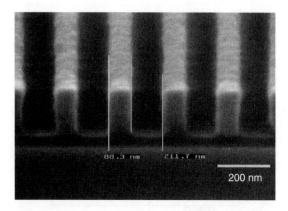

Figure 10.12 PSM: 100 nm lines printed with 193 nm light source. Reproduced from Fritze *et al.* (2003) by permission of IEEE

Figure 10.11 Two schemes for fabrication of rim PSMs: double exposure self-aligned on the left; standard single exposure on the right side. Both processes result in identical mask plates

mask plate and the chrome acts as a self-aligned mask for the shifters. The shifters are then etched, followed by chrome undercutting in a second etching step.

Process flow for PSM fabrication by single exposure

- Chrome deposition
- Shifter deposition
- Photoresist application
- Pattern generation
- Shifter etching
- Chrome etching and underetching
- Photoresist stripping

Chrome undercutting in both methods results in exactly the same degree of dimensional control. The difference is in mask inspection and repair: in the self-aligned method, the chrome pattern can be inspected and repaired before shifter fabrication. Lack of inspection and repair for PSMs has been one main factor holding back their adoption. PSM has been used in production since the 180 nm generation. As shown in Figure 10.12, it is possible to print 100 nm lines with 193 nm wavelength.

10.6.2 Off-axis illumination (OAI)

Normally the illuminating light hits the mask perpendicularly but in off-axis illumination (OAI) the light hits the mask tilted (Figure 10.13). This has the effect of shifting the diffraction pattern so that only the undiffracted zeroth order and the first diffracted order are transmitted (and two components are needed to form an image). While in normal illumination the zeroth order is at the middle of the aperture, and the first orders at the edges, giving pitch as λ/NA, in OAI the pitch is halved because the zeroth and first order are twice that far apart. Theoretically OAI can therefore provide a pitch that is half the standard illumination pitch. Various pupil designs for OAI are shown in Figure 10.14. One of the drawbacks of OAI is that parallel and perpendicular lines will be different, and in fact each mask should have a pupil optimized for it.

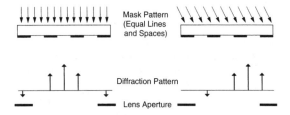

Figure 10.13 Off-axis illumination (OAI). Reproduced from Mack (2007), by permission of John Wiley & Sons, Ltd

Figure 10.14 Conventional, annular and quadrupole apertures

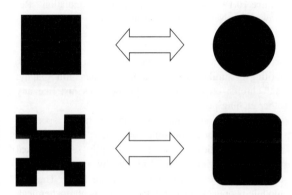

Figure 10.15 Standard mask (top left) and its resulting resist pattern (top right); optical proximity corrected mask (bottom left) and its resist pattern (bottom right)

10.6.3 Optical proximity correction (OPC)

Isolated and dense lines print differently (Figure 10.3) because diffraction in arrays is different from that of an isolated structure. This effect becomes more significant when feature sizes are scaled down. It can be combated by designing the structures to take into account the neighboring structures. This is an inverse problem: we know the desired end result, and we have to work backward from the correct resist pattern to a mask pattern which will give us that. In Figure 10.15 this is represented by arrows pointing in both directions: the mask on the left produces the pattern on the right, or, alternatively, to get the pattern on the right, we have to design the mask to be as shown on the left.

10.6.4 Subresolution assist features (SRAF)

Subresolution assist features (SRAFs) are lines that are too small to be printed in the lithography process, but they will still affect the diffraction patterns, enabling better control of feature shapes (Figure 10.16). SRAFs around an

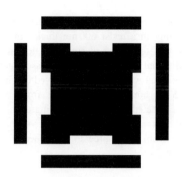

Figure 10.16 Subresolution assist features (SRAFs)

isolated line make it look more like an array line, making the difference between the two smaller, thus improving process latitude. On the negative side, it takes a lot more shots from the pattern generator to print all those extra small pixels.

10.7 Non-Optical Extension of Optical Lithography

Because optical lithography tools are expensive, there is a constant need to find ways to produce narrow lines in a more cost-effective way. One such technique is resist trimming. The minimum resist line is first produced by optical lithography, and isotropic plasma etching of the photoresist is then performed (Figure 10.17). The resist line gets narrower and thinner. This method is most suitable when narrow lines can be used as a starting point. In a university lab, a 1 μm minimum line can be

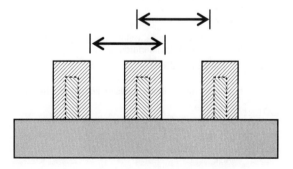

Figure 10.17 Resist trimming: resist lines made narrower by isotropic etching of the resist in oxygen plasma. In both cases the resolution (line + space) remains constant

narrowed to 200 nm by 400 nm lateral narrowing on both sides (resulting in a resist 600 nm thick). In modern IC fabrication this technique is employed to narrow down critical individual lines: the microprocessors gates. Lines of 50 nm original width can be narrowed down to 20 nm: that is, 15 nm horizontal narrowing from both sides. Pitch is not affected, and device packing density remains unchanged, but transistor speed improves because the transit time from the source to drain becomes smaller, see Equation 26.1.

10.8 Lithography Simulation

Lithographic pattern formation starts with the designer's layout file which is turned into a physical mask plate in a mask shop. This mask is inserted into the exposure tool, where it modifies illumination from the light source. After complex photochemistry steps in the photoresist, development creates patterns in the resist. This information flow (Figure 10.18) has many points where errors can occur and where dimensions are not accurately transferred. Some of these are data errors related to formats used in drawing and mask writing, and some are physical and chemical errors related to both mask writing and exposure resolution and to etching tolerances.

Note that the mask writing process has a similar information flow and similar error sources: the mask

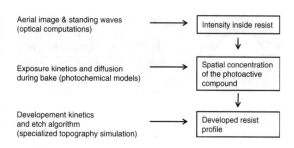

Figure 10.19 Modules of lithography simulation. Redrawn after Neureuther and Mack (1997)

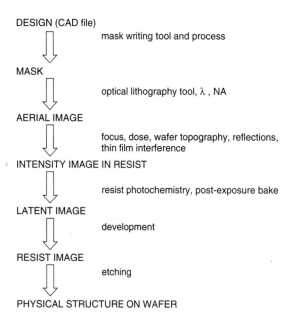

Figure 10.18 Lithography information flow. Adapted from Brunner (1997)

writer has finite resolution; the photoresist used in mask writing is similar to resists used in optical lithography; and chrome etching has its non-idealities just like any other etching process.

Lithography simulation is a self-contained specialty within microfabrication simulations. It is partly physical simulation (optical modeling) and partly semiempirical (development simulation). Lithography simulators have three basic functions as shown in Figure 10.19. The first module is optical modeling, the second is photochemical, time-dependent, diffusion modeling, and the third module is an etch simulator specifically built for resists. The development of novolak resist in alkaline developer is an etching reaction, and it uses models similar to etching, but because its application field is very specific, higher accuracy is possible. These steps have been modeled with good success even though an understanding of many of the basic mechanisms in resist exposure and development is yet to be revealed.

SAMPLE 2D is a lithography and topography simulator containing a concise optical lithography model (Figure 10.20). Lithography simulation input parameters include light source data like wavelength, exposure dose, NA and coherence; resist thickness and Dill parameters A, B and C; wafer and resist refractive indices; and development rate parameters. SAMPLE 2D is able to predict resist profiles with standing waves. PROLITH is an advanced lithography simulator especially for deep-UV lithography.

10.9 Lithography Triangles

Optical lithography has been amazingly persistent, and it has been scaled down more than anybody could ever imagine. Contenders for optical lithography will be discussed in Chapter 38. Some of them excel in certain aspects, but optical lithography has all the elements it takes: light sources, resist, masks, metrology, design software,

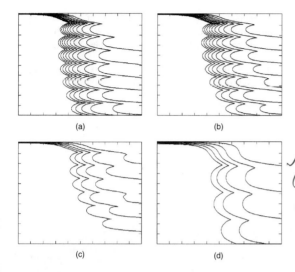

Figure 10.20 SAMPLE 2D simulation of resist exposure and development; nominal linewidth is 1.0 μm (only right hand side is shown because the structure is symmetric): (a) exposure dose 100 mJ/cm², development time 65 s; (b) dose 80 mJ/cm², development 75 s, leads to sloped profile; and (c) dose 70 mJ/cm², development 70 s, leads to incomplete development. In (d) conditions are identical to (c) but resist thickness is only 0.5 μm

etc. It takes time to build an infrastructure for a new lithography technology.

Critical dimension, CD, is basically a 2D concept. However, the resist sidewall is seldom 90°, so CD should really be able to describe the shape of a trapezoidal object. Real resist lines are also rough to some extent, and line edge roughness (LER) is used to describe this.

Fantastically narrow lines are printed optically but there are tradeoffs. These are shown by two "magic triangles" in Figure 10.21. Magic triangles tie together properties that are difficult to optimize simultaneously.

10.10 Exercises

1. Produce a graphical presentation of projection lithography resolution vs. depth of focus!
2. Estimate the contrasts of resists in Figure 10.3.
3. If subresolution assist features are half the feature size, how is e-beam mask writing time affected?
4. How much will the swing ratio be reduced if the top antireflection coating can reduce air/resist reflections by 20%? How much will the swing ratio be reduced if absorbance increases from 0.5 to 1/μm?
5. Calculate some good and bad resist thicknesses for novolak resist at 365 nm exposure!
6. What is the resist thickness and linewidth in Figure 10.6?
7. How does resist trimming work if the original linewidth is 0.5 μm and resist thickness 1 μm? What if the original linewidth is 100 nm and resist thickness 150 nm?
8. How thick are phase shifters if a 193 nm laser is used for exposure?

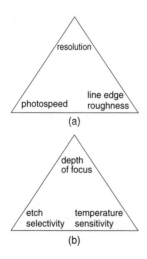

Figure 10.21 Magic triangles in lithography. Adapted from Zell (2006)

References and Related Reading

Ausschnitt, C. P. *et al.* (1997) Advanced DUV photolithography in a pilot line environment, *IBM J. Res. Dev.*, **41**, 21.

Brunner, T. (1997) Pushing the limits of lithography for IC production, International Electron Devices Meeting, p. 9.

Fritze, M. *et al.* (2003) Enhanced resolution for future fabrication, *IEEE Circuits Devices Mag.*, **1**, 43.

Gu, A. and A. Zakhor (2008) Optical proximity correction with linear regression, *IEEE Trans. Semicond. Manuf.*, **21**, 263–271.

Holmes, S. J. *et al.* (1997) Manufacturing with DUV lithography, *IBM J. Res. Dev.*, **41**, 7.

Lin, B. J. (2006) The ending of optical lithography and the prospects of its successors, *Microelectron. Eng.*, **83**, 604–613.

Mack, C. (2007) **Fundamental Principles of Optical Lithography**, John Wiley & Sons, Ltd.

McCallum, M., M. Kameyama and S. Owa (2006) Practical development and implementation of 193nm immersion lithography, *Microelectron. Eng.*, **83**, 640–642.

Neureuther, A. R. and C. A. Mack (1997) Optical lithography modeling, in P. Rai-Choudhury (ed.) **Handbook of Microlithography, Micromachining and Microfabrication**, SPIE.

Peace, R. F. and S. Y. Chou (2008) Lithography and other patterning techniques for future electronics, *Proc. IEEE*, **96**, 248–270.

Peterson, B. *et al.* (1996) Approaches to reducing edge roughness and substrate poisoning of ESCAP photoresists, *Semicond. Fabtech*, **8**, 183.

Rai-Choudhury, P. (ed.) (1997) **Handbook of Microlithography, Micromachining and Microfabrication**, SPIE.

Schellenberg, F. M. (2005) A history of resolution enhancement technology, *Opt. Rev*., **12**, 83–89.

Zell, Th. (2006) Present and future of 193nm lithography, *Microelectron. Eng*., **83**, 624–633.

11

Etching

The pattern transfer process consists of two steps: lithographic resist patterning and the subsequent etching of the underlying material. Resist protects the areas where the material needs to remain, and open areas are etched.

In the etching process material is chemically and/or physically attacked and eroded in the unprotected areas. Some materials are spontaneously chemically etched, like silicon by fluorine, aluminum by chlorine, or oxide by hydrofluoric acid. Sometimes physical processes are needed to assist in etching, like ion bombardment in chlorine etching of aluminum oxide and fluorine etching of silicon dioxide. In those cases the resulting etch rate is a synergistic sum of both chemical and physical processes.

All materials can be etched by energetic ions, and this applies to the resist mask and the underlying film, too. But in practical processes it is important to achieve selectivity: that is, high etch rate ratio between two materials. In the ideal case etching would stop when film clears, but in practice some underlying material loss is almost inevitable (Figure 11.1). Resist is also consumed in the process, and the sidewall of etched structure is not necessarily perfectly vertical. It calls for engineering judgment to decide which degree of profile control and selectivity are acceptable.

In lithography rework is easy: if the resist pattern is found to be faulty in inspection, the resist is stripped and lithography repeated, but once the pattern has been transferred into solid material by etching, rework is much more difficult, and usually impossible.

11.1 Etch Mechanisms

Etching is often divided into two classes, wet etching and plasma etching. Wet etching equipment consists of a heated quartz tank or bath plasma etching equipment consists of a vacuum chamber with a RF generator and a gas system.

The basic reactions in etching are given by

Wet etching:

solid + liquid etchant ⇒ soluble products

$$Si\ (s) + 2\ OH^- + 2\ H_2O \Rightarrow Si(OH)_2(O^-)_2\ (aq) + 2\ H_2\ (g) \quad (11.1)$$

Plasma etching:

solid + gaseous etchant ⇒ volatile products

$$SiO_2\ (s) + CF_4\ (g) \Rightarrow SiF_4\ (g) + CO_2\ (g) \quad (11.2)$$

There are several processes that must take place for etching to proceed:

- transport of etchants to surface (flow and diffusion)
- surface processes (adsorption, reaction, desorption)
- removal of product species (diffusion and flow).

These basic steps are the same for both wet etching and plasma etching, but clearly diffusion in the gas phase is

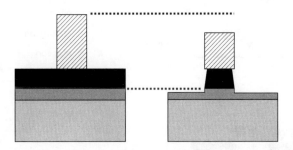

Figure 11.1 Etching with photoresist mask: thin film has been etched, with some etching of the underlying material and resist, too. There is some undercutting and the profile is not perfectly vertical

more rapid than in the liquid phase. In plasmas there are many possibilities to provide extra energy to the process, in the form of accelerated ions. The chemical component of etching can be enhanced by heating, in both wet and plasma etching.

A comparison of the basic processes of plasma etching in Figure 11.2 to those of CVD (Figure 5.6) reveals obvious similarities. Gases are fed in by forced convection, some species diffuse through the boundary layer to the surface and react there, desorb and diffuse back to the main flow stream, and get pumped away. Note that utilization of source gases is never perfect: some source gas molecules "fly by" the wafer in the main flow. A very small fraction of molecules are ionized, and these ions enhance surface reactions. Plasma etcher can be run as a deposition tool, and PECVD reactor as an etcher, which is beneficial for chamber cleaning.

If etching is too slow, any of the above steps can be the cause of the problem: inadequate inflow of etchant, slow diffusion through a thick boundary layer, slow chemical reaction on the surface, low volatility (or solubility) of products. Redeposition of products on the surface can also take place, and sometimes byproducts, like hydrogen gas evolution (as in Equation 11.1), are so strong that they prevent etchant from reaching the surface.

Etch rates are typically 100–1000 nm/min, for both wet and plasma processes. The lower limit comes from manufacturing economics and the upper limit from resist degradation, thermal runout and damage considerations. Silicon etching is exceptional: rates up to 20 μm/min are available in both wet etching (HF:HNO$_3$) and in plasma etching (DRIE) in SF$_6$.

11.1.1 Footnote on terminology

The term dry etching, as an opposite to wet etching, is often used as a synonym for plasma etching, but there are dry methods which do not involve plasmas, e.g. XeF$_2$ etching of silicon. We use the terms plasma etching and RIE interchangeably. RIE, Reactive Ion Etching is a misnomer: ions in RIE are in a minor role, most etching is done by excited neutrals. Ion etching, also known as ion milling, is a completely different kind of an etching process, to be discussed in Section 11.11.

11.2 Etching Profiles

The isotropic etching front proceeds as a spherical wave from all points open to the etchant. Because the etch front proceeds under the mask (resulting in undercut), isotropic etching cannot be used to make fine features (Figure 11.3). In fully isotropic etching lateral extent (undercutting) is identical to vertical etched depth.

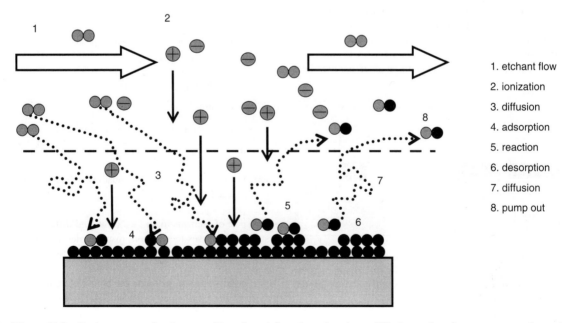

Figure 11.2 Basic processes in plasma etching: forced flow, boundary layer diffusion and surface processes enhanced by ion bombardment

Etching 129

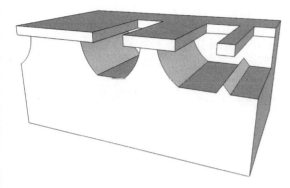

Figure 11.3 Undercutting in isotropic etching: wide lines are narrowed but narrow lines are completely undercut and released

Etch bias, defined as the difference between etched feature size and mask size, is then twice the etched depth.

Isotropic profile is the most commonly encountered etch profile. Most wet etchants result in an isotropic profile, and it is encountered also in plasma and dry etching. In plasma etching, the degree of isotropy can be controlled by the etching parameters, from fully isotropic to fully anisotropic (which may not be easy !).

Undercut can be compensated by making the initial mask feature larger than the desired width, for light-field structures, and smaller for dark-field structures. This approach works quite well for isolated structures, but in dense arrays its utility is limited.

Wet etching profiles are seldom perfectly isotropic, and both deep slopes and gently sloping sidewall profiles are possible. The main parameters affecting the slope are the same as those governing the other main features of etching: etchant concentration and temperature. Silicon dioxide etching can produce steep slopes in NH_4F:HF at a ratio of 7:1 at 25 °C, but a ratio of 30:1 at 55 °C leads to a gentle slope. Gentle slopes may be desirable for step coverage in subsequent deposition steps. NH_4F:HF mixture is known as buffered HF (BHF) because the consumption of fluoride ions by the etching reaction is compensated, buffered, by ionization of NH_4F into ammonia and fluoride.

Undercutting is sometimes desirable and even necessary. In order to fabricate free-standing beams and plates, isotropic etching is a must: the beams are released when the underlying material is completely removed (Figure 11.3). Such released beams and plates are essential building blocks in surface micromechanics, for instance the bolometer of Figure 2.1 has been released by isotropic undercutting.

Anisotropic etching results in a vertical or almost vertical profile, which is suitable for fine structure fabrication. Isotropic and anisotropic profiles are compared in Figure 11.4. In plasma etching the profile can be tailored by process conditions: pressure, power, temperature and gas flows. Bombardment supplies energy to horizontal surfaces and drives vertical etching. Sidewalls do not experience ion bombardment, and therefore the etch rate in the lateral direction is reduced. Low-pressure operation favors anisotropy because bombardment is more directional, but low-pressure operation requires either a bigger pump or reduced flow rate, in which case the rate is lower. Other effects of ion bombardment include ion-induced desorption, ion-induced damage and ion-activated chemical reactions. Increased ion energy

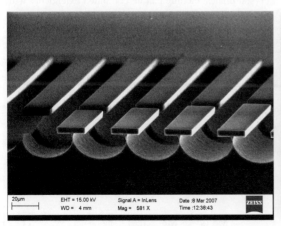 

Figure 11.4 Isotropic (left) and anisotropic etch profiles (right)

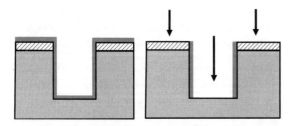

Figure 11.5 Mechanism of anisotropy: all surfaces are passivated by a thin film, but directional ion bombardment will clear films from a horizontal surface while leaving passivation film on the sidewalls, enabling etching to proceed vertically only

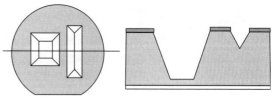

Figure 11.6 Anisotropic crystal plane-dependent wet etching of silicon: top view and cross-sectional view of a membrane and a V-groove etched on (100) silicon wafer

can be helpful for increasing vertical etch rate, but the masking material is also etched faster if power is high.

The other main mechanism for anisotropy is sidewall passivation. Passivating films, for example $(CF_2)_n$-type fluorocarbon films (from CHF_3 or C_4F_8 gases), deposit on all surfaces, but ion bombardment removes them from horizontal surfaces, and sidewalls remain passivated (Figure 11.5). Operation at cryogenic temperatures (e.g., $-120\,°C$) leads to reduced chemical reaction rates according to the Arrhenius equation. Suppression of spontaneous reactions means that only ion bombardment-driven reactions take place, and they are anisotropic. Cryogenic etching of silicon in SF_6/O_2 etching of silicon results in SiO_xF_y reaction products that are not very volatile, and they will redeposit on the wafer. And just as with fluoropolymer films, ion bombardment will remove this redeposited film from horizontal surfaces, leaving the sidewalls passivated.

11.3 Anisotropic Wet Etching

Isotropy, or homogeneity of space in all directions, is sometimes useful as we can neglect directions. Anisotropic processes are spatially directional, but there are two completely different usages of the term anisotropic etching: anisotropic wet etching and anisotropic plasma etching. Anisotropic wet etching could be termed crystal plane anisotropy, because the different silicon crystal planes have different etch rates.

Potassium hydroxide (KOH) is the prototypical anisotropic wet etchant for silicon. In KOH silicon (100) crystal planes are etched 200 times faster than (111) planes. KOH etching at about $80\,°C$ cannot be done with photoresist mask. Instead, resist is used to etch silicon dioxide or silicon nitride, and after resist stripping silicon etching is done in KOH with oxide or nitride mask. Initially the shape is determined by the fast etching crystal planes (100) and (110) but in the end etching terminates when only slow etching (111) planes are present. When the slow etching (111) planes meet, etching will terminate. The etched depth is determined by the mask opening and the angle between the (111) and (100) planes, $54.7°$. This angle is very characteristic of silicon MEMS structures. Some simple geometries are depicted in Figure 11.6, namely V-grooves and membranes. Note that it is important that the structures are aligned to the crystal planes (indicated by the wafer flat).

A square-shaped mask pattern will lead to a square-shaped well, or, if thought of differently, to a silicon membrane that can be used in pressure sensors as the active bending element. If etching is continued until silicon is completely removed, a nitride membrane results. If the mask opening is small, the slow etching (111) planes will meet, and an inverse pyramid shape is formed (as seen in Figure 1.14a).

Because anisotropic wet etching is controlled by the crystal planes, etching can be continued extendedly without loss of control. Through-wafer etching is possible, enabling, among others, microrocket thrusters (Figure 11.7). The thruster is made of two identical wafers which have been bonded together. The variety of shapes that can be made by crystal plane-dependent etching is astonishingly large, as will be seen in Chapters 20 and 30.

11.4 Wet Etching

Wet etching mechanisms fall into two major categories: metal etching by electron transfer, Equation 11.3, and insulator etching by acid–base reaction, Equation 11.4:

$$Me\ (s) \Rightarrow Me^{n+}\ (aq) + ne^- \qquad (11.3)$$

$$SiO_2 + 6\ HF \Rightarrow H_2SiOF_6\ (aq) + 2\ H_2O \qquad (11.4)$$

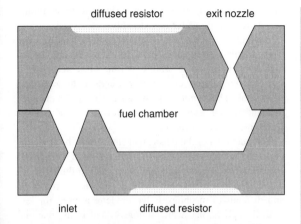

Figure 11.7 Anisotropic wet etching of silicon: two identical silicon wafers bonded together to form a microrocket fuel chamber and inlet and exit nozzles (with diffused resistor heater). Adapted from Mukerjee et al. (2000)

Table 11.1 Wet etchants for photoresist masked etching (mostly room temperature)

SiO_2	$NH_4F:HF$ (7:1),[a] 35 °C
SiO_2	NH_4F: $CH_3COOH:C_2H_6O_2$ (ethylene glycol). H_2O (14.32.4.50)
Poly-Si	$HF:HNO_3:H_2O$ (6:10:40)
Al	$H_3PO_4:HNO_3:H_2O$ (80:4:16), water can be changed to acetic acid
Mo	$H_3PO_4:HNO_3:H_2O$ (80:4:16)
W, TiW	$H_2O_2:H_2O$ (1:1)
Cr	$Ce(NH_4)NO_3:HNO_3:H_2O$ (1:1:1)
Cu	$HNO_3:H_2O$ (1:1)
Ni	$HNO_3:CH_3COOH:H_2SO_4$ (5:5:2)
Ti	$HF:H_2O_2$ (1:1)
Au	$KI:I_2:H_2O$, $KCN:H_2O$
Pt, Au	$HNO_3:HCl$ (1:3), "aqua regia," H_2O dilution may be used

[a] Called BHF, for buffered HF, and also known as BOE, for buffered oxide etchant.

The rates of wet chemical reactions can have two behaviors which we encountered earlier in the CVD reaction mechanisms:

1. The surface reaction is slow and it determines the rate.
2. The surface reaction is fast and the rate is determined by etchant availability (transport of reactant by convection and diffusion).

Surface reaction-limited processes exhibit activation energies of 30–90 kJ/mol. The rate increases with increasing etchant concentration and is insensitive to stirring. Different crystal planes can have different surface reaction rates and this leads to anisotropy, for example in KOH and TMAH etching of silicon. Aluminum etching in H_3PO_4 is also surface reaction limited: Al_2O_3 dissolution is the rate determining step, with 54 kJ/mol activation energy.

Transport-controlled reactions are characterized by activation energies of 4–25 kJ/mol. Their rate increases with agitation and stirring because more reactant is being brought within the vicinity of the surface. Furthermore, all crystal planes etch at the same rate, which is natural. Silicon etching in $HF:HNO_3$ mixture is limited by HF diffusion through the product layer. The activation energy is 17 kJ/mol.

Wet etching processes are available for most materials, with diamond and GaN being the most notable exceptions. Table 11.1 lists a number of etch processes for photoresist masked etching. As a general rule, lower temperature and more dilute solutions can be tried if the photoresist mask does not survive the etching conditions. It must be remembered that other factors will change too, for example sidewall profile.

Oxide etch rate goes down linearly with decreasing HF concentration. However, aluminum etch rate goes up when HF concentration decreases: 49% HF etches aluminum at 38 nm/min, but 10:1 diluted HF results in a 320 nm/min rate. This is because water has an active role in aluminum surface oxidation. Buffering agents and other additives can dramatically change etch rates and selectivities, as shown in Table 29.2. Wet etch processes for non-masked etching are listed in Table 11.2. These processes remove films without making any patterns.

Wet etching is an indispensable tool in defect analysis: microstructural defects like stacking faults and pinholes can be made visible by wet etching. Sirtl, Secco, Wright, Dash and Sailor are etchants for delineating defects. In reverse engineering and failure analysis thin films are removed selectively by isotropic etching (wet or dry) to reveal layer by layer the wanted structures.

Table 11.2 Wet etchants for unmasked etching

SiO_2	DHF, dilute HF (1%), for removing native oxide (about 10 nm/min)
SiO_2, PSG	HF (49%) sacrificial layer removal (>1 μm/min)
Si_3N_4	HF (49%) layer removal (4 nm/min)
Si_3N_4	H_3PO_4 (concentrated, 180 °C)
Si	$HF:HNO_3:H_2O$ (6:10:40) (thinning etch, up to 20 μm/min)
Cu	HCl

Wet etching processes are easy in theory, but in practice wet etching is difficult:

1. Reaction products may affect the etching reaction, for example hydrogen evolves when silicon is etched by hydroxide (KOH, for instance) and this hydrogen can prevent etchant from reaching the surface.
2. Etching products catalyze or inhibit the reaction ($HF:HNO_3$ etching of silicon).
3. The etching reaction is sensitive to stirring (mass or heat transport limited).
4. The etching reaction is exothermic and temperature rises during etching (for these reactions, stirring decreases the etch rate because it removes heat).
5. Evaporation of etchant leads to concentration changes during etching.

11.4.1 Wet etching tools

Wet etching comes in three major variants: tank immersion (bath) (Figure 11.8), spray tool and single wafer processor. The tank is for example a quartz vessel with heating and temperature control. It is filled with water and chemicals and the wafers are immersed in liquid for a required time and then transferred to similar tanks for rinsing. Spray tools handle a cassette (or cassettes) but, instead of immersion, liquid is sprayed from stationary nozzles on rotating wafer cassette(s). After the first spraying the process continues with either another chemical or water spray and nitrogen drying, in the same vessel. Fresh mixing of chemicals and lower liquid volumes are spray tool advantages over tanks. Single wafer tools are akin to photoresist spinners, and in a sense they are spray tools,

Figure 11.8 Wet etching tank. Courtesy VTT

too. However, processing acts on a single wafer at a time, and works on the wafer topside only.

Heating wet etching tanks uniformly is no easy task because highly reactive and corrosive chemicals are used at high temperatures (e.g., 180 °C boiling nitric acid to etch nitride). Temperature uniformity depends on flow patterns in the tank. This is not trivial because stirring can enhance reactant supply, reaction product removal, or heat removal from an exothermic reaction.

The materials of the tanks and heaters must be compatible with the process, in chemical, thermal and mechanical respects. Teflon and quartz are often used in the most demanding applications, but both are expensive materials and difficult to machine. Polypropylene is used for less critical applications, while stainless steel is the material for solvent tanks.

11.5 Plasma Etching (RIE)

Vertical walls and highly accurate reproduction of photoresist dimensions translate to closely spaced structures. High packing density of devices is possible by anisotropic plasma etching. When undercut becomes significant relative to linewidth, wet etching faces serious problems. In IC fabrication this led to the adoption of plasma etching at about 3 µm linewidths. With anisotropy, that is vertical sidewalls, undercut compensation schemes became unnecessary, and all the resolving power of lithography tools could be used to increase device packing density.

Plasma etching is done in a vacuum chamber by reactive gases excited by RF fields (Figure 11.9). Both excited and ionized species are important for plasma etching. Excited molecules like CF_4^* are very reactive, and ionic species like CF_3^+ are accelerated by the RF field and impart energy directionally to the surface. Plasma etching is thus a combination of chemical (reactive) and physical (bombardment) processes.

Deep reactive ion etching (DRIE) is an extension of RIE to make deep structures with high etch rate. DRIE reactors have two power generators, one for generating a high-density plasma (10^{12}–13 ions/cm^3) and another for biasing the wafer electrode, see Figure 33.2. This arrangement enables high powers to be used for active species generation, but independent control of ion bombardment via the second power source. DRIE will be discussed in detail in Chapter 21.

11.5.1 Plasma etch chemistries

In a plasma discharge a number of different mechanisms for gas phase reactions are operative. The discharge generates both ions and excited neutrals, and both are important

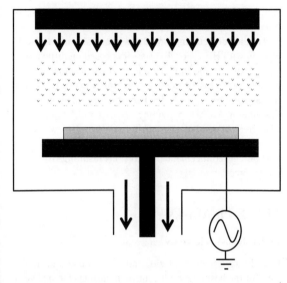

Figure 11.9 Plasma etching system (RIE): gases are introduced through the top electrode; wafers are on the powered bottom electrode

Table 11.3 Typical etch gases

Fluorine	Chlorine	Bromine	Stabilizers	Scavengers
CF_4	Cl_2	HBr	He	O_2
SF_6	BCl_3		Ar	
CHF_3	$SiCl_4$		N_2	
NF_3	$CHCl_3$			
C_2F_6				
C_4F_8				
XeF_2				

Table 11.3 lists typical plasma etch gases. These include not only the etchant gases but also scavengers and stabilizers. SF_6- and CF_4-based processes have typically 10–40% oxygen added to them. Oxygen has several roles: it reacts with SF_n and CF_n fragments, and keeps fluorine concentration high by preventing fluorine recombination with the fragments. Oxygen reacts with resist, which contributes to sidewall polymer formation and therefore improved anisotropy. Oxygen also increases the resist etch rate, so that selectivity can in fact be worse even though the silicon etch rate is enhanced.

11.5.2 Plasma etch mechanisms

Chemical bonds need to be broken for etching to take place. Bond energies, therefore, give indications of possible etching reactions. Reactions that lead to bonds stronger than the Si–Si bond will etch silicon, and if the products have stronger bonds than Si–O, silicon dioxide will be etched (Table 11.4). These simple predictions are experimentally confirmed: fluorine, chlorine and bromine will etch silicon because silicon–halogen bonds are stronger than silicon–silicon bonds. Only the Si–F bond is stronger than the Si–O bond and therefore only fluorine is predicted to etch oxide. But because of ion bombardment, oxide is slightly etched also in chlorine and bromine plasmas, but to a much lesser extent than in fluorine plasmas.

In practice, the volatility of reaction products (i.e., high vapor pressure) is used as a criterion for etchant selection. Boiling points of reaction products can be used to estimate volatility (Tables 11.5 and 11.6). (Note that boiling points are usually for 1 atm pressure, not for reduced pressures, but they are very useful as quick estimates for volatility.)

for etching, that is

Ionization: $e^- + Ar \Rightarrow Ar^+ + 2\,e^-$

Excitation: $e^- + O_2 \Rightarrow O_2^* + e^-$

Dissociation: $e^- + SF_6 \Rightarrow e^- + SF_5^* + F^*$

The most abundant species in the plasma reactor is the source gas. Etch reaction products are the next most abundant, and they may represent a few percent or 10% of all moieties. Excited neutrals may be present at a few percent, but ions are just a very minor component, for example one in 100 000. They are, however, often important for the mechanism.

Plasma etching is based on reaction product volatility. Silicon is easily etched by halogens: the fluorides (SiF_4), chlorides ($SiCl_4$) and bromides ($SiBr_4$) of silicon are volatile at room temperature at millitorr pressures. No ion bombardment is needed for etching because the reactions are thermodynamically favored and the role of ion bombardment is used to induce directionality. Silicon nitride (Si_3N_4) is etched by fluorine, producing SiF_4 and NF_3. Aluminum is spontaneously etched by Cl_2, but the surface of aluminum is always protected by a native aluminum oxide layer a few nanometers thick, and aluminum etching can only commence after this oxide has been removed. Ion bombardment is essential for aluminum oxide etching.

Table 11.4 Bond energies (kJ/mol)

C–O	1080	Si–F	550
Si–O	470	Si–Cl	403
Si–Si	227	Si–Br	370

Table 11.5 Etch product boiling points (T_{bp}, °C; d = decomposition, s = sublimation)

SiF_4	−90	$SiCl_4$	−70	CO_2	−56
NF_3	−206	$AlCl_3$	190	PH_3	−133
WF_6	2.5	$GaCl_3$	78	AsH_3	−116
WOF_4	110	$TiCl_4$	−25		
TaF_5	96.8	$WOCl_4$	211	$SiBr_2$	5.4
MoF_6	17.5	WCl_6	275		
$MoOF_4$	98	$InCl_2$	235		
NbF_5	72	$MoCl_5$	194		
GaF_3	800s	$PtCl_4$	370d		
		$PbCl4$	−15		
		$Cr(CO)_6$	110d		

Table 11.6 Non-etchable reaction products (T_{bp}, °C; d = decomposition, s = sublimation)

$CuCl_2$	620	TiF_4	>400
CuF_2	950d	PbF_2	855
$CrCl_2$	824	CrF_2	1100
AlF_3	1290s	TiF_3	1200

Reaction products like WOF_4 (from CF_4 and O_2 etching of tungsten) and $AlCl_3$ (Cl_2 etching of aluminum) have boiling points around 200 °C and are volatile enough for practical etching, but AlF_3 or CrF_2 have boiling points about 1000 °C and therefore fluorine is not a suitable etchant for these materials. Ion bombardment enhances the removal of material and can be used to drive reactions which might otherwise not be suitable for etching. Such reactions are, however, prone to residues. Gallium fluoride is involatile but chlorides are volatile, and therefore GaAs and GaN etching uses chlorine.

The exact plasma etch mechanisms remain unknown in many cases. It has been shown that damaged single crystal tungsten is etched much faster than perfect crystal. Silicon etch rate has been shown to be synergistic, with both ion bombardment and chemical components: etching with argon ion bombardment alone results in low rate, and similarly for XeF_2 gas etching, but for simultaneous Ar^+/XeF_2 process the etch rate increases by a factor of 10–100.

11.6 Isotropic Dry Etching

There are plasma-less dry etching methods: XeF_2 and ClF_3 gases will spontaneously break down and release free fluorine, which aggressively etches silicon. XeF_2 is a crystalline solid which sublimes to produce XeF_2 vapor with a vapor pressure of a few torr. It is very reactive with water and forms hydrofluoric acid (HF) which presents hazards to operators.

Another dry etching process without plasmas is oxide etching with HF vapor. Because it is based on diffusion of vapors, there is no directionality and the resulting profile is isotropic. This process, too, requires extensive safety precautions because of HF. Isotropic dry etching is very desirable in MEMS, as will be discussed in Section 29.6. The great benefit of dry etching compared to wet etching is the elimination of the drying step and the surface tension-caused mechanical distortions resulting from capillary forces.

11.7 Etch Masks

11.7.1 Resist selectivity

Usually a vertical resist sidewall is desirable and necessary for the best linewidth control in plasma etching. Most often the resist is, however, slightly sloped, for example 86° or 88° (positive slope), or even negatively sloped (retrograde). Fluorine plasmas are "milder" than chlorine plasmas, and resist selectivity is better, for example 5:1 or 10:1, while in chlorine plasmas 3:1 or 5:1 is more typical. In severe cases 1:1 is seen. This then means that only rather thin layers can be etched.

In wet etching resist selectivity is often infinite in practice, but adhesion of resist becomes an issue. This is dependent on priming, feature size, resist thickness and the chemical character of the resist. Some resists are specifically tailored for wet processing (etching and/or electroplating). Generally, thicker resists are mechanically more stable. If there is any loss of adhesion between the resist and substrate, etching will easily penetrate along the interface, leading to complete resist delamination, or just to a sloped sidewall, which may be beneficial in subsequent deposition steps.

11.7.2 Etching with a hard mask

Many wet and dry etching processes utilize hard masks because resists are simply not tolerant enough under harsh etch conditions. Harsh can mean aggressive chlorine plasmas, very long etch times or hot acids or bases like 80 °C 25% KOH for silicon anisotropic etching. In deep submicron processes resist thickness has to be scaled down for maximum lithographic resolution, but these thin resists are not always suitable as etch masks.

Photoresists are materials that combine photoactivity and mechanical/thermal/chemical stability; obviously photoactivity is the property that cannot be sacrificed. In order to find optimum materials as etch masks, the concept of hard mask has been devised. The hard mask material is etched with photoresist mask, photoresist is then stripped and the etch process is performed using the hard mask only. The hard mask material can be optimized to suit the application, irrespective of photoresist.

In silicon etching in KOH, silicon dioxide or silicon nitride hard masks are standard materials. When glass wafers (or thick oxides) are etched, nickel, chromium, polysilicon and amorphous silicon are suitable masking materials for concentrated HF (49%). Silicon carbide (PECVD SiC) and tantalum pentoxide (Ta_2O_5) are excellent hard masks for many wet and dry etching processes.

In DRIE etched depths can be 500 μm (through the wafer) and resists cannot be used as masks. Many materials are used as hard masks, such as silicon dioxide and aluminum. Aluminum nitride (AlN) and aluminum oxide (Al_2O_3) are very resistant in many plasmas, but can be easily wet etched by KOH and even dilute NaOH photoresist developer. This fact can sometimes make processing much faster and easier, compared to other hard masks that are very stable materials (which is why they were chosen in the first place).

RIE processes that use Cl_2 chemistry use metals like chromium or nickel as etch masks. However, the use of metal masks poses a problem in plasma etching. Even though the mask is chemically inert, it is always etched somewhat under ion bombardment. Redeposition of these non-volatile ion-etched species on the surfaces leads to non-etchable spots. This is called micromasking. In the case of perfect anisotropy micromasking leads to the formation of dense forest of pillars or grass.

Polysilicon gate etching can be done with an oxide hard mask. Because poly etching is highly selective against gate oxide, it is also highly selective against oxide hard mask, therefore a very thin oxide hard mask is enough, and very thin photoresist can be used to etch this hard mask. Elimination of carbon (i.e., elimination of photoresist) from the reaction brings about a major selectivity improvement: the selectivity between poly and oxide can be as high as 300:1 compared to 30:1 with resist mask, while keeping plasma parameters (RF power, pressure and gas flows) constant. In the presence of carbon CO is formed because it is energetically favorable (Table 11.4) and the source of carbon for CO formation is the photoresist, hence the lower oxide selectivity. In the absence of carbon/resist, no CO is formed.

11.8 Non-Masked Etching

Plasma etching replaced wet etching because of less undercut and better linewidth control. But this argument

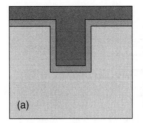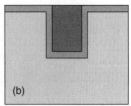

Figure 11.10 Etchback: (a) the trench wall is coated with a thin Ti/TiN layer and filled with CVD tungsten; (b) tungsten etchback will result in a planar surface and filled vertical plug

applies to patterning etching only: there are plenty of applications where etching is not used for making patterns, therefore there is neither photoresist nor hard mask.

Etchback is a process where extra material is removed, with the whole wafer exposed to plasma. In tungsten etchback, tungsten is removed from planar areas, leaving only the contact/via plugs filled (Figure 11.10). Etchback is used in multilevel metallization to make tungsten plugs (see Chapter 28 for more on this).

Non-masked wet etching removes for instance diffusion mask oxide after diffusion, and similarly, mask nitrides are removed in H_3PO_4 or concentrated HF (which would immediately attack resist). Dilute HF removes native oxides. High rate maskless silicon etching is needed in wafer thinning.

11.9 Multistep and Multilayer Etching

Etching a single layer structure can be accomplished in a single step, but multistep etching can be used for improved process control. In polysilicon gate etching a three-step process is typical. The first step removes native oxide in a short, non-selective process. Bulk etching is then done, but before the end point is reached, the process is switched to a low-rate, high-selectivity recipe. Note that the underlying oxide loss is a sum of four different factors: (1) polysilicon film (non)uniformity; (2) polysilicon etch process (non)uniformity; (3) poly:oxide selectivity; and (4) overetch time.

11.9.1 Multilayer etching

Thin-film functionalities are enhanced by multilayer structures. This is bad news for etch engineers, because there is no guarantee that the materials behave at all similarly in etching.

It seldom happens that both (or all) layers can be etched with the same process parameters, and it may well be

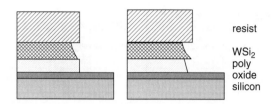

Figure 11.11 Double layer plasma etching: simultaneous optimization of both films is not easy – photoresist is still in place

that completely different etch chemistries must be used. In two-step double layer etching an end point signal must be obtained so that etching can be stopped, or else etch chemistry must provide high selectivity. High selectivity, however, is not always beneficial: if TiN on top of aluminum is etched in fluorine plasma, etching will definitely stop once the underlying aluminum is met, but the aluminum surface will turn to AlF$_3$, which is a very stable material, and initiation of the aluminum etch step is endangered. Etching of polycide gate (WSi$_2$/poly) is shown in Figure 11.11 with potential profile issues.

Process flow for WSi$_2$ /polysilicon (polycide) etching

1. WSi$_2$ etching: Cl$_2$/He/O$_2$ for WSi$_2$
2. Poly etching: Cl$_2$/HBr for poly
3. Poly end point step: HBr/He/O$_2$ for etching last 20 nm of poly
4. Overetch step: HBr/He/O$_2$ optimized for high oxide selectivity

Etching of the bottom layer has all the usual requirements about rate, selectivity and profile, and the extra requirement of not etching the top layer. Of course the acceptable degree of undercut in either of the layers calls for engineering judgment. Problems with film stacks which require different etch chemistries (chlorine vs. fluorine) has led to multichamber etch reactors, with each chamber reserved for one material and/or specific etch chemistry. This will be discussed in Chapter 31.

11.10 Etch Processes for Common Materials

11.10.1 Silicon

Fluorine, chlorine and bromine processes are standard for silicon etching, resulting in reaction products SiF$_4$, SiCl$_4$ and SiBr$_4$, respectively. Single crystal silicon, polysilicon and amorphous silicon can all be etched with similar etch processes, which is not necessarily true for wet etching.

Fluorine processes are safer to use, but seldom fully anisotropic. Chlorine processes which have less spontaneous chemical etching and more ion enhancement result in vertical sidewalls, and the same applies to bromine processes. These two gases are, however, highly toxic, and the equipment for Cl$_2$ or HBr etching must be equipped with loadlocks. Loadlocks complicate system operation but simultaneously improve reproducibility since the reaction chamber is not exposed to room air and humidity.

Silicon isotropic wet etching is often done in HF:HNO$_3$:CH$_3$COOH (sometimes water is used instead of acetic acid). Depending on the ratio of nitric acid to hydrofluoric acid, the etch rate can be modified from a few hundred nanometers per minute to tens of micrometers per minute. A common mixture is 1:3:8 which etches silicon at about 3 μm/min at room temperature. Hydroxyl ions (OH$^-$) form according to

$$HNO_3 + H_2O + HNO_2 \rightarrow 2\,HNO_2 + 2\,OH^- + 2\,h^+ \quad (11.5)$$

Hydroxyl ions react with silicon to form silicon dioxide, Equation 11.6. Holes (h$^+$) are needed in the reaction to balance charges. HF finally etches the oxide. The overall reaction is given by Equation 11.7:

$$Si + 2\,OH^- + 2\,h^+ \rightarrow SiO_2 + H_2 \quad (11.6)$$

$$Si + HNO_3 + 6\,HF \rightarrow H_2SiF_6 + HNO_2 + H_2O + H_2 \quad (11.7)$$

The reaction is problematic for many reasons. First of all, it is autocatalytic: nitric acid (HNO$_3$) produces nitrous acid (HNO$_2$) which reacts with nitric acid to produce more nitrous acid. The reaction is also very exothermic, releasing a lot of heat which accelerates the reaction. If the etchant is stirred, heat is carried away, reducing the etch rate. Because the reaction needs holes, its rate depends on silicon doping level. The rate is high for highly doped silicon but decreases rapidly when doping concentration falls below 10^{17} cm^{-3}. This has applications in MEMS, like etching highly doped epitaxial layers, see Figures 22.7 and 30.7.

11.10.2 Silicon dioxide

Silicon dioxide plasma etching is driven by ion bombardment. Isotropic plasma etching of oxide is therefore difficult, but a high enough radical concentration will result in reasonable isotropic etch rates. These processes are, however, not selective against silicon. Any fluorine-containing gas can be used as an etchant for oxide, CF_4 or SF_6 for example. But both gases etch silicon too, and they are suitable for non-selective etching only.

CHF_3, C_2F_6 and C_4F_8 are used as oxide etch gases when selectivity against silicon is required. They provide fluorine and carbon for etching (SiF_4, CO_2 etch products) and CF_2^* radicals which are polymer precursors. Polymerization takes place on silicon surfaces, whereas on oxide surfaces $(CF_2)_n$ polymerization does not take place due to oxygen supply: ion bombardment-induced reactions on oxide result in CO_2 formation.

A three-step oxide etch process consists of a bulk etching step, an end point step which is highly selective (and polymerizing), followed by a low-power step that removes polymeric residues: a few extra nanometers of silicon are lost in the low-power etch step but the wafer cleaning that follows will be much easier.

Wet etching of oxide with HF-based etchants offers high selectivity against silicon, and if the only role of oxide etching is to remove oxide, not to make patterns, it is often advisable to use wet etching, while plasma etching excels in pattern etching. A major problem in wet etching of oxides is that different CVD oxides etch at widely different rates. Therefore thermal oxide etch rate is usually given as a reference.

11.10.3 Silicon nitride

Nitride etching has aspects of both silicon and oxide etching. SF_6- and CF_4-based plasma processes etch nitride quickly and isotropically, and without selectivity against silicon. They do, however, show some selectivity against oxide, for instance 2:1. CHF_3-based processes, on the other hand, etch nitride and provide selectivity against silicon. In fact, CHF_3 and C_4F_8-based oxide etch processes usually perform well as nitride etch processes too, and result in anisotropic profiles, unlike SF_6- and CF_4-based processes.

The wet etchant for Si_3N_4 is boiling concentrated phosphoric acid. Photoresist cannot tolerate such etching conditions. Instead, oxide is used as an etch mask: CVD oxide is deposited on top of nitride, and oxide is patterned by photoresist and etched with HF. After resist stripping oxide acts as a mask for nitride etching. When CF_4 plasma was found to etch nitride, manufacturers were willing to invest in plasma etching, even though it was immature technology and not very production worthy, just because the wet etching process was so difficult.

11.10.4 Aluminum

Aluminum has a native oxide Al_2O_3 which is very difficult to etch. Chlorine and chlorine-containing gases are used, with $AlCl_3$ as the main etch product. Multistep etching is needed to etch aluminum: in the first few seconds high power is used to ion-etch native Al_2O_3 away; power is then reduced to etch the bulk of aluminum. Aluminum is spontaneously etched in Cl_2, and a polymerizing agent is needed to passivate the sidewalls for anisotropic profile; $CHCl_3$ and CH_4 are often used. In some low-pressure reactors Cl_2/BCl_3 gases without polymer-forming gases will result in clean, anisotropic profiles. Nitrogen or argon is often added to stabilize the plasma and to improve photoresist selectivity.

If linewidths are large, phosphoric acid-based wet etching is very simple. Above 3 μm linewidths H_3PO_4 etchant will perform nicely, especially if the film is so thin that undercutting can be neglected.

11.10.5 Copper

Copper is not plasma etched in current microfabrication processes. It is a difficult material to etch because neither fluorides (CuF_2) nor chlorides ($CuCl_2$) are volatile at room temperature. Increased temperature will help, but even at 100–200 °C the rate is still low and photoresist is severely attacked. If aluminum is alloyed with copper (to improve electromigration resistance), aluminum etching will be difficult for the same reason. Al–0.5% Cu is still fairly easy to etch but Al–4% Cu leaves a residue of copper chlorides that is difficult to remove.

Wet etching of copper can be done by a number of acidic solutions. The criteria are really resist stability, metal thickness and other factors. Copper patterns can be made by two other techniques instead of etching: electroplating into a resist mold (Figures 5.11 and 5.12), or by depositing in a groove and polishing away the excess copper (Figure 16.1).

11.10.6 Refractory metals and silicides

Tungsten etching is similar to silicon in many respects. In fluorine plasmas the reaction product is WF_6; in oxygen–halogen plasmas, WOF_4 or $WOCl_4$. Tungsten hexafluoride has a boiling point of 17 °C and an isotropic etching profile easily results. Oxyfluorides and oxychlorides are less volatile and ion bombardment is needed to remove them completely, which translates to better anisotropy. Molybdenum, too, is etched by both chlorine and fluorine plasmas, with or without oxygen. For titanium etching, chlorine etching is preferred, but fluorine etching is possible, and for TiW (30 at. % Ti) SF_6 is a typical choice. Tantalum and niobium are etched similarly. Silicides WSi_2, $MoSi_2$ and $TaSi_2$ are etched with both fluorine and chlorine-based processes which are similar to silicon and refractory metal etch processes, for example SF_6/O_2 for tungsten silicide or Cl_2/O_2 for $MoSi_2$.

Refractory metals are wet etched by hydrogen peroxide-based solutions. These are often harsh on photoresists, and therefore etching must be done at room temperature and in dilute solution.

11.11 Ion Beam Etching

Accelerated ions will etch all materials. This technology is known as ion beam etching (ion milling). It is, however, difficult to find suitable non-eroding masking materials: if all materials can be etched by ion bombardment, this applies to masking materials as well. Typical ion milling rates by argon ions are about 20 nm/min for Si, SiO_2, SiC, Si_3N_4 and resist, and 20–100 nm/min for metals. Ion milling is therefore much slower than plasma etching. One of the benefits of ion milling is inclined etching: the wafer can be tilted relative to the ion beam, and inclined structures can be made. Many solid state laser and magnetic materials (of the type $Gd_3Ga_5O_{12}$, gadolinium gallium garnet) are etched by ion beam etching.

11.12 Etch Process Characteristics

11.12.1 Linewidth and profile

Linewidth is also known as CD, for critical dimension, in the IC industry. Linewidth measurement checks any deviation from design values. A deviation of 10% is acceptable for digital devices, but the error budget has to be divided between lithography and etching.

The sidewall profile of the finished feature has important implications for subsequent process steps: the step coverage of the next deposition process depends on it. The profile can be measured from top view optical or SEM measurements, but destructive cross-sectional SEM pictures are considered the ultimate profiles.

Line edges are seldom abrupt, and judgment must be used to locate a line edge properly. Real lines do not have perfectly vertical sidewalls but sloped, or even retrograde, walls (Figure 11.12), with edge roughness which can be a significant fraction of the linewidth for narrow lines. Multiple scans must be made to average over edge roughness. Substrate and film roughness add noise to stylus measurements, and, for soft materials, stylus penetration can be a

Figure 11.12 Line profiles: left, ideal vertical wall; middle, retrograde wall; right, positively sloped wall with rough edge

problem. Linewidth can also be measured electrically, as was discussed in Chapter 2.

11.12.2 Selectivity

Selectivity is a measure of etch rate ratios (ERRs). It can be defined between the film and substrate and between the film and masking material. Selectivities range from 1:1 to 100:1 in typical plasma etching processes. Resist selectivities range from 1:1 to 10:1 in plasma etching (with 100:1 possible). In wet etching resist selectivity is often so good that it does not need to be considered, but resist adhesion loss and peel-off are severe limitations. When polymeric films are etched, selectivity and photoresist stripping are problematic: resist is polymeric material, too, and selectivity between two similar materials is difficult to achieve.

Etch stop is the term used for etching processes where the selectivity is so high that etching essentially stops when the underlying material is reached. This will be discussed in more detail in Chapter 20 because it has important implications in bulk MEMS.

11.12.3 Etch rate and etch time

Etch rate should be determined by plotting etched depth against etch time for a number of data points, and taking the slope as the rate. There are many mechanisms which make etch rate determination less straightforward than expected. There may be etch initiation lag: for example, a thin native oxide layer will prevent etching initially. This is especially true for silicon and aluminum. Then there are etch slowdown effects: etch rate in deep narrow structures is not identical to flat open areas, and as etching continues and the hole becomes deeper, etch rate will slow down. This issue is covered in more detail in Chapter 21.

Etch time seems like a simple concept: film thickness divided by etch rate. Add corrections for both film thickness and etch process non-uniformity, say 5% for each, and etch time results. Overetch required to clear film on

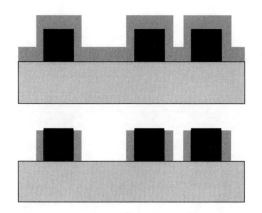

Figure 11.13 Spacer formation: top, thin-film deposition over a step; bottom, anisotropic etching to clear film from horizontal surfaces leaves spacers at feature edges

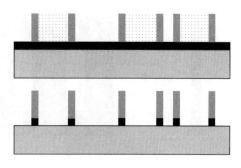

Figure 11.14 Spacer lithography: after the formation of spacers the original pattern is etched away, and the spacers act as etch masks for the next etching step

planar samples amounts then to 10–20% depending on process details.

But when the films to be etched run over topography, the situation changes dramatically. Film thickness at the edge of a step will be the sum of the film thickness and step height. If anisotropic etching is stopped at the end point calculated from planar film thickness, residue equal to original step height remains at the edge (Figure 11.13). Long overetch will eventually remove this residue but this puts a high demand on etch selectivity between the two materials.

Sometimes it is desirable to leave this residue in place, and utilize it in the fabrication process. It is then termed spacer. Spacers have important applications in both MOS and bipolar transistor fabrication, as will be seen in Chapters 26 and 27. Note that it is essential for spacer formation that etching is anisotropic: in isotropic etching sideways etching would remove the material at the step edge.

If the steps are made of conducting material and the spacer is dielectric (e.g. CVD oxide), the spacer can be left in place. But if the steps are made of dielectric material and conductive lines go over them, the conductive spacers will short all the lines. Extended overetch is then needed to ensure complete spacer removal.

11.12.4 Spacer lithography

The concept of spacer lithography (also know as sidewall lithography and double patterning) was originally introduced in 1984 when the end of optical lithography was speculated. It was used to fabricate MOS ICs in an industrial setting, and now it is again being considered for extending the life of optical lithography.

In spacer lithography the sidewalls form the final structure; the initial structure will be selectively etched away after spacer formation (Figure 11.14). The width of the structure is determined by the film thickness, which is easily controlled even in the tens of nanometers range. Of course additional lithography steps are needed to complete the patterns, but they are not critical lithography steps.

11.13 Selecting Etch Processes

When narrow lines need to be made, the obvious choice is plasma etching. But in many applications the choice of wet vs. plasma etching is a question of convenience: certain equipment or an etch bath is available or some suitable masking material is handy. When spherical etch profiles are required, or when undercutting is needed, isotropic etching must be used. Isotropic etching of silicon can easily be done by SF_6 plasmas, and isotropic etching of silicon dioxide or glass by HF, both at fairly high rates, even tens of microns per minute.

The microbolometer of Figure 11.15 requires two lithography and three etching steps in its fabrication. The first lithography step defines the resistor pattern; anisotropic plasma etching defines the resistor. The second lithography step defines openings in the oxide, and the same photoresist serves as a mask for both anisotropic etching of oxide in CHF_3 plasma and isotropic

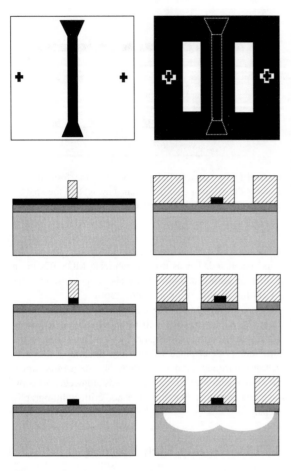

Figure 11.15 Bolometer fabrication process: left, resistor lithography and etching; right, second lithography, oxide etching and silicon isotropic etching

silicon etching in SF$_6$ plasma. A SEM micrograph of the finished device is shown in Figure 11.16.

11.14 Exercises

1. What would you use as plasma etch gases and etch masks for etching the following materials: diamond
    - TiN
    - SiC
    - InP
    - GaN
    - GaAs
    - PbZrTiO$_3$
    - BCB (benzo cyclobutadiene)?

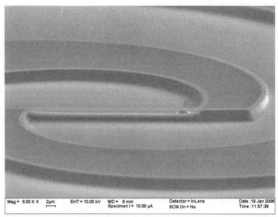

Figure 11.16 Spiral antenna microbolometer: silicon is isotropically etched to release the narrow resistor. SEM courtesy Leif Grönberg, VTT

2. Silicon is etched in plasma according to the reaction Si (s) + 2 Cl$_2$ (g) $\Rightarrow$ SiCl$_4$ (g). What is the theoretical maximum etch rate of a 200 mm diameter silicon wafer when chlorine flow is 100 sccm (standard cubic centimeters per minute)?

3. Polysilicon etched depth in chlorine plasma is given in the table below. Determine the etch rate.

Time (s)	Depth (nm)
20	70
40	170
60	320
80	415

4. The etch rate of <100> silicon in 20% TMAH is given below. What is the activation energy?

Arrhenius

Temperature (°C)	Rate (μm/h)
60	29
70	36
80	62
90	87

5. How much underlying oxide is lost when a tungsten film 500 nm thick is etched from a sample that has 300 nm steps on it? Tungsten:oxide selectivity is 10:1.

6. How much dimensional error does chromium wet etching introduce to (a) 1× photomasks; (b) 5× reticles?
7. What problems will be met in trying to compensate wet etching undercut by drawing lines on the mask broader?
8. If oxide film thickness is 500 nm, oxide non-uniformity ±5% and oxide etching non-uniformity similarly ±5%, plot the underlying silicon loss as a function of oxide:silicon etch selectivity!
9. Draw cross-sections of silicon etching with oxide hard mask for the following cases:
   (a) KOH etching
   (b) Cl_1 etching
   (c) $HF:HNO_3$ etching
   (d) TMAH etching
   (e) SF_6 etching
   (f) H_3PO_4 etching.
10. How can the following structures be etched?

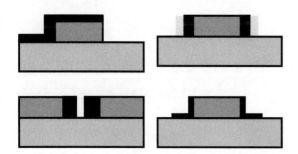

11. What is the difference in making inside vs. outside spacers by anisotropic etching?

References and Related Reading

Arana, L. R. *et al.* (2007) Isotropic etching of silicon in fluorine gas for MEMS micromachining, *J. Micromech. Microeng.*, **17**, 384–392.

Armacost, M. *et al.* (1999) Plasma-etching processes for ULSI semiconductor circuits, *IBM J. Res. Dev.*, **43**, 39.

Bell, F. H. and O. Joubert (1996) Polysilicon gate etching in high density plasmas, *J. Vac. Sci. Techol.*, **B14**, 3473.

Clawson, A. R. (2001) Guide to references on III-V semiconductor chemical etching, *Mater. Sci. Eng.*, **31**, 1–438.

Kiihamäki, J. *et al.* (2000) Depth and profile control in plasma etched MEMS structures, *Sens. Actuators*, **82**, 234–238.

Lang, W (1996) Silicon microstructuring technology, *Mater. Sci. Eng.*, **R17**, 1–55.

Mukerjee., E. V. *et al.* (2000) Vaporizing liquid microthruster, *Sens. Actuators*, **83**, 231–236.

Oehrlein, G. S. and J. F. Rembetski (1992) Plasma-based dry etching techniques in the silicon integrated circuit technology, *IBM J. Res. Dev.*, **36**, 140.

Sainiemi, L. and S. Franssila (2008) RIE, in D. Li (ed.) **Encyclopedia of Micro and Nanofluidcs**, Springer.

Walker, P. and W. H. Tarn (eds) (1991) **Handbook of Metal Etchants**, CRC Press.

Williams, K. R. and R. S. Muller (1996) Etch rates for micromachining processes, *J. Microelectromech. Syst.*, **5**, 256–269.

Williams, K. R. and R. S. Muller (2003) Etch rates for micromachining processes II, *J. Microelectromech. Syst.*, **12**, 761–778.

12

Wafer Cleaning and Surface Preparation

Microfabrication takes place under highly controlled conditions: all materials for cleanroom construction, processing equipment and wafer handling tools are carefully selected to eliminate particles and to reduce atomic contamination. Water, gases and chemicals are purified and filtered. These are, however, passive precautions, and active wafer cleaning and surface conditioning must be undertaken before practically every major process step. These preparatory steps can account for up to 30% of all process steps.

Wafer cleaning is about removing particles and unwanted atoms and films, but it is also about leaving the surface in a known and controlled condition. This means damage removal, surface chemistry tailoring and proper hydrophobicity/hydrophilicity. As shown in Figure 12.1, ammonia/peroxide cleaning will render a silicon surface oxidized and hydrophilic while HF treatment results in a hydrogen-terminated hydrophobic surface.

In some cases surface preparation involves intentionally depositing films on the surface rather than removing material, as the term "cleaning" would hint. Many cleaning chemistries result in thin-film deposition, for example sulfuric and nitric acids will oxidize silicon surfaces, resulting in films 1 or 2 nanometers thick. The next process step after cleaning should commence immediately, before particle deposition and surface chemical reactions take place.

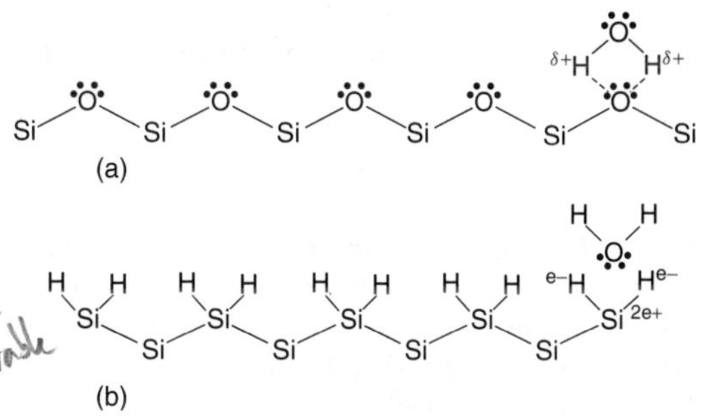

12.1 Classes of Contamination

Air cleanliness in an advanced cleanroom is so good that airborne particles are no longer the main contamination source. The human contribution has also been reduced significantly with correct gowning and working procedures, or by factory automation. These matters are dealt in more detail in Chapter 35 on cleanrooms.

The main sources of contamination are the fabrication processes themselves: films flaking from deposition chamber walls, vapors rising from wet benches, resist debris on wafer edges. Because of device size downscaling, contamination becomes evermore critical. Finer patterns demand control of finer particles, ultrathin films cannot tolerate unwanted atoms, and low leakage currents necessitate low metal contamination levels. The purity of starting materials is important: liquid chemicals can have impurity concentrations measured in parts per trillion (ppt) and sputtering target purities are for example 99.999%. Similar "5Nine" purities are typical for many process gases, but some applications need 99.99999% (7N) purity.

Contamination comes in various forms, which have different sources, effects on device and cleaning methods. The main classes of contamination are:

- particles
- metals
- organics
- native oxide.

Adsorbed water could be called contamination, too. Baking at elevated temperature removes water, and baking is

Figure 12.1 Surfaces treatments: (a) hydrophilic silicon surface after ammonia/peroxide cleaning attracts water; (b) hydrophobic silicon surface after HF cleaning repels water. Reproduced from Hattori (1998)

Introduction to Microfabrication, Second Edition Sami Franssila
© 2010 John Wiley & Sons, Ltd

a standard preparation step in many processes. Water can be very persistent, and not all water is removed before a 400 °C bake.

Particles greater than about a third of minimum linewidth are potential causes of fatal flaws, for example shorts between conductors and even smaller particles can cause pinholes during film growth. The minimum dimension is really the thickness of the thinnest film, not the narrowest line. Particles are also a major concern in wafer-bonding (the subject of Chapter 17).

Metal contamination cannot be avoided as long as machine parts are made of metals, so it has to be controlled by cleaning. Metal contamination on surfaces can spread into the silicon bulk, and metal precipitates in the bulk act as charge carrier traps. If the metals segregate into the oxide during oxidation, they can prevent, retard or degrade oxide film growth, and result in poor-quality oxides.

Organics often work through prevention of the cleaning process. This has to do with the generally hydrophobic nature of organics and repellency of water-based cleaning solutions. Organic materials in cleanrooms include photoresists, HMDS, various solvents (isopropyl alcohol (IPA), acetone, MEK, PGMEA), wafer boxes and chemical containers.

Native oxide films grow readily on silicon, aluminum and titanium, for instance. Growth is not instantaneous, however, and proper surface finishing can protect the surfaces for extended periods of time. Hydrophobic (HF-last) cleaning chemistry results in hydrogen-terminated surfaces (Figure 12.1) that can survive a week without native oxide forming, if the oxygen concentration in the ambient is low enough (e.g., <0.1 ppm). In normal cleanroom air, 0.5 nm native oxide film will grow in a few hours.

Native oxides degrade contacts, prevent epitaxial growth, cause defects at the substrate/epi interface and prevent solid state reactions like silicide formation. HF etching is a typical last step before oxidation and also before epitaxy. However, in epitaxy an additional in situ cleaning just prior to deposition is also done (Figure 34.7).

12.2 Chemical Wet Cleaning

Acid, base and solvent wet cleaning are the main methods of cleaning. Chemical wet cleaning is a simple, yet powerful method with high throughput. Wet benches are cheap and reliable tools but chemical consumption can become very high because the chemicals need to be changed regularly, several times a day, otherwise there is a risk of redeposition of contaminants from the cleaning solutions back to the wafers. Obviously, very high-purity chemicals must be used, otherwise the cleaning baths would become contamination sources themselves.

Practical cleaning processes use many different chemicals in sequence, each designed to attack different contamination: in a classic RCA clean the first ammonia/peroxide clean removes particles and organics, the hydrogen chloride/peroxide step is effective against metallic contamination and the HF step removes any oxide. Table 12.1 presents the most common wet cleaning chemicals. The water in cleaning solutions is de-ionized water (DI water), also known as ultrapure water (UPW).

The two main mechanims for wet cleaning are:

1. Dissolution/decomposition.
2. Etching.

They have a very important distinction as far as surface roughness is concerned: etching processes tend to make surfaces rougher. This effect can be seen in subsequent process steps: LPCVD polysilicon deposited on thermal oxide coming directly from a furnace has larger grain size than poly that is deposited on freshly cleaned oxide. Increased surface roughness from peroxide cleaning solution leads to a larger number of nuclei, therefore smaller grain size. If the cleaned wafer is annealed, the surface will become smoother, and subsequently grain size will be larger.

Ammonia/peroxide solution works by oxidizing the silicon surface, and subsequently etching the oxide away:

$$2\ H_2O_2 \Rightarrow 2\ HO_2^- + 2\ H^+$$

peroxide disproportionation (12.1)

$$Si + 2\ HO_2^- \Rightarrow SiO_2 + 2\ OH^-$$

silicon oxidation (12.2)

$$Si + 2\ H_2O_2 \Rightarrow SiO_2 + 2\ H_2O$$

total reaction for oxidation (12.3)

$$SiO_2 + OH^- \Rightarrow HSiO_3\ (aq)$$

oxide etching (12.4)

Silicon etch rate in ammonia/peroxide is about 0.5 nm/min, and a typical 10 min clean thus results in about 5 nm silicon etching. This leads to undercutting and removal of the particles. For modern CMOS fabrication, however, 5 nm silicon loss is unacceptably large.

The compositions given above are the traditional ones, and recently there has been a trend toward more dilute and lower temperature cleaning solutions. For example, RCA-1 clean of composition 0.1:1:5 is also used, and even ammonia water without peroxide has been used. Similarly, dilute HCl clean can replace RCA-2 in some applications.

Table 12.1 Wet cleaning solutions

Name/alias	Chemical composition	Temp./time
RCA-1 (SC-1, APM ammonia-peroxide)	$NH_4OH:H_2O_2:H_2O$ (1:1:5)	70–85 °C/10–20 min
RCA-2 (SC-2, HPM, hydrochloric acid-peroxide)	$HCl:H_2O_2:H_2O$ (1:1:6)	70–85 °C/10–20 min
SPM sulfuric acid peroxide mixture, Piranha sulfuric peroxide mixture, or. Piranha	$H_2SO_4:H_2O_2$ (4:1)	120 °C/10–20 min
DHF (dilute HF)	$HF:H_2O$ (1:100–1000)	room temp./30 s

Standard chemicals come in the following concentrations:
HCl	37%
H_2SO_4	96%
H_2O_2	30%
NH_4OH	29%
HF	49%

Preparation: The order of mixing and heating is important. Water is the first component, acids are added later, to prevent excessive heat generation. Hydrogen peroxide is added to a heated solution just prior to use, to minimize thermal decomposition.
Bath life: RCA clean is often used just once. Alternatively, peroxide can be added to prolong bath life (dubbed "spiking"). SPM bath life is similarly elongated by adding more peroxide.
Disposal: HF requires a separate disposal system because its health effects are different from other mineral acids, which may all be collected in the same container (volume and disposal rate limited). Sometimes acids which contain heavy metals must be collected separately (e.g., titanium or cobalt containing salicide etchants).

Cleaning bare silicon wafers is straightforward and the results can be measured at ease. Complexity increases when more materials are present: the cleaning method has to work on all these materials with similar efficiency. Measurements, too, become more difficult because, for example, optically transparent and opaque materials reflect laser light differently, and minimum detectable particle size gets bigger. Various limitations kick in when the first metal is deposited: acid cleans are then forbidden because acids dissolve and corrode metals.

With deep trenches and other highly 3D structures wetting ability becomes critical, and drying becomes more difficult too. The use of surfactants can improve wetting but they introduce problems of their own, namely possible organic residues. The measurement of surface properties (sidewall properties, really) in deep narrow trenches is a formidable task, even in the research phase where time and effort can be used to measure a few data points. Monitoring in a manufacturing environment is even more difficult because measurement speed becomes an issue.

Chemical consumption in wet benches is a major environmental concern. With larger wafer sizes, larger tanks have to be used, with increasing volumes of expensive high-purity liquids, which are dangerous to handle and which have to be disposed of under controlled conditions. CMOS fabrication on 200 mm wafers consumes thousands of liters of ultrapure water and tens of kilograms of liquid chemicals are required. Hundreds of liters of acid waste are produced. Rinse water can be recycled, and acid recovery and reuse are also common practices. Spin processing (Figure 12.2) is more economical in chemical consumption.

12.3 Physical Wet Cleaning

Physical cleaning is mostly about particle removal. But the difference between chemical and physical wet cleaning is not clear cut: often chemical cleaning is enhanced by the application of physical forces, for example ultrasonic energy or spraying. Three methods of physical wet removal of particles are widely used:

1. Brush scrubbing.
2. Jet scrubbing/nozzle.
3. Ultrasonic/megasonic.

In brush scrubbing, nylon or PVA brushes move water close to the wafer surface and induce flow that brushes away the particles. This is especially effective when lots of particles or large particles are present on the wafer. Therefore, brush scrubbing is very useful in removing particles after wafer scribing or chemical mechanical polishing (Chapter 16).

In jet scrubbing, high-pressure water is sprayed on the wafer. The removal mechanism is similar to brush scrubbing. Increasing pressure improves cleaning efficiency, but electrostatic charging can damage thin films.

In sonic cleaning, shock waves supply sound energy that helps in particle removal. Ultrasonic agitation

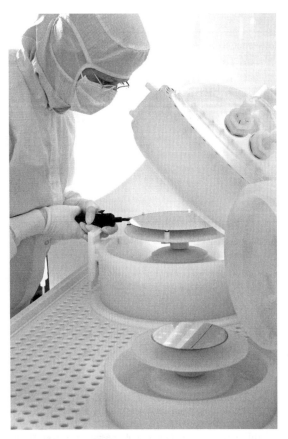

Figure 12.2 Wet cleaning: spin processor. Courtesy VTT

Ultrasonic (20–40 kHz) is also beneficial in the wet removal of photoresist. However, cavitation (exploding bubbles) may damage the wafers. Above 1 MHz this is not an issue, and the method is termed "megasonics." However, the removal efficiency is not very good for particles smaller than 300 nm.

12.4 Rinsing and Drying

Rinsing in DI water and drying must be considered as essential parts of any wet cleaning process. As a general strategy the wafer should be kept wet all along the cleaning process and minimize the number of times when wafers are drawn from liquid to air. When drying is required, there are a number of methods available: spinning, nitrogen blowing, vapor drying, lamp drying, vacuum drying; dry wafers can also emerge by slow removal from hot DI water. Spinning techniques are prone to charging and particle adherence, which are inherent in high-speed spinning equipment. Various isopropyl alcohol (IPA) drying methods rely on the low surface tension and good wettability of IPA. In Marangoni drying, the wafer is drawn from water into an IPA–nitrogen atmosphere, and water is pulled back, leaving a dry surface. IPA drying methods must be considered for chemical consumption, hot vapors and solvent accumulation.

12.5 Dry Cleaning

Gases, vapors and plasmas can be used to clean wafers (Table 12.2). Because it is easy to integrate process modules with similar pressure and temperature regimes, it is attractive to integrate dry cleaning into a plasma and vacuum tool, for example plasma etchers, sputters, PECVD, RTP and single wafer epitaxial reactors.

Compared to wet cleaning, dry cleaning has two very advantageous features: there are no surface tension effects in small structures and no drying is necessary.

UV ozone has been tried for organics removal, UV Cl_2 for metal removal and HF vapor for native oxides. Argon and H_2 plasmas have also been utilized, in sputtering systems, to improve contact by etching oxide just prior to metal deposition. Dry cleaning has a central role in epitaxial systems where the utmost surface cleanliness is mandatory. Thin oxides can be desorbed by a hydrogen bake. The exact temperatures depend on surface termination: hydrogen-terminated surfaces can be baked at temperatures as low as 700 °C to reveal a perfect surface for epitaxy. To date, however, dry cleaning has remained a special method, especially because it is difficult to remove particle contamination with dry methods.

12.6 Particle Removal

Particle contamination is dangerous in lithography, but lithography is rather insensitive to metal ion contamination. Deposition processes are sensitive to small particles that can "grow" in size during deposition: the film encapsulates the particle. This may eliminate the particle as an

Table 12.2 Dry cleaning agents

Vapors	Anhydrous HF
Gases	H_2, HCl
Ions	Ar+
Atoms	Si
Photons	UV (with Cl_2 or O_3)
Plasmas	CF_4
Aerosols	CO_2

Table 12.3 Sources of particles

- Chemical reactions in deposition and etching
- Moving parts in tools: robot arms, valves, doors
- Static parts: wafer holders, cassettes, O-rings
- Vacuum: pumping, venting, condensation
- Gases, chemicals, water
- Cleanroom personnel

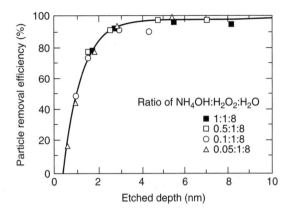

Figure 12.4 Etching as a method for particle removal: about 4 nm undercut etch is enough to remove most particles. Ammonia dilution as a parameter. Reproduced from Hattori (1998)

electrical or chemical contaminant but it will be a problem in lithography and bonding.

Fabrication processes themselves are major sources of particles. Table 12.3 lists some materials and mechanisms which contribute to particle contamination.

An electrical double layer forms on the particle surface in an ionic solution. The zeta potential is the potential at the fixed charge layer/diffuse charge layer interface (typically a few nanometers from the surface). It can be negative or positive, leading to either attraction or repulsion of particles toward the surface, depending on the wafer surface charge. The zeta potential of the particle is independent of particle size but depends on the electrolyte pH: in acidic conditions the zeta potential is typically positive, and in alkaline solution it is negative, as shown in Figure 12.3. The wafer surface will be oxidized in RCA-1, and it is negatively charged (Figure 12.1). When etching undercuts the particle, like charges repel each other and redeposition on the wafer is unlikely.

Particle removal efficiency in RCA depends on etch undercutting. As shown in Figure 12.4, it is really the etched depth that matters: all the different ammonia/peroxide compositions result in the same particle removal efficiency when the same amount of silicon is etched.

Cleaning solution composition and time have to be optimized with respect to both cleaning efficiency and roughness increase. Decomposition of cleaning solutions and impurities can also catalyse surface reactions leading to increased roughness.

12.6.1 Wafer particle measurements

Particle measurements on wafers down to 100 nm size range can be performed by laser scattering equipment (Figure 2.6). A laser illuminates the wafer surface, and forward scattered (Mie scattering) and reflected light are measured by a (hemispherical) detector.

Scatterometric particle sizes are calibrated against contamination standards which have polystyrene latex spheres (PSLs) of certified sizes on them. These PSLs are nearly spherical, have tight size distribution and have a known refractive index of about 1.6. The number of particles is better calibrated against etched features with known light scattering properties and known positions on the wafer. Such standards can be cleaned and reused, whereas contamination standards cannot.

Because real particles are not spheres with known optical constants, particle sizes cannot strictly be measured by light scattering. Latex sphere equivalent (LSE) size should be reported. Mirror polished unpatterned wafers are good for basic studies, but real wafers present a number of problems. Because forward scattered light is reflected by the wafer before reaching the detector, thin films on the wafer must be taken into account. On the oxide, particle calibration needs to be done for each film thickness. On metallized wafers, surface roughness leads to a decreased signal-to-noise ratio, therefore small particles cannot be detected. Correlating a scattering

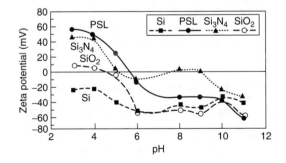

Figure 12.3 Zeta potential: pH influences particle adhesion and removal (PSL = polystyrene latex). Reproduced from Hattori (1998)

event to a physical particle is usually difficult, even though scatterometry produces a map of the wafer. If particles can be seen in a SEM, chemical identification is possible by either EDX or EMPA analysis. This can be important for particle source identification.

Most of the discussion above has centered on front-side cleaning and surface preparation, but wafer back-side cleanliness is becoming increasingly important, too. Big particles on the back side prevent wafer leveling for lithography, and make thermal contact to chucks poor.

12.7 Organics Removal

There are many sources of organic contamination in the cleanroom. Table 12.4 lists some of the most usual ones.

RCA-1 ammonia-peroxide solution is very good at removing organics because of the oxidizing power of peroxide. If a lot of organics need to be removed, sulphuric acid-peroxide (SPM) is used (as in photoresist stripping !). SPM cleaning leaves difficult-to-remove sulfur residues, and the RCA-1 step is often carried out immediately after SPM to turn sulfides into soluble sulfates. Organic residues often lead to surface roughening. Organic film protects part of the surface for some time during cleaning, while etching proceeds elsewhere.

Because sulfuric acid constitutes an environmental concern, and a safety hazard, other candidates have been sought for organics removal. Ozonated DI water with 10–100 ppm ozone has proven to be very effective. Furthermore, it is a room temperature process, as opposed to 120 °C SPM. The ultimate cleaning method for organic contamination is thermal oxidation: no organic compound can tolerate 1000 °C in an oxygen atmosphere. This provides a reference surface for analytical methods but of course is not a practical cleaning process.

12.7.1 Measurement of organic contamination

Organic contamination can be conveniently measured by FTIR (Fourier Transform Infrared Spectroscopy), which

Table 12.4 Sources of organic contamination

- Liquid chemicals and vapors used in fabrication processes: HMDS, IPA, acetone
- Gases, for example according to the reaction $n\ CF_4 \Rightarrow (CF_2)_n + 2n\ F^*$
- Organic films (resist, spin-on polymers)
- Wafer holders and boxes
- Vacuum systems: pump oils, O-rings
- Cleanroom materials: sealants

identifies not only elements but also chemical bonds. XPS can also identify chemical bonds, which is often important in understanding the origin of the contamination.

Molecular surface contamination can be measured by thermal desorption spectroscopy (TDS). TDS consists of a furnace connected to a mass spectrometer, and desorption of contaminants is monitored as a function of furnace temperature. Silicon surface condition has also been clarified by TDS: at 340 °C, water desorbs; at 400 °C, the hydrogen-terminated silicon surface undergoes the reaction $SiH_2 \Rightarrow SiH + 1/2\ H_2$; and at 500 °C, $SiH \Rightarrow Si + 1/2\ H_2$.

12.8 Metal Removal

There are numerous sources of metals, even though other materials like silicon, Teflon, SiC and quartz are extensively used in making process equipment and wafer handling tools. Table 12.5 lists some common sources of unwanted metals.

12.8.1 Device effects of metal contamination

Metal contaminants degrade the performance of electronic devices in various ways depending on their chemical and physical nature, that is, reactivity with silicon and silicon dioxide, and diffusion. Harmfulness of metal atoms comes from the fact that they have their energy levels deep in the forbidden energy gap of silicon.

Non-electronic devices are less sensitive to metal contamination, but metals cannot be completely ignored: metal contamination causes stacking faults in oxidation, and metals can catalyze peroxide decomposition, which leads to reduced particle cleaning efficiency in RCA-1.

12.8.2 Metal removal

The acidic solutions $HCl-H_2O_2$ and $H_2SO_4-H_2O_2$ are the main methods for metal removal. Dilute HF, which removes a thin oxide layer, will additionally remove some metallic contaminants. Ammonia solutions

Table 12.5 Sources of metal contamination

- Tool materials (shutter blades, collimators, chucks)
- System components (pipes, valves)
- Wafer handling (tweezers, robot arms, wafer holders)
- Chemicals (some resist developers contain NaOH as an essential constituent)
- Human contribution (sodium from sweat, heavy metals from cosmetics)

(RCA-1) can also form complexes with metals and remove some metals.

The cleaning efficiencies of HCl–H$_2$O$_2$ and HF are very different, though. Both can reduce Fe and Ni levels below the detection limit but HF is much more effective in removing Al, and HCl–H$_2$O$_2$ in removing Cu. Dilute HF needs to be specified because various workers use different concentrations. For aluminum removal 0.1% HF is sufficient, but below that amount the removal efficiency rapidly deteriorates. HCl concentration in HCl–H$_2$O$_2$ has to be at least 5% for it to remove iron.

The wet chemicals themselves contain metallic impurities, and at the 10 ppb level their deposition on the wafer surface is of some concern. For example, iron at 10 ppb in RCA-1 solution results in a surface concentration of 2×10^{12} atoms/cm^2. The use of higher purity chemicals helps to reduce the effect but it is an expensive way. If a RCA-1 bath is used several times, contamination from previous batches remains in the solution. RCA-1 must be accompanied by a clean that removes metals efficiently.

Newer cleaning solutions include HF:H$_2$O$_2$ which has both oxidizing and metal removal capabilities. It can be used at room temperature as opposed to 70 °C typical of RCA cleans. HF:H$_2$O$_2$ seems to increase surface roughness, so cleaning time needs to be optimized.

12.8.3 Measurement of metallic contamination

Metal surface concentrations range from 10^{10} to 10^{14} atoms/cm^2 depending on general contamination control strategies and particular process steps. Total reflection X-ray fluorescence (TXRF) uses a grazing incident angle to probe only the very top layer of the wafer, at nanometer thickness. It is most sensitive for medium-mass atoms and less sensitive toward both ends of the mass range. The detection limit of TXRF is about 10^{10} atoms/cm^2. TXRF is a non-destructive method that can be done on whole wafers, and it can be used to generate maps of contamination. This is sometimes very useful for determining the source of contamination.

In VPD (Vapor Phase Decomposition) and WSA (Wafer Surface Analysis) methods, surface impurities are first collected in the growing thermal oxide, which is then dissolved in HF. This concentrate is analyzed by graphite furnace atomic absorption spectroscopy (GFAAS), which is sensitive to picogram quantities, or detection limits of 10^9–10^{10}/cm^2, or by an inductively coupled plasma mass spectrometer (ICP-MS) which has lower detection limits by an order of magnitude.

Metallic contaminants can be measured by their effects on charge carriers. Minority lifetime will be degraded by contamination. Surface photovoltage (SPV) and microwave photoconductivity decay (µPC) methods provide this information. Quantitative iron determination is possible in p-type silicon by the SPV method because of a special feature of Fe–B pairs: at 200 °C they dissociate, and the interstitial Fe$^+$ is a much more effective recombination center.

12.9 Contact Angle

Removal of native oxide is easily seen by the behavior of water on the wafer surface: oxide is hydrophilic and water spreads evenly, but silicon is hydrophobic and water droplets form. Quantitative measurement of contact angle is used to characterize surface hydrophilicity/hydrophobicity. Hydrophilic surfaces have contact angles (CAs) < 90°, hydrophobic surfaces >90° (Figure 12.5). In practice, distinct droplets form on surfaces with contact angles of 50°–90° while uniform wetting takes place when CA < 10°.

It is important that cleaning solutions are matched to surface properties: aqueous solutions will spread and dissolve contamination on hydrophilic surfaces, but their effect is erratic on hydrophobic surfaces. For example, organic residues from photoresist may render the surface hydrophobic, and two-step cleaning is required, the first step to remove organic contamination and the second step to remove particles, for instance. Ammonia/peroxide mixture is the standard solution for making silicon surfaces hydrophilic.

HF-last clean results in CA ~ 70°, and water droplets form on silicon. Water sometimes remains on the wafer after rinsing, resulting in watermarks during drying. These can be minimized by tailoring the contact angle to either high or low values.

Hydrophobic surfaces with CA > 90° are typical of polymeric surfaces: methyl-terminated surfaces (like PDMS) have CA ~ 110° and fluoropolymers (like Teflon) CA ~ 120°.

High contact angles indicate low surface energies and vice versa. Self-assembled monolayers (SAMs) (Figure 5.14) follow the general rule that the longer the carbon

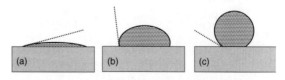

Figure 12.5 Examples of contact angles of water droplets: (a) on a hydrophilic surface, 20°; (b) on a hydrophobic surface, 95°; (c) on an ultrahydrophobic surface, 150°

chain in the polymer, the more hydrophobic the surface. The maximum contact angle possible on planar surfaces is about 120°. If higher contact angles are desired (so-called super or ultrahydrophobic surfaces), micro- and nanostructures coated by fluoropolymers can be used. The water droplets then rest on top of nanostructures like a fakir on a bed of nails. Contact angles approaching 180° can be achieved.

When contact angles are discussed, water contact angles are most often quoted (Table 12.6). It is important to remember that other liquids have other contact angles, and even salt concentration in water can drastically change contact angle: biological buffer solution on silicon has a contact angle of 35°, while DI water has 70°. Similarly, a surface coated with fluoropolymer DDMS is highly hydrophobic, water contact angle 103°, but the hexadecane contact angle is 38°. Contact angle can change during device operation due to adsorption of material on the surface: for example, protein adsorption can turn a hydrophobic surface into very hydrophilic one.

The values of contact angles depend on surface treatment details and the values in Table 12.6 are to be used as rough guides only. First of all, contact angle measurements are accurate only ±2%. Surface roughness can also change contact angle.

What is more, contact angles change as time goes by: native oxide formation on silicon will mean that a freshly HF-dipped wafer has a 70° contact angle, but once 1 nm of native oxide has been formed, the contact angle is reduced. Oxygen plasma modified polymer surfaces become hydrophilic but revert back to their original, more hydrophobic, structure. The timescale of this is very much dependent on materials and processes: sometimes it takes less than an hour or it can take months.

12.10 Surface Preparation

Conditioning the surface to a known state is of paramount importance. HMDS priming before lithography is a prototypical example: it leaves the wafer slightly hydrophobic due to the methyl-terminated surface (Figure 9.2). Resists are polymers, and a slightly hydrophobic surface makes good contact with resists. HMDS prevents atmospheric moisture from condensing on the wafer, improving adhesion. This equalizes wafers: irrespective of underlying film, and irrespective of storage time, the resist will see the same surface.

Sometimes surface roughening is the goal, for instance if it is necessary to prevent sticking (to be discussed in more detail in Chapter 29). Many etching processes lead to roughening, like RCA-1, and even minor roughening can prevent bonding. Elimination of particles is essential in wafer bonding (Chapter 17). Surface smoothness is also critical. RCA-1 clean is a compromise because there is some roughening, but good particle removal and hydrophilic surface finish.

Plasmas are used as a cleaning method, but it is more generally to be considered a surface modification method. Plasmas can etch surfaces, and this cleans the surface. Plasmas are energetic, and as a result the surface has broken bonds which tend to be reactive. This is advantageous in many applications: bonding benefits from readily available bonds, and the same applies to adhesion.

Oxygen plasma can be used to grow oxides on silicon and aluminum. These oxides are only 1 or 2 nanometers thick, but they can protect the underlying material. For instance, the $AlCl_3$ etch product from Cl_2 plasma etching of aluminum reacts with atmospheric water vapor to produce HCl, which corrodes aluminum. Therefore oxygen plasma treatment after aluminum plasma etching will stabilize the surface.

Sometimes the treatment only modifies the outermost atomic layer, replacing some atoms with other, more suitable, atoms or molecules. Polymer PDMS surface is methyl terminated (Figure 12.6) but oxygen plasma treatment will result in some methyl groups being replaced by hydroxyl groups. Hydroxyl groups are

Table 12.6 Water contact angles for various surfaces and treatments

Surface	Angle
Ammonia/peroxide cleaned silicon	5°
Oxygen plasma treated SU-8	5°–40°
Sulfuric acid cleaned silicon	10°
RCA-1 + RCA-2 cleaned silicon	10°
KOH etched silicon	25°
Thermal oxide	45°
Native oxide	45°
Oxygen plasma treated PDMS	50°
HMDS coated silicon	60°
HF dipped silicon	70°
Polyimide	75°
Native SU-8	80°
Native polystyrene	90°
Native PDMS	108°
ECT (eicosanethiol)	110°
Fluoropolymer	120°
Microstructure + PDMS	150°
Nanostructure + fluoropolymer	170°

Note that all the values are approximate and depend on surface treatment details and duration, and on time delay.

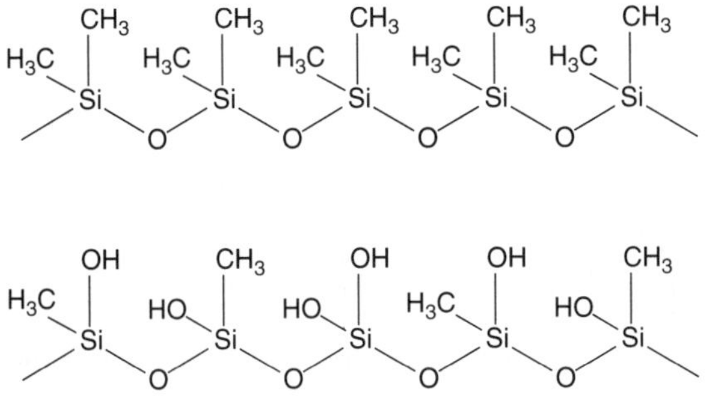

Figure 12.6 Native PDMS surface with methyl groups (top); oxidized hydrophilic PDMS surface with hydroxyl and methyl groups (bottom)

much more reactive (which is important in bonding) and are hydrophilic, while the methyl-terminated surface is hydrophobic.

Not only plasmas but also gases can be used for surface conditioning: hot HCl gas etches silicon, revealing undamaged silicon beneath the original surface (Equation 6.3). Thermal oxidation (Chapter 13) can also be used as a cleaning method: oxidation and oxide etching in HF will reveal a new silicon surface that is very clean.

CVD processes can also be used to deposit thin protective layers. A copper surface can be stabilized by short exposure to silane and subsequently to ammonia, forming a $CuSi_xN_y$ layer. This layer will prevent electromigration at the copper surface (see Chapter 36 for more on reliability).

Deposition of self-assembled monolayers (SAMs) is used extensively to modify surfaces (Figure 5.14): by suitably selecting the terminating group of the SAM, a wide range of contact angles can be obtained. SAMs can be used as adhesion layers too: gold as a noble metal has poor adhesion, but gold and sulfur form strong bonds, therefore SAM with sulfur termination will improve gold adherence (Figure 17.10).

There are also equipment solutions to surface cleanliness: if processes are performed immediately after each other, or if the wafers are kept under a vacuum or nitrogen, there is no time and no chance for contamination and particles to deposit on the wafers. This will be discussed in Chapter 31.

Wettability and cleaning are surface issues: in SAMs the effects are due to a single molecular layer only. In many other cases it is also only the very surface that matters. In Figure 12.7 silicon nanograss is shown. On the left hand side it is coated by ~1 nm thick silicon dioxide and the contact angle is close to zero; on the right hand

Figure 12.7 Nanometer thick layers of oxide (left) and fluoropolymer (right) drastically change contact angle. Reproduced from Jokinen *et al.* (2008) by permission of Wiley-VCH

side a few nanometers of fluoropolymer is deposited, and the contact angle is ~170°.

12.11 Exercises

1. Translate surface iron contamination of 10^{10} atoms/cm^2 into a number of monolayers!
2. If there is one monolayer coverage of organic contamination on a wafer, how much of that is counted as carbon atoms/cm^2? Select one common organic chemical used in microfabrication as an example.
3. How can the structure of Figure 12.7 be made?
4. Calculate the chemical and DI water consumption over 24 hours for the following cleaning cycle
   SPM
   DIW rinse
   RCA-1
   DIW rinse
   DHF
   DIW rinse
   RCA-2
   DIW rinse 1
   DIW rinse 2
   when a tank for 25 wafers of 200 mm diameter is used. Assume a 4 h changing interval for RCA cleans and 24 h bath life for SPM and DHF.
5. What happens to particle contamination in (a) wet etching and (b) plasma etching?
6. If an Olympic swimming pool is full of UPW, how many droplets of sweat can be dissolved before Na$^+$ and Cl$^-$ exceed the specification level of 0.01 ppb?

Make rough estimates of sweat salt concentration based on your own experiences.

7. Which cleaning chemicals are allowed and which are forbidden for the following materials, and why:
   (a) aluminum
   (b) gold
   (c) copper
   (d) aluminum oxide
   (e) polyimide
   (f) silicon nitride?

References and Related Reading

Grannemann, E (1994) Film interface control in integrated processing systems, *J. Vac. Sci. Technol.*, **12**, 2741.

Hattori, T. (ed.) (1998) **Ultraclean Surface Processing of Silicon Wafers**, Springer.

Jokinen, V., L. Sainiemi and S. Franssila (2008) Complex droplets on chemically modified silicon nanograss, *Adv. Mater.*, **20**, 3453–3456.

Kern, W. (1990) The evolution of silicon wafer cleaning technology, *J. Electrochem. Soc.*, **137**, 1887.

Kinoshita, K. and T. Hara (2008) A cleaning process for fine particles of silicon nitride, *J. Electrochem. Soc.*, **155**, H642–H647.

Kitajima, H. and Y. Shiramizu (1997) Requirements for contamination control in the gigabit era, *IEEE Trans. Semicond. Manuf.*, **10**, 267.

Li, X.-M., D. Reinhoudt and M. Crego-Calama (2007) What do we need for a superhydrophobic surface? A review on the recent progress in the preparation of superhydrophobic surfaces, *Chem. Soc. Rev.*, **36**, 1350–1368.

Middleman, S. and A. K. Hochberg (1993) **Process Engineering Analysis in Semiconductor Device Fabrication**, McGraw-Hill.

Ohmi, T. *et al.* (1992) Dependence of thin-oxide film quality on surface microroughness, *IEEE Trans. Electron Devices*, **39**, 537.

Okorn-Schmidt, H. (1999) Characterization of silicon surface preparation processes for advanced gate dielectrics, *IBM J. Res. Dev.*, **43**, 351.

Reinhardt, K. A. and W. Kern (2007) **Handbook of Silicon Wafer Cleaning Technology**, 2nd edn, William Andrew.

Schroder, D. K. (1998) **Semiconductor Material and Device Characterization**, 2nd edn, John Wiley & Sons, Inc.

Zhang, F. *et al.* (2000) The removal of deformed submicron particles from silicon wafers by spin rinse and megasonics, *J. Electron. Mater.*, **29**, 199.

13

Thermal Oxidation

Silicon dioxide, SiO_2, is probably a more important material in silicon technology than silicon itself: whereas GaAs and Ge have higher electron mobilities than silicon, and potentially faster transistors, they do not have native oxides which protect their surfaces, and neither do stable, thick oxides exist. Silicon dioxide has functions as passivation layers, capacitor dielectrics and electric isolation layers in finished devices. Thin oxides (1–20 nm) are used for example as capacitor dielectrics (including the CMOS transistor) and as tunnel electrodes in flash memories and chemical sensors, while thick oxides (100–1000 nm) serve as diffusion and etch masks and as electrical isolation layers. Thermal oxidation is often done repeatedly during device processing: for example, the first oxide may be used as an etch mask, and it is removed after etching, but another thermal oxidation is done immediately. This new oxide is again of high quality.

13.1 Thermal Oxidation Process

Silicon is easily oxidized: a native oxide of nanometer thickness grows on a silicon surface in a couple of hours or days, depending on the surface conditions, and similar thin oxides form easily in oxygen plasma or in oxidizing wet treatment. These oxides are, however, limited in their thickness and they are not stoichiometric SiO_2. Two basic schemes are used in thermal oxidation: wet and dry oxidation, respectively.

$$Si\ (s) + 2\ H_2O\ (g) \Rightarrow SiO_2\ (s) + 2\ H_2\ (g) \quad (13.1)$$

$$Si\ (s) + O_2\ (g) \Rightarrow SiO_2\ (s) \quad (13.2)$$

Thermal oxidation is a slow process: 1 h of dry oxidation at 900 °C produces oxide about 30 nm thick and 1 h of wet oxidation about 130 nm. Exact values are dependent on silicon crystal orientation: the oxidation rate of <111> is somewhat higher than that of <100> silicon; highly doped silicon oxidizes faster than lightly doped material.

Thin gate oxides, flash-memory tunnel oxides and DRAM capacitor oxides are grown in dry oxygen at 850–950 °C. In addition to better thickness control due to the lower rate, dry oxides exhibit lower interface charges and trap states which are important for transistor operation. Thin oxides also have many auxiliary and sacrificial roles: a thin oxide under nitride relieves stresses caused by the nitride film. Thicker oxides are used for device isolation and as masking layers for ion implantation, diffusion and etching steps. They are usually 100–1000 nm thick and grown by wet oxidation. Typical process temperatures are 950–1100 °C, and process times are hours or tens of hours.

Even though silicon is easily oxidized, oxidation furnaces are by no means cheap or simple equipment. They involve high-temperature parts which must be non-contaminating and which must tolerate oxidizing conditions and hot cleaning gases. Exacting temperature control and fine mechanics are also required. In Figure 13.1 a batch of silicon wafers is being loaded into a horizontal furnace.

When silicon and oxygen react to form SiO_2, the resulting oxide is roughly twice the volume of silicon it replaces: for a SiO_2 layer of thickness D, silicon thickness consumed is $0.45D$ as can be calculated from molar volumes:

density of silicon is 2.3 g/cm^3

mass 28 g/mol   molar volume 12.17 cm^3

density of SiO_2 is 2.2 g/cm^3

mass 60 g/mol   molar volume 27.27 cm^3

The original silicon surface is somewhat below the oxide midpoint. This volume increase leads to restrictions in

Introduction to Microfabrication, Second Edition   Sami Franssila
© 2010 John Wiley & Sons, Ltd

154 Introduction to Microfabrication

Figure 13.1 Wafers being loaded into a horizontal oxidation furnace. Courtesy VTT

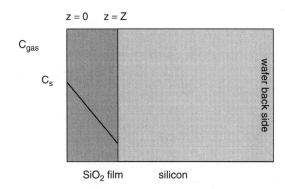

Figure 13.2 Model of thermal oxidation: oxygen diffuses through SiO$_2$ film and reacts at the SiO$_2$/Si interface. Concentration of oxygen inside oxide decreases linearly

the oxidation of structured surfaces, because stresses can become excessively large in the corners of the structures. On the other hand, the fact that oxidation consumes silicon can be used as a cleaning method: thin oxide is grown and immediately removed by HF etching, to reveal a perfect silicon surface.

Oxide thickness is usually measured by optical methods: either by ellipsometry or by reflectometry. Thermal oxides can be grown with very tight specifications, for an oxide 10 nm thick the uniformity is 1%, or equal to one atomic diameter. For thermal oxides a refractive index value of $n = 1.46$ is usually used, but for oxides thinner than 10 nm this is not really accurate. A quick and easy way to gauge oxide thickness, especially for intermediate thicknesses of 50–500 nm, is by its color. An oxide color chart can be found in Appendix C.

Various electrical measurements are also used: breakdown voltage is one of many. High-quality silicon dioxide can sustain 10 MV/cm, even 12 MV/cm. Oxide defects and electrical quality are closely connected; this topic will be discussed further in Chapter 36.

13.2 Deal–Grove Oxidation Model

A model for oxide growth has been developed by Deal and Grove. It is a phenomenological macroscopic model which does not assume anything about the atomistic mechanisms of oxidation. Oxygen diffusion through the growing oxide and chemical reaction at silicon/oxide interface are modeled by the classical Fick diffusion equation and chemical rate equation. Model geometry and boundary conditions are shown in Figure 13.2.

Oxidation is modeled as if the boundaries were stationary (which is a reasonable assumption because oxidation is slow). The diffusion equation for oxygen is

$$\frac{dC}{dt} = 0 = D\frac{d^2C}{dz^2} \qquad (13.3)$$

where C is oxygen molar concentration (mol/m^3), subject to boundary conditions

$$C = C_s \quad z = 0 \quad \text{at the SiO}_2 \text{ surface} \qquad (13.4)$$

$$-D\frac{dC}{dz} = R \quad z = Z \quad \text{at the SiO}_2/\text{Si interface} \qquad (13.5)$$

where R is the reaction rate at the interface (in units of mol/m^2-s).

The latter equation specifies that all oxygen reaching the interface will react there to form oxide: there will be no build-up of unreacted oxygen inside oxide or silicon.

For a reaction like Si (s) + O$_2$ (g) $\Rightarrow$ SiO$_2$ (s) the rate is assumed to be first order, that is $R = kC$, directly related to the concentration of reactive species, C, and characterized by a rate constant k. We can then rewrite the second boundary condition as

$$-D\frac{dC}{dz} = kC \quad \text{at } z = Z \qquad (13.6)$$

A solution that satisfies these conditions is

$$C = C_s - \left(\frac{kC_s}{kZ + D}\right) \qquad (13.7)$$

The rate (at the SiO$_2$/Si interface $z = Z$) is then

$$R = kC(Z) = \frac{kDC_s}{kZ + D} \quad (13.8)$$

To calculate the thickness growth rate, we must convert molar concentration to volume through density:

$$RM_{SiO_2} = \rho_{SiO_2} \frac{dZ}{dt} \quad (13.9)$$

where the molar volume of SiO$_2$ is $v = M_{SiO_2}/r_{SiO_2}$ (60 g/mol ÷ 2.2 g/cm^3 = 27.3 cm^3/mol).

When we solve for $Z(t)$ from the rate equation, we get

$$\frac{dZ}{dt} = \frac{kDC_s v}{kZ + D} \text{ subject to } Z = 0 \text{ at } t = 0 \quad (13.10)$$

This leads to the oxide thickness equation

$$t = \frac{Z}{KC_s v} + \frac{Z^2}{2DC_s v} \quad (13.11)$$

When thin oxides are considered, we can ignore the second term, and the rate is then simply

$$Z = kC_s t \quad (13.12)$$

or growth is linear in time and linearly related to rate constant k.

For thick oxides, we can ignore the first term and get

$$Z = \sqrt{2DC_s \, vt} \quad (13.13)$$

or growth is parabolic, related to diffusion length $\sqrt{Dt}$. The thicknesses of wet and dry thermal oxides obtained by a 1D simulator are shown in Figure 13.3.

The Deal–Grove model thus predicts linear oxidation rate initially, followed by a parabolic behavior for thicker oxides. The linear regime covers the initial stages of oxidation, with some success. The model works much better for thick oxides, and theory and experiment agree that doubling oxide thickness requires quadrupling oxidation time in the parabolic regime (this can be used as a quick estimate for oxidation time once one process is known and fixed). Alternatively, if thickness and time are known for one combination, others can be easily calculated from Equation 13.13: a 200 min dry oxidation results in oxide 165 nm thick, then 50 min corresponds to about 80 nm. It should be noted that the demarcation between the linear and parabolic regimes is not clear cut, and this introduces uncertainty into the simple calculation.

Dry oxidation is slower than wet oxidation even though diffusion of oxygen molecules through silicon

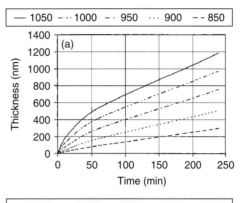
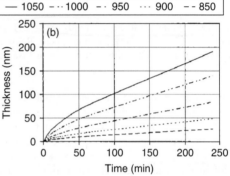

Figure 13.3 Oxidation of <100> silicon at temperatures between 850 and 1050 °C, wet and dry. Maximum practical oxide thickness is 1 or 2 micrometers, because of the decrease in parabolic oxidation rate

dioxide is faster than diffusion of water molecules. The explanation lies in the water solubility of silicon dioxide: it is four orders of magnitude larger that oxygen solubility, and therefore the concentration of the oxidant in oxide is much greater.

Generally, when thin oxides are required, a lower temperature is used. It is not, however, possible to reduce temperature without affecting oxide quality. Therefore one solution is to reduce the partial pressure of oxygen: for instance, in the experiment of Figure 13.4 the gas mixture is 10% oxygen, 90% nitrogen. If thick oxides are needed, higher temperature and higher partial pressure of oxygen are used. HIPOX, for high-pressure oxidation, can reduce oxidation time significantly.

Oxidation rate depends slightly on silicon crystal orientation. Silicon of <111> orientation oxidizes faster than <100> (and all other orientations fall between these two). This is speculated to arise from the number of silicon atoms and bonds at the oxidation front. For thick oxides, diffusion of oxygen through oxide determines the rate, and

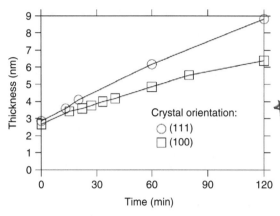

Figure 13.4 Initial oxidation of (100) and (111) silicon at 850 °C, in 10% O_2 + 90% N_2 atmosphere. Reproduced from Niskanen *et al.* (2009), by permission of Elsevier

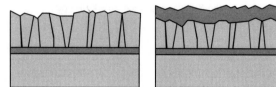

Figure 13.5 Polysilicon is rough and consists of grains of different orientations, which oxidize at slightly different rates, leading to rough oxide with non-uniform thickness

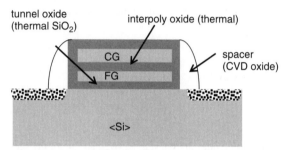

Figure 13.6 Flash memory cell: interpoly oxide between floating and control gates is limited to about 10 nm because poly oxides are of inferior quality

clearly this cannot depend on silicon crystal orientation, but in the linear regime the rate constant k is different for <111> and <100>. This is seen in Figure 13.4: <111> oxidation is faster than <100>. Note also that a thin oxide is present even at zero oxidation time. It is difficult to eliminate oxygen altogether: rinsing water has dissolved oxygen in it, and wafers are loaded into the furnace under atmospheric conditions. These topics will be discussed again in conjunction with CMOS gate oxidation in Chapter 26. The initial stages of oxidation are surprisingly poorly known, in spite of decades of research and substantial manufacturing experience. Most models cannot explain the very first nanometers of oxidation.

13.3 Oxidation of Polysilicon

Polysilicon oxidation presents a number of complications compared to single crystal silicon oxidation. Polysilicon consists of grains of many orientations which have different oxidation rates. Polysilicon grains are most often of (110) orientation and the oxidation rate of undoped poly is somewhere between the (100) and (111) silicon oxidation rates. In polycrystalline materials there are two different diffusion paths, through the bulk and along grain boundaries, requiring more advanced models. Grains also grow during oxidation, further complicating analysis.

The polysilicon surface is rough (Figure 7.4) and oxide quality will be inferior to single crystal oxides. Roughness and different oxidation rates of different grains lead to non-uniform poly oxide thickness (Figure 13.5). While single crystal oxides can tolerate 10–12 MV/cm electric fields, poly oxides break at 3–5 MV/cm.

Poly oxides are intensively used in EPROM/flash memories (Figure 13.6). Tunnel oxide is clearly thermally oxidized single crystal silicon, but the floating gate (FG) is poly and interpoly oxide between the floating gate and control gate (CG) is therefore thermally oxidized polysilicon. The fact that it is difficult to scale down poly oxides is partly limiting flash memory scaling.

13.4 Oxide Structure

Thermally grown silicon dioxide is glassy and exhibits only short-range order, in contrast to quartz, which is crystalline SiO_2. The basic unit of silica structure is SiO_4, as shown in Figure 13.7.

In a perfect arrangement, such as crystalline quartz, all oxygen atoms bond to two silicon atoms (oxygen has valence 2, silicon valence 4) but at the silicon/oxide interface some bonds are not made, leaving unbonded charged oxygen atoms (Figure 13.8), making oxide less stable than quartz. This is also reflected in their

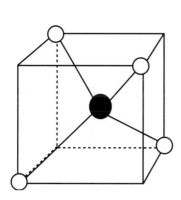

Figure 13.7 Basic structure of silica: a silicon atom tetrahedrally bonds to four oxygen atoms

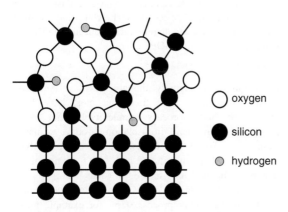

Figure 13.8 The structure of the silicon/silicon dioxide interface: single crystalline silicon and amorphous oxide. There are dangling bonds (not having their full valence) and some have hydrogen atoms bonded to them

properties: quartz density is 2.65 g/cm^3, silicon oxide density 2.2 g/cm^3; Young's modulus is 107 GPa for quartz and 87 GPa for oxide.

Incompletely oxidized silicon atoms are positively charged, and this is known as fixed oxide charge, Q_f. It is located in the oxide in the first few nanometers above the silicon surface. There is also interface trapped charge Q_{it} which can be either positive or negative. The defect density at the silicon/oxide interface can be made very low by proper cleaning and well-controlled oxidation. Both Q_f and Q_{it} densities are 10^9–10^{11}/cm^2, which should be compared to the silicon surface density of 10^{15} atoms/cm^2. Thus only one atom in 10^4 or one in 10^6 has incomplete bonds.

Thermal oxidation is often complemented by a post-oxidation anneal (POA) in nitrogen (in the very same furnace; oxygen gas is switched to nitrogen). This step densifies the film and anneals out some defects. It also causes diffusion of dopants and it has to be included in calculation the doping profiles. Hydrogen anneal is used to passivate dangling bonds: hydrogen attaches to the free valence of the silicon and eliminates further charge trapping. However, high electric fields can easily accelerate electrons to such energies that hydrogen atoms become mobile during device operation.

If post-oxidation anneal is done in ammonia (NH$_3$) or nitrous oxide (N$_2$O), surface nitridation will take place (while nitrogen anneal is an inert anneal). Nitridation results are very much dependent on anneal details, but the whole oxide is not turned into nitride: only the surface layer is affected. Nitridation, too, has many effects: the dielectric constant goes up (nitride $\varepsilon_r \approx 7$ vs. $\varepsilon_r \approx 4$ for oxide) which increases capacitance; it also improves oxide tolerance to damage caused by hot electrons injected into the oxide. Nitrided oxide reduces boron diffusion through it, which is extremely important for thin CMOS gate oxides. Ammonia anneal also introduces hydrogen. Oxidation of nitrided oxide is a way to reduce hydrogen concentration, and the resulting film is known as ONO, for oxidized nitrided oxide.

Dopant atoms have a major impact on oxidation: high doping level will increase oxidation rate even by 200% for thin oxides at low temperatures. High phosphorus concentration in silicon is associated with high concentration of vacancies, and because the volume of silicon dioxide is larger than the volume of silicon it replaces, oxidation consumes vacancies. Highly doped phosphorus therefore provides plenty of vacancies for oxidation to proceed. Boron behaves differently: it is incorporated into the growing oxide, weakening its bond structure and thus enabling faster diffusion through it.

In doped polysilicon, dopants precipitate at grain boundaries. Boron doping leads to minor oxidation rate enhancement, and phosphorus doping to clearly increased oxidation rate, via increased vacancy concentration, just as in the case of single crystal material.

13.5 Local Oxidation of Silicon

When local oxidation of silicon is needed, a silicon nitride mask is used. Nitride will prevent oxygen diffusion, and areas under the nitride will not be oxidized. This is known as LOCOS, for local oxidation of silicon. LOCOS is pictured in Figure 13.9.

158   Introduction to Microfabrication

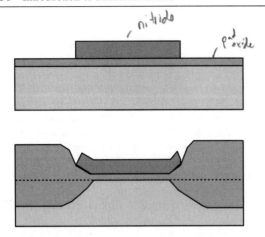

Figure 13.9 LOCOS: top, before oxidation, thin-pad oxide and patterned nitride; bottom, after oxidation, no oxidation under nitride but "bird's beak" at nitride edge. Note that about half of the oxide will be below the original silicon surface

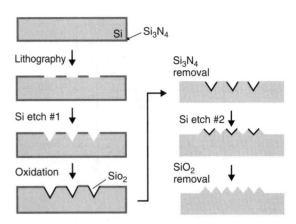

Figure 13.10 LOCOS for a sawtooth structure in (100) silicon. Nitride first acts as a silicon etching mask and then as an oxidation mask. Second KOH etching and HF oxide removal result in sawtooth relief. Reproduced from Ribbing *et al.* (2003), copyright IOP

LOCOS process flow

- Thermal oxidation (pad oxide)
- LPCVD nitride deposition
- Lithography
- Nitride etching
- Photoresist strip
- Cleaning
- Wet oxidation

LOCOS variables are pad oxide thickness (10–50 nm), LPCVD nitride thickness (100–200 nm) and oxidation temperature. Pad oxide serves as a stress-relief layer, and it diminishes the stress-induced dislocations that thick nitride exerts in silicon. Nitride acts as a diffusion barrier for oxygen diffusion and as a mechanical stiffener: the thicker the nitride, the smaller the oxide growth under the mask. This lateral extension is known as bird's beak, because of its visual appearance. Thinner pad oxide would help to minimize bird's beak but at the expense of silicon damage from nitride stress.

The LOCOS oxide surface will be above the original silicon surface. If we desire to have the oxide surface level with the original surface, we can start the LOCOS process by etching into silicon, with the silicon etched depth approximately half the desired oxide thickness, which then will result in approximately equal surface height for oxide and silicon.

LOCOS isolation has been used for 30 years because of its simplicity. It has been scaled to much smaller linewidths than anybody thought possible. Numerous modifications have been tried, but most have failed because the added process complexity has not offered enough improvement in isolation. Today, transistor isolation is based on shallow trench isolation (STI), which uses RIE of silicon and CVD filling of the trenches. This will be discussed in Chapter 26.

The fact that nitride acts as an oxidation mask can be combined with other process steps, for example nitride also acts as an etch mask for KOH etching of silicon. In Figure 13.10 a LOCOS-type KOH etching process is shown for wafers of (100) orientation. KOH etching is done using a nitride mask, resulting in inverse pyramid shapes and trenches. The wafer is cleaned and the newly formed silicon (111) sidewalls are oxidized. Nitride is then etched away, and a second silicon KOH etching step is performed. This leads to "period doubling" of trenches.

13.6 Stress and Pattern Effects in Oxidation

Oxide volume is greater than the volume of the silicon it replaces. Oxides are therefore under compressive stresses, and this causes a number of pattern-dependent phenomena which can be either beneficial or disadvantageous. Typical stress values are on the order of 300 MPa. Somewhere between 975 and 1000 °C oxide exhibits viscous flow. Oxidation above that temperature will result in reduced stress and wafer bow. Below that temperature, oxide needs to be treated as an elastic material with appropriate elastic

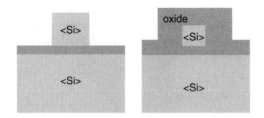

Figure 13.11 SOI structure (left) after plasma etching. After low-temperature thermal oxidation (right) unoxidized silicon remains because stresses build up. Redrawn from Heidemayer *et al.* (2000)

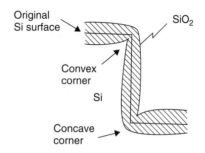

Figure 13.12 Cross-section of an oxidized silicon step with oxide thinning at both the convex (top) and concave (bottom) corner. Reproduced from Minh and Ono (1999) by permission of AIP

constants. Scaling of LOCOS to smaller linewidths meets an inevitable limit at submicron dimensions: stresses in the growing oxide prevent full oxidation of narrow gaps.

Thermal oxidation of small silicon wires (Figure 13.11) shows a self-limiting effect due to high stresses and this has been utilized in making nanostructures. This is illustrated in the SOI nanowire process.

Process flow for silicon nanowires

- SOI wafer with device silicon 21 nm thick
- Lithography
- Silicon etching
- Photoresist stripping and wafer cleaning
- Thermal oxidation

Thermal oxidation proceeds for a while, but then a self-limiting effect sets in: a critical stress, which stops oxidation, is about 2.6 GPa at 850 °C. After the self-limiting oxide thickness has been grown, no further oxidation takes place. If oxidation is carried out at higher temperature, say 1000 °C, this stress can be overcome, and the whole structure will be oxidized.

Stresses are also responsible for non-uniform oxidation in convex and concave corners as shown in Figure 13.12. Uneven oxide thickness causes problems for reliability because electric field strength is different in the corners and planar areas. Etched trenches have concave corners, and therefore thermal oxidation leads to non-uniform thickness which may be detrimental to for example DRAM trench capacitors, because the capacitor dielectric has a weak spot at the corner. Etch processes can be tailored to some extent for smoother bottom profiles, but this is a limited option because the top corner needs rounding, too. Oxide and nitride can be deposited by conformal CVD, but in very deep trenches the conformality may not be adequate. Sacrificial thermal oxidation can be used to

smooth corners. Second thermal oxidation then provides the actual thin dielectric film, which serves for example as DRAM capacitor dielectric. Anisotropic wet etching of silicon produces V-grooves, and oxide at the apex of the V-groove is highly stressed. This oxide has a much higher wet etch rate in HF than ordinary oxide. This has been employed to make nanoscopic holes, as shown in Figure 13.13.

13.6.1 Oxidation sharpening

Sharp tips are used as AFM probes and as field emitters in vacuum microelectronic devices, for high resolution in the former application and for low operating voltage in

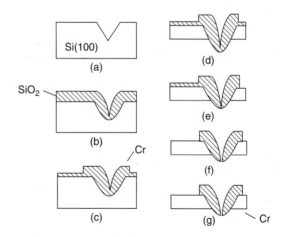

Figure 13.13 Oxide thinning at the apex is used as a method to fabricate nanoscopic holes: the apex can be etched open while leaving oxide elsewhere because oxide is thin and stressed at the apex. Reproduced from Minh and Ono (1999) by permission of AIP

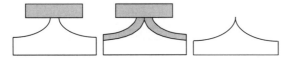

Figure 13.14 Silicon tip fabrication: left, isotropic silicon etching with an oxide mask; middle, thermal oxidation; right, silicon tip recovery by HF etching

the latter. Such tips can be fabricated by isotropic etching, but the final part of tip release is difficult: the mask will fall off. Thermal oxidation can help as shown in Figure 13.14: after initial isotropic (or KOH anisotropic) etching, the final sharpening takes place during oxidation. Mask removal is done by isotropic etching, but this is a non-critical, non-patterning etch. Thermal oxidation process control is also much tighter than shape control in an etch process.

13.7 Simulation of Oxidation

Oxidation simulation, together with diffusion simulation, is the backbone of all process integration simulators. Thermal oxidation is well understood and can be accurately modeled. However, the atomistic mechanisms of thin oxides (and early stages of oxidation in general) are still under intensive study.

Oxidation simulation requires as input:

- wafer orientation <100>/<111>
- doping level
- temperature
- time
- oxidizing ambient wet/dry.

Additional model parameters such as oxygen partial pressure (1 atm as default), high concentration effects and viscous/elastic models can be used instead of default models.

The Deal–Grove model (Equation 13.1) is the default model for wet oxidation and for thick oxides in general. It is not, however, applicable to thin dry oxides. A power-law model from Nicollian and Brews (2002) can be used for this regime. Oxidation is modeled as

$$x_{ox} = a \left(\frac{t}{t_0} \right)^b \tag{13.14}$$

Simulators produce results that are accurate within experimental error for 1D oxidation. Additionally, simulators can account for segregation, the distribution of dopants at the oxide/silicon interface.

13.7.1 Segregation

Dopants which are initially in the silicon are redistributed between silicon and the growing oxide during oxide growth. As shown in Figure 13.15, boron is depleted and phosphorus accumulated in the silicon side of the interface. Segregation has a major effect on device properties: if the dopant is mostly incorporated in the oxide and depleted in the silicon near the interface, inversion of silicon doping type may occur. Segregation proceeds as long as the chemical potentials of the dopants differ in the oxide and silicon. The equilibrium segregation coefficient, m, is defined as the ratio of dopant in silicon to that in oxide.

Metal atoms experience segregation just like the dopants: for example, Al and Ca are segregated preferentially into the oxide (and cause oxide quality problems) whereas Ni and Cu diffuse into the bulk (and cause defects which act as lifetime killers).

13.8 Thermal Oxides vs. other Oxides

Thermal oxidation of silicon produces high-quality oxide (in terms of uniformity, density, electric breakdown field, etc.) but thermal oxidation is a high-temperature process. CVD oxides are sometimes indispensable because they allow lower process temperatures, but CVD oxides have, however, lower breakthrough voltage, higher wet etch rate in HF and rougher surfaces than thermal oxides. Sometimes CVD oxides have to be used because thermal oxides cannot be used. One such situation concerns very thick oxides: thermal oxidation of micrometer layers takes hours or days while CVD oxides of identical thicknesses can be done in a matter of minutes or an hour, depending on the particular equipment.

Very few materials can tolerate oxidizing ambients at about 1000 °C. Metals cannot usually withstand such conditions. Tantalum is an exception: thermal oxidation of tantalum can produce device-quality Ta_2O_5 thin films in a standard oxidation furnace (for reasons of contamination, a separate oxidation tube needs to be reserved for such an

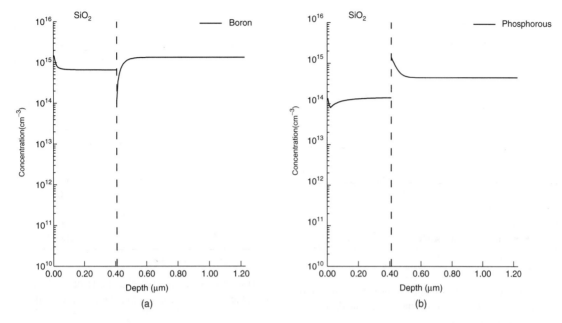

Figure 13.15 Segregation of dopant at silicon/oxide interface during wet oxidation (1000 °C, 60 min): (a) boron-doped wafer shows dopant loss at interface; (b) phosphorus-doped wafer shows accumulation of dopant at the interface. Substrate resistivity is 10 ohm-cm in both cases

exotic process). Silicides will thermally oxidize to form SiO_2, with the exception of $TiSi_2$, which will turn into TiO_2. Tungsten polycide gates (WSi_2/poly) can be oxidized similarly to polysilicon gates. Making the silicide silicon rich, $WSi_{2.2}$, will ensure proper oxidation. Silicon carbide, SiC, can be oxidized to produce SiO_2 with standard silicon oxidation processes but the rate is very low compared to silicon oxidation.

In addition to thermal oxidation, there are other oxidation processes in microfabrication, namely chemical oxidation in acidic solutions, as discussed in the previous chapter; for example, nitric acid will oxidize silicon and form a SiO_2 layer 1 nm thick. This has been used as a depth profiling method: after each oxidation cycle oxide is removed and the surface analyzed, and with continued oxidation and HF etching cycles depth information is obtained, for example on dopant distribution. Anodic oxidation of silicon produces silicon dioxide, and similarly aluminum is anodically oxidized in acidic solutions (to alumina, Al_2O_3) which can be used for example as a tunnel dielectric in chemical sensors. Plasma is often used to grow thin oxides, for example aluminum oxide formed by oxygen plasma treatment is 2 nm thick and useful as a tunnel oxide in Josephson junction superconducting devices. Similar plasma oxide is used as a passivation layer after chlorine plasma etching of aluminum.

13.9 Exercises

1. A wet oxidation of 250 minutes results in oxide 1 μm thick. How long will it take to grow 10 μm oxide under the same conditions? How long will it take to grow oxide 0.1 μm thick?
2. Holes are etched in thermal oxide 1 μm thick. The wafer is then given 1 hour of wet oxidation at 1000 °C. All oxide is then etched away. What is the resulting step height in silicon?
3. Recessed LOCOS has the final oxide surface at the same level as the original silicon surface. Its fabrication starts by etching through nitride and oxide into silicon as shown below. Calculate the required etched depth and choose proper oxidation conditions which will give the final oxide thickness. Draw a cross-sectional figure of the resulting structure.

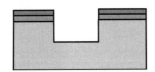

4. Explain where thermal oxide will grow in the structure shown below.

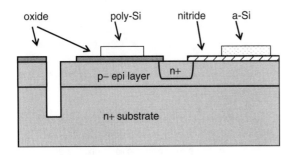

5. Draw the following structure after thermal oxidation. Assume 500 nm height for the oxide islands and new oxide 500 nm thick.

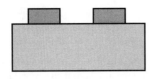

6. What difference do the following materials have in 1000 °C dry oxidation:
   (a) lightly doped single crystalline silicon
   (b) highly phosphorus-doped single crystal silicon
   (c) phosphorus-doped polysilicon
   (d) undoped amorphous silicon
   (e) lightly doped epitaxial silicon?

Simulation exercises:

7. The Deal–Grove oxidation model is not valid for thin oxides. Experimental data for dry oxidation is shown below. Check how your simulator works for thin oxides!

Time (min)	850 °C	1000 °C
20	6 nm	26 nm
40	8 nm	42 nm
60	11 nm	56 nm
80	13 nm	68 nm

Data from Massoud *et al.* (1985a).

8. Phosphorus-doped polysilicon (20–80 ohm/sq) oxidation produces 50 nm oxide in 30 min of dry oxidation at 1000 °C. At 900 °C dry oxidation results in 10 nm oxide. How do these values compare to single crystal silicon oxidation?

9. High-pressure oxidation (HIPOX) increases oxidation rates. Data for dry oxidation at 900 °C is as follows.

Pressure (atm)	Time (min)	Thickness (nm)
10	30	40
10	60	65
10	120	100
20	30	55
20	60	100
20	120	180

Data from Lie *et al.* (1982).

How does your simulator handle HIPOX oxides?

10. What is the segregation behavior of n-type dopants As, P, Sb?

References and Related Reading

Aït-Kaki, A. *et al.* (2002) Characterization of sub-micrometre silicon films (Si-LPCVD) heavily in situ boron-doped and submitted to treatments of dry oxidation, *Semicond. Sci. Technol.*, **17**, 983–992.

Chabal, Y. J. (ed) (2001) **Fundamental Aspects of Silicon Oxidation**, Springer.

Green, M. L. *et al.* (1999) Understanding the limits of ultrathin SiO_2 and Si-O-N gate dielectrics for sub-50 nm CMOS, *Microelectron. Eng.*, **48**, 25.

Heidemayer, H. *et al.* (2000) Self-limiting and pattern dependent oxidation of silicon dots fabricated on silicon-on-insulator material, *J. Appl. Phys.*, **87**, 4580.

Kameda, N. *et al.* (2007) High quality gate dielectric film on poly-silicon grown at room temperature using UV light excited ozone, *J. Electrochem. Soc.*, **154**, H769–H772.

Krzeminski, C. *et al.* (2007) Silicon dry oxidation kinetics at low temperature in the nanometric range: modeling and experiment, *J. Appl. Phys.*, **101**, 064908.

Lie, L. N., R. R. Razouk and B. E. Deal (1982) High pressure oxidation of silicon in dry oxygen, *J. Electrochem. Soc.*, **129**, 2828.

Massoud, H. Z., J. D. Plummer and E. A. Irene (1985a) Thermal oxidation of silicon in dry oxygen growth-rate enhancement in the thin regime, Part I, *J. Electrochem. Soc.*, **132**, 2685.

Massoud, H. Z., J. D. Plummer and E. A. Irene (1985b) Thermal oxidation of silicon in dry oxygen: growth-rate enhancement in the thin regime, Part II, *J. Electrochem. Soc.*, **132**, 2693.

Minh, P. N. and T. Ono (1999) Non-uniform silicon oxidation and application for the fabrication of aperture for near-field scanning optical microscopy, *Appl. Phys. Lett.*, **75**, 4076.

Nicollian, E. H. and J. R. Brews (2002) **MOS Physics and Technology**, John Wiley & Sons, Inc.

Niskanen, A. *et al.* (2009) Ultrathin tunnel insulator films on silicon for electrochemiluminescence studies, *Thin Solid Films*, **517**, 5779–5782.

Plummer, J. D., M. D. Deal and P. B. Griffin (2000) **Silicon VLSI Technology**, Prentice Hall.

Ribbing, C., Cederström B. and Lundqvist M. (2003) Microfabrication of saw-tooth refractive x-ray lenses in low-Z materials, *J. Micromech. Microeng.*, **13**, 714–720.

Shimidzu, H. (1997) Behavior of metal-induced oxide charge during thermal oxidation in silicon wafers, *J. Electrochem. Soc.*, **144**, 4335.

Suryanarayana, P. *et al.* (1989) Electrical properties of thermal oxides grown over doped polysilicon thin films, *J. Vac. Sci. Technol.*, **B7**, 599.

Vollkopf, A. *et al.* (2001) Technology to reduce the aperture size of microfabricated silicon dioxide aperture tips, *J. Electrochem. Soc.*, **148**, G587.

Watanabe, T., K. Tatsumura and I. Ohdomari (2006) New linear-parabolic rate equation for thermal oxidation of silicon, *Phys. Rev. Lett.*, **96**, 196102.

Zhang, C. and K. Najafi (2004) Fabrication of thick silicon dioxide layers for thermal isolation, *J. Micromech. Microeng.*, **14**, 769–774.

14

Diffusion

The power of silicon microelectronics technology stems from the ability to tailor dopant concentrations over eight orders of magnitude by introducing suitable n- or p-type dopants into the silicon. The upper limit is set by the solid solubility of the dopants (about 10^{21} cm^{-3}), the lower limit (about 10^{12} cm^{-3}) by impurities which result from the silicon crystal growth. This enables a wealth of microstructures and devices, witnessed by the multiplicity of diode, transistor, thyristor and other semiconductor device designs. In silicon IC technology dopant diffusion is such a key step that the country of origin of semiconductor devices is defined as the country where diffusions were made.

Dopants can be introduced into silicon by five different methods:

- during crystal growth
- by neutron transmutation doping (NTD)
- during epitaxy
- by diffusion
- by ion implantation.

The first two techniques are applied to whole ingots, and epitaxy results in a uniformly doped layer all over the wafer. Diffusion and ion implantation are techniques to locally vary the dopant concentration, and they are the topics of this and the following chapter.

14.1 Diffusion Process

Diffusion is the movement of atoms along concentration gradients. Atoms from high-concentration areas move to areas of lower concentration (and if we wait long enough, there will be no concentration gradients). In microtechnology diffusion is a technique to introduce and drive boron, phosphorus and other dopant atoms into the silicon lattice.

Thermal diffusion is a high-temperature process: diffusion temperatures for the common dopants are in the range of 900–1200 °C. Diffusion furnaces are identical to oxidation furnaces, and diffusion is a batch process where long process times are compensated by a huge loads, 100 or even 200 wafers, in a batch.

Thermal diffusion can be done from the gas phase. In gas phase doping the wafers are put in a furnace and a suitable doping gas, $POCl_3$ for phosphorus doping, or BBr_3 for boron doping, is introduced. The wafers are exposed to dopant atom vapors and doped (Figure 14.1). The alternative technique is diffusion from doped thin films. For example, boron-doped polysilicon, phosphorus-doped silica glass (PSG) or doped spin-on glass is deposited on the wafer, which is then put into a furnace. Dopants from the doped film diffuse into the silicon. The junction depth (x_j) is the depth where diffused dopant concentration equals substrate dopant concentration. Solid solubilities of dopants in silicon are shown in Figure 14.2. If the dopant concentration exceeds the solubility limit, the dopant will precipitate, and it does not contribute to electrical conductivity.

If only certain areas need be doped, oxide is used as a diffusion mask. Even though the dopants do not diffuse through the oxide, they do modify it to the extent that diffusion mask oxides are practically always etched away after diffusion and new oxide is grown. Oxide mask thicknesses are typically around 500 nm, but might have to be thicker if it is absolutely required that no dopants penetrate the mask. This might be the case if very high-resistivity starting wafers are used: their initial doping level is very low, and a very small amount of dopant will easily counterdope them.

Typical diffusion process at at 1000 °C for a few hours will result in roughly 1 µm diffusion depth. Much longer

Introduction to Microfabrication, Second Edition   Sami Franssila
© 2010 John Wiley & Sons, Ltd

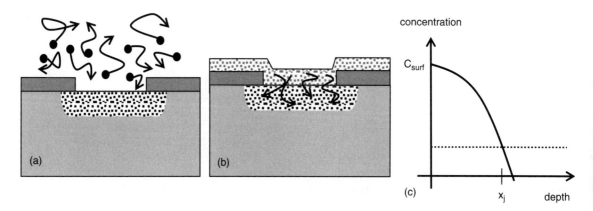

Figure 14.1 Thermal diffusion: (a) gas phase diffusion with oxide mask; (b) diffusion from doped thin film with oxide mask; (c) dopant profile and junction depth x_j

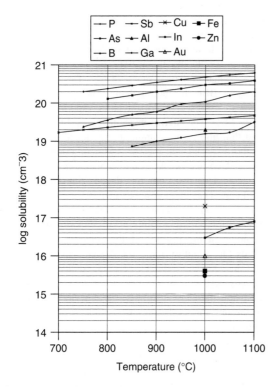

Figure 14.2 Solid solubilities of the most important dopants and impurities in silicon technology

introduced on the wafer, and during drive-in they will diffuse deeper. In gas phase diffusion this means that $POCl_3$ gas is switched off during drive-in, and nitrogen or oxygen introduced. Ion implantation, where ionized dopants are accelerated into silicon, can be considered as a pre-deposition step for diffusion. Diffusion is therefore the general term for doping processes, irrespective of the actual mechanism of dopant introduction.

Doping can be performed many times over, and silicon doping type may change from p- to n-type and back again, depending on the process sequence. The device shown in Figure 14.3, a UV photodiode, is made in a modified npn–bipolar process. UV photons are absorbed in the top p^+ diffusion layer. The process flow will only show the diffusion aspects of the device.

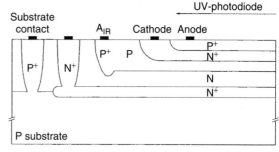

Figure 14.3 UV photodiode with shallow p^+ anode diffusion. The structure is based on an npn–bipolar transistor. Reproduced from Zimmermann (1999) by permission of Springer. Note that lateral and vertical dimensions in the figure are not true: diffusion is isotropic and for example 5 μm vertical diffusion leads to 5 μm lateral (sideways) diffusion on both sides!

diffusions are also common: CMOS well diffusion could be 16 hours at 1150 °C, for 5 μm depth.

Diffusion is often carried out in two steps: pre-deposition (or pre-diffusion) and drive-in. In pre-deposition a known and limited number of dopants are

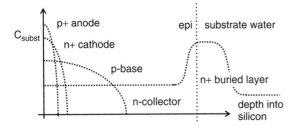

Figure 14.4 UV photodiode doping profile along the anode of Figure 14.3

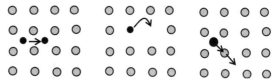

Figure 14.5 Diffusion mechanisms: left, interstitial; middle, substitutional/vacancy; right, interstitialcy

Process flow for UV photodiode (lithography, etch and oxidation steps omitted)

- p-type substrate wafer
- n^+ buried layer diffusion
- n epitaxial layer deposition
- p^+ substrate contact diffusion
- n^+ diffusion to contact buried layer
- p^+ base contact enhancement diffusion (under A_{IR})
- p base diffusion
- n^+ cathode diffusion
- p^+ anode diffusion

The area directly underneath the anode changes its doping type three times: it is originally n-type epilayer, doped by PH_3 gas during epitaxy. Base diffusion changes it to p-type when the boron concentration exceeds the phosphorus concentration in the epilayer; n-cathode diffusion turns it back to n-type because the phosphorus concentration is higher than the boron concentration; and finally the surface anode diffusion with the highest boron concentration of all results in p^+ silicon. Figure 14.4 shows the dopant depth profile underneath the anode of Figure 14.3.

14.2 Diffusion Mechanisms

Fairly simple mathematical models can describe concentration profiles in solids, but at the atomistic level diffusion remains to be fully explained. This has consequences for simulators, because mechanisms are not fully known, therefore modeling remains inaccurate.

Dopant atoms move with the help of point defects: they jump to vacancies and interstitials. Substitutional dopants are fairly stable without point defects. Vacancies are always present through thermal equilibrium processes: vacancies are thermodynamic defects, and their nature is different from for example dislocations and stacking faults, which are "frozen." Vacancies as a fraction of all sites can be estimated by

$$f = e^{(-E_a/kT)} \quad (14.1)$$

For 1 eV activation energy it gives about 0.01% vacant sites at 1000 °C (1273 K).

Here we outline some mechanisms for diffusion (Figure 14.5). In interstitial diffusion atoms jump from one interstitial site to another interstitial site which is always available. This is the diffusion mechanism for small atoms, like sodium and lithium. Substitutional/vacancy diffusion necessitates that an empty lattice site is available next to the diffusing atom. At high temperatures substitutional sites are thermally created. Antimony and arsenic demonstrate substitutional mechanisms. The interstitialcy mechanism is related to the substitutional mechanism: self-interstitial atoms move to lattice sites, and knock dopants out to interstitial sites, and from there they move to lattice sites. Boron and phosphorous are expected to diffuse via interstitialcy mechanism, but there are still some open questions even in diffusion of the most widely used dopants.

The substitutional and interstitialcy mechanisms with activation energies of about 3.5–4 eV are the most important for doping in silicon technology. Boron, phosphorus and arsenic as well as antimony, indium and gallium all have activation energies in this range. Therefore, doping by diffusion must take place at high temperature.

Many metallic impurities diffuse with the interstitial mechanism (E_a 1–1.5 eV), and nickel for instance can diffuse through the wafer (500 µm) at 750 °C in one hour.

14.3 Doping of Polysilicon

Polysilicon can be doped almost like single crystal silicon. Again, the difference arises from the grain boundaries. Grain boundaries attract and lock dopants, and a proportionately smaller concentration of dopant ends up in the grains. Therefore polysilicon resistivity is never as low

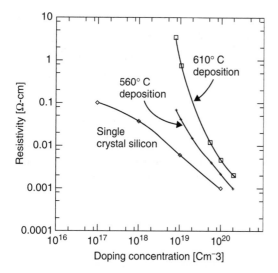

Figure 14.6 Silicon doping. Single crystal silicon resistivity is lower than poly for same doping concentration. True poly (610 °C) and a-Si (560 °C) resistivities also differ. From French (2002) by permission of Elsevier

Table 14.1 D_0 and E_a values for boron and phosphorus

	Boron	Phosphorus
D_0 (cm^2/s)	0.76	3.85
E_a (eV)	3.46	3.66

as that of single crystal (Figure 14.3). This is true for in situ doping during poly deposition at all temperatures, as well as diffusion doping and ion implantation. Another feature seen in Figure 14.6 is the effect of poly deposition temperature: at 570 °C "poly" is amorphous, and at 610 °C poly is truly polycrystalline. They crystallize differently and end up with different grain sizes, hence different resistivities. The lowest possible n-type poly doping is achieved with phosphorus, resulting in 400 μ ohm-cm resistivity, while for boron-doped poly the low limit is about 2000 μ ohm-cm.

14.4 Doping Profiles in Diffusion

Concentration-dependent diffusion flux is described by Fick's first law:

$$j = -D\left(\frac{\partial N}{\partial x}\right) \quad (14.2)$$

where D is the diffusion coefficient (cm^2/s) and N is concentration (in cm^{-3}). The unit of flux is atoms/s-cm^2.

Diffusion coefficients can be presented by

$$D = D_0\, e^{(-E_a/kT)} \quad (14.3)$$

where D_0 is the frequency factor (related to lattice vibrations, $10^{13} - 10^{14}$ Hz), E_a is the activation energy (related to the energy barrier that the dopant must overcome), k is Boltzmann's constant, $k = 1.38 \times 10^{-23}$ J/K or 8.62×10^{-5} eV/K, and T is temperature in kelvin.

Values of D_0 and E_a for boron and phosphorus are given in Table 14.1. From those values the boron diffusion coefficient D can be calculates as 4×10^{-15} cm^2/s at 950 °C and as 4.7×10^{-14} cm^2/s at 1050 °C.

The characteristic diffusion length is given by

$$x \approx 2\sqrt{Dt} \quad (14.4)$$

so that 1050 °C boron diffusion for 1 h corresponds to roughly 0.26 μm diffusion depth. This distance is a characteristic length scale only: diffusion profiles are gently sloping and there is no clear cut-off depth.

The sheet resistance of doped layers in the general case is given by Equation 14.5a, and is approximated for a box profile (assuming constant dopant level up to junction depth, and then zero dopant) by Equation 14.5b. The box profile can be quite a reasonable assumption: for example, the arsenic profile of Figure 2.12 is fairly box-like.

$$\frac{1}{R_s} = \int_0^{x_j} q\mu(N(x) - N_b)\, dx \quad (14.5a)$$

$$\frac{1}{R_s} = q\mu x_j N(x) \quad (14.5b)$$

where q is elementary charge, μ the mobility, $N(x)$ the dopant concentration, N_b the background concentration and x_j the junction depth. Mobilities of n- and p-type silicon are about 1400 cm^2/V-s and 500 cm^2/V-s, respectively, at low concentrations ($<10^{15}$ cm^{-3}), and about 50 cm^2/V-s at high concentrations ($>10^{19}$ cm^{-3}), irrespective of dopant. Sheet resistances of doped areas are typically from 10 Ω/sq to 10 kΩ/sq.

14.4.1 Infinite dopant supply (constant surface concentration of dopant)

Infinite dopant supply corresponds to gas phase doping where new dopant is constantly being injected into the diffusion tube. A heavily doped thin film (doped polysilicon or PSG or BSG CVD oxide) can be approximated as an infinite source when diffusion times and temperatures are moderate. The concentration profile of the dopant in silicon is given by the complementary error function (erfc):

$$N(x, t) = N_0 \, \text{erfc}\left(\frac{x}{4Dt}\right) \quad (14.6)$$

where N_0 is the dopant concentration (1/cm^3) in the surface layer, x the depth (cm), t the time (s) and D the diffusion constant at given temperature (cm^2/s). Longer doping times will lead to deeper diffusions but the surface concentration is unchanged.

14.4.2 Limited dopant supply (constant dopant amount)

Limited dopant supply describes the case of pre-deposition: a small amount of dopants is introduced in the first step. In the second step they diffuse deeper but because no new dopants are available, peak concentration at the surface will get lower as time goes by. Ion implantation, to be discussed in the next chapter, can be used to introduce a controlled number of dopants into silicon. The concentration profile of limited source diffusion is Gaussian:

$$N(x, t) = \frac{Q_0}{\sqrt{\pi Dt}} \exp\left(-\frac{x^2}{4Dt}\right) \quad (14.7)$$

where Q_0 is the total amount of dopant on the surface (1/cm^2). The junction depth x_j is given by

$$x_j = \sqrt{4Dt} \times \ln\left(\frac{Q_0}{C_{\text{subs}}\sqrt{\pi Dt}}\right) \quad (14.8)$$

Equation 14.8 cannot be solved in analytical form for diffusion time. An approximate solution for diffusion time can be obtained by a graphical solution: calculate x_j for a few diffusion times, plot the results, and estimate the junction depth from the graph. Simulators should be used for more accurate estimates.

14.4.3 Diffusion profile measurement

Diffusion profiles are measured either physically or electrically. The standard physical measurement is secondary ion mass spectrometry (SIMS). The dynamic range of SIMS is up to eight orders of magnitude, that is dopant concentrations down to 10^{14} cm^{-3} can be detected (silicon atom density is 5×10^{22} cm^{-3}). Spreading resistance profiling (SRP) measurement (Figure 2.10) measures resistance with probes at the surface, and then beveling or anodic oxidation is done in order to have access to dopants deeper inside the silicon. SRP data needs some major calculations before dopant profiles are obtained. Both SIMS and SRP are sample destructive methods. Simple four-point probe measurement gives sheet resistance, which is good for monitoring diffusion.

14.5 Diffusion Applications

Gas phase doping by POCl$_3$ gas for n-type and BBr$_3$ gas for p-type was used in the early years of semiconductor manufacturing for steps where a high degree of control was required, for example bipolar base diffusion. Solid source doping was used when high dopant concentration (near or at the solid solubility limit) was required, for example in bipolar emitters and MOS source/drain. Solid source doping has the problem that it is often very difficult to remove the dopant source material after diffusion, and residues may be left.

Resistors can be made easily if diffusion is part of the process. Polysilicon, with its higher resistivity, is better suited for making high-resistance resistors.

Bipolar devices and power transistors (Figure 1.14) rely on diffusion as the key element in their fabrication. As shown for MOS-controlled thyristor (Figure 14.7), many diffusions, some of them very deep, are needed.

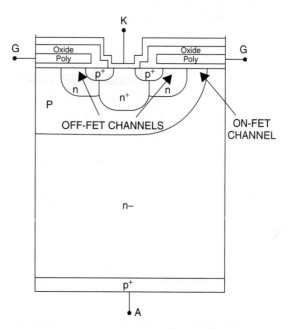

Figure 14.7 MOS-controlled thyristor. Reproduced from Sankara Narayanan et al. (2004), copyright © 2004 by permission of Elsevier

There are concentration and electric field effects which make actual device diffusions more complex than the simple Fick models predict. In the emitter push effect phosphorus diffusion enhances boron diffusion (see Figure 14.8). Boron diffusion alone would result in a profile predicted by simple theory, but boron diffusion under a phosphorus-doped region is much faster. This is explained by self-interstitial generation in the phosphorus diffusion process, and these interstitials enhance boron diffusion.

In oxidation-enhanced diffusion (OED) the vacancies generated by volume changes associated with thermal oxidation lead to enhanced diffusion underneath the oxide. This is pictured in Figure 14.9. Simulators can handle the emitter push effect, OED and high dopant concentration effects among other subtleties. We will see in the next chapter that ion implantation produces vacancies, and these vacancies will similarly enhance diffusion.

14.6 Simulation of Diffusion

All high-temperature process steps contribute to diffusion, therefore diffusion is the omnipresent process to be simulated in the front-end of the process. There can easily be tens of steps that contribute to dopant profiles. Segregation effects during oxidation and dopant outdiffusion from free surfaces add to the computational and modeling loads.

The simulation of phosphorus diffusion needs to consider at least five species:

- phosphorus (P)
- vacancies (v)
- interstitials (i)
- phosphorus–vacancy pairs (P–v)
- phosphorus–interstitial pairs (P–i).

Vacancies and interstitials are not permanent species like phosphorus atoms, and we must account for the annihilation of point defects, for example vacancy meets interstitial and annihilation takes place (v + i = nil). Point defects can also form pairs like v–v. To make the situation even more difficult to analyze, many of the species are charged: diffusion models have to account for equilibrium processes like $P^- + v^0 \Leftrightarrow Pv^-$ (charged phosphorus–vacancy pair) or $P^- + i^0 \Leftrightarrow Pi^-$. Clustering and precipitation of dopants lead to inactivation. These phenomena are especially important when concentrations are near the solid solubility limit.

A standard simulator requires as inputs for diffusion simulation:

- wafer orientation <100>/<111
- wafer doping level/resistivity
- dopant type
- concentration of dopant (gas phase/solid phase/implanted)
- temperature
- ambient (oxidizing/inert/reducing).

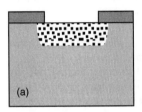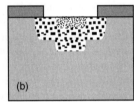

Figure 14.8 Emitter push effect: (a) unimpeded boron diffusion; (b) phosphorus diffusion pushes boron diffusion deeper under the same conditions

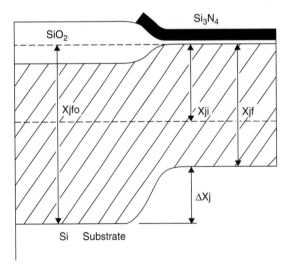

Figure 14.9 Oxidation-enhanced diffusion (OED): vacancy injection during oxidation enhances dopant diffusion under oxide. Reproduced from Taniguchi et al. (1980) by permission of Electrochemical Society Inc.

The doping profiles of Figure 14.10 have been calculated with the 1D simulator ICECREM.

Longer diffusion time of course leads to deeper diffusion, but the two cases of limited and infinite supply have differences. In the case of limited dopant supply the surface concentration decreases the longer the diffusion time. The area under the depth-concentration curve is constant: the number of dopant atoms is constant. In the infinite supply case the surface concentration remains constant: there is an ample supply of new dopant atoms.

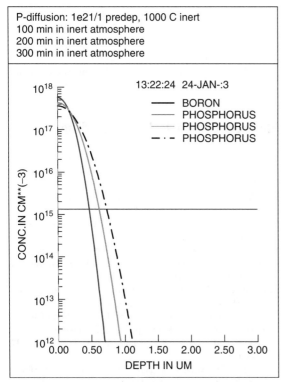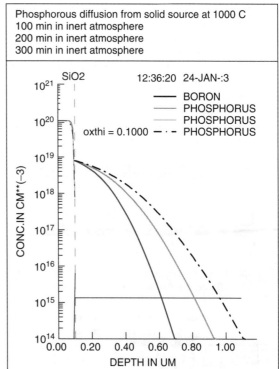

Figure 14.10 Diffusion at 1000 °C, for 100, 200 and 300 min in inert atmosphere for boron-doped substrate: (a) diffusion from a limited source, implanted dose 10^{13} cm^{-2}; (b) diffusion from phosphorus-doped oxide film (with 10^{20} cm^{-3} phosphorus concentration, approximating an infinite source)

14.7 Diffusion at Large

Diffusion is of course a much larger concept than described above: it includes many processes where concentration gradients are reduced by atom movements; for example, in microfluidics two parallel liquid flows will mix due to diffusion, and hydrogen in a fuel cell is spread uniformly by a gas diffusion layer, an intricate web of nanotubes or microfibers.

Diffusion is inevitable in all high-temperature steps, but it can be minimized by minimizing process time. In rapid thermal annealing (RTA; or RTP for Rapid Thermal Processing) wafers are heated rapidly by powerful lamps, and $2\sqrt{Dt}$ is brought down by annealing for very short times at high temperatures: a furnace diffusion of 950 °C for 30 min, might be replaced by a RTA process of 1050 °C for 10 s.

Many processes in addition to diffusion are temperature dependent: grain growth, interfacial chemical reactions, surface oxidation, etc., all experience higher rates at higher temperatures. Also, unwanted processes like impurity diffusion are accelerated. Metallic impurities have diffusion coefficients that lead to 500 μm diffusion distances during typical heat treatments. This means that back-side contamination can reach the wafer front, and destroy device functionality there. Therefore wafer cleaning before any high-temperature step is very important.

In solar cell fabrication, cost reduction pressures are high. Diffusion steps have been simplified and shortened. P-type diffusion can be done by spin coating phosphoric acid on the wafer, and then performing 950 °C diffusion for 15 min. This resembles in some sense the result of incomplete cleaning: whatever is left on the surface will be driven inside the silicon by subsequent annealing.

14.8 Exercises

1. What is the diffusion time required to form a pn junction at 1 μm depth at 1000 °C when boron pre-deposition is 10^{14} cm^{-2} and a phosphorus-doped wafer (10^{15} cm^{-3}) is used.

2. Polysilicon sheet resistance is 50 ohm/sq. What is the polysilicon thickness?
3. What is the sheet resistance of diffusion after the anneal shown in Figure 2.12?
4. If deep n-type diffusions are needed, which n-type dopant should be used?
5. How far will metallic impurities diffuse during thermal oxidation?
6. Explain the order of the fabrication steps for the spreading resistance thermometer shown below.

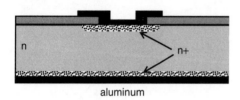

7. Explain step by step the fabrication process for the triple diffused bipolar transistor shown below.

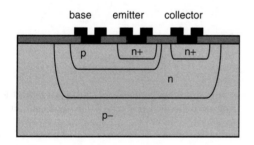

8. Explain the order of diffusion steps in the fabrication of the MOS-controlled thyristor of Figure 14.7.

Simulator exercises:

9. Boron-doped oxide film (200 nm thick, concentration 10^{21} cm^{-3}) is deposited on a phosphorus-doped wafer (10^{15} cm^{-3} phosphorus concentration). What is the junction depth doping after a 300 min, 1100 °C diffusion step?
10. What is the magnitude of the emitter push effect?
11. What is the magnitude of OED? Run some simulations to find which process parameters are important.

References and Related Reading

Baliga, J.B. (2001) The future of power semiconductor device technology, *Proc. IEEE*, **89**, 822 (special issue on power electronics technology).

French, P.J. (2002) Polysilicon: a versatile material for microsystems, *Sens. Actuators*, **A99**, 3–12.

Ghandhi, S.K. (1994) **VLSI Fabrication Principles**, 2nd edn, John Wiley & Sons, Inc.

Kim, D.S. *et al.* (2006) Development of a phosphorus spray diffusion system for low-cost silicon solar cells, *J. Electrochem. Soc.*, **153**, A1391–A1396.

MRS Bulletin (2000) Defects and diffusion in silicon technology, *MRS Bull.*, **25** (June), special issue.

Naganawa, M. *et al.* (2008) Accurate determination of the intrinsic diffusivities of boron, phosphorus, and arsenic in silicon: the influence of SiO$_2$ films, *Jpn. J. Appl. Phys.*, **47**, 6205–6207.

Plummer, J.D., M.D. Deal and P.B. Griffin (2000) **Silicon VLSI Technology**, Prentice Hall.

Sankara Narayanan, E.M. *et al..* (2004) Progress in MOS-controlled bipolar devices and edge termination technologies, *Microelectron. J.*, **35**, 235–248.

Taniguchi, K. *et al.* (1980) Oxidation enhanced diffusion of boron and phosphorus in (100) silicon, *J. Electrochem. Soc.*, **127**, 2243.

Zimmermann, H. (1999) **Integrated Silicon Optoelectronics**, Springer, p. 36.

15

Ion Implantation

Ion implantation (I/I) is a process where accelerated ions hit the silicon wafer, penetrate into silicon, slow down by collisional, stochastic processes, and come to rest within femtoseconds. Implantation today is the main method of introducing dopants into silicon, and in CMOS fabrication, it has replaced thermal diffusion almost completely.

Typical ion implantation energies are 10–200 keV, and doses are 10^{11} to 10^{16} ions/cm^{-2}. Because implantation depths are of the order of hundreds of nanometers, the corresponding concentrations are ca. 10^{15} cm^{-3} to 10^{20} cm^{-3}. In 1 μm CMOS technology source/drain diffusions are made by 5×10^{15} cm^{-2} ion implant doses, and the depth is ca. 200 nm, which translates to ca. 25 Ohm/sq. For more advanced CMOS technologies S/D sheet resistances are rapidly increasing because S/D depths are scaled down.

Lateral confinement of implanted dopants is better than in diffusion: sideways spreading under the mask is considerably less, as a rule of thumb, it is one-third of vertical range. This is especially important in advanced CMOS where extremely small dimensions are fabricated. In thermal diffusion the depth and the lateral extension are identical in the first approximation.

Photoresist can mask ion implantation, an obvious advantage iover thermal diffusion which requires an oxide mask. However, implantation is always connected with a high temperature anneal step. Implantation damages the silicon crystal, and in order to recover defect-free single crystalline state, this damage has to be annealed away. Second, dopants must be activated, that is they have to be in substitutional lattice sites. Activation of dopants and damage removal both take place simultaneously, but the process cannot always be optimized for both, as will be discussed in Chapter 26.

15.1 The Implantation Process

Ion implantation is depicted in Figure 15.1a. Ions penetrate into silicon and into the mask, too. The mask has to be thick enough so that it will block ions. Photoresist, oxide, nitride and polysilicon are typically used as mask materials. The higher the implantation energy, the deeper the ions will penetrate, and the lighter the ion, the deeper it will go. The range of ions in matter (R_p) is statistically distributed and rather broad. This is described by straggle, ΔR, in Figure 15.1b.

Implanted ions scatter stochastically, traveling a distance R (range). However, we are more interested in projected range, R_p, the depth ions reach underneath the silicon surface (Figure 15.2). Straggle ΔR is the deviation in range, the width of the depth distribution. As a rule of thumb, the concentration at $R_p \pm 2\Delta R$ is 10% peak concentration, and at $R_p \pm 3\Delta R$ it is 1% of peak concentration. Also of interest is lateral straggle, R_L, or the deviation from the incident direction.

Ions are decelerated in the lattice by nuclear and electronic stopping, that is by collisions with atomic nuclei of atomic number Z and mass M, and by collisions with the electronic cloud, respectively. Under a number of simplifying assumptions (about the nature of material, interaction potentials, energy independence of various variables, etc.) the Linhard solution to nuclear stopping for projectile (M_1, Z_1) hitting a wafer of (M_2, Z_2) is

$$S_n = 2.8 \times 10^{-15} \frac{Z_1 Z_2}{Z} \frac{M_1}{(M_1 + M_2)} \text{ eV cm}^2 \quad (15.1)$$

where Z is the reduced atomic number, $Z = (Z_1^{2/3} + Z_2^{2/3})^{3/2}$. Nuclear energy loss is independent of ion energy in this approximation. Electronic stopping is

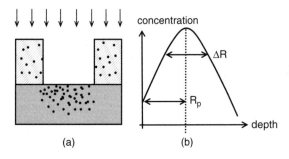

Figure 15.1 (a) Implantation: mask layer blocks selected areas; (b) dopant concentration profile inside silicon, with projected range R_p and straggle ΔR

Table 15.1 Energy loss of implanted ions in silicon

Nuclear energy loss in silicon (independent of energy) in keV/μm:

Boron	92
Phosphorus	447
Arsenic	1160

Electronic energy loss in silicon in keV/μm:

E (keV)	Boron	Phosphorus	Arsenic
10	65	88	90
50	145	196	200
100	205	277	283
200	290	391	401

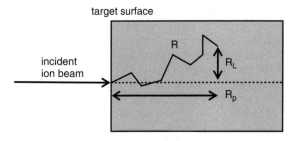

Figure 15.2 Key concepts for implanted ions: R: range is the length of ion travel; R_p is the projected range, and R_L lateral straggle

proportional to the square root of energy:

$$S_e = 3.3 \times 10^{-17}(Z_1 + Z_2)\sqrt{\frac{E}{M_1}} \text{ eV cm}^2 \quad (15.2)$$

Total energy loss is calculated as

$$\frac{dE}{dx} = -(S_n + S_e)N \quad (15.3)$$

where N is silicon atom density, 5×10^{22} cm^{-3}.

Table 15.1 lists the nuclear and electronic energy losses in silicon. Equation 15.3 can now be used to estimate ranges. As a rough guide, the projected range is 90% of the range.

The combined energy loss from nuclear and electronic stopping for 100 keV phosphorus is thus 724 keV/μm. Equation 15.3 then gives a range of about 140 nm and a projected range of about 126 nm. Boron ions are light and they penetrate deep into silicon, with 100–500 nm as typical implanted depth, while heavy arsenic ions penetrate only 10–50 nm when standard implant energies of 10–200 keV are used.

Figure 15.3 shows dopant concentration profiles: the high peak concentration is usually below the surface (while in diffusion it is always at the surface) and there is a long tail deep into the silicon. Note that not only does higher energy result in deeper projected range, but also the peak concentration will be lower. Thinking visually of Figure 15.3, the areas under the concentration vs. depth curves must be identical for identical doses: the same numbers of dopant atoms are just distributed differently.

Masking layers for ion implantation have to be substantially thicker than projected ranges, to ensure that the ions in the tail of distribution do not penetrate the mask. Mask thickness is especially important when working with high-resistivity substrates: it does not take many dopant atoms to change dopant type.

Dose control in ion implantation is superior to thermal diffusion. Another benefit of ion implantation compared to thermal diffusion is doping through oxide. When energy is high enough, a sizable proportion of ions penetrate through the oxide and dope silicon, as shown in Figure 15.4. Both 50 keV and 150 keV boron ions will penetrate through the oxide into silicon. However, 50 keV arsenic ions are completely stopped and only a very small dose of 150 keV arsenic ions will reach silicon.

15.1.1 Measurements for implantation

Implanted wafers can be measured by a four-point probe for sheet resistance. It is a natural control measurement for doping. It is, however, a fairly slow feedback loop because the wafer has to be cleaned and annealed before 4PP measurement. Sheet resistance measurement sees only electrically active dopants, and annealing is therefore not just an auxiliary step for measurement but an essential part of ion implantation doping. What is more, the wafer

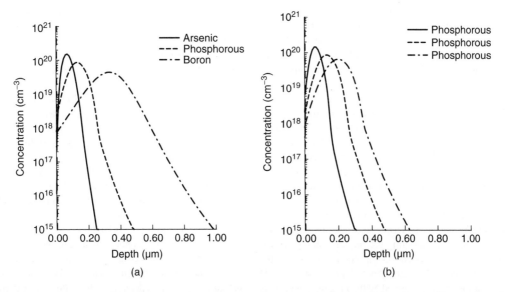

Figure 15.3 (a) The 50 keV implantation of arsenic, phosphorus and boron: the lighter ions will penetrate deeper. (b) Phosphorus implantation with 50, 100 and 150 keV energies

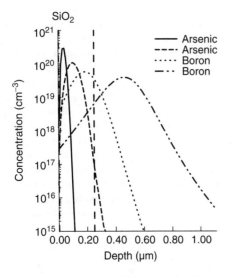

Figure 15.4 Implantation through oxide: 250 nm oxide will block 50 keV arsenic ions and most of 150 keV arsenic, too, but both 50 keV and 150 keV boron ions penetrate the oxide and dope silicon underneath

has to be discarded after 4PP measurement because it is a contact measurement and likely to cause contamination.

Alternatively, the dose can be monitored by modulated photoreflectance (also known as the thermal wave method). A modulated laser beam heats the wafer and the periodic heat waves are monitored by another, small-power laser. Dissipation lengths are correlated to implant damage and hence to dose. This is a fast, non-contact measurement which needs no wafer preparation and can be done even on photoresist patterned wafers.

If the dopant depth distribution needs to measured, the wafer destructive methods of spreading resistance profiling (SRP) and SIMS are available (Chapter 2). Both measurements are long and tedious and used mostly in the research phase of process development. Point defects created by implantation cannot be seen by physical analysis, but extended defects like dislocations can be seen by a TEM. Amorphization can be measured by a TEM or by XRD.

15.2 Implant Applications

Ion implantation emerged in the 1970s when accurate low-dose doping was required for CMOS threshold voltage control. Implantation excelled over thermal diffusion for this, and it opened up the way for implantation to start penetrating the doping business. Whenever shallow-doped structures are needed, ion implantation is used. This applies to piezoresistors on AFM cantilevers as well as advanced CMOS source/drain diffusions.

Actually, diffusion is never displaced by implantation because some diffusion takes place during implant activation anneal. And in many applications, like CMOS wells, implantation and thermal diffusion are used in tandem: the accurate dose control of implantation is used to perform pre-diffusion, which determines the number of dopants, and long thermal drive-in, which determines the final dopant profile. This is necessary because implantation depths are shallow (<1 μm) and wells used to be rather deep, in the 5 μm range.

Self-aligned polysilicon gate was one of the great innovations of CMOS fabrication in the 1970s. Previously, source and drain (S/D) were formed by thermal diffusion, using an oxide mask. After oxide mask removal and new thermal oxidation, aluminum gate was deposited and aligned on S/D, leading to inevitable alignment errors, as shown in the left column of Figure 15.5. In the self-aligned gate process (right column of Figure 15.5) the polysilicon gate is patterned and acts as a mask for S/D doping. Suitably selecting the implant energy so that the thin oxide is penetrated but the thick poly is not, the channel area under the gate is not doped, and the S/D areas are automatically aligned with the poly gate. This reduces overlap capacitances and resistances. Aluminum cannot be used in the self-aligned gate process because the annealing step after implantation requires temperatures about 1000 °C.

Doping polysilicon is one of the applications where implantation offers multiple choices. Most manufacturers prefer to deposit undoped LPCVD poly because of better process control. Traditionally, poly doping was by POCl$_3$ gas phase doping, resulting in heavily doped n-type polysilicon. This is an application where thermal diffusion excels: namely, high doping concentration (e.g., 10^{20} cm^{-3}) blanket deposition. In the self-aligned process the poly becomes phosphorus doped during NMOS S/D implantation, and no separate poly doping step is needed. Similarly, when PMOS S/D are implanted by boron, PMOS gate poly becomes p-type as well. Of course such a dual gate structure could be done by two lithography and two doping steps before poly etching, but the self-aligned process eliminates two lithography steps. In another approach, n$^+$ doped LPCVD poly is deposited, but lithography is done next, and PMOS areas are implanted by a heavy boron dose that is so high that dopant type inversion n$^+ \rightarrow$ p$^+$ occurs.

Photoresist masking works nicely for implants: it is easy to spin-coat thick enough resists to block ions. There are, however, a few issues to tackle. Accelerated ions break bonds due to their high energy. This can lead to resist carbonization, especially if high doses are used. The wafers also heat up during implantation because accelerated ions carry a lot of energy. This heating will further bake the resist and change its structure. Both of these processes make stripping of implanted resist difficult.

Channeling is a phenomenon where ions are channeled between silicon crystal planes, rather like light in optical fibers. This effect is more pronounced for light ions, and especially for <100> crystal orientation which has a very open structure (Figure 4.6). In order to avoid channeling, many implanters tilt the wafers by 7°, so that the ions hit the wafer obliquely, avoiding channeling. However, this leads to device asymmetry (Figure 15.6). Lateral diffusion during damage removal anneal is becoming very small in advanced devices, and this makes the problem more serious than before.

Another way to reduce channeling is to implant through thin oxide (e.g., 10 nm). Thin oxide randomizes incoming ions, reducing channeling. These oxides serve another

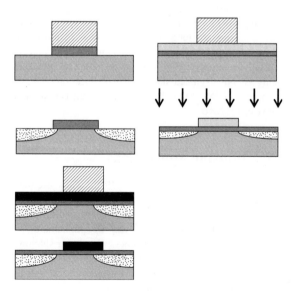

Figure 15.5 Non-self-aligned aluminum gate MOS process (left) with thermally diffused S/D vs. ion-implanted, self-aligned, polysilicon gate MOS process (right)

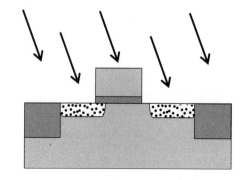

Figure 15.6 Doping asymmetry due to tilted implantation

function, too: implantation is a high-energy process, and accelerated ions can sputter metal atoms from implanter hardware. Thin oxide prevents these metal atoms from penetrating the silicon. In the post-implantation clean, this thin oxide and the metals on it can easily be removed by a HF dip.

15.3 Implant Damage and Damage Annealing

Nuclear stopping displaces atoms from the silicon lattice: a 100 keV arsenic ion displaces about 2000 silicon atoms along its trajectory. Damage creation depends on:

- implant species (heavy ions produce more damage)
- energy (more energy, more damage)
- dose (above about 10^{14} cm^{-2} extended damage set in)
- current (high current leads to overlapping collision cascades).

At low doses (below 10^{14} cm^{-2}) the predominant damage type is point defects such as vacancies and interstitials, or clusters of point defects. At high doses extended defects are created. Dislocation loops are created in the crystalline silicon just next to the amorphous/crystalline interface. These are known as end-of-range (EOR) defects. If the concentration of dopants is above the solid solubility limit, dopants precipitate.

Boron does not cause appreciable amorphization, irrespective of dose, because it is a light-mass ion. High-dose phosphorus and arsenic implants can amorphize silicon, but if amorphization is needed without doping, germanium, a group IV element isoelectronic with silicon, can be used. The critical dose for amorphization is about 10^{14} cm^{-2}. Note that channeling is therefore not an issue for high implant doses because the amorphized surface layer will block the crystal channels.

The annealing temperature must be so high that atoms move. But contrary to thermal diffusion, atoms do not need to move significant distances. The dopant atoms could in theory find a suitable lattice site within an atomic distance. This is not possible in practice, though, because the defects created by implantation enhance diffusion.

Activation refers to dopant atoms being electrically active after annealing. They then occupy lattice sites in the crystal and act as donors or acceptors. A high concentration of active dopants is needed for low resistance. Dopant atoms above the solid solubility limit form precipitates or are found as interstitial atoms, and they do not contribute to electrical conductivity. The interplay

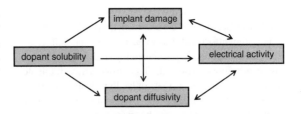

Figure 15.7 Implantation–diffusion interaction matrix. Redrawn from Jones (1993)

between damage and doping is shown graphically in Figure 15.7.

15.4 Tools for Ion Implantation

Ion implantation acceleration voltages used to range from 10 to 200 kV but low-energy implanters (1 kV minimum) and high-energy implanters (max. 3 MV) exist. Low-energy implants are needed to fabricate shallow S/D junctions (below 100 nm) in deep submicron CMOS. High-energy implanters (HEIs) implant deep into silicon, 1 micrometer or even deeper. The ability to fabricate retrograde profiles, that is to have low concentration at the surface and high concentration deep down, exactly the opposite to thermal diffusion, offers some interesting possibilities, for example as a replacement for buried layers and epitaxy.

Medium-current implanters (MCIs) are 1 µA to 5 mA, 20–200 keV, single wafer machines. High-current implanters (HCIs) are batch machines, with for example 13 wafers on a rotating wheel.

The minimum energy in a HCI is 80 keV. Extraction beam current scales as $V^{3/2}$ which explains why a low-voltage HCI is not practical. This scaling means difficulties for low-energy, high-dose implantations which are needed for advanced CMOS S/D implants.

Implant currents can be anything from 1 µA to 30 mA, and with doses from 10^{11} to 10^{16} cm^{-2} implantation times are seconds or minutes. Beam currents are limited if photoresist is used as a mask: currents that are too high will damage the resist, and removal of the resist becomes difficult. Cooled wafer stations can be used to minimize resist damage. Implanter throughput can be up to 500 wafers per hour for low-dose applications.

Scaling down ion energy involves a number of techniques. One of the oldest is to implant molecular ions instead of simple ions: BF_2^+ has a mass of 49 vs. 11 for B^+, and its range is about one-fifth the boron range to a first approximation. Carborane molecular ions ($C_2B_{10}H_{12}^+$) provide 10 times more boron ions per molecule, and energy can be high, yet range will be small. Replacement of boron ions with BF_2^+ or carborane is not straightforward, however, because the behavior of fluorine or carbon during annealing and further processing needs to be assessed. True low-energy implanters must accept the fact that only low beam current is available. In the limit of 1 kV, the sputtering of surface atoms becomes important: because low implant energy equals low penetration depth, every atom layer removed from the surface will affect the final implant profile.

15.4.1 Implanter design and operation

Implantation requires ions, and these are generated in ion sources which are plasma discharges. Dopants have be vaporized or be in a gaseous state before ionization. Dopant gases include PH_3, AsH_3, BF_3, but evaporation of solids in a furnace can also be used, and almost all elements in the periodic table can be implanted. However, the efficiency of solid sources is low and switching between ions slow. Ions are extracted from the source by voltage and enter the selection magnet (Figure 15.8).

Ion selection is based on a mass spectrometric separation according to radius of curvature r in a magnetic field B balanced by the centrifugal force:

$$|F| = |q(v \times B)| = m|v|^2/r = qV \qquad (15.4)$$

where m is mass and q is charge, which can be solved for $B = \sqrt{(2mV/qr^2)}$. By adjusting the magnetic field of the selection magnet, the ion of the desired mass is selected. The magnet selection can be fooled by similar ion masses, termed mass contamination. Doubly charged molybdenum ions Mo^{+2} can pass along with BF_2^+ ions (molybdenum is a common construction material for vacuum equipment). The $^{11}BHF^+$ ion behaves like $^{31}P^+$ for the selection magnet. This situation might emerge when PH_3 gas is used after BF_3 gas and some residual gas remains in the ion source. Energy purity refers to the spread of ion energies in the beam and consequently their range in silicon.

The acceleration tube must be kept under high vacuum in order to steer the beam toward the wafer in a collisionless fashion. After acceleration either electromagnetic or mechanical scanning spreads the beam over the wafer. Implantation is an inherently slow process because of the scanning nature of operation. Alternative implantation techniques which work in parallel mode have been devised: plasma immersion ion implantation (PIII) is a process where the wafer is immersed in plasma, and biased. Very high dose rates are possible, but energy purity is sacrificed because the selection magnet has been eliminated from the system. PIII may have uses in large-area applications like flat-panel displays because of its high throughput.

Wafers will be charged when ions are implanted. Current flows from the beam to the wafer holder, and it passes any oxides on its way. Also, beam non-uniformity between wafer centre and edge can cause lateral currents. Charging is compensated by flooding: electron gun-generated electrons hit the wafer and neutralize charges. This approach is prone to overcompensation and problems with electron charging. Plasma discharge, which produces an ion density an order of magnitude higher than the beam, is used in neutralization. Charge neutrality is inherent in the plasma system.

Implant dose is monitored during implantation by Faraday cup current measurement.

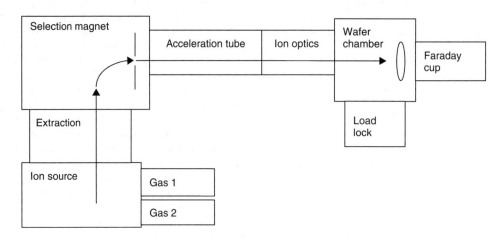

Figure 15.8 The main elements of an implanter. Adapted from Current (1996) by permission of AIP

15.4.2 Safety aspects

Ion implanters pose a number of safety issues that have to be tackled. The obvious one is the high voltage that is present inside the machines. The second issue is X-rays, which are produced as ions decelerate. Lead radiation protection is routinely used around the parts where X-rays are generated. If hydrogen is implanted, as in the Smart-cut process (Chapter 22), nuclear reactions are possible at fairly low energies of 150 keV and gamma rays are then generated.

Implant gases AsH$_3$, PH$_3$ and BF$_3$ are extremely toxic. Hydride detectors are placed inside the system to sniff for leaks. The operation and maintenance of an implanter are therefore for highly trained staff only. More safety issue discussion can be found in connection with cleanrooms in Chapter 35.

15.5 Ion Implantation Simulation

Implantation simulation must make a critical first choice on how to treat matter: amorphous matter is easy to model, but silicon really is single crystalline. Many simulators use single crystal silicon material parameters, but ignore the actual crystal structure.

Monte Carlo (MC) simulation offers many advantages over semianalytical implantation simulations because it can truly take silicon crystal structure into account. MC simulations not only can predict ranges and straggle, but enable physically based damage prediction, including amorphization. MC simulations are, of course, much more computationally intensive than semianalytic ones. SRIM (Simulation of Ranges of Ions in Matter) is one widely used MC simulator for implantation and other ion beam processes.

The input for a prototypical semianalytical implantation simulation includes:

- wafer orientation
- dopant concentration
- ion species
- energy
- dose.

The accuracy of simulation is very good in the surface and peak concentration regimes, but worse in the tail of the distribution (Figure 15.9). This is partly due to ion channeling, which is not readily implemented in semianalytical simulators. For heavier elements discrepancies can come from the amorphization treatment: single crystal may be used initially but, as the dose increases, the simulator adopts amorphous silicon material parameters for further

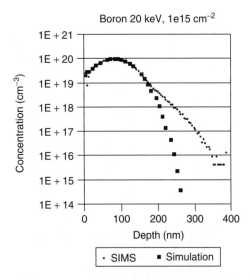

Figure 15.9 Boron implantation into silicon, 20 keV, 10^{15} cm^{-2}. SIMS measured data shown by small diamonds, ICECREM simulation by large squares. The discrepancy in the tail results partly from ion channeling and partly from model deficiencies. SIMS data courtesy Jari Likonen, VTT

calculations. This approach is depicted in Figure 15.10 for increasing phosphorus doses.

15.6 Implantation Further

Implantation is used everywhere where doping is needed, from resistor fabrication (Chapters 25 and 30) to CMOS (Chapter 26) to advanced bipolar doping (Chapter 27). Implantation enables heavily doped shallow junctions for MOS S/D. Implantation is also a key technology in SOI wafer fabrication, with both oxygen and hydrogen implantation serving special roles, as will be described in Chapter 22. The thermal processes for annealing implant damage will be presented in Chapter 32.

15.7 Exercises

1. What will be the implant time for a wafer of 200 mm diameter when arsenic ions are implanted with a dose of 10^{15} cm^{-2} and implant current of 100 µA?
2. How much will wafer temperature rise during the implantation of arsenic ions of energy 100 keV and dose 10^{15} cm^{-2} with a current of 1 mA on a 200 m wafer? Make simplifying assumptions as needed.
3. Under the conditions of 10^{15} cm^{-2} phosphorus implant dose, 200 mm wafer size, PH$_3$ bottle volume 3 liters

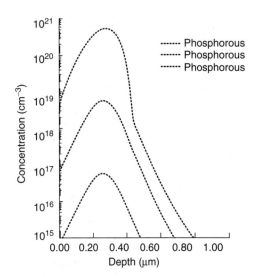

Figure 15.10 Different phosphorus doses compared: 10^{12} cm^{-2}, 10^{14} cm^{-2} and 10^{16} cm^{-2} at 200 keV. The shape is different for 10^{16} cm^{-2} because it is above the amorphization limit, and different stopping parameters are applied for the amorphized region

(STP), how many wafers can be implanted? If the ion current is 1 mA, what is the interval for bottle changing?

4. What is the range of 20 keV ^{11}B$^+$, ^{49}BF$_2^+$ and carborane C$_2$B$_{10}$H$_{12}^+$ ions?
5. What needs to be considered if boron implantation is replaced by BF$_2^+$ implantation?
6. What is the range of 100 keV germanium implantation?
7. How deep will boron, phosphorus and arsenic of 100 keV penetrate into a photoresist mask?
8. Translate implant dose 10^{16} cm^{-2} into film thickness assuming the thin film is made of implanted atoms only.

Simulator exercises:

9. How thick an oxide layer is needed to mask boron implantation? Present your results as a function of boron energy.
10. Check by simulator the range of 100 keV phosphorus ions and compare it to the simple estimate discussed in the text.

References and Related Reading

Brodie, I. and J.J. Muray (1982) **The Physics of Microfabrication**, Springer.
Chanson, E. *et al.* (1997) Ion beams in silicon processing and characterization, *J. Appl. Phys.*, **81**, 6513–6561.
Current, M. (1996) Ion implantation for silicon device manufacturing: a vacuum perspective, *J. Vac. Sci. Technol.*, **A14**, 1115.
Fair, R.B. (1996) Conventional and rapid thermal processes, in C.Y. Cheng and S.M. Sze, **ULSI Technology**, McGraw-Hill.
Jones, K.S. (1993) Extended defects from ion implantation and annealing, in R.B. Fair, **Rapid Thermal Processing: Science and Technology**, Academic Press.
Kawaguchi, M.N., J.S. Papanu and E.G. Pavel (2006) Low temperature, ion-enhanced, implanted photoresist removal, *J. Vac. Sci. Technol.*, **B24**, 651–656.
LeCoeur, F. *et al.* (2000) Ion implantation by plasma immersion: interest, limitations and perspectives, *Surf. Coating Technol.*, **125**, 71.
Mok, K.R.C. *et al.* (2005) Ion-beam amorphization of semiconductors: a physical model based on the amorphous pocket population, *J. Appl. Phys.*, **98**, 046104.
Pelaz, L. *et al.* (2010) Simulation of pn-junctions: present and future challenges for technologies beyond 32 nm, *J. Vac. Sci. Technol.*, **B28**, p. C1A1
Pelletier, J. and A. Anders (2005) Plasma-based ion implantation and deposition: a review of physics, technology, and applications, *IEEE Trans. Plasma Sci.*, **33**, 1944–1959.
Rubin, L. and J. Poate (June/July 2003) Ion implantation in silicon technology, *The Industrial Physicist*, p. 12
Suzuki, K. (2009) Extended Lindhard–Scharf–Schiott theory for ion implantation profiles expressed with Pearson function, *Jpn. J. Appl. Phys.*, **48**, 046510.
White, N.R. (1996) Moore's law: implications for ion implant equipment – an equipment designer's perspective, Proceedings of the 11th International Conference on Ion Implantation Technology, Austin, TX, p. 355.
Zechner, C. and V. Moroz (2008) Simulation of doping profile formation: historical evolution, and present strengths and weaknesses, *J. Vac. Sci. Technol.*, **B26**, 273–280.

16

CMP: Chemical–Mechanical Polishing

Polishing is a key technology in silicon wafer manufacturing where final polishing yields wafers with RMS roughness as small as 0.1 nm, but it emerged elsewhere in microfabrication only in the late 1980's. It has several different uses:

- polishing of deposited films
- planarization of topography (step height reduction)
- removal of hard-to-etch materials

Polishing in microfabrication is a decendant of glass polishing which is an established technology since 400 years. Abrasive particles are dispersed in a suitable liquid to create a slurry, which is fed in between a polishing pad and the piece to be polished. In the case of a blanket wafer surface irregularities are smoothed out.

In the case of a wafer with topography, the elevated structures are preferentially removed since the local pressure is highest there. This leads to planarization. Both mechanical force and chemical etching are needed for high rate polishing, and the technology has been named CMP: chemical–mechanical polishing.

Grinding may look similar to CMP, but the two are quite different. In grinding abrasive particles of 1–100 μm size are mounted in resin, and micrometer-sized chunks of material are removed by crack propagation and brittle fracture. Grinding is fast, but also very coarse: the substrate is damaged due to mechanical forces acting on the microstructures. This subsurface damage extends even 20 μm deep. Grinding is used when hundreds of micrometers need to be removed, as in wafer thinning. CMP is used to remove micrometers only. In CMP abrasive particles of 10–300 nm are dispersed in a slurry. The mechanism is different from grinding, as CMP works in the atomic regime. Atomic bonds are weakened or broken, and removal is based on chemical reactions between the slurry and the surface and the mechanical effect of the abrasive particles. Surface roughness after CMP is of the order of nanometers, orders of magnitude better than with grinding.

The first application of CMP in microfabrication was in multilevel metallization: oxide and tungsten were polished in order to planarize wafer topography, easing the next lithography step. As shown in Figure 16.1, two sequences result in identical final structures.

Copper cannot be plasma etched, but copper polishing is reasonably easy. Oxide is plasma etched, and copper is deposited into the grooves, with extra copper removed by CMP. Copper minimum linewidth is determined by oxide plasma etching, and no copper etching is needed. The difference between copper and tungsten is that tungsten can be plasma etched (Figure 11.10).

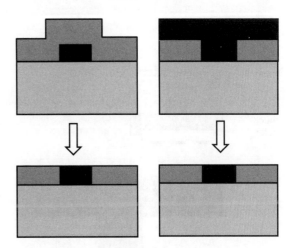

Figure 16.1 Metal deposition, metal etching, oxide deposition, oxide polishing (left); oxide deposition, oxide etching, metal deposition, metal polishing (right)

Introduction to Microfabrication, Second Edition Sami Franssila
© 2010 John Wiley & Sons, Ltd

16.1 CMP Process and Tool

The CMP tool consists of a solid, extremely flat platen, on which the polishing pad is glued (Figure 16.2). The wafer chuck, which holds the wafer upside down, is situated on a spindle. A slurry introduction mechanism feeds the slurry onto the pad. Both platen and spindle are rotated, and the linear velocity (used in Preston's equation) is the sum of two velocities.

There are four major elements in a CMP process:

- topography
- materials
- polishing pad
- slurry.

The machine parameters that can be used to control them are listed in Table 16.1.

Local polishing pressure is down force divided by contact area. For a flat wafer the pressure is low because load is evenly distributed over the whole geometrical area, but on a structured wafer the effective contact area is only a fraction of wafer area, and local pressure is much higher (Figure 16.3). The polishing rate is thus not constant: when the contact area is small, local pressure is high and the polishing rate is high. As polishing continues, the steps are reduced and contact area increases, leading to a decrease in polishing rate.

Table 16.1 CMP tool parameters and process responses

Platen rotation	10–100 rpm
Velocity	10–100 cm/s
Down force (applied pressure)	10–50 kPa
Slurry supply rate	50–500 ml/min
Polish rate	100–1000 nm/min
Selectivity	1:1 to 100:1
Uniformity across wafer	10%
Wafer-to-wafer repeatability	10%

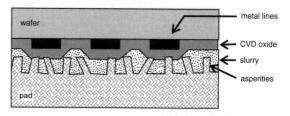

Figure 16.3 Close-up of CMP set-up: the wafer, upside down, is pressed against the pad with slurry in between. Pad asperities make contact with the wafer

Structure height obviously affects CMP, but pattern density is also important because it determines the effective contact area: denser patterns are polished at a lower rate due to lower pressure. Polishing a single material is easier than polishing stacks of materials, or structures with different materials present simultaneously. The mechanical properties of the wafer itself must also be considered: if it is bowed, pressure will be different at the center and the edges, leading to non-uniform polishing. Pressure can be applied through the chuck to the wafer back side: this will equalize center–edge differences and compensate for wafer bow.

The pad should be rigid so that it uniformly polishes the wafer. However, such a rigid pad would have to be aligned and kept in alignment with the wafer surface at all times. Therefore pads are often stacks of hard and soft materials which conform to the wafer topography to some extent. Pads are porous polymeric materials (with 30–50 μm pore size) which are consumed in the process and must be reconditioned regularly. Polyurethane is commonly used for pads. Pads are very much proprietary, and are usually referred to by their trade names, rather than by any chemical or physical property.

Slurries incorporate both mechanical elements via abrasive particle size and hardness, and chemical effects via reactivity and pH of the fluid. Typical abrasive particles are silica (SiO_2), alumina (Al_2O_3) and ceria

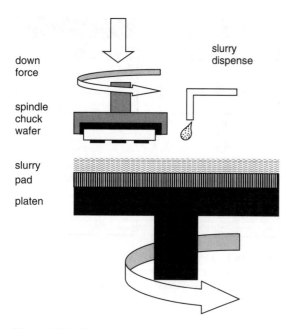

Figure 16.2 Schematic structure of rotary CMP equipment. Wafer is held face down in the spindle chuck

(cerium oxide, CeO_2). The abrasive particle size distribution is related to smoothness: monodispersed slurry leads to smoother surfaces.

The fluids in CMP are alkaline and acidic solutions, and the abrasive particle concentration can be for example a few percent. Copper is polished by 2% NH_4OH with 2.5 wt % alumina particles.

Slurries are a cause of concern for post-CMP: particles must be cleaned away after polishing. Like pads, slurries are often proprietary, and the information given is often restricted to pH value, base liquid (for instance, NH_4OH based) and abrasive particle size. Slurries can be buffered against consumption in the process (cf. etching in buffered HF). At the end of CMP a soft polishing step is often done: no slurry is used, just water. This step does not remove solid material but it is effective in washing away abrasive particles and corrosive chemicals.

Pad type, compressibility, hardness and elastic modulus, conditioning, pore size and ageing can be considered variables, too. Because there is a chemical component in CMP, temperature will have an effect on polishing results. CMP process factors resemble those encountered in etching: overpolishing (cf. overetching), selectivity and pattern density effects (see Chapter 21 for etch-related pattern density effects).

Plasma etching and CMP resemble each other also in the sense that both depend on the interaction between chemical and physical processes: in etching ion bombardment removes reaction products from the surface; in CMP mechanical abrasion removes surface layers which have been modified chemically, for instance by oxidative slurries.

Polish rate can be limited by the transport of reactants, or by surface processes, just like in etching or CVD. This can be found out by varying the input variables: if the rate is unaffected by a change in a variable, it cannot be the rate controlling factor. Another similarity to etching is pattern dependency: small pattern density leads to higher rates. Pattern size effect is, however, the opposite: in CMP small patterns are polished faster, but in etching small patterns will be etched slower than large ones.

16.2 Mechanics of CMP

There are three modes in polishing, depending on the degree of contact between the pad and the wafer, Figure 16.4. In direct contact (boundary lubrication) mode the pad makes contact with the wafer, resulting in high and constant friction because there is no lubrication from the slurry. Polish rate is very high. In rolling contact mode (mixed lubrication mode) slurry particles occasionally roll on the wafer surface. In non-contact

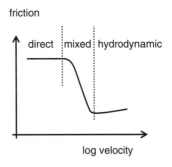

Figure 16.4 Stribeck diagram of CMP: three different lubrication modes

mode (hydrodynamic lubrication mode) slurry particles are accelerated hydrodynamically and impart energy to the wafer surface, weakening the surface so that chemical attack can occur. Hydrodynamic lubrication takes place at high velocities where the load is borne by the fluid, and the system is well lubricated. The friction force between the pad and the wafer is very different in these modes and it is classified in a Stribeck diagram.

The penetration depth of the abrasive particles into the substrate is very small indeed: this is the reason for smooth surfaces with no visible grooves or scratches. The penetration depth is given by

$$R_s = \frac{3}{4} d \times \sqrt[3]{\left(\frac{P}{2kE}\right)^2} \qquad (16.1)$$

where d is the abrasive particle diameter (for example, 100 nm), k is the filling factor of abrasive particles (for instance, 50%), P is local pressure (not down force, which is 10–50 kPa) and E is Young's modulus of the surface being polished. Penetration depths are on the order of a nanometer, which is similar to surface roughness after polishing, as would be expected. Increasing pressure will lead to deeper penetration but also to higher removal rate. Sometimes the abrasive particles agglomerate into huge chunks, which leads to much larger penetration depths, and microscratches tens of nanometers deep will result.

16.2.1 Preston model

Polish rates were measured experimentally by Preston (in 1927) and found to obey

$$R = \frac{\Delta H}{\Delta t} = K_p \times P \times \frac{\Delta s}{\Delta t} \qquad (16.2)$$

where ΔH is the change in the height of the surface, P the pad pressure, K_p the Preston coefficient and $(\Delta s/\Delta t)$ the linear velocity of the pad relative to the wafer.

Figure 16.5 Copper polish rate as a function of velocity (15 kPa pressure). Reproduced from Steigerwald *et al.* (1997) by permission of John Wiley & Sons, Ltd

Experimental results (Figure 16.5) show a fairly good fit to Preston's equation, especially in the low-pressure/low-velocity regime, that is in the direct contact mode.

The Preston coefficient is related to the elastic properties of the material, and it can be approximated by

$$K_p = \frac{1}{2E} \quad (16.3)$$

where E is Young's modulus. With Young's moduli in the range of 100 GPa for many inorganic and metallic solids, values of K_p are on the order of 10^{-11} Pa^{-1}. Applied pressures are on the order of 10 kPa and velocities on the order of 0.10 m/s, which lead to polish rates on the order of 10 nm/s or 600 nm/min, which is the correct order of magnitude. This estimate is, however, not accurate enough to be of predictive use. But it does explain many basic features of polishing, for instance the fact that tungsten is polished at a lower rate than oxide.

16.3 Chemistry of CMP

In CMP there are two components: in addition to the mechanical pressure, chemical modifications and etching take place. For instance, a tungsten surface is turned into tungsten oxide according to the reaction

$$W + 6\ Fe(CN)_6^{3-} + 3\ H_2O$$
$$\rightarrow WO_3 + 6\ Fe(CN)_6^{4-} + 6\ H^+ \quad (16.4)$$

Tungsten oxide has two important roles: it is a protective layer, and in the valleys it protects tungsten from further chemical attack. But it is mechanically weaker and more brittle material than tungsten, and in the high points it can be removed by mechanical abrasion. The same mechanism is at work in copper polishing: Cu_2O is removed by mechanical action while copper is not. For hard materials like tungsten and tantalum the mechanical effects are usually important, whereas for soft materials like aluminum and polymers the chemical effects often dominate.

When WO_3 is removed by polishing, the underlying metal is etched according to

$$W + 6\ Fe(CN)_6^{3-} + 4\ H_2O$$
$$\rightarrow WO_4^{2-}\ (aq) + 6\ Fe(CN)_6^{4-} + 8\ H^+ \quad (16.5)$$

Corresponding reactions in copper polishing are

$$Cu \Leftrightarrow Cu^{2+} + 2\ e^- \quad (16.6)$$
$$2\ Cu^{2+} + H_2O + 2\ e^- \Leftrightarrow Cu_2O + 2\ H^+ \quad (16.7)$$

Copper polishing is carried out with slurries like $Fe(NO_3)_3$ and H_2O_2. Hydrogen peroxide oxidizes copper, which enhances the removal rate. Typical rates are 100–1000 nm/min, selectivity to oxide ranges from 40:1 to 200:1 and residual step height 100–300 nm. Copper polishing uniformities can be 10–15%, which is among the worst uniformities of any microfabrication process.

Aluminum polishing can be done in acidic solutions, for instance phosphoric acid (pH about 3–4) with alumina abrasive. Aluminum CMP proceeds by aluminum oxidation and mechanical removal of the oxide, not unlike copper and tungsten polishing. Selectivity to oxide can be 100:1.

Oxide polishing slurries are ammonia or KOH based, for instance 1–2% NH_4OH in DI water, with up to 30% silica abrasives of 50–100 nm. Oxide polishing slurries are mildly alkaline, with pH values of about 11. The oxide polishing mechanism depends on surface modification of the oxide: leaching of oxide by the slurry softens the top layer, and the mechanical abrasion rate goes up.

CMP slurries do etch without mechanical polishing, just like fluorine will etch silicon without plasma, but in both cases it is the interaction between chemical and mechanical processes that leads to the desired total process: slurry etch rates of 10 nm/min are typical, but CMP removal rates of 500 nm/min are standard.

16.4 Non-Idealities in CMP

CMP is an interplay between many process factors. Pressure, velocity, slurry composition, etc., can be varied for optimization, but device design cannot usually be changed

CMP: Chemical–Mechanical Polishing

(even though sometimes dummy patterns are made in order to eliminate pattern density effects). Polish stop layers add process complexity too, but improved process control can balance the cost. Polish selectivities are not dissimilar from etch selectivities: they range from 1:1 to 200:1. Copper to tantalum selectivities are so high that measurements are difficult. Oxide to nitride selectivities can be 50:1, and this is useful in shallow trench isolation, which will be discussed in Chapter 26.

Because of finite selectivity some underlying layer loss is unavoidable. This is termed erosion and it is pictured in Figure 16.6. Another non-ideality is the dishing. It is caused by two factors: the pad conforms to some extent to the structures on the wafer, and softer material is polished faster than the surrounding hard material. Recess etching is a chemical effect. Recess in CMP can be as low as a few tens of nanometers, and in this respect CMP is superior to etchback.

Copper dishing is strongly feature size dependent, but rather insensitive to pattern density. Oxide erosion, on the other hand, is strongly pattern density dependent, but feature size independent.

On the practical side, slurry cost is a major problem. Slurries are consumables with very low utilization: in some processes it is estimated that only 2% of slurry actually participates in the process, the rest being swept away by platen rotation. Various solutions to this problem are being investigated: structured pads with grooves and channels of various shapes retain the slurry better, and also result in more uniform slurry distribution, leading to better uniformity. Another solution is to use a fixed abrasive: the abrasive particles are attached to the pad, and the slurry is replaced by particle-free chemicals.

Temperature is not constant during CMP: friction leads easily to a 10 °C temperature rise which is detrimental to reproducibility and uniformity. Rates of chemical reactions go up as expected, and this temperature rise can easily double the removal rate. Pad hardness decreases as temperature goes up, which leads to more asperities in contact with the wafer and reduced local contact pressure. This effect, is, however, not significant compared to chemical rate increase.

16.5 Monitoring CMP Processes

Top view microscopy, either optical or with a SEM, can be used for gross checking of CMP. Stains from slurry residues, scratches, layer peeling and other coarse problems can be identified. Scanning probe methods, mechanical stylus and the AFM are widely used to study nanoscale phenomena. Submicron resolution is needed because many CMP effects are strongly feature size dependent. Many optical, electrochemical, mechanical, thermal and acoustic methods are being developed to monitor CMP in real time.

16.5.1 Post-CMP cleaning

The introduction of CMP was obviously resisted by many people because the very idea of bringing myriads of particles intentionally onto the wafer was against all accepted cleanroom and manufacturing policies. Post-CMP cleaning was, and remains, a topic of paramount importance. Brush cleaning and other physical cleaning techniques are good for removing large particles, but the smaller particles pose problems. RCA-1 cleaning is efficient in particle removal, but it is of limited use on metallized wafers. In addition to the particle problem, there is metal contamination: potassium hydroxide is a common slurry liquid, and copper residues may be embedded in PSG, which is a soft material. HF etching can remove a thin top layer of PSG and reduce the amount of copper. In order to minimize the spread of particle and chemical contamination, the CMP section is usually separated from the rest of the cleanroom, and DI water is drained immediately after use, even though used DI water is normally recycled.

16.6 Applications of CMP

From a single application in multilevel metallization, CMP has expanded into a major technology which is used not only in IC fabrication but also in optics, MEMS, superconducting devices, micromagnetics and other areas. Polishing processes have been developed for tens of new materials, including but not limited to oxides of all sorts, noble metals, III–V semiconductors, Ge, SiGe and SiC, and polymers.

Conformal deposition processes replicate underlying topography dutifully and step height is unchanged after

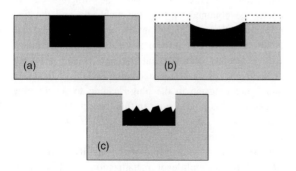

Figure 16.6 (a) Ideal CMP result; (b) erosion and dishing; (c) plug recess (chemical attack)

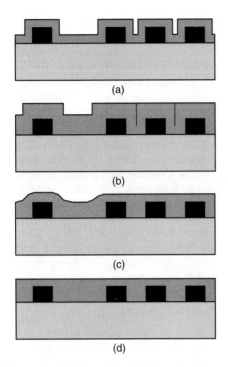

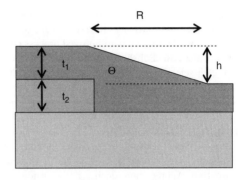

Figure 16.8 Planarization relaxation distance R

Figure 16.7 Planarization: (a) thin conformal deposition, no planarization; (b) thick conformal deposition, gap filling but no planarization, (c) local planarization by spin-on film; (d) global planarization by CMP of thick conformal deposition

conformal deposition, as shown in Figure 16.7a. Thick conformal deposition completely fills small gaps between lines but there is no step height change in large spaces, Figure 16.7b. Spin-on dielectrics flow over topography, resulting in smoothing but incomplete planarization, Figure 16.7c. CMP of thick conformal film is the closest we can get to global planarity, Figure 16.7d.

Planarization length (Figure 16.8) is defined by

$$R = \frac{h}{\tan \Theta} \quad (16.8)$$

Planarization lengths are in the range of micrometers or tens of micrometers in the maximum.

Polishing rate and planarization rate are two different concepts. Polishing rate is applicable to one material. Planarization rate is the rate of decrease in step height: the high peaks are polished, which decreases step height, but some material is removed from the valleys too, which decreases planarization rate. Toward the end of the process planarization rate drops to zero, even though the overall polishing rate is still finite.

Selectivity in CMP bears a close resemblance to etching: we need to know the polishing rates of the top and bottom films in order to calculate, for instance, substrate loss during overpolishing. As in etching, it is sometimes beneficial to have the same 1:1 selectivity between films, but most often it is desirable to remove one film relatively rapidly, and to have high selectivity against the bottom film, which can then be processed in a separate step.

Oxide polishing is the oldest and most widely practiced CMP process. Its main application is planarization in multilevel metallization in advanced ICs, where it provides a planar surface which makes subsequent lithography and deposition steps easy. One problem with oxide polishing is the lack of an end point: there is no clear end for polishing. This is called blind polishing. The opposite is stopped polishing, where for instance a nitride layer acts as a polish stop (cf. etch stop layer) but selectivities are not necessarily very high.

Tungsten polishing is another CMP process that was adopted rapidly. Contact holes and via holes are filled by CVD tungsten, which is then removed from planar areas, leaving just the contact plug filled with metal (Figure 16.1). The same structure can of course be obtained by tungsten etchback, and the first implementations of the tungsten plug process did use etchback. CMP has proven to be better than etchback with respect to plug loss: at the etching end point the etchable area decreases dramatically, and the etchant will attack the tungsten in the plug, leading to severe plug recess. CMP is much better in this respect, but naturally process optimization with either technology can bring about improvements.

CMP is used whenever global planarity is required. In addition to multilevel metallization for ICs, other applications have sprung up. In superconducting quantum interference devices (SQUIDs) CMP planarization of

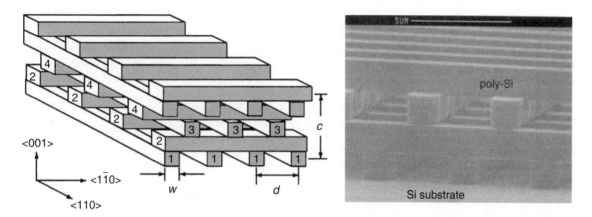

Figure 16.9 The infrared wavelength selective photonic lattice has been made with the help of CMP: oxide deposition, oxide trench etching, polysilicon LPCVD and polysilicon CMP have been repeated five times to create the lattice. As the last step all oxide has been etched away in HF. Reproduced from Lin *et al*. (1998) by permission of Nature

PECVD oxide is performed before metallization to eliminate step coverage problems and conductor cross-section variation to ensure high and constant current density, up to 10^7 A/m^2, an order of magnitude higher than in IC metallization.

Photonic crystals (photonic band gap materials) are artificial lattices where electromagnetic wave propagation is selectively restricted due to forbidden energy levels. There are many ways to fabricated photonic lattices and CMP is just one approach. Grooves are etched in the oxide and filled by CVD polysilicon. Poly is polished and another CVD oxide is deposited. Lithography and oxide etching are followed by another cycle of poly CVD and CMP and the process is continued until the desired number of layers has been made. Oxide is finally etched away to create the air gaps. The resulting structure is shown schematically and in the SEM micrograph in Figure 16.9.

Wafer bonding (Chapter 17) depends on surface smoothness (and flatness). Bonding two silicon wafers with 0.1 nm RMS roughness is easy, but bonding two CVD oxide layers with 3 nm RMS roughness is impossible. CMP can bring smoothness of deposited films to a level comparable to silicon wafers (Figure 16.10), enabling bonding of many new materials. Polishing is also used in layer transfer applications (Chapter 22):

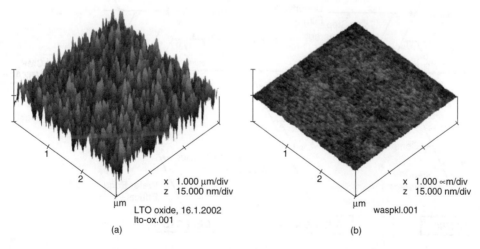

Figure 16.10 AFM scans of PECVD oxide: (a) as-deposited film peak-to-valley height is 26 nm, with RMS roughness of 3.3 nm; (b) after CMP peak-to-valley height is 2 nm and RMS roughness is 0.2 nm. Courtesy Kimmo Henttinen, VTT

thin layers of semiconductor are cut from the substrate and bonded to another wafer. The cut surface is fairly smooth but polishing is needed to put it on par with silicon wafers.

Polycrystalline films are generally rough, and thicker films rougher than thinner. One application of CMP is copper polishing in GMR magnetic head fabrication: thin films a nanometer thick need to be deposited on an electroplated copper layer a micrometer thick. The roughness of plated copper can be 100 nm, and this needs to be brought to nanometer level. CMP of copper can achieve this.

16.7 CMP as a Whole

CMP is a relatively young microfabrication technique, and one with very complex phenomenology. As shown in Figure 16.11, there are issues related to chip design (density of patterns, pitch, die size), basic mechanical issues like material hardness as well as practical mechanical issues like platen rotation. Materials issues are also very much involved in the choice and processing of dielectric thin films, like film composition and stress, and gap fill, which is also a design issue.

16.8 Exercises

1. What is the Preston coefficient for copper on theoretical grounds? What is the experimental value of the Preston coefficient? Use data from Figure 16.5.
2. How do the polish rates of tungsten, silicon dioxide and polymers compare?
3. How do polish rate and planarization rate measurements differ from each other?
4. If a titanium layer 20 nm thick is used as a polish stop underneath tungsten 500 nm thick, and film thickness non-uniformities are ±5% and CMP non-uniformity is ±10%, what must polish selectivity be?
5. Work out a step-by-step fabrication process for the photonic crystal shown in Figure 16.9. Include film thicknesses, too.

References and Related Reading

Hernandez, J. *et al.* (1999) Chemical mechanical polishing of Al and SiO_2 thin films: the role of consumables, *J. Electrochem. Soc.*, **146**, 4647.

Jindal, A. *et al.* (2003) Chemical mechanical polishing of dielectric films using mixed abrasive slurries, *J. Electrochem. Soc.*, **150**, G314.

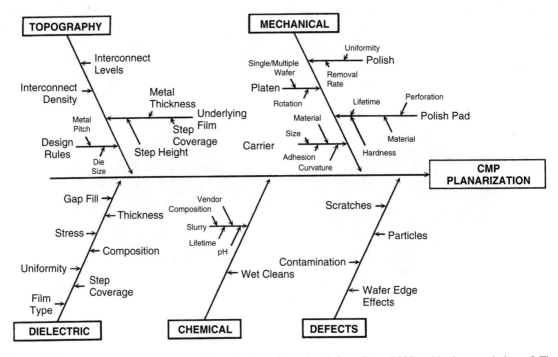

Figure 16.11 Fishbone diagram of CMP planarization. Reproduced from Rao (1993) with the permission of The McGraw-Hill Group of Companies

Lin, S. Y., *et al.* (1998) A three-dimensional photonic crystal operating at infrared wavelengths, *Nature*, **394**, 251.

Rao, G. P. (1993) **Multilevel Interconnect Technology**, McGraw-Hill.

Steigerwald, J. M., S. P. Murarka and R. J. Gutman (1997) **Chemical Mechanical Planarization of Microelectronic Materials**, John Wiley & Sons, Inc.

Stine, B. E. *et al.* (1998) Rapid characterization and modeling of pattern-dependent variation in chemical-mechanical polishing, *IEEE Trans. Semicond. Manuf.*, **11**, 129.

Suni, I. I. and B. Du (2005) Cu planarization for ULSI processing by electrochemical methods: a review, *IEEE Trans. Semicond. Manuf.*, **18**, 341–349.

Tas, D. K. *et al.* (2005) Online end point detection in CMP using SPRT of wavelet decomposed sensor data, *IEEE Trans. Semicond. Manuf.*, **18**, 440–447.

Wrschka, P. *et al.* (2000) Chemical mechanical planarization of copper damascene structures, *J. Electrochem. Soc.*, **147**, 706.

Yasseen, A. A. *et al.* (1997) Chemical-mechanical polishing for polysilicon surface micromachining, *J. Electrochem. Soc.*, **144**, 236.

Zantye, P. B., A. Kumar and A. K. Sikder (2004) Chemical mechanical planarization for microelectronics applications, *Mater. Sci. Eng.*, **R45**, 89–220.

Zhang, F. *et al.* (1999) Particle adhesion and removal in chemical mechanical polishing and post-CMP cleaning, *J. Electrochem. Soc.*, **146**, 2665.

Zhong, Z. W., Z. F. Wang and B. M. P. Zirajutheen (2005) Chemical mechanical polishing of polycarbonate and poly methyl methacrylate substrates, *Microelectron. Eng.*, **81**, 117–124.

17

Bonding

Bonding mates two wafers. This can be done at wafer manufacturer, to create more versatile starting wafers. It can be done as part of wafer processing, just like any other process step, and there is no need to limit it to two wafers: bonding can be continued with more and more wafers added. It can also be done at the end of the process to encapsulate the finished structures. Bonding is also essential in layer transfer: thin slices of material can be detached from substrate wafers and bonded to other wafers.

Bonding together two wafers with different properties (crystal orientations, doping levels, doping types), or wafers of two different materials, opens up new possibilities for the device engineer. Bonding dissimilar wafers is, however, difficult because lattice mismatches and thermal expansion differences create stresses and surface chemistries that are not always suitable for bond formation.

Bonding of structured wafers is used to create channels, cavities and gaps, for example for microfluidic channels, capacitive pressure sensors and RF switches. The gaps formed by bonding can be 500 μm, wafer thickness, as in Figure 17.1, or as small as submicron, as in Figure 17.2, and anything in between. Gases trapped inside closed cavities need careful consideration because bonding temperatures can be quite high, up to 1200 °C.

There are other uses of the bonding technologies in microtechnology: wire bonding is about attaching leads to finished chips, and flip-chip bonding is about attaching fully processed chips to suitable substrates and carriers. Some of the techniques are common with wafer bonding, like thermocompression bonding, but in this chapter we will concentrate on wafer-level bonding applications only.

17.1 Bonding Basics

Two wafers can be joined by a number of methods, but two main classes can be distinguished:

- direct bonding (also known as fusion bonding or thermal bonding)
- bonding with intermediate layers (glasses, metals, adhesives).

At least theoretically, a wafer of any material can be bonded at room temperature to another wafer of any material via van der Waals intermolecular forces. This bonding requires that the bonding surfaces are sufficiently smooth, flat, clean and the surface chemistry is suitable for bond formation. A strong bond can then develop across the bonding interface. Two flat and smooth wafers with 0.3 nm RMS roughness do not actually have a very large contact area with each other: 0.3 nm RMS roughness translates to perhaps 3 nm peak-to-peak heights, and the weak van der Waals forces only work at distances of 0.5 nm.

In order to bring the wafers into intimate contact, several strategies are available: one of the wafers can conform to the other, and this can be done by heating. It is easy for polymers, but requires about 500–600 °C for glasses, while silicon wafers require about 1000 °C before viscous

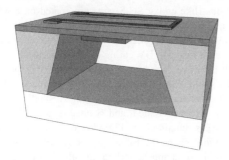

Figure 17.1 Wet etched silicon microreactor bonded to a glass wafer. Heater resistor on nitride membrane, catalyst metal inside channel

Introduction to Microfabrication, Second Edition   Sami Franssila
© 2010 John Wiley & Sons, Ltd

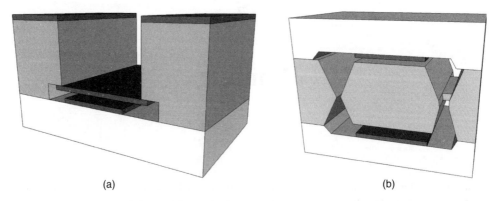

Figure 17.2 Narrow gap bonding: left, tunable aluminum plate capacitor by silicon–glass bonding, adapted from Etxeberria and Gracia (2007); right, capacitive accelerometer, bonding of a silicon wafer between two glass wafers

flow takes place. A second alternative is wetting: suitable treatment will allow one material to cover the surface, for example solder flow upon heating or solvent treatment of a polymer wafer. A third method is to apply external force to ensure intimate contact.

When dissimilar materials with different thermal expansion coefficients are bonded, high stresses can emerge. This prevents many bonding processes. In anodic bonding silicon and glass CTEs have to be similar, and this can be done for some glasses but not for all of them (more on glass properties in Chapter 19). Quartz-to-silicon bonding remains much more difficult because there is no possibility to tailor the quartz CTE.

Direct bonding usually involves two identical wafers: silicon to silicon, glass to glass, PMMA to PMMA, etc. It results in identical chemical bonds at the interface as inside the material itself, and if breakage occurs, it takes place inside the wafers, not at the bond interface. The bond strength is then the same as the tensile strength of the material. The bonded wafers can be processed further as if they were one wafer.

If the temperatures involved are low enough, CTE-induced problems are minimal and dissimilar wafers can be bonded. There is constant progress toward lower and lower bonding temperatures, that is for lower temperatures without sacrificing bond strength. Direct bonding is preferred, because all the additional materials at the interfaces increase the possibilities for unwanted reactions, during either processing or device use.

Silicon–glass bonding is a special case of the direct bonding process. Usually different materials cannot be direct bonded. Silicon and glass are bonded permanently and hermetically by the application of elevated temperature and high voltage in a process called anodic bonding.

The applied voltage creates a strong electrostatic field which pulls the wafers together. The sodium and oxygen ions in the glass become mobile at about 300–500 °C, and the mobile oxygen ions move toward the silicon interface, where they react with silicon and form strong Si–O bonds.

Indirect bonding uses three classes of materials as "glues": metals, glasses and polymers. Depending on the intermediate material, the force and temperature needed to effectuate bonding are very different: with polymer adhesives mild pressure and temperatures around 100–200 °C are sufficient, but with metal intermediates pressures of several megapascals and temperatures of 300–400 °C are typical. Glass frit bonding requires similar temperatures but less pressure.

Adhesive bonding can be used to bond anything to anything. Gluing is a form of adhesive bonding but in microfabrication more advanced methods are used. Polymer films (like epoxy SU-8) are spin coated on a wafer, patterned lithographically. This way the thickness of the adhesive is well controlled and narrow lines can be made. Thermal curing and UV curing can both be used.

Self-adhesive bonding is found with the polymer PDMS. It is an elastomer, and its softness allows intimate contact. The bond strength is small, and the bond is easily debonded, and bonded again, as long as the surfaces remain clean. PDMS is much used to create microfluidic channels. Either planar PDMS is attached as a roof to channels etched in silicon or glass, or structured PDMS is bonded to a planar silicon or glass wafer. Such bonding is simple because no alignment is needed. The great benefit of this process is that it is a room temperature process. This is important if, for instance, proteins are handled: they denaturate quickly

Table 17.1 Bonding process steps

Particle removal
Surface chemistry modification
Vacuum pumping (optional)
Wafer alignment (optional)
Room temperature joining
Application of force/heat/voltage
Wafer thinning (optional)

Figure 17.3 Silicon-on-insulator (SOI) wafer fabrication by bonding an oxidized wafer to a bare silicon wafer: left, surface preparation; middle, room temperature joining and annealing for bond improvement; right, top wafer thinning

above 40 °C. Bonding of PDMS to glass and to PDMS is much practiced in microfluidics, as will be discussed in Chapter 18.

The basic requirements for good wafer bonding are as follows:

- The materials being bonded form chemical bonds across their interface.
- High stresses are avoided.
- No interface bubbles develop.

The driving force for bonding can be temperature, pressure, electric field or a combination of these. Irrespective of the details of various bonding processes, the general outline is the same. The basic steps of bonding are listed in Table 17.1.

If closed cavities are formed, the behavior of gases needs to be analyzed. Gases are trapped inside cavities when wafers are mated, but additional gases are released from reactions (especially during high-temperature bonding) and outgassing from cavity materials contributes to the final cavity atmosphere, too. To a first approximation these gases are treated by the ideal gas law, for example to calculate cavity pressure during annealing.

17.2 Fusion Bonding Blanket Silicon Wafers

When bonding is done to produce more advanced starting wafers, it is usually about blanket wafers (wafers with prefabricated cavities will be discussed in Chapter 30).

Silicon direct bonding to an oxidized silicon wafer results in strong silicon–oxygen bonds. The resulting material now has an oxide layer in between two silicon wafers, and this property is very useful: it is the basis of silicon-on-insulator (SOI) wafers. The basic steps of SOI wafer formation by bonding are shown in Figure 17.3. This type of SOI is known as BESOI, for bond etchback SOI, or simply as BSOI. There are other techniques for SOI formation, to be discussed later on in this chapter and in Chapter 22. The SOI top layer, known as the device wafer, is electrically insulated from the bottom wafer, known as the handle wafer. This opens up many possibilities to optimize device properties in the device layer, and separate device and handle processing from each other, because the oxide acts as electrical insulation and as an etch stop layer.

Fusion bonding, like all bonding processes, begins with a cleaning step. RCA-1 cleaning with ammonia/peroxide mixture takes care of two requirements at the same time: it is effective in particle removal and it leaves the surface in a hydrophilic condition, with silanol groups (Si–OH). RCA-1 cleaned surfaces are smooth, and roughness remains smaller than 0.5 nm. CVD oxides are usually not smooth enough for bonding, and CMP must be done to achieve surface roughness below the 1 nm required for successful bonding (see Figure 16.10).

Surface energy (γ) is the energy required to break a bond and to create two new surfaces. It can be estimated from bond strengths (E_{bond}) and bond densities (d_{bond}):

$$\gamma = \frac{1}{2} E_{bond} d_{bond} \qquad (17.1)$$

The factor 1/2 comes from the fact that when a bond is broken, two surfaces are created. Two wafers in close contact are bonded via water molecules, Figure 17.4. We can get an estimate for surface energies from silicon atom surface density, about 10^{15} cm^{-2}, and hydrogen bond energies, 25–40 kJ/mol, which translate to about 200–350 mJ/m^2. Measured values for room temperature hydrogen-bonded silicon wafers are between 50 and 80 mJ/m^2. This indicates that less than maybe 20–30% of area is in contact with hydrogen bonds. This is understandable because the wafer surfaces are neither perfectly flat nor smooth but have local roughness and waviness, and hydrogen bonds have short range. The saturation value of surface energy after mild thermal treatment or extended time has been measured to be about 250 mJ/m^2, in very good agreement with the simple bond strength calculation.

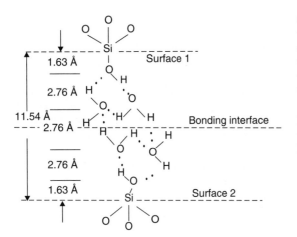

Figure 17.4 Bonding of hydrophilic silicon surfaces via intermediate water molecules. Reproduced from Tong and Gösele (1999) by permission of John Wiley & Sons, Inc.

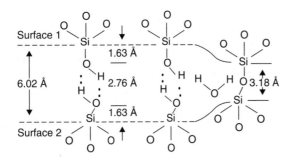

Figure 17.5 Water removal and siloxane bond formation. Reproduced from Tong and Gösele (1999) by permission of John Wiley & Sons, Inc.

The reaction that takes place during annealing (for silicon–silicon fusion bonding is typically about 1000 °C) is siloxane bond (Si–O–Si) formation (Figure 17.5)

$$\text{Si–OH} + \text{HO–Si} \Rightarrow \text{Si–O–Si} + H_2O \quad (17.2)$$

Siloxane bonds are much stronger than hydrogen bonds, and measured surface energies are about 1300 mJ/m^2. This surface energy is almost constant from 150 to 800 °C.

However, surface energies calculated from Si–O bond energies (4.5 eV/bond or 430 kJ/mol) translate to about 3000 mJ/m^2. This discrepancy is due to the fact that the surfaces are not yet fully bonded but have some areas which bond via hydrogen bonds only, and somewhere above 1000 °C thermal oxide becomes viscous and flows, which increases contact area and leads to higher surface energy. Surface energies of 3000 mJ/m^2 are not encountered in experiments, however, because wafer breakage will take place inside the silicon because Si–Si bonds are weaker than the Si–O bonds.

Silicon-to-silicon bonding without oxides is possible. Hydrogen bonds are involved, again. Surface treatment with HF leaves the wafers hydrophobic with either H or F termination. Because HF does not remove particles like ammonia/peroxide solution, hydrophobic bonding is inferior to hydrophilic bonding. Silicon-to-silicon bonding without oxides can yield abrupt pn junctions when p- and n-type wafers are bonded.

17.2.1 Low-temperature direct bonding

Low-temperature bonding is needed for instance when there are structures on the wafer, and these structures do not tolerate standard fusion bonding temperatures, or because two dissimilar materials are being bonded and low temperature will keep CTE problems at bay.

Plasma activation is the main method to allow low-temperature bonding. It is not fully clear why plasma surface treatment is so beneficial. There are a number of processes that may contribute: ion bombardment breaks bonds at the surface, leaving surface atoms in reactive states; UV from plasma discharge also breaks Si–O and Si–H bonds; the surface is oxidized in oxygen plasma, but the oxide is porous and diffusion through it beneficial for water removal, Equation 17.2.

Plasma activation will enable bonding at 400 °C, which is low enough for fusion bonding to compete with anodic bonding. Bonding has been done at 200 °C, and even room temperature has been reported, but no such process has become standard procedure. Low-temperature bonding is even more sensitive to surface roughness than standard fusion bonding, and wafers with 0.1 nm RMS roughness are used.

17.3 Anodic Bonding

Anodic bonding of silicon to glass (also known as field-assisted thermal bonding, FATB) is the oldest bonding technique in microfabrication. It has many features which make it easy: glass is a soft material which will conform at 400–500 °C bonding temperatures, sealing structures and irregularities of up to 50 nm hermetically. Native oxides, and thin grown or deposited oxides, do not prevent bonding. Anodic bonding can be visually checked through the glass side: bonded surfaces look black, and non-bonding areas are lighter.

Not all glasses are amenable to anodic bonding (see Chapter 19 for a discussion on glass properties). Thermal mismatch between silicon and glass needs to be considered at two temperatures: the bonding temperature and room temperature/operating temperature of the device. Glasses have higher coefficients of thermal expansion than silicon, but a match at two temperatures is approximately met with glasses like Schott 8339 and 8329 and Corning 7070 and 7740 (Pyrex). The CTE of 7740 is almost constant at $3.3 \times 10^{-6}/°C$ from room temperature to 450 °C, and that of silicon increases from 2.5×10^{-6} to $4 \times 10^{-6}/°C$.

When glass is heated to about 400 °C, sodium oxide (Na_2O) decomposes into sodium and oxygen ions. The bonding process uses 300 to 1000 V applied to the glass wafer (Figure 17.6). Sodium ions (Na^+) move toward the glass top surface and oxygen ions (O^{2-}) toward the silicon wafer. This will create a depletion layer and electrostatic force pulls the glass and silicon wafer together. The resulting electrostatic forces are very strong: if the thickness of the depletion region is 1 μm, the field is on the order of 500 MV/m ($E = 500$ V/1 μm) and the electrostatic force is proportional to E^2.

Oxygen ions react at the glass/silicon interface according to

$$Si + 2\ O^{2-} \Rightarrow SiO_2 + 4\ e^- \qquad (17.3)$$

and sodium ions are neutralized at the cathode. If higher temperatures are used, more sodium atoms are ionized and participate in the process, and will diffuse faster, so the depletion width is greater, leading to stronger bonds.

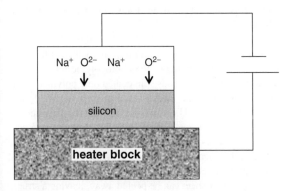

Figure 17.6 Anodic bonding: at about 350 °C glass ionizes partly, and the voltage of about 400 V drives oxygen ions toward the glass/silicon interface, creating a depletion region and a strong electrostatic field which pulls the wafers together

Bonding initiation is by applying pressure at the wafer center, but if bonding is done in vacuum, it is possible to bond without an initiation point. Typical bonding times are tens of minutes. It consists of alignment, chamber evacuation, temperature ramping, actual bonding and cooling. Total time is for example 30–60 minutes. This is fairly long for a single wafer operation, and special wafer holders have been designed so that wafer loading and unloading can be done while another wafer is being bonded.

There is usually the option for trading bonding temperature, voltage and time for each other. Microscope glass slides have been anodically bonded to silicon at 150 °C (typically microscope slides have CTEs of 7–9 ppm/°C). One hour bonding was done, which is quite long, using a fairly high voltage of 500 V.

Anodic bonding leads to a hermetic bond seam. Vacuum cavities, or others with controlled atmosphere, can be sealed. A "collar" around the cavity is necessary, but there are no standardized design rules for this. The size of bondable area can be very small. For instance, anodic bonding has been used to make pillar frits: silicon pillars 5 μm in diameter (Figure 21.9) can be bonded to glass.

Oxidizable metal films like aluminum can be sealed between glass and silicon if the films are thin enough (<300 nm). Metals like gold or chromium will prevent bond formation because either they do not oxidize (like gold) or their oxides are conductive (CrO). Signal lines out of a bonded structure can be done by diffused lines in the silicon wafer. The resistance of diffused wires will be high, but the surface remain planar, and these wires tolerate all possible processing steps. This method is also suitable for fusion bonded wafers.

Anodic bonding of multilayer structures is also possible: glass/silicon/glass systems (Figure 17.2, right) can be made in a single bonding step. Heating uniformity is important, and double side heating is usually employed. Contacting the middle wafer electrically requires special jigs.

17.4 Metallic Bonding

Thermocompression bonding (TCB) applies pressure and heat simultaneously to the samples. This is the standard wire bonding technique for attaching gold or aluminum leads to ICs. Gold is suitable because it is a noble metal: there are no gold oxides on the surface preventing TCB, and the softness of gold (low yield point) is also advantageous. Typical pressures and temperatures for wafer-level TCB are in the range of 1–10 MPa at 300–400 °C. Bonding times are then minutes or tens of minutes. An example of gold–gold bonding can be seen in Figure 30.8:

the two wafers forming the microphone have been bonded together by 300 °C, 15 min Au–Au bonding. Wafer-level TCB is difficult because uniform pressure is needed over a large area.

Another metallic bonding method is eutectic bonding. It is based on the fact that eutectic alloys have low melting points: for example, Au–Si eutectic alloy (19 at. % Si) melts at 363 °C even though the pure gold and silicon melting points are 1063 and 1421 °C, respectively. When gold dissolves into silicon, the eutectic point shifts to higher temperature, stopping the reaction. If silicon wafers with gold thin films are bonded, there is an ample supply of silicon, but if thin films are used, their thicknesses need to be designed so that eutectic composition is achieved. Any oxides on silicon will have to be carefully removed, otherwise Au–Si alloying will not take place.

There are tens of technologically important eutectic compositions that have been tested for bonding and many are useful in microsystem bonding. Aluminum–germanium and gold–tin systems are common. While Au–Si and Au–Sn are easy and well established, gold is a contamination source in silicon electronics. And even if there were no electronics on the wafers, gold would spread around in process tools, and therefore lab policy may prevent gold being used in equipment.

17.5 Adhesive Bonding

Adhesive bonding with a polymeric intermediate layer offers several advantages for bonding:

- temperatures from 20 °C to a few hundred degrees
- tolerant to (some) particle contamination
- bonding of structured wafers
- low-cost simple process.

Many negative resists have good bonding capabilities: SU-8, polyimides and benzocyclobutene have all been successfully used for adhesive bonding. Because polymers are soft materials they conform to particles, and voids are not a similar problem as with stiffer materials like silicon. The main problem with adhesive bonding is limited long-term stability and limited thermal range, with 100–400 °C maximum depending on polymer. Because of the low temperatures and benign processes, CMOS wafers can be used as substrates. A mirror array with individually addressable pixel elements steered by electronics in the bottom wafer is shown in Figure 17.7.

Prototypical steps in adhesive bonding are as follows:

1. Surface cleaning (optional adhesion promoter application).

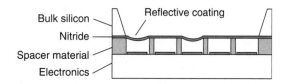

Figure 17.7 Aluminum mirror on nitride membrane is addressed pixel-wise by electronics in the bottom wafer. Photoresist serves the roles of both spacer and adhesive. Reproduced from Sakarya *et al.* (2002) by permission of Elsevier

2. Spin coating of polymer.
3. Initial curing (solvent bake).
4. Evacuating vacuum and joining the wafers.
5. Final curing of the polymer: pressure/heat/UV.

Initial curing is important to remove excess volatile material and thus reduce interface bubbles later on. The final curing temperature has to be above the glass transition temperature T_g of the polymer, otherwise no bonding will take place. Adhesive bonding, and bonding of polymeric materials in general, will be further discussed in Chapter 18.

17.6 Layer Transfer and Temporary Bonding

In layer transfer methods thin slices of material are cut from a donor wafer and bonded to a handle wafer. This cutting can be accomplished by weakening the wafer with ion-implanted hydrogen and then mechanically rupturing the top layer. SOI wafer fabrication by an ion cutting method (Smart-cut) will be discussed in more detail in Chapter 22.

Layer transfer can also be based on thin films. As an example of transferring a thin film, microchannel fabrication by transfer bonding of the photopolymer Ormocer is discussed. Ormocer film is spin coated on a polymer sheet (e.g., PDMS or PET) and bonded to a patterned Ormocer channel wafer. UV exposure through the transparent polymer sheet causes a reaction at the photopolymer interface. If the polymer sheet is inert, the bond strength between the sheet and the photopolymer is less than the bond strength between two photopolymer layers, and the carrier polymer sheet can be peeled away, leaving a structure made completely out of photopolymer. The process is shown in Figure 17.8.

As in all bonding, surface preparation is the key: the receiving surface must be able to form chemical bonds with the transferred layer. In the case of gold

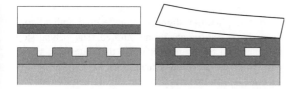

Figure 17.8 Channel formation by transfer bonding: photopolymer on a polymer sheet is bonded to a channel wafer and detached after curing the photopolymer by UV light

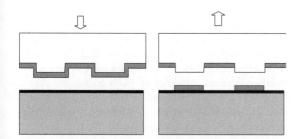

Figure 17.9 Transfer of gold film from a PDMS stamp: the silicon surface is functionalized by thiol groups, and gold–sulphur bonds are stronger than gold–PDMS bonds

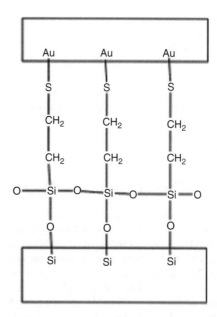

Figure 17.10 Gold–thiol bonds

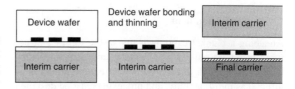

Figure 17.11 Temporary bonding: device wafer and interim carrier bonded by adhesive; device wafer thinned by grinding/etching/polishing; adhesive bonding to final carrier; detachment of interim carrier

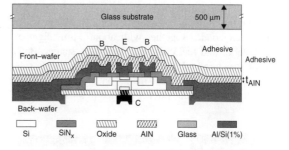

Figure 17.12 Bipolar transistor with AlN heat spreader layer transferred to glass, with silicon wafer removed. Reproduced from La Spina *et al.* (2008), copyright © 2008 by permission of Elsevier

transfer, Figure 17.9, the receiving surface is coated by a thiol–SAM: a monomolecular layer with a sulfur atom at the end of the carbon chain, because gold forms a strong bond with sulfur (Figure 17.10) and this bond is stronger than the van der Waals hydrogen bonds which hold gold to PDMS.

Temporary bonding is used to assist handling very thin wafers, or to make very thin wafers: the wafer to be thinned is bonded to a carrier, thinned down and possibly processed further, while being bonded to the carrier. Then, at a suitable later stage, the temporary bond is detached, and the thin wafer may be bonded to its final carrier, as shown in Figure 17.11. This is common in 3D packaging applications. The final carrier is chosen for example because it has high thermal conductivity (used for laser heat sinks) or because it is a dielectric and eliminates parasitic capacitances (as in RF devices).

Bonding and layer transfer have many applications. In power transistors heating is a major issue, as is insulation voltage. While silicon is a good thermal conductor, it is a semiconductor and leakage currents will increase with temperature. Glass would solve this issue, but it is a thermal insulator. A solution is to deposit a thermally conducting but electrically insulating interlayer before bonding the power transistor to the glass. Aluminum nitride can be used (Figure 17.12). For this application the AlN deposition process has to be optimized so that thermal conductivity is maximized while the piezoelectric effect is minimized.

Temporary bonding is a key element in 3D integration: fully processed wafers are joined together temporarily, thinned and metallized, for greater functionality. More examples will be shown in Chapter 39. Flexible, wearable electronics have been demonstrated by laminating thinned silicon chips on metallized polymers. One example is shown in Figure 39.13.

17.7 Bonding of Structured Wafers

Blanket wafer bonding requires smoothness and flatness, and if one (or both) of the wafers have been structured, they may be compromised from a bonding point of view. If etching has been done to make microchannels (Figure 17.1), the etch mask has to be removed somehow. If this removal process, for example wet or plasma etching, introduces microscale roughness, bonding properties may be compromised. On a wafer scale, metal deposition may introduce stresses which will bend the wafer, making bonding difficult or even impossible.

17.7.1 Bond alignment

Attaching a blanket capping wafer to the top of a structured wafer is easy. If both wafers have structures they need to be aligned to each other. Perfect alignment is as difficult as in lithography, and processes must be designed to work with less than perfect alignment (Figure 17.13).

Glass and PDMS are very good with respect to bond alignment: transparency makes optical alignment straightforward. Anodic bonding alignment resembles standard lithography: the glass wafer with its metal patterns can be aligned to the bottom silicon wafer (photomasks are glass plates with metal patterns).

In the microturbine (Figure 1.17) five silicon wafers have been bonded together with about 1 μm alignment to create air and fuel channels, rotors and stators. Bonding of structured silicon wafers requires a double-sided alignment tool. Alignment marks on the first wafer are registered, the second wafer is aligned to those marks and wafers are then brought into contact. Alternatively, infrared through-wafer alignment can be used with silicon, because it is transparent in the infrared. The critical step is to maintain the alignment while the wafers are transferred to the bonding equipment. This is accomplished by a special fixture that fits both the aligner and the bonder, therefore wafers need not be handled after alignment.

17.7.2 Gas entrapment

Silicon and glass are impermeable to gases, therefore the pressure inside a cavity needs to be understood. This is especially important when thin membranes are made, because they will deflect due to any pressure difference between the cavity and the outside world (Figure 17.14). With polymer devices the issues are quite different and mostly concerned with keeping gases outside the cavity, that is hermeticity of the bond. In both cases gases desorb from the structures themselves during bond anneal. Polymers are more prone to outgassing, but then, on the other hand, polymer bonding temperatures are much lower.

When wafers with cavities are joined together, three possibilities for the cavity atmosphere are commonly encountered: air, nitrogen and vacuum. When vacuum-bonded wafers are cooled down and taken to atmospheric pressure, the membrane will be deflected down. In fusion bonding oxygen has reacted with silicon and formed oxide (this is dry oxidation, Equation 13.2). Nitrogen remains, resulting in a pressure of about 0.8 atm, since air is roughly 80% nitrogen, 20% oxygen. Joining the wafers under a nitrogen atmosphere will result in equal pressure after bond annealing. The ideal gas law is a good approximation for gas pressures inside cavities.

The pressure inside microcavities can be calculated from membrane bending and Equation 17.4, which gives circular membrane centerpoint deflection (w) under applied pressure:

$$w = 0.662a \times \sqrt[3]{\frac{\Delta P \times a}{Et}} \qquad (17.4)$$

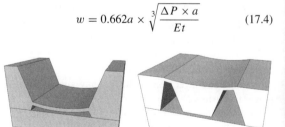

Figure 17.13 Bond alignment: none, perfect and misaligned

Figure 17.14 Deflection of membrane due to vacuum cavity: atmospheric pressure bends the membrane. Note that the design on the right retains parallelism of the gap

where a is membrane radius, t the thickness and E Young's modulus. Alternatively, the chips can be placed in a vacuum chamber which is pumped down. The flat-membrane condition is equated to gas pressure inside the cavity.

However, gases are not just trapped inside cavities and desorbed from walls – some gases are generated in the bonding process itself. The direct bonding reaction products of silicon are hydrogen and water. These gases behave differently: hydrogen dissolves into silicon dioxide, and if the oxide is thicker than 50 nm it can absorb all the hydrogen, but in the case of thin oxide, hydrogen bubbles may evolve. Water oxidizes silicon (wet oxidation !), because the temperature of bond annealing is similar to the oxidation temperatures.

If CVD oxide is used in bonding, it will release gases when annealed: remember that hydrogen is always embedded in CVD oxides. These oxides have to be annealed above the bonding temperature before bonding. This is even more pronounced when polymeric materials are present: polymers can absorb quite considerable amounts of water, and this will be released upon heating.

In anodic bonding oxygen is diffusing toward the interface, and oxygen gas accumulates in the cavity. Bonding pressure needs some attention when anodic bonding is done on wafers with cavities. At millitorr pressures a glow discharge can be initiated in the cavity. Therefore either a good vacuum or atmospheric pressure is desirable.

17.7.3 Mechanical strength

Even when bonding is successful from the surface chemistry/bond strength points of view, there may be problems: it is not guaranteed that the cover wafer is strong enough for the span intended (Figure 17.15). Membranes will deflect because of a pressure difference, but this is elastic, and when pressures are equalized, flat membranes will result. But with thin membranes at elevated temperatures there is the danger (and possibility) of irreversible plastic deformation. First of all, thin membranes are mechanically not very rigid, and, second, yield stresses are much

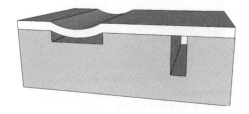

Figure 17.15 Sagging over a long span

reduced at elevated temperatures. See Figure 19.15 for a discussion on irreversible membrane deflection.

Span and gap are related: if narrow gaps with long spans need to be made, bending of the wafers can result in closure of the gap. This can be circumvented by using stiffer materials (higher Young's modulus) or thicker wafers. The design also affects strength: circles are sturdier than long lines, for instance. The ratio of span to gap can be for instance 1000:1 (millimeter span for micrometer gap) for anodically bonded sturdy designs.

17.8 Bond Quality Measurements

Cleanliness is paramount in wafer bonding: particles at the bond interface will prevent bonding locally. Anodic bonding can easily be observed through the glass side, but if the wafers are not transparent, infrared optical measurement through the wafer is possible. For silicon this translates to 1.1 µm wavelength and above. The height of voids can be inferred from interferometric rings, with $\lambda/4$ as the minimum detectable height, or about 0.28 µm for silicon.

Acoustic microscopy (SAM, Scanning Acoustic Microscopy) can be used to check voids in the finished wafer stack non-destructively. The wafer is immersed in water and high-frequency ultrasound is aimed at it. Theoretical resolution is down to micrometers, but full wafer scanning is often desirable, and scanning time can be traded for resolution so for example a 1 min scan with 100 µm resolution is done.

Bond strength can be measured by a simple crack opening method: a razor blade is inserted between the wafers and the crack opening length is measured. Bond strength γ is given by

$$\gamma = \frac{3Ed^3 t_b^2}{32L^4} \qquad (17.5)$$

where E is Young's modulus, d the wafer thickness, t_b the blade thickness and L the crack length. This method is quick and easy with no wafer preparation, but because of the L^4 dependence on crack length, any uncertainties in crack length determination will lead to erratic results.

Debonding the wafers and visual or microscopic examination reveal bond interface quality. Bond strength can also be checked by pull tests. In a pull test the two wafers are torn apart, and the debonding force is measured. This is better than the crack method because it measures bond strength all over the wafer, not just at the edge (where the bond might be of inferior quality). The problem with a bond test is that the bond may be stronger than the attachment of the pulling set-up, and only a lower limit is obtained.

Anodic bonding bond strengths are 20–30 MPa, similar to the best polymer-to-polymer bonds. The usual polymer bonds are 1–10 MPa and PDMS self-adhesive bonds about 0.5 MPa. Bond tightness can be measured by gas leakage. When patterned and etched silicon wafers have been direct bonded to form cavities, etched depths of 6 nm can be sealed gas-tight, but 9 nm grooves will result in leakage. Higher anneal temperature will seal slightly better. Anodic bonding is much more flexible: even 50 nm grooves can be sealed in a gas-tight manner.

17.9 Bonding for Packaging

Bonding is also a typical method to create zero-level packages for MEMS devices. A cavity with controlled ambient is needed for several reasons: damping of mechanical resonators depends on pressure; clogging of movable structures by particles must be prevented; and surfaces have to be protected from atmospheric water vapor condensation. With all the sensitive, delicate micromechanical parts covered by a capping wafer, subsequent operations of dicing, encapsulation, mounting, etc., can be generic, whereas packaging of unprotected chips with vibrating beams or bending plates would have to be developed for each and every design separately. The alternative method for cavity formation is deposition. This will be discussed in Chapter 30. Deposition avoids one of the main drawbacks of bonding, which is the extra cost of an additional wafer.

Glass-frit bonding is an old established technique for fairly low-temperature hermetic bonding. Hermeticity equals low leakage and long-term stability of the cavity atmosphere. The glass frit consists of a powder of low-melting-point lead glass and polymeric binder. It is screen printed on a wafer (making it amenable to large linewidths only) and baked to remove binder. The glass frit is then pressed against the wafer to be bonded and heated to about 400 °C. The glass frit is quite thick, for example 15 μm, which makes it insensitive to particles and microstructures. It is easy to have metallization run

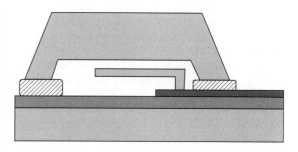

Figure 17.16 MEMS resonator enclosed by a silicon cap wafer and glass-frit bonding: metallization runs under the glass frit from a vacuum cavity. Glass frit conforms to some extent to height differences

Table 17.2 Bonding techniques compared

Method	Advantages	Disadvantages
Direct Si bonding	Strong bond	Smooth and flat surface needed
	Hermetic sealing	Very high temperature >1000 °C
Low-T Si direct	Hermetic at low T	Very flat and smooth surface needed
	CMOS compatible	Sensitive to surface preparation
Fusion, glass	Strong bond	650 °C needed for Pyrex
	Hermetic seal	Wafer deformation a problem
Fusion, polymer	Strong bond	Depends on T_g
	Identical materials	Flow of material may block cavities
Anodic	Strong bond	High voltage and discharge danger
	Hermetic	Sodium accumulation on cathode
	Seals small irregularities	Requires CTE matched glass
Thermocompression	Non-flat surface OK	High forces needed
	Hermetic	Difficult on wafer level
Eutectic	Hermetic	Flat surface needed
	Strong bond	Requires special metals
Glass frit	Hermetic	Large-area bond seam
	Covers metallization	Lead may be contamination source
Adhesive	Versatile	Non-hermetic
	Covers metallization	Low-temperature uses only
	Insensitive to particles	Reactions with fluids

under the glass frit, thus enabling electrical contacts to hermetic cavities (Figure 17.16). The poor thickness control prevents the use of glass-frit bonding in applications where bonding defines device-critical vertical dimensions. Glass-frit bonding is utilized in many bulk micromechanical applications like pressure sensors and accelerometers.

Adhesive bonding can be used in much the same way as glass-frit bonding; for example, structures resembling exactly Figure 17.15 can be made by adhesive bonding. However, the polymer seal is not hermetic, and the long-term reliability of adhesively bonded structures has to be assessed for each application. All polymers are permeable to some degree, even though Teflon-like polymers have permeabilities 10 000 times less than that of PDMS, but metals have permeabilities another 10 000 times less than the densest polymers.

17.10 Bonding at Large

Because there are so many different bonding techniques available, some bonding solution is almost invariably applicable. All methods have their pros and cons, and they depend on device requirements. Table 17.2 lists some important issues of major bonding techniques.

17.11 Exercises

1. Which materials can be present on a wafer undergoing silicon–silicon fusion bonding?
2. Calculate the gas pressure inside an anodically bonded cavity when bonding has been done at ambient pressure and 400 °C temperature!
3. Outline the fabrication sequence for the tunable capacitor of Figure 17.2! Which process steps contribute to gap size?
4. Design a fabrication process for the accelerometer of Figure 17.2 (right).
5. What is the resolution of a 160 MHz acoustic measurement of voids?
6. Which measurements could reveal the role of sodium ion depletion in anodic bonding?
7. How much will a silicon membrane 0.5 µm thick and 1 mm in diameter bend due to 0.2 atm pressure difference? What about a 5 µm membrane?
8. Calculate the bond strength from Equation 17.5 for two silicon wafers when a blade of 50 µm has opened a crack of 8 mm.

References and Related Reading

Berthold, A. et al. (2000) Glass-to-glass anodic bonding with standard IC technology thin films as intermediate layers, *Sens. Actuators*, **82**, 224.

Christiansen, S. H., R. Singh and U. Gösele (2006) Wafer direct bonding: from advanced substrate engineering to future applications in micro/nanoelectronics, *Proc. IEEE*, **94**, 2060–2106.

Dziuban, J. A. (2006) **Bonding in Microsystem Technology**, Springer.

Etxeberria, J. A. and F. J. Gracia (2007) Tunable MEMS volume capacitors for high voltage applications, *Microelectron. Eng.*, **84**, 1393–1397.

Henttinen, K. et al. (2000) Mechanically induced Si layer transfer in hydrogen-implanted Si wafers, *Appl. Phys. Lett.*, **76**, 2370.

Huff, M. A. et al. (1993) Design of sealed cavity microstructures formed by silicon wafer bonding, *J. Microelectromech. Syst.*, **2**, 74.

Jourdain, A. et al. (2002) Investigation of the hermeticity of BCB-sealed cavities for housing (RF-)MEMS devices, Proceedings of IEEE MEMS 2002, p. 677.

La Spina, L. et al. (2008) Aluminum nitride for heatspreading in RF IC's, *Solid-State Electron.*, **52**, 1359–1363.

Lee, B. et al. (2003) A study on wafer level vacuum packaging for MEMS devices, *J. Micromech. Microeng.*, **13**, 663.

Mack, S. et al. (1997) Analysis of bonding-related gas enclosure in micromachined cavities sealed by silicon wafer bonding, *J. Electrochem. Soc.*, **144**, 1106.

Niklaus, F. et al. (2006) Adhesive wafer bonding, *J. Appl. Phys.*, **99**, 031101.

Sakarya, S. et al. (2002) Technology of reflective membranes for spatial light modulators, *Sens. Actuators*, **A97–98**, 468.

Tong, Q.-Y. and U. Gösele (1999) **Semiconductor Wafer Bonding**, John Wiley & Sons, Inc.

Tsau, C. T., S. M. Spearing and M. A. Schmidt (2002) Fabrication of wafer-level thermocompression bonds, *J. Microelectromech. Syst.*, **11**, 641–647.

Vallin, Ü., K. Jonsson and U. Lindberg (2005) Adhesion quantification methods for wafer bonding, *Mater. Sci. Eng.* **R50**, 109–165.

18

Polymer Microprocessing

Polymer microprocessing employs a wide variety of methods, some of them scaled-down versions of macroscopic polymer processing, like injection molding and casting, and some borrowed from microfabrication and applied to novel materials like CVD of parylene, plasma deposition of Teflon and lithography of epoxy. Polymers are essential in optical lithography and newer patterning methods like embossing and nanoimprint lithography (NIL) also make patterns in polymers. Most of the time photoresist or NIL polymer is stripped away after it has served as etch, implant or plating mask, but it is also possible to leave the polymer structure as a permanent part of the finished device. Such structures are central in this chapter. In Figure 18.1a epoxy pillars 300 μm tall, 30 μm in diameter made by optical lithography are shown, and in Figure 18.1b imprinted epoxy grating with 60 nm feature size is shown.

Polymers are soft materials (have small Young's modulus) and this property is utilized when flexible structures or sensitive mechanical elements are needed. Polymers can have exceptional combinations of properties, like optical transparency and elasticity (PDMS); or a refractive index of 1.34 and low loss tangent in RF (amorphous fluoropolymer); or low thermal expansion and excellent solvent tolerance (polyimides). But then some other properties might be disadvantageous: water absorption is high and solvent tolerance often poor. Because of the wide range of polymers existing, it is often possible to find a material with a suitable mix of properties. Polymers are cheap, which is important when disposable devices are made. The main drawback of polymers is their limited stability (mechanical softness, limited thermal range, porosity) and this applies to both device processing and operation.

Polymer processing methods can be divided into two gross categories: direct fabrication and replication. Direct fabrication results in the final structure, as in lithographic patterning. Replication methods rely on a master or mold, which is used to force the polymer into the desired shape. These two methods can result in exactly the same final structures, as shown in Figure 1.9. There are many different implementations, and the terminology is somewhat diffuse. Embossing and imprinting both use a 3D stamp to force the structure into softened polymer, while molding (also spelled moulding) and casting both describe processes where polymer is flowing (because either it is molten or the monomer is fluid) and filling a master/mold. This difference between imprinting and molding is depicted in Figure 18.2.

Various hybrid methods exist, like UV embossing, where liquid photopolymer is stamped and cured by UV radiation. Molding techniques can be applied not only to polymers, but also to inorganic materials, like steel powder. These will be discussed in Chapter 23 on special processes.

18.1 Polymer Materials

Polymer molecules are long chains of repeating monomer units. In polyethylene (PE) the C_2H_4 monomer unit is repeated thousands of times, resulting in molecular weights (MWs) of for example 30 000 or 300 000. Polymerization processes do not produce exact chain lengths but result usually in rather wide distributions, and in addition to chemical formula, it is important to know MW. Many polymer properties are a function of MW, for example viscosity, which is important in micromolding and stamping processes.

The bonds between carbon atoms in the chain are strong, but the chains interact only via weak hydrogen bonds. Polymers are held together also by mechanical entanglement of the chains. Because of the weak hydrogen bonds, polyethylene easily softens when heated: the polymer chains slide past each other.

Figure 18.1 (a) fluidic sieve with 300 μm high, 30 μm diameter pillars made by photolithography, courtesy Santeri Tuomikoski, Aalto University; (b) optical grating, 60 nm linewidth made by nanoimprint lithography, reproduced from Martinsen (2008) by permission of Elsevier. Both structures are made of SU-8 polymer

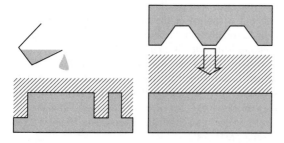

Figure 18.2 Molding (left): material flow into mould master. Imprinting/embossing (right): the stamp is pressed against solid (but softened) polymer

In polyvinyl chloride (PVC) the repeating unit is $CHCl-CH_2$, and in polytetrafluoroethylene (PTFE) C_2F_4 units are repeated. Parylene-C has the repeating unit consisting of a benzene ring with chlorine. Such aromatic benzene rings render polymers thermally stable. BCB (benzocyclobutadiene) is another aromatic thermally stable polymer widely used in microsystems. Epoxy SU-8 consists of aromatic rings and eight epoxy groups which will result in tight crosslinking upon heating. The bonds are strong covalent bonds which are not weakened by heating, therefore cross-linked SU-8 does not flow. The chemical formulas of common microfabrication polymers are shown in Figure 18.3.

Polymers are available in many formats: they are available as sheets and wafers, with thicknesses ranging from a few micrometers to millimeters. They serve as substrates for processing. Some are available as solutions and can be applied by spin coating, while others can be synthesized from monomers in liquid phase, vacuum and plasma processes, and applied as thin films not unlike sputtered or PECVD films. Some polymers are available in many formats: polymethylmethacrylate (PMMA) is used in wafer format for hot embossing, and as spin-coated thin film it serves as electron beam resist. More commonly PMMA is known as Plexiglas. Similarly, polyimides are used as substrates 50 μm thick for flexible electronics under the trade name Kapton, and as high thermal stability spin-coated negative resists, as well as spin-coated non-photoactive films.

Silicon is also present in many polymers, and because of its chemical similarity to carbon (both are column IV elements that form four bonds), there are interesting silicon-based polymers, for example siloxanes. Poly(dimethyl)siloxane (PDMS) has the basic repeating unit of $-(O-Si(CH_3)_2-O-)$. Its backbone consists of silicon and oxygen, but the side chains are organic groups. Similar polymers with different organic groups have been synthesized with interesting properties ranging between polymers and SiO_2. These materials can sometimes combine the best of both worlds: for example, low dielectric constant and improved thermal stability. One such hybrid polymer is Ormocer (for organically modified ceramic) which can be processed like negative resist, but which results in oxide-like material, for instance a low contact angle compared to polymers, and much higher temperature stability than most polymers.

Polymers can be divided into two major classes: thermoplasts and thermosets (or duroplasts). Thermoplast molecules have only weak bonds between the polymer chains and, when heated, the polymer flows (like polyethylene). Thermoplasts can be repeatedly heated and cooled, with no permanent change. Many common

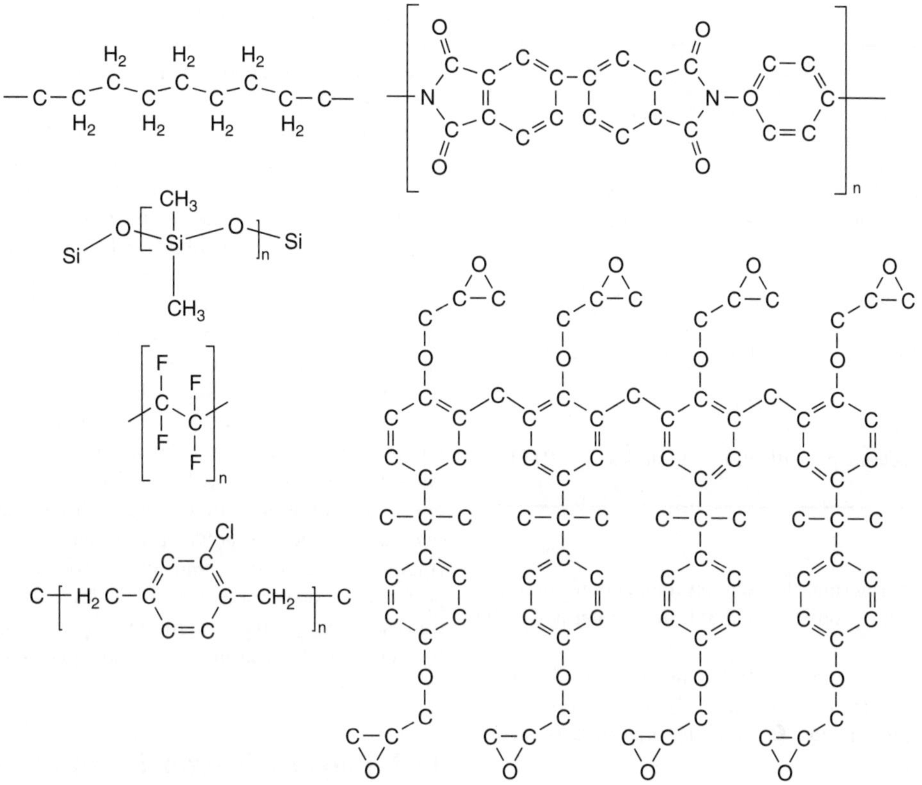

Figure 18.3 Polymer chemical structures. Left column: PE, PDMS, PFTE, parylene. Right column: polyimide, SU-8 epoxy

polymers are thermoplasts: PE, polycarbonate (PC), PMMA, COC (Cyclic Olefin Copolymer), PET, PEEK, PVC and PS.

Thermosets are polymers that react upon heating and form strong covalent bonds between the chains. Thermosets will flow for a while when heated for the first time, but then crosslinking and hardening take place. The resulting crosslinked material will not flow even above T_g because of the rigidity of their structure. Thermosets include photoresists, epoxies (like SU-8), PTFE, polyimides, polyesters, parylene and BCB.

Elastomers form a third group. In these materials the bonds between the chains are flexible and the material returns to its shape even after 1000% elongation. Rubbers are typical elastomers. In microfabrication PDMS and Viton are common elastomers.

There are other classifications, too. Some polymers are amorphous. They are glassy, in the same sense as glasses, they exhibit no long-range order and they are optically transparent. Amorphous polymers include PMMA, PC, COC, PI, PDMS and PVC. Semicrystalline polymers show local crystallinity in the midst of amorphous areas. Sometimes they are called crystalline polymers for short, but there are really no crystalline polymers. In semicrystalline polymers the crystal/amorphous interfaces diffract and reflect light, and the resulting material is opaque. Semicrystalline polymers include PP, PET, PEEK, PVDF and PTFE.

Polymers are often called by their trade names, increasing confusion. Teflon is PTFE, Topas and Zeonor are COC, PMMA is Plexiglass and PC Lexan. Kapton is polyimide and Mylar is polyester. Viton and CYTOP are fluoropolymers, AZ resists are novolak resins and Sylgard 184 is PDMS.

Polymer properties vary a lot, and Table 18.1 gives just a glimpse of typical values. It indicates the major properties of polymers in general. Table 18.2 lists in a little more detail the properties of common microfabrication polymers. In both tables the values do not cater for all blends and varieties; for instance, imides are a large class of polymers with wide-ranging properties. Temperature limits are partly set by the glass transition temperature T_g,

Table 18.1 Typical polymer properties

	Value	Comments
Density (kg/m^3)	~1000	Teflon ~2000
Dielectric constant	2–4	SiO$_2$, $\varepsilon \sim 4$
Volume resistivity (ohm-cm)	10^{14}–10^{17}	Similar to SiO$_2$ and Si$_3$N$_4$
Index of refraction	1.35–1.65	Resists ~1.6
Dielectric strength (MV/cm)	2–3	SiO$_2$ ~10
Young's modulus (GPa)	1–5	PDMS ~0.01, SiO$_2$ ~76
Tensile strength (MPa)	10–100	Imides even higher
Residual stress (MPa)	10–30	PDMS very low
Water contact angle (deg)	70–100	Imides 50°, PTFE 114°

and partly by thermal degradation temperature, or melting. Additional properties and selection criteria will be discussed below.

It must stressed that different manufacturers offer polymers of the same basic formula with different properties, and any values in this chapter are indicative of major trends only.

Note that MW and polymer processing history (baking time and temperature, and hence crosslinking density) have considerable effects on many of the above parameters.

Other properties which may be of importance in certain applications include the dielectric loss tangent, which ranges from 0.0002 for PTFE to 0.04 for PMMA. Shrinkage upon crosslinking is a source of stresses, and it can be considerable (5–10%).

Polymer selection involves further considerations: chemical stability against acids, bases and solvents, for example. PDMS has excellent chemical resistance to weak acids and alkaline solutions but it is unsuitable for most organic solvents. SU-8, parylene, PTFE and polyimides are very resistant to most solvents, acids and bases. PTFE is exceptional in its chemical stability and low water absorption (<0.01%), while for PDMS it is 0.1% and for PMMA and polyimides a few percent.

Polymers are generally electrical and thermal insulators, and can serve many of the same functions as silicon dioxide and silicon nitride: as intermetal insulators, gate dielectrics, passivation films and free-standing membranes. Semiconducting polymers like polythienylene vinylene (PVT) and PEDOT can be used in transistor channels and conducting polymer polyanilin (PANI) has been used for "metallization" of transistors. The resistivity of a conducting polymer can be as low as 10 000 μohm-cm, a thousand times higher than metals, but only a hundred times the resistivity of polysilicon. As will be discussed in Chapter 26, polymers can serve every function needed in a transistor (Figure 26.22). The main advantages of polymer electronics are flexibility (Figure 18.4) and low-temperature processing, but their performance does not match silicon devices, and polymer transistors are geared toward very different applications like embedded electronics, enabling cheap intelligence in products.

18.2 Polymer Thermal Properties

Because polymer MWs have wide distributions, polymer properties change gradually, not abruptly. The glass transition temperature (T_g) is an important parameter. Below T_g polymers are brittle and glass-like, but above T_g thermoplasts become soft (viscosity decreases) and they will flow and can be molded. Baking above T_g is also a stress reduction method: above T_g the polymer will relax stresses because it becomes softer and is able to accommodate stresses. Baking photoresist above T_g will mean that it will flow and end up as a hemispherical dot (see Figure 18.30 below). While a non-vertical resist sidewall profile is generally regarded as a processing

Table 18.2 Properties of common microfabrication polymers

	T_g (°C)	T_{deg} (°C)	CTE (ppm/°C)	Thermal conductivity (W/K-m)	UV transparency
PMMA	100	200	70	0.2	Opaque
PC	150	230	65	0.2	>350 nm
PDMS	−130	400	300	0.15	>240 nm
SU-8	240	340	100	0.2	>350 nm
Polyimide	400	620	3–50	0.2	Opaque
Parylene	150	290	35	0.1	>300 nm
Teflon	130	330	100	0.1	opaque

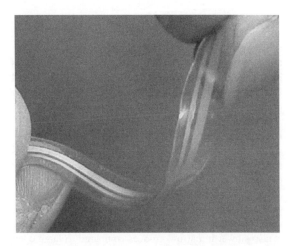

Figure 18.4 Flexible glucose sensor. Reproduced from Kudo et al. (2006), copyright © 2006 by permission of Elsevier

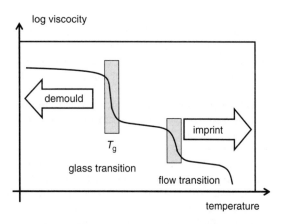

Figure 18.5 Thermoplast polymer viscosity vs. temperature diagram. Adapted from Schift (2008)

problem, these dots can be used for microlenses. Elastomers are weakly crosslinked materials with values of T_g below room temperature, and therefore they are soft and easily deformed by pressure, but when pressure is removed, they return to their original shape. PDMS is the most widely used elastomer in microfabrication, with T_g of $-130\,°C$.

Temperature is a key variable in polymer microprocessing because polymer viscosity depends strongly on temperature. The different temperature regimes of a polymer are shown in Figure 18.5. Viscosity, which describes the mechanical response to a shearing force, is plotted vs. temperature. The first major transition occurs at T_g. Polymer becomes viscous, and can be embossed/imprinted, even though embossing is usually done at higher temperature, around the flow transition, in order to speed up the process. At higher temperatures the polymer melts; this is the regime of injection molding. Not all polymers flow and melt: some will decompose when heated.

Polymer dimensional stability is poor compared to inorganic materials. Thermal expansion is huge, typically 50–100 ppm/°C, compared to 3 ppm/°C for silicon and 23 ppm/°C for aluminum. This leads to many problems: even mild temperatures, like the 120 °C encountered in lithography bake processes, can cause large thermal stresses. And the shrinkage of polymer after it has been molded at elevated temperatures has to be accounted for. In CDs and DVDs these two factors are somewhat easier than in microfabrication in general, because the molten polymer is fed in from the middle, and shrinkage is radially symmetric.

Polymers are low-density materials and always porous to some extent. Water vapor absorption varies a lot: for some fluoropolymers the coefficient of permeability is 10^{-15} mol-m/m^2-s-Pa, for polyimide two orders magnitudes more, and yet another two orders of magnitude more for elastomers. So while polymer packaging of microsystems is attractive because of ease, it is not applicable to cases where hermetic sealing is needed. Permeability is sometimes useful: in cell culture devices oxygen can diffuse through the PDMS roof, and no separate fluidic channels are needed to provide oxygen. Metallization of polymers is difficult because desorption of water from pores will poison the vacuum, and this water vapor will oxidize metals, resulting in increased resistivity.

Polymers can also be filled with other materials, to modify their properties. Microfabrication makes no exception: for electrical conductivity, carbon black has traditionally been used, and carbon nanotubes and silver nanoparticles are now used in microtechnology. Glass spheres and glass fibers have been used in the macro world, and the same applies in the micro world. Young's modulus can be tailored by an order of magnitude by fillers. Magnetic nanoparticles can be used to make polymers react to magnetic fields and it is also possible to retain photoactivity when the nanoparticle size is small relative to exposure wavelengths (and their density low enough).

18.3 Thick-Resist Lithography

Thick can mean very different thicknesses to different people. In IC fabrication resist thickness is from a hundred nanometers to a few micrometers. In MEMS and thin-film magnetic recording head (TFH) fabrication, thick can mean anything from 5 to 200 µm, and in X-ray lithography thick extends into the millimeter regime. As with thin resists, a number of sometimes conflicting requirements are imposed on thick resists, such as resolution, sidewall profile, sensitivity (photospeed), thermal stability, adhesion and shelf life.

Thick-resist processing has a few extra factors that need attention, compared to standard resists. Rapid solvent evaporation has to be prevented because rapid and large shrinkage leads to defective and non-uniform films as the surface layer dries and encapsulates the material beneath. One solution is a closed spinner bowl which creates a saturated solvent vapor atmosphere. This buys extra time to ensure uniform resist spreading before viscosity increases so much that flow stops. Solvent evaporates during final spinning to some extent, but for thick resists it is advantageous to perform an additional slow spinning step at the end, to further dry the resist.

With thick resists the edge bead is sizable, and it will prevent contact exposure because the mask touches the edge bead at wafer edges. Edge bead removal is beneficial also for cleanliness: the extra resist material can flake off and cause particle problems.

Soft bake can be done in a convection oven, on a hotplate or by infrared heaters. Solvent removal is diffusion limited and long baking times are required for thick resists; for 100 µm thick resists, bake times are about an hour. While oven bakes are usually done at one temperature, hotplates allow easy temperature ramping. During the bake the wafers should be perfectly horizontal because viscosity is smaller at elevated temperatures and therefore resist can easily flow. A minor bevel can cause major resist thickness non-uniformity across the wafer. Positive novolak resists need an additional rehydration period after baking: water is an essential compound in exposure (Figure 9.7).

In novolak–DNQ resists, exposed DNQ still absorbs some light, and thick layers of such resist will need very high exposure doses, that is long exposure times. The practical thickness limit of novolak–DNQ positive resists is 50–100 µm, which translates to exposures of tens of minutes. In negative resists based on a photoacid generator, the reactive acid molecule is produced throughout the volume of the resist, and absorbance is not affected by the exposure. Therefore even layers a millimeter thick can be exposed in reasonable times. All polymers absorb to some extent below 350 nm, and in order to avoid absorbance by the base resin, wavelengths below 350 nm should be filtered out. Exposure is supposed to interact with the photoactive compound, not with the base resin.

It is usually not the exposure but the mechanical strength of the final resist structures that limits the thicknesses of negative resist structures. This involves also the properties of liquids used in development and rinsing and drying. Capillary forces can be so strong that drying water can pull neighboring resist lines into contact. This will be discussed in more detail in Chapter 29, because such drying-induced stiction problems are central in surface micromechanics.

In order to make deep trenches or high pillars, collimated light is needed: if incoming light is not collimated, sizable exposure will take place underneath the mask. This same reason explains why thick resists are not exposed by projection mask aligners.

Thick-resist development is normally done by immersion. Development is diffusion limited and development time increases as a function of thickness. Development is also slower in narrow grooves compared to large open areas. Development times are tens of minutes.

For negative photoresists post-exposure bake (PEB) is required to finish the crosslinking reaction. For positive novolak resists PEB is not normally required. It is important that the temperature is high enough to complete crosslinking, but not too high because crosslinking creates stresses (which can only be controlled by crosslinking density). Stresses also appear because of thermal expansion coefficient differences between the substrate and resist.

Even though resists are soft materials and thin relative to wafer thickness, visible wafer bowing can result from resist thermal stresses.

Polymer structures of high aspect ratio in for example SU-8 resist (Figure 18.1a) resemble DRIE structures in appearance (Figure 21.9). Both technologies are capable of 10:1 aspect ratios easily, 40:1 ratios with effort and 100:1 ratios have been reported in special cases. In both SU-8 and DRIE the sidewall profile is seldom 90°, and an acceptable sidewall angle is a question of engineering judgment. For example, in mirror and interferometer applications, or with parallel-plate capacitors, perfect verticality is desired, while in fluidics or mechanical applications deviations can be tolerated to some extent. The smoothness of resist sidewalls can be as low as 10 nm, which is better than that of DRIE sidewalls.

Multilayer structures can be made by repeated exposure–bake–spin coat–expose–bake cycles, leaving the development step till the very end. This is shown in Figure 18.6 both schematically and in the SEM micrograph. The structure is used in tissue engineering.

18.3.1 Thick resists as structural materials

There are many microfluidic applications for thick-resist structures. The choice of microfluidic channel wall

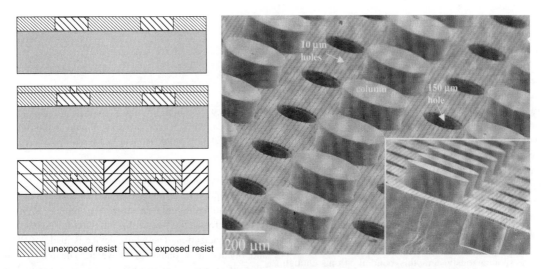

Figure 18.6 Three-level SU-8 structure: three spin coatings and three exposures with different masks, and one development at the end. Reproduced from Mata *et al.* (2006) by permission of IOP

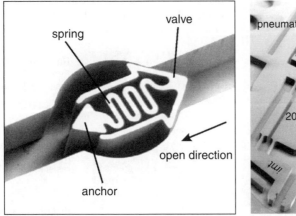
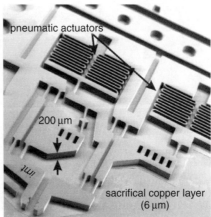

Figure 18.7 SU-8 microvalve (left); SU-8 pneumatic actuator (right). Reproduced from Seidemann *et al.* (2002), copyright © 2002 by permission of Elsevier

material is mostly about surface chemistry, surface charge for electro-osmotic flow, adsorption of analytes and contact angle for wetting and capillarity. Transparency and lack of autofluoresence for optical detection, integration of electrodes for conductivity detection and bondability are also important.

Polymers are suitable for mechanical devices, too. Spring constants can be smaller by making long, slender beams, and this is certainly done in MEMS, but the alternative is to use softer materials, like polymers (see Equation 29.2). In Figure 18.7 mechanical elements made of SU-8 epoxy polymer are shown for a passive microfluidic valve and pneumatic actuators. The SU-8 structures are free to move because a copper thin film underneath the polymer has been etched away.

18.4 Molding Techniques

18.4.1 Replica molding in PDMS

Molding of PDMS is easy and therefore widely used. A simple SU-8 process can be used as a mold master. PDMS prepolymer is mixed with a crosslinking agent (e.g., in a 10:1 ratio) and degassed. It is poured over the mold master and degassed again to ensure bubble removal. Curing can be anything from 65 °C for 10 hours to 80 °C for 2 hours, depending on process details. PDMS will demold easily because of its inertness and flexibility (it is an elastomer).

PDMS devices are not only easy to fabricate, but also good for aqueous fluids; they are optically transparent down to 250 nm and mechanically flexible. A simple PDMS device is shown in Figure 18.8. The PDMS piece is bonded to a flat glass wafer, forming a fluidic actuator/valve. However, because the PDMS CTE is about 300 ppm/°C, PDMS is not suitable for applications that require accurate pattern positioning.

One of the great advantages of PDMS is its self-adhesiveness: a clean PDMS piece will adhere to a clean surface spontaneously, and this bond is strong enough for many applications. It is great for R&D but long-term stability of the bond is not good. If a stronger bond is

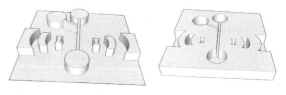

Figure 18.9 Optofluidic chip: SU-8 master and its PDMS cast replica. Microlenses focus light onto a microfluidic channel. Adapted from Seo and Lee (2004)

needed, oxygen plasma activation of PDMS can be used (more on polymer bonding later in this chapter).

An optofluidic chip is shown in Figure 18.9: microlenses focus light onto the microfluidic channel, for improved detection. The whole device is made of PDMS and bonded to a planar glass wafer. The lenses come "free of charge": they are formed in the same molding process as the channels.

More complex mold masters enable more complex shapes. Figure 18.10 shows a hemispherical SU-8 mold master. It is made by standard lithography, but after development SU-8 is baked above its glass transition

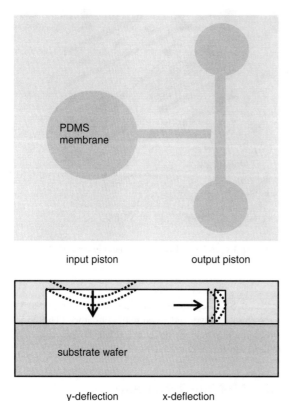

Figure 18.8 PDMS vertical input piston induces lateral deformation in output piston, closing a microchannel. Adapted from Lee *et al.* (2007)

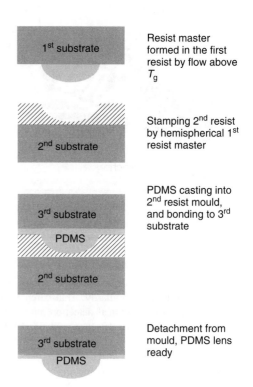

Figure 18.10 PDMS microlenses by resist flow, stamping and transfer bonding. Adapted from Chen *et al.* (2005)

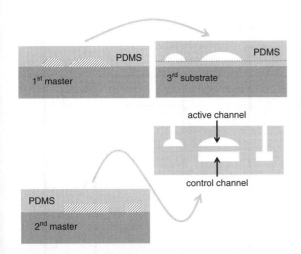

Figure 18.11 Microfluidic chip with three PDMS layers: pneumatic pressure in control channel deforms the thin PDMS membrane, closing the active liquid channel. Fluidic inlets are made by piercing. Adapted from Lin and Su (2008)

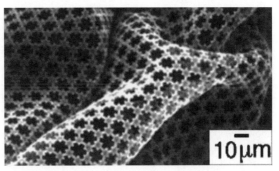

Figure 18.12 Polyurethane net by micromolding in capillaries. Reproduced from Kim *et al.* (1996), copyright 1996 American Chemical Society

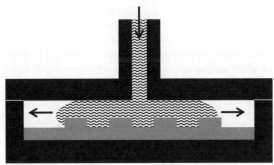

Figure 18.13 Injection molding: high-pressure injection of molten polymer into a mold

temperature, which will make is flow. Figure 18.11 shows a fluidic system: two SU-8 masters are used, one with the standard SU-8 process and the other by resist flow. Three PDMS steps are required: two casting steps and one spinning step for thin PDMS membrane formation. The actuator channel does not need to close completely, but the analyte channel with its hemispherical shape will be fully closed.

18.4.2 Micromolding in capillaries (MIMIC)

Micromolding in capillaries takes advantage of microfabricated mold masters: liquid precursor fills microfluidic channels in the master by capillary forces. The precursor is then cured, and the master is removed. This method is applicable to passive structures (like Figure 18.12) and active polymer transistor components alike. Not only polymers but other materials that can be made soluble can be processed, for example nanoparticles in suspension, metal powders, etc. Micromolding is also applicable to multiple layer thicknesses in a single molding step.

18.4.3 Injection molding

Injection molding set-up is pictured in Figure 18.13. Injection molding is applied for micrometer dimensions in mass manufacturing: molten plastic is injected into a mold insert to fabricate CDs, DVDs and Blu-ray discs (with 1, 0.6 and 0.3 μm feature sizes, respectively). However, from a general microfabrication point of view, these are easy applications because aspect ratios are about 0.2 only, pattern density is quite uniform and pattern sizes are quite similar. Because the molds are expensive (they have to tolerate molten polymers, often around 300 °C), long production series are compulsory to bring the cost down.

The main parameters of injection molding are the temperature of the injected polymer, mold temperature, injection speed and pressure, and holding time and pressure. Micro injection molding requires some modifications relative to large-scale systems: for instance, materials usage needs to be rethought, otherwise only a fraction of polymer ends up in the finished parts. In mold fabrication technologies, materials, dimensions and surface roughness need to be assessed. Polymer properties must be understood, for example viscosity changes rapidly in micromolds because the large surface-to-volume ratio leads to rapid cooling of the polymer melt.

Compared to all other replication methods, injection molding is the fastest: it takes only a few seconds per piece. Injection molding is used in microfluidics for disposable devices where cost minimization is essential, but some very complex injection-molded microfluidics applications exist, too.

18.5 Hot Embossing

Hot embossing involves pressing a master against a polymer at a temperature around T_f, typically 50–100 °C above the polymer glass transition temperature T_g, see Figure 18.14. Popular hot embossing polymer PMMA has $T_g \sim 100$ °C and polycarbonate (PC) has $T_g \sim 150$ °C. The embossing force is on the order of 100–1000 kPa and hold time is on the order of minutes. De-embossing takes place after cooling below the glass transition temperature, therefore the embossed shape is retained. In theory the features are true replicas of the master, and any size patterns can be made, from nanometer size to macroscopic (but nanometer master fabrication is very expensive !). Note that in order to emboss large areas, uniform pressure needs to be applied. In addition to precision mechanics, a soft compliance layer helps Figure 18.14.

The hot embossing process is visualized in the time–temperature–pressure graph in Figure 18.15. The process starts by heating the wafer (and the master) to T_2, which is for example 80 °C above the polymer T_g. Pressure of 100–1000 kPa is then applied to press the master against the softened polymer. Polymer will flow and fill the cavities in the master, forming an inverse replica of the master. After a few minutes the temperature is ramped down, and

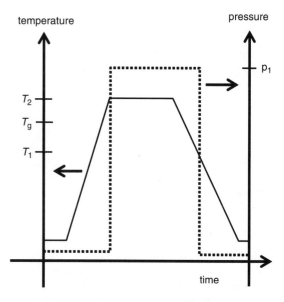

Figure 18.15 Pressure–temperature diagram of hot embossing: heat to T_2 (50–100 °C > T_g) apply pressure (100–1000 kPa), cool down to T_1 (e.g., 20 °C below T_g), and remove pressure

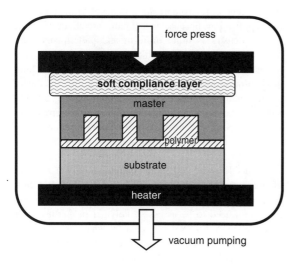

Figure 18.14 Thermal imprinter/hot embossing tool: soft compliance layer assures conformal large-area contact between master and substrate

below T_g (at T_1, which is 80 °C for PMMA) force is removed. The polymer has solidified into the form of the master, and because the temperature is now below T_g, it will retain its shape upon de-embossing.

Hot embossing has three major issues: filling of structures by polymer, reproduction fidelity and de-embossing. Polymeric materials have CTEs on the order of 50–100 ppm/°C, whereas silicon has a CTE of 2.6 ppm/°C and nickel, a typical electroplated master material, 13 ppm/°C. Thermal cycling is mandatory in hot embossing but it should be minimized around T_g to avoid thermal mismatch cracking.

Hot embossing results in microparts with low internal stresses because stresses have time to relax in the slow process. Its slowness is compensated in R&D by the simplicity of the experimental set-up: changing masters and materials is quick and easy, and the various technologies for master fabrication enable cheap masters. Hot embossing is applicable to large areas, with 200 mm wafer size embossing tools available. The slowness is inherent in hot embossing because the polymer and master are heated by conduction, and heating is an integral part of the process and not easily decoupled from other process parameters.

The thickness of hot embossed structures can be varied enormously, from nanometers to millimeters (Figure 18.16). There is no resolution limit, and

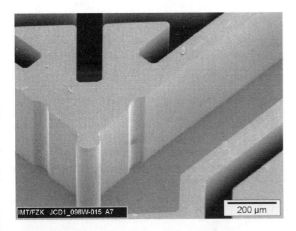

Figure 18.16 An embossed optical bench 400 μm high. Courtesy Mathias Worgull, Forschungszentrum Karlsruhe

embossing can replicate structures down to 10 nm size, making the master become the limiting factor. Aspect ratios of embossed structures can be as high as 20:1 and up to 50:1 when special release coatings have been applied. Making needle-like structures by embossing is difficult, however, because filling such long needles by polymer is often inadequate.

18.6 Nanoimprint Lithography

Nanoimprint lithography (NIL) is hot embossing applied to nanofabrication. The smallest imprinted features have been 5 nm wide. This indicates that any nanoscopic irregularities in hot embossing masters will be replicated. This was in fact observed long ago, but NIL only emerged in the late 1990s. In lithography application, the removal of a residual layer at the bottom of features (Figure 1.9) becomes a necessity, and the thickness contrast h_r/h_f must be high enough in order to reasonably make patterns.

At the macroscopic level the NIL process depends critically on stamp contact with the substrate. Using a soft compliance layer between the stamp and the upper platen, small deviations from planarity can be compensated. Thermal NIL works on 200 mm wafers. On a microscopic level air evacuation from cavities is important, and a vacuum can help in this. The squeeze flow of polymer is obviously very important. Polymer viscosities are on the order of 10^3–10^7 Pa-s and imprinting pressures around 1000 N/cm². Temperature and pressure can be traded for each other, and low pressure can be compensated by longer process time. But as discussed above, structure size affects the time requirement, and therefore time is not a completely free variable.

NIL is a very simple process for making nanostructures: if master fabrication can be subcontracted, the NIL equipment price is only a fraction of the optical lithography tool capable of identical linewidths: from $100 000 for a NIL lab system, to $1–3 million for an advanced system. There is a twist, however: in optical lithography tools alignment is a standard feature, while the cheaper NIL systems offer no alignment or only rudimentary alignment. And if alignment accuracy is 500 nm and resolution 50 nm, most the high-resolution capability is left unused. But if a single layer pattern is needed, imprint lithography is very cost effective. Hard disks have been suggested as an application and optical devices like gratings can also be done without alignment.

18.6.1 Imprinting theory

An idealized description of the embossing/imprinting process is shown in Figure 18.17. A downward force F presses the stamp against a polymer with viscosity η. Polymer height h changes over time as $h(t)$. Stefan's law, Equation 18.1, governs the flow of polymer between solid parallel plates. That is,

$$F = \frac{3\pi R^4}{2h(t)^3} \frac{dh}{dt} \eta \qquad (18.1)$$

Interpretation of Stefan's equation tells us that achieving a thin residual layer is exceedingly difficult because

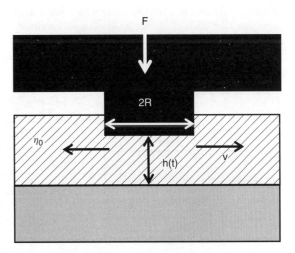

Figure 18.17 Factors contributing to Stefan's law: viscosity, force, stamp protrusion size, residual thickness

of the $1/h^3$ term. This is a critical difference between traditional hot embossing and NIL is in layer thicknesses. In embossing the polymer sheet is for example 2 mm thick, and 50 μm structures are embossed. Residual thickness is 1950 μm. In NIL original polymer thickness is for example 300 nm and target residual thickness 50 nm. In Stefan's equation h is large and dh/dt is small.

Another observation is that in fact it is easier to make small structures than large ones, because the required force increases as R^4. Low viscosity is beneficial because then lower force can be used, therefore elevated temperature is practically always used, even though cold embossing does exist. A further point to note is that structure filling is nonlinear in time: initially polymer flow will fill the cavities in the master fast, but filling becomes slower as time goes on.

A fundamental principle of mass conservation illustrates why the imprinting of uneven areal density structures is difficult. In Equation 18.2 mass conservation is used to calculate final heights. The volume of polymer is unchanged (nothing disappears during imprinting) and it is only surface heights that change. The situation is pictured in Figure 18.18. The residual thickness h_f depends on the areal density (A_0/A_f) of patterns:

$$A_0 h_0 = A_f h_f + (A_0 - A_f) h_e \quad (18.2)$$

18.6.2 Practical NIL

Several non-idealities make NIL more difficult. Non-uniform pattern height results when a rigid stamp (or low imprint pressure) does not allow large patterns to form, while nanostructures are rapidly filled. A submicrometer grating is imprinted in a fraction of a second, while large areas can take minutes. In the case of high imprinting force the stamp may bend and the smaller features penetrate deeper into the polymer, resulting in thicker

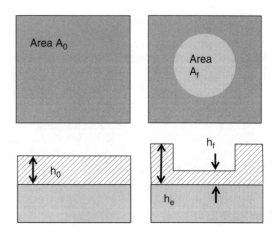

Figure 18.18 Mass is conserved in imprinting, and residual thickness h_f and final height h_e can be calculated once the embossed area is known

residual layers for the large structures. Incomplete nanopattern filling can also result because of the volume constancy requirement: large and small areas side by side both affect pattern height.

There are many variants of NIL. In UV-NIL, photoactive polymer is used. In one version liquid photopolymer is used, and the process is basically a room temperature process. The structure is cured by UV light (Figure 18.19). Compared to thermal imprint, optical/UV-NIL has certain strong points: UV curing can be much faster than thermal ramps, an obvious productivity feature. And if temperature ramps can be eliminated, polymer shrinkage problems can also be minimized. On the other hand, thermal NIL masters can be made of silicon, nickel or practically anything. Optical methods obviously require transparent masters, which limits the choices. For accurate

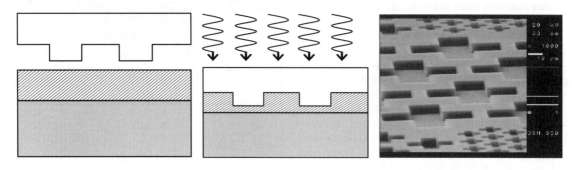

Figure 18.19 UV-NIL: stamping of photopolymer with a transparent master stamp and curing by UV exposure. Reproduced from Aura *et al.* (2008) by permission of Elsevier

nanofabrication fused silica masters are used. In the micrometer range PDMS is an excellent master material. There are also equipment issues: optical paths have to be designed, and if both heat and light are needed, the design becomes even more difficult.

Thermal NIL can be done on 200 mm wafers (but the writing time and cost of a 200 mm stamp with nanostructures are astronomical), therefore many imprint methods "expose" smaller areas. This is similar to step-and-repeat optical lithography (Figure 10.2): stamp size is for example 25×25 mm, and multiple stampings are needed to fill the wafer. The method is sometimes called step-and-stamp, and as S-FIL, for step-and-flash imprint lithography, in the case of UV-NIL.

18.7 Masters for Replication

Masters and molds can be made by a number of technologies:

- photolithography of SU-8
- photolithography and silicon etching
- photolithography and metal electroplating
- discharge machining
- laser machining
- mechanical milling
- natural objects.

Photoresists are common mold masters, especially SU-8, which is often used as a master for PDMS casting. The benefits of SU-8 are in the simplicity of making masters in various sizes, both laterally and in the z-direction, from micrometers to millimeters. Additionally, since SU-8 is a polymer, its thermal expansion is similar to other polymers, which reduces thermal stresses.

Silicon is an excellent master mold material in many cases: its machining accuracy is good, the surface finish is good (especially in wet etching) and with DRIE practically any shape can be made. Wet etching is more limited in shapes, but the 54.7° angle is good for detachment, see Figure 18.20.

Thomas Alva Edison used sputtered gold seed layer, wax mask and gold electroplating to fabricate phonograph masters. The technology entered production in 1901 and it could replicate 125 μm pitch (200 grooves/inch) structures 25 μm thick into phonograph records. Electroplating is still a major method for mold master fabrication, with nickel as the metal of choice. In microfluidic applications the dimensions are similar to Edison's, and in fact traditional machine tools can be used to fabricate the masters, but often the surface finish is too rough and pattern complexity makes machining throughput low, though it is a useful method for making a few test devices.

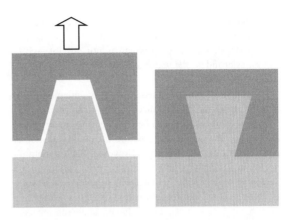

Figure 18.20 Mold master should have positive slope for easy detachment; negative (retrograde) profile does not allow detachment

An important issue in all replication methods is the detachment (also known as de-embossing or release). In fact, it is often equal in importance to replication itself. There are a number of requirements that are essential and beneficial. First of all, the master must have positively sloped walls (Figure 18.20). Release is not possible with retrograde walls.

Another requirement is surface smoothness: the smoother the surface, the easier the release. Microfabricated master molds are very smooth, especially photoresist mold (and metal mold made using resist molds) and anisotropically wet etched silicon molds. Plasma etched molds are somewhat rougher. Milled and machined molds are clearly much rougher.

Chemical bond formation between master and replica should be avoided. This is especially important in molding processes where monomers are chemically reactive, not just physically flowing. Diamond is a good choice for a mold because of its inertness and abrasion resistance. Another way is to deposit a sacrificial layer on the mold master, and release the structures by etching this sacrificial layer away.

Alternatively, the mold can be coated by low-surface-energy material. Teflon-like films and fluorinated SAMs are obvious candidates, because of their extremely low friction. Alternatively, the surface of the polymer can be treated with fluoropolymer and then stamped.

Often polymers can be used as masters and molds for other polymers. PDMS has been used to emboss polycarbonate, and PDMS is also an excellent transparent material for UV embossing masters. Of course, thermal and other limitations apply, but clearly the choices are many. The polymer masters are soft, while the silicon and metal masters are rigid. Depending on process details, either property can be beneficially utilized, for example a soft master will conform and make intimate contact, allowing large-area stamping which is not possible with a rigid master unless the wafer and master are very flat and carefully leveled.

Molding can be continued to further generations: instead of using the molded piece itself, it can be used as a new mold. This process can be continued for many generations in certain applications before the quality of molded pieces becomes unacceptable. However, each generation results in a reverse polarity structure of its parent, so it is necessary to decide beforehand which generation is going to be used. If the master fabrication process is expensive, as in X-ray LIGA, it is useful to have a slave master made of the expensive original, and thus limit the costs.

Master fabrication can sometimes benefit from existing structures: both butterfly wings and lotus leaves have been used as masters for replication, the former for optical effects, the latter for superhydrophobicity. In order to use the master repeatedly, it has to be transferred to some more stable material, but there are many choices.

18.8 Processing on Polymers

Polymer etching is not often done because lithography, imprinting and casting produce polymer microstructures easily. Thermosets must sometimes be etched because molding is not possible. Polymers are etched by oxygen-containing gases, with or without additional gases, for example O_2/CF_4, or O_2/Ar. Non-erodible masks like aluminum and chromium are used in deep etching. But as in silicon etching, achieving anisotropy is difficult because of the spontaneous chemical reactions. In most cases the rates are limited to about 1 μm/min. Wet etching is usually not an option for polymers. In theory solvent etching could be done by suitable selection of polymer and solvent. Sometimes NaOH etching is done, not for patterning, but rather for surface treatment.

Polymers can be polished by CMP and this has been used in the creation of thermal insulation: a deeply etched cavity wafer is coated by BCB, which is planarized by CMP. Devices are then processed on BCB. Thermal isolation is often done by etching away the silicon wafer, leaving only a thin membrane. This approach has problems with mechanical strength. BCB has a high operating temperature, but at about 400 °C it is still low compared to inorganic materials.

18.8.1 Deposition of polymers

The adhesion of polymers on thin films and thin films on polymers is generally poor because polymers do not readily form bonds. Adhesion then depends on mechanical interlocking (surface roughness) and surface cleanliness. Surface cleaning can be done prior to metal deposition by many techniques, for example baking, wet treatment, plasma treatment. These have the effect of removing contaminants and absorbed water vapor, and etching also induces surface roughening, which is beneficial for adhesion. Use of adhesion promoter layers underneath the main metal can be made as usual, for example titanium or chromium. Evaporated films have poor adhesion on polymers, but sputtering fares somewhat better. Sputtered atoms have kinetic energy and impinge on the polymer surface, and some of them are implanted inside the polymer, creating a mixed layer which holds the substrate and film together. If layer thickness is measured on a polymer film, the result will differ from an inorganic substrate, because some of the deposited atoms will find their way inside the polymer, due to both their kinetic energy and polymer porosity.

Adhesion can be improved by nano- and microstructuring the surfaces. Figure 18.21 shows one technique for improved polymer adhesion. It utilizes an isotropic etch profile to create undercut structures which will lock polymer structures, just as a retrograde profile will prevent molded piece detachment from the mold master (Figure 18.20).

One technique to secure mechanical interlocking is to cast polymer over T-shaped or mushroom-shaped metals (Figure 18.22). Overplating (Figure 5.13) is performed on a standard resist pattern. The T-shape of the plated metal now securely keeps the cast PDMS in place, and the structure tolerates further processing, in this case through-wafer DRIE. The resulting membrane is flexible due to PDMS and can be magnetically actuated due to the metallic coil structure. The standard approach does not work well, even with the adhesion layer, because the large thermal expansion of PDMS during resist baking will lead to metal detachment.

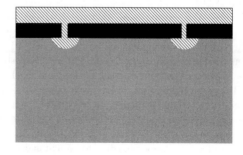

Figure 18.21 Isotropically etched holes anchor the polymer mechanically

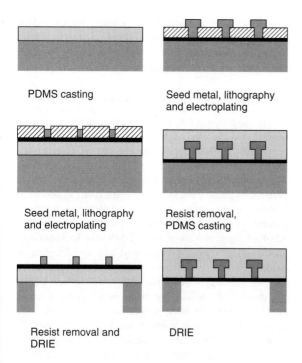

Figure 18.22 Flexible PDMS membrane for micropump. Left, standard approach of metal plating on top of PDMS; right, PDMS casting on overplated metal. Adapted from Yin *et al.* (2007)

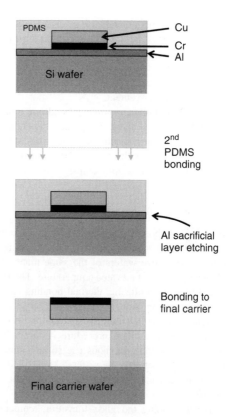

Figure 18.23 Metallized flexible PDMS antenna. Adapted from Tiercelin *et al.* (2006)

Resist spinning on polymers requires careful attention to solvents used in resists. A reaction between polymer and solvent can render the whole process unusable. Therefore alternative methods like shadow masks are often used (see Chapter 23) because neither lithography nor etching is required. This is beneficial because the typical 120 °C temperature of resist baking can already result in thermal stresses that will delaminate the metal from the polymer.

Thin-film metals have higher resistivity than bulk materials, and thin-film metals on polymers have even higher resistivities than those on silicon or glass. On PMMA or other fairly non-porous polymer metal, the resistivity can be for example 10% or 100% higher than on reference oxide, but on porous polymer like PDMS the resistivity can be even 1000% higher than on oxide. Post-deposition annealing for resistivity reduction is not an option with most polymers.

In Figure 18.23 PDMS is used to make highly flexible antennas. The process uses aluminum as a sacrificial layer, copper as the antenna conductor and chromium as an adhesion layer and barrier between the copper and oxide. PDMS is spin coated on top and cured. Another PDMS piece is then bonded on top of the first PDMS, and TMAH is used to etch aluminum. PDMS bonding to a copper carrier wafer finalizes the process. Unlike the process shown in Figure 18.22, there are no process steps above room temperature once PDMS has been cured, so the large CTE of PDMS does not come into play.

PECVD films can be deposited on polymers, temperature permitting. It is often possible to lower the deposition temperature and get reasonable film quality, even though lower temperature invariably correlates with poorer film quality. ALD alumina (Al_2O_3) is usually deposited at around 220 °C, but it can be done at 80 °C which enables deposition on polymers. However, its properties are different; for instance, it etches much faster in wet etching than the standard film.

18.9 Polymer Bonding

The basics of bonding were presented in Chapter 17. In this chapter we will explore applications specific to polymers. As in all bonding processes, a few universal requirements apply: the surfaces to be bonded must be clean of particles, have proper bond-forming surface chemistry, and the substrates must be flat and smooth to ensure intimate contact between the wafers to be bonded. Adequate bond strength at the lowest possible temperature is desired, and with polymers there is little room for annealing for bond strength improvement.

Bond strengths of polymer–polymer bonds are typically in the 1–10 MPa range, and differ greatly depending on process details. The best results have been obtained for COC solvent bonding, with over 30 MPa, which is adequate for liquid chromatography chips.

18.9.1 Thermal bonding

The term thermal bonding is often used to describe a process similar to fusion bonding or direct bonding in silicon technology: two wafers of the same material are bonded by application of elevated temperature. The terminology is somewhat confusing: thermal bonding can also refer to thermocompression bonding (where pressure is an important element). Vacuum-assisted thermal bonding is to be considered standard procedure in direct bonding: most fusion bonding methods use some vacuum to reduce gas bubbles between the wafers. At elevated temperatures polymers may degrade due to reactions with oxygen, therefore a vacuum is useful.

In polymer bonding the glass transition temperature plays a critical role: above T_g the materials become soft, which compensates some non-flatness and non-smoothness. In the case of two identical materials, interface bond strength should be identical to material bulk strength. Bonding identical polymers is easier than dissimilar materials because thermal mismatch is eliminated. One of the advantages of fusion bonding is that the final structure is made of a single material, and this is beneficial in microfluidics as all the walls of the system have identical contact angle, surface charge and adsorption.

Most polymer bonding processes operate between 100 and 200 °C: for example, 95–165 °C for PMMA (the values differ both because of experimental details like pressure and because of polymer molecular weight differences), 250 °C for BCB, 130–230 °C for parylene ($T_g = 90$ °C, decomposition 290 °C), 160 °C for CYTOP fluoropolymer and around 200 °C for polyimides. Typical bonding times are a few hours. This is rather slow and alternative methods are often motivated by faster bonding processes. Pressures of 500–1000 N are usual.

Optimizing thermal bonding processes is difficult: differences between polymers mean that temperature–pressure relations have to be found for each polymer. Bonding is also dependent on chip design: bonding shallow, wide channels is much more critical to roof sagging than bonding narrow, deep channels. A general purpose process has to be able to do both simultaneously, or else designers must be given restrictions on allowed structures.

18.9.2 Bonding by surface treatments

PDMS elastomer is soft and it is easy to achieve intimate contact. This is one reason for its adhesiveness. But because the PDMS surface is mostly methyl groups, they are not very reactive, and the bond is easily detached. This reversibility is advantageous sometimes, but if more permanent bonds are needed, oxygen plasma treatment of PDMS will turn some of the methyl groups to hydroxyl groups (–OH), which are much more reactive (Figure 12.6). Oxygen plasma-treated PDMS can be bonded to another PDMS piece permanently, and also to glass or oxide because their surfaces are also hydroxyl terminated. The bond strengths are on the order of 0.5 MPa, which is not much, but adequate for many applications. Contact angle measurement after plasma treatment can be used as a quick monitoring method: small contact angles, or hydrophilic surfaces, are essential to strong bonds.

One technique to ensure intimate contact between boding surfaces is to make them soft by solvent treatment. The solvent should be selected to match the polymer, and properly selected solvent will soften only the very top layer of polymer, for example 50 nm in the case of DMSO and PMMA. This soft layer is sufficient for polymer chains to become mobile and make contact. A thicker solvated layer is in fact detrimental: softened polymer flow will clog microchannels. UV treatment can also be used to modify surface chemistry by breaking bonds (and by cleaning the surface). It can even be that T_g of the surface layer is lowered so much that bonding can be carried out below bulk T_g.

18.9.3 Localized bonding

Laser welding of polymers is possible but it must be applied to bond dissimilar wafers: the top wafer has to be transparent, but the bottom wafer has to be absorbing. Laser bonding benefits include minimized thermal distortions because the heating is very local.

Ultrasonics heats up the polymer pieces, and softening will help make a good contact between the wafers. The best results are obtained if there are structures designed to guide ultrasonic energy, which makes the method sensitive to chip design. Microwave welding relies on the heating of metal thin films, which will then soften the polymers, leading to bonding. The obvious drawback is the need to deposit and pattern metallizations on polymers.

18.9.4 Adhesive bonding

The advantage of adhesive bonding is its universality: any two wafers can be joined together by an adhesive, irrespective of their surface chemistries, thermal expansion coefficients or other properties. Bond strengths in epoxy-based adhesive bonding can reach 20 MPa which is much higher than most polymer–polymer direct or thermal bonding methods. However, the foreign material between the wafers is sometimes undesirable, and direct bonding must then be used. For example, the adhesive may interact with the analytes in microfludic systems. Thermal stability of adhesive bonds is limited, and their long-term behavior is poorly known.

Adhesives work by many different mechanisms, and not all of them are suitable for microsystems. Some adhesives solidify by drying, releasing gases which will be trapped at the bonding interface (unless one or both of the wafers are permeable to gases). Thermoplastic adhesives are applied hot, and are soft and make good contact. They solidify upon cooling, forming bonds, and are very general purpose adhesives. Epoxies are thermosets that experience hardening either by UV light or by heating. UV adhesives like SU-8 are good because the bonding temperature is low, but there is the limitation that one of the wafers has to be transparent. BCB is a thermoset polymer which can be cured by heating (to about 200–300 °C).

Thermoplastic adhesives tolerate a maximum of 300 °C, while some thermoset adhesives can tolerate even 450 °C. They are generally also more resistant to chemicals.

In many applications it is advantageous to have all the walls of a fluidic channel made of one material only. A simple full wafer, single photomask process for full SU-8 channels is illustrated in Figure 18.24 (left). The first layer, blanket exposed, serves as the floor; the second layer, exposed through a mask, defines the walls; and the third layer, on a Pyrex wafer, serves as the roof. After joining the wafers, the SU-8 on Pyrex is exposed through the Pyrex wafer to bond it to cured SU-8.

However, because of the large thermal expansion mismatch between SU-8 and silicon and glass, large SU-8 structures are prone to cracking. Therefore it is advisable to break up large SU-8 areas into smaller areas. As shown in Figure 18.24 (right), two photmasks are needed in this version. The bottom SU-8 is exposed through mask #1 but not developed. The second SU-8 is spin coated and exposed with the second photomask. Both layers are then developed. The third layer is spin coated on top of the Pyrex wafer as before.

In a similar fashion, three layers of SU-8 have been used to fabricate an electrospray nozzle (Figure 18.25). Additionally, the structure has been released from the supporting glass wafer by HF etching.

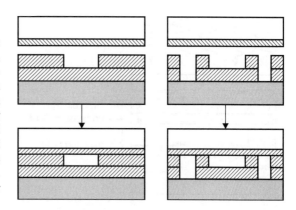

Figure 18.24 SU-8 microfluidic channels by SU-8 adhesive bonding: left, simple channel by one-photomask process; right, two-photomask process with stress relief auxiliary channels. Adapted from Tuomikoski *et al.* (2005)

Figure 18.25 Electrospray tip made of three-layer SU-8: after two layers (left); and after bonding (right). Reproduced from Tuomikoski *et al.* (2005) by permission of Wiley-VCH

Another benefit of adhesive bonding is the ease of creating patterns: either photoactive epoxy like SU-8 is defined by lithography, or non-photoactive glue is patterned by screen printing. Adhesive bonding can create overhanging structures which cannot be made by lithography or imprinting processes.

The drawback of adhesive bonding is that the "glue" must usually be flowable, and there is the danger that it will flow into microchannels and cavities already made. This can be minimized by working with thinner glues, but then the advantage of adhesive bonding starts to disappear: thick adhesive compensates wafer non-flatness, but with thin adhesive the usual flatness requirement reemerges. Thin adhesive is also less forgiving of particle contamination. Dry resists (laminates) can be used to close structures, as will be discussed at the end of this chapter.

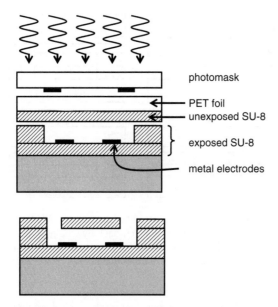

Figure 18.26 Microfluidic channel with electrodes, lithography with auxiliary carrier film: SU-8 on PET foil, which is detached after SU-8 exposure. Adapted from Abgrall *et al.* (2006).

18.10 Polymer Devices

The rest of this chapter discusses examples of polymer microdevices. Various aspects include the interaction of materials properties with fabrication and operation, and a comparison of different techniques and materials.

Dry film resists are useful because of their simplicity and ability to cross trenches (recall Figure 9.8). The drawback is poor resolution: dry resists are suitable for structures in the tens of micrometers range and larger. Sometimes standard resist and an auxiliary film can serve the same purpose, as shown in Figure 18.26: resist is spin coated on PET foil for example, and this stack is then inverted on top of the structure. Exposure through the PET film hardens the resist, and PET film is peeled off.

The Braille actuator shown in Figure 18.27 uses three layers of PDMS. It is very similar to the microvalve shown in Figure 18.11: both devices use pneumatic pressure to move the PDMS membrane up and down.

An in-plane fluidic valve (diode) has also been made of PDMS (Figure 18.28). A flap valve by design, the large vertical PDMS flap deflects to the right, but is prevented from deflecting to the left, creating a fluidic diode.

PDMS permeability to oxygen is a benefit in cell growth chambers, but in many applications this permeability is a drawback: drug stability is compromised if oxygen permeates through the device and makes contact with the drug, and in PCR DNA amplification at 95 °C operating temperature water will be vaporized and liquid volume will decrease over time. Of course other materials can be used instead of PDMS, or else additional layers can be added to the structure: for example, in Figure 18.29 a low-permeability Mylar film is bonded to PDMS to reduce vapor losses.

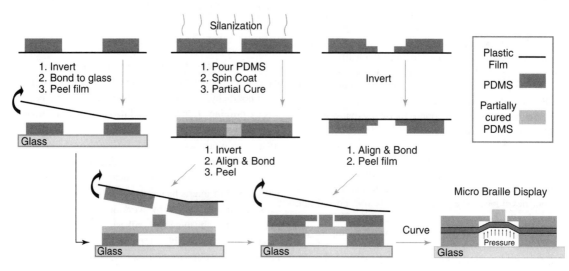

Figure 18.27 Fabrication steps of a Braille actuator: bonding of three PDMS pieces. Reproduced from Moraes *et al.* (2009) by permission of IOP

Polymer Microprocessing   221

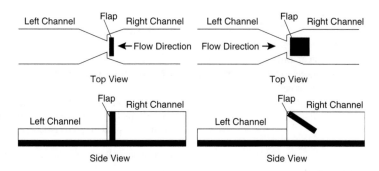

Figure 18.28 PDMS fluidic diode: left-to-right flow is allowed, right-to-left flow is blocked. Adapted from Adams *et al.* (2005)

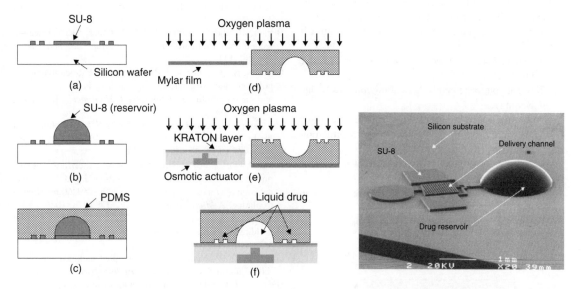

Figure 18.29 Drug delivery device: SU-8 mold for PDMS casting, Mylar vapor barrier. Reproduced from Su *et al.* (2002), copyright 2002, by permission of IEEE

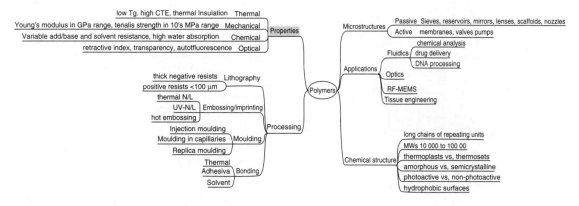

Figure 18.30 Polymer microprocessing and devices

18.11 Polymer Overview

Polymers are a vast group of materials, with widely different chemical structures, physical properties, processing methods and device applications. The mind map of Figure 18.30 lists some important aspects of polymers for microfabrication.

18.12 Exercises

1. If SU-8 structures 50 μm thick need to be exposed, what is the smallest feature size that can be fabricated?
2. What degree of positional repeatability can be achieved in stamping if the polymer stamp CTE is 60 ppm/°C and the cleanroom temperature control is ±1 °C?
3. Design a fabrication process for the SU-8 microvalve of Figure 18.7.
4. How can you make hemispherical microlenses by hot embossing?
5. Draw a time–height plot to show how feature filling evolves in imprinting/embossing: (a) for submicrometer features; (b) for 20 μm features.
6. What ratios of original NIL resist height to residual thickness are feasible?
7. Calculate silicon wafer bow due to stress from an SU-8 layer 200 μm thick.
8. Explain the fabrication process of the fluidic diode of Figure 18.28.
9. Design a fabrication process for the three-layer PDMS valve shown below.

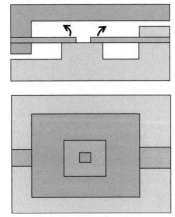

10. Invent a process for making superhydrophobic surfaces by NIL using real lotus leaves as a starting point.
11. Explain step by step the fabrication of an SU-8 channel with electrodes as in Figure 18.26.

References and Related Reading

Abgrall, P. *et al.* (2006) Novel fabrication method of flexible and monolithic 3D microfluidic structures using lamination of SU-8 films, *J. Microelectromech. Syst.*, **16**, 113–121.

Adams, M.L. *et al.* (2005) Polydimethylsiloxane based microfluidic diode, *J. Micromech. Microeng.*, **15**, 1517–1521.

Aura, S. *et al.* (2008) Novel hybrid material for microfluidic devices, *Sens. Actuators*, **B132**, 397–403.

Beck, M. *et al.* (2002) Improving stamps for 10 nm level wafer scale nanoimprint lithography, *Microelectron. Eng.*, **61–62**, 441–448.

Becker, H. and C. Gärtner (2008) Polymer microfabrication technologies for microfluidic systems, *Anal. Bioanal. Chem.*, **390**, 89–111.

Chen, K.-S., K. Lin and F.-H. Ko (2005) Fabrication of 3D polymer microstructures using electron beam lithography and nanoimprinting technologies, *J. Micromech. Microeng.*, **15**, 1894–1903.

del Campo, A. and C. Greiner (2007) SU-8: a photoresist for high-aspect-ratio and 3D submicron lithography, *J. Micromech. Microeng.*, **17**, R81–R95.

Dixit, P. *et al.* (2007) Fabrication and characterization of fine pitch on-chip copper interconnects for advanced wafer level packaging by a high aspect ratio through AZ9260 resist electroplating, *J. Micromech. Microeng.*, **17**, 1078–1086.

Giboz, J., T. Copponnex and P. Mele (2007) Microinjection molding of thermoplastic polymers: a review, *J. Micromech. Microeng.*, **17**, R96–R109.

Guo, J.L. (2007) Nanoimprint lithography: methods and material requirements, *Adv. Mater.*, **19**, 495–513.

Haas, K.-H. (2004): Hybrid Inorganic-organic polymers based on organically modified Si-alkoxides, *Adv.Eng.Mater.* pp. 571–582

Heckele, M. and W.K. Schomburg (2004) Review on micro molding of thermoplastic polymers: *J. Micromech. Microeng.*, **14**, R1–R14.

Kang, W.-J. *et al.* (2006) Novel exposure methods based on reflection and refraction effects in the field of SU-8 lithography, *J. Micromech. Microeng.*, **16**, 821–831.

Kim, E., Y. Xia and G.M. Whitesides (1996) Micromolding in capillaries: applications in materials science, *J. Am. Chem. Soc.*, **118**, 5722–5731.

Koukharenko, E. *et al.* (2005) A comparative study of different thick photoresists for MEMS applications, *J. Mater. Sci.: Mater. Electron.*, **16**, 741–747.

Kudo, H. *et al.* (2006) A flexible and wearable glucose sensor based on functional polymers with Soft-MEMS techniques, *Biosens. Bioelectron.*, **22**, 558–562.

Lee, D.-W. and Y.-S. Choi (2008) A novel pressure sensor with a PDMS diaphragm, *Microelectron. Eng.*, **85**, 1054–1058.

Lee, S.J. *et al.* (2007) Characterization of laterally deformable elastomer membranes for microfluidics, *J. Micromech. Microeng.*, **17**, 843.

Lin, B.-C. and Y.-C. Su (2008) On-demand liquid-in-liquid droplet metering and fusion utilizing pneumatically actuated membrane valves, *J. Micromech. Microeng.*, **18**, 115005.

Loechel, B. (2000) Thick-layer resists for surface micromachining, *J. Micromech. Microeng.*, **10**, 108.

Luharuka, R. *et al.* (2006) Improved manufacturability and characterization of a corrugated Parylene diaphragm pressure transducer, *J. Micromech. Microeng.*, **16**, 1468–1474.

Mata, A., A.J. Fleischman and S. Roy (2006) Fabrication of multi-layer SU-8 microstructures, *J. Micromech. Microeng.*, **16**, 276–284.

Meng, E., P.-Y. Li and Y.-C. Tai (2008) Plasma removal of Parylene C, *J. Micromech. Microeng.*, **18**, 045004.

Moraes, C. *et al.* (2009) Solving the shrinkage-induced PDMS alignment registration issue in multilayer soft lithography, *J. Microelectromech. Syst.*, **19**, 065015.

Prakash, A.R. *et al.* (2006) Small volume PCR in PDMS biochips with integrated fluid control and vapour barrier, *Sens. Actuators*, **B113**, 398–409.

Pranov, H. *et al.* (2006) On the injection molding of nanostructured polymer surfaces, *Polym. Eng. Sci.*, **46**, 160–171.

Samel, B., M.K. Chowdhury and G. Stemme (2007) The fabrication of microfluidic structures by means of full-wafer adhesive bonding using a poly(dimethylsiloxane) catalyst, *J. Micromech. Microeng.*, **17**, 1710–1714.

Schift, H. (2008) Nanoimprint lithography: an old story in modern times? A review, *J. Vac. Sci. Technol.*, **B26**, 458–480.

Seidemann, V., S. Bütefisch and S. Büttgenbach (2002) Fabrication and investigation of in-plane compliant SU8 structures for MEMS and their application to micro valves and micro grippers, *Sens. Actuators*, **A97–98**, 457–461.

Seo, J. and L.P. Lee (2004) Disposable integrated microfluidics with self-aligned planar microlenses, *Sens. Actuators*, **B99**, 615–622.

Stephan, K. *et al.* (2007) Fast prototyping using a dry film photoresist: microfabrication of soft-lithography masters for microfluidic structures, *J. Micromech. Microeng.*, **17**, N69–N74.

Su, Y.-C., L. Lin and A.P. Pisano (2002) A water-powered osmotic microactuator, *J. Microelectromech. Syst.*, **11**, 736.

Sun, Y. and Y.C. Kwok (2006) Polymeric microfluidic system for DNA analysis: a review, *Anal. Chim. Acta*, **556**, 80–96.

Tiercelin, N. *et al.* (2006) Polydimethylsiloxane membranes for millimeter-wave planar ultra flexible antennas, *J. Micromech. Microeng.*, **16**, 2389–2395.

Tsao, C.-W. and D.L. DeVoe (2009) Bonding of thermoplastic polymer microfluidics, *Microfluid Nanofluid*, **6**, 1–16.

Tuomikoski, S. and S. Franssila (2005) Free-standing SU-8 microfluidic chips by adhesive bonding and release etching, *Sens. Actuators*, **A120**, 408–415.

Tuomikoski, S. *et al.* (2005) Fabrication of enclosed SU-8 tips for electrospray ionization-mass spectrometry, *Electrophoresis*, **26**, 4691–4702.

Worgull, M. *et al.* (2006) Modeling and optimization of the hot embossing process for micro- and nanocomponent fabrication, *J. Microlithogr. Microfabr. Microsyst.*, **5**, 011005.

Yin, H.-L. *et al.* (2007) A novel electromagnetic elastomer membrane actuator with a semi-embedded coil, *Sens. Actuators*, **A139**, 194–202.

19

Glass Microprocessing

The term glass is used as a shorthand for many different materials, but the common silicate glasses are amorphous silica with alkali metal oxides. Glass used for beverage bottles contains 72.5% SiO_2, 13% Na_2O, 9.3% CaO, with the remainder consisting of K_2O, MgO, Al_2O_3 and Fe_2O_3. This basic formula has remained pretty much unchanged since antiquity.

Borosilicate glasses are important in microsystems because their coefficients of thermal expansion can be matched with silicon, important for stress-free bonding. Each manufacturer has slightly different compositions, but Pyrex 7740 borosilicate glass suitable for anodic bonding has a composition of 80.3% SiO_2, 12.2% B_2O_3, 4.0% Na_2O, 2.8% Al_2O_3, 0.4% K_2O and 0.3% CaO. Other anodically bondable glass might replace Na_2O for Li_2O and include some Fe_2O_3 and TiO_2. The wide variability of glass properties can be seen in Table 19.1 where the properties of different borosilicate glasses do vary a lot.

There are at least 3000 varieties of glass. While silicon properties can be tailored by doping and crystal orientation, glasses can be tailored by a huge repertoire of methods. First of all, glass composition can be varied, and sometimes very small changes in composition lead to drastically different properties: changing Na_2O content from 15.75% to 15.5%, Al_2O_3 content from 3.4% to 2.2% and increasing MgO from 3.0% to 3.2% make glass biosoluble. Pyrex's thermal expansion can be tailored by its composition, but it is also affected by the cooling rate used during solidification. Tempering can be used to introduce compressive stress in the surface layers of glass, blocking the propagation of cracks and greatly improving mechanical strength.

The emphasis in this chapter will be on Pyrex-type glasses that are available in wafer format and widely used in MEMS and microfluidics. Flat-panel display processing on large glass panes is discussed in Chapter 37 and only briefly touched upon here.

19.1 Structure and Properties of Glasses

Glasses are metastable liquids without long-range order. Glass networks are based on Si–O bonds (Figure 19.1). These bonds form tetrahedral basic units (Figure 13.18). When dopant atoms are incorporated into a silicon dioxide network, they can take either substitutional or interstitial positions. Boron can take the position of a silicon atom in the network and form oxide (B_2O_3). Sodium, potassium and lead are interstitial network modifiers which bond to one silicon atom only, because they have a valence of one. Therefore they do not contribute to networked structures.

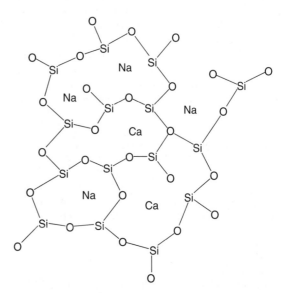

Figure 19.1 The structure of glass: Si–O network with metal atoms dispersed (the fourth bond of silicon is out-of-plane)

Introduction to Microfabrication, Second Edition Sami Franssila
© 2010 John Wiley & Sons, Ltd

As temperature rises, the electrical behavior of glass is not too different from that of semiconductors. Thermally generated charge carriers are important in many applications, but in glasses these thermally generated charge carriers are sodium ions, not holes or electrons. In Pyrex glass the room temperature resistivity is 10^{14} ohm-cm, at 250 °C it is 10^8 ohm-cm, but at 350 °C only 10^6 ohm-cm. Anodic bonding is also based on sodium ionic conduction (at 300–500 °C).

Glasses do not experience sharp melting but gradual softening. As a simplification the glass transition temperature is often used to characterize glasses: at T_g glass behavior changes from a brittle–elastic solid to viscous melt. Room temperature viscosities of glasses are 10^{18} Pa-s, and T_g is the temperature when viscosity equals $10^{12.3}$ Pa-s. The usable temperature range of glass wafers is limited by softening, which takes place at about 820 °C for Pyrex glasses. In practice the processing temperatures are limited to about 600 °C because even minor dimensional and shape changes, for example increased warp, may render bonding impossible and lithography out of focus. Table 19.1 lists some important properties of borosilicate glasses, and compares them to quartz glasses.

The proper term for pure amorphous SiO_2 is fused silica even though it is often called "quartz." It is very different from glasses: it does not exhibit temperature-dependent resistivity because there are no metal oxides which dissociate and release metal and oxygen ions.

One major microfabrication difference between soda lime glass and fused silica is optical transmission: the elimination of metals leads to superior optical transmission. Fused silica is transparent from NIR to deep UV while soda lime transmission in UV is strongly attenuated (Figure 8.9). This means that photomasks for DUV lithography must be made on "quartz" but above 400 nm exposure wavelength the much cheaper soda lime glass can be used. The coefficient of thermal expansion for soda lime glass is 10 ppm/°C (cf. 2.6 ppm/°C for Si) and as a photomask material soda lime glass is limited to applications above 3 μm linewidths where dimensional control requirements are lax (recall Exercise 9.4). Fused silica with its extremely low CTE retains its dimensions much better under intense UV illumination.

Quartz is pure crystalline silica. Because quartz is crystalline, it can be anisotropically wet etched by HF while glasses are etched isotropically because of their amorphous structure. And because of crystallinity, quartz properties are anisotropic, for example thermal conductivity parallel and perpendicular to the z-axis differs by a factor of two, and Young's modulus parallel to the z-axis is 97 GPa but 76 GPa perpendicular to it. These properties have been used to make micromechanical devices.

CVD oxides are sometimes called glasses, for example undoped silica glass (USG) and phosphorus-doped silica glass (PSG). These are not glasses proper because they do not contain metal oxides. They are more akin to silica in chemical composition. Spin-on glasses come in two main varieties, silicate (inorganic) and siloxane (inorganic–organic hybrid), but neither is glass in the traditional sense. All these "glasses" share some properties with glasses, like amorphous structure, electrical and thermal insulation, and optical transparency, but not ionic conductivity, or sodium contamination danger.

19.2 Glass Substrates

Glass can be cut into many shapes, and it is available in silicon-compatible formats: round wafers, even with wafer flat, even though glass is amorphous and flat in silicon is used to identify crystal orientation. Flats are useful when bonding glass to silicon wafers with flats.

The glass wafer fabrication process is not too different from the silicon wafering process. The main steps are:

1. Shape cut
2. Edge grinding
3. Lapping
4. Polishing
5. Cleaning
6. Inspection
7. Packaging

An example of glass wafer specification is shown in Table 19.2. Because there are many glass varieties, the specs are never as detailed or standardized as with silicon.

Table 19.1 Quartz vs. borosilicate glasses

	Quartz glasses	Borosilicates
Density (g/cm^3)	2.0–2.2	2.25–2.45
Tensile strength (MPa)	70–120	80–150
Compressive strength (MPa)	1600–2000	600–1000
Young's modulus (GPa)	62–75	65–85
CTE (10^{-6}/°C)	0.53	3–6
Resistivity (ohm-cm)	10^{18}–10^{19}	10^{14}–10^{18}
Dielectric constant ε_r	3.7–3.9	4.5–8
Thermal conductivity (W/m-K)	4.8	2.0–3.8
Glass transition[a] (°C)	1300	600

[a] Defined as temperature where viscosity $\eta = 10^{12.3}$ Pa-s.
Source: Hülsenberg, D., A. Harnicsch and A. Bismarck (2008)

Table 19.2 Glass wafer specifications

Diameter	100.0 ± 0.3 mm
Thickness	500 ± 25 µm
Bow/warp	<10 µm
Top side	Polished
Back side	Polished
Roughness	<1.5 nm RMS
TTV[a]	<10 µm
Primary flat	32.5 mm
Particles	<5 µm
Scratch and dig	60–40 according to MIL-PRF-13830
Edge exclusion	6 mm
Packaging	Under class 1000 according to Federal Standard 209

[a] Total Thickness Variation.

Glass wafer thicknesses range from 50 µm to 10 mm and in general glass wafers can be found with a very wide range of specs. In many respects glass wafer specs are similar to silicon wafer specs, for instance wafers with tighter TTV can be bought, or wafers with less than 0.5 nm RMS roughness (surface roughness is an important factor for bondability). Glass wafers are available up to 300 mm in diameter, just like silicon. Thick glass wafers are standard in the millimeter range. Glass is commonly available also as square or rectangular sheets, ranging in size from microscope slides (76 × 25 mm) to flat-panel display "mother panes" (2160 × 2460 mm).

19.3 General Processing Issues with Glasses

Some problems of glass substrates are related to their processing in silicon-oriented labs. Even though glass wafers are round like silicon, have flats like silicon and are available in the same thicknesses as silicon, complications can still arise, especially in automated tools. The detection of the presence and movements of wafers in processing equipment is based on either optical or capacitive sensors, and these are fooled by transparent dielectric wafers. Amorphous silicon or polysilicon deposition on the wafer back side can be used as a preventive measure, but the role of this extra film needs to be considered for all process steps and tools.

Glasses contain alkali metals which are serious contaminants to silicon transistors. There are various ways of dealing with this: use fused silica instead of glass (which will make bonding very difficult); limit glass processing to specific equipment and/or cleanroom areas; or take special precautions, like depositing barriers on glass to prevent sodium diffusion. This is a standard starting step in flat-panel display manufacturing, and it works: the thin-film transistors in displays are silicon devices which work well for extended periods.

In this section we will briefly discuss various processing issues and then devote more attention to two technologies, etching and bonding. Glasses are also machined by various non-standard techniques (non-standard from a microfabrication point of view), powder blasting for example. This will be discussed in Chapter 23. Various grinding, milling and drilling methods will be discussed in Chapter 24.

19.3.1 Surface preparation

Glass can be cleaned with RCA cleans just like silicon wafers; in fact, the RCA clean was invented for glass cleaning in TV tube manufacturing. The surface of the glass is chemically identical to oxidized silicon, with Si–OH as the typical surface termination. Surface functionalization techniques developed for one can easily be transferred to the other, like silane coatings for hydrophobicity.

19.3.2 Lithography

The exposure of resist on a glass wafer requires more energy than exposure on silicon because there is no reflected light. An additional 50% of the exposure dose might be used as a starting point. Reflections from etched structures or thin films need to be accounted for, because they can cause unwanted exposure (see Figure 10.9).

Fabrication on glass substrates offers intriguing ways for self-alignment by back-side exposure. A bottom gate thin-film transistor (TFT) is shown in Figure 19.2. After chromium bottom gate lithography, etching and stripping, a stack of PECVD oxide (gate oxide), a-Si:H (channel) and nitride (passivation) are deposited. Photoresist is applied on the top side, but exposure is made from the back side, with the Cr gate blocking light (photomasks are glass plates with chromium patterns on them!). Resist is then developed and nitride etched. After resist stripping and wafer cleaning, chromium is deposited. During annealing, chromium silicide will form on the a-Si layers, but not on the nitride (recall Figure 7.15).

19.3.3 Deposition on glass

Basically, deposition on glass is identical to deposition on silicon, with only minor differences. Sputtering and evaporation, and PECVD, can be done as on silicon. With thermal CVD temperature limitations arise: for example,

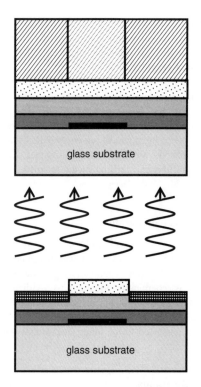

Figure 19.2 Top: Cr gate has been patterned on the glass substrate, and PECVD gate oxide, a-Si:H channel and nitride stopper layers have been deposited. Bottom: resist on the top side is exposed through the back side, making S/D metallization self-aligned to the gate. Redrawn after Hirano et al. (1996)

polysilicon deposition by LPCVD around 600 °C is at the limit of mechanical strength of glass wafers. Another issue is contamination: very few laboratories allow polysilicon deposition on glass wafers.

Film stresses are always present. When processing thick SU-8 resist on thin microscope glass, considerable bending can occur, even rendering the whole device useless. Thicker glass would help, but sometimes optical thickness considerations require thin substrates.

Because glass is used in optical devices, transparent conducting oxides (TCOs) are often needed, and indium-doped tin oxide (ITO), fluorine-doped tin oxide (FTO) and the like are used. These films can be deposited by vacuum techniques like sputtering, evaporation and (PE)CVD, but in large-area devices like flat-panel displays (FPDs) and solar cells, pyrolysis and various wet processes are also used.

Another big optical application is windowpane coating for self-cleaning and optical and thermal properties.

Titanium dioxide in its crystalline anatase phase is a photocatalyst that can break down organic dirt on windows, which is then flushed away by rain. Thin-film multilayer filters are deposited on glass, for example by tailoring infrared reflectivity for minimizing thermal losses in windows or improving windshield optics. Metallic and dielectric layers with thicknesses from 10 nm onward are sputtered by the square meter in the glass industry for building and automotive applications. Over 90% of the sputtering target industry (by weight) is geared up for these applications.

19.3.4 Glass polishing

While silicon microtechnology adopted polishing by CMP only recently, glass polishing for microscope and telescope lenses is 400-year-old technology. In fact, early CMP equipment was directly borrowed from the glass polishing industry. Melted glass as well as spin-coated glass can be polished to create smooth surfaces, for either ease of further processing, or optical properties.

19.3.5 Glass embossing, molding and blowing

Glass softens upon heating and embossing can be done to create glass microstructures. The principles are the same as discussed for polymer embossing in Sections 18.3–18.5. First of all, embossing must be done above T_g. For glasses the transition temperatures are around 400–800 °C. Second, reaction between master and glass must be avoided. Because of the higher temperatures involved in glass embossing, this is more difficult than with polymers, but non-oxidizing materials like TiN and TiAlN have been successfully used. For de-embossing, the surface quality of the master is important, and microfabricated silicon and nickel promise smooth-enough surfaces.

Glass molding is an old technique for lens fabrication: instead of polishing each lens individually, molten glass is poured over a mold, and lenses are formed. The basic idea and lens shapes are identical to polymer lenses (Figure 18.10), but of course the mold material has to be high-temperature resistant. The method is best suited for cheap mass-produced lenses.

The silicon mold can be made by RIE, enabling delicate shapes that are not possible with glass etching. Figure 19.3a shows the starting situation: the silicon mold has been filled with borofloat glass (at temperatures around 800 °C). Three possible process options then emerge: in Figure 19.3b glass etching and polishing end up with thick glass structures embedded in silicon. These might be used as thermal isolation islands. In

to thermal oxide etching. Hydrofluoric acid etches silica according to Equations 19.1 and 19.2. The resulting etch profile is isotropic (Figure 19.5):

$$SiO_2 \text{ (s)} + 4 \text{ HF (aq)} \rightarrow SiF_4 \text{ (ad)} + 2 \text{ H}_2\text{O} \quad (19.1)$$

$$SiF_4 \text{ (ad)} + 2 \text{ HF (aq)} \rightarrow H_2SiF_6 \text{ (aq)} \quad (19.2)$$

The insoluble SiF_4 is turned to soluble H_2SiF_6. Pyrex and other glasses are much more difficult to etch because insoluble fluorides are formed, for example AlF_3 and CaF_2. These etch products slow down the etch rate but are removed by undercutting.

The Pyrex etch rate in hydrofluoric acid (49%) is about 8 µm/min at room temperature. Etching 300 µm can be done in 40 min, and the resulting surface roughness is about 20 nm. A microfluidic nozzle made by double-sided etching of a Pyrex wafer 500 µm thick is shown in Figure 19.5. Amorphous LPCVD silicon was used as the etch mask. The addition of some HCl to HF improves smoothness: the insoluble metal fluorides are dissolved by chlorine, and surface roughness below 10 nm can be obtained in deep etching. HCl has only a minor effect on etch rate and the most common mask materials behave similarly with regard to HF and 10:1 HF:HCl mixture. But remember that, because each manufacturer has a proprietary glass composition, glass etch rates are variable. High aluminum oxide content will reduce the etch rate, and the same applies to magnesium and calcium.

When glass or fused silica wafers are etched, the etched depth can be anything up to 500 µm, and the issue of etch mask stability becomes paramount. This applies to both

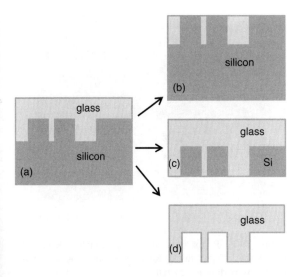

Figure 19.3 (a) Silicon mold made by RIE, filled by molten glass with three alternative process continuation options: (b) glass etching/polishing to planarize; (c) silicon etching/polishing to planarize; and (d) complete silicon etching

Figure 19.4 Polysilicon micromotor before glass coating and after glass casting and polishing. Reproduced from Yasseen *et al.* (1999), copyright 1999, by permission of IEEE

Figure 19.3c silicon etching and polishing result in embedded silicon islands inside glass. These might be used as conductive paths or as thermal sinks. Finally, in Figure 19.3d all the silicon has been etched away to create glass microstructures of high aspect ratio. One example is shown in Figure 19.4, namely a polysilicon mold, with and without molded glass.

19.4 Glass Etching

19.4.1 Wet etching

Etching of glass follows the same basic ideas as etching of silicon dioxide. Fused silica etching is very similar

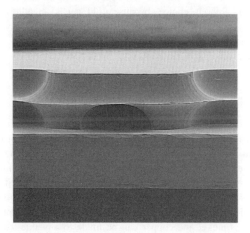

Figure 19.5 Wet etching Pyrex wafer from both sides: a-Si mask, 49% HF. Reproduced from Saarela *et al.* (2009) by permission of IOP

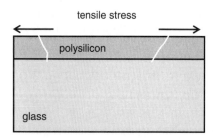

Figure 19.6 Tensile stresses in thin films tend to open cracks which allow etchant to penetrate through the mask

wet and dry etching, but for different reasons. Photoresist masks fail by adhesion loss. Concentrated HF is a very aggressive etchant, and it etches rapidly under the mask. If there is, for example, a water-rich interface layer between the wafer and resist, etchant will penetrate very fast laterally. Faster lateral than vertical etching is often seen in isotropic etching. In wet etching, mask mechanical strength is important, especially when large undercut develops, and long overhangs are formed.

Common hard masks for glass etching include metals (Ni, Cr, Au) and CVD films: PECVD a-Si and SiC and LPCVD a-Si. In theory all these are resistant to HF, but often fail in practice. The main explanation is tensile film stress. Any cracks in tensile-stressed film tend to enlarge as shown in Figure 19.6, and the etchant will eventually penetrate through the mask. The surface of the mask matters: if it is hydrophobic, it lasts longer because it will reject aqueous etchant, while if it is hydrophilic the surface will attract etchant.

Double layer films are often used as etch masks: Cr/Au is an obvious choice, because gold needs an adhesion promotion layer. But it is useful to deposit multiple layers of gold, as the later layers will seal some of the cracks in the preceding films, improving mask quality.

19.4.2 RIE

Oxide etching is driven by ion bombardment, therefore anisotropy is good, but mask selectivity is low. The problem of mask erosion really arises when deep etching is done. Non-erodible masks for fluorine plasmas include many metals (Ni, Cr, Al, Cu), oxides and nitrides (Al_2O_3, AlN). If these are deposited by PVD or CVD, the practical thickness limit is a few micrometers. Electroplated metals nickel and copper can be made to any thickness. The same applies to thick resists. Thick-resist SU-8 has been found to be quite a good mask for glass DRIE. The stability of SU-8 results in the problem of resist stripping

Figure 19.7 Pyrex glass DRIE: 80 μm etched depth with nickel mask 5 μm thick. Reproduced from Kolari *et al.* (2008) by permission of IOP

afterward. A profile of plasma-etched Pyrex is shown in Figure 19.7. The sidewall angle is steep, but not vertical.

Oxide and glass RIE is critically dependent on the exact composition of the material. In both plasma and wet etching pure silica (SiO_2) is easier to etch than glass with metal oxides (Na_2O, CaO, K_2O, MgO, Al_2O_3, Fe_2O_3). In plasma etching the lower volatility of the metal fluorides results in a lower rate and larger risk of residues.

Silicon is etched in fluorine plasmas but if the silicon is thick enough, and selectivity reasonable, deep glass etching can be done with a silicon mask. Complete silicon wafers with through-wafer holes have been used as etch masks for glass DRIE. Both anodic bonding and simple mechanical clamping have been employed.

One RIE etch application of glass/fused silica is making masters for UV NIL (see Figure 18.19). There the problem is not etched depth or mask erosion but feature size control, sidewall angle control and smoothness.

It should be remembered that there are good alternatives to glass etching. One technique is to make a silicon mold and cast molten glass on it, and remove the silicon (Figure 19.3), leaving glass microstructures whose dimensional accuracy is determined by the silicon DRIE process. In the case of larger structures drilling, powder blasting and spark-assisted etching can be used. There are also laser processes to pattern glass, and some of these are described in Chapter 24.

19.5 Glass Bonding

All the usual requirements for successful bonding apply to glass bonding: flat, smooth surfaces, cleaned of particles and with proper bond forming surface chemistry, are needed. Various surface cleaning and activation processes have been tried, including RCA-1, sulfuric acid, nitric acid, KOH, acetone and other solvents. Because the surface chemistry of glasses is similar to oxidized silicon, RCA cleans (ammonia/peroxide, see Chapter 12) are most common.

Thermal fusion bonding takes place above the glass transition temperature, which translates to about 600 °C for Pyrex glasses. Bonding times range from 1 to 6 hours, with extended cooldown. Specialty glasses may have much lower or higher values of T_g, but fusion bonding must be between two identical wafers. Fused silica bonding can be done around 1000 °C, but because fused silica does not soften like glasses, bonding is very sensitive to flatness and smoothness deviations.

Anodic bonding of glass and silicon (Section 17.3) is a very special case of direct bonding, and it requires special properties from the glass, namely suitable sodium concentration and matching CTE. A glass wafer is connected as a cathode and a silicon wafer as an anode. Current increases rapidly at the initiation of bonding because the contact area increases, and then decreases exponentially as oxygen ions react at the interface to form SiO_2 and the oxide becomes thicker (Figure 19.8). When the current has dropped to 10% of its peak value, bonding is termed finished.

Bonding glass to glass anodically is possible if thin films are deposited on one of the wafers. Silicon nitride, silicon carbide, polysilicon and a-Si can be used. When a voltage is applied, sodium ions migrate toward the interface, which acts as a diffusion barrier for sodium (silicon dioxide does not work because it is not a barrier for sodium). A depletion layer is formed and electrostatic force pulls the wafers strongly together, just as in standard anodic bonding. Oxygen ions diffuse toward the interface and react there to form Si–O bonds. The interlayer film should be free of hydrogen, because otherwise oxygen

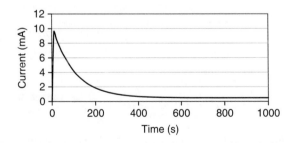

Figure 19.8 Anodic bonding current vs. time: initial current peak, followed by a long tailout

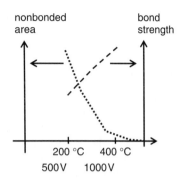

Figure 19.9 Non-bonded area and bond strength as a function of temperature and voltage, for anodic bonding of two glass wafers with polysilicon interlayer. Adapted from Wei *et al.* (2003)

will react with the hydrogen and not with silicon. Other outgassing products from PECVD films are also problematic. Bondable area and bond strength increase with voltage and temperature as expected (Figure 19.9), but often it is desirable to have reasonable bond strength at low temperature.

Glass–glass bonding can be done in a liquid environment. Two cleaned wafers with a proper surface finish are brought together under liquid, and capillary forces draw the wafers together (this is something that must be avoided in release etching of micromechanical free-standing structures, as will be discussed in Chapter 29). If the surfaces are clean and smooth, fairly strong bonds can form even at room temperature, and annealing can be done to increase bond strength.

Various wet cleaning strategies have been developed for liquid bonding. Sulfuric acid (with and without peroxide) cleaning has been shown to work, with or without RCA-1 (ammonia/peroxide) activation. The actual bonding takes place under a DI water stream

Another bonding technique uses spin coating of a sodium silicate layer, a few tens of nanometers thick, on top of one of the wafers, followed by a 90 °C anneal for 1 h. Bond strengths similar to silicon–silicon fusion bonding are possible, showing that Si–O–Si bonds have been formed at high density across the interface.

Adhesive bonding is an option in many microfluidic devices. Again, the principles described in Chapter 17 apply here: adhesive bonding is a low-temperature process which is insensitive to some particle contamination, and it is quite general purpose.

Because glasses are transparent, UV curing adhesives are handy. But adhesive bonding introduces additional

19.6 Glass Devices

Capillary electrophoresis chips are the most studied microfluidic devices. They can be made of polymers but glass devices are also widely used. Most often they are made simply by etching fluidic channels in one wafer (in the case of polymers, by casting or embossing) and bonding it to a blanket wafer which has the fluidic inlets. In case the capping wafer is the same material, thermal or fusion bonding is performed, or a PDMS cover can be attached (permanently or intermittently). The chips do not usually have metal electrodes integrated; rather Eppendorf tubes are attached to fluidic inlets and platinum wires stuck in. This works well for high-voltage electro-osmotic pumping, but if electrochemical detection is used, the detection electrodes need to be integrated in the fluidic channels (recall Figure 18.26).

In the capillary electrophoresis (CE) device shown in Figure 19.10 the fluidic channels are HF etched in the top wafer, and fluidic inlets are drilled in the top wafer, too. Electrodes are fabricated in the bottom wafer. In order to have a flat surface for fusion bonding, the metal electrodes are recessed (Figure 19.11). A glass wafer is etched to the same depth as the designed metal thickness. After etching of the glass, the resist is dried by baking, and used again, as a lift-off mask for metal deposition. Next, an acetone wash removes photoresist, and together with it the metal that was deposited on top (see Chapter 23 for more on lift-off). The resulting metal surface is fairly close to the

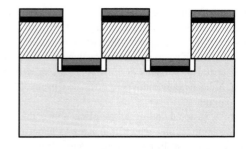

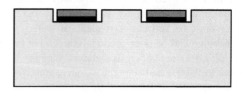

Figure 19.11 Embedded metal electrodes: photoresist masked etching of glass wafer, metal deposition without removing the resist, followed by lift-off. Metal thickness is designed to match the etched depth. Redrawn from Keynton (2005)

original glass surface, except for minor inaccuracies in both etched depth and metal thickness.

Glass–glass fusion bonding is also used in the microfluidic nebulizer shown in Figure 19.12. The nebulizer has heater electrodes that vaporize incoming liquid and mix it with nebulizer gas, shooting out a hot vapor jet from a nozzle. This vapor jet is ionized and the ions analyzed by a mass spectrometer. The same device can be made of silicon but glass is preferred because it is thermally insulating: in order to vaporize high-boiling-point analytes,

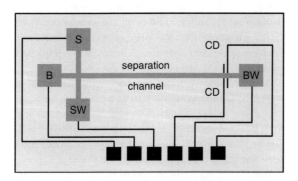

Figure 19.10 Capillary electrophoresis (CE) chip: fluidic channels (gray) etched in top glass wafer and electrodes and pads (black) in bottom glass wafer. Fluidic reservoirs for buffer (B), sample (S), sample waste (SW) and buffer waste (BW) have high-voltage electrodes. Low-voltage electrodes are for conductivity detection (CD)

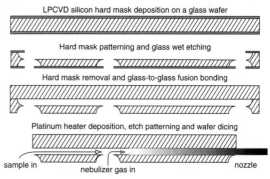

Figure 19.12 Fusion bonded glass–glass fluidic nebulizer chip. Reproduced from Saarela et al. (2007), Royal Society of Chemistry

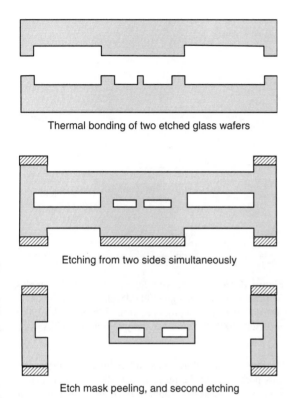

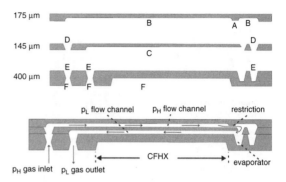

Figure 19.14 Cryogenic cooler with three wafers fusion bonded. Reproduced from Lerou *et al.* (2006) by permission of IOP

Figure 19.13 Glass–glass bonded PCR reactor. Redrawn from Easley *et al.* (2007)

the chip operates at 400 °C; in a glass chip this hot zone is local, but in a silicon chip heat spreads all over.

When fast temperature ramping is required, it is essential to minimize thermal mass. This can be achieved by etching away excessive glass. In the PCR reactor of Figure 19.13 this has been done after bonding. The channels end up being suspended, with thin roofs and floors. Timed etching was used, so ultimate thinness cannot be achieved.

Glass is the material of choice for not only high-temperature devices like the nebulizer and the PCR reactor, but also for cryogenic devices, and because of the thermal insulation. The cryogenic cooler of Figure 19.14 is fabricated from three glass wafers, requiring seven photomasks. The top wafer is 175 μm thick, and flow channels 50 μm deep are fabricated by HF etching. The middle wafer is 145 μm thick, and again the flow channels are HF etched, but the through-wafer vias are made by powder blasting. In the bottom wafer all the structures are powder blasted because they are large, deep and non-critical, for example the bottom wafer is thinned to improve thermal isolation. Fusion bonding of the three wafers is done in a single step.

Glass blowing is an old technique that has been given a microfabrication twist. A thin glass sheet is bonded in air to a cylindrical cavity formed by DRIE (Figure 19.15). When the device is heated to 850 °C, the trapped air expands, blowing the thin, softened glass sheet into a spherical bubble. Again in the spirit of traditional glass blowing, the structure is annealed at 560 °C for 30 min to remove stresses and slowly cooled below the strain point, 510 °C for Pyrex. In further processing the device is filled with rubidium chloride and barium, for atomic clock action.

The DNA processing chip of Figure 19.16 consists of three glass wafers and a PDMS layer. The top glass wafer holds the fluidic channels, the second glass wafer the thermal elements, namely heaters and temperature sensors. The PDMS layer acts as a movable membrane valve and is actuated by vacuum/air pressure in the bottom manifold wafer. Microfluidic chips like this offer benefits in DNA processing in many ways: small sample volumes (nanoliters) are important for reagent cost minimization, small size is also important for rapid temperature ramping, and the integration of many analysis steps on one device eliminates manual labor, saves time and improves reproducibility. The PCR chip of Figure 25.7 is similarly of glass–polymer hybrid design.

19.7 Specialty Glasses

The variety of glasses is vast, so we offer here only a short glimpse of some specialty glasses. Vycor is porous glass. Various annealing steps cause phase separation, and etching in HCl and NaOH removes practically everything else except silica (with a few percent sodium borate). By

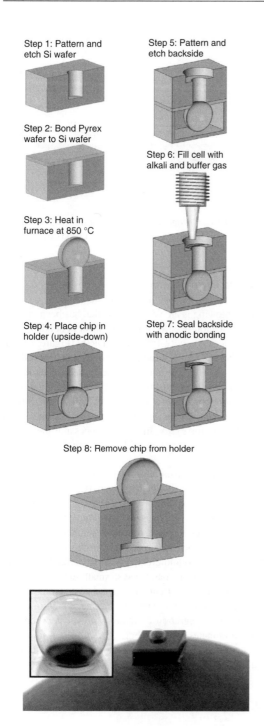

Figure 19.15 Blown glass sphere atomic clock. Reproduced from Eklund et al. (2008), copyright © 2008 by permission of Elsevier

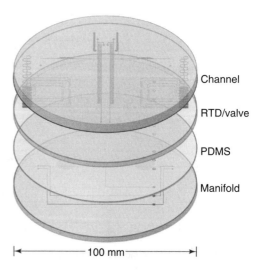

Figure 19.16 Bioprocessor for nanoliter DNA samples, fabricated on a glass/glass/PDMS/glass wafer stack of 100 mm diameter. Reproduced from Blazej et al. (2006), copyright 2006 National Academy of Sciences USA

tailoring the annealing steps and etching conditions, various pore sizes in the range of 10–100 nm are available. Such porous glass has a huge surface area which is useful in many biochips.

Photostructurable glasses (e.g. Foturan) can be exposed by UV light, and after annealing the exposed areas can be anisotropically etched in HF. These glasses contain about 70% silica, 10% Li_2O, and the usual alkali metal oxides plus small amounts (<1%) of cerium, silver and antimony and tin oxides. When hit by UV radiation cerium ions absorb the light and turn Ce^{3+} into Ce^{4+} and release an electron, which reacts with Ag^+ to form metallic silver. During the thermal treatment atomic silver forms nuclei and lithium metasilicate crystallizes. This crystallized phase can be etched in 10% HF with 20:1 selectivity to amorphous glass. The etch rate can be up to 20 μm/min. Etched depth is controlled by UV exposure dose. For example, 8 J/m^2 is used to expose through the wafer, while 3 J/m^2 exposes half the wafer thickness.

The problem is that crystal size is quite large, in the range of 10 micrometers, which means that small structures cannot be made and sidewalls will be rough. T_g values of photostructurable glasses are around 500 °C with CTEs ranging from 7 to 10 ppm/°C, which means that bonding with silicon is out of the question. Figure 19.17 shows a MEMS combustor. The middle wafer is made of photostructurable glass, including the movable piston. De-

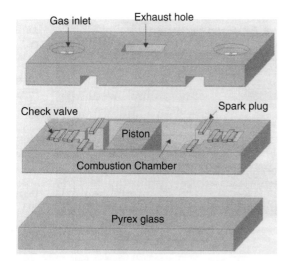

Figure 19.17 MEMS combustor and reciprocating device. Reproduced from Lee *et al.* (2002) by permission of IOP

spite gas leakage between the movable piston and frame, 6 mm movement has been achieved within 0.1 s.

19.8 Exercises

1. Design a process for molding glass microlenses.
2. List the processes and phenomena that become important for borosilicate glass as temperature goes up.
3. Estimate the device dimensions for the PCR reactor of Figure 19.13. How thin a glass is there after etching?
4. How could the glass melting process of Figure 19.3 be used to fabricate a thermally insulated PCR reactor?
5. In the cryogenic cooler of Figure 19.14 some structures are wet etched, others powder blasted. Which ones and why?
6. Design an alternative version of the nebulizer of Figure 19.12 but with the heater inside the channel!

References and Further Reading

Bien, D. C. S. *et al.* Characterization of masking materials for deep glass etching, *J. Micromech. Microeng.*, **13**, S34.

Blazej, R., Kumaresan, P. and Mathies, R. A. (2006) Microfabricated bioprocessor for integrated nanoliter-scale Sanger DNA sequencing, *Proc. Natl Acad. Sci.*, **103**, 7240–7245.

Danel, J. S. and G. Delapierre (1991) Quartz: a material for microdevices, *J. Micromech. Microeng.*, **1**, 187–198.

Easley, C. J., J. A. C. Humphrey and J. P. Landers (2007) Thermal isolation of microchip reaction chambers for rapid non-contact DNA amplification, *J. Micromech. Microeng.*, **17**, 1758–1766.

Eklund, E. J. *et al.* (2008) Glass-blown spherical microcells for chip-scale atomic devices, *Sens. Actuators*, **A143**, 175–180.

Hirano, N. *et al.* (1996) A 33cm diagonal high-resolution TFT-LCD with fully self-aligned a-Si TFT, *IEICE Trans. Electron.*, **E79**, 1103.

Hülsenberg, D., A. Harnicsch and A. Bismarck (2008) **Microstructuring of Glasses**, Springer.

Iliescu, C., B. Chen and J. Miao (2008) On the wet etching of Pyrex glass, *Sens. Actuators*, **A143**, 154–161.

Keynton, R. S. (2004) Design and development of microfabricated capillary electrophoresis devices with electrochemical detection, *Anal. Chim. Acta*, **507**, 95–105.

Kim, M.-S. *et al.* (2005) Fabrication of microchip electrophoresis devices and effects of channel surface properties on separation efficiency, *Sens. Actuators*, **B107**, 818–824.

Kolari, K., V. Saarela and S. Franssila (2008) Deep plasma etching of glass for fluidic devices with different mask materials, *J. Micromech. Microeng.*, **18**, 064010.

Lee, D. H. *et al.* (2002) Fabrication and test of a MEMS combustor and reciprocating device, *J. Micromech. Microeng.*, **12**, 26–34.

Lee, J.-H. *et al.* (2008) A simple and effective fabrication method for various 3D microstructures: backside 3D diffuser lithography, *J. Micromech. Microeng.*, **18**, 125015.

Lerou, P. P. P. M. *et al.* (2006) Fabrication of a micro cryogenic cold stage using MEMS-technology, *J. Micromech. Microeng.*, **16**, 1919–1925.

Saarela, V. *et al.* (2007) Glass microfabricated nebulizer chip for mass spectrometry, *Lab Chip*, **7**, 644–646.

Saarela, V. *et al.* (2009) Microfluidic heated gas jet shape analysis by temperature scanning, *J. Micromech. Microeng.*, **19**, 055001.

Wang, H. Y. *et al.* (1997) Low temperature bonding for microfabrication of chemical analysis devices, *Sens. Actuators*, **B45**, 199–207.

Wei, J. *et al.* (2003) Low temperature glass-to-glass wafer bonding, IEEE Trans. Adv. Packag., **26**, 289–294.

Yasseen, A. A., J. D. Cawley and M. Mehregany (1999) Thick glass film technology for polysilicon surface micromachining, *J. Microelectromech. Syst.*, **8**, 172–179.

20

Anisotropic Wet Etching

Microsystems technology relies on anisotropic wet etching of silicon for many major applications. Bulk micromechanics, the machining of silicon by anisotropic wet etching, depends on silicon crystal plane-dependent etching, and many surface micromechanical and SOI devices make use of silicon wet etching for auxiliary structures, even though the main device features are defined by plasma etching. Because <100> silicon is the workhorse of microsystems, our discussion concentrates on it. Both <110> and <111> etching will be briefly discussed.

Etched grooves, trenches and wells exemplify the basic features of crystal plane-dependent etching. They can be used as sample wells and flow channels in microfluidics, or as optical fiber alignment fixtures. Other basic structures are membranes (diaphragms), beams and cantilevers. Mechanical devices like pressure sensors, resonators and AFM cantilevers rely on these basic elements. Through-wafer structures include nozzles and orifices, for example for ink jets or micropipettes.

20.1 Basic Structures on <100> Silicon

Anisotropic etching relies on aligning the structures with wafer crystal planes. The primary flat, which is along the [110] direction, is used as a reference. Rectangular structures with concave corners are easily made, with four (111) sidewalls and the (100) plane as the bottom (Figure 20.1).

Self-limiting depth is the depth were the slow etching (111) planes meet. The angle between the (100) and (111) planes is 54.7° and the self-limiting depth (d) for mask opening W_{mask} is calculated from

$$d = \frac{W_{mask}}{\sqrt{2}} \qquad (20.1)$$

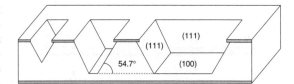

Figure 20.1 Anisotropic wet-etched profiles in <100> wafer. The sloped sidewalls are the slow etching (111) planes; the horizontal planes are (100). Etching will terminate if the slow etching (111) planes meet

A pattern 742 μm wide on the mask will then result in a through-wafer "zero-sized" hole, for a wafer 525 μm thick.

Typical mask films for anisotropic wet etching, thermal oxide and LPCVD nitride are deposited on both sides of the wafer. This is useful in protecting one side while processing the other, but it can also be utilized to speed up processing: this was used in microthruster (Figure 11.7) etching from both sides simultaneously. In the AFM cantilever–tip structure (Figure 20.2) the tip is etched first, followed by oxide etch and p^{++} diffusion. Another oxidation is done for backside mask and front side protection.

20.2 Etchants

A number of alkaline etchants have been tried for crystal plane-dependent etching but KOH has emerged as the main etchant. Typical etch rates are about 1 μm/min which translates to 6 hours for through-wafer etching of 380 μm wafers. KOH poses a contamination hazard for CMOS work, therefore CMOS-compatible etchants are desirable. Tetramethyl ammonium hydroxide, $(CH_3)_4NOH$, usually known as TMAH, is such a compound. In fact, both NaOH and TMAH are used as photoresist developers,

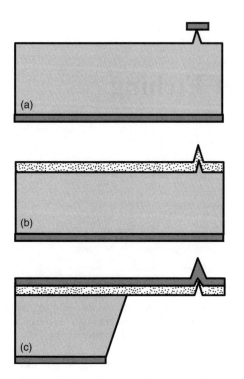

Figure 20.2 AFM cantilever and tip: (a) oxide-masked etching of tip; (b) p^{++} boron doping on front side; (c) Thermal oxidation and KOH etching from back side, stopping on p^{++} layer

in diluted concentrations and at room temperature, so the contamination danger can be handled with proper working procedures. Organic amines have also been used for anisotropic etching, most notably ethylene diamine $(NH_2)(CH_2)_2NH_2$ mixture with pyrocathecol and water, known as EDP or EPW. Hydrazine (N_4H_2) has also been tried. Both amines pose occupational safety and health hazards, and thus are not widely used. Ammonia has been shown to etch silicon reasonably well, but the stability of ammonia etch baths during extended etching needs special attention. Etch rates of 1.5 μm/min at 70 °C have been demonstrated, with high selectivity against oxide and aluminum masks, and very smooth surfaces of 2.4 nm RMS roughness, whereas typical KOH-etched surfaces exhibit 5–10 nm RMS roughness.

Even though all the alkaline etchants share the same basic features of etching (100) crystal planes quickly and (111) planes slowly, the actual selectivity between the crystal planes needs careful attention. KOH exhibits 200:1 selectivity between (100) and (111), whereas selectivity is only 30:1 in TMAH. Exact selectivities are dependent on etchant concentration and temperature. But when other crystal planes are considered, even more differences pop up: when planes like (110) and high-index planes like (311) are studied, the differences multiply. Figure 20.3 shows the etch rates for KOH. It is important for etch geometry which are the (local) maximum etch rate planes. Table 20.1 lists etch rate relative to (110).

Isopropanol (IPA) addition to KOH will change the relative etch rates of crystal planes and, depending on exact conditions, either the (100) or (110) planes will be the maximum etch rate planes. Other surfactants have also been used to modify crystal plane selectivities.

Because etch times are rather long, etchant stability over hours and days needs special attention. When silicon is etched in TMAH, the etched silicon dissolves

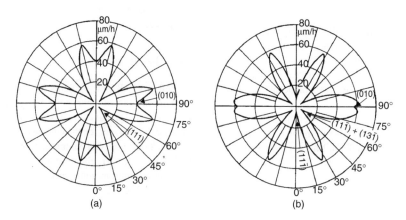

Figure 20.3 Etch rates in different crystal directions in 50% KOH at 78 °C: (a) <100> Si: fast, but not maximum etching in (010) direction; (b) <110> Si: (010) near maximum etch rate. Reproduced from Seidel *et al.* (1990a) by permission of Electrochemical Society Inc.

Table 20.1 Etch rates normalized to (110) crystal plane in 30% KOH

(110)	1.00	(100)	0.548
(210)	1.072	(211)	0.906
(221)	0.491	(310)	1.00
(311)	0.987	(320)	1.06
(331)	0.797	(530)	1.069
(540)	1.039	(111)	0.004

Source: Sato, K. *et al.* (1998).

into the etchant solution and changes its properties, especially selectivities. In order to stabilize the etchant, some silicon is dissolved into TMAH before etching, in a concentration higher than that caused by etched silicon. Pyrocathecol is employed in EDP for similar reasons: the decomposition of ethylene diamine releases small amounts of pyrocathecol, which changes the etchant composition, but if pyrocathecol is added in large amounts to begin with, the decomposition has negligible effect.

20.3 Etch Masks and Protective Coatings

Silicon dioxide and silicon nitride are the common masking materials for anisotropic wet etching. Oxide etch rate in KOH is a few nanometers per minute, while TMAH and EDP etch oxide hardly at all. Nitride is more resistant than oxide in both etchants. Silicon etch rate and mask etch rate depend on temperature and concentration but some general guidelines can be given. An oxide thickness of 2 μm is needed for through-wafer etching in KOH whereas 200 nm is sufficient in TMAH or EDP. Plasma oxides etch fastest and thermal oxides slowest. LPCVD nitride films of 200 nm are thick enough for almost any applications, but PECVD nitrides have widely varying properties.

As a practical issue it should be noted that thermal oxide and LPCVD nitride are furnace processes and film is grown/deposited on both sides of the wafer so that the wafer back side is protected. This is important when deep etching is done. PECVD deposition is usually on the front side of the wafer only.

All silicon etchants etch aluminum, which means that either aluminum deposition has to be done after silicon etching, or aluminum has to be protected during silicon etching. In some cases aluminum has been replaced by another metal, like gold. Some relief can be achieved by saturating TMAH solution with silicon, but typically only very short alkaline etchings are done after metallization.

20.3.1 Peeling masks/nested masks

Photoresist coating over severe topography can be eliminated by double masking (also known as peeling masks or nested masks): two different mask materials are patterned on a planar wafer, before the first silicon etching step (Figure 20.4). Mask 1 is discarded after the first silicon etching step, and silicon etching continues

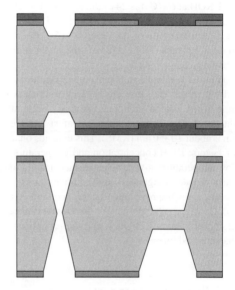

Figure 20.4 Peeling mask/nested mask: nitride (medium gray) deposition and patterning; CVD oxide (dark gray) deposition and patterning; first silicon etching; oxide etching in HF; second silicon etching with nitride mask

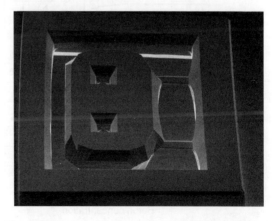

Figure 20.5 Symmetrical bulk silicon micromachined accelerometer: wafer 380 μm thick has been etched through from both sides simultaneously. SEM courtesy VTI Technologies

with the second mask. Combinations of resist, oxide and nitride are common. An accelerometer fabricated by peeling masks and double-sided symmetric wet etching is shown in Figure 20.5.

20.4 Etch Rate and Etch Stop

Silicon etch rate follows the Arrhenius exponential law: for 30% KOH at 60 °C it is 24 μm/h, at 80 °C it is 79 μm/h and at 100 °C it reaches 225 μm/h. KOH, TMAH and EDP practical etch rates are of the order of 0.5–1 μm/min for (100) crystal planes. Table 20.2 lists the major properties of alkaline etchants for silicon.

In addition to silicon etch rate, other factors must be considered: mask etch rate and surface roughness often increase when silicon etch rate increases. The surface roughness of deep (almost through-wafer) KOH etched silicon is about 10–50 nm (RMS). This depends on a number of factors: KOH concentration and temperature (high and low temperatures give rougher surfaces), agitation (ultrasonics reduces roughness) and additives (arsenic and antimony oxides improve smoothness, and surfactants do so too). However, other properties also change: surfactants affect crystal plane selectivities, and the process needs to be characterized anew.

Table 20.2 Alkaline anisotropic etchants: some main features of etchants

Etchant	KOH	TMAH	EDP/EPW
Rate (at 80 °C) (μm/min)	1	0.5	1 (at 115 °C)
Typical concentration	40%	25%	80%
Selectivity (100):(111)	200:1	30:1	35:1
Selectivity Si:SiO$_2$	200:1	2000:1	10 000:1
Selectivity Si:Si$_3$N$_4$	2000:1	2000:1	10 000:1
Etch stop factor (10^{20}cm^{-3})	10	100	50

Etch stop is an idealization; infinite selectivities are not encountered in the real world. High selectivity is termed etch stop when selectivity is so high that etch timing becomes non-critical. Etch stop can happen through various mechanisms. The etch rate of boron-doped silicon decreases rapidly when doping levels exceed 10^{19} cm^{-3}, as shown in Figure 20.6. The exact mechanism is unknown but high stresses in heavily doped film may play

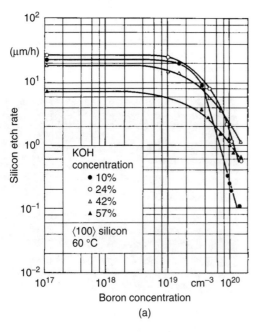
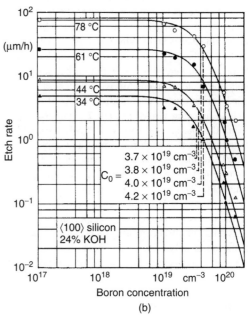

Figure 20.6 The p^{++} etch stop: (a) with KOH concentration as a parameter; (b) with etch temperature for 24% KOH as a parameter. Reproduced from Seidel *et al.* (1990b) by permission of Electrochemical Society Inc.

a part. This property is frequently used in bulk MEMS, as a way to fabricate simple mechanical structures. It is, however, not possible to fabricate electrical devices on such highly doped material. For instance, piezoresistors cannot be made because the p^{++} etch stop doping level is higher than the piezoresistor doping level. The stresses in p^{++} doped structures make then mechanically inferior to lightly doped material. Furthermore, slips are introduced in silicon because of high stresses, and this makes bonding of highly doped wafers difficult.

On the positive side, more shapes can be made. Simple through-wafer etching can produce only square and rectangular nozzles, but with the boron etch stop technique all shapes become possible. As shown in Figure 20.7, circular nozzles can be defined by lithography and boron ion implantation. Features can also be placed close to each other. KOH etching stops at p^{++} regions, but continues in the areas that were protected by photoresist during implantation. In this application back side to front side alignment is non-critical: the nozzle shape is defined by the front-side implantation mask only.

20.4.1 Electrochemical etch stop

When a silicon wafer is an anode in an alkaline etching solution, biased positively, above the passivation potential the surface will be oxidized, which stops silicon dissolution. The n-type layer of a pn structure can similarly be protected (Figure 20.8). Positive potential, above the passivation potential, is applied to the n-type layer. Etching of p-type silicon continues until the diode is destroyed, and n-type silicon is then passivated. Electrochemical etch stop can be achieved with lightly doped material, and it is therefore possible to fabricate a number of devices on such material.

20.5 Front-Side Processed Structures

Initially the etched shapes are determined by the fast etching planes, usually the (100) planes. But in the end it is

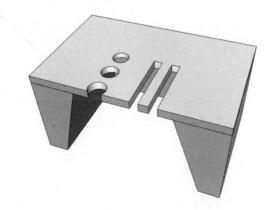

Figure 20.7 p^{++} etch stop: dots and rectangles were masked during doping and are etched in KOH, while p^{++} doped areas form a silicon membrane

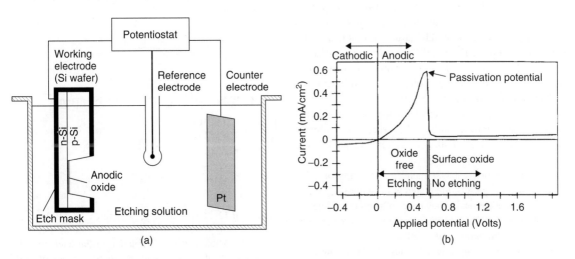

Figure 20.8 (a) Electrochemical cell for silicon electrochemical etching in KOH: p-type silicon etched; n-silicon passivated by anodic oxide. Reproduced from Wong *et al.* (1992) by permission of Electrochemical Society Inc. (b) Passivation potential and anodic oxidation regime. Reproduced from Collins (1997) by permission of IEEE

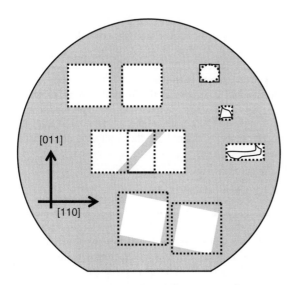

Figure 20.9 The largest rectangle bounded by [110] orientations along and perpendicular to flat will be etched: original mask openings in white, final etched shapes shown by dotted lines. Oxide shown gray. Free-standing oxide bridge will be formed in the middle geometry. Defects of arbitrary shape will end up as rectangles

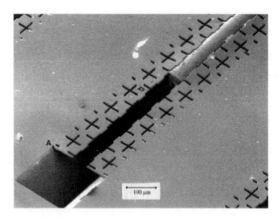

Figure 20.10 Silicon nitride microclips for optical fiber positioning. Reproduced from Bostock *et al.* (1998) by permission of IOP

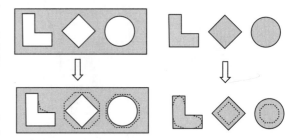

Figure 20.11 Etched shape difference with mask polarity: upper row shows the mask patterns (oxide mask shown gray); lower row shows structures after a short anisotropic wet etch

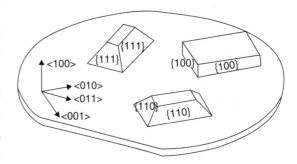

Figure 20.12 Orientation of structures on (100) wafer. Alignment to wafer flat leads to 54.7° angles and {111} sidewalls. Alignment of 45° relative to flat leads to {110} walls and {100} vertical walls result when rates of {110} relative to {100} fulfill conditions 20.2 and 20.3. Reproduced from Powell and Harrison (2001) by permission of IOP

the slow etching planes, usually the (111) planes, that determine the shape. Mask patterns are usually aligned to major [110] crystal axes (Figure 20.9), but due to either wafer miscut or alignment error, patterns can be off-axis. Anisotropic wet etching carves out the shape that is determined by the largest rectangle bounded by the pattern. This can be used to fabricate free-standing structures by undercutting, as in the middle structure of Figure 20.9. The nitride clips of Figure 20.10 are formed similarly.

Note that inverse polarity structures (opposite mask oxide vs. silicon openings) behave rather differently in anisotropic wet etching (Figure 20.11).

Cantilevers and bridges can be made by front-side micromachining by undercutting. The structures are aligned not to the main axes of silicon, but for example 45° off, so that fast etching planes appear.

Sensor resistors, catalyst metals, AFM tips or other microstructures must be integrated with cantilevers. Whatever structures are made, they have to be processed before the silicon release etch because topology and topography do not allow lithography after release.

If the structures are aligned along the [100] direction (45° relative to wafer flat), instead of the usual flat direction [110], new possibilities arise. The basic situation is shown in Figure 20.12.

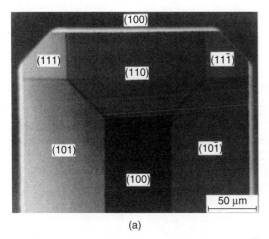 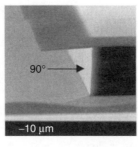 

Figure 20.13 (a) An etched well with 45° slanted sidewalls in <100> wafer by 45° degree off-orientation. Reproduced from Strandman *et al.* (1995) by permission of IEEE. (b) Mesas with 90° angles in <100> wafer, before and after etch mask removal. Reproduced from Vazsonyi *et al.* (2003) by permission of IOP

For instance, 45° walls suitable for fiber coupling mirrors and 90° sidewall mesas can be made. These structures depend on the relative etch rates of (100) and (110) planes according to

$$\text{rate}\{100\}/\text{rate}\{110\} < 1/\sqrt{2} \quad 90° \text{ walls} \quad (20.2)$$

$$\text{rate}\{100\}/\text{rate}\{110\} > \sqrt{2} \quad 45° \text{ walls} \quad (20.3)$$

Condition (20.2) leads to vertical walls which are (100) planes, and condition (20.3) leads to 45° walls which are (110) walls. This is shown in Figure 20.13. Also shown are vertical (100) walls. However, severe undercut is unavoidable in order to make vertical walls in <100> silicon.

KOH etchant of 25–50% fulfills condition (20.2), and KOH–IPA solution is an example of condition (20.3). When the rate condition is close to the limit values, as with <25% TMAH, inadequate stirring or some other disturbance can lead to unexpected changes in final shapes.

20.6 Convex Corner Etching

The etch rate of (100) planes is high relative to (111) planes. When simple concave shapes are etched, the fast etching planes will disappear and the slow etching (111) planes will dominate in the final structure. The fastest etching planes, usually (110) and some high-index planes like (311), are not present in the simple rectangular wells, channels and nozzles, which have only concave 90° inside corners. Convex corners reveal these high-etch-rate planes, and etching under the mask takes place. The time evolution of a convex corner shape is shown in Figure 20.14. The etched shape is initially determined by the fast etching planes, but the slow etching (111) planes will finally be limiting the structures.

Depending on initial mask size and etch time, either flat mesas or sharp silicon needles will be produced (Figure 20.15).

Nitride bridges, membranes and cantilevers, as well as p^{++} doped silicon structures, are made by corner undercutting (Figure 20.16).

20.6.1 Corner compensation

We noted above that convex corners are dominated by (311) planes (Figure 20.15). In many designs it would be very useful to have sharp corners. This is possible with a little extra effort in mask design by adding compensation structures, shown in Figure 20.17. The fast etching planes start to erode at convex corners. But the final convex corner is protected by this sacrificial structure so that after the compensation structure has been etched away, a rectangular corner remains.

Timing is the difficult part: if etching is stopped too early, a peak remains on the corner. Overetching leads to a structure with an undercut corner, similar to the non-compensated case, but with less undercut. Even though this method looks perfect in two dimensions, it leaves some small <311> surfaces in three dimensions. Another shortcoming of this method is that it can take a lot of space to form these compensation structures.

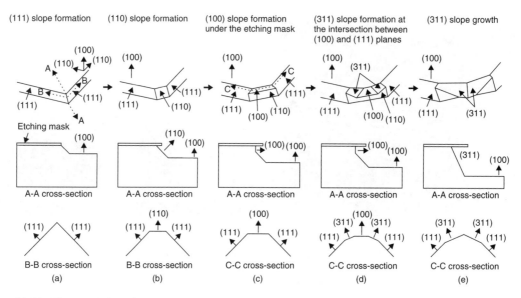

Figure 20.14 Time evolution of convex corner undercutting. Reproduced from Shikada *et al.* (2001) by permission of Springer

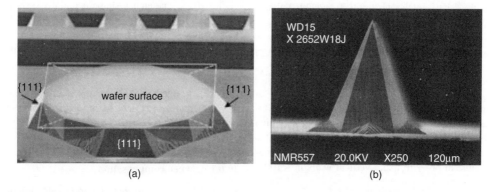

Figure 20.15 Left: etching a silicon mesa. Reproduced from Schröder *et al.* (2001), copyright 2001, by permission of IEEE. Right: silicon pyramid. Reproduced from Wilke *et al.* (2005) by permission of IOP

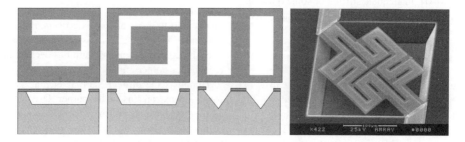

Figure 20.16 Cantilever and bridge structures by front-side etching. Underetching from convex corners is used, with structures aligned to [110] main axes on a wafer. Simple rectangular holes along [110] axes result in V-grooves only. Reproduced from Ma *et al.* (2009) by permission of IOP

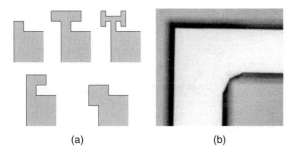

Figure 20.17 (a) Different designs for corner compensation. (b) Optical microscope image of a compensated corner after etching. Courtesy Ville Voipio and Seppo Marttila, Aalto University

20.7 Membrane Fabrication

There are two basic membrane (diaphragm) structures: either the membrane is made of a deposited film, or it is made of single crystal silicon. In the first case etching is quite simple: all silicon is removed, and the thin film remains. There are two main considerations for the membrane material: it has to be (slightly) tensile stressed because a compressively stressed film would buckle, and too high a tensile-stressed film would crack. The film has also to be resistant to alkaline etchants. Silicon nitride fulfills both requirements, and it is widely used. It is also electrically (and thermally) insulating so that resistors can be readily deposited on it, and it is optically transparent. (We use the word membrane for a continuous film: some workers in the fluidics field assume the membrane to be permeable, as in filters. We call such structures perforated membranes.)

Silicon membrane fabrication pictured in Figure 20.18a relies on timed etching but this is a very unsatisfactory approach, especially if thin membranes are needed. Something like 40 μm is the thinnest that can reasonably be made by timed etching in a manufacturing environment.

Etch stop techniques (pn junction, p^{++}) offer better thickness control, and SOI wafers are also amenable to very good membrane thickness control. Etch stops based on p^{++} and electrochemical techniques differ with respect to the final membrane properties: the p^{++} membrane is highly doped and highly stressed (because boron is a small atom and the lattice has contracted due to the high concentration of boron), so it cannot be used for electrical devices and it is no good for mechanical devices either. It can be used for passive structures, however. Standard p^{++} etch stop has two variants: either the p^{++} layer is made by diffusion (or implantation) or it is an epitaxial layer. Because

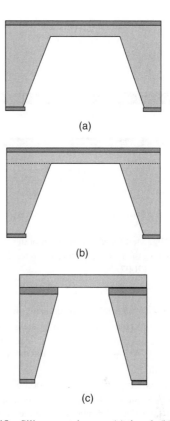

Figure 20.18 Silicon membranes: (a) timed; (b) etch stop; (c) SOI membranes

the doping levels required for etch stop are very high indeed, diffusion p^{++} is limited to very thin membranes.

Alternatively, Si:Ge:B etch stop can be used: germanium, a big atom, is introduced to compensate for the small boron atoms. Electronically nothing changes because germanium is isoelectronic with silicon. Si:Ge:B etch stop layers are always grown by epitaxy. Usually a structural layer with freely chosen doping is epitaxially grown on top of the Si:Ge:B etch stop layer. This gives a lot of freedom, but epitaxy is an expensive process.

The pn junction etch stop has the same variants as p^{++} as far as doping methods are concerned, that is diffusion and epitaxy, therefore the same thickness limits. Additionally, the n-layer has to be electrically contacted, and this contact has to be protected from alkaline silicon etchant. Holders of various designs have been invented, with the drawback that part of the wafer front side is used for sealing the holder, leading to silicon real-estate loss, sometimes up to 20% fewer chips than in free etching.

SOI wafers offer an elegant but somewhat expensive way of making membrane structures. The buried oxide

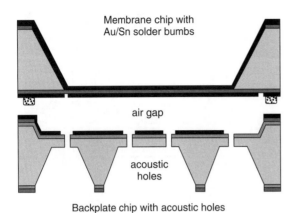

Figure 20.19 Metallized nitride membrane to be bonded to a backplate chip to make a microphone. Adapted from Kwon and Lee (2007)

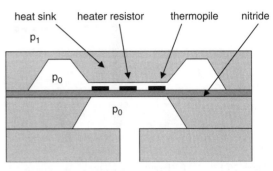

Figure 20.20 Thermal pressure sensor: flexible silicon heat sink will bend when pressure is applied, and the narrower gap enhances heat conduction from the heater resistor to thermopile detector

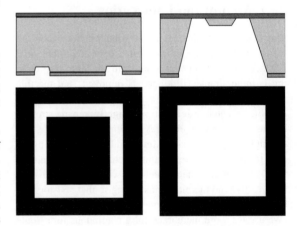

Figure 20.21 Nitride membrane with silicon heat spreader island, with the two photomasks used in its fabrication. Adapted from Briand et al. (2000)

of SOI acts as an etch stop layer, leaving the SOI device layer untouched by the etch process. Bonded SOI device layer thicknesses are usually specified at about 10%, so that a 10 µm membrane with ±1 µm thickness variation results.

Microphones are pressure sensors too, and many structural features of the two are identical. The microphone of Figure 20.19 consists of a metallized silicon nitride membrane and an acoustic backplate. The air gap is defined by wet etching (9 µm) plus Au/Sn solder thickness (3 µm). Acoustic holes are defined by DRIE (150 µm deep, 60 µm in diameter). Thin and large membranes and small gaps equal sensitive microphones, but small gaps are prone to stiction during fabrication and problematic because of condensing water.

The thermal pressure sensor shown in Figure 20.20 has a number of important points on membrane structures. The top wafer has two thin silicon supports and a large silicon mass (boss) etched in it. The device responds to pressure because of the thin support beams, and it will move downward retaining parallelism with the lower wafer because the large boss does not bend. Comparison of thin membrane and thick bossed membrane was shown in Figure 17.14. Second, the thick boss will act as a heat sink (due to the excellent thermal conductivity of silicon). The central wafer has a nitride membrane. The heater resistor and thermopile detector are patterned on it. The fact that nitride is an electrical and thermal insulator is crucial. The heat produced by the resistor is lost to the gas, and when pressure reduces the gap, heat transfer to the boss is enhanced. Temperature is sensed by the thermopile detector. Similar bossed design is also used in capacitive pressure sensors.

A silicon boss can be made for thermal uniformity reasons in a micro hotplate sensor (Figure 20.21). The first mask defines the silicon island size (which depends on both mask dimensions and initial wet etch depth) and the second mask defines the size of the membrane (together with wafer thickness).

Membrane devices can be made in three basic ways:

1. Membrane first, thin films afterward.
2. Thin films first, membrane etching at the end.

3. Partial wafer etching first, then thin films, then final silicon removal at the end.

The first method is easy in the sense that the harsh KOH (or TMAH) etching step is already done, and the chemical stability of thin films need not be considered. The drawback is that the wafer becomes rather fragile when the membrane is made, and the larger the membrane, the bigger the concern. This applies to processing in general (spin coating, high temperature steps, etc.) and thin-film stresses in particular: high stresses may damage the membrane. The "membrane-first" approach was used to make the membrane chip of the microphone in Figure 20.19. Electrode metallization (Cr/Ni/Au) and solder bonding metals (Au/Sn) were deposited afterward.

The membrane-last approach necessitates some sort of protection of the structures processed on top of the wafer. This protection may be polymer coating, but photoresist is not chemically tolerant. Instead, parylene, fluoropolymers, or black wax can be used. Other protective layers include the same materials that are used as etch masks, namely oxide and nitride. Instead of protective films, a single wafer holder can protect the top side during backside KOH/TMAH etching. This approach leads inevitably to some loss of effective area on the wafer front. But the holder can provide other functions, like electrical contacts in electrochemical etching or the etch stop technique, which can compensate the costs. The membrane-last approach was used in the micro hotplate device of Figure 20.22. However, the chemically sensitive film was deposited after membrane formation because it does not tolerate KOH, unlike platinum, oxide and nitride.

The mixed approach tries to combine the best of both worlds: mechanical stability offered by a silicon "backbone" during processing, and shorter final etch time, which means lesser protection requirements for the front-side devices. Then, for instance, simple PECVD oxide can serve as a protective coating.

20.8 Through-Wafer Structures

A nozzle is a basic through-wafer structure. It can be done by single-sided lithography and etching: the nozzle size is determined by the mask size, wafer thickness and silicon crystal geometry. The condition for zero nozzle orifice on a <100> wafer is given by Equation 20.1. This simple process is very sensitive to wafer thickness variations and it is not practical for making small nozzles.

In the microthruster of Figure 11.7 the size of the nozzle depends critically on etch timing: in order to achieve inlet shapes as drawn, etching must be stopped immediately when the top and bottom etching fronts meet in the middle of the wafer, otherwise the structures start widening as shown in Figure 20.23. So, even though <100> silicon usually results in 54.7° angles, vertical walls can be obtained in <100> etching. The SEM micrograph

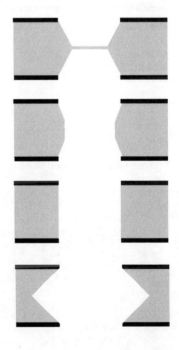

Figure 20.23 Time evolution of etching through <100> silicon from two sides simultaneously

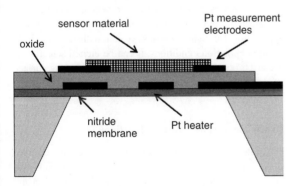

Figure 20.22 Membrane-last micro hotplate sensor

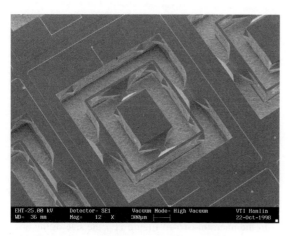

Figure 20.24 Accelerometer by wet etching of <100> silicon. Courtesy VTI Technologies

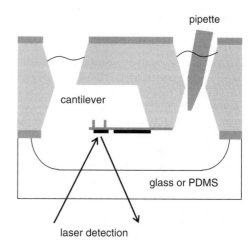

Figure 20.26 Cantilever biosensor. Adapted from Yue et al. (2004)

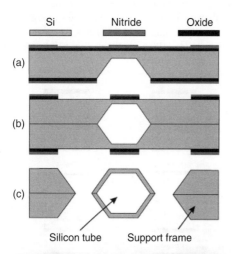

Figure 20.25 Hollow silicon tube for densitometer, double masking with thermal oxide and LPCVD nitride. Reproduced from Najmzadeh et al. (2007) by permission of IOP

of Figure 20.24 shows what a wet-etched accelerometer looks like when etching is stopped at the right moment.

Through-wafer etching is often done in steps: etching some structures, doing further processing and continuing etching later on. In the densitometer of Figure 20.25 fusion bonding has been done on partially etched wafers. Wet etching was then continued, to create hollow tubes for liquid density measurements.

In the cantilever biosensor of Figure 20.26, etching from two sides simultaneously is used to increase the apparent etch rate. The etched shapes are not critical to device operation, and the functionality is in the gold-coated LPCVD nitride cantilever: antibodies are attached to the gold-coated cantilever via gold–sulphur (thiol) bonds. When antigens bind to antibodies, surface stress increases and the cantilever bends. This bending is detected optically. Silicon is very suitable for this application for many reasons: it can withstand LPCVD nitride deposition conditions, it can be etched selectively, it can be bonded to glass (and to PDMS) and its surface is suitable for fluidics.

20.8.1 Membranes with perforations

Perforated nitride membranes can be made by double-sided lithography and anisotropic wet etching. Top-side lithography determines the size for perforations. Plasma etching of nitride exposes silicon to wet etchant. Inverted pyramids will form in silicon, and etching will grind to a halt. But because (111) planes have a small but finite etch rate, the inverted pyramids will eventually merge,

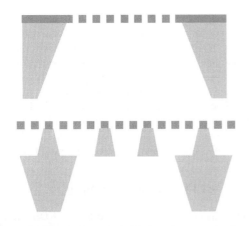

Figure 20.27 Perforated nitride membranes: simple and doubly supported designs. Adapted from Kuiper et al. (2000)

Figure 20.28 Rectangular groove bottoms in KOH–IPA etching of <110> silicon. Reproduced from Dwivedi et al. (2000) by permission of Elsevier

and etching then continues. This could be helped by designing filter openings not along the main crystal axes, but slanted. However, this strategy might still not work: hydrogen evolution may prevent etchant from entering small cavities. Back-side etching is therefore needed, too. Nitride is a very advantageous material for this application because of its very high etch tolerance in KOH and TMAH. Nitride is also mechanically a good material, but large membranes and high pressures pose problems. The formation of dual support bars (Figure 20.27) enables large-area membranes with high aperture ratios (high percentage of open area).

20.9 <110> Etching

Silicon of <110> orientation offers an interesting possibility anisotropically to wet-etch perfectly vertical walls when the mask is aligned so that slow etching (111) planes form the sidewalls. However, just as in the case of <100> silicon etching, the relative rates of different crystal planes can be changed by etchant concentration and temperature. It is possible to find conditions where a square bottom profile can be achieved, for instance KOH (23% wt)–H_2O–isopropanol (10–15% wt) at 85 °C or 30% KOH at 70 °C give the etch profile shown in Figure 20.28.

Under other etch conditions a self-limiting shape, a U-groove, is encountered, for instance with 40% KOH at 70 °C (Figure 20.29). U-grooves are self-limiting just like V-grooves on (100) wafers, when planes that etch slower than (110) appear.

Etching will proceed until the six slow etching (111) planes meet. The self-limiting depth D for U-grooves

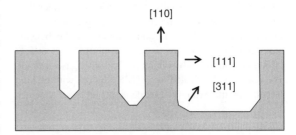

Figure 20.29 Etching of <110> silicon: slow etching (111) planes form vertical sidewalls. Depending on etchant concentration, composition and temperature, slow etching planes start to limit the groove

for initial mask opening sizes a and b (Figure 20.30) is given by

$$D = \frac{(a + \sqrt{2} \times b)}{2 \times \sqrt{6}} \quad (20.4)$$

A major limitation of vertical walled structures on (110) silicon is that only diamond-shaped structures will have all four walls vertical. Diamonds have 70.5° and 109.5° angles. Rectangular shapes will turn into hexagons, but diamonds oriented with crystal axes will retain their shape in the etching process.

There is no fundamental limitation as to size of wet-etched structures. In practice minimum size may be limited by lithography or by crystal plane etch selectivity. With 100 nm initial size and 100:1 etch selectivity, a

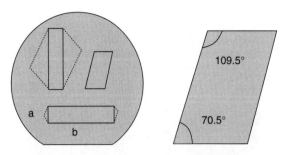

Figure 20.30 <110> etched shapes: solid lines indicate mask openings, dotted lines final etched shapes. Diamonds oriented along major crystal axes retain their shape

100 nm undercut appears after etching 10 μm deep, but finite crystal plane selectivity can be designed in, and in fact structures with aspect ratios greater than 100:1 have been made.

Chemical microreactors form a broad class of microfabricated devices, usually with various chamber/flow channel geometries. The hydrogen separation device shown in Figure 20.31 is one example of the benefits that microfabrication has to offer. Hydrogen can diffuse through palladium–silver thin film, but other gases cannot. The reactor is fabricated on <110> silicon, with etching stopping on the oxide. The oxide is etched away to reveal

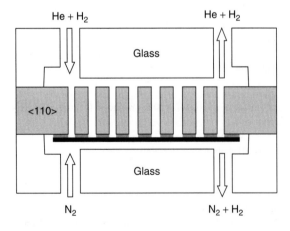

Figure 20.31 A microreactor for hydrogen separation. KOH etching of <110> silicon stops on oxide, HF etching removes oxide, and palladium membrane remains. Pd allows hydrogen diffusion but blocks all other gases. Adapted from Tong (2003)

the Pd–Ag membrane. The flow channels are formed by anodic bonding. Defect-free thin metal membranes can be made reproducibly because fabrication takes place in a cleanroom, and because the silicon surface is extremely flat and smooth. Higher separation selectivity between hydrogen and other gases is possible because thin, yet defect-free, membranes do not leak. Moreover, the membranes tolerate high pressures because silicon etching can be used to leave silicon support struts as desired. This enables higher pressures and higher gas fluxes.

20.10 <111> Silicon Etching

<111> silicon wafers cannot be etched in KOH because (111) planes are the slow etching planes. If, however, initial trenches are opened by plasma etching, other crystal planes will be exposed. The depth of the structure is determined by the initial plasma etch step because the bottoms are (111) planes just like the wafer surface and they do not etch further in KOH.

The sixfold symmetry which was seen in the vertex view of silicon crystal (Figure 4.6) is evident in <111> wafers (Figure 20.32). Triangular and hexagonal patterns will retain their shapes if oriented along the proper crystal planes. The sidewalls will be either 70.5° or 90°. Structures with concave corners will result in hexagonal final shapes as (111) planes meet.

Free-standing thin-film structures can be made by etching an initial release hole, and then continuing with anisotropic wet etching. Complete undercutting leads to free-standing structures not unlike those made on (100) silicon. However, lateral undercutting in some directions is fairly large, as shown in Figure 20.33.

If free-standing silicon bridges and beams need to be made (Figure 20.34), multiple etch and deposition steps are needed.

Process flow for <111> silicon microbridge

1. Oxide mask
2. First silicon RIE
3. CVD oxide
4. Oxide RIE
5. Second silicon RIE
6. Silicon wet etching
7. Oxide removal

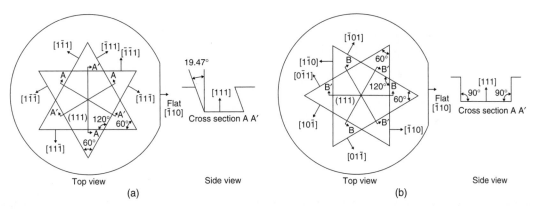

Figure 20.32 <111> silicon crystal planes. Note the hexagonal symmetry. Not all walls are bound by slow etching (111) planes. Reproduced from Park *et al.* (1999) by permission of Institute of Pure and Applied Physics

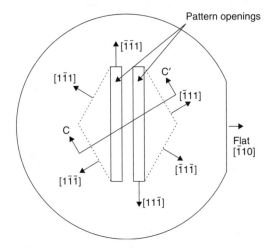

Figure 20.33 Etching of <111> silicon bridge: two rectangular pattern openings are undercut, and etching will proceed until slow etching (111) planes are met. Undercutting to the left and right of the bridge is large compared to bridge width. Reproduced from Park *et al.* (1999) by permission of Institute of Pure and Applied Physics

This microbridge can be of any doping type, while (100) silicon bridges are made either by p^{++} or pn etch stop or by timed etching. The bridge thickness of a (111) bridge is determined by the first silicon RIE step, and release gap thickness by the second silicon RIE step. The depths of RIE steps are not very accurate but because the bridge roof and ceiling are slow etching (111) planes, surface quality is excellent.

20.11 Comparison of <100>, <110> and <111> Etching

If an initial trench has been etched in the wafer by anisotropic plasma etching (i.e., vertical sidewalls, Figure 20.35), anisotropic wet etching will proceed until slow etching (111) planes are encountered. On a (100) wafer this will result in a rhombohedral structure with 54.7° angles. On a (110) wafer the flat bottom will be further etched and, depending on relative etch rates in the etchant in question, either a flat bottom remains or a U-groove sets in. On (111) wafers either vertical or slanted walls will result, depending on pattern orientation.

Anisotropic wet etching of silicon has served the MEMS community well for decades and is still going strong. There are of course applications where (D)RIE excels, and wet etching is no substitute, but the opposite is also true in certain applications. Batch processing with excellent throughput, uniformity and surface quality is a strong point of wet etching. Exact slope angles determined by silicon crystal planes cannot be matched by DRIE: obtaining perfectly vertical walls by DRIE is only approximate and often a wet etch step is done to reveal the vertical crystal planes, as will be shown in Figure 21.24. Wet etching is also used to smooth surfaces after DRIE.

Integrating wet etching into a process flow requires a number of issues to be tackled: mask oxide thickness is on the limit of practical thermal oxides for through-wafer etching with KOH, but then TMAH can be used. However, the two are not direct substitutes for each other. Use of a silicon nitride etch mask usually solves masking

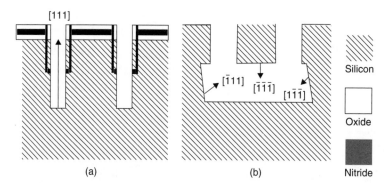

Figure 20.34 Silicon bridges in (111) silicon. The first RIE defines silicon bridge thickness. A spacer is formed before the second RIE step which defines the release gap. The spacer protects the bridge during undercutting etch in KOH. Reproduced from Park *et al.* (1999) by permission of Institute of Pure and Applied Physics

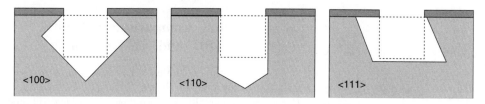

Figure 20.35 RIE vertical walled starting shape (dotted), followed by KOH wet etching: top, <100>; middle, <110>; bottom, <111> wafers

problems. DRIE will be discussed in the following chapter and Chapter 30 discusses MEMS applications at length.

20.12 Exercises

1. What is the activation energy of <110> silicon etching in 30% KOH?

30	40	50	60	70	80	90	100 °C
4.7	9.8	19.4	37	68	121	209	350 µm/h

2. KOH etch rates (µm/h) for <100> silicon are given below. Calculate the activation energies for the different KOH concentrations!

	40 °C	60 °C	80 °C	100 °C
20%	7.09	26.7	86.3	246
30%	6.48	24.4	79	225
40%	5.28	19.9	64.4	183
50%	3.77	14.2	45.9	131

   Data from Lang (1996).

3. Silicon <100> wet etch rate in 25% KOH at 90 °C has been measured to be 2.5 µm/min, and activation energy was determined to be 0.61 eV (59 kJ/mol). If membranes 40 µm thick are made in wafers 380 µm thick, and etch bath temperature is controlled to ±1 °C, how does this affect membrane thickness control?

4. In Exercise 1.5 pressure sensor membrane deflection was calculated. Compare this to the surface roughness of a KOH etched wafer.

5. Micromechanical pressure sensor chips have diaphragms 40 µm thick which are 1 × 1 mm in area. How many such chips can be made on
   (a) 380 µm thick 3 inch wafers?
   (b) 525 µm thick 100 mm wafers?
   (c) 675 µm thick 150 mm wafers?

6. Nozzles are fabricated by etching anisotropically through a <100> silicon wafer 380 µm thick. A mask pattern 540 µm wide is used. Calculate the size of holes produced by an ideal process. Then calculate the effect of the following real-world uncertainties:
   (a) wafer thickness variation 380 ± 5 µm
   (b) total thickness variation (TTV) of 1 µm
   (c) <100>:<111> crystal plane selectivity 33:1 vs. 30:1.

7. If a piezoresistive pressure sensor membrane is made in an epitaxial layer, and diaphragm etching is stopped by a pn junction etch stop, how do the following affect membrane thickness:
   (a) wafer thickness
   (b) wafer TTV
   (c) epitaxial layer thickness?
8. Detail all the fabrication steps of the AFM cantilever-tip of Figure 20.2.
9. Detail the process sequence for making the device shown in Figure 20.7.
10. What is the angle between the (111) and (311) planes shown in Figure 20.29?
11. Design "corner compensation" structures that will result in a circular hole in a wet etched <100> wafer.
12. Design a process and mask set for the fabrication of silicon bridges on (110) wafers.
13. Calculate the aperture ratios (percentage of open area) in the two filter designs of Figure 20.27.
14. The deflection of a circular membrane under pressure is given by $h = 0.666(r^4 \Delta p/Et)^{1/3}$, where r is diaphragm radius, t thickness and E Young's modulus. What deflection corresponds to 25 mtorr pressure difference? What is the corresponding capacitance change?
15. How small an opening can you make by wet etching in (a) bulk wafers and (b) SOI wafers for the structure shown below? Assume top-side etched depth to be 10 μm.

16. How is nozzle shape control different below if etching is done (a) from both sides simultaneously, (b) from the top side first, then thermally oxidized and etched from the back side?

17. How would you fabricate the optical fiber positioning device shown below? Draw photomasks and cross-sections after each silicon etch step.

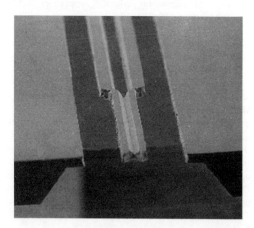

Reproduced from Hoffman and Voges (2002) by permission of IOP

18. Design a process and calculate the mask dimensions for a 500 × 500 μm nitride membrane with a 300 × 300 μm silicon island, 20 μm thick, for the heat spreader structure of Figure 20.21.
19. Estimate the dimensions of structures in the accelerometer of Figure 20.5!

References and Related Reading

Bostock, R.M. *et al.* (1998) Silicon nitride microclips for the kinematic location of optic fibres in silicon V-shaped grooves, *J. Micromech. Microeng.*, **8**, 343–360.

Briand, D. *et al.* (2000) Design and fabrication of high-temperature micro-hotplates for drop-coated gas sensors, *Sens. Actuators*, **B68**, 223–233.

Brida, S. *et al.* (2000) Microstructures etched in doped TMAH solutions, *Microelectron. Eng.*, **53**, 547–551.

Collins, S.C. (1997) Etch stop techniques for micromachining, *J. Electrochem. Soc.*, **144**, 2242.

Dwivedi, V.K. *et al.* (2000) Fabrication of very smooth walls and bottoms of silicon microchannels for heat dissipation of semiconductor devices, *Microelectron. J.*, **31**, 405.

Elwenspoek, M. and H. Jansen (1998) **Silicon micromachining**, Cambridge University Press.

Fang, W. (1998) Design of bulk micromachined suspensions, *J. Micromech. Microeng.*, **8**, 263–271.

Hannemann, B. and J. Fruhauf (1998) New and extended possibilities of orientation dependent etching in microtechnics, Proceedings of IEEE MEMS'98, p. 234.

Hoffmann, M. and E. Voges (2002) Bulk silicon micromachining for MEMS in optical communication systems, *J. Micromech. Microeng.*, **12**, 349.

Kovacs, G.T.A. et al. (1998) Bulk micromachining of silicon, *Proc. IEEE*, **86**, 1543.

Kuiper, S. et al. (2000) Wet and dry etching techniques for the release of sub-micrometre perforated membranes, *J. Microelectromech. Syst.*, **10**, 171–174.

Kwon, H.-S. and K.-C. Lee (2007) Double-chip condenser microphone for rigid backplate using DRIE and wafer bonding technology, *Sens. Actuators*, **A138**, 81–86.

Lang, W. (1996) Silicon microstructuring technology, *Mater. Sci. Eng.*, **R17**, 1–55.

Lindroos, V. et al. (2010) **Handbook of Silicon Based MEMS Materials and Technologies**, Elsevier.

Ma, T., Y. Liu and T. Li (2009) A <100> direction front-etched membrane structure for a micro-bolometer, *J. Micromech. Microeng.*, **19**, 035022.

Najmzadeh, M., S. Haasl and P. Enoksson (2007) A silicon straight tube fluid density sensor, *J. Micromech. Microeng.*, **17**, 1657–1663.

Oosterbrook, R.E. et al. (2000) Etching methodologies in <111> -oriented silicon wafers, *J. Microelectromech. Syst.*, **9**, 390.

Pal, P., K. Sato and S. Chandra (2007) Fabrication techniques of convex corners in a (100)-silicon wafer using bulk micromachining: a review, *J. Micromech. Microeng.*, **17**, R111–R133.

Park, S. et al. (1999) Mesa-supported, single-crystal microstructures fabricated by the surface/bulk micromachining process, *Jpn. J. Appl. Phys.*, **38**, 4244.

Powell, O. and H. Harrison (2001) Anisotropic etching of {100} and {110} planes in (100) silicon, *J. Micromech. Microeng.*, **11**, 217.

Sasaki, M. et al. (2000) Anisotropically etched Si mold for solid polymer dye microcavity laser, *Jpn. J. Appl. Phys.*, **39**, 7145.

Sato, K. et al. (1998) Characterization of orientation-dependent etching properties of single-crystal silicon: effects of KOH concentration, *Sens. Actuators*, **A64**, 87–93.

Schröder, H. et al. (2001) Convex corner undercutting of {100} silicon in anisotropic KOH etching: the new step-flow model of 3-D structuring and first simulation results, *J. Microelectromech. Syst.*, **10**, 88–97.

Seidel, H. et al. (1990a) Anisotropic etching of crystalline silicon in alkaline solutions I, *J. Electrochem. Soc.*, **137**, 3612.

Seidel, H. et al. (1990b) Anisotropic etching of crystalline silicon in alkaline solutions II, *J. Electrochem. Soc.*, **137**, 3626.

Shikida, M. et al. (2000) Differences in anisotropic etching properties of KOH and TMAH solutions, *Sens. Actuators*, **80**, 179.

Shikida, M. et al. (2001) A new explanation of mask undercut in anisotropic silicon etching: saddle point in etching rate diagram, Proceedings of Transducers'01, p. 648.

Strandman, C. et al. (1995) Fabrication of 45° degree mirrors together with well-defined V-grooves using wet anisotropic etching of silicon, *J. Microelectromech. Syst.*, **4**, 214.

Tong, H.D. (2003) Microfabrication of palladium–silver alloy membranes for hydrogen separation, *J. Microelectromech. Syst.*, **12**, 622–629.

Vazsonyi, E. et al. (2003) Anisotropic etching of silicon in a two-component alkaline solution, *J. Micromech. Microeng.*, **13**, 165.

Wilke, N. et al. (2005) Process optimization and characterization of silicon microneedles fabricated by wet etch technology, Microelectron. J., **36**, 650–656.

Wong, S.S. et al. (1992) An etch stop utilizing selective etching of n-type silicon by pulsed potential anodization, *J. Microelectromech. Syst.*, **1**, 187.

Yue, M. et al. (2004) A 2-D Microcantilever array for multiplexed biomolecular analysis, *J. Microelectromech. Syst.*, **13**, 290–299.

21

Deep Reactive Ion Etching

Reactive ion etching (RIE) (which we will use as a synonym for plasma etching) opens up many possibilities which cannot be done by wet etching. Practically any size and shape can be made by RIE. All sorts of mechanical devices, like springs, beams and plates, are fabricated by plasma etching, as are vertical capacitors and mirrors, in-plane and out-of-plane microneedles and nozzles, and fluidic filters and separation devices.

DRIE is an extension of RIE for deep etching, but depth is very different from application to application. For example, in SOI MEMS there is a need to etch the SOI device layer, which can be 5–50 μm, but in many cases also the handle wafer needs to be etched, that is 300–500 μm. While typical RIE rates are on the order of 0.1–1 μm/min, DRIE rates are 2–20 μm/min. Even higher rates have been reported, but then other criteria have been sacrificed: sidewall profile control is poor and high rates are only available for designs which have small etchable area.

This chapter concentrates on silicon DRIE processes, with some basic structures discussed. Comparison is often made to anisotropic wet etching because sometimes DRIE and KOH can both be used, and selection has to be reasoned. More advanced device examples will be found in Chapter 30 on MEMS integration.

21.1 RIE Process Capabilities

Small linewidths can be made by RIE. If lithographers have been able to print something on the wafer, etch engineers have always been able to etch those patterns. RIE was adopted in IC fabrication at about 3 μm linewidths, and it has been scaled down to current generation 45 nm processes. RIE is therefore suitable for nanostructure fabrication.

Structures of high aspect ratio can be made (Figure 21.1). Pillars and holes can be very deep and narrow, with 100:1 record structures used in DRAM trench capacitors (10 μm deep, 100 nm wide). In MEMS through-wafer structures 10 μm wide and 400 μm deep have been made. Making structures of high aspect ratio is by no means easy, and etch rate slows down in such deep cavities, as will be discussed later in this chapter.

Combining the narrow gap and high aspect ratio capabilities, RIE enables comb-drive devices. Comb drives are widely used as sensors and actuators in microsystems. Basically a comb drive is a vertical capacitor with moving and fixed capacitor plates. The principle is shown in Figure 21.2 and a SEM micrograph of it is shown in Figure 21.3. All the capacitor fingers are undercut, but every other one is connected to a fixed anchor and every other to a movable plate (more on this in Chapter 29 on surface micromachining).

A comb drive can be used as a capacitive sensor, with its capacitance given by Equation 21.1 and force by Equation 21.2. DRIE is important in many ways, because it can be used to fabricate narrow gaps, spacing d is small and the number of finger pairs, n, can be made large and, because deep structures can be etched, height h and area A will be large:

$$C = \frac{n\varepsilon A}{d} \tag{21.1}$$

$$F = \frac{1}{2}CV^2 = \frac{n\varepsilon A}{2d}V^2 \tag{21.2}$$

In sensor applications the gap between the capacitor plates changes due to some external force, for example acceleration (in accelerometers and gyroscopes) or gravity (inclinometers). High sensitivity is possible because of large capacitor area. When a comb drive is used as an electrostatic actuator, the narrow gap means low actuation voltages, and large area equals large forces. Long actuation distances of tens of micrometers can be made, and this is used for example in the laterally moving optical attenuator of Figure 21.4. DRIE is essential also in fabricating

Introduction to Microfabrication, Second Edition   Sami Franssila
© 2010 John Wiley & Sons, Ltd

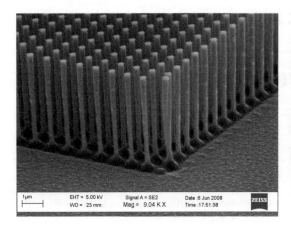
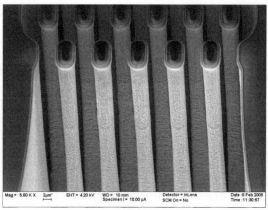

Figure 21.1 Nanopillars (left) and micropillars by DRIE (right). Courtesy Nikolai Chekurov Aalto University, and Kai Kolari, VTT

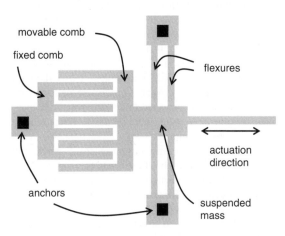

Figure 21.2 Comb-drive actuator

the vertical mirror which is inserted in the optical path. A variable optical attenuator (VOA) has been made by two 45° mirrors, one fixed, one movable (Figure 21.4). DRIE is essential for both the vertical wall and 45° angle.

The third major difference to wet etching is shape freedom: RIE will truthfully replicate any shape that has been patterned, irrespective of wafer crystal orientation. This opens up endless opportunities for device design. In making watch springs and other mechanical structures, the capability to fabricate variable cross-sections opens up new possibilities to fine-tune mechanical features, unlike traditional machining (Figure 21.5).

RIE produces vertical walls. This is true of anisotropic wet etching of <110> silicon as well, but there are limitations in wet etching: it is only possible along certain crystal planes, and cannot for example make two mirrors that are 90° to each other. With RIE, any shapes can

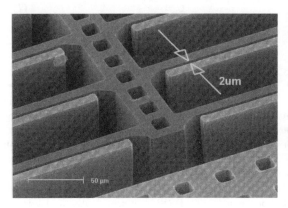

Figure 21.3 SEM micrograph of a comb-drive actuator (left) and vertical mirror at the end of the actuator (right). Reproduced from Acar and Shkel (2005) by permission of IOP and from Yun *et al.* (2006), copyright 2006, by permission of Elsevier

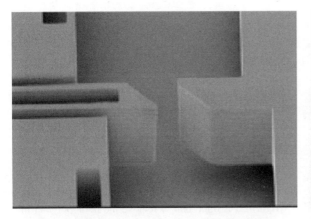 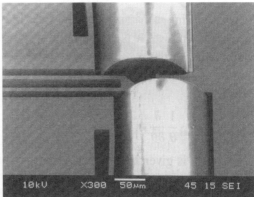

Figure 21.4 Variable optical attenuator (VOA) with fixed and movable 45° vertical mirrors (left); optical fibers inserted into VOA (right). Reproduced from Kim and Kim (2005) by permission of IOP

 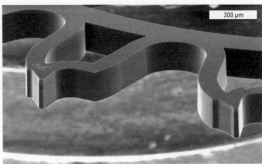

Figure 21.5 Mechanical parts for watches. Courtesy CSEM

be made, as the gyroscope of Figure 21.6 reveals. RIE walls are seldom exactly 90°, but for example 90 ± 2°, which may, or may not, be critical. In the case of vertical capacitors (and the gyroscope is one) and mirrors, 90° sidewalls are a must. Negative angles (>90°), also called retrograde, are problematic in many cases, for example in micromolding and casting, see Figure 18.20.

Micro-optics offers a wealth of examples where the special qualities of DRIE are utilized. The shape freedom is used to fabricate curving and spiraling waveguides (Figure 21.7). Sometimes the radius of curvature is as small as 5 μm, so the small feature size capability of DRIE is important. There are many applications where subwavelength sizes are needed, for example in antireflection "coatings" by artificial refractive index structures (Figure 21.7).

Another important issue in micro-optics is the sidewall quality. Most applications would prefer perfectly vertical sidewalls, but sometimes slanted walls or continuous surface profiles are needed. Any roughness on the sidewalls is usually detrimental to optical performance, and DRIE is sometimes followed by additional smoothing steps (Section 21.6 below).

In order to increase device packing density, the front surface area should be reserved for active devices only. This is especially important for optical devices like solar cells and CMOS camera chips. One solution is through-silicon vias (TSVs), making contacts on the back of the wafer, and having vias through the wafer to devices on the front side. DRIE and high aspect ratios are involved, but the etched depth is typically only 30–100 μm, depending on whether etching is done from the front or back side. In all cases it is important that via profile is positively sloping, or at least not retrograde, because the vias need to be filled by a conductor later on. TSV will be discussed in Chapter 39.

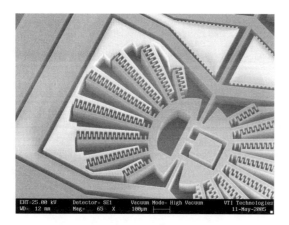

Figure 21.6 Gyroscope. Courtesy VTI Technologies

21.2 RIE Process Physics and Chemistry

Fluorine, chlorine and bromine etch silicon. Fluorine is chosen for MEMS DRIE because of its etch rate: etch rates an order of magnitude higher are possible with SF_6 than with Cl_2 or HBr. More fluorine can be generated at the same power level, and fluorine is very reactive. Fluorine, however, does not easily result in a perfectly anisotropic profile, but in many MEMS applications near-vertical profiles can be tolerated. In DRAM trench capacitor etching profile control is of paramount importance, and Cl_2 and HBr are used, and the etched depth is small enough so that a slower rate can be tolerated. Two major variants of MEMS DRIE processes exist: Bosch and cryogenic. Both take advantage of sidewall passivation mechanisms (Figure 11.5), but differ in many details.

In the Bosch process (named after the company which developed it) SF_6 and C_4F_8 gases are pulsed: an SF_6 pulse etches a few micrometers of silicon, but etching is not completely anisotropic. A C_4F_8 pulse is then applied, and a fluoropolymer protective film is deposited all over the wafer. The next SF_6 etching pulse removes the polymer film from the trench bottom by ion-assisted etching, but the sidewalls do not experience ion bombardment and remain protected (though are slightly etched by fluorine radicals). After removing the protective film from the trench bottom, SF_6 etching of silicon can continue. The next C_4F_8 pulse deposits a new protective film and then another SF_6 pulse is fed into the reactor. The pulsed operation leads to an undulating sidewall (shown schematically in Figure 21.8 and in the SEM micrograph of Figure 21.9 (right)). It is important to remove the fluoropolymer film completely at the end of the etching process because it can interfere with the following steps, for example by preventing wet cleaning via its hydrophobicity.

In cryogenic deep etching continuous SF_6/O_2 flow is used at $-120\,^\circ$C. Etching proceeds vertically because a film of SiO_xF_z from the reaction products is deposited on the cooled feature walls, and it is removed from horizontal surfaces by ion bombardment, leaving the sidewalls protected. Lateral etching by spontaneous chemical reaction is also suppressed by low temperature as predicted by the Arrhenius law (Equation 1.1). The smooth sidewall of the cryogenic process is contrasted to the undulating Bosch profile in the SEM micrographs of Figure 21.9.

Both the Bosch and cryogenic processes are low-pressure, high-density plasma processes. Various process parameters can be used to fine-tune DRIE processes, as shown in Figure 21.10. Process inputs like pressure, RF power, flow rate and temperature influence ion density, energy and angular distribution; radical density, surface reaction probability and desorption probability are also determined by machine parameters. In the end we are not interested in ion angular distributions, but

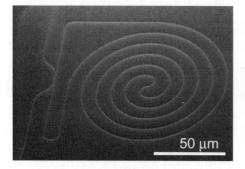

Figure 21.7 RIE in micro-optics: left, spiraling waveguides, reproduced from Xu *et al.* (2009) by permission of Optical Society of America; right, end-of-ridge waveguide antireflective facets, reproduced from Schmid *et al.* (2007) by permission of Optical Society of America

and more directional ion bombardment. But if high flow rates are needed (and they are needed to provide enough etchant), high pumping capacity is mandatory. Temperature increase has to be prevented because the resist is degraded when temperatures approach and exceed 100 °C. Passivation film thickness (either from C_4F_8 source gas or from SiO_xF_y reaction products) is good for anisotropy, but the passivation film slows down etching, and there is a danger of overpassivation, which leads to residues, and in the extreme to silicon nanograss, often known as black silicon (because it is an efficient absorber of light).

21.3 Deep Etching

Because DRIE etch rates are high, through-wafer etching is a standard step. This can be used to make holes, nozzles, membranes, cantilevers and others. Figure 21.11 shows a through-wafer etched triangular aligner for optical fiber. Its size is designed to match the fiber size, obviously, but the flexible fins allow for minor dimensional variations and ensure intimate contact. Holders of silicon nitride for fiber alignment made by wet etching were shown in Figure 20.10, but these were for in-plane fiber alignment, and wet etching could not be used to make a through-wafer version.

In Figure 17.2a a variable capacitor was shown. DRIE through-wafer etching releases the aluminum membrane and allows it to move. Aluminum is used as an etch mask, and in fluorine plasmas very high etch selectivity is obtained against aluminum because AlF_3 is involatile. If the variable capacitor were made by wet etching, the process flow would need to be completely different because aluminum does not tolerate KOH or TMAH. Making the aluminum membrane would then need to be done after silicon etching.

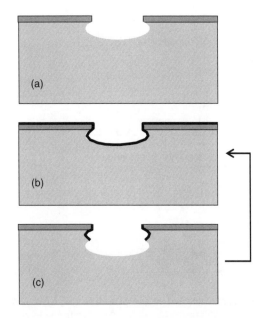

Figure 21.8 Bosch process, etch and passivation pulses repeated: (a) SF_6 isotropic etch step; (b) C_4F_8 passivation layer deposition; (c) next etch step

in wafer-level measurable quantities, like etch rate, selectivities, sidewall angle and the like.

Some general guidelines can be given regarding process optimization, but process recipes are very much reactor dependent, and not easily transferred from one etcher to another. Higher RF power results in higher etch rate, as expected, but the etch rates of oxide and resist also increase, lowering selectivities, so some balance must be obtained. Lower pressure is good for anisotropy because at low pressure there are fewer ion collisions

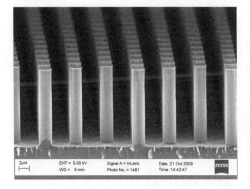 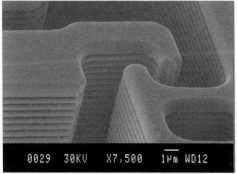

Figure 21.9 Silicon microstructures by cryogenic (left) and Bosch processes (right). Courtesy Ali Shah, Aalto University and Mikael Sterner and Joachim Oberhammer, KTH Royal Institute of Technology

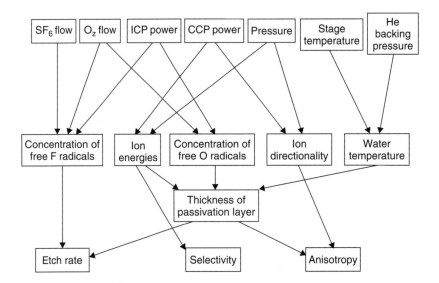

Figure 21.10 The interdependence of reactor parameters to plasma parameters and etch responses on a wafer. Courtesy Lauri Sainiemi, Aalto University

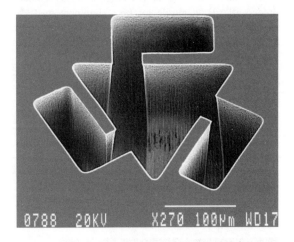

Figure 21.11 A holder for optical fiber. Reproduced from Holm *et al.* (2000), copyright 2000, by permission of Elsevier

Both the thin SOI device wafer and the thick handle wafer can be etched by DRIE. In the torsion mirror of Figure 21.12, all thin-film deposition and patterning steps on the SOI device layer are finished, and device layer silicon is then etched. Back-side lithography and handle wafer etching then commence. In this case all the critical structures are in the device layer, and back-side etching is non-critical, so both wet etching and DRIE can be used. BOX etching finally releases the torsion bars and combs and allows them to move.

21.4 Combining Anisotropic and Isotropic DRIE

As discussed in Chapter 11, anisotropic etching is the special case and isotropic profile is the default profile. The high rate offered by DRIE is beneficial also in isotropic etching. The combination of anisotropic and isotropic etching steps can be used to make free-standing single crystal silicon structures with vertical walls (Figure 21.13). The method relies on anisotropy/isotropy control in many steps.

Process flow for suspended silicon bridges

- Oxide mask
- DRIE for the trench
- Oxide CVD
- Oxide RIE: bottom cleared, sidewalls remain
- Isotropic silicon etching

This method has many limitations, the most important being that the width of the released structure and its thickness are related, and the method is best suited for narrow and thick structures. The underside of the structure is not well defined, which may be a limitation in some applications.

Silicon bridges with excellent dimensional control can be achieved when SOI wafers are used (Figure 21.14).

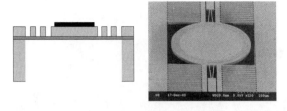

Figure 21.12 Schematic comb-drive torsional analog mirror: left, before SOI BOX etching; right, SEM micrograph after HF etching. Reproduced from Tsou *et al.* (2005) by permission of IOP

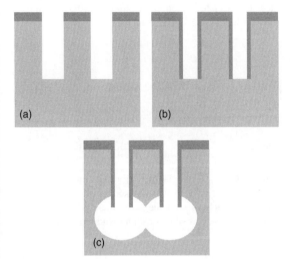

Figure 21.13 Suspended silicon bridge fabrication: (a) oxide-masked DRIE; (b) CVD oxide deposition and anisotropic oxide etching; (c) isotropic etching of silicon

Thermal oxide masks the DRIE of SOI device layer etching. CVD oxide is then deposited. Next, anisotropic RIE of oxide leaves sidewalls protected, but CVD oxide on the bottom is etched. So far the bulk silicon and SOI processes are identical, but then oxide etching continues and BOX is etched, too. The oxide mask on device silicon is also consumed, so the three oxide thicknesses have to be carefully designed. After etching through BOX, the isotropic silicon etch process step releases the bridge. The bridge is protected by thermal mask oxide from the top, by CVD oxide on sidewalls and by BOX from the bottom, and the bridge dimensions are preserved during the isotropic silicon etch step. The stress test structure of Figure 2.2 was made with this process. Bridge doping is a free parameter, while in many wet-etched versions p^{++} or pn etch stop techniques necessitate certain doping levels.

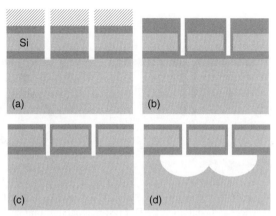

Figure 21.14 Microbridge in SOI device layer: (a) mask oxide, SOI device silicon and BOX anisotropic etching with resist mask; (b) PECVD oxide; (c) oxide anisotropic etching; (d) SOI handle silicon isotropic etching. Redrawn from Sainiemi *et al.* (2009)

The same method has been used to create buried microchannels (Figure 21.15). Anisotropic silicon etching is used to etch a narrow trench, which is protected by CVD oxide. Isotropic silicon etching creates a circular channel. Conformal deposition closes the channel. Because thin-film deposition thicknesses are practically limited to a few micrometers, the initial trench opening has to be in the micrometer range.

21.5 Microneedles and Nozzles

In-plane microneedles can be made very long because silicon is strong. A length of a few millimeters is not unusual. Out-of-plane microneedles are limited to the wafer thickness, in practice a few hundred micrometers. There are needles and nozzles for biomedical applications (blood extraction and drug injection), for chemical applications (for sample droplet dispensing, spraying and vaporization), printers (ink jets, DNA array spotting), and for cooling, aerosol generation and cell probing.

The in-plane fluidic microneedle shown in Figure 21.16 makes use of DRIE's special capabilities, as it could not be fabricated by wet etching. The process flow is as follows. The buried channel is formed as described above using both anisotropic and isotropic DRIE, and conformal CVD. The fluidic reservoir part (along cut line AA′) is very wide. It will be coated by CVD but not closed. Needle shape is determined by the next DRIE step. Chip size is defined on the same mask. Needle thickness is determined by the back-side DRIE release etch.

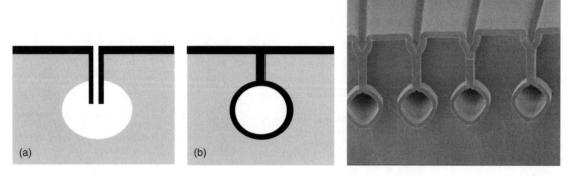

Figure 21.15 Buried microchannels: (a) anisotropic DRIE, sidewall spacer formation and isotropic DRIE; (b) removal of spacer and conformal CVD. SEM micrograph from de Boer *et al*. (2000) by permission of IEEE

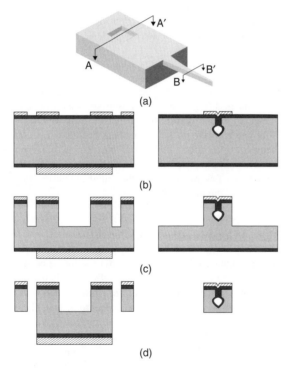

Figure 21.16 In-plane microneedle: buried microchannel, Reproduced from Paik *et al*. (2004), copyright 2004, by permission of Elsevier

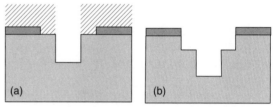

Figure 21.17 Nested mask: (a) oxide lithography and etching, followed by new resist spinning and lithography and silicon etching with the resist mask; (b) resist mask removal and second silicon etching with oxide mask

and is then stripped away. The oxide revealed underneath will act as a mask for the second, deeper DRIE step.

There are various ways of making nozzles with the through-wafer fluidic channel. In this section we will compare many different versions, with regard to lithography–etching interplay: the number of process steps differs, and masking and alignment requirements vary. In every case, we are talking about arrays of nozzles, because here DRIE excels over wet etching. In wet etching nozzle density is limited to about 4/mm^2 while DRIE nozzle arrays can have densities of 100/mm^2 easily, and 1000/mm^2 with some process tuning. The most common limiting factor is the through-wafer etching at high aspect ratio: 10 μm holes equal 40:1 aspect ratios, and most applications will settle for larger channels (even though the nozzles can be made very small).

The first approach uses three front-side lithography and DRIE steps (Figure 21.18). If the first DRIE etch (determining nozzle height) is shallow, resist spinning over topography can reasonably be done. The second lithography defines the nozzle itself. The third patterning

Deep structures can be etched easily, but lithography over deep steps is difficult. Double masking (or peeling or nested masks) can be used just as in wet etching (Figure 20.4). For example, oxide is grown and patterned, and a second pattern of resist is done on top (Figure 21.17). The resist serves as a mask during the first shallow DRIE step

Deep Reactive Ion Etching 263

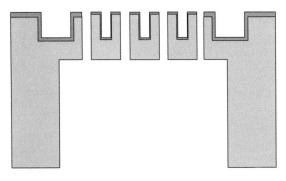

Figure 21.18 Nozzle array with common fluid reservoir: three lithography and three DRIE steps. Redrawn from Yu *et al.* (2009)

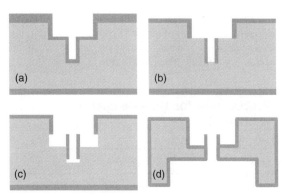

Figure 21.20 Oxide nozzle for patch clamp cell probing. After two-step DRIE with peeling masks: (a) thermal oxidation; (b) anisotropic oxide RIE; (c) anisotropic and isotropic silicon RIE; (d) back-side silicon DRIE and thermal oxidation. Adapted from Lehnert *et al.* (2002)

on the wafer back side is non-critical for both lithography and DRIE: the structures are large, and alignment requirements are lax. All the nozzles will be squirting the same fluid because of the common reservoir but the density of nozzles is very high.

The DRIE requirements for the nozzles of Figure 21.19 are much more demanding than in the previous example. For 100 μm nozzles aspect ratios of 4:1 are needed in through-wafer DRIE. On the other hand, only two lithography steps are needed, and both of them are done on a planar surface. Alignment is critical to make the top and bottom channel openings coincide.

In both the above cases simple silicon nozzles sufficed, but if for instance a dielectric nozzle is needed, the process becomes a little more complex (Figure 21.20). This device is used in patch clamp cell membrane electrical studies, and the device has to have gigaohm resistance. A peeling mask process similar to Figure 21.17 is used to create the initial top pattern. The wafer is then thermally oxidized, and oxide RIE is performed. Oxide on the front surface is thicker than elsewhere, because there was mask oxide to begin with. The oxide on the sidewalls is preserved because oxide RIE is anisotropic, but the bottoms of etched features are cleared of oxide. Silicon DRIE is then continued, to form an oxide nozzle (Figure 21.21). Another oxidation step ensures dielectric isolation everywhere.

The side-opened microneedle (Figure 1.13) provides a sharp tip for effective piercing of the epithelium, but because the fluidic channel is not at the tip, it will not

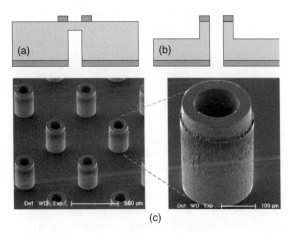

Figure 21.19 Silicon nozzle array: (a) back-side DRIE; (b) front-side DRIE; (c) SEM micrographs of nozzles. Reproduced from Deng *et al.* (2006), copyright 2006, by permission of Elsevier

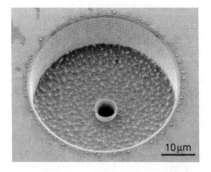

Figure 21.21 Oxide nozzles of previous figure as seen by a SEM. Reproduced from Lehnert *et al.* (2002), copyright 2002, American Institute of Physics

be easily clogged. The fabrication process is depicted in Figure 21.22. Front-to-back alignment is needed, but because device dimensions are fairly large, 100 μm, alignment is not difficult. The process flow is as follows:

Process flow for the side-opened microneedle:

Step	Comment
Thermal oxidation	Both sides of wafer, protects front
Back-side lithography	Non-critical lithography
Oxide RIE on back	Etch mask pattern
DRIE from backside	Fluid access channel
Oxide etching in HF	Oxide removed from both sides
Wet oxidation	Front etch mask and etch stop in channel
Top-side lithography	Needle lateral shape defined
Oxide RIE on top	Etch mask pattern, no back-side protection
DRIE sequence on top side	First isotropic shallow etch, 30 μm
	Second anisotropic medium etch, 100 μm
	Third isotropic narrowing etch, until oxide
Oxide etch	HF etching opens channels and clears tip

21.6 Sidewall Quality

As a first approximation DRIE sidewalls are vertical (Figure 21.9) and this is important in many applications. Vertical capacitors and mirrors necessitate perfectly vertical walls. In an optical etalon (Exercise 30.13) the vertical mirrors should be perfectly parallel, otherwise the light reflected in between will rapidly leak out. There are many possible non-idealities. The first is a non-vertical but smooth sidewall, with a deviation of a few degrees from the vertical (Figure 11.4). A similar negative (retrograde) slope is also common (Figure 21.1 (left)) and they can be tuned to some extent by changing the process parameters. Other sidewall non-idealities include Bosch process undulation (Figure 21.9b) and roughness and bowing, which are shown schematically in Figure 21.23. The periodic undulation (also known as

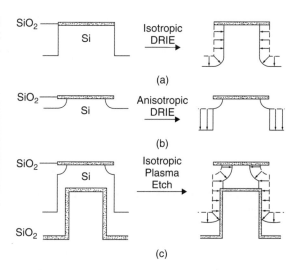

Figure 21.22 Microneedle with a sharp tip for piercing and side-opened liquid channel. See also SEM micrograph of the finished needle in Figure 1.13. Reproduced from Griss and Stemme (2003), copyright 2003, by permission of IEEE

scalloping) in the Bosch process makes it unsuitable for many optical applications. There are ways to minimize undulation for example by using shorter pulses, which unfortunately results is a slower etch rate. The sidewall angle may also change, so undulation minimization will affect many other process responses.

High ion energy is useful because it can sputter away material, but it is non-selective etching. It may be beneficial in the early stages of etching to have a non-selective etch step which will remove everything, for example native oxide. But too high an ion energy starts to affect the shape. Sharp corners will be rounded by ion bombardment and deflected high-energy ions will lead to trenching at the feature bottoms, Figure 21.23d. Deflected ions in general will result in deviations about verticality, for instance barreling (bowing) of the profile (Figure 21.23b).

The sidewalls of DRIE etched walls are quite rough, with 50–100 nm RMS roughness typical. Optical scattering losses increase rapidly when roughness exceeds about 5% of wavelength (which corresponds to 20–30 nm for visible wavelengths). A combined plasma–wet process can provide benefits: wet smoothing after dry etching. A short KOH/TMAH/EDP etch step etches of course in crystal plane-dependent fashion and if suitable crystal planes are found, the surface quality will be equal to wet etching. These walls are very smooth and precisely oriented, as witnessed by their use as a master for polymer optical

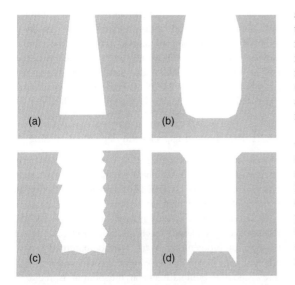

Figure 21.23 DRIE sidewall non-idealities: (a) retrograde profile; (b) barreling (bowing); (c) roughness; (d) faceting (top) and trenching (bottom)

an isotropic low-power plasma etch is done to remove the damaged surface layer. A factor of two increase in mechanical strength can be achieved. The role of lapping in removing grinding damage is analogous. Hydrogen annealing, which is used in Smart-cut SOI wafer fabrication (Section 22.5), smoothes out small irregularities, and it is a useful post-DRIE treatment too; for example, smoothing requires a 1000–1100 °C anneal for 10 minutes.

Post-etch smoothing can take many forms. Anisotropic wet etching obviously works better for rectangular shapes (on <100> silicon) and hexagonal shapes (on <111> silicon) than for circular shapes. Thermal oxidation plus HF wet etching is also possible (Figure 21.25). If the process is continued by LPCVD polysilicon deposition and poly oxidation, smoothness improves and simultaneously the etched hole can be made smaller. All these steps increase process complexity, so there has to be a strong motivation for such multistep smoothing/narrowing.

21.7 Pattern Size and Pattern Density Effects

Etch rates are very much dependent on particular process details, including etch reactor geometry, RF power coupling scheme, wafer temperature and gas flow rate. They also depend on device layout. First of all, the etch rate depends on etchable area: more area, slower rate, and vice versa. Second, the etch rate of small features is slower than that of large features. Third, etch rate slows down. This happens in high aspect ratio structures. All these phenomena necessitate a more detailed analysis of how the interactions of chemical and physical processes really work in etching.

The loading effect, or area-dependent reaction rate, is a common phenomenon in chemical reactions. For a process optimized for certain etchable area, the flow may not be high enough to supply reactants to keep an identical etch rate when area is increased. This is a major problem for ASIC manufacturers who face hundreds of different designs.

The loading effect is very general and operates in all etching processes. It manifests itself when reactions are under a mass transport/diffusion-limited regime. If the flow of reactants is insufficient, the rate cannot be maintained. On the other hand, if the reaction is controlled by surface reactions (i.e., there is ample supply of reactants, and the etching process is limited by surface reactions), the process does not exhibit loading effects.

Loading effects operate at various scales:

- In batch reactors the etchable area changes because the number of wafers changes.

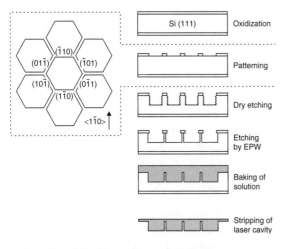

Figure 21.24 Bosch process ripple removed by wet etching in EPW, simultaneously ensuring vertical, crystal plane-defined walls. Reproduced from Sasaki *et al.* (2000) by permission of Institute of Pure and Applied Physics

device casting (Figure 21.24). This typically translates to 10–30 nm RMS roughness.

Smoothing is also important for mechanical strength: a surface with cracks and hillocks is mechanically weak because the irregularities act as starting points for fracture. A two-step dry–dry etching will help: after the DRIE step

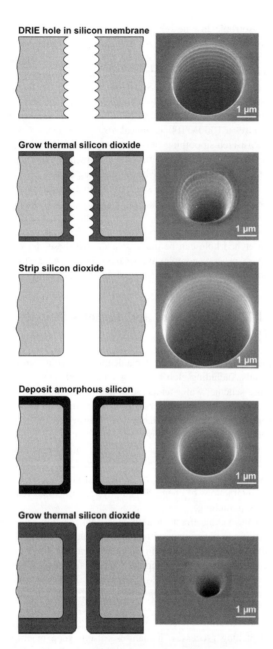

Figure 21.25 Reducing ripple in the Bosch process by postprocessing. Reproduced from Matthews and Judy (2006), copyright 2006, by permission of IEEE

- In single wafer reactors different chip designs have different etchable areas.
- The local pattern density on a chip is different in every design (microloading).

Microloading manifests itself as an etch depth difference between isolated and array features: there is more material to be etched in arrays, therefore the rate is slower. Identical trench widths will result in different etched depths depending on pattern density in the neighborhood (Figure 21.26). Other pattern dependences discussed below are deceptively similar, yet different.

21.7.1 RIE lag and aspect ratio-dependent etching

Plasma etching of structures of 1:1 aspect ratio is fairly straightforward, but somewhere around 2:1 aspect ratio a phenomenon known as RIE lag manifests itself: smaller features etch slower than larger features (Figure 21.26).

Gas conductance in deep narrow holes is low and the reactants simply cannot reach the bottom effectively (or the reaction products removed). Ion bombardment is also affected: ions experience sidewall collisions in deep structures, and the bombardment at the bottom is reduced. These effects lead to a reduced etch rate in deep structures of high aspect ratio. RIE lag can be seen from a single experiment with a test structure that contains many different linewidths. RIE lag is not related to RIE reactors; it is present in all plasma etching systems irrespective of actual reactor design.

Aspect ratio-dependent etching (ARDE) is a dynamic effect: the etch rate slows down as etching proceeds, for every linewidth! The basic mechanism for RIE lag and ARDE is the same. But in order to see ARDE, many wafers have to be etched, with different etch times. To appreciate the difference, see Figure 21.27 (left) where etch rate is plotted vs. etch time. As etching proceeds, the etch rate slows down. This is true for $2\,\mu m$, $5\,\mu m$ and all other linewidths alike. ARDE is therefore seen as a downward-sloping line in Figure 21.27 (right). RIE lag, on the other hand, is the vertical displacement of the two lines: wider lines are higher up on the chart.

DRIE is fairly straightforward for structures with aspect ratios of 10:1 while 20:1 is much more demanding. And even though 40:1 has been demonstrated in the lab, it is not to be considered a standard fabrication step. For

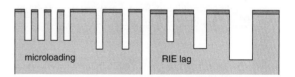

Figure 21.26 Microloading vs. RIE lag. Microloading results in different etched depths for identical lines, with lines in low pattern density areas resulting in deeper structures. RIE lag: smaller etch rate for narrower lines

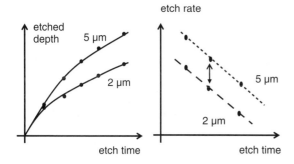

Figure 21.27 Raw data of etched depth vs. etch time for 2 μm and 5 μm features (left); plotted as rate vs. time (right), RIE lag is the vertical distance between the lines and ARDE is the slope

380 μm wafers these numbers translate to feature sizes about 40, 20 and 10 μm. In bonded SOI wafers, device layer thicknesses range from 5 μm upward. Feature size is then limited by lithography and undercutting of the pulsed (Bosch) process, rather than by aspect ratio effects.

Another charging-related non-ideality in DRIE is the notching effect (or footing effect): when the silicon etching end point is reached, the underlying oxide (either oxide on the back side of a bulk wafer, or BOX) becomes charged. This charging leads to repellency of incoming ions, and they are deflected sideways, enhancing lateral etching near the silicon/oxide interface (Figure 21.28). Note that RIE lag has an effect on notching: the larger features have experienced longer overetching, therefore the notching effect has had more time to operate. Charge can accumulate on isolated conductors, and the oxide beneath these conductors can be damaged by this charge accumulation. Not only plasma etching, but all plasma processes, PECVD and sputtering are potential sources of oxide damage.

21.8 Etch Residues and Damage

Etch anisotropy relies on passivation films, and this means that we intentionally deposit material on the wafer; not all of it is beneficial. Too much passivation results in a reduced etch rate, and really heavy passivation equals etch stop. This can happen locally, so that some areas are not attacked by the etchant, leading to roughness and in extreme cases pillar-like structures. Similar pillar-like residues result from particles on the surface. In the case of a completely anisotropic process, even nanoscopic particles act as etch masks (Figure 21.29).

Similar to overpassivation, redeposition can lead to roughness and pillar-like structures. If the etch product is only marginally volatile, it can be redeposited on the wafer surface and act as a mask. It is also possible that the mask material is redeposited: high ion energy can detach atoms from an inert mask (remember that ion beams can etch anything) and if redeposited on the wafer, they will certainly act as masks; after all, mask material is chosen to be non-etchable.

Often etch products are incorporated into a sidewall passivation film. They can also react with photoresists. Photoresist removal after aggressive plasma etching is therefore difficult. Aggressive etching will also etch the resist mask, and line narrowing can take place. The opposite effect can also happen (but not often!): so much

Figure 21.28 Notching effect. Reproduced from Chekurov *et al.* (2007) by permission of IOP

Figure 21.29 Nanoparticle-masked etching. Reproduced from Sainiemi *et al.* (2007) by permission of IOP

material (etch products, redeposited non-etchable material, passivation film, resist debris) is deposited on the resist sidewall that in fact linewidth increases!

RIE lag is not the only lag that is encountered in etching. Etch initiation lag is also typical. If etched depth is plotted vs. etch time, the line should go through the origin, but often there is a time lag before etching starts. Native oxides is one explanation, and other residues can also be present on the wafer surface. The "first-wafer effect" is often seen in RIE: the etch rate only stabilizes after a few wafers have been etched. There are both thermal and chemical reasons for this. RIE is an energetic process and the system heats up during RIE. And even if the etch reactor is clean to begin with, passivation films will soon cover its inside surfaces. If the reactor is used for very different materials, using different gases, like SF_6-based silicon etching and Cl-based GaAs etching, the etch residues in the chamber will foul the process.

21.9 DRIE vs. Wet Etching

Both DRIE and anisotropic wet etching have their advantages, as in Tables 21.1 and 21.2, and in many applications both etching techniques are needed. The decision in favor of either technique depends not only on technological factors like etched shape, sidewall angle or surface quality, but on practical issues like etch rate, back-side protection or equipment availability.

The ink jet example of Figure 21.30 shows how many different etch techniques are utilized in one device: the manifold etching through the wafer is done by TMAH anisotropic wet etching, the critical inlet channel is defined by DRIE, and the chamber is made hemispherical by isotropic plasma etching. The nozzle

Table 21.1 DRIE main features

- Any shape can be made
- Tightly spaced structures can be made
- Vertical structures of high aspect ratio are possible
- Difficulty in making silicon membranes unless SOI wafers used
- Photoresist masking is possible
- Single-sided processing, no back-side protection needed
- 1–3 hours for through-wafer etching in single wafer operation
- 1–3 days to etch a batch of 25 wafers through-the-wafer

Table 21.2 Anisotropic wet etching main features

- Very accurate dimensional control by crystal plane-dependent etching
- Structural shapes limited by crystal plane-dependent etching
- Accurate 45°, 54.7°, 70.5° or 90° sidewall
- Smooth and well-defined surfaces
- About 4–8 hours for through-wafer etching for a single wafer
- About 4–8 hours for through-wafer etching for a batch of 25 wafers
- Etches both sides, protection needed on back side
- Etches both sides, symmetric structures can be made in a single etch step
- Aggressive to metals and many other materials, protective layers needed
- Limited selection of mask materials, thick oxide and LPCVD nitride standard
- Many etch stop mechanisms available: boron p^{++}, pn junction, SOI BOX

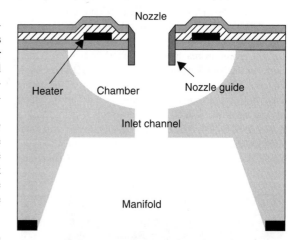

Figure 21.30 Ink jet etching features: chamber is etched by isotropic plasma etching, anisotropic plasma etching for inlet channel, anisotropic spacer etching for nozzle guides and anisotropic TMAH wet etching for the manifold. Reproduced from Shin *et al.* (2003) by permission of IEEE

guides are spacers formed by plasma etching. In a flame ionization detector (Figure 1.10), the nozzle chamber is KOH etched in silicon, and micronozzles and sample gas channels are etched in silicon by DRIE. Pyrex glass wafers are isotropically wet etched in HF.

21.10 Exercises

1. What must SF_6 gas flow at least be in a DRIE reactor if the silicon etch rate is 10 μm/min, wafer size 150 mm and etchable area 20%?
2. Determine the DRIE single crystal silicon etch rate from the following trench etching data:

Etch time	Etched depth (μm)		
(min)	80 μm wide	40 μm wide	12 μm wide
20	109	104	85
40	205	193	156
60	292	278	215

3. How much etch non-uniformity can native oxide cause in silicon (D)RIE?
4. Design a fabrication process for the cell filter shown below.

Reproduced from Prince et al. (2007), by permission of *Proc. IMechE*

5. Compare silicon tip fabrication by anisotropic wet etching, plasma etching and thermal oxidation.
6. Draw top view mask layouts to show how the microneedle of Figure 1.13 and 21.22 is made.
7. Explain the dimensions and film thicknesses and quantify the narrowing of the 3 μm hole shown in Figure 21.25.
8. How could you make a DRIE version of the filter shown in Figure 20.27? Could you improve the aperture ratio (percentage of open area)?
9. What areal densities of nozzles could be made by the processes of Figures 21.18, 21.19 and 21.20?
10. Design a fabrication process for the variable optical attenuator of Figure 21.4.
11. Explain step by step the fabrication of the comb-driven torsional mirror of Figure 21.12!
12. Explain in detail the process steps of buried channel formation (Figure 21.15).

References and Related Reading

Acar, C. and A.M. Shkel (2005) Structurally decoupled micromachined gyroscopes with post-release capacitance enhancement, *J. Micromech. Microeng.*, **15**, 1092–1101.

Agarwal, R., S. Samson and S. Bhansali (2007) Fabrication of vertical mirrors using plasma etch and KOH:IPA polishing, *J. Micromech. Microeng.*, **17**, 26–35.

Chekurov, N. et al. (2007) Atomic layer deposition enhanced rapid dry fabrication of micromechanical devices with cryogenic deep reactive ion etching, *J. Micromech. Microeng.*, **17**, 1731–1736.

Chen, K.-S. et al. (2002) Effect of process parameters on the surface morphology and mechanical performance of silicon structures after deep reactive ion etching (DRIE), *J. Microelectromech. Syst.*, **11**, 264.

de Boer, M.J. et al. (2000) Micromachining of buried micro channels in silicon, *J. Microelectromech. Syst.*, **9**, 94.

Deng, W. et al. (2006) Increase of electrospray throughput using multiplexed microfabricated sources for the scalable generation of monodisperse droplets, *J. Aerosol Sci.*, **37**, 696–714.

Gottscho, R.A. et al. (1992) Microscopic uniformity in plasma etching, *J. Vac. Sci. Technol.*, **B10**, 2133–2147.

Griss, P. and Stemme, G. (2003) Side-opened out-of-plane microneedles for microfluidic transdermal liquid transfer, *J. Microelectromech. Syst.*, **12**, 296.

Holm, J. et al. (2000) Through-etched silicon carriers for passive alignment of optical fibers to surface-active optoelectronic components, *Sens. Actuators*, **82**, 245–248.

Jansen, H.V. et al. (2009) Black silicon method X: a review on high speed and selective plasma etching of silicon with profile control: an in-depth comparison between Bosch and cryostat DRIE processes as a roadmap to next generation equipment, *J. Micromech. Microeng.*, **19**, 033001.

Kiihamäki, J. and S. Franssila (1999) Pattern shape effects and artefacts in deep silicon etching, *J. Vac. Sci. Technol.*, **A17**, 2280.

Kim, C.-H. and Y.-K. Kim (2005) MEMS variable optical attenuator using a translation motion of 45° tilted vertical mirror, *J. Micromech. Microeng.*, **15**, 1466–1475.

Lee, M.-C.M and M.C. Wu (2006) Thermal annealing in hydrogen for 3-D profile transformation on silicon-on-insulator and sidewall roughness reduction, *J. Microelectromech. Syst.*, **15**, 338–343.

Lehnert, T. et al. (2002) Realization of hollow SiO_2 micronozzle for electrical measurements on living cells, *Appl. Phys. Lett.*, **81**, 5063–5065.

MacDonald, N.C. (1996) SCREAM MicroElectroMechanical Systems, *Microelectron. Eng.*, **32**, 49.

Matthews, B. and J.W. Judy (2006) Design and fabrication of a micromachined planar patch-clamp substrate with integrated microfluidics for single-cell measurements, *J. Microelectromech. Syst.*, **15**, 214–222.

Paik, S.-J. et al. (2004) In-plane single-crystal-silicon microneedles for minimally invasive microfluid systems, *Sens. Actuators*, **A114**, 276–284.

Prince, M. et al. (2007) The development of a novel Bio-MEMS filtration chip for the separation of specific cells in fluid suspension, *Proc. IMechE, Part H: J. Eng. Med.*, **221**, 113–128.

Ranganathan, N. et al. (2008) The development of a tapered silicon micro-micromachining process for 3D microsystems packaging, *J. Micromech. Microeng.*, **18**, 115028.

Sainiemi, L. et al. (2007) Rapid fabrication of high aspect ratio silicon nanopillars for chemical analysis, *Nanotechnology*, **18**, 505303.

Sainiemi, L. et al. (2009) Fabrication of thermal microbridge actuators and characterization of their electrical and mechanical responses, *Sens. Actuators*, **A149**, 305–314.

Sasaki, M. et al. (2000) Anisotropically etched Si mold for solid polymer dye microcavity laser, *Jpn. J. Appl. Phys.*, **39**, 7145.

Schmid, J.H. et al. (2007) Gradient-index antireflective subwavelength structures for planar waveguide facets, *Opt. Lett.*, **32**, 1794–1796.

Shin, S.J. et al. (2003) Firing frequency improvement of back shooting inkjet printhead by thermal management, Proceedings of Transducers'03, p. 380.

Sterner, M., N. Roxhed, G. Stemme and J. Oberhammer (2007) Mechanically tri-stable SPDT metal-contact MEMS switch embedded in 3D transmission line, Proceedings of the 37th European Microwave Conference, 2007, pp. 1225–1228.

Tsou, C. et al. (2005) A novel self-aligned vertical electrostatic combdrives actuator for scanning micromirrors, *J. Microelectromech. Syst.*, **15**, 855–860.

Xu, D.-X. et al. (2009) Spiral cavity Si wire resonators as label-free biosensors, OSA Integrated Photonics and Nanophotonics Research and Applications (IPNRA), paper IMB2.

Yu, L.M. et al. (2009) A microfabricated electrode with hollow microneedles for ECG measurement, *Sens. Actuators*, **A151**, 17–22.

Yun, S.-S., S.-K. You and J.-H. Lee (2006) Fabrication of vertical optical plane using DRIE and KOH crystalline etching of (110) silicon wafer, *Sens. Actuators*, **A128**, 387–394.

22

Wafer Engineering

Silicon crystal growth influences wafer processing in various ways: for example, the quality of thin thermal oxide depends on microscopic voids formed during crystal pulling, and oxygen content and distribution inside wafers are essential for trapping metallic impurities. In spite of continuous developments, wafers never become "perfect" because each new generation of devices sets different demands on wafers, for example larger and thicker wafers are needed, which changes crystal pulling conditions and thermal balances, leading to different dominating defect mechanisms. These wafers will be processed at lower temperatures, using novel processes, which will change the relative importance of impurities and defects.

When silicon-on-insulator (SOI) wafers are made, different techniques result in different possible thicknesses for buried oxide layer and the top device silicon. Defects in different SOI technologies are different, and have varying degrees of impact on devices. In epitaxy multiple layers enhance possibilities for optimizing various device design targets, but bulk wafers are always being considered as an option because of the higher price of epi (and SOI).

In addition to silicon-to-silicon bonding, there are other possible materials that can be combined with silicon in order to achieve the best of both worlds. Silicon-on-glass leads to transparency, while indium phosphide-on-silicon leads to high-speed electronics, and silicon-on-diamond is the ultimate thermal conduction solution for high-power devices. Wafer bonding is the key technology to make these heterogeneous substrates.

22.1 Silicon Crystals

Vacancies and interstitials will always be present even in the best possible material because they are created by thermal equilibrium processes (there are also vacancies and interstitials resulting from the crystal pulling process).

Vacancies and interstitials diffuse like dopant atoms. They also aggregate at temperatures of 900–1050 °C. Vacancy clusters are voids, and aggregates of atoms are precipitates, for example dopant atoms or oxygen above the solubility limit will form precipitates (see Figure 14.2 for solubility vs. temperature). Silicon interstitials can also cluster, forming dislocation loops (also known as L-pits). Processes which cause volume changes, like thermal oxidation, are prone to produce defects. Oxidation-induced stacking faults (OISFs) are one class of such defects. On the other hand, the volume changes associated with oxidation inject vacancies into silicon, and these affect diffusing species. The temperature differences between wafer center and edge can generate dislocations, so the material that was perfect to begin with is modified critically if improper processing is carried out.

When a wafer is heated non-uniformly in a high-temperature step, for example during epitaxy or oxidation, temperature differences between wafer center and edge can easily lead to thermal stresses above silicon yield strength. These stresses will relax by forming defects, for example slip lines. The order of magnitude of thermally generated stress σ can be gauged by

$$\sigma = \varepsilon E = \alpha L \times \Delta T \times E \qquad (22.1)$$

where strain ε depends on the silicon coefficient of thermal expansion, α, temperature difference ΔT and wafer size, L. Silicon yield strength (or critical shear stress) is strongly temperature dependent: at 850 °C it is about 50 MPa, at 1000 °C only on the order of 10 MPa, and about 1 MPa at 1200 °C.

Oxygen is always present in CZ silicon because of the silica crucible (Figure 22.1). All dissolved oxygen does not remain in the melt or get incorporated into the ingot: most of it is evaporated as SiO. Oxygen at 5–20 ppma (according to ASTM standard F121-83) will end up in the crystal.

Introduction to Microfabrication, Second Edition   Sami Franssila
© 2010 John Wiley & Sons, Ltd

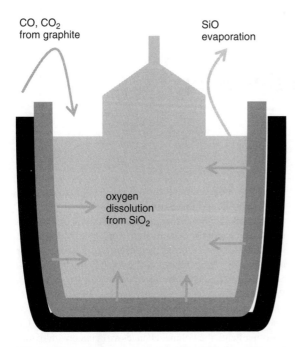

Figure 22.1 Oxygen from fused silica (SiO_2) crucible and carbon from graphite holder

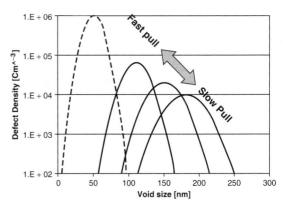

Figure 22.2 Vacancy cluster (COP) size vs. density at different pull rates. Reproduced from Aminzadeh et al. (2002), copyright 2002, by permission of IEEE

The silica crucible is not mechanically strong enough at about 1400–1500 °C temperatures, and a graphite susceptor provides mechanical strength. The silica crucible reacts with the graphite holder according to the equation $SiO_2 + 3\,C \Rightarrow SiC + 2\,CO$. This carbon monoxide gas is the source of carbon which is always present in CZ crystals, at concentrations about 10^{16} cm^{-3}, below 1 ppm.

Oxygen is initially dissolved in interstitial sites but can precipitate during thermal treatments. These precipitates, often known simply as O_2P, are important in many respects, for example O_2P are beneficial for wafer mechanical strength because they will prevent slip lines from spreading in the wafer. Carbon impurities act as nucleation sites and centers for oxygen precipitation. Oxygen precipitates are prime examples of bulk micro defects (BMDs). Oxygen concentration can have a drastic effect on KOH wet etching of silicon: oxygen precipitates cause local stresses which change the relative etch rates of crystal planes. This will de discussed in Chapter 30 (see Figure 30.4). Very small and uniform oxygen precipitates can be achieved with low oxygen concentrations, for example 7 ppma. This may, however, be too low for ICs, and wafer selection for CMOS-MEMS integration is difficult.

Microvoids are clusters of vacancies formed inside the ingot during crystal pulling. They are known, for historical reasons, as COPs (Crystal-Originated Particles). Typical COP sizes are 50–200 nm, and they are found in concentrations of 10^4–10^6 cm^{-3} (Figure 22.2). When wafers are cut and polished, COPs end up at wafer surface after alkaline cleaning step during wafer processing. A COP is detected by laser scattering because it reflects light almost like a particle (advanced multiangle scatterometry tools can distinguish COPs from particles). Vacancy clusters were therefore classified as particles, and were given the name COP. It was the fact that the number of COPs did not decrease in cleaning (and it could in fact increase!) that lead to a reassessment of their nature.

COP formation during crystal pulling depends on V/G ratio: pull rate to temperature gradient at crystallization interface. When V/G is high, usually at high pull rates, COPs (microvoids) are formed throughout the crystal. At medium pull rates oxidation induced stacking faults (OISF) appear at periphery. Yet lower pull rates lead to L-pits appear outside the OISF ring and at very low pull rates OISF disappears and L-pits are everywhere. Because pull rates tend to be smaller for larger diameter crystals, larger wafers have quite different defect profile from smaller crystals.

22.2 Gettering

Defects are not passive objects: they change and morph during wafer processing, especially during high-temperature steps. For example, metallic impurities diffuse, agglomerate and trap charge carriers. Metallic impurities come from EGS polysilicon, the silica crucible, the graphite crucible and heaters and other hot parts of the crystal growth system. The segregation coefficients of most metals (Table 4.2) are very small, and the crystal is purified relative to melt. Metals are, however, fast diffusers in silicon, and they react with other defects, forming clusters. Metals affect electronic devices by creating trapping centers in silicon midgap, reducing minority carrier lifetimes and lowering mobility. Metals can also precipitate at the Si/SiO$_2$ interface and reduce oxide quality. The allowed iron level in advanced silicon wafers is limited to 10^{10} cm^{-3}.

In order to contain the impurities, and keep the active transistor area defect-free, a multizone approach to wafers has been devised. The wafer is designed to have different zones with different roles. The top surface is the prime quality area, with minimized particles, minimal roughness and ultimate flatness. Under the surface is the defect-free region, known as the denuded zone (DZ). The rest of the wafer can be used as a getter, a sink for collecting and holding impurities. The back side of the wafer can also be made to act as a getter. These zones are depicted in Figure 22.3.

This multizone approach is clearly not suitable for devices that extend through the wafer, like power transistors, solar cells or bulk MEMS, because of the vertical non-uniformity it introduces. If both ICs and MEMS devices are made on the same wafer, it is beneficial to have small, uniform oxygen precipitates, as a compromise which satisfies to some extent the demands of both internal gettering and vertical uniformity.

Oxygen is depleted in the top silicon layer by annealing in hydrogen. Oxygen outdiffuses from the surface, and an oxygen-depleted region develops. This denuded zone thus has low oxygen concentration and minimized oxygen-induced defects. It is formed in three steps:

1. The outdiffusion step (1100–1200 °C; 1–4 h): oxygen diffuses out of the surface region, leaving <5 ppma oxygen.
2. The nucleation step at 600 °C: SiO$_x$ formed homogeneously throughout the wafer volume.
3. SiO$_x$ precipitate growth and gettering (950–1200 °C, 4–16 h).

Denuded zone depth depends strongly on device requirements and can range from 10 to 40 μm.

Similar treatments can be done to create a bulk wafer which resembles an epitaxial wafer: a p-type wafer is annealed in hydrogen at 1200 °C, leading to boron and oxygen outdiffusion. A reduction of boron concentration by an order of magnitude can be achieved, essentially forming p$^-$ on a p$^+$ type structure (similar to Figure 4.15). The denuded zone is formed simultaneously. Also, vacancy clusters (COPs) are annealed out and the wafer surface is smoothed during the hydrogen anneal.

Gettering of impurities can be done either inside silicon (intrinsic gettering, IG) or at the wafer back side (external gettering, EG). In both cases four essential steps must take place: (1) release of metals from active circuit region; (2) transport of these metals to the gettering zone; (3) capture; and (4) retention for the rest of the process.

Extrinsic gettering on the wafer back side can be achieved by a number of techniques: a damaged layer from sand blasting or laser damage, polysilicon thin film, phosphorus doping and ion implantation damage are possible. The number of gettering sites increases in these steps, or metal diffusion is modified, as in the case of phosphorus. Ion implantation damage is annealed away during high-temperature steps, and its gettering ability reduced, but new back-side implants can be made. Extrinsic gettering can be added to a process flow before critical high-temperature steps. Solar cell fabrication, which is very cost sensitive, uses gettering to ensure cleanliness in a few critical steps only.

External gettering becomes difficult with growing wafer size and even completely unusable. Large-diameter wafers need to be thicker for mechanical strength, and diffusion distances for impurities increase. Simultaneously, process temperatures are getting lower (see Chapter 26 for more details), which further decreases the efficacy of back-side gettering. And finally, double-sided polishing and the need to keep the wafer back side flat and clean eliminate many external gettering options.

Intrinsic gettering is closely related to bulk microdefects (BMDs) and the thermal cycles that the wafer will experience during processing. Oxygen precipitates act as precipitation sites for other impurities, creating an impurity gradient which drives impurities toward designed precipitation sites. Wafer oxygen concentration is thus critical for intrinsic gettering. By and large, intrinsic gettering is determined when wafer processing begins. On the other hand, oxygen precipitates mechanically strengthen the wafer, which is even more important for large-diameter wafers.

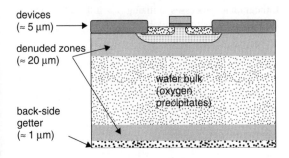

Figure 22.3 Wafer cross-section with denuded zone (not to scale)

22.3 Wafer Mechanical Specifications

Wafer thickness refers to centerpoint thickness (Table 22.1). It is difficult to produce tight thickness specifications because some wafering steps are batch processes for many wafers at a time and some are single wafer steps, therefore variations are inevitable. Wafer thicknesses are compromises between material usage and mechanical strength. Mechanical strength is especially important in high-temperature steps as many mechanical properties (for instance, yield strength) are strongly temperature dependent. MEMS devices which extend through the whole wafer require exacting thickness control. Anisotropic wet etching results in slanted 54.7° sidewalls which waste area, and more so for thicker wafers. In DRIE thick wafers lead to longer etch times.

Specialty wafers with practically any thicknesses are available but thin-wafer handling is very difficult. Mechanical stability increases with thickness, and thickness has to increase with wafer size. Through-wafer MEMS have not been done on 300 mm so far, and 200 mm is on the fringe, too. Thicker wafers are readily available in thicknesses of 1–1.5 mm.

Wafers bend and warp at high temperatures, either because their front and back sides are heated non-uniformly (e.g., lamp heating from the top side only), or because films deposited on one side of the wafer introduce stress. This is sometimes compensated by the fact that both sides of the wafer are coated, as in thermal oxidation, but when the film on either side is etched, asymmetry is created and bowing develops. Even though epitaxial films are assumed to be zero-stress films, doping leads to bowing. Because boron atoms are smaller than silicon atoms, the resulting film will be tensile stressed (and large arsenic atoms result in compressive stresses): the higher the doping level, the more the stress. Boron doping of 10^{18} cm^{-3} corresponds to 10^{-4} lattice mismatch, which leads to bowing by tens of micrometers. This makes wafer bonding, for instance, much more difficult.

Bow and warp relate to shape deformations of free, unclamped wafers (Figure 22.4). Wafers can be concave

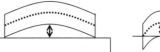

Figure 22.4 Bow (left): deviation of median surface from reference plane; warp (right): range of median surface distances from reference plane

or convex, or undulating. Bow may be eliminated by clamping, that is forcing the wafer flat on a chuck. Warp is the difference between the maximum and minimum distances of the median surface. Warp is a bulk property, in contrast to flatness, which is a surface property. Warp and bow can develop during high-temperature process steps, or result from ingot sawing and lapping operations. The presence of excessive bow and warp will affect lithographic performance via depth-of-focus problems.

Wafer surface topography can be divided into a few distinct scales: roughness is atomic/nanoscale, flatness is chip-scale and bow and warp are wafer-scale phenomena. Smoothness and flatness are essential parameters for fusion bonding: wafers with 0.1 nm roughness are preferred for fusion bonding. Anodic bonding is more forgiving to surface roughness, and wafers with 0.5 nm roughness are fine for anodic bonding.

Flatness is characterized by total indicator reading (TIR). It is a front-side reference measurement. TIR is defined as the sum of the maximum positive and negative deviations from a reference plane. If this reference plane is chosen to coincide with the focal plane of the mask aligner, focal plane deviation (FPD) is defined as the largest deviation, positive or negative, from this plane (Figure 22.5). TIR and FPD are measured for clamped wafers, while bow and warp are measured for unclamped (free) wafers.

Flatness is measured over an area that is relevant to the lithography process and chip size. It directly impacts linewidth variation through lithographic depth of focus. Lithographic processes utilizing 1× full wafer imaging systems are sensitive to global flatness, whereas step-and-

Table 22.1 Standard wafer sizes and thicknesses (centerpoint thickness)

3 inch	380 μm	
100 mm	525 μm	(380 μm for MEMS; thinner wafers exist)
150 mm	625 μm	(380 μm for MEMS; 250 μm minimum)
200 mm	725 μm	(500 μm MEMS)
300 mm	770 μm	

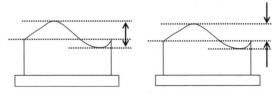

Figure 22.5 Left, total indicator reading (TIR); right, focal plane deviation (FPD)

repeat imaging systems are sensitive to local site flatness, over an exposure area which is less than 10 cm square.

22.4 Epitaxial Wafers

Epitaxial wafers offer extreme purity: carbon and oxygen, which are always present in CZ wafers, are practically absent in epitaxial layers. If there are COPs in bulk material, they will be buried by the epitaxial layer, reducing a 100 nm void to a few nanometers. Epitaxial layers are not defect-free, however; stacking faults created in epitaxial growth are the largest yield limiters in epitaxy.

Whereas CZ wafers have cylindrical distribution of doping and defects because of the rotation during crystal pulling, epitaxial deposition is highly uniform. Epi doping uniformity is typically below 4% and thickness uniformity about 1%. Epitaxial deposition is very reproducible, for both resistivity and thickness. And while resistivity in an ingot has lengthwise gradient, and bulk wafers therefore have slightly different resistivities, epitaxial wafers have identical resistivity.

Minimum thickness by CVD homoepitaxy is about 0.5 μm, and the maximum thickness is determined by economics of epitaxial growth, not by physics and chemistry. Practical maximum epitaxial thicknesses are about 100 μm for certain power semiconductor devices. Epitaxial wafers have applications in almost all areas of microfabrication (Table 22.2), but the wafer cost limits their use to more expensive applications only. In CMOS, one-third of wafer usage is epi, two thirds bulk.

Because p^{++} etch stop material is too heavily doped for electrical devices and mechanically poor due to tensile stresses from high boron concentration, alternatives have been explored. An advanced etch stop structure relies on a double epitaxial layer structure: an etch stop layer and a device layer. The first epilayer is heavily boron doped, but in order to minimize mechanical stresses from boron doping, the film is compensated by germanium (10^{21} cm^{-3} germanium, 10^{20} cm^{-3} boron) (Figure 22.6). The boron atom is smaller than the silicon one, and germanium one is larger, which prevents stresses from volume mismatch building up. Germanium is a column IV element (beneath silicon in the periodic table), therefore isoelectronic with silicon, so no electrical effects are introduced. The second layer, lightly doped, is deposited on top of the Si:Ge:B etch stop layer. This second layer is the actual device layer, and we can choose device doping level freely.

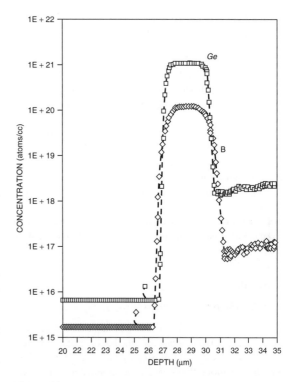

Figure 22.6 SIMS doping profile of Si:Ge:B epitaxial etch stop layer. Courtesy Okmetic

22.5 SOI Wafers

Three SOI techniques will be discussed in this chapter:

- Bonded SOI (also known as BESOI, for bond and etch-back SOI)

Table 22.2 Epitaxial wafer applications

Technology	Subst.	Epi	ρ (ohm-cm)	Thick (μm)	Motivation
CMOS	p$^+$	p	5–10	5–20	Latch-up
Power MOS	n$^+$	n	5–10	10–20	Conductivity
Analog bipolar	p$^+$	p	1–20	10–100	Speed
MEMS	p	n	1–10	7–150	Etch stop
MEMS	p	p^{++}	0.005	3–5	Etch stop

- Smart-cut (ion cut layer transfer)
- SIMOX (implanted oxygen).

Each has its characteristic SOI device layer thickness as well as typical buried oxide (BOX) thickness. They also differ in defect density and in the applicability of cavity structure formation. Bonded SOI is best suited for thick device layer applications in MEMS. BOX thickness is quite freely chosen. SIMOX and Smart-cut are used for thin SOI device layers in CMOS. In SIMOX both BOX and device silicon thicknesses are limited, while in Smart-cut silicon thickness is limited but BOX thickness is a free variable. SOI specifications and applications are listed in Table 22.3.

Intermediate SOI device layer thicknesses are difficult to make and usually involve an epitaxial step, as will be discussed below.

22.5.1 Bonded SOI

22.5.1.1 The bonding process

Bonding is a straightforward way to make SOI structures. The bonded SOI technique uses the bonding of two wafers (one or both oxidized) followed by thinning (Figure 17.3). Bonding without thinning has its applications too, see Figures 1.17 and 30.16.

Wafer bonding allows independent optimization of the top device layer and the supporting handle wafer. The handle wafer is chosen for mechanical support, thermal compatibility, micromachining, doping level or some other property. The device layer can have material, crystal orientation, doping level or thickness tailored to the particular device design, irrespective of handle wafer properties.

Thinning of the device wafer involves grinding, polishing and etching, much like in silicon wafering (Table 4.3). Thinning down to 10 μm thickness is reasonably easy, and 5 μm can be done. For layers thinner than that, special techniques are required: either real-time thickness monitoring during final polishing, or etch stop layers.

Figure 22.7 shows a process for thin device layer bonded SOI. Epitaxial layers with different etching properties have to be grown on the device wafer before bonding. Grinding (or etching) removes the bulk of the

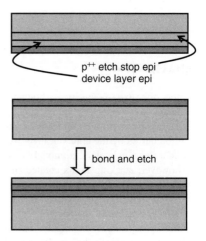

Figure 22.7 Thin bonded SOI: top carrier wafer with two epitaxial layers and thermal oxide is bonded to an oxidized handle wafer. KOH etching of carrier wafer until p^{++} etch stop layer, and HF:HNO$_3$ etch to remove p^{++} silicon

silicon, and selective etching (KOH/TMAH) removes the remaining material until the p^{++} etch stop layer is met. High boron doping is used but a second epitaxial layer is grown on it. The highly doped etch stop layer can then be removed by, for example, 1–3–8 etchant (HF, HNO$_3$ and CH$_3$COOH in the volume ratio of 1:3:8) which does not etch lightly doped material. Etch stop layers enable fabrication of device silicon layers 100 nm thick with ±5–10 nm variation.

Bonding two oxidized wafers instead of one bare and one oxidized wafer is useful when thick BOX is required. The thickest BOXs made of thermal oxides are 4 μm (2 μm of oxide on both wafers) thick. Bonding of wafers with thick deposited (CVD) oxides has been actively studied, but the films are generally not smooth enough for good bonding. If CMP is used to polish the surface (Figure 16.10), process cost rapidly increases.

Edge processing is needed in BESOI, as shown in Figure 22.8. Wafer edges are rounded because sharp edges are crack-prone. Some 3 mm of device layer silicon is removed from the edges, which means that it is easy to identify the SOI device side.

Table 22.3 SOI wafer applications

Application	Device layer	Buried oxide	SOI technology
CMOS	10–200 nm	200–400 nm	Smart-cut, SIMOX
Bipolar	1–10 μm	0.1–1.0 μm	Various
MEMS	5–50 μm	0.5–4 μm	Bonded SOI
Power IC	1–100 μm	1–4 μm	Bonded SOI

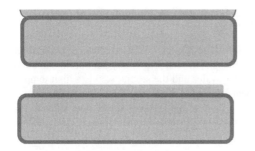

Figure 22.8 Bonded SOI edge treatment: top, after thinning; bottom, after edge treatment

In hydrophilic bonding there are two possibilities for the pair to be bonded: a silicon wafer and an oxidized wafer, or two oxidized wafers. The latter results in reduced bond strength, just 70–80% of the former, but the resulting structure is symmetric with respect to interfaces. In SOI wafer specifications it is stated which wafer has thermal oxide on it.

22.5.1.2 Bonding chemistry and physics

The water released during the formation of Si–O–Si bonds will oxidize silicon further ($Si + 2\ H_2O \Rightarrow SiO_2 + 2\ H_2$; wet oxidation!). The effect of this oxide is the more important, the thinner the oxide on the wafers; if wafers with thick oxides are bonded, water diffusion will be slow and additional oxidation minuscule. A combination of thin (or native) oxide wafer and thick oxide wafer is a compromise: oxidation will proceed according to the aforementioned equation, strengthening the bond, and hydrogen can dissolve in the oxide, preventing the build-up of gas bubbles at the interface.

While hydrophilic bonding is mainstream technology, hydrophobic bonding is of interest in some applications. For example, forming pn diodes by bonding requires bonding without interfacial oxide. HF-last cleaned wafers have a hydrogen-terminated surface (Figure 12.1) and bonding proceeds as described by

$$\equiv S-H + H-Si \equiv\ \Rightarrow\ \equiv Si-Si \equiv\ +H_2 \quad (22.2)$$

In the case of hydrophobic surfaces (–Si–H terminated) roughness is on the order of 0.5 nm and their bonding properties are worse than those of hydrophilic surfaces. Hydrogen bonds between HF units are weak. Hydrogen bubble prevention is very important in hydrophobic bonding. Hydrogen will diffuse along the bonding interface, and will not dissolve into bulk silicon below 500 °C. The bond energies of hydrophobic bonding are much lower

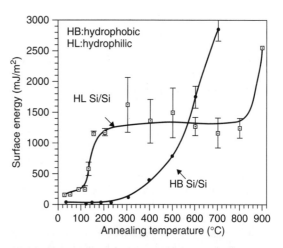

Figure 22.9 Surface energies for hydrophilic (HL) and hydrophobic (HB) bonding. Reproduced from Tong and Gösele (1999) by permission of John Wiley & Sons, Inc.

than those of hydrophilic bonding at low temperatures (as shown in Figure 22.9), but they can be improved by annealing.

22.5.2 Smart-cut™ ion cut layer transfer

In the Smart-cut method hydrogen is implanted into an oxidized wafer (called the donor wafer) which is subsequently bonded to a handle wafer. The donor wafer is then split along the mechanically weak region that was created by hydrogen implantation (Figure 22.10). The hydrogen implantation method has been patented, under the name Smart-cut, and wafers manufactured with the method are marketed as Unibond.

Hydrogen bubble-induced layer splitting is based on hydrogen implantation. Gas bubbles form at the depth of maximum hydrogen concentration. These bubbles lead to mechanical weakening of the silicon material and microcracks lead to cleavage of the implanted layer when a suitable thermal treatment or mechanical pressure is applied. The donor wafer provides mechanical strength during thermal or mechanical treatment, and the whole wafer can be split. In the case of a thin silicon layer, hydrogen bubbles can burst the thin top layer.

Process flow: Smart-cut

- Thermal oxidation of donor wafer
- H^+ implantation into donor wafer (60 kV, $6 \times 10^{16}\ cm^{-2}$)

- Hydrophilic bonding at room temperature
- Anneal at 400–600 °C to split the wafers
- High-temperature anneal at 1100 °C for 2 h strengthens the chemical bonds
- Final polishing or hydrogen anneal

The hydrogen dose required for bubble formation is $3\text{--}10 \times 10^{16}\,\text{cm}^{-2}$. The thickness of the splitting layer is related to the H^+ energy, which can accurately and easily be controlled. CMP or hydrogen annealing is necessary to eliminate the microroughness of the SOI device layer, even though the layer thickness just after splitting is uniform to a few nanometers.

An alternative way to detachment is mechanical force. Water jets or pressurized gas can be used. The bonding energy at the bonding interface is much higher than that in the H-implanted region which is embrittled. Thus, even at room temperature, the H-implanted layer can be peeled off from the donor wafer.

Smart-cut is a quite generic technique: implantation and mechanical splitting have been applied to transfer thin layers of many precious materials onto cheaper handle wafers. In the case of a silicon handle, the bonding process is always the same, that is the familiar silicon-to-oxide bonding. The device layer material may impose some limitations, but for example germanium-on-insulator (GeOI) and strained silicon-on-insulator (SSOI) have been made in straightforward fashion.

22.5.3 Separation by implantation of oxygen (SIMOX)

In SIMOX technology, SOI structure is realized in two main steps. The first step is oxygen ion implantation into a silicon wafer and the second step is a high-temperature anneal during which the implanted oxygen atoms form a buried oxide layer inside the silicon. Process conditions are:

- oxygen dose $2 \times 10^{18}\,\text{cm}^{-2}$
- oxygen energy 150–200 keV
- wafer temperature 550–650 °C

Good uniformity of both the silicon device layer and BOX thickness can be achieved. SIMOX is best suited for thin-silicon/thin-BOX applications. Typically both layers are on the order of 100 nm thick. Increasing oxygen ion energy would lead to deeper penetration, but at the cost of more expensive equipment.

SIMOX material exhibits inherent defect problems: the device silicon layer is damaged by the implantation process and it cannot be fully recovered during annealing. SIMOX dislocation densities can be $10^6/\text{cm}^2$, orders of magnitude more than in bulk silicon. Implantation time poses another limitation: the required doses are two orders of magnitude higher than those in common usage in CMOS manufacturing.

An improvement of SIMOX quality can be achieved by internal oxidation (ITOX), shown in Figure 22.11. During

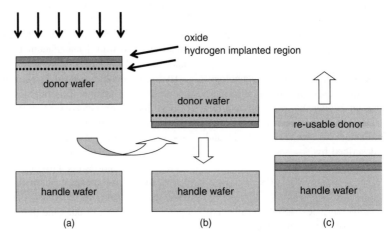

Figure 22.10 Smart-cut: (a) H^+ implantation into an oxidized donor wafer; (b) donor wafer is bonded to a handle wafer; (c) cleavage along the weak region created by implantation results in a SOI wafer

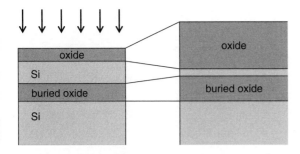

Figure 22.11 SIMOX SOI wafer fabrication: oxygen implantation through oxide to create buried oxide inside silicon, followed by thermal oxidation (called ITOX, for internal oxidation) which anneals silicon and thickens and densifies buried oxide

ITOX the device silicon layer is annealed and simultaneously BOX is densified. Oxidation will consume silicon and make the device silicon layer thinner. SIMOX BOX electrical breakdown voltages are 3–5 MV/cm, but can be increased to 8 MV/cm by ITOX. Thermal oxides exhibit ≥ 10 MV/cm breakdown voltages.

Hydrogen implantation in Smart-cut also produces damage, but hydrogen is a light ion, and the damage can be more readily repaired. From a productivity point of view, hydrogen implantation dose (and thus time) in Smart-cut is 10–100 times less than in SIMOX.

One interesting option with SIMOX is patterned (or partial) isolation: oxide masked implantation will result in BOX only at desired locations (Figure 22.12). This can mean local patterns, or larger scale patterns: for example,

Figure 22.12 Masked implant for SIMOX: BOX is formed only locally. Reproduced from Cheng *et al.* (2005), copyright 2005, Elsevier

the DRAM cell array is done on bulk, while the peripheral circuits are made on SOI.

22.6 Bonding Mechanics

Because of local roughness and global waviness, the two wafers will not touch fully. It is possible to estimate the dimensions of cavities which can be closed in the bonding process. The same equations govern both random cavities from wafer irregularities as well as micromachined cavities.

Gap closing is a function of wafer thickness (t), wafer mechanical strength determined by Young's modulus (E) and Poisson's ratio v, and surface energy γ (about 100 mJ/m^2 for room temperature silicon–silicon bonding) (Figure 22.13).

Cavities of radius R ($R > 2t$, $R \gg h$) will be closed if the distance between the wafers, h, is smaller than that given by Equation 22.3, and for cavities of radius R ($R < 2t$, $R \gg h$) it is given by Equation 22.4:

$$h \leq \frac{R^2}{\sqrt{\frac{2Et^3}{3\gamma(1-v^2)}}} \quad (22.3)$$

$$h \leq 3.5\sqrt{\frac{R\gamma(1-v^2)}{E}} \quad (22.4)$$

For dissimilar materials (different E) more complex formulas have to be used.

Particles between wafers cause non-bonding areas (voids) because wafers cannot conform abruptly to particles (Figure 22.14). The radius of non-bonding area is given by

$$R = \sqrt[4]{\frac{2Et^3}{3\gamma(1-v^2)}}\sqrt{h} \quad (22.5)$$

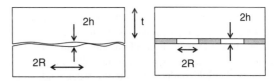

Figure 22.13 Geometry for analyzing closing of cavities for the case height $h \ll$ radius R. Wafer thickness is t

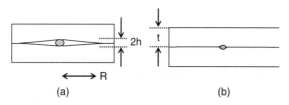

Figure 22.14 Particle-caused void in bonding: left, a large particle leads to a non-bonded area much larger than the particle itself; right, wafers conform to particles below critical size

Below a critical particle size h_{crit} the wafers are able to conform to particles, and void size is practically identical to particle size. This critical size is given by

$$h_{\text{crit}} = 5\sqrt{\frac{t\gamma}{E}} \qquad (22.6)$$

22.7 Advanced Wafers

Bonding wafers with more layers on them leads to more complex wafers. This approach can be continued with more wafers (and more grinding, etching and CMP steps), but of course each additional wafer incurs extra costs. One such application is shown in Figure 39.10; there, two buried conductor layers of WSi_2 are employed.

Bonding of dissimilar materials is made easier if surface quality is good, and low-temperature annealing is sufficient to achieve adequate bond strength. Then it is possible to bond for example silicon to III–V semiconductors like InP or GaAs. The process shown in Figure 22.15 uses epitaxial InGaAs etch stop layer to protect the InP device layer during thinning. Plasma activation of InP enables low-temperature bonding. Of course, all the usual requirements of smoothness and flatness must be fulfilled, too.

22.7.1 Silicon-on-diamond

One of the drawbacks (in many but not all applications) of SOI wafers is thermal insulation due to BOX. The alternative dielectrics silicon nitride or aluminum oxide have a thermal conductivity an order of magnitude higher. The ultimate thermal conduction material is diamond, with a thermal conductivity a thousand times higher than silicon dioxide, yet it is an electrical insulator. The wafers are dubbed SOD, for silicon-on-diamond (Figure 22.16). SOD wafer fabrication starts by CVD diamond deposition (Equation 5.11) on a hydrogen-implanted, Smart-cut-like wafer. As discussed in Chapter 5, it is not really diamond,

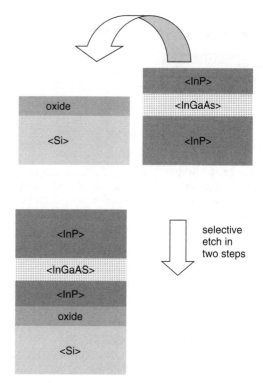

Figure 22.15 Heterogeneous bonding: InP to silicon bonding using InGaAs etch stop layer. Redrawn from Arokiaraj *et al.* (2006)

but diamond-like carbon (DLC) and its thermal conductivity is less than that of diamond gems, yet better than that of copper, for films a few micrometers thick. In order to improve the mechanical strength, a thick polysilicon can be deposited on it. Or, if the device processing only tolerates low-temperature processing, copper electroplating instead of polysilicon can be used, to utilize the high thermal conductivity of copper. The cleavage of the carrier wafer at 450 °C completes the process. Strong Si–C bonds are formed at the interface, according to the reaction

$$\text{Si-OH} + \text{H-C} \rightarrow \text{Si-C} + \text{H}_2\text{O} \qquad (22.7)$$

During the annealing steps this water will oxidize the silicon wafer. Hydrogen released in wet oxidation will be incorporated in the DLC film, which has hydrogen in it to begin with.

22.8 Variety of Wafers

A wafer manufacturer has thousands of specifications. Ingots are different, and the wafering process, especially the

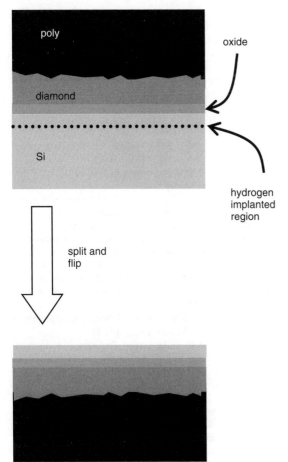

Figure 22.16 Silicon-on-diamond (SOD): diamond and copper deposition on hydrogen-implanted handle wafer, with splitting in Smart-cut fashion

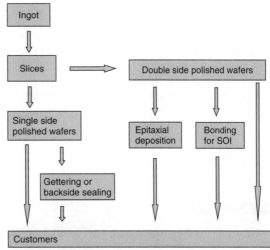

Figure 22.17 From ingot to customer

thermal treatments, adds to the variety. DSP wafers, epitaxial wafers and SOI wafers increase this variety further, as shown in Figure 22.17.

In CMOS bulk wafers command about 60% of the market, epitaxial wafers about 30% and SOI less than 10%. These values are in terms of dollars; in terms of area the shares of epi and SOI are smaller. In MEMS DSP wafers, SOI and epi are more prominent, but overall the MEMS wafer market is just a few percent of the total wafer market. In addition to those prime wafer markets, there are other markets, for solar silicon, test wafers, reclaimed wafers and non-polished wafers. The latter three are important for non-device applications as monitors for individual process steps, or for equipment testing and development. In 2010, 450 mm wafers are available only as test wafers, with basically the mechanical features in place, namely diameter and thickness. These wafers are used by equipment makers in developing reactors, robotics, and associated tools and techniques.

22.9 Exercises

1. If the CZ ingot neck is 2 mm in diameter, what is the maximum ingot weight that can be pulled before silicon yields catastrophically?
2. If COP density in ingot is 10^5 cm^{-3}, what is COP density on the wafer surface?
3. What is the maximum device silicon thickness in (a) SIMOX and (b) Smart-cut if a 200 keV implanter is used?
4. If ultrathin SOI needs to be made, Smart-cut device silicon has to be thinned down, in the extreme case of 6 nm MOS transistors to, for example, 2 nm. Analyze the strengths and weaknesses of doing this 50 nm → 2 nm thinning by the following techniques:
   (a) wet etching in TMAH
   (b) plasma etching is Cl_2 or SF_6
   (c) CMP
   (d) thermal oxidation + HF wet etching.
5. How thick a silicon dioxide layer will be formed inside silicon when the implant dose is 2×10^{18} cm^{-2} in SIMOX?
6. If hydrogen ion currents are in the milliampere range, how many wafers per hour can be implanted for Smart-cut?

7. What is the critical particle radius for 100 mm silicon wafer bonding?
8. What is the non-bonded area caused by a 0.3 μm particle on 150 mm wafers? If 150 mm wafers are specified to have 50 particles of size 0.3 μm, what fraction of wafer area will be unbonded?
9. Analyze metallic bonding by copper as shown below. What are its strengths and limitations?

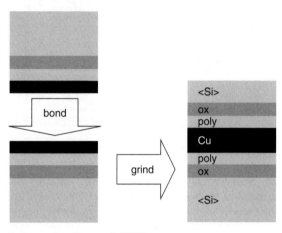

Redrawn after Chen et al. (2005)

References and Further Reading

Alexe, M. and U. Gösele (2004) **Wafer Bonding**, Springer.
Aminzadeh, M. et al. (2002) Pseudo epi – cost reduction approach and a paradigm shift in substrate material, *IEEE Trans. Semicond. Manuf.*, **15**, 486–492.
Arokiaraj, J., S. Vicknesh and A. Ramam (2006) Integration of indium phosphide thin film structures on silicon substrates by direct wafer bonding, *J. Phys.: Conf. Ser.*, **34**, 404–409.
Borghesi, A. et al. (1995) Oxygen precipitation in silicon, *J. Appl. Phys.*, **77**, 4169.
Celler, G.K. and S. Cristoloveanu (2003) Frontiers in silicon-on-insulator, *J. Appl. Phys.*, **93**, 4955–4978.
Chen, K.N. et al. (2005) Process development and bonding quality investigations of silicon layer stacking based on copper wafer bonding, *Appl. Phys. Lett.*, 87, 031909
Cheng, X. et al. (2005) Patterned silicon-on-insulator technology for RF Power LDMOSFET, *Microelectron. Eng.*, **81**, 150–155.
Dornberger, E., D. Temmler and W. van Ammon (2002) Defects in silicon crystals and their impact on DRAM device characteristics, *J. Electrochem. Soc.*, **149**, G226–G231.
Falster, R., V.V. Voronkov and F. Quast (2000) On the properties of the intrinsic point defects in silicon: a perspective from crystal growth and wafer processing, *Phys. Stat. Solidi (b)*, **222**, 219–244.
Ghyselen, B. et al. (2004) Engineering strained silicon on insulator wafers with the Smart Cut™ technology, *Solid-State Electron.*, **48**, 1285–1296.
Gösele, U. et al. (1999) Wafer bonding for microsystems technologies, *Sens. Actuators*, **74**, 161–168.
Graff, K. (2000) **Metal Impurities in Silicon-Device Fabrication**, 2nd edn, Springer.
Matsumura, A. (2003) Technological innovation in low-dose SIMOX wafers fabricated by an internal thermal oxidation (ITOX) process, *Microelectron. Eng.*, **66**, 400–414.
Miller, D.C. et al. (2007) Characteristics of a commercially available silicon-on-insulator MEMS material, *Sens. Actuators*, **A138**, 130–144.
Muller, T. et al. (2000) Assessment of silicon wafer material for the fabrication of integrated circuits sensors, *J. Electrochem. Soc.*, **147**, 1604–1611.
Müssig, H.-J. et al. (2001) Can Si(113) wafers be an alternative to Si(001)? *Microelectron. Eng.*, **56**, 195.
Pettinato, J.S. and D. Pillai (2005) Technology decisions to minimize 450-mm wafer size transition risk, *IEEE Trans. Semicond. Manuf.*, **18**, 501–509.
Sama, S. et al. (2001) Investigation of Czochralski silicon growth with different interstitial oxygen concentrations and point defect populations, *J. Electrochem. Soc.*, **148**, G517.
Shimura, F. (ed.) (1994) **Semiconductors and Semimetals: Oxygen in silicon**, Willardson.
Tang, Z. et al. (2009) Effect of nanoscale surface topography on low temperature direct wafer bonding process with UV activation, *Sens. Actuators*, **A151**, 81–86.
Taraschi, G., A.J. Pitera and E.A. Fitzgerald (2004) Strained Si, SiGe, and Ge on-insulator: review of wafer bonding fabrication techniques, *Solid-State Electron.*, **48**, 1297–1305.
Tong, Q.-Y. and U. Gösele (1999) **Semiconductor Wafer Bonding**, John Wiley & Sons, Inc.
Varma, C.M. (1997) Hydrogen-implant induced exfoliation of silicon and other crystal, *Appl. Phys. Lett.*, **71**, 3519.
von Ammon, W., E. Dornberger and P.O. Hansson (1999) Bulk properties of very large diameter silicon single crystals, *J. Cryst. Growth*, **198/199**, 390–390.

23

Special Processes and Materials

This chapter is different from the previous ones: instead of one topic, it is a collection of various specialty techniques used in wafer processing. Some of the techniques are materials specific, some are excellent in one limited application and some require non-standard equipment. Many of them are quite simple, but on the other hand they may not offer full microfabrication benefits. All these techniques can be applied on a wafer scale, just like other microfabrication processes, and this is the difference from techniques to be presented in the following chapter: those are serial methods for (slowly) writing one pattern at a time.

23.1 Substrates other than Silicon

A plethora of materials have been used in addition to silicon. Some of these have already been discussed: glasses, fused silica and crystalline quartz (Chapter 19), and polymers, for instance PET and Kapton (Chapter 18). Other substrate materials used in microfabrication include:

- steel
- nickel
- alumina
- AlN
- ZnO
- GaN
- SiC
- sapphire
- PCBs
- LTCCs.

Steel is cheap and available in large areas, which is good for solar cells; nickel can act as an electrode in a fuel cell; alumina and aluminum nitride have excellent microwave properties; ZnO is a transparent conducting glass; GaN can be used for UV detection; SiC is excellent in high-temperature, harsh-environment cases; sapphire is hard and transparent and can support silicon epitaxy; LTCCs (Low-Temperature Co-fired Ceramics) are ceramics that can be processed easily. PCBs (Printed Circuit Boards) are polymer plates with thick copper metallization ready-made on them. So for larger linewidth applications which require metals on a thermally insulating substrate, PCBs offer a quick starting point.

Compared to round and standardized silicon wafers, these exotic substrates have a number of issues that need to be addressed. Square and rectangular shapes are not well suited to photoresist spinning, but of course laminate resist and spray resist are options. Non-standard sizes complicate processing in equipment designed for silicon processing. Multicrystalline silicon solar cells are most often 5 and 6 inch squares, and they are much thinner than silicon wafers, for example 200 μm. Handling thin wafers is more delicate than with thicker ones. There are also very thick substrates, and for example thermal equilibrium of a thick insulating substrate in a single wafer tool may take a lot longer than with a thin silicon wafer of high thermal conductivity. And some substrates are 10 times heavier than silicon of the same diameter, which may be an issue for handler robots.

Silicon and glass are flat and smooth, which does not apply to many other substrates. For instance, polished metals might have a surface roughness of 100 nm, a hundred times more than silicon. Thin films of 100 nm thickness cannot really cover such a mountainous substrate very well. Some of the substrates are porous, for example all polymers are somewhat porous. This leads to problems in vacuum systems: water vapor (and other chemicals) are slowly released from pores, leading to vacuum poisoning and subsequently to poor metal quality.

23.2 Pattern Generation

The general purpose pattern generation methods discussed in Chapter 8 are applicable to all possible pattern shapes: rectangles, circles, wedges, spirals, etc. There are, however, ways to make patterns in much simpler ways, but these methods are usually capable of producing just one shape only, for example lines, dots or circles.

23.2.1 Holographic lithography

Holographic lithography, also known as interference lithography, uses intersecting laser beams to create line gratings, dot arrays and 3D meshes. It is very easy to create regular patterns by interfering laser beams, with minimum linewidths in the 100 nm range for visible wavelength lasers. It is also possible to vary the line-to-space ratio, and to produce, for example, 200 nm lines with 200 nm spaces, or with 1000 nm spaces. 3D shapes can also be made (Figure 23.1).

23.2.2 Microstereolithography

Rapid prototyping of 3D objects can be accomplished by the photohardening of negative-resist types of polymers. There are several different ways of doing this:

- physical mask and blanket exposure of 2D layers
- virtual mask and blanket exposure of 2D layers
- $x - y$ scanning laser and mechanical z-stage control
- two-photon laser polymerization

(the latter two are serial microfabrication methods and belong to Chapter 24, really).

In all the approaches the 3D design is broken down to 2D layers which will be hardened layer by layer starting from the bottom layer. After a layer has been exposed, the z-controller lowers the stage, and new photopolymer is allowed to cover the previous layer (Figure 23.2). In the masked versions each layer requires a mask, and clearly the physical mask approach is not amenable to any but the simplest of designs. Virtual masks are patterns generated by micromirrors or video projectors that use software to create exposure pattern on-the-fly.

Exposure times are similar to lithography (and no wonder: similar photoactive polymers are used), with a few seconds for layer thicknesses of tens of micrometers. Structures with up to a thousand layers have been made, in a few hours. Minimum feature sizes depend on mask technology but tens of micrometers and larger are usual. If direct laser writing is used, even micrometer sizes are possible. The layer-by-layer structure of the finished objects is clearly seen in Figure 23.3.

In addition to polymers, metallic and ceramic microstructures can be made by stereolithography. Metallic or ceramic powders, with solvents and binders, are hardened just like photoactive polymers. And it is also possible to make metallic microstructures in a roundabout way: a polymeric 3D structure is made by stereolithography, and it is filled by electroplating. Polymer is then removed. Compared to lithographic resist molds, stereolithography enables retrograde shapes and hollow structures to be made.

23.2.3 Block copolymer lithography

Block copolymers (BCPs) consist of blocks that are bonded together covalently, but free to rotate and bend in space. If the two blocks are chosen to repel each other, for example by having hydrophobic and hydrophilic ends in the chains, the block copolymer will self-organize into a regular pattern.

The natural size scale for block (co)polymer structures is the size of macromolecules, 5–50 nm. This is very

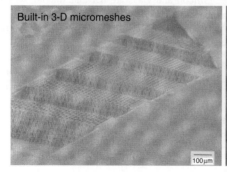
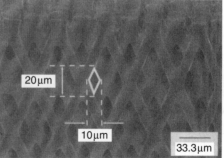

Figure 23.1 Built-in 3D meshes by three-laser interference lithography. Reproduced from Sato *et al.* (2006) by permission of IOP

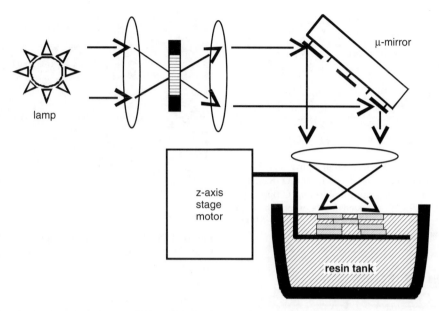

Figure 23.2 Mask projection stereolithography: lamp, optical system, micromirror array virtual mask, resin tank and z-stage. Adapted from Kang *et al.* (2005)

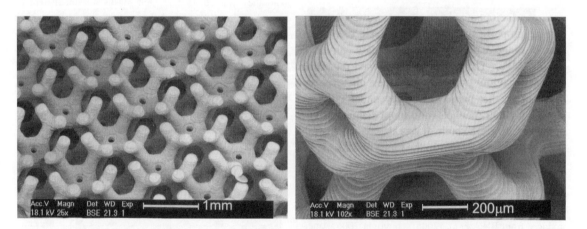

Figure 23.3 Example 3D objects made by microstereolithography: the magnification on the right clearly reveals the layer-by-layer structure. Reproduced from Stampfl *et al.* (2008) by permission of IOP

attractive for making nanostructures. There are a number of drawbacks, however: the phase separation process is very slow; and a 24 h bake might be needed to form the pattern. Also, while local order can be very good, long-range order of BCP patterns is often less perfect. This might be acceptable for some applications, but fatal in many others.

Depending on the particular polymers, many shapes are possible. For example, dot and hole arrays and line gratings can be made (Figure 23.4). It is essential that one of the polymers can be selectively etched relative to the other, to create a pattern that can be used as an etch or plating mask.

BCP patterns are molecule size and also molecule thick, 5–50 nm, which means that they are very thin and not very good as etch masks. In Figure 23.5 the BCP has been used as a mask for SiO_2 etching, and the oxide then acts as a hard mask for SOI silicon layer etching.

286 Introduction to Microfabrication

Figure 23.4 Block copolymer lithography: left, PS–PB BCP self-organized lines, reproduced from S.O. Kim *et al.* (2003) by permission of Nature Publishing Group; right, PS–b–PMMA BCP self-organized dots between lithographically defined lines, reproduced from Park *et al.* (2003), copyright 2003, Elsevier

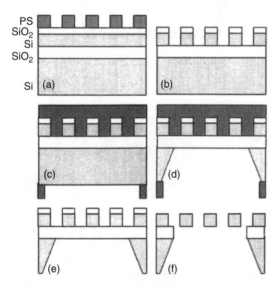

Figure 23.5 Membrane filter fabrication by BCP lithography: PS BCP nanostructures, other processing standard. Reproduced from Black *et al.* (2006), copyright 2006, American Institute of Physics

23.2.4 Colloidal bead lithography

Colloidal lithography is also known by the names of nanobead patterning and micro- and nanoparticle assembly. All the same, the method is based on micro- or nanobeads assembling into a regular hexagonal array upon drying (Figure 23.6). The basic arrangement for nanobead lithography is very easy: spinning or otherwise spreading a nanobead-containing solution on a wafer, and letting it dry in a controlled manner so that the

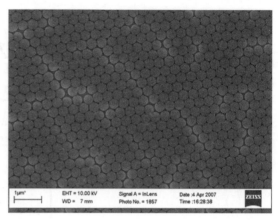

Figure 23.6 Bead lithography: single layer of polystyrene beads, after drying. Courtesy Kestas Grigoras, Aalto University

beads will self-organize. Both single layer structures and multilayers can be done, see Figure 23.3.

The beads act like photoresist patterns, for example, and they can be used as etch masks. Or metal can be evaporated in the spaces between the beads, Figure 23.7. These spaces between the beads are triangular, but annealing and ball-up (surface energy minimization) will lead to round metallic nanostructures. Alternatively, the bead array can be buried completely by a thick polymer, and then the beads dissolved, leaving an array of empty spaces, known as inverse opals.

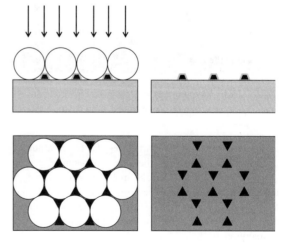

Figure 23.7 Evaporated metal through bead pattern, and after bead dissolution

23.3 Patterning

23.3.1 Microcontact printing (μCP)

Microcontact printing (μCP) is a microlithographic version of ink-and-stamp patterning: a soft polymeric stamp (most often made of PDMS) is wetted by "ink," for example alkanethiol $CH_3(CH_2)_{15}SH$ or octadecyltrichlorosilane OTS (Figure 5.14), and the wetted stamp is pressed against a gold surface (Figure 23.8). A reaction between thiol and gold leaves a self-assembled monolayer (SAM) pattern on the wafer because the bonds between the ink and surface are stronger than the bonds between the stamp and ink.

Microcontact printing is an example of so-called soft lithographies, which include a number of methods where soft stamps and masters are utilized, as opposed to hard masks of chromium-on-glass type. The claims that soft lithography techniques work without cleanrooms is not true: any technique that intends to produce micron and submicron features requires cleanliness matching those pattern sizes. But if 100 μm lines are made, then micrometer particles obviously are not of great concern and relaxed cleanliness requirements apply.

SAMs are usually only 2–3 nm thick, and their usefulness as a plating, etch or lift-off mask needs to be improved, even though 20–30 nm etched depths have been demonstrated, but this is clearly not enough for the majority of applications.

The stamps are soft and elastic, which ensures an intimate contact with the substrate, and allows for minor irregularities. Soft materials are sensitive to printing force: too much pressure will deform the soft stamp and the patterns will be distorted. Stiffer materials offer higher resolution, but worse contact. Hybrid stamps with a hard backplate/soft cushion and a stiff stamping surface have been devised in order to have the best of both worlds (Figure 23.9).

Of course the patterns need not be flat: in fact pyramid-shaped tips are useful (Figure 23.10). If a very small

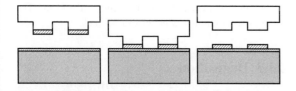

Figure 23.8 Microcontact printing on a gold coated surface: left, alkanethiol-inked PDMS stamp and gold-coated wafer; middle, alkanethiol stamped against the gold surface; right, alkanethiol pattern on wafer, PDMS stamp lifted

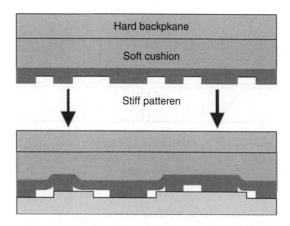

Figure 23.9 Composite stamp. Adapted from Michel et al. (2001)

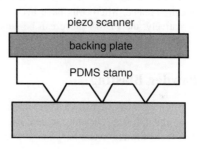

Figure 23.10 PDMS pyramid stamper. Adapted from Huo et al. (2008)

pattern is needed, the tip is gently pressed, but if larger patterns are needed, more force can be applied and the soft PDMS will be pressed against the substrate, forming a larger pattern.

The inks must balance between two opposing trends: shorter molecules form better SAMs, but also diffuse faster. Longer molecules result in imperfect SAMs, but experience less diffusion, therefore they allow more accurate pattern edges.

Contact area plays an important role: light-field structures, with small contact area, are easier because the separation force is small. Structures with aspect ratios around 1:1 and structures with uniform fill factors, such as periodic gratings, are easy. If aspect ratios of structures to be stamped differ from unity or from each other considerably, stamping becomes problematic. Structures of low aspect ratio, say 0.1, are subject to sagging, and those with high aspect ratio are subject to lateral instability and collapse (Figure 23.11).

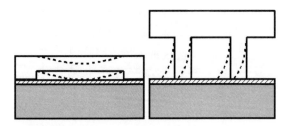

Figure 23.11 Problems with soft stamps: when spacing is sparse, sagging can take place. In stamps of high aspect ratio, lateral collapse can take place

23.3.2 Stamping non-planar objects

PDMS is flexible, and this opens up special applications: patterns can be contact printed on curved surfaces. Gratings on optical fibers have been realized. Similarly, a round object can be rolled over a PDMS stamp and a spiral structure created. Microcoils have been made this way. Alternatively, the PDMS piece can be curved, and used as mold.

23.4 Powder Blasting

In powder blasting abrasive particles are ejected toward the wafer at about 100 m/s, leading to removal rates on the order of tens of micrometers per minute. Material removal is based on mechanical crack propagation. This is the same as grinding in wafer thinning, and therefore the methods share the same strengths and limitations: a very high rate of material removal, but rough and damaged surfaces, with the damage extending micrometers inside the material. A powder blasting system is shown schematically in Figure 23.12. In order to powder-blast a full wafer, a scanning system has to be implemented.

When abrasive particles of tens of micrometers are used, the rate is high, up to 1 mm/min, but surface roughness is similar to particle size. When smaller particles are used, the rate is lower but submicrometer roughness can be obtained. A powder-blasted profile is sloped, and while powder blasting can be used to make through-wafer holes, these are not holes with high aspect ratios. Powder blasting is extensively used to blast holes in glass wafers because both wet and dry etching of deep holes is difficult.

Masking in powder blasting requires something that is not eroded by abrasive particles. Hard materials like perforated steel and thick electroplated metals like nickel are candidates. Selectivities on the order of 50:1 are possible with metal masks. Photoresists cannot be used: they are too brittle and will crack. Soft, elastomeric materials

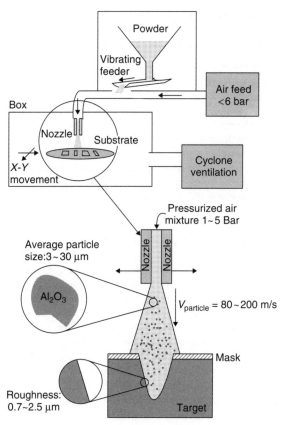

Figure 23.12 Powder blasting. Reproduced from Cui (2005) with kind permission of Springer Science and Business Media

like PDMS absorb energy from the particles, and are deformed, not broken. Selectivities of 100:1 to 1000:1 are possible with elastomeric masks. The minimum feature sizes that can be made depend not only on mask size, but also on particle size used, with minimum linewidth roughly 10 times the particle size.

One special application of powder blasting is inclined blasting (Figure 23.13). Because the particle jet is directed, the sample can be tilted relative to the beam. This enables free-standing structures to be made.

23.5 Deposition

23.5.1 Lift-off metallization

Lift-off is metallization with sacrificial resist: after lithography, metal deposition is done on the resist pattern, followed by resist dissolution in solvent and

Special Processes and Materials 289

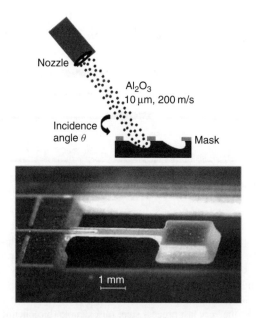

Figure 23.13 Inclined angle powder blasting, quartz paddle resonator. Courtesy Martin Gijs, EPFL

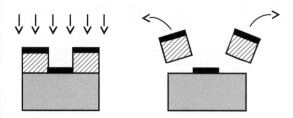

Figure 23.14 Lift-off process: left, metal deposition on resist pattern; right, resist dissolution and metal lift-off

lift-off, with all the metal that is not in contact with substrate being removed (Figure 23.14).

Lift-off metallization should have poor step coverage, and therefore the method of choice is evaporation, even though sputtering can be used too. The deposition process has, however, photoresist-imposed limitations: it must take place under about 150 °C temperature because of resist thermal stability. There is always some deposition on the sidewalls, too, but if the films are thin, they are discontinuous, and resist dissolution can take place.

Special techniques have been devised to make lift-off easier. One way is to use negative resists which typically have negative sidewall angles (Figure 9.10). Another technique is to use a two-layer resist (Figure 23.15). The bottom resist, called the lift-off resist (LOR), can be thick because it has no role in defining linewidth. LOR's main

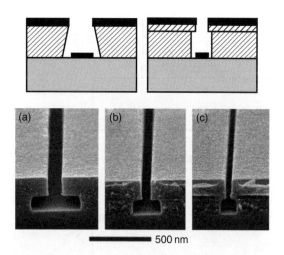

Figure 23.15 Negative resist and two-layer lift-off resist. SEM micrographs reproduced from Chen et al. (2004), copyright 2004, Elsevier

role is to undercut the top resist in development, and it is designed for a high development rate (and good adhesion and thermal stability). The top resist is thin to enable high-resolution patterning.

Lift-off is used in magnetic recording head (GMR heads) fabrication: lead metallization has to make contact with the GMR element as closely as possible. This self-alignment is shown in Figure 23.16 and described in the process flow below.

Process flow for GMR sensor metallization

- GMR sensor deposition
- Lithography of lift-off resist
- Ion beam etching of GMR layers
- Deposition of hard bias metal
- Deposition of lead metal
- Resist stripping (= lift-off)

Ion beam etching (ion milling) is used to etch the GMR sensor (Figure 7.19) because there are so many different materials that finding RIE gases would be impossible, and therefore argon ion milling is the straightforward way. After milling the GMR head, the same resist acts as the metallization lift-off mask: hard bias metal (CoPtCr) and lead metallization (Au for resistance minimization) are sputtered, and lifted off. Because one resist does it all, the metallization will be automatically aligned to the GMR sensor.

290 Introduction to Microfabrication

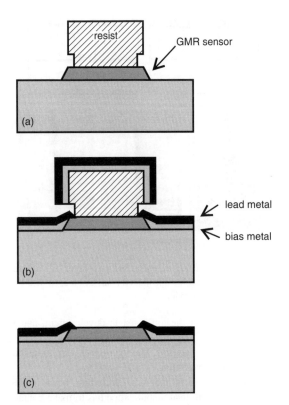

Figure 23.16 GMR head fabrication: (a) ion beam etching of GMR sensor with two-layer lift-off resist profile; (b) two-layer metallization; (c) lift-off. Adapted from Childress and Fontana (2005)

Lift-off is very general: all metals and their alloys can be patterned with the same basic process. Lift-off is suited for hard-to-etch metals like gold and platinum. A GMR head is also suitable for ion beam etching because the total thickness is only about 40 nm and the slow rate is therefore not a major handicap.

Lift-off is, however, a special technique, and not an alternative to etching in general. If an etching process exists, it is universally used, and lift-off is reserved for those special applications where etching is difficult or impossible. Because lift-off calls for poor step coverage, it is difficult to make lift-off metallization over topography.

23.5.2 Shadow masks

Shadow masks (also known as stencil masks) are mechanical aperture plates. Shadow mask patterning is basically lift-off with a mechanical mask instead of a resist mask. The shadow mask is aligned with and temporarily attached to a substrate, and this stack is then positioned in

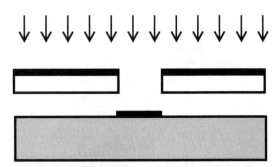

Figure 23.17 Deposition with a shadow mask

the deposition system (Figure 23.17). Basically any deposition technique can be used, but line-of-sight methods like evaporation lead to the smallest penumbra under the mask. The smaller gap between the stencil and the wafer minimizes unwanted deposition, too.

Some materials are so sensitive that their deposition has to be the very last process step, for example (bio)chemical sensor films. The application of photoresist on these films is not possible, and solvent dissolution cannot be used, ruling out lift-off. Another reason for using stencil masks is that wafer topology does not allow photoresist spinning: for example, there are through-wafer holes, or freestanding cantilevers. And the shadow mask saves a lot of process steps: resist spinning, baking, exposure, development, rinsing and drying are eliminated.

If the shadow mask and wafer can be aligned with each other in a mask aligner or a bond aligner, micrometer alignment accuracy is possible; often, however, shadow masks are only used for non-critical applications where manual $\pm 10\,\mu$m alignment is enough. Minimum linewidths that are possible with shadow masks are in the 10 nm range. In practice, however, such small apertures are quickly clogged by the deposited film, and the sidewall profiles of the deposited structures are far from ideal. But in theory any hole larger than atomic size can be used as a shadow mask. One special limitation of shadow masks is the impossibility of donut-shaped structures. There are, however, tricks around this: by using very narrow bridges and relying on deposition penumbra under the mask, donut shapes can be patterned. The general solution is to perform shadow mask deposition twice, with "half-masks."

23.5.3 Screen printing

Screen printing is an old technique which has emerged as the major technique in solar cell metallization. A perforated screen holding the metallization pattern is

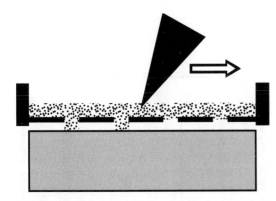

Figure 23.18 Screen printing: a squeegee pushes metal paste through the gridded screen

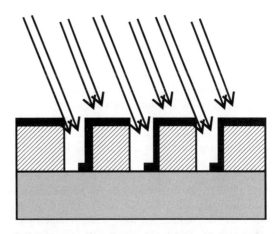

Figure 23.19 Inclined angle evaporation leads to structures smaller than the lithographic feature size

stretched over the wafer, and paste is applied to the screen. A squeegee scans the screen, pushing paste through the holes in it (Figure 23.18). Additional drying and firing steps are then required to finish metallization, but this is similar to other thin-film techniques: the film is seldom in final form immediately after deposition. Screen printing is cheaper than vacuum deposition methods, and it can produce thick metal wires, for example 10 μm, reducing resistive losses. Limitations are similar to shadow masking: for example, donut-shaped patterns cannot be made.

Other applications of screen printing in microfabrication include adhesive and glass-frit bonding: the adhesive or glass frit is defined by screen printing. In all these applications rather wide lines are satisfactory, in the range of tens or hundreds of micrometers.

23.5.4 Inclined angle evaporation

The directionality of evaporation, its line-of-sight deposition geometry, is favorable for lift-off, and if this is combined with a tilted wafer, very small structures can be made (Figure 23.19). Some of the smallest ever MOSFETs have been demonstrated by oblique angle evaporation.

23.5.5 Glancing angle deposition

Columnar growth (Figure 7.2) is taken to extremes in glancing angle deposition (GLAD). The wafer is tilted to about 80° relative to the arriving atoms, and the atoms arrive obliquely. With low surface mobility and a shadowing effect from already deposited atoms, tall, narrow crystals will grow. If the initial surface is smooth, random columns will form, and if there are patterns, they will act as shadowing elements and GLAD film deposits on top of those (Figure 23.20).

GLAD films are highly porous but electrically conductive. They can be used as top electrodes in chemical sensors: fluids can penetrate through the GLAD film, and interact with the sensor film. Alternatives to GLAD top electrodes include thin percolated metal films – films that are so thin that they are physically discontinuous but electrically continuous. Percolated films have much higher resistance than the much thicker GLAD films. Normal films can also be used, but deposition, lithography and etching of those films on top of soft polymeric sensor films is undesirable.

23.6 Porous Silicon

Silicon is not etched in HF. If, however, silicon is made an anode in an electrochemical etching set-up (Figure 23.21), etch rates of about 1 μm/min are observed. Hydrofluoric acid, with or without ethanol and/or water, is used as an electrolyte. Platinum is the standard cathode. Depending on current density, silicon can be etched in two rather different modes: pore formation and electropolishing (Figure 23.22). Both electropolishing and pore formation take place in the anodic regime. In pore formation etching proceeds anisotropically downward leaving a silicon "skeleton" with up to 80% empty space.

The reactions that take place in HF electrolyte are given by

$$\text{Si} + 6\,\text{HF} \Longrightarrow \text{H}_2\text{SiF}_6 + \text{H}_2 + 2\,\text{H}^+ + 2\,\text{e}^- \quad (23.1)$$
(pore formation, low current density)

$$\text{Si} + 6\,\text{HF} \Rightarrow \text{H}_2\text{SiF}_6 + 4\,\text{H}^+ + 4\,\text{e}^- \quad (23.2)$$
(electropolishing, high current density)

Pore formation starts at the wafer surface from a defect or an intentional initial pit. Electronic holes from the bulk silicon are transported to the surface, and they react at the defect or pit. Further etching occurs at the newly formed pore tips because they attract more holes due to higher electric field strength, and the process leads to a uniform porous layer depth as the holes are consumed by the growing tips and other surfaces are depleted of holes. This etching mode takes place under low hole concentration, and it is limited by hole diffusion, not by mass transfer in the electrolyte cell. If the hole density increases, some holes reach the surface and react there, leading to surface smoothing. This is the electropolishing regime and ionic transfer from the electrolyte plays a role. These two regimes are pictured in Figure 23.22.

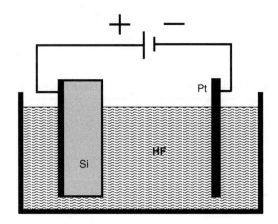

Figure 23.21 Electrochemical etching set-up for making porous silicon

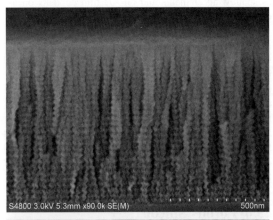

Figure 23.20 GLAD films: left, random crystals; right, seeded crystals. Courtesy Marianna Kemell, University of Helsinki

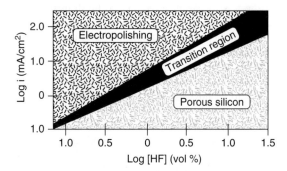

Figure 23.22 Regimes of silicon anodic etching in HF: porous silicon formation and electropolishing. Reproduced from Collins (1997) by permission of Electrochemical Society Inc.

In p-type silicon holes are naturally present, and in n-type silicon they can be generated by illumination. A very wide range of pore sizes from 2 nm to 20 µm can be etched by varying electrolyte concentration, current density and illumination. Pore size dependence on silicon resistivity is shown in Figure 23.23. As a rough guide for n-type silicon, the pore diameter in micrometers is half the resistivity in ohm centimeters: for 1 µm pores in 2 ohm-cm, and 20 nm pores in 0.04 ohm-cm material.

If the initial pits are prepared by lithography and etching, regular arrays of pores can be made (Figure 23.24). Because etch rate depends on etching current, current modulation can be used to increase etch rate, periodically, leading to interesting 3D shapes.

Special Processes and Materials   293

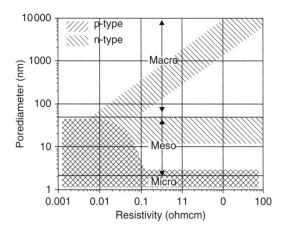

Figure 23.23 Pore size ranges of electrochemically etched silicon: macroporous, mesoporous and microporous regimes. Reproduced from Lehmann (1995) by permission of IEEE

There are a couple of drawbacks in electrochemical etching (and deposition): electrical contact has to be made to the wafer back side, and this contact has to tolerate the etchant. Concentrated HF (49%) is often employed, which seriously limits the choice of metals. Alternatively, a wafer holder can be used to protect the wafer back side, and any metal is good. However, such a holder takes up area on the wafer front, reducing the number of usable chips.

Porous silicon is single crystalline silicon even though it is a sponge-like network rather than continuous material. Epitaxial deposition on porous silicon is possible, and other thin films can be deposited too. Depending on the process, either pores will be filled or porous material will be covered with a continuous surface of thin-film material.

Conformal CVD into macroporous grooves is no different from CVD into etched grooves of similar dimensions

23.6.1 Sacrificial structures using porous silicon

The electrochemical etch rate of n-type silicon (10–20 ohm-cm) in HF electrolyte is very low compared to p-type silicon or low-resistivity n-type silicon (about 0.01 ohm-cm). Doping (by diffusion or epitaxy) can therefore be used to create local porous silicon patterns. Alternatively, protective etch masks can be used as in any etching. Photoresist, silicon nitride, amorphous silicon and silicon carbide are candidates; silicon dioxide cannot be used because of the HF electrolyte, and photoresist is limited to cases with dilute HF.

Free-standing microstructures can be made by depositing films on top of porous silicon, patterning them and then removing the porous silicon (Figure 23.25). Porous silicon presents a curious case where etch selectivity can be obtained between silicon and silicon: porous silicon etching proceeds rapidly because the sidewalls between the pores can be as thin as a few nanometers, whereas solid silicon is attacked from the top surface only. Etch selectivity can be as high as 100 000:1. The material for the free-standing structure can be for instance silicon nitride, but epitaxial silicon can also be used.

Porous silicon is a mechanically weak material, and it can be destroyed by capillary forces during drying. It can also be destroyed by gas bubbles: KOH etching releases hydrogen (Equation 11.1) and if gas evolution is rapid the bubbles can burst porous structures. For this reason dilute KOH, 0.1–1%, is used, rather than the 20–50% typical of silicon anisotropic etching.

In a modification of the above scheme, a free-standing structure can be made from bulk single crystal silicon. A p-type silicon substrate (10 ohm-cm) is used. A deep, heavy p-diffusion is done, followed by a shallow

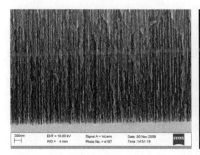 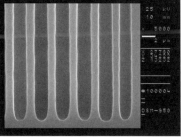 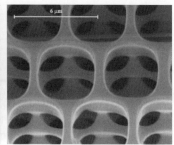

Figure 23.24 Porous silicon: (a) random initial pits; (b) lithographically defined initial pits and constant current etching, reproduced from Grigoras *et al.* (2001); (c) lithographically defined and modulated etching, reproduced from Trifonov *et al.* (2008), copyright 2008, Elsevier

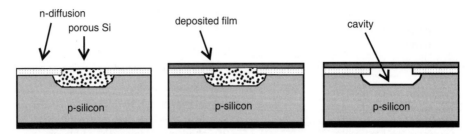

Figure 23.25 Fabrication of a free-standing bridge on a p-type substrate: left, n-diffusion of selected areas, followed by electrochemical etching; middle, bridge material deposition; right, removal of porous silicon in dilute KOH resulting in a bridge over a cavity. Adapted from Hedrich *et al.* (2000).

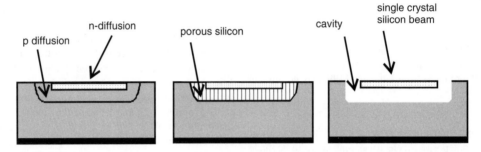

Figure 23.26 Single crystal silicon bridges by porous silicon sacrificial layer: left, p-type and n-type diffusions; middle, electrochemical etching; right, KOH etching of porous silicon. Adapted from Lee *et al.* (2000).

n-diffusion (Figure 23.26). Electrochemical etching attacks the p-diffusion and transforms it into porous silicon. The n-type silicon is intact in electrochemical etching. A short KOH etching removes the porous silicon, leaving an n-type single crystal silicon microbridge.

23.7 Molding with Lost Mold

Hard-to-etch materials can be made into patterns by the following methods:

- ion beam etching (argon ion milling)
- selective deposition
- blanket deposition and polishing
- molding and mold dissolution.

Ion beam milling is a brute force method. It suffers from very slow rates. Selective deposition depends critically on chemical surface processes which are hard to control. Highly conformal deposition combined with polishing is quite universal, but conformal deposition processes are rare.

Molding is rather a universal process because so many different ways of transporting the material are available. The reverse of the final pattern is fabricated in silicon or resist, filled with the desired material, and the mold removed. AFM tips of silicon nitride are made this way: inverse pyramids are etched into silicon, filled by LPCVD nitride, and silicon is etched away. The diamond structures shown in Figure 23.27 are made similarly: CVD diamond deposition plus silicon wafer etching.

When the mold is completely removed, shape freedom is unlimited. These processes are variously called dissolved wafer, lost mold or disposable mold processes. Because there are no detachment problems (Figure 18.20), the most pressing issue is filling of the mould. If the material to be molded can fill retrograde features, then any shape can be made.

Etch selectivity between silicon and the molded material limits the use of this method: the usual silicon etchants, hot concentrated KOH or HF:HNO$_3$ mixtures, are very aggressive solutions. Alternatively, silicon can be removed by isotropic SF$_6$ plasma etching or by XeF$_2$ or ClF$_3$ dry etching. A number of devices have been made with silicon molds: for

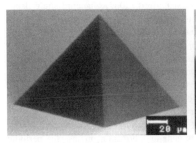  

Figure 23.27 Diamond microstructures made with silicon wafer disposable molds. Reproduced from Björkman *et al.* (1999) by permission of Elsevier

Figure 23.28 Silicon mold (top) and cast PDMS replica (bottom): the whole silicon wafer has been etched away to release the PDMS "mushroom", courtesy Lauri Sainiemi, Aalto University

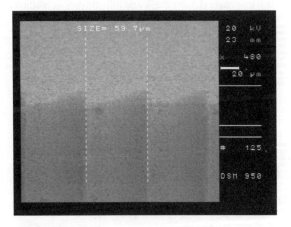

Figure 23.29 Steel powder molded by a PDMS mould. Courtesy Pia Suvanto, Aalto University

instance, PZT ultrasonic transducers and parylene needles, in addition to hollow diamond channels. In Figure 23.28 the PDMS "mushrooms" have been made by etching a mold in silicon in a fashion similar to Figure 21.15, filling with PDMS and then completely removing the silicon wafer.

Resist molds are easy to make and electroplating into resist is widely used. Resist molds are, however, usable only once. But resist can be used as a master to make PDMS parts, which can be used as disposable molds. In Figure 23.29 stainless steel powder (with solvent and binder) has been cast into PDMS mold and cured. After detachment, it is further annealed to improve its mechanical properties.

23.8 Exercises

1. How does shadow mask thickness affect the resulting linewidth and profile?
2. What is the gap effect on feature size and shape in shadow mask patterning?
3. When new dielectrics are tested, it is usual to deposit the top electrode through a shadow mask because of speed and simplicity. How much error will be introduced to dielectric constant ε_r if shadow mask dimensional control is $100 \pm 5\,\mu m$?
4. It has recently been proposed to use shadow masks in ion implantation. Explore the issues that need to be addressed for such an approach.

5. Explain in detail how donut shapes can be made using shadow masks.
6. PDMS with CTE of 300 ppm/°C is made by molding over a 100 mm silicon wafer. By how much will features shrink? What positional accuracy can be achieved?
7. Explain in detail how the hollow diamond tubes shown in Figure 23.27 are made!
8. What dimensional limitations apply to structures made by the porous silicon sacrificial layer technique?
9. Estimate the feature sizes that can be made by inclined angle evaporation in Figure 23.19!
10. How could shadow mask and lift-off types of methods work for liquid phase patterning?
11. How can the porosity of porous silicon be measured by weighing?
12. What percentage of silicon atoms are surface atoms if porous silicon has a 5 nm pore size and 3 nm wall thickness?
13. Explain the difference (in materials, applications, processes, temperatures, etc.) between the etching and lift-off method.

References and Related Reading

Björkman, H. *et al.* (1999) Diamond replicas from microstructured silicon masters, *Sens. Actuators*, **73**, 24.

Black, C.T. *et al.* (2006) Highly porous silicon membrane fabrication using polymer self-assembly, *J. Vac. Sci. Technol.*, **B24**, 3188–3191.

Chen, Y., K. Peng and Z. Cui (2004) A lift-off process for high resolution patterns using PMMA/LOR resist stack, *Microelectron. Eng.*, **73–74**, 278–281.

Childress, J.R. and R.E. Fontana (2005) Magnetic recording read head sensor technology, *C. R. Physique*, **6**, 997–1012.

Collins, S.C. (1997) Etch stop techniques for micromachining, *J. Electrochem. Soc.*, **144**, 2242.

Cui, Z. (2005) **Micro-Nanofabrication**, Higher Education Press/Springer.

Ekkels, P. *et al.* (2009) Evaluation of platinum as a structural thin film material for RF-MEMS devices, *J. Micromech. Microeng.*, **19**, 065010.

Graff, M. *et al.* (2004) Microstenciling: a generic technology for microscale patterning of vapor deposited materials, *J. Microelectromech. Syst.*, **13**, 956–962.

Grigoras, K., A.J. Niskanen and S. Franssila (2001) Plasma etched initial pits for electrochemically etched macroporous silicon structures, *J. Micromech. Microeng.*, **11**, 371–375.

Hawkeye, M.H. and M.J. Brett (2007) Glancing angle deposition: Fabrication, properties, and applications of micro- and nanostructured thin films, *J. Vac. Sci. Technol. A*, **25**, 1317–1335.

Hedrich, F., S. Billat and W. Lang (2000) Structuring of membrane sensors using sacrificial porous silicon, *Sens. Actuators*, **84**, 315.

Huie, J.C. (2003) Guided molecular self-assembly: a review of recent efforts, *Smart Mater. Struct.*, **12**, 264–271.

Hulteen, J.C. and R.P. Van Duyne (1995) Nanosphere lithography: a materials general fabrication process for periodic particle array surfaces, *J. Vac. Sci. Technol.*, **A13**, 1553–1558.

Huo, F. *et al.* (2008) Polymer pen lithography, *Science*, **321**, 1658–1660.

Kang, H.-W., J.-W. Rhie and D.-W. Cho (2009) Development of bi-pore scaffold using indirect solid freeform fabrication based on microstereolithography technology, *Microelectron. Eng.*, **86**, 941–944.

Kim, H.-C. and W.D. Hinsberg (2008) Surface patterns from block copolymer self-assembly, *J. Vac. Sci. Technol.*, **A26**, 1369–1382.

Kim, S.O. *et al.* (2003) Epitaxial self-assembly of block copolymers on lithographically defined nanopatterned substrates, *Nature*, **424**, 411–414.

Kim, G.M., M.A.F. van den Boogaart and J. Brugger (2003) Fabrication and application of a full wafer size micro/nanostencil for multiple length-scale surface patterning, *Microelectron. Eng.*, **67–68**, 609–614.

Lee, C.-S., J.-D. Lee and C.-H. Han (2000) A new wide-dimensional freestanding microstructure fabrication technology using laterally formed porous silicon as a sacrificial layer, *Sens. Actuators*, **84**, 181.

Lehmann, V. (1995) Porous silicon – a new material for MEMS, Proceedings of IEEE MEMS'95, p. 1.

Michel, B. *et al.* (2001) Printing meets lithography: soft approaches to high-resolution patterning, *IBM J. Res. Dev.*, **45**, 697–719.

Park, C., J. Yoon and E.L. Thomas (2003) Enabling nanotechnology with self assembled block copolymer patterns, *Polymer*, **44**, 6725–6760.

Pawlowski, A.-G., A. Sayah and M.A.M. Gijs (2005) Accurate masking technology for high-resolution powder blasting, *J. Micromech. Microeng.*, **15**, S60–S64.

Perl, A., D.N. Reinhoudt and J. Huskens (2009) Microcontact printing: limitations and achievements, *Adv. Mater.*, **21**, 2257–2268.

Sato, H. *et al.* (2006) An all SU-8 microfluidic chip with built-in 3D fine microstructures, *J. Microelectromech. Syst.*, **16**, 2318–2322.

Sayah, A. *et al.* (2005) Elastomer mask for powder blasting microfabrication, *Sens. Actuators*, **A125**, 84–90.

Shibata, T. *et al.* (2002) Stencil mask ion implantation technology, *IEEE Trans. Semicond. Manuf.*, **15**, 183–188.

Stampfl, J. *et al.* (2008) Photopolymers with tunable mechanical properties processed by laser-based high-resolution stereolithography, *J. Micromech. Microeng.*, **18**, 125014.

Sun, C. *et al.* (2005) Projection micro-stereolithography using digital micro-mirror dynamic mask, *Sens. Actuators*, **A121**, 113–120.

Syms, R.R.A. *et al.* (2009) Silicon microcontact printing engines, *J. Micromech. Microeng.*, **19**, 025027.

Trifonov, T. *et al.* (2008) Macroporous silicon: a versatile material for 3D structure fabrication, *Sens. Actuators*, **A141**, 662–669.

24

Serial Microprocessing

Microfabrication manufacturability stems largely from the fact that millions or even billions of identical structures are made simultaneously, by lithography, imprinting, embossing or molding, and then etched, implanted, metallized or otherwise further processed in parallel fashion. In this chapter we will discuss different breeds of processes: single features are made, slowly and painstakingly, but with great precision. Pixel by pixel, single feature processing is slow, but there are major benefits: the cost and waiting time for the photomask/master/mold is eliminated, and new designs can be implemented on-the-fly by changing the code that drives the writing tool. The cycle time of experimenting becomes much faster than with mask/master/mold-based techniques. These processes are used in research and development, and whenever a limited number chips are enough. And of course they are used when no alternatives exist, as in photomask writing.

There are two main varieties of serial microprocessing tools: beam based and tip based. The beam spot size and the tip radius vary greatly, but all rely on scanning the sample surface to create the pattern. Focused ion beams (FIBs), focused electron beams (FEBs) and laser beams are all used, with or without additional gases (Figure 24.1). The same beams can also be used to expose resist, but in this chapter we are interested in fabricating the microstructures directly in the final material, sidestepping the intermediate polymer step. The tip-based methods have also been used to write resist patterns, but again we are interested in direct etching/deposition of the final structures.

Vacuum, atmospheric and liquid phase systems are all in use. Ion and electron beams are of course vacuum techniques, but lasers and tips can operate in all environments. However, there are no universal techniques: all are more or less specialized in some materials and some environments.

24.1 Focused Ion Beam (FIB) Processing

Ion beams are energetic, as we know from ion implantation (Chapter 15). The processes that occur include not only implantation but also amorphization and sputtering, that is ion beam etching. Secondary electrons are also emitted (Figure 24.2).

In order to achieve high enough beam current, a liquid gallium ion source is commonly used in FIB systems (and in SIMS analysis systems, too). In serial microprocessing ion energies are similar to ion implantation (e.g., 50 keV), but they are focused to very small spot sizes, for example 10 nm at 10 A/cm^2 current density. Sputtering, which is also present in implantation to a minor extent, can be the dominant phenomenon in FIB. The ion beam-induced secondary electron emission is most useful and in fact FIB systems act as electron microscopes, too.

Ion beam etching was mentioned in Chapter 11. In ion beam etching (as in ion implantation) a broad beam system is used. The mechanism is the same for focused beams but now we do not need a mask because etching only takes place where the beam hits. This is used in repairing photomasks: chromium pinspots, protrusions and bridges are etched away (Figure 8.10). The small spot size of FIB is essential in this application. The imaging capability of secondary electrons is also essential.

The FIB etching process is slow, but the rate can be enhanced by adding a reactive gas, for example fluorine-containing gases like trifluoroacetamide ($C_2H_2F_3NO$) for silicon etching. The same trick is used in broad beam etching where the method is called CAIBE (for chemically assisted ion beam etching) and also RIBE (for reactive ion beam etching). In addition to the slow rate, another major problem with ion beam etching is the high energy which erodes sharp corners, leading to rounding (Figure 24.3).

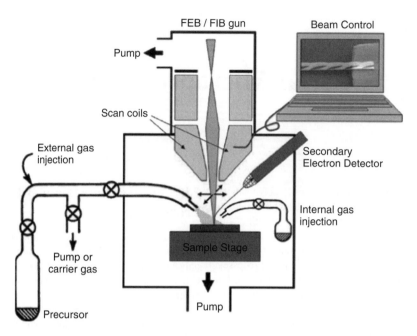

Figure 24.1 Serial writing by focused beam system. Reproduced from Utke *et al.* (2008), copyright 2008, American Institute of Physics

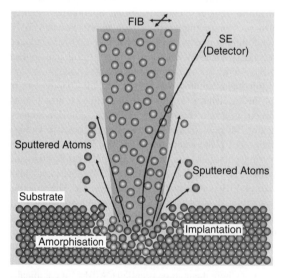

Figure 24.2 Atomistic processes in FIB. Reproduced from Utke *et al.* (2008), copyright 2008, American Institute of Physics

Figure 24.3 Focused ion beam, chemically enhanced etching of silicon pillars. Courtesy Antti Peltonen, Aalto University

This high fluence of ions can be used to locally dope silicon, just as in ion implantation. But because the doses are very high locally, ion concentration can be orders of magnitude higher than in ion implantation, for example peak concentration 10% vs. 0.1%. This has been used to create a gallium-doped layer 30 nm thick, which can be used as an etch mask for subsequent DRIE steps (Figure 24.4).

Ion beam deposition of tungsten according to Equation 24.1 is used to deposit metal. Gaseous tungsten

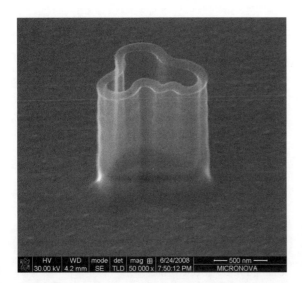

Figure 24.4 Gallium-doped etch mask for DRIE. Design by Alvar Aalto, processing by Nikolai Chekurov, Aalto University

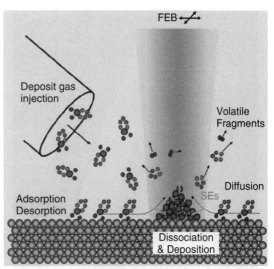

Figure 24.5 FEB deposition. Reproduced from Utke et al. (2008), copyright 2008, American Institute of Physics

hexacarbonyl source gas is broken down by the gallium ion beam and tungsten is deposited, releasing carbon monoxide:

$$W(CO)_6 \text{ (g)} \rightarrow W \text{ (s)} + 6 \text{ CO (g)} \quad (24.1)$$

The resulting tungsten thin film is not of very high quality: it contains a lot of residual carbon, and its resistivity can be 100 µohm-cm vs. 10 µohm-cm for CVD tungsten. It is useful for instance when a defective chip needs to be rewired so that it can be measured.

One application where the high resistivity is not a concern is mask repair. The role of the deposited metal is to block light, and the electrical properties are irrelevant. FIB-deposited tungsten and carbon are used to repair photomask defects.

24.2 Focused Electron Beam (FEB) Processing

Electrons do not dislodge atoms, but they can induce reactions. With suitable gases both deposition and etching are possible (Figure 24.5). Electrons interact with molecules both in the gas phase and on the surface.

One application for FEB is mask repair. Missing chrome can be repaired by deposition. For instance, platinum organometallic source gases can be used to deposit opaque layers which are similar to chrome, both optically and mechanically, that is they can tolerate mask cleaning. The repair of extra chrome is done by electron beam-enhanced etching. The etch gas is injected in the vicinity of the beam spot, at very small flow rate because the e-beam requires a high vacuum. The etch gas is selected to produce volatile products just as in RIE. Molybdenum silicide mask repair is easier than chrome mask repair because there are more volatile molybdenum compounds. Mask metals are rather thin, in the range of 100 nm, so the slow etch rates can be tolerated. But accuracy requirements for advanced masks are formidable: linewidth errors of only 10–15 nm are tolerated on the mask (and a factor of four smaller at wafer level due to reduction optics).

FEB can also be used to create 3D objects: the focal point is moved in space, and new material is deposited at that focal point. Precursor gases similar to those of Equation 24.1 are used to deposit delicate 3D structures, like the nanotip shown in Figure 24.6.

24.3 Laser Direct Writing

Laser beam writing can be done in room air whereas electron and ion beams always require high vacuum. Laser processing is therefore scalable to larger areas, as discussed already in mask making (Chapter 8). Laser beam spot size is larger than electron or ion beam spot size, and submicron spot size is difficult to achieve. And when spot size is made smaller, writing time increases according to Equation 8.2: writing with 1 µm spot takes 100 times

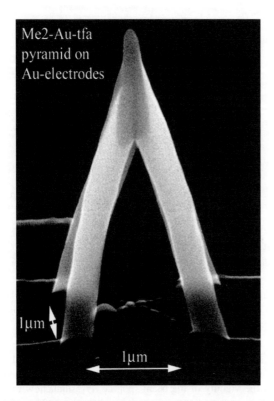

Figure 24.6 SEM micrograph of a gold nanotip made by FEB. Reproduced from Utke *et al.* (2008), copyright 2008, American Institute of Physics

longer than with 10 μm spot. In the following discussion typical linewidths are in the 10–100 μm range. This includes many solar cell, flat panel display and micromachining applications like via drilling and wafer scribing.

Laser processes fall into three major categories: ablation, photolytic and photothermal. In ablation material is explosively evaporated when intense laser beam is absorbed by the material. This hints at one limitation of laser processing: laser wavelength and material absorbance need to match. It is less of a problem with newer picosecond and femtosecond lasers because non-linear effects take place and practically all materials can be ablated. However, nanosecond lasers have higher pulse energies and therefore higher machining speeds.

In photolytic processes laser light induced reactions are important. In laser assisted CVD (LCVD or LACVD) breakdown of source gases is a photolytic effect. Metals can be made in processes resembling FIB metallization, Equation 24.1. Similar to FIB-deposition, metals made by LACVD are not very good in terms of resistivity.

In photothermal processes the localized heat produced by the laser drives the processes. Laser annealing and sintering are thermal effects. When laser pulses are short, the heat supplied by laser does not have time to diffuse: $x \approx \sqrt{Dt}$, with thermal diffusivities of 0.1–1 cm^2/s for silicon result in submicrometer diffusion distances for nanosecond pulses. Therefore laser thermal processes can be done on sensitive substrates: for example amorphous silicon crystallization on glass and polymer substrates is possible.

Lasers used in direct writing are usually diode pumped solid state (DPSS) lasers, like Nd:YAG ($\lambda = 1064$ nm) and Ti:sapphire (680–1100 nm). Laser parameters of interest include not only wavelength but also energy (μJ/pulse), power (10–100 W), pulse temporal duration (ms to fs), pulse repetition rate (from a few Hz to tens of MHz) and fluence (0.1–10 J/cm^2). Beam shape is also important: when Gaussian beams are used, there can be ablation at beam center, and thermal processes at edges, leading to heat affected zones.

Excimer lasers form the other major class of micromachining lasers. They are suitable for large area processing, for example square centimeter areas are treated. Masked ablation is often done with excimer lasers. This resembles optical lithography and in fact same excimer lasers KrF (248 nm) and ArF (193 nm) are used in lithography, too (Chapter 10). Laser fluence is, however, much higher. Beam shaping enables larger area flat beam profile (known as top hat). Carbon dioxide laser of 10.6 μm wavelength which is extensively used in metal machining is seldom used in micromachining, but for example PMMA and some other polymers can be processed with CO$_2$ lasers.

The case of ablating indium tin oxide (ITO) thin film on top of glass shows many important issues in laser processing: Nd:YLF laser with 1047 nm wavelength is used. In this wavelength range ITO absorbs strongly but glass is transparent, and ablation without damage to substrate is possible. At second harmonic (523.5 nm) and third harmonic (349 nm) wavelengths both ITO and glass have low absorption, and much higher fluencies are required for ablation. But at 262 nm (4th harmonic) absorption in ITO is strong, and heating is limited to a very thin top layer, leading to excellent ablation. Every harmonics generation, however, reduces pulse energy to half, which leads to lower overall removal rate. Therefore if the fundamental wavelength works, it is preferred.

All materials can be ablated, but not all materials are volatile, meaning that some materials will start condensing, forming particles and redepositing on the wafer. This depends on laser pulse duration, with shorter pulses generally vaporizing the sample more completely, and resulting in less residue. Atmosphere is also important: ablated material may react and form for example oxide particles. This depends on material reactivity, and it can be minimized

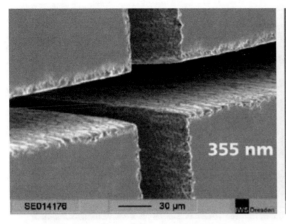 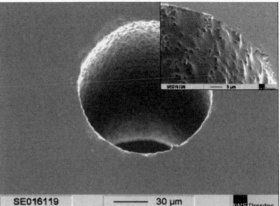

Figure 24.7 Silicon wafer dicing by laser: a) scribe lines; b) vias, from ref. Otani by permission of Springer

by working under vacuum, which makes systems more complex. A protective polymer or oxide layer is often used on silicon during ablation. After ablation, this layer is removed and the particles with it.

24.3.1 Laser scribing/drilling/etching

Just as in etching, there are three main cases in laser machining: removing thin films selectively against underlying material, cutting a limited depth into substrate (sometimes called blind via) which requires good rate control, and scribing and drilling through-wafer (Figure 24.7) which calls for high rate. In laser ablation of silicon 100–1000 nm of material is removed per pulse and 10 μm/pulse has been demonstrated.

Silicon removal rate of $10\,mm^3$/second translates to 300 mm/s of 100 μm deep, 10 μm wide trench. In many applications sidewall profile is important, for example flow channels and nozzles. Laser ablated grooves and holes are seldom vertical walled. Gaussian beam shape leads to gaussian groove profile, as shown in Figure 24.8. A through-wafer hole might have 50 μm entrance diameter, and 20 μm exit diameter on wafer backside.

Lasers are used in dicing wafers. There are a number of benefits: irregular shapes can be cut easily, brittle substrates pose no problems and water is eliminated. This is important for MEMS structures with moving parts which might stick upon drying. Speed is also faster than dicing with a mechanical saw. Laser dicing is especially attractive for thin substratres (<200 μm) because such wafers are fragile and because laser dicing time is reduced for thinner wafers, unlike mechanical dicing.

Just as in electron beam mask writing (Figure 8.3) lines are made up of circular spots, and certain overlap is needed to have straight edges, and not "pearl-necklace".

But there are applications like electrical isolation in solar cells where separation is enough, and line quality irrelevant. The edges of laser cut lines have burr, or residue wall (Figure 24.8). Using more pulses and larger overlap between the pulses will minimize burr but careful cleaning is needed if bonding is to be done.

24.3.2 Laser annealing

Laser annealing (sometimes known as ELA, for excimer laser annealing) is used in flat panel display fabrication: amorphous silicon on glass plates is crystallized by laser. The high temperature needed for crystallization is limited to a top micrometer layer by using nanosecond laser pulses, and therefore the glass substrate is not excessively heated up. A problem with ELA is orientation dependence: the grains are not randomly oriented, but aligned along the laser scan direction. Therefore transistors fabricated in different orientations will have slightly different properties.

In photomask repair laser annealing can be used: metal-containing thin film is spin coated on the mask, and the laser locally anneals the film to become opaque. Untreated film can be selectively etched away.

Writing and erasing CDs and DVDs is also a laser annealing process: GST film (Germanium Antimony Telluride) is turned to amorphous state by rapid heating and cooling, and to crystalline state by slow heating and cooling, resulting in reflectivity change by a factor of two between the states.

24.3.3 Laser processing of solar cells

Solar cells make use of laser processing in many instances: it has for example replaced lithography and etching in many steps. Process flow for buried contact solar

Figure 24.8 Profile of grooves in silicon by 800 nm wavelength, 150 fs pulses: a) burr (residue) with 20 pulses; b) burr-free, with 100 pulses, from ref. Crawford by permission of Springer

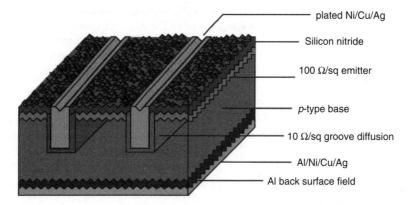

Figure 24.9 Buried collector solar cell in crystalline silicon, from ref. Neuhaus (creative commons article)

cell is given below, and the resulting cell is shown in Figure 24.9. This cell has the following features: lithography is completely eliminated and metallization is done in the laser cut grooves by electrochemical techniques. The 20 μm deep and 30 μm wide grooves need to be smoothed and cleaned after laser processing by wet chemical etching. Laser grooves enable close spacing of metallization, and even though they are narrow, they offer low resistance because they are deep.

Process flow for buried contact solar cell from ref. Neuhaus

- P-type silicon wafer
- Texturing etch in KOH
- Phosphorous oxide (P_2O_5) deposition on front
- PECVD nitride on front
- Laser grooving
- Groove damage etching
- $POCl_3$ gas phase diffusion & P_2O_5 drive in
- Aluminum evaporation on back
- Rear contact diffusion
- Electroless nickel plating into grooves
- Sintering
- Electroless copper and silver metallization
- Laser edge isolation

Thin film solar cell processing uses lasers, too. The thin film layers of a thin film solar cell are shown in Figure 24.10. In this cell the light enters the cell through glass and zinc oxide, and is absorbed by the amorphous

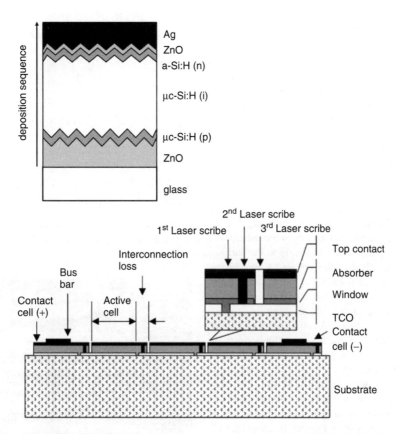

Figure 24.10 Thin film solar cell with ZnO as transparent conducting oxide (TCO): a) thin film layers, from ref. Repmann by permission of Elsevier; b) three laser cuts, from ref. Chopra by permission of John Wiley & Sons

silicon layer, while the current is collected on top by the thick silver conductor. Cell processing consists of seven main steps, including three laser patterning steps, as shown below:

Process flow for μc-Si thin film solar cell (from ref. Repmann)

1. first conductor (ZnO) deposition
2. ZnO texture etching in HCl
3. ZnO patterning by laser
4. PECVD of Si:H p–i–n structure
5. silicon patterning by laser
6. ZnO/Ag back contact sputtering
7. back contact laser patterning

The first laser cut patterns ZnO (transparent conducting oxide, TCO) front electrode (which was deposited first); the second one cuts through the active semiconductor material (μc-Si:H) and the third cut defines the back metallization (which was deposited last). ZnO is textured to enhance light trapping. Similar texture etch process is used in silicon solar cells, Figure 1.14a.

Lasers can also be used to sinter/anneal screen printed metallization. Metal-to-silicon contact is made by forcing metal through silicon nitride by a laser annealing step. Contact hole lithography and etching are then eliminated. Laser doping is also possible: contact arrays in solar cells are made by local drive-in by laser pulses (Figure 37.6).

24.4 AFM Patterning

AFM tip radius of curvatures are in the 10 nm range, and this then is the range where minimum feature sizes lie. AFM tips can be used in various modes of localized processing:

- subtractive: removal of material

306   Introduction to Microfabrication

- additive: deposition of material
- material modification.

In subtractive mode etching, electrochemical etching or discharge machining is used to remove material. By applying a voltage between the AFM tip and the substrate, anodic oxidation of silicon is also possible. In additive mode the STM (Scanning Tunneling Microscope), the vacuum brother of the AFM, has been used to move and position individual atoms. While this is the ultimate in patterning accuracy, it is hopelessly slow: it takes minutes to move each atom.

The dip pen is an AFM tip which writes with liquid ink, with speeds of 10 μm/s. There are three steps in dip pen writing: a water meniscus forms between the tip and the surface, ink is transported from the tip to the surface, and finally ink is attached to the surface. A typical ink would be thiol SAM, which has active sulfur atoms ready to react with a gold surface. The tip has to be occasionally replenished, by dipping into the ink reservoir. In order to increase dip pen writing speed, ink can be supplied continuously, as in a fountain pen. Such a microfluidic fountain pen design in shown in Figure 24.11. Its fabrication is left as an exercise in Chapter 30.

Thermomechanical actuation of AFM tips has been used to press dimples into polymer, as a future memory technology. This process is similar to hot embossing, as can be witnessed by comparing Figures 18.15 and 24.12. Thermomechanical memory, as well as other AFM-based techniques, can be made to run in parallel, by fabricating

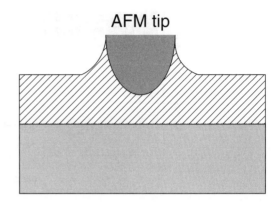

Figure 24.12 Polymer imprinting by a hot AFM tip. Adapted from Vettiger *et al.* (2002)

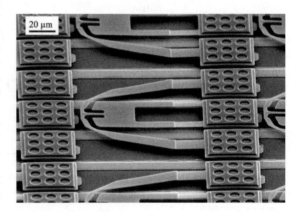

Figure 24.13 AFM tip array for thermomechanical data storage. Reproduced from Despont *et al.* (2004), copyright 2004, by permission of IEEE

arrays of AFM tips with, for instance, thousands of tips (Figure 24.13).

24.5 Ink Jetting

Ink jet printers routinely produce dots in the 20–100 μm range at kilohertz frequencies. Ink-jetted lines can be used as molding masters for PDMS fluidic channels as such: the typical thickness of 10 μm is quite suitable for microfluidics. Layer-by-layer ink jetting can produce real 3D structures in much the same way as microstereolithography: heights up to 400 μm have been demonstrated with ceramic powders.

Picoliters and tens of picoliters are typical droplet volumes. This may sound small but it results in droplets tens

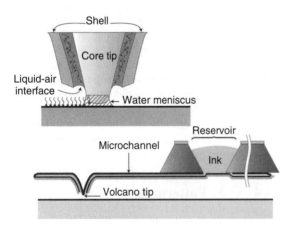

Figure 24.11 Dip pen writing: top, detail of tip showing thiol–SAM attachment on a gold surface; bottom, fountain pen principle for dip pen. Reproduced from Salaita *et al.* (2007) by permission of Nature Publishing Group

of micrometers wide on the substrate. Droplet spreading on the surface depends on surface energy and ink surface tension. Surface tension is not a free variable, because ink jetting requires a certain surface tension (30–300 mN/m). The surface energy of the substrate can be modified by the same methods as always, for example HMDS or OTS coating. Adhesion on the surface is delicate: it depends on smoothness/roughness, as well as contact angle; a large flat droplet evaporates faster than a hemispherical droplet. Higher contact angle (lower surface energy) surfaces result in narrower lines, but are subject to occasional drop bulging. Additionally, heated surfaces can be used, to control spreading and evaporation.

Many classes of materials are amenable to ink jetting. Metals can be ink jetted using metal nanoparticles, metal salts or organometallic compounds as precursors; oxides can be made from metal halides by oxidation after ink jetting; and various polymers can be used. Nanoparticle inks use very small particles (5–200 nm) to avoid clogging, even though ink jet nozzles are tens of micrometers. Metal salt concentrations are for example 10–20%. Dilute polymer solutions must be used because of ink jet viscosity requirements. On the other hand, running the ink jet chip hot enables ink jetting of molten waxes, for instance. Inks typically consist of many components: water and solvents, binders (if ceramic particles are printed), dispersion agents (preventing nanoparticle coalescence) and adhesion promoters. Piezo jets are preferred for complex fluid mixtures because the high temperatures in thermal ink jets may degrade some ink components.

Polymer transistors have been made using ink jet printing for selected features, like the channel polymer deposition and silver metallization for source and drain in the device of Figure 24.14. Sometimes silicon microtechnology elements are included, like thermal oxidation in the transistor of Figure 24.14. Because this device is processed on silicon, heat treatments pose no limitations. This is usually not the case in all-polymer electronics: temperatures must be kept low, and the annealing temperature is limited to for example 150 or 200 °C. Silver conductivity will then be for example only 50% of bulk conductivity. Gold nanoparticle inks have achieved 70% but in many experiments the result is 1% of bulk conductivity. Adhesion is as important in polymer devices as in any others, and before the conducting polymer PQT-12 for the transistor channel is ink jetted, OTS SAM was applied to the wafer to improve adhesion.

Traditional techniques are routinely mixed with ink jetting. For instance, wax can be ink jetted on metal and used as an etch mask. Embossing/imprinting has been used to create some of the smallest features made by ink jetting.

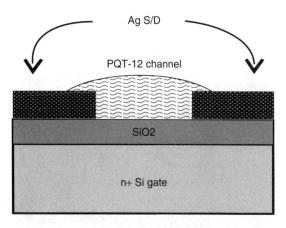

Figure 24.14 Polymer transistor with ink jet printed silver source/drain and ink jet printed conductive polymer channel. Adapted from Doggart *et al.* (2009)

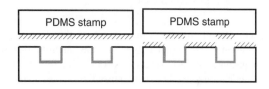

Figure 24.15 Planar SAM-coated PDMS stamp has been printed against embossed substrate, leaving high parts coated by hydrophobic SAM and recesses hydrophilic

A droplet lands on the embossed surface and adheres to feature bottoms which are hydrophilic, while the upper surfaces have been rendered hydrophobic by microcontact printing (Figure 24.15).

Another mixed technology approach is used in ink jet printing color filters for displays. Lithographically defined polyimide forms fences around each subpixel. This prevents ink-jetted droplets from spreading over neighboring pixels. This is necessary because subpixel sizes are on the limit of ink jet resolution. Additionally, plasma treatment has been done to render the imide hydrophobic, so that the ink will adhere to the feature bottom only. The brute force method would require three lithography steps, one for each color.

24.6 Mechanical Structuring

Practically all traditional machining techniques, cutting, grinding, drilling, turning, dicing, etc., have their microscopic counterparts. The main strength of these techniques is their simplicity and applicability to

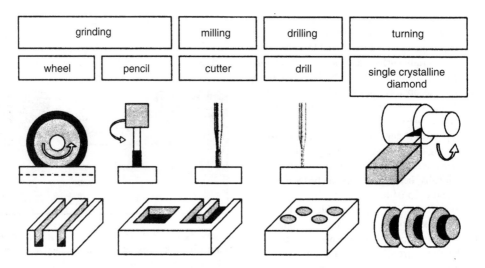

Figure 24.16 Mechanical structuring. Reproduced from Hülsenberg *et al.* (2008) with kind permission of Springer Science and Business Media

practically all materials, but of course the dimensions are usually not comparable to true microfabricated dimensions. Surface quality is often inferior to microfabrication, and this may pose limitations, even in applications where dimensions would be sufficiently small. This is the case with microfluidic molds: a 100 μm linewidth is fine for many applications but a 0.3 μm roughness makes demolding difficult. Yet another limitation is shape: cutting and milling can make complex shapes but many others are limited to certain shapes only, as shown in Figure 24.16.

Wafer dicing is a grinding process with a wheel. It can produce vertical walls, and the thinnest blades are quite thin, 20–60 μm; the placing accuracy is very good, so quite delicate microstructures can be made. Dicing saws have 1–10 cm/s feed rates, and if coarse microstructures are sufficient, a dicing saw is an excellent tool.

When dicing saw structures are combined with wet etching, even more advanced structures can be realized (Figure 24.17). Vertical (100) walls will etch just like horizontal (100) walls, and all the familiar shapes will develop as etching proceeds.

Drilling is often needed in fluidics, to create inlet holes. While this works fine in research, making tens of holes, it is much more difficult in production. Drill bits wear down rapidly and need to be replaced in rapid succession. If not replaced, the wearing down will affect hole shape. And while 100 μm drill bits are available, much larger ones are usually used, for example 250 or 500 μm. Glass wafers with 1000 drilled holes are available, but mechanical drilling cannot go much further. Laser drilling and DRIE are options when a large number of holes with high aspect ratio are needed.

Cutting by pencil cutters gives shape freedom, but as a serial technique it is slow and limited to about 200 μm minimum dimensions. Microfabrication offers some improvement in cutting pencils: CVD diamond coatings add to the working life of the pencils. Water microjets can also be used to cut structures of about 200 μm. And in both methods, adding abrasive particles speeds cutting up.

24.7 Chemical and Chemomechanical Machining Scaled Down

24.7.1 Micro electric discharge machining (μEDM)

In electric discharge machining (EDM) a conductive piece is machined by applying a voltage between the piece and a conductive electrode, in an insulating fluid. A high enough voltage leads to dielectric breakdown of the fluid, causing a spark. This leads to local heating, melting and vaporization. The only important material property is electrical conductivity, and hard materials like tungsten, which are otherwise difficult to machine, are readily amenable to EDM.

μEDM is a miniaturized version of EDM, and it is capable of drilling holes in the tens of micrometers range (Figure 24.18). Surface roughness is a few micrometers, but this can be improved by inclusion of an electropolishing step (with a conducting electrolyte), to obtain submicrometer roughness. Even though the

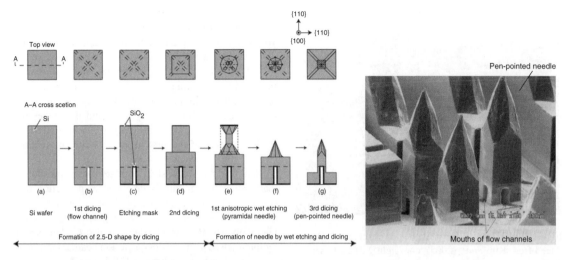

Figure 24.17 Microneedles by multiple dicing saw cuts and anisotropic wet etching: the vertical sidewalls of (100) wafers formed by dicing are also (100) planes, and will etch accordingly. Minimum diced dimension is 30 μm. Reproduced from Shikida *et al.* (2006) by permission of IOP

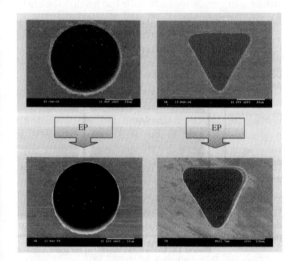

Figure 24.18 μEDM with electropolishing post-treatment. Reproduced from Hung *et al.* (2006) by permission of IOP

resolution of μEDM would suffice for mold master fabrication for many microfluidic devices, surface roughness is often excessive, making detachment of a molded piece difficult.

Machining rates can be micrometers per second. High aspect ratios are difficult to make, and wall verticality is also problematic: due to wear of the electrode, the shape of the wall changes as machining proceeds.

24.7.2 Micro electrochemical machining (μECM)

Micro electrochemical machining (electrochemical etching and deposition) uses an inert microelectrode, for example platinum, that is not consumed in the process. Therefore it can be used to make much smaller structures than μEDM (μECM electrodes are often made by μEDM!). And because there is no tool wear, structure dimensions are better preserved, and higher aspect ratios can be made.

In μECM local electrolysis or electroplating takes place in the space between the electrode and the working piece. Voltages are a few volts and frequencies are in the kilohertz range. Electrode gap control in the 10 μm range is essential for reproducible machining. Typical feature dimensions are in the tens and hundreds of micrometers (Figure 24.19).

Steel can be machined in sulfuric acid, and copper can be plated in $CuSO_4/H_2SO_4$ electrolyte, at 5 μm/s. When a high rate is used, however, hydrogen evolution leads to a rough surface. The rate can be further increased by having multiple electrodes, as in Figure 24.20.

24.7.3 Spark-assisted machining

Various spark-assisted methods have been developed over the years. They rely on the high temperature created by local discharge in the electrolyte. The mechanism of material removal is believed to be thermal melting, even though chemical etching is assumed to be in action, too. SACE, for spark-assisted chemical engraving, is suitable

Figure 24.19 Stainless steel 60 μm structures. Reproduced from Kim *et al.* (2005), copyright 2005, Elsevier

Figure 24.20 Multiple electrode ECM: 45 μm diameter, 200 μm deep holes in copper. Reproduced from Kim *et al.* (2005), copyright 2005, Elsevier

for non-conductive substrates. Very high machining rates of 10–100 μm/s are possible.

There are several parameters which affect the results:

- electrolyte parameters
- tool electrode parameters
- power supply parameters
- workpiece parameters.

The electrochemical discharge process is a complex phenomenon which depends on a multitude of factors, for instance electrolyte electrical and thermal conductivity and surface tension, gas generation, and workpiece thermal conductivity and local temperature.

24.8 Conclusions

Serial writing is like writing with a pen: anything can be written, even if you change your mind at the last minute, and it is very fast for writing short notices. But writing a book with a pen is very slow, and if you need a copy of your book, the writing time for the second copy is almost identical to that for the first.

Parallel writing, optical lithography, imprinting, molding are like a printing press: it is very cost efficient to produce huge volumes of identical works. But the cost of making the first copy is very high, and if changes need to be implemented, the starting cost of the second version is close to that of the first because of the non-recurring costs of masks/masters/molds.

Some techniques exist in both serial and parallel formats. Stereolithography is practiced in projection mode (Figure 23.2), but it can also be done in serial mode: instead of masks, direct laser writing is used to solidify photopolymer. Powder blasting can also be done serially by microjets. This is analogous to ion beams: ion implantation is a masked process but FIB is serial.

In large-area applications direct processing is advantageous because the mask price goes up with area, and resist coating large rectangular areas is difficult. Even in smaller area applications where the line length to be drawn is small, laser processing may win, because resists, developers, etchants and other consumables are eliminated.

The main criteria of the patterning process, namely linewidth, patterning speed and alignment, are difficult to achieve simultaneously. AFM, FIB and electron beams excel in linewidth and alignment is good too, but the writing speed is very slow. Nanoimprint and injection molding are strong in linewidth and speed, but poor in alignment. Laser beams are a good compromise when ultimate linewidths are not required, but optical lithography is currently the dominant solution to this optimization problem.

24.9 Exercises

1. What problems does redeposition in laser grooving cause, and how can they be overcome?
2. What differences does photomask repair with ion beams, electron beams and laser beams have?
3. How do laser cutting and dicing saw compare in chip dicing?
4. How could a multiple head ink jet system speed up deposition?
5. Consider the question of using an ink jet as a spray etcher to locally etch away material.
6. Explain in detail how the microneedles of Figure 24.17 are made!
7. How could local thermally driven processing be done by resistive heating instead of laser beam heating?

References and Further Reading

Calvert, P. (2001) Inkjet printing for materials and devices, *Chem. Mater.*, **13**, 3299–3305.

Chopra, K.L., P.D. Paulson and V. Dutta (2004) Thin-film solar cells: an overview, *Prog. Photovolt: Res. Appl.*, **12**, 69–92.

Crawford, T.H.R., A. Borowiec, and H.K. Haugen (2005) Femtosecond laser micromachining of grooves in silicon with 800nm pulses, *Appl. Phys.*, **A80**, 1717–1724.

Despont, M. et al. (2004) Wafer-scale microdevice transfer/interconnect: its application in an AFM-based data-storage system, *J. Microelectromech. Syst.*, **13**, 895–901.

Doggart, J., Y. Wu and S. Zhu (2009) Inkjet printing narrow electrodes with <50 µm line width and channel length for organic thin-film transistors, *Appl. Phys. Lett.*, **94**, 163503.

Edinger, K. et al. (2004) Electron-beam-based photomask repair, *J. Vac. Sci. Technol.*, **B22**, 2902–2906.

Gierak, J. et al. (2005) Exploration of the ultimate patterning potential achievable with focused ion beams, *Microelectron. Eng.*, **78–79**, 266–278.

Harder, N.-P. et al. (2009) Laser-processed high efficiency silicon RISE-EWT solar cells and characterization, *Phys. Stat. Solidi*, **C6**, 736–743.

Henry, M., P.M. Harrison and J. Wendland (2007) Laser direct write of active thin films on glass for industrial flat panel display manufacture, *J. Laser Micro/Nanoeng.*, **2**, 49–56.

Hülsenberg, D., A. Harnicsch and A. Bismarck (2008) **Microstructuring of Glasses**, Springer

Hung, J.-C. et al. (2006) Micro-hole machining using micro-EDM combined with electropolishing, *J. Micromech. Microeng.*, **16**, 1480–1486.

Kim, B.H. et al. (2005) Micro electrochemical machining of 3D micro structure using dilute sulfuric acid, *CIRP Ann. – Manuf. Technol.*, **54**, 191–194.

Kim, K.-H. et al. (2008) Direct delivery and submicrometer patterning of DNA by a nanofountain probe, *Adv. Mater.*, **20**, 330–334.

Kumagai, M. et al. (2007) Advanced dicing technology for semiconductor wafer – stealth dicing, *IEEE Trans. Semicond. Manuf.*, **20**, 259–265.

Neuhaus, D.-H. and A. Münzer (2007) Industrial silicon wafer solar cells, *Adv. OptoElectron.*, 24521.

Ogane, A. et al. (2009) Laser-doping technique using ultraviolet laser for shallow doping in crystalline silicon solar cell fabrication, *Jpn. J. Appl. Phys.*, **48**, 071201.

Repmann, T. et al. (2006) Microcrystalline silicon thin film solar modules on glass, *Sol. Energy Mater. Sol. Cells*, **90**, 3047–3053.

Reyntjens, S. and R. Puers (2001) A review of focused ion beam applications in microsystem technology, *J. Micromech. Microeng.*, **11**, 287–300.

Salaita, K., Y. Wang and C.A. Mirkin (2007) Applications of dip-pen nanolithography, *Nature Nanotechnol.*, **2**, 145–155.

Shikida, M., T. Hasada and K. Sato (2006) Fabrication of densely arrayed micro-needles with flow channels by mechanical dicing and anisotropic wet etching, *J. Micromech. Microeng.*, **16**, 1740–1747.

Straub, M. et al. (2004) Complex-shaped three-dimensional microstructures and photonic crystals generated in a polysiloxane polymer by two-photon microstereolithography, *Opt. Mater.*, **27**, 359–364.

Sun, Y. (2002) Laser link cutting for memory chip repair, *Proc. IEEE*, **90**, 1627–1636.

Sun, Y. et al. (2006) Low-pressure, high-temperature thermal bonding of polymeric microfluidic devices and their applications for electrophoretic separation, *J. Micromech. Microeng.*, **16**, 1681–1688.

Tekin, E., P.J. Smith and U. S. Schubert (2008) Inkjet printing as a deposition and patterning tool for polymers and inorganic particles, *Soft Matter*, **4**, 703–713.

Tseng, A.A. (2004) Recent developments in micromilling using focused ion beam technology, *J. Micromech. Microeng.*, **14**, R15–R34.

Utke, I., P. Hoffmann and J. Melngailis (2008) Gas-assisted focused electron beam and ion beam processing and fabrication, *J. Vac. Sci. Technol.*, **B26**, 1197–1276.

Vettiger, P. et al. (2002) The "Millipede" – nanotechnology entering data storage, *IEEE Trans. Nanotechnol.*, **1**, 39–55.

Wüthrich, R. (2008) **Micro-machining Using Electrochemical Discharge Phenomena**, William Andrew.

25

Process Integration

Process integration is the task of putting together individual process steps to create functional devices. This necessitates interfacing device design and processing, knowledge of process capability and device operation, understanding materials interactions, being prepared for equipment limitations – all aspects of microfabrication.

Process integration is about questions like the following.

Wafer selection:
- Are wafer mechanical specifications important, or electrical, or both?
- Do we need transparency or thermal insulation?
- Can single-sided polished wafers be used, or should DSP wafers be used?

Materials compatibility:
- What is the maximum temperature the materials can tolerate?
- Will thermal expansion coefficient mismatches create stresses?
- Will the polymer parts withstand the chemical treatments that follow?

Process–device interactions:
- How do thermal treatments add to diffusion profiles?
- Are etch profiles important for device performance?
- Does the stress-relief anneal affect structures already fabricated?

Equipment and process capability:
- What is the surface roughness allowed for good bonding?
- What is step coverage of sputtered films in contact holes?
- Can thick bonded wafer stacks be inserted into wafer boats?

Design rules:
- What is the minimum allowed linewidth?
- What is the minimum allowed spacing between lines?
- Do we need to design dummy patterns to equalize area usage?

Mask considerations:
- Which photomasks are critical, which are non-critical?
- Does etch undercutting need to be compensated on the mask?
- Can test structures be fitted in dicing lanes?

Order of process steps:
- Should front-side processing be completed before back-side processing?
- What processes can be done after through-wafer etching?
- Can any steps be done after thin-membrane formation?

Reliability:
- Do current densities in wiring need to be limited?
- How do stresses build up when more layers are deposited?
- What is the breakdown voltage of thin oxides?

We will start by discussing the general features of wafers in different process steps: sometimes processes act on both sides and sometimes on one side only. We will then go through a few examples: a solar cell, fluidic sieves, a DNA amplification chip (PCR) and an integrated passive chip (RCL chip). The solar cell is a bulk wafer process with critical properties defined oxidation and diffusion steps. The filter case involves comparing many different technologies which all result in more or less the same functionality, but via very different processes. Etching and bonding are the key technologies. In the PCR chip polymer processing and bonding are key elements. In the RCL chip alignment and etch selectivities, as well as multiple thin-film depositions, are important. General issues will be discussed along with the examples.

Introduction to Microfabrication, Second Edition Sami Franssila
© 2010 John Wiley & Sons, Ltd

25.1 The Two Sides of the Wafer

Processes come in two forms: directional and diffuse (beam-like and immersion, if you wish). The former include processes where beams of atoms, photons, electrons, ions or fluid sprays impinge on the wafer (like lithography, evaporation and implantation); the latter are immersion processes where wafers are surrounded by vapors, gases or liquids (like wet etching or oxidation). In order to prevent immersion processes acting on the whole wafer, both sides of the wafer have to be protected by masking layers. Sometimes back-side protection comes free of charge, when for instance thermal oxidation oxidizes both sides of the wafer. Additionally, directional processes can also be blanked by absorbers, collimators or stencil masks, which are not on the wafer but above it (Figure 25.1).

CVD, PECVD and plasma etching processes can be either directional or diffusive: if wafers are loaded upright in a wafer boat (Figure 11.8), deposition/etching takes place on both surfaces, but if wafers are loaded flat, or clamped, on an electrode (Figure 13.1), only the top side is processed (Table 25.1), with some unintentional spillover over the edge.

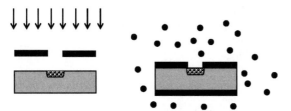

Figure 25.1 Left: directional process blanked by a stencil above the wafer. Right: diffuse process blanked by a masking layer on the wafer

Table 25.1 Double-sided and single-sided processes

Double sided	Single sided
Furnaces, oxidation	Sputtering
Furnaces, CVD	Evaporation/MBE
Furnaces, diffusion	Ion implantation
Furnaces, annealing	PECVD
Furnaces, RTA	Lithography
Wet etching and cleaning in a tank	CVD epitaxy
Spray processing	CMP
Barrel plasma etching/stripping	RIE/plasma etching
Resist stripping in a tank	Spin processing

In most equipment inserting the wafers into the reactor upside down is allowed, but potential damage to the patterns on the front by transport mechanisms, clamping or chucking must be considered. Temperature permitting, photoresist is a quick fix that protects the front side. Sometimes a film that was deposited on both sides is first patterned on the back, while the front side is covered. Processing must be tailored so that both sides of the wafer are under controlled conditions at all times.

Three kinds of processes take place on the wafer back side:

1. Patterning.
2. Blanket processing (doping, growth, deposition).
3. Unintentional processes.

Thin films on the wafer back side are often of poor quality because most processes are optimized for the front side only. If single-sided polished wafers are used, backside roughness (Figure 4.12) prevents proper film growth. Sometimes back-side films result from front-side processing spillovers: photoresist will cover the wafer edge erratically, and some resist will be deposited on the wafer back side; or, alternatively, material from the wafer chuck or transport system will adhere to the wafer back.

Thermal oxidation oxidizes both sides of the wafer, which may, or may not, be advantageous. Oxide on the back side can be a useful protective layer, for example to prevent diffusion in the next step. LPCVD nitride similarly covers both sides.

Diffusion from the gas phase will dope both sides of the wafer. Again, oxide or nitride films can prevent unwanted diffusion. Doping by implantation and from thin-film sources (e.g., PSG or BSG) are single-sided processes.

Epitaxy presents a special case of back-side effects on the front side: if a lightly doped epilayer is grown on a highly doped substrate wafer, evaporated dopant from the substrate will mingle with the source gases and affect epilayer doping. Therefore, CVD oxide is used as a backside capping layer to prevent dopant outdiffusion from the substrate.

Blanket processing involves growth and deposition of films either simultaneously or sequentially on both sides. Thermal diffusion can be done either way, with an oxide film to prevent diffusion on the protected side. Ion implantation doping is inherently one sided. Applications of blanket processing include doping to improve back-side metal contact or etch mask formation.

Rather thick stacks of films can build up on the wafer back side. If both sides of the wafer are coated, the stresses are initially hidden, but when the film on either side is processed, stress is able to deform the wafer. Stresses can cause flaking and rupture, which generate particles. For these reasons back-side films are sometimes removed even though no device reason would necessitate it.

25.2 Device Example 1: Solar Cell

The simple solar cell (Figure 25.2) process described below features many important interactions between process steps which arise when complete processes are put together. Volume manufacturing of solar cells uses additional techniques to reduce costs (some of them are discussed in Chapters 24 and 37) but the integration aspects remain similar.

Process flow for solar cell

- Wafer selection
- Front-end processing
  - wafer cleaning
  - thermal oxidation
  - photoresist spinning on front
  - back-side oxide etching
  - resist stripping
  - wafer cleaning
  - p^+ backside diffusion
  - front-side oxide etching
  - wafer cleaning
  - n-diffusion
- Back-end processing:
  - resist spinning on front
  - aluminum sputtering on back side
  - resist stripping
  - wafer cleaning
  - PECVD nitride deposition on front side
  - lithography for contact holes
  - RIE of front-side nitride
  - resist stripping
  - wafer cleaning
  - aluminum deposition on front side
  - resist spinning on back side to protect aluminum
  - lithography for front aluminum
  - aluminum etching
  - photoresist stripping
  - wafer cleaning
  - contact improvement anneal

All processes begin with substrate selection: p-type silicon is chosen, and the pn junction will be made by n-diffusion. If an n-type wafer were chosen, opposite diffusions would need to be made. It is advantageous to do back-side p^+ diffusion before the pn junction.

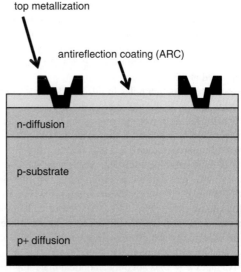

Figure 25.2 Solar cell cross-section

The heavy p^+ diffusion (e.g., boron concentration of 10^{19} cm^{-3}) on the back is unaffected by the light n-diffusion (e.g., 10^{16} cm^{-3}) on the front because of the huge difference in doping levels. If n-diffusion were done first, another oxidation would be needed to protect the lightly n-doped front side during heavy p^+ back-side diffusion. Wafer orientation is not important in this application.

The process is divided into two very different parts: front end and back end. The front end contains high-temperature steps where silicon is oxidized, diffused, implanted and annealed at temperatures around 1000 °C. Cleaning steps are especially important before high-temperature steps because whatever contamination is present will diffuse rapidly at high temperatures. Thermal oxide is superior in quality to CVD oxides, and if thermal oxidation can be used, it usually is. It is, however, a slow process step with a duration of hours, even tens of hours. (PE)CVD systems are fast, which may favor CVD oxides in the case thick oxides are needed. Sometimes the temperature forbids thermal oxidation, and CVD and PECVD remain the only options.

After first-metal deposition (back-side metallization for this solar cell) the process temperatures must be limited to about 450 °C to stabilize the silicon–metal interface. This rules out many deposition processes for the antireflective coating (ARC), for example thermal oxide, TEOS CVD oxide or LPCVD nitride.

Back-side metallization is done before front-side ARC and metal. This is because the front side is more important for device operation, and we would not like to clamp the wafer face down in a sputtering system after front-side processing is completed. PECVD nitride ARC is deposited at 300 °C.

We now have to etch holes in this nitride to make contact with silicon. If the top metal were the same size as the contact holes, perfect alignment and zero undercut etching would be needed for the metal to cover the hole completely. Because such processes do not exist, the top metal is designed to be somewhat wider than the contact hole, to make sure that minor misalignment or linewidth loss in etching will not result in structures where some silicon (in n-diffusion) would be exposed to ambient air. If this were the case, cell performance would rapidly deteriorate as humidity and other environmental agents would make contact with the pn diode. Nitride ARC (with a refractive index $n \approx 2$) is not only an optical matching layer between air ($n = 1$) and silicon ($n \approx 4$), but also serves scratch resistance, moisture and ion barrier functions.

Top surface metallization should be as narrow as possible, because it blocks part of the incoming sunlight. It should, however, be capable of carrying considerable currents. Therefore, metallization of high aspect ratio should be used. Electroplated copper is a good choice, but for research devices sputtered aluminum will do.

The back end and front end are independent of each other: metallization can be changed from sputtered aluminum to electroplated copper to screen-printed silver paste without changing the front end. Similarly, diffusion conditions can be changed and there is no need to modify the back end because of that.

25.3 Device Example 2: Microfluidic Sieves

Microfluidic sieves are low-pass filters which block particles larger than their specified pass size. There are many designs that can be used to realize such devices, and four of them are pictured in Figure 25.3: the etched hole sieve, pillar sieve, bonded Weir and sacrificial layer sieve.

In the etched hole sieve pass size is simply determined by lithography and etch capability. Silicon provides mechanical strength. In another version the holes are etched in silicon nitride but the nitride membrane is thin compared to silicon, which can be made into practically any thickness, and one figure of merit for a sieve is the pressure it can withstand. Another figure of merit for a sieve is the aperture ratio, or the percentage of holes. It is

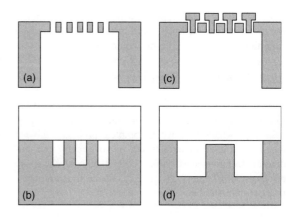

Figure 25.3 Fluidic sieves: (a) etched hole; (b) pillar; (c) sacrificial; (d) Weir

limited to about 25% for a lithographically defined sieve (recall Figure 9.13). Lithographic size limitations can be somewhat circumvented by conformal thin-film deposition after etching: it makes the holes smaller. The backside access hole can be etched by DRIE or KOH/TMAH, depending on the application.

The pillar sieve calls for DRIE and bonding. The lithography requirement is similar to that for the etched sieve, but etching requirements are more demanding because pillars of high aspect ratio must be made. Narrowing the gap between the pillars by conformal deposition is not an option: the deposited film will cover the tops of the pillars too, and this will make cover bonding difficult. The pillar sieve is more expensive than the etched hole sieve: DRIE is an expensive step, and a second wafer is needed for capping. The benefits come from the shape freedom of DRIE, and the fact that the flow channels can be made with pillars parallel to the flow, while in the etched hole sieve the flow is through the filter, which makes clogging a problem.

The sacrificial layer sieve relies on thin-film deposition for pass size definition. Holes are etched in silicon and covered by a conformal thin film, for example oxide. Polysilicon is then deposited and patterned. Oxide is then etched away, and very narrow holes created. In theory any thickness that can be controllably deposited can be used. Sacrificial filters with 10 nm gaps are feasible from a fabrication point of view, but fluid dynamics, clogging and other issues have to be handled, too. Even though the pass size can be made really small, the aperture ratio remains very small.

In the Weir sieve the shallow etch depth determines the pass size, and the deep groove etching defines the flow channels. Shallow etches in the micrometer range are easy, and shallower ones could be made. However, the anodic bonding process and glass structural stability determine how the shallow passages will remain open (as discussed in Chapter 22). Auxiliary pillars can be made to support the glass roof, with minimal effect on the fluidics and no effect on pass size. In the Weir sieve sacrificial techniques can also be used: instead of etching silicon to the desired depth, thermal oxidation and HF–wet etching could be used. This is beneficial for two reasons: pass size can easily be checked by ellipsometry after oxidation, and oxide thickness control is much better than etching depth control. The surface quality of the silicon after HF oxide removal is also superior to an etched silicon surface.

25.4 Wafer Selection

Wafer selection and process design go hand in hand. Bulk silicon wafers are the default choice for practically any microdevice. The use of more special wafers must always be thoroughly justified. Czochralski (CZ) wafers are also the default choice, and float zone (FZ) wafers are reserved for special applications only, for instance when extremely high resistivity is needed.

<100> wafers are used unless otherwise specified. MOS transistors are made on <100> for silicon/oxide interface quality: fewer trapped charge and interface defects are generated in the oxidation of <100> silicon than <111> silicon. For MEMS, anisotropic etching of <100> silicon is standard technology. In bipolar technology <111> is used. When both MOS and bipolar transistors are on the same chip (BiCMOS), <100> wafers are used because oxide for the MOS part is more critical than <111> special features of the bipolar part.

Wafers of crystal orientations other than <100> have some special applications. Wafers of <110> orientation enable wet-etched 90° sidewalls, which has a competitive advantage over DRIE in some special cases. Recently, <110> wafers have been explored as alternatives to <100> in advanced CMOS because they offer 50% higher hole mobility than <100>, thus boosting PMOS transistor performance, but electron mobility is less than with <100> wafers. Similarly, <113> wafers have been studied for future CMOS, because of a better silicon/oxide interface.

CZ wafers are available over a wide range of dopant density, or, in other words, over a wide range of resistivities. Typical CZ resistivities are listed in Table 25.2.

If epitaxial wafers are used, then process design offers both greater freedom, because some bulk effects can be

Table 25.2 CZ silicon resistivity ranges

p-type	Boron	0.01–60 ohm-cm
n-type	Phosphorus	0.01–40 ohm-cm
n-type	Antimony	0.008–10 ohm-cm
n-type	Arsenic	0.002–20 ohm-cm

ignored, though it also introduces some limitations, and incurs extra wafer costs. SOI wafers usually require full process rethinking, in order to realize their full potential in reducing the number of process steps or enhanced device performance.

Wet etching in MEMS calls for wafers cut exactly to the [100] normal vector. Whereas the standard cut is specified ±1°, MEMS wafers can have ±0.02° specification. Miscut <111> wafers have terraces and kinks (Figure 4.10) which are preferred sites for deposited atoms to attach to, an important issue in epitaxy and thin-film deposition in general. A large miscut of 4° changes the apparent lattice constant of silicon and offers possibilities to grow epitaxial oxides like <Y_2O_3> on silicon.

Wafer thickness increases with diameter to improve mechanical strength. Mechanical strength is important, especially during the high-temperature steps of oxidation, diffusion and epitaxy. Above 1000 °C, thermally generated stresses may cause slip dislocations on uneven cooling. Thick wafers are also generally more robust to handle.

In many applications thin wafers are needed. Solar cells would be cheaper if they used less silicon, wet-etched bulk MEMS devices with 54.7° angle require less area in through-wafer etching, and in power transistors resistive losses are minimized by using thin wafers. Wafer thicknesses down to 200 µm are readily available but they require special attention during processing. Wafers can also be thinned down to final thickness after all device processing is done. This improves flexibility of the silicon dies and helps packaging in applications like smart cards.

For surface micromachining almost any wafer is good because by definition the structures will be made in the thin films deposited on top of the wafer. Most people use silicon wafers, but there are surface micromachined devices fabricated on quartz, glass, GaAs, polyimide and other substrates. For test runs reclaimed wafers could be used. For CMOS transistors prime-quality wafers must be used.

Non-silicon substrates commonly used include fused silica and glass. They are transparent, thermally and electrically insulating, and have surface chemistry similar to silicon. They are also available in shapes and thicknesses identical to silicon. Processing non-silicon substrates need not be problematic, but if silicon and glass are processed in the same fabrication facility, contamination issues arise.

25.5 Masks and Lithography

The lithography tool must be specified early on in process design because, with the tool, the exposure wavelength, mask size, wafer size and chip size become fixed. The exposure wavelength sets limits on photoresist selection, mask plate material and resolution. In 1× exposure tools the mask size is somewhat larger than the wafer size (e.g., 5 inches for 100 mm wafers; 7 inches for 150 mm wafers). With the 1× aligner, chip size is limited by wafer size and edge exclusion. With step-and-repeat lithography tools chip size is limited by the exposure field size which is about 8 cm^2 maximum. Optimization is needed to fit many small chips in the field, or, alternatively, stitching is needed to make larger chips.

Photoresist polarity, negative or positive, needs to be selected before mask making. It is possible to draw the design in one polarity, and to invert polarity computationally in the mask making process, but once the physical mask plates have been drawn, the mask and resist are tied together. Minimum linewidth affects the choice of mask material: soda lime is acceptable for micrometer dimensions, but in submicrometer projection lithography with a 365, 248 or 193 nm wavelength, fused silica is used.

Not all lithography steps are equal; some are more critical than others. Critical levels determine device functionality in a critical way, for example the CMOS gate mask determines the gate length, which affects transistor speed and leakage current. It is possible to mix two lithographic techniques: this approach is known as mix-and-match. For instance, in 0.50 μm CMOS technology the critical levels might be exposed by a 365 nm 5× stepper and the non-critical levels by a 1× tool. This approach is investment related: some additional work from mix-and-match (in, for example, an alignment scheme) is traded for major savings in equipment purchase prices.

Undercutting in wet etching can be compensated by biasing the photomask. The patterns on the mask are made wider by the amount of etch undercutting for light-field structures, and narrower for dark-field structures. Mask biasing can be done in a global fashion: for example, all structures on an aluminum level can be biased wider by, say, twice the designed aluminum film thickness. For 3 μm nominal linewidth this translates to patterns 5 μm wide assuming 1 μm aluminum thickness, and thus 1 μm etch undercutting per side. If the resolution of the lithography tool is 6 μm (capable of printing 3 μm lines with 3 μm spaces), mask biasing cannot be done, because 1 μm spaces would need to be resolved. Mask biasing wastes silicon real estate, and the resolving power of the lithography tool is not fully utilized for increasing device packing density.

On a 1× mask there are usually three elements: device chips, test structures and alignment marks (Figure 1.20). The area usage between these elements depends on process and device maturity. In early phase development more area is spent on test structures, but in volume manufacturing device chips take up practically all the area, with test structures embedded in the scribe lines between the chips. Test structures include both device-specific and process-specific measurements. The latter are identical in all runs using the same process, and they are used for collecting information on process performance, stability, drifts and variation, for statistical process control (SPC).

If the first process step is diffusion or implantation, there will be nothing visible (or something barely visible) on the wafer, and the second lithography step – the first alignment – cannot be done. Therefore it is common practice to etch special alignment marks into the silicon at the very beginning of the process. This is called zero level. Planarization later in the process may smear alignment marks, and it might be that in some process steps the alignment marks must be protected in order to maintain them.

25.6 Design Rules

Design rules are statements about allowed structures, with regard to linewidths and spacings, overlap and layer-to-layer positioning. These are often referred to as layout rules, as opposed to electrical design rules, which include information about sheet resistances, contact resistances, current density limitations, etc. Layer thickness design rules are needed in capacitor design: oxide thickness determines capacitance density, both when the oxide is used as the capacitor dielectric as such and when it is used as a sacrificial layer in the fabrication of an air gap capacitor. In addition to design rules, device models (for transistors, resistors, capacitors) are higher level abstractions of the process for circuit designers. Design rules and models are always process specific. They are also company specific: for example, 0.13 μm CMOS processes from different suppliers have different sets of rules and models even though the basic dimensions and devices are similar.

Device interactions come in many guises and are device and process specific. Transistors need to be isolated from each other, and this isolation takes up space. Inductor coils must be placed rather far away from each other because of magnetic field coupling over distance. It is also important to understand and to limit structures that can be placed between two coils as these can couple into the magnetic field.

25.6.1 Single level layout rules

Layout design rules are formal geometric rules which relieve the designer from the details of the fabrication process. The process engineer has distilled the physical capabilities and limitations of the fabrication process into the design rules, with the aim of making the process more robust. Sometimes breaking the rules leads to zero yield, sometimes more subtle effects are encountered. The design rules are often divided into compulsory and advisory rules, the latter being hints of known good practice.

Minimum size and spacing are basic layout rules. Three elements contribute to them:

1. Lithographic process capability.
2. Structure widening in subsequent process steps.
3. Device interactions.

Lithographic capability involves the optical tool, photomask quality, resist properties and resist thickness. If the lines are not accurate on the mask, the design width cannot be obtained on the wafer. Breaking the minimum line and space rules will lead to catastrophic failures.

Very often minimum space is different from minimum linewidth. For one thing, lithographic resolution (pitch) is not usually divided equally between a line and space: it is typical that, for example, the 0.5 μm linewidth process has 0.5 μm minimum line and 0.7 μm minimum space. Sometimes processes are specified by half-pitch: the previous process would then be classified as a 0.6 μm process.

The final structure width is determined by process step properties. Diffusion is an isotropic process and a 3 μm diffusion depth leads to sideways spreading of about 3 μm, too. Similarly, isotropic etch undercutting necessitates similar design concerns: namely, equal spacing of 10 μm width; grooves 5 μm deep would result in touching of the neighboring grooves.

Some processes behave differently for different sizes: for example, DRIE rate and sputtering step coverage are different in small holes compared to large ones. The most blatant design rule is to allow only one hole size. Areas of low pattern density tend to etch faster (loading effect) and polish slower. It is then useful to design dummy features. These are used to fill area, to equalize things. In embossing/NIL the force needed goes up as the fourth power of the structure radius, and it makes sense to break up large areas into small dummy lines and spaces if possible.

Stresses build up in large structures, and it is advisable to break large uniform areas into smaller segments, if possible. This is true for wide aluminum metallization and large SU-8 fluidic structures alike. Stress minimization is also important in injection molding and other high-pressure processes, and sturdy dummy patterns can help delicate microstructures to survive.

25.6.2 Alignment and layer-to-layer rules

When structures on two different layers need to coincide, overlap rules must be invoked.

Overlap rules make sure that the layers that need to touch will do so irrespective of process variation, and structures that must not touch will stay separated. The alignment of structures on different levels depends on three factors:

1. Lithography tool alignment performance.
2. Pattern placement accuracy.
3. Alignment sequence.

It is not advisable to place two structures on different mask levels exactly on top of each other because misalignment (and resist linewidth variability and etch undercut) will always introduce some uncertainty into edge position. If the underlying structure is the same width as the upper level structure, misalignment will lead to the formation of a severe crevasse. This is pictured in Figure 25.4 for contact holes. When the contact hole is etched into CVD oxide, a misaligned contact exposes the underlying oxide, which will also be etched. The subsequent metal sputtering and/or CVD process will have difficulties in filling the crevasse. In order to make sure that the contact hole will touch the resistor, the resistor contacting area is made larger to accommodate any misalignment. This is termed collar or border or dogbone. It wastes area but it is necessary for process robustness.

Different mask levels may have different linewidth rules. One mask level may contain critical structures, and narrow lines are allowed, but other levels may have only non-critical structures. For example, pads for wire bonding are, say, 50×50 μm or 100×100 μm and the design rules are then much more relaxed, with for instance a 5 μm minimum overlap rule, while a 0.3 μm overlap rule is used for the 1 μm minimum linewidth critical levels.

Tool alignment performance is roughly one-third or one-fifth of minimum linewidth. If a tool with a 3 μm minimum capability (typical of 1× proximity mask aligner) is used to print contact holes 3 μm wide, an alignment tolerance of 1 μm needs to be designed in.

The second contribution to alignment accuracy between levels comes from pattern placement on the mask: the masks for two different layers are two separate physical

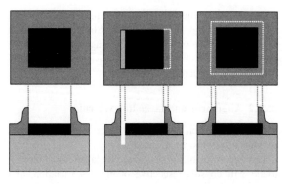

Figure 25.4 Top views and cross-sectional views of contact holes with different alignment cases: left, perfect alignment of contact hole with underlying metal; middle, misaligned contact without misalignment allowance etches into underlying oxide; right, oversized misalignment relative to metal

objects and the exact position of structures on the mask plate is subject to its own statistical variation. If image placement error on the mask is one-tenth of minimum linewidth, its contribution is

$$\sqrt{x_1^2 + x_2^2} \approx \sqrt{2x}$$

if mask errors are identical on both plates. This translates to about 14%, usually less than the contribution from misalignment.

Alignment sequence is the third factor. Layers are not aligned to the previous layer but to some important layer: for instance, in making resistors both contact holes and metallization are aligned to the resistor; after all, the whole idea of the structure is to make metal-to-resistor contact. If metal were aligned to the contact hole, we would have to account for two tool misalignment tolerances: one for contact hole-to-resistor alignment and another for contact hole-to-metal alignment. Assuming a Gaussian distribution, this leads to an alignment tolerance of $\delta\sqrt{n}$, where n is the number of alignments involved.

Automatic checking of design rules is a standard procedure for advanced chips. Design rule checking (DRC) includes both individual level checks (dimensional rules) as well as layer-to-layer checks (overlap rules, positioning rules).

25.7 Resistors

Resistors are simple elements, but they offer good insights into many aspects of process integration. First of all, resistors can be made by widely different technologies:

- diffused bulk silicon
- SOI silicon
- polysilicon
- metals.

They can be used for many different applications:

- heater resistors
- precision analog resistors
- load resistors in SRAM
- piezoresistors in MEMS
- thermal sensors.

Resistance is determined by linewidth (W), line length (L), thickness (T) and resistivity (ρ), and usually sheet resistance R_s ($\equiv \rho/T$) is used. High resistance values call for thin resistors, long and narrow lines or high-resistivity material. In practise different resistances are obtained by changing L and W, keeping ρ and T unchanged. This is because thin-film resistivity is thickness dependent, which would necessitate new characterization of the material if different thickness were to be used.

Resistors for analog ICs are usually made of polysilicon or metal alloys and compounds. Poly resistors have sheet resistance values between 50 and 5000 ohm/sq for typical film thicknesses of hundreds of nanometers. Tantalum nitride can be used when low resistance values are needed: 1–10 ohm/sq (100–200 µohm-cm resistivity). Alloys like SiCr enable high-resistance resistors to be made by sputtering, with 2000–20 000 µohm-cm resistivity. NiCr has a resistivity of 100 µohm-cm. In advanced analog device processes there are two different materials for resistors, that is two polysilicon layers with different doping levels, to enable a wide range of resistances.

Resistor linewidths are seldom the minimum linewidths that are available in the process, but rather large, in order to improve absolute value control. Long, straight resistors complicate circuit topology and meandering resistors are usually employed. However, corners do not contribute to resistance equally with the linear parts.

When isotropic etching is used in the resistor process, etch undercutting of the resistor and contact holes work in opposite directions: the resistor is narrowed by etch undercutting, whereas the contact holes become wider. These processes add up and the overlap rule has to accommodate that. The resistor process is outlined in Figure 25.5. In a similar fashion, contact holes and metal etching work in opposite directions. In general, the overlap rules for plasma-etched processes are much tighter than those for wet-etched processes. Plasma etching increases device packing density not only by its ability to make narrower lines, but also through smaller overlap requirements.

Different resistor applications have different requirements: analog components are more demanding than

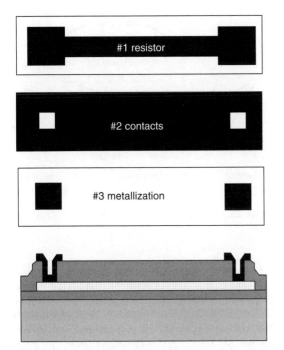

Figure 25.5 Thin-film resistor: photomasks and cross-sectional view of finished device. The resistor has collars in the ends to ensure that the contact holes in the oxide will always land on the resistor; and metallization has an overlap relative to contact holes so that no resistor will be exposed. Metal is, however, aligned to the resistor, not the contact hole

digital ones, for instance in digital MOS transistors 10% linewidth variation will not affect on/off action, but it changes the resistance of a resistor by 10%. In many cases absolute values of resistances (or capacitances) are not used, rather the ratios of two resistances (or capacitances). The deposition process may be non-uniform across the wafer, but locally within the chip area uniformity is usually very good.

Diffused resistors are simple in the sense that they often come "free of charge": if diffusions are made during wafer processing, resistors can be made. In addition to simplicity, diffused resistors can tolerate high temperatures during processing. They can also tolerate high temperatures during device operation, and they have been used in microrockets as heaters. In the design rules for diffused resistors the lateral extent of diffusion must be accounted for: a diffused resistor 5 µm wide and 2 µm deep ends up 9 µm wide.

Metal resistors are used as heaters in various chemical microsystems and sensors. Platinum has a linear temperature coefficient of resistivity (TCR) and it is easy to make a thermometer along with the heater. But depending on the required temperature range, many other metals can be used. Glass wafers are useful for thermal isolation, and then metal resistors must be used, because glasses do not stand poly deposition temperatures.

In micro hotplate chemical sensors (Figures 20.21, 20.22), in bolometers (Figures 11.15, 11.16) and in thermal flow sensors, resistors must be placed on thermally isolated membranes in order to reduce the heating requirements or to improve sensitivity. Resistor film stresses must then be considered: silicon nitride or alumina membranes are often very thin (to reduce thermal conductivity) and if stresses that are too high are present in heater metallization, the membrane will be stressed. It is also important to consider whether the resistor is made first, and release etching afterward, or perhaps shadow mask patterning is used to make the resistor after membrane formation. It is easier to process on a solid wafer than on a thin membrane, but it may be that the release etch also etches the resistor material. The alternatives are then either to protect the resistor, or to find another material.

Diffused bulk resistors (both piezo- and standard) behave differently from SOI and polysilicon resistors: at elevated operating temperatures the pn junctions leak and set an upper limit to diffused bulk resistors of about 150 °C (application dependent). Because there are no pn junctions in SOI or polysilicon resistors, they can operate at much higher temperatures, for example 300 °C.

Polysilicon can be "true" polysilicon, LPCVD deposited at about 625 °C, or amorphous silicon (a-Si), deposited about 570 °C. This difference pops up when the films are doped by ion implantation and annealed: the originally amorphous films will have lower resistivity, for example 10^{19} cm^{-3} boron will result in 1 ohm-cm for polycrystalline material but 0.1 ohm-cm for the amorphous film (and 0.01 ohm-cm for single crystal silicon). This can be explained by the relative roles of grains and grain boundaries: in the originally amorphous material a greater portion of dopants will be incorporated into grains, but in the polycrystalline material more dopant is trapped at grain boundaries early on, and those dopants do not effectively contribute to conductivity.

TCR is also dependent on the original state of the film: a-Si has a smaller TCR than poly: boron doping with 2×10^{18} cm^{-3} results in TCRs of $-1\%/°$C for amorphous material vs. $-2\%/°$C for polycrystalline material. Small grains and grain boundaries have negative TCRs, while large grains and single crystals have positive TCRs. This can be utilized to fabricate resistors with zero TCR, as shown in Figure 25.6.

Making electrical contact to high-resistivity poly is not straightforward, and the customary procedure is to

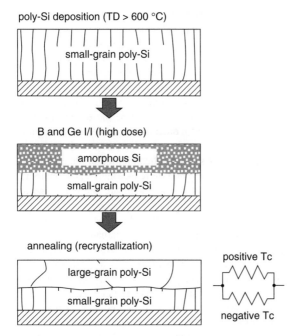

Figure 25.6 Fabrication of zero-TCR polysilicon resistor: boron and germanium ion implantation, amorphization and subsequent annealing result in large-grained poly with positive TCR, which compensates the negative TCR of small grained poly. Reproduced from Washio *et al.* (2003), copyright 2003, by permission of IEEE

make a contact enhancement implantation (which adds one lithography step to the process). This, however, sets limits on resistor length: it has to be long enough so that the heavily doped contact region does not contribute to resistance. In a short resistor the heavily doped region acts as a solid phase diffusion source, and dopes the whole resistor!

25.8 Device Example 3: PCR Reactor

The PCR (Polymerase Chain Reaction) chip is a DNA amplification device. Nucleotides and primers are added to a microfluidic chamber that is repeatedly ramped between 60 and 95 °C to produce copies of the original DNA. One design for a PCR chip is shown in Figure 25.7. As far as process design goes, the first issue is wafer selection. Glass is chosen because this is a thermal reactor, and a thermally insulating substrate will reduce heating power requirements and enable faster thermal ramps. Platinum is chosen for heater resistors because it is easy to fabricate temperature sensors from platinum, due to its linear

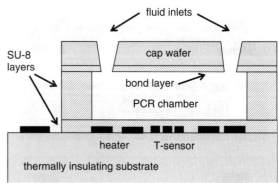

Figure 25.7 PCR reactor. Redrawn after El-Ali *et al.* (2004)

TCR. Because platinum is a noble metal, its adhesion is poor. Therefore 10 nm of titanium is sputtered before a platinum layer 200 nm thick. The resistors are covered by spin-coated SU-8 epoxy (5 µm thick) to prevent metal-to-DNA contact. Epoxy SU-8 is also chosen as the chamber material. SU-8 has excellent chemical stability (for a polymer) and low adsorption of analytes. It is also thermally stable (glass transition temperature in excess of 200 °C) and therefore suitable for PCR operating temperatures. The second SU-8 layer determines the chamber height, which can be 400 µm.

The cover lid is the biggest challenge. One approach is to use PDMS sheet. PDMS is transparent, which is good for fluorescence measurements and visual monitoring of chamber filling (and clogging), but there are a number of problems: it is permeable to water vapor and air, and loss of liquid during operation is expected (Figure 18.29). Some monomers can leak out PDMS during 95 °C operation. PDMS is mechanically soft, and the device is rather large (1 cm^2), so a roof made of PDMS is not mechanically very stable. It is, however, simple to implement and reversibly bondable, which is an advantage in the initial tests. A more stable roof can be made by spin coating a PDMS layer on a carrier wafer, and transfer bond the thin PDMS to a (non-permeable) glass wafer that has fluidic access holes etched (or drilled) in it. This bond is made permanent by oxygen plasma treatment before bonding. This two-layer structure is then bonded to SU-8 without plasma treatment, for reversibility. The thin PDMS is manually pierced to open the fluidic ports. For further development of the device a more manufacturing-friendly way will probably be implemented.

Another approach would be to use SU-8 for the roof. A glass wafer with fluidic ports is laminated by SU-8 dry resist, and this is bonded to the base chip. Fluidic ports can be exposed in SU-8 through the glass wafer, which will also cure the SU-8 and form the bond. The accuracy of lithography through the wafer is not good, but fluidic ports are hundreds of micrometers in diameter, and their lithography is non-critical. The SU-8-to-SU-8 bond is permanent. Now all the walls of the reactor are made of the same material, which is beneficial: the same contact angle will ensure uniform liquid wetting, and if there is analyte adsorption, it should be identical everywhere. SU-8 is optically not as good as PDMS, but optical detection can be done at longer wavelengths (above 550 nm).

25.9 Device Example 4: Integrated Passive Chip

An integrated passive chip (RCL chip) with two different resistors, a capacitor and an inductor coil is shown in Figure 25.8. A fused silica wafer is chosen to eliminate parasitic substrate capacitances. A high-resistivity silicon wafer could be used, but some parasitics would then have to be tolerated. In order to improve inductor behavior, it could be possible to remove the silicon by back-side through-wafer etching, but this is a considerable cost issue because through-wafer DRIE is an expensive process step. A glass wafer is not an option because the high-quality LPCVD nitride capacitor dielectric requires a deposition temperature of about 800 °C. Molybdenum is used for low-resistivity resistors ($\rho \approx 10\,\mu\text{ohm-cm}$), SiCr for high-resistivity resistors ($\rho \approx 2000\,\mu\text{ohm-cm}$), molybdenum and aluminum for capacitor electrodes and gold coils for inductors. LPCVD nitride is used for the capacitor dielectric, and three layers of CVD oxide insulate the devices from each other.

Process flow for RCL chip (cleaning steps omitted)

- Wafer selection
- Molybdenum deposition
- Lithography 1: molybdenum resistor and capacitor bottom plate
- Molybdenum etching and resist stripping
- Nitride deposition by LPCVD
- CVD oxide 1 deposition
- Sputtering of SiCr high-resistivity resistor
- Lithography 2: SiCr resistor pattern
- SiCr etching and resist stripping
- CVD oxide 2 deposition
- Lithography 3: contact holes to molybdenum
- Plasma etching of CVD oxide 2/CVD oxide 1/nitride and resist stripping
- Lithography 4: contact holes to SiCr resistor and to capacitor top
- Wet etching of CVD oxide 2/CVD oxide 1 and resist stripping
- Aluminum deposition
- Lithography 5: aluminum pattern
- Aluminum etching and resist stripping
- CVD oxide 3 deposition
- Lithography 6: contact holes to aluminum
- Etching of CVD oxide 3 and resist stripping
- Lithography 7: inductor coil pattern
- Gold electroplating and resist stripping

The first lithography step defines the molybdenum resistor and capacitor bottom plate. Because resistors

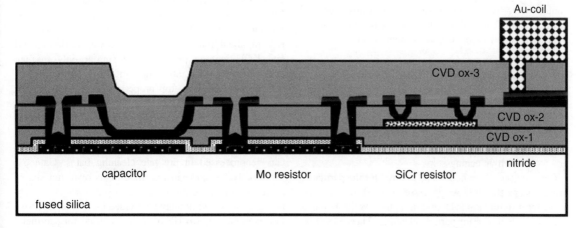

Figure 25.8 RCL chip on a fused silica substrate: four metallic layers (Mo, Al, SiCr, Au) and four insulator layers (a LPCVD nitride and three CVD oxides) are used. Adapted from VTT Microelectronics Annual Review 2000

and capacitors do not use aggressively scaled linewidths, and because the films are thin (50–200 nm), both wet etching and plasma etching can be used. The second lithography defines the SiCr resistor. SiCr is etched in chlorine plasma, but there is a danger of chromium residues (recall Table 11.6).

In the third lithography step contacts to the capacitor bottom molybdenum electrode and molybdenum resistor are made. Plasma etching is chosen because etching through both oxide and nitride is needed, and nitride wet etching is difficult. Selectivity against molybdenum is somewhat of a concern because oxide, nitride and molybdenum are attacked by fluorine plasmas. Partial thinning of molybdenum does not affect device performance, however.

In the fourth lithography step the capacitor area and contacts to the SiCr resistor are defined. Wet etching in HF is now used because high selectivity is needed: if the capacitor nitride is etched, the capacitance will change. There will be undercut because of wet etching, and the contacts to SiCr especially will be subject to extended overetching because oxide there is thinner than in the capacitor area. Underetching is acceptable because contact hole size is not very important and capacitors are large-area devices, and the minor undercut can be designed in. Of course it is possible to fabricate the capacitor and SiCr contacts separately, with dedicated photomasks and lithography and etching steps, but because the structures are non-critical, it makes sense to eliminate extra process steps.

Sputtered aluminum serves as the capacitor top electrode, and it also makes contacts to the resistors. In the RCL chip with CVD oxide ~0.5 µm thick and 3 µm minimum linewidth (1:6 aspect ratio), sputtering step coverage is reasonably good in both plasma-etched and wet-etched contact holes. Aluminum serves as a seed layer for gold electroplating. The gold plating must be done outside the main cleanroom, to avoid gold contamination.

25.10 Contamination Budget

Wafer cleaning can be viewed as an important stabilization tool: surfaces will be in a known state after wafer cleaning. Cleaning steps are the most numerous of all process steps: most other major steps are both preceded and followed by cleaning steps.

Cleaning processes need to be tailored for the particular process steps that follow: processes have different tolerances for different kinds of contamination. Wafer bonding is a major challenge for particle cleaning. Thermal oxidation is very sensitive to metal contamination because metals diffuse rapidly at elevated temperatures and degrade

oxide quality. Epitaxy requires crystal information and is extremely sensitive to native oxides or other surface layers.

The processes generate contamination themselves: ion implantation and sputtering where energetic ion bombardment is present produce metallic contamination, by sputtering metals from shield plates; deposition processes generate films, and particles form when films on reactor walls flake; and lithography chemicals (HMDS, photoresist) are major sources of organic contamination, as is plasma etching where $(CF_2)_n$ types of etch gas fragments and photoresist debris are abundant.

Contamination is partly a materials selection problem: some materials are allowed and some are forbidden. This can be either device related, or tool related: in the RCL example (Figure 25.8) a separate LPCVD nitride tube must be used for nitride-on-molybdenum deposition, and another LPCVD tube is reserved for non-metal processes. CMP is carried out in a separate cleanroom, because of the danger of excessive particle contamination.

Cleaning strategies are also process integration issues. Iron contamination increases oxide defect density and results in a lower oxide breakdown voltage. Use of p-type wafers differs from n-doped wafers because some iron is held immobile in Fe–B pairs. Contamination is strongly oxide thickness dependent, and a pre-oxidation cleaning strategy must be designed accordingly. Use of ultrapure chemicals in a 20 nm gate oxide process is a waste of money but an absolute must when the gate oxide thickness is below 10 nm.

Photoresist developers are hydroxides, and NaOH-based developers were once the mainstay, also in MOS fabrication, but organic developers like TMAH do not pose alkali contamination risks. MEMS fabrication with KOH etching tends to be strictly separated from all MOS activities. If MEMS fabrication is done in a MOS fabrication lab, TMAH etchant is used to eliminate the risk of alkali ion contamination. However, the TMAH and KOH etching processes are similar only in their gross features, and all details of rates, crystal plane selectivities and etch stop properties need to be redone, as discussed in Chapter 20.

Wet cleaning baths must also be dedicated to certain processes only. In CMOS pre-gate cleaning is very critical, and only wafers which are very clean to begin with can be processed in pre-gate cleaning baths. Gate oxide usually has an oxidation tube of its own, not shared even with other front-end oxidation processes. Wet etching baths may additionally be divided into no-resist/resist. For example, of two HF baths one is used for sacrificial oxide removal, the other for pattern etching. Photoresist stripping can similarly be divided into metallized and non-metallized wafers (non-metallized meaning that the wafer has not yet been metallized during the process).

25.11 Thermal Processes

25.11.1 Film modification

Dielectric films deposited at 300 °C by PECVD can be annealed at any desired temperature, other materials permitting. Metal films have limitations both because of the presence of metal/silicon interfaces, and because of oxidation danger. Sputtering, evaporation and electrochemical deposition are basically room temperature processes, and even mild thermal treatments, at and below 400 °C, can modify film properties dramatically. Electroless copper can have a resistivity of 4 μohm-cm as-deposited, but a 400 °C anneal in N_2/H_2 can bring it down to 2 μohm-cm. This results from grain growth and void annihilation. Grain growth is proportional to the square root of anneal time, indicative of a diffusion-limited process (cf. thermal oxidation).

CVD films and spin-coated films are often porous and unstable. PECVD films may contain up to 30 at. % hydrogen, which can be annealed out. Inert anneal at 900 °C will densify (PE)CVD oxide film, with for example a thickness reduction of 10%. This densification is seen as HF wet etch rate reduction and CMP polish rate reduction. There is room for high-temperature annealed (PE)CVD oxides because thermal oxide thicknesses are limited by the diffusion-controlled parabolic growth law, whereas (PE)CVD film thickness scales linearly with deposition time. PECVD of film 2 μm thick plus annealing can be completed in about 2 hours, whereas thermal oxidation would require 2 days. Thick oxides (>1 μm) are needed as mask oxides in MEMS and in optical devices as waveguides.

Deposited films may need stoichiometry tailoring, and, for oxide films, an oxygen anneal can result in more stoichiometric films. Sputter- and MOCVD-deposited Ta_2O_5 films are often annealed at 700 °C in oxygen. This causes crystallization and oxygen deficiency is compensated. The dielectric constant ε of amorphous Ta_2O_5 is about 25 whereas ε for crystalline Ta_2O_5 is about 35.

Annealing temperature can be used to tailor stresses: a long low-temperature anneal of a-Si (deposited at 580 °C) will result in a slightly compressively stressed film, while a high-temperature anneal will result in tensile stress (Figure 25.9).

25.11.2 Surface modification

Silicon nitride is the standard masking material for localized thermal oxidation of silicon (LOCOS). The surface

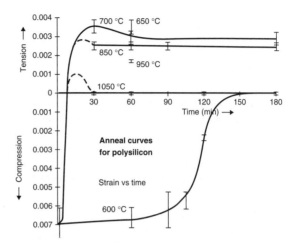

Figure 25.9 Deposited a-Si at 580 °C is compressively stressed; annealing can reduce the stress and even turn it to tensile stress. Reproduced from Guckel et al. (1988) by permission of IEEE

of the nitride will react with oxygen, even though oxygen cannot diffuse through the nitride. This modified surface layer is termed oxynitride. Its thickness is limited to a few nanometers. Somewhat similar, extremely etch-resistant material can be deposited by PECVD, using a process that mixes elements of oxide and nitride processes.

Molecular nitrogen (N_2) is unreactive and often employed in place of argon when inert gas is needed. When wafers are loaded into an oxidation furnace, nitrogen is used as a curtain gas. It is, however, possible that nitride a few atomic layers thick is formed because the temperatures are fairly high.

Intentional nitridation is usually done with ammonia. Oxide can be nitrided in NH_3. Oxynitride film has a higher dielectric constant than pure oxides and better electrical quality. Films like this are known as NO, ONO and RONO, for nitrided oxide, oxidized nitrided oxide and re-oxidized nitrided oxide. These films are standard CMOS gate dielectrics in deep submicron technologies where oxide thickness is below 10 nm.

The most commonly encountered unintentional surface modification is oxidation: some residual oxygen or moisture in a furnace atmosphere will lead to oxidation. Copper annealing in a moist atmosphere will result in copper oxide, and 5 ppm of water vapor is enough to lead to titanium oxidation, which will disturb titanium silicide formation. Oxidation is sometimes done to protect the surface: for example, aluminum oxide is chemically much more stable than aluminum, and it is preferable to oxidize the aluminum surface. Room temperature plasma oxidation (e.g., a RIE step with oxygen) will do the job.

25.11.3 Thermal budget

The thermal budget concept is central to front-end process integration. Diffusion of dopants takes place in all high-temperature steps: in addition to diffusion itself, it manifests itself during epitaxy, oxidation, densification anneal and implant damage annealing. The final doping profile is the sum of diffusion in all these steps. Effective Dt, which is a measure of diffusion distance, is calculated as

$$(Dt)_{\text{eff}} = \sum D_n t_n \qquad (25.1)$$

where the D are diffusivities under appropriate conditions and the t are the times for the high-temperature steps.

In an aluminum gate CMOS process source/drain (S/D) diffusions are done before gate oxidation, and dopants will thus diffuse further during gate oxide growth. In a self-aligned polygate process, gate oxide growth is done before S/D formation, therefore shallower junctions are possible because there are fewer high-temperature steps after doping.

A thermal budget sets limits on possible process steps. PSG and BPSG are called glasses, and glasses flow at elevated temperatures. Flow (also called reflow) was once a standard technique to make topography smoother in CMOS processes in linewidth generations above 1 μm. Of course, it was only applicable after polysilicon, not after metal deposition, because it was done at about 950–1000 °C, depending on the boron and phosphorus content. With more advanced processes the thermal budget had to be limited to make shallower junctions (reduce dopant diffusion) and glass flow became non-usable. In order to reduce thermal loads, rapid thermal processing (RTP, also called rapid thermal annealing, RTA) has been introduced: it combines very short anneal times with very high temperatures, in the extreme millisecond anneal times.

Dopant segregation must be taken into account when designing a fabrication process. Segregation of dopants between silicon and oxide can seriously deplete the interface of dopants, but this segregation is dependent on the annealing/oxidation atmosphere: wet oxidation, dry oxidation, inert anneal in nitrogen or reducing anneal in hydrogen-rich ambient can behave differently.

Ion implantation annealing has two different elements: activation of dopants and damage removal. Activation energies for these processes are different and, depending on temperature, damage removal can either be accomplished in a few seconds, or take hours. Typical temperatures are 950–1150 °C. Transient-enhanced diffusion has major implications for diffusion profiles, as will be discussed in connection with shallow junctions in Chapter 26.

25.12 Metallization

All electrical devices need at least one level of metallization in order to connect to the outside world, and so do most mechanical, thermal, fluidic and biodevices because electrical sensing and actuation are widely used.

In order to be able to probe devices, or to wire-bond them, bonding pads must be made. For both probing and bonding, soft metals are better: aluminum and gold are the standard choices. Pads made of chromium, molybdenum or tungsten may survive processing better than aluminum or gold, but they cannot be used as bonding pads.

Metal-to-semiconductor contacts come in two basic varieties (Figure 25.10): ohmic (resistive) or diode-like (Schottky). Even ohmic contacts have some diode character because the metal and semiconductor work functions are never exactly equal. If the semiconductor doping level is below 10^{19} cm^{-3}, charge carriers will have to overcome the barrier (which is proportional to the difference of metal work function and semiconductor electron affinity, $\varphi_{\text{metal}} - \chi_{\text{semiconductor}}$) by thermionic emission. In heavily doped semiconductors the situation is different: charge carriers can tunnel through the barrier because it is thin. Barrier thickness is related to depletion width in the semiconductor (which is proportional to $1/N_D$).

The silicon doping level needs to be in excess of 10^{19} cm^{-3} for good ohmic contact. This often leads to one extra doping step to increase contact hole doping.

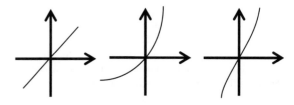

Figure 25.10 Metal–semiconductor contact I–V curves: left, ohmic; middle, diode-like (Schottky); bottom, real metal–semiconductor contact

Aluminum is the most widely used ohmic contact metal to silicon. Aluminum, which is a p-type dopant for silicon, can also be used to make ohmic contact to lightly doped p-type silicon: during the contact anneal (in N_2/H_2, the forming gas, at 450 °C), aluminum will dope the top surface of silicon, and good contact is made. Schottky contacts to silicon are usually made with PtSi.

25.13 Passivation and Packaging

Final passivation provides protection against the environment. There are mechanical elements to passivation, like scratch resistance, chemical aspects like moisture resistance and gettering, and physical effects, like the prevention of sodium diffusion. Electronic chips can be fairly easily passivated, but MEMS chips with movable parts are much more demanding to protect. A hermetic package for a MEMS resonator costs easily $5 or more, while a non-hermetic package costs perhaps 50 cents. So there is a big incentive to understand exactly the vacuum requirements for each device.

The standard passivation materials are PSG and PECVD nitride, either alone or as a two-layer stack (Figure 5.19). Phosphorus doping of CVD oxide film is beneficial for sodium ion gettering, but too much phosphorus makes the oxide hygroscopic, so there is a delicate balance. Usually the phosphorus content is about 5% wt. Nitride provides mechanical strength and chemical resistance, but this chemical stability translates to plasma etching for bonding pad opening, whereas oxide passivation can be etched in HF-based solutions (not, however, without difficulty, because HF–water solutions attack aluminum; acetic acid + NHF_4 (or ethylene glycol + NHF_4) are known as pad etches or PSG etch). Packaging introduces more materials into the system. This is problematic because more materials compatibility issues must be addressed. Each new materials interface is a potential reliability problem, with diffusion, reaction, thermal expansion mismatch and other issues that need to be addressed.

25.14 Exercises

1. How many lithography steps are needed to fabricate the solar cell shown in Figure 1.14? How many critical alignments are there?
2. Design a fabrication process for the JFET shown below.

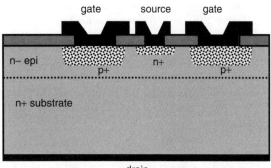

3. Redo Exercise 9.4, this time for 5× step-and-repeat lithography and quartz masks.
4. List step by step the fabrication process for the platinum silicide Schottky diode shown below. Platinum silicide is formed by metal/silicon reaction.

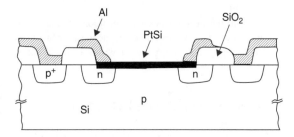

Source: Chen, C.K. *et al.* (1991) by permission of IEEE

5. Draw the photomask set required to fabricate the RCL chip of Figure 25.8. Include design features such as overlaps.
6. In an integrated passive chip (Figure 25.8):
   (a) What is the nitride thickness if the areal capacitance density is 4 nF/mm^2 and nitride $\varepsilon = 7$?
   (b) Why is the first contact etching by plasma and the second by wet etching?
   (c) SiCr thin-film resistor resistivity is 2000 µohm-cm. Design a 5 kohm resistor!
7. Capacitor nitride deposition uniformity across the wafer is ±1%, and across the batch ±2%. The top electrode area is defined by etching CVD oxide (thickness and etch non-uniformity ±5%) against the capacitor nitride. If the oxide thickness is 200 nm and nitride thickness 10 nm, plot capacitance variation as a function of oxide:nitride etch selectivity.
8. Design a process where the capacitor top metal is deposited directly after the capacitor dielectric.

9. Can you reduce the pass size of an etched sieve (Figure 25.3a) if KOH etching is used?
10. Design a process for a sacrificial sieve (Figure 25.3c) and include the mask layout figures showing which parts will stop the polysilicon from falling apart.
11. What is the aperture ratio of hole, pillar and sacrificial sieves if all use the same lithography process?
12. Explain the fabrication steps for the capacitor shown below.

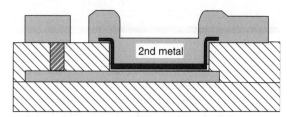

Reproduced from Washio *et al.* (2003) by permission of IEEE

13. Explain the design rules for the bipolar transistor of Exercise 14.7. Assume that the deepest diffusion is about 5 μm and minimum linewidth 1 μm. What is the transistor area?

References and Related Reading

Andersson, H., W. van det Wijngaart and G. Stemme (2001) Micromachined filter-chamber array with passive valves for biochemical analysis, *Electrophoresis*, **22**, 249–257.

Barlian, A. A. *et al.* (2007) Design and characterization of microfabricated piezoresistive floating element-based shear stress sensors, *Sens. Actuators*, **A134**, 77–87.

Brody, J. P. *et al.* (1996) A planar microfabricated fluid filter, *Sens. Actuators*, **A54**, 704–708.

Chen, C. K. *et al.* (1991) Ultraviolet, visible, and infrared response of PtSi Schottky-barrier detectors operated in the front-illuminated mode, *IEEE Trans. Electron Devices*, **38**, 1094.

Chu, W.-H. *et al.* (1999) Silicon membrane nanofilters from sacrificial oxide removal, *J. Microelectromech. Syst.*, **8**, 34.

El-Ali, J. *et al.* (2004) Simulation and experimental validation of a SU-8 based PCR thermocycler chip with integrated heaters and temperature sensor, *Sens. Actuators*, **A110**, 3–10.

Guckel, H. *et al.* (1988) Fine-grained polysilicon films with build-in tensile strain, *IEEE Trans. Electron Devices*, **35**, 800.

Honer, K. A. and G. T. A. Kovacs (2001) Integration of sputtered silicon microstructures with pre-fabricated CMOS circuitry, *Sens. Actuators*, **A91**, 386–397.

Leslie, T. *et al.* (1994) Photolithography overview of 64 Mbit production, *Microelectron. Eng.*, **25**, 67.

Lin, H. *et al.* (2008) Test structures for the characterization of MEMS and CMOS integration technology, *IEEE Trans. Semicond. Manuf.*, **21**, 140–147.

Neuhaus, D.-H. and A. Münzer (2007) Industrial silicon wafer solar cells, *Adv. OptoElectron.*, 24521.

Ono, M. *et al.* (2002) A complementary-metal-oxide-semiconductor-field-effect transistor-compatible atomic force microscopy tip fabrication process and integrated atomic force microscopy cantilevers fabricated with this process, *Ultramicroscopy*, **91**, 9–20.

Washio, K. (2003) SiGe HBT and BiCMOS technologies for optical transmission and wireless communication systems, *IEEE Trans. Electron Devices*, **5**, 656.

26

MOS Transistor Fabrication

CMOS is and remains the most voluminous microfabricated device by a wide margin. Many of the process steps of microfabrication were developed originally for CMOS fabrication and later adapted to other microdevices. Linewidth scaling has been driven almost exclusively by CMOS. Technology generations have followed at the pace of roughly 30% linewidth reduction every three years (generations are described by the minimum linewidths or half-pitches). The path from 0.8 μm CMOS in the late 1980s to 45 nm CMOS 20 years later equals eight generations, involving a lot of ingenuity in device structures, materials and fabrication methods. This chapter discusses some generic issues of MOS transistor fabrication, but not the latest generation of advanced devices. These will be covered in Chapter 38.

In the 1950s there was an argument between whether single crystal silicon (SCS) or polycrystalline silicon should be used for transistors. Polycrystalline silicon was more readily available, and single crystal growth difficult. However, the rapid advances in single crystal growth led to its adoption for transistors. Pixel driver transistors in flat-panel displays (i.e., LCDs) are MOS transistors made of amorphous or polycrystalline silicon, known as thin-film transistors (TFTs). TFTs made of polycrystalline silicon are inferior compared to SCS transistors, but adequate for display applications. More recently, polymers have emerged as materials for MOSFETs. Again, performance is not even close to that of SCS transistors, but the applications are in intelligent price tags or in disposable diagnostic devices.

26.1 Polysilicon Gate CMOS

Early MOS processes used aluminum for gates, thermal diffusions for source and drain, and a single level of aluminum for metallization. Then came the polysilicon

Table 26.1 Aluminum gate vs. polygate

	Al gate	Polygate
Linewidths	>5 μm	<10 μm
Doping	Thermal diffusion	Ion implantation
Isolation	pn junction	Oxide (LOCOS)
Gate material	Aluminum	Doped polysilicon
Gate process	Non-self-aligned	Self-aligned
Etching	Wet/isotropic	Plasma/anisotropic

gate CMOS, which exhibited most of the essential process steps that have characterized CMOS from the mid 1970s till today: CMOS is an oxide-isolated, ion-implanted, plasma-etched, self-aligned polysilicon gate process (Table 26.1). CMOS linewidths were in the 5 μm range in the mid 1970s but the basic design was very successful and it has gone through many generations up to the 65 nm processes in the 2000s, and only very recently have new paradigms emerged, with deposited oxides replacing thermal oxide, and metal gate (TiN, Ta, W) replacing polygate.

CMOS transistors are made infront-end: by oxidation, diffusion and ion implantation. Back-end processes create wiring to interconnect the transistors to each other. Over the years the number of metal levels has increased to 10. Back-end is about metal and dielectric deposition, etching and CMP. In today's chips back-end contributes to more process steps than front-end. Chapter 28 is devoted to multilevel metallization.

Contact hole fabrication defines the division between front-end and back-end: after the metal/silicon interface has been formed, process temperatures are limited to about 450 °C.

The CMOS process can be further divided into modules: wells, isolation, gate, contact, metallization

330 Introduction to Microfabrication

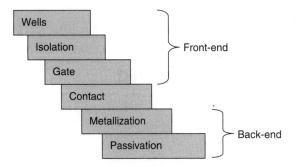

Figure 26.1 Main modules of a CMOS process

and passivation. These are further divided into process sequences, where the detailed recipes for ion implantation doses and energies are specified, or oxidation times and atmospheres given. The main modules of CMOS fabrication are pictured in Figure 26.1.

26.2 Polysilicon Gate CMOS: 10 μm to 1 μm Generations

26.2.1 Wafer selection

Process integration begins with wafer selection: n-type silicon, of 4 ohm-cm (phosphorus concentration about 1.5×10^{15} cm^{-3}), is chosen as the starting material. This means that NMOS transistors will be made in p-well, and PMOS transistors directly in the substrate. The choice of p-type starting material would lead to a reversed configuration. Simple polygate CMOS is shown in Figure 26.2. Note that this and the following figures are highly schematic, so, for instance, the lateral spreading of implants during oxidation or etch profiles are not drawn realistically.

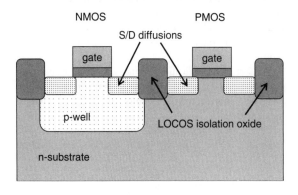

Figure 26.2 Prototypical CMOS

26.2.2 Isolation

LOCOS isolation is used. Wafers are cleaned, a thermal oxide (pad oxide) of 40 nm is grown in dry oxygen, followed by LPCVD nitride deposition (100 nm). The first lithography step defines the transistor active areas. Nitride will cover these transistor active areas, and it will be etched away from areas that will become isolation oxide. Nitride etching in CF$_4$ plasma stops on pad oxide. By stopping the etch at the oxide, the silicon surface is not damaged and cleaning of the wafer will be easy.

Figure 26.3 shows a top view of the photomask, together with a cross-sectional view of the CMOS device (showing also parts that will be made later on).

After photoresist strip, arsenic is implanted with an energy of 50 keV and dose 10^{12} cm^{-2}. No lithography

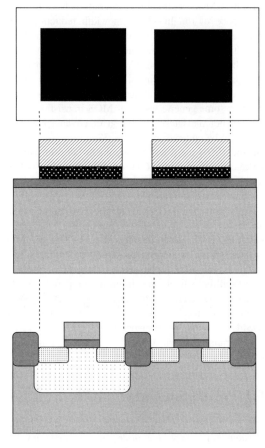

Figure 26.3 Mask 1: nitride plasma etching, stopping on pad oxide, with resist still in place. Mask 1 is also called the active area mask because the transistors will eventually be made in the area protected by nitride, as seen in the bottom figure

is required (Figure 26.4 top). The implant penetrates the thin pad oxide but not the thicker nitride. Later on, when LOCOS oxidation will be done, the arsenic implant will be buried under the LOCOS oxide. This field oxide implant improves isolation between neighboring transistors.

The second lithography step defines p-well areas (Figure 26.4). Boron is implanted with a dose of $2 \times 10^{13}\,\text{cm}^{-2}$ and energy 40 keV. Drive-in diffusion is performed next: the implanted boron will be driven deeper by a two-step process; a short oxidation step (50 min at 950 °C, dry oxidation) is followed by a 500 min, 1150 °C diffusion in nitrogen. The oxidation will result in oxide about 30 nm thick. Note that diffusion spreads laterally: the final size of the diffused area is significantly bigger than the area defined by mask 2.

LOCOS oxidation then follows: 6 h at 1050 °C, wet oxidation. This will result in oxide about 1.2 μm thick. The p-well is diffused to a depth of about 4 μm. After oxidation, the nitride/oxide stack is removed: nitride in CF_4 plasma and oxide in HF.

26.2.3 Gate module

The third lithography step (Figure 26.5) is used to tailor the threshold voltage of PMOS transistors. The threshold voltage (V_{th}) where the transistor turns on is proportional to the square root of doping concentration in the channel, and this can be modified by implantation. This is true for NMOS and PMOS alike, but in a simple CMOS process NMOS V_{th} control is not done. A dose of $1 \times 10^{12}\,\text{cm}^{-2}$ of boron is implanted with an energy of 50 keV.

Gate pre-cleaning then commences. An RCA-1, HF, RCA-2 cleaning sequence is used (but other combinations are used, too). Dry oxidation at 1050 °C for 65 min produces gate oxide about 80 nm thick. The threshold-implanted boron diffuses further during gate oxidation.

Polysilicon, about 500 nm thick, is deposited undoped. A separate $POCl_3$ gas phase doping step is performed after deposition, and the resulting poly sheet resistance is about 30 ohm/sq. Both NMOS and PMOS gates are made of the same material, namely phosphorus-doped poly.

The fourth photomask (Figure 26.6) defines the polysilicon gates. Gate poly etching is done either with fluorine

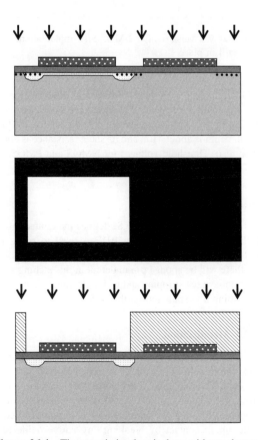

Figure 26.4 The arsenic implant is done without photoresist, with energy so low that oxide is penetrated but not the oxide/nitride stack (top figure). Mask 2: boron ion implantation for the p-well. The implant energy is high enough to penetrate the oxide/nitride stack.

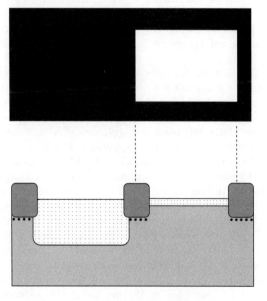

Figure 26.5 Mask 3: PMOS threshold voltage implant. Arsenic field stop implant is seen under field oxide

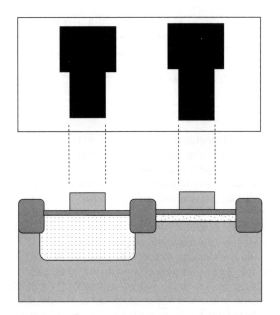

Figure 26.6 Mask 4: after polysilicon etching and resist stripping

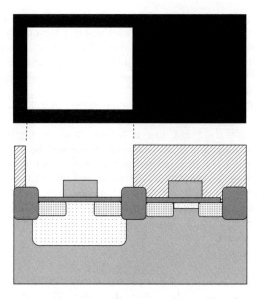

Figure 26.7 Mask 6: after NMOS S/D implantation, with resist still in place

(SF_6 or CF_4) or with chlorine-based plasmas. The selectivity requirement is not very demanding because gate oxide is fairly thick, so the process can be optimized for sidewall profile, rate and/or uniformity. After photoresist stripping and cleaning, a mild oxidation step (900 °C, 10 min, dry oxidation) is performed, and about 50 nm of oxide grown on polysilicon. This removes plasma etch damage and regrows gate oxide on the source/drain areas a little.

The fifth mask defines PMOS S/D implants. In fact this fifth photomask is actually the same mask as mask 3, the PMOS threshold voltage mask: it defines the PMOS transistor area. This time it protects the NMOS areas from PMOS S/D boron ion implantation. A high dose, 2×10^{15} cm^{-2}, of boron is implanted at 40 keV.

The sixth mask (Figure 26.7) is for NMOS high-dose S/D implants. Phosphorus is implanted at 50 keV energy and 5×10^{15} cm^{-2} dose. This is a non-critical mask: it is rather large in size and LOCOS oxide all around active areas means that minor misalignment does not matter.

Insulator deposition follows. Phosphorus-doped CVD oxide (PSG) of about 1 μm thickness is deposited. PSG is a glassy material and above its glass transition temperature (about 1050 °C) it will flow, resulting in beneficial smoothing of the top surface. This is the last high-temperature step, and dopant profiles are now "frozen."

Junction depths of both PMOS and NMOS transistors are about 1 μm, with S/D sheet resistances of about 30 ohm/sq for NMOS and about 90 ohm/sq for PMOS; p-well depth is about 4 μm and its sheet resistance is about 4 kohm/sq. Threshold voltages for NMOS and PMOS are about 1.3 V and −1.5 V, respectively.

26.2.4 Contact

The seventh mask (Figure 26.8) defines the contact holes in the oxide. Wet etching in BHF is used to open the contacts. Contact hole design rules must take into account that there will be about 1 μm undercut in this etching step. After photoresist stripping and wafer cleaning, about 1 μm of aluminum is sputtered on the wafers.

26.2.5 Metallization

The eighth mask (Figure 26.9) defines aluminum patterns. Aluminum (1 μm thick) is etched in H_3PO_4-based wet etch. Overlap rules must make sure that the metal covers the contact completely. After stripping and wafer cleaning, a forming gas anneal at 450 °C improves silicon-to-aluminum contact by breaking any native oxide that might be at the interface. The hydrogen atoms in the forming gas will also bond to dangling bonds, stabilizing CVD oxide.

A passivation layer of silicon oxynitride is deposited by PECVD. Mask 9 defines bonding pad openings, and

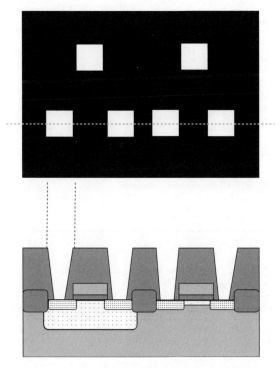

Figure 26.8 Mask 7: contact holes after resist stripping. The cross-sectional drawing is across the S/D contact holes, therefore the polygate contacts do not show

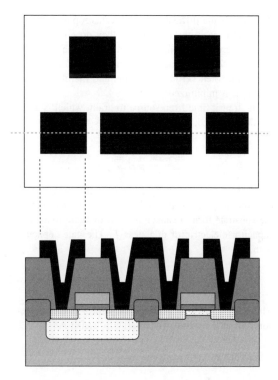

Figure 26.9 Mask 8: after aluminum etching and resist stripping

plasma etching of oxynitride opens those pads. Wafer-level processing is now complete.

The wafers will be tested electrically, at wafer level, and non-functional chips will be inked. Dicing will separate the chips, and functional chips will proceed to encapsulation and packaging. Many tests cannot be performed at the wafer level and more characterization will take place on packaged chips. The cost of testing can be very high if the chips need to be tested for a multitude of parameters.

26.2.6 CMOS process variations

A prototypical 5 μm CMOS process was described above. There are many variations between different CMOS manufacturers: implant doses and diffusion times differ, oxide thicknesses and junction depths vary, mask compensations can be used, etc. More variety enters the picture if, for example, analog CMOS is made. Then some of the doping steps will be used to make resistors, and extra lithography masks may be needed. In more advanced analog CMOS processes an extra polysilicon layer is added for resistor and capacitor fabrication. EEPROM processes also need extra polysilicon, for the floating gate. A couple of masks have been added in each CMOS generation, with some 35 masks in modern devices. Bipolar transistors can be added to the CMOS process, to create BiCMOS, a topic that will be discussed in Chapter 27 on bipolar technologies. This needs one to four extra masks.

26.3 MOS Transistor Scaling

As linewidths were scaled from 5 μm to about 1 μm, plasma etching replaced wet etching in all patterning etch steps, poly, oxide and metal alike. Oxidation and diffusion times were scaled down in order to make shallower junctions. We will now discuss some issues relevant to the scaling of CMOS, from both the device and fabrication points of view.

26.3.1 Transistor scaling

CMOS transistor scaling is most often discussed from a lithographic, linewidth scaling point of view. But vertical scaling is equally important. S/D diffusions must be made shallower because they must not extend sideways under the gate. If the diffusions touch, catastrophic failure occurs, but even in the case where they do not touch, they degrade device performance via increased leakage current and parasitic capacitances. Sideways diffusion is kept to a minimum when vertical diffusion, and therefore junction depth x_j, is minimized.

The transit time from source to drain, which is a proxy for device speed, can be calculated as

$$\tau = \frac{L}{v} = \frac{L}{\mu E} = \frac{L^2}{\mu V_{ds}} \quad (26.1)$$

where L is channel length, v the velocity and μ the electron mobility in electric field $E = V_{ds}/L$. The gate and the substrate form a capacitor, with the gate oxide as the capacitor dielectric of thickness T. The gate capacitance is then

$$C = \frac{\varepsilon W L}{T} \quad (26.2)$$

where W is the width of the gate and ε is the dielectric constant of the oxide. The charge in transit is

$$Q = -C_g(V_{gs} - V_{th}) = -\frac{\varepsilon W L}{T}(V_{gs} - V_{th}) \quad (26.3)$$

and the current is

$$I_{ds} = \frac{Q}{\tau} = \frac{\mu \varepsilon W}{LT}(V_{gs} - V_{th})V_{ds} \quad (26.4)$$

where V_{gs} is the gate/source voltage, V_{th} is the threshold voltage where the gate starts to control the charge carriers, and V_{ds} is the drain/source voltage.

Scaling down transistor lateral dimensions L and W, and the vertical dimension, oxide thickness T, by a factor n ($n > 1$), leads to the following new dimensions:

$$L' = L/n \quad W' = W/n \quad T' = T/n \quad (26.5)$$

For many CMOS generations the operating voltage was kept constant at 5 V, but the electric field cannot be increased without limit because of dielectric breakdown and hot electron considerations, which necessitate lower operating voltage, V', given by $V' = V/n$. Using the shorthand $V \equiv V_{gs} - V_{th}$, we can write the physical parameters for the scaled devices (Table 26.2).

Scaling is mostly beneficial: the transistor area is scaled as $1/n^2$, transistor speed increases as $1/n$, switching power decreases as $1/n^2$, and switching energy decreases as $1/n^3$. Power density (P'/A') remains constant. Junction depth scaling x_j has been mostly in line with oxide thickness scaling, but more recently it has been difficult to maintain the pace. This is because ion implantation damage

Table 26.2 MOS scaling by a constant factor n (>1)

$$\tau' = \frac{1}{\mu}\frac{L^2}{n^2}\left(\frac{V}{n}\right)^{-1} = \frac{1}{\mu}\frac{L^2}{V}\frac{\tau}{n} = \frac{\tau}{n}$$

$$V' = \frac{V}{n} \quad C' = \frac{C}{n} \quad I' = \frac{I}{n}$$

$$P'_{switch} = \frac{C'V'^2}{2\tau'} = \frac{P_{switch}}{n^2}$$

$$E'_{switch} = \frac{C'V'^2}{2} = \frac{E_{switch}}{n^3}$$

$$P'_{dc} = I'V' = \frac{P_{dc}}{n^2}$$

$$A' = L'W' = \frac{LW}{n^2} = \frac{A}{n^2}$$

Table 26.3 Front-end scaling (about 1980–1995), supply voltage constant at 5 V

Generation	3 µm	2 µm	1.5 µm	1 µm	0.7 µm	0.5 µm
T_{ox} (nm)	70	40	30	25	20	14
x_j (nm)	600	400	300	250	200	150
Gate delay (ps)	800	350	250	200	160	90

Table 26.4 Front-end scaling (about 1995–2005), supply voltage scaling

Generation	0.35 µm	0.25 µm	0.18 µm	0.13 µm
T_{ox} (nm)	8	6	4.5	4
Supply (V)	3.3	2.5	1.8	1.5
V_{th} (V)	0.65	0.6	0.5	0.45

necessitates high-temperature annealing, which inevitably leads to diffusion, however shallow the original implantation profile. Table 26.3 lists real-world CMOS scaling data from the 1980s and 1990s, and Table 26.4 gives more recent trends.

Note that gate oxide thickness is roughly linewidth divided by about 50 and junction depth is approximately linewidth divided by 5. Linewidth scaling is just one factor in packing density increase: process and device cleverness can contribute amazingly large area reductions.

26.3.2 Front-end simulation

CMOS front end simulation involves diffusion profile and oxide thickness calculations, which are fed into device simulators to obtain transistor properties like

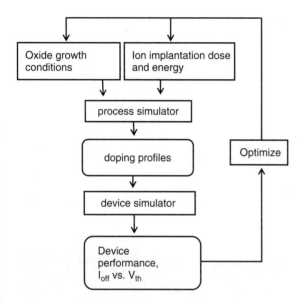

Figure 26.10 Front-end process development loop depends heavily on process simulation

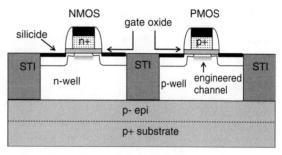

Figure 26.11 Deep submicron CMOS: 200 nm gate length, 5 nm gate oxide, 70 nm junction depth; n^+ poly for NMOS and p^+ poly for PMOS. Shallow trench isolation on epitaxial n^+/p^+ wafer

threshold voltages and current-voltage characteristics. 1D, 2D and 3D simulations are used, depending on application. If a 1D process simulator is used, it feeds 1D device simulation, and similarly 2D for 2D and 3D for 3D. This process development loop is pictured in Figure 26.10.

26.4 CMOS from 0.8 μm to 65 nm

The 5 μm CMOS process presented above has the main features of any modern CMOS process. Even though the basic features of a 65 nm CMOS process certainly differ from the nine-mask 5 μm process, the basic idea has remained unchanged. We will not discuss changes generation by generation, but rather look at some important trends in processes and structures themselves.

Several new design ideas, structures and materials have been invented to keep polygate CMOS scaling going to submicron dimensions. A schematic 0.25 μm CMOS is shown in Figure 26.11. It includes for instance the following new features which will be discussed shortly:

- Lithography: i-line step-and-repeat 5× reduction lithography with $\lambda = 365$ nm
- Device structures: spacers and lightly doped drain (LDD) implants
- New materials: silicides
- New processes: CVD-W plugs with etchback.

Deep submicron (0.35, 0.25, 0.18 and 0.13 μm) generations have taken advantage of many more new techniques and materials:

- DUV lithography with $\lambda = 248$ nm
- SiO_2 gate oxide replaced by nitrided oxides
- The p^+ gate for PMOS and the n^+ gate for NMOS
- Tilted and halo implants for S/D engineering
- RTA junction annealing.

26.4.1 Wafer selection

CMOS process integration begins, like all other processes, with wafer selection (Table 26.5). Note that the tightening of wafer specifications goes hand in hand with wafer size via the linewidth: 300 mm wafer specs are tighter because 90 nm linewidth devices with gate oxides 2 nm thick are made on 300 mm wafers, whereas 0.5 to 0.8 μm linewidths and 10 nm oxides are typical of 150 mm wafers.

26.4.2 Wells and isolation

Wells are the deepest diffusions in CMOS, and they must be fabricated early on in the process. There are several ways of making the wells, depending on starting wafer choice and device design requirements: n-well, p-well and twin-well processes are all possible.

The twin-well process requires two lithography steps, but both NMOS and PMOS doping levels can be optimized independently.

LOCOS isolation served CMOS fabrication for 30 years, and it has been scaled to much smaller linewidths than was previously thought possible. Below half-micron technologies, LOCOS was finally replaced: first, bird's peak lateral extent wastes area (Figure 13.9); second,

Table 26.5 Wafer specifications for CMOS

Specs	100 mm	125 mm	150 mm	200 mm	300 mm
Thickness (μm)	525 ± 20	625 ± 20	675 ± 20	725 ± 20	775 ± 25
TTV (μm)	3	3	2	1.5	1
Warp (μm)	20–30	18–35	20–30	10–30	10–20
Flatness (μm)	<3	<2	<1	0.5–1	0.5–0.8
Oxygen (ppma)	20	17	15	14	12
OISF (cm^{-2})	100–200	100	<10	None	None
Particles (per wafer)	10@0.3 μm	10@0.3 μm	5–10@0.3 μm 100@0.2 μm	10–100@0.16 μm 20–30@0.2 μm	50–100@0.12 μm 10–20@0.16 μm 5–10@0.20 μm
Metals (atoms/cm^2)	10^{12}	10^{11}	10^{11}	5×10^{10}	10^9

field oxide growth in narrow spaces is suppressed by compressive stresses, that is oxide does not grow to full thickness in narrow spaces. The main isolation method in the deep submicron technologies is shallow trench isolation (STI) (deep trench isolation of bipolar transistors will be described in Chapter 28). A schematic STI process is described below and shown in Figure 26.12.

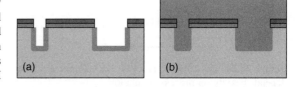

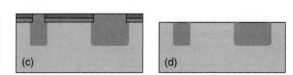

Figure 26.12 Shallow trench isolation (STI): trench filling and polishing. See text for details

Process flow for shallow trench isolation (STI)

- Pad oxide (thermal)
- Pad nitride (LPCVD)
- Lithography
- Etching nitride/oxide/silicon
- Resist strip and cleaning
- Liner oxide (thermal) (Figure 26.12a)
- CVD oxide deposition (Figure 26.12b)
- CMP planarization of the oxide (Figure 26.12c)
- Nitride etching
- Oxide etching (Figure 26.12d)

Dimensions for STI are scaled down with each technology generation, but for 0.25 μm CMOS the following values are representative:

- pad oxide thickness 40 nm
- nitride thickness 100 nm
- narrow trench width 250 nm
- trench depth 300 nm
- liner oxide thickness 30 nm
- CVD oxide thickness 500 nm.

There are tens of variations of STI, but all of them have to fulfill certain common criteria. The liner oxide is a thermal oxide. Its role is to minimize silicon/oxide interface defects (as discussed in Section 13.4 on oxide structure). The CVD oxide process must be able to fill conformally both narrow and wide spaces and it must be thicker than the trench (see Figure 16.7 for step coverage, gap filling and the requirements for planarization). CMP planarization has also to be able to polish narrow and large areas at the same rate. If the large-area polish rate is higher, planarization will only work for the narrow gaps. Instead of CMP, various etchback processes have also been tried, but they have the pattern size and pattern density effects similar to or worse than CMP, and the results are therefore no better. In addition to CVD, spin coating of dielectrics can also be used to fill trenches.

26.5 Gate Module

The gate module is critical for transistor action. Gate oxide thickness, channel doping, gate length and S/D doping profiles determine critical transistor parameters such as threshold voltage, switching speed, leakage current and noise.

26.5.1 Gate oxide

Making thin gate oxides is a major wafer cleaning challenge: 100 nm particles are permissible in 0.35 μm technology from a linewidth point of view, but compared to oxide thicknesses of less than 10 nm, they are not allowed. Atomic contamination also becomes more crucial as film thicknesses are scaled down. Metals and organics can be removed from the wafers by cleaning, but for very thin oxides impurities in the gas phase also matter: residual water vapor at a concentration level of 20 ppm in the oxidation tube will dramatically enhance the dry oxidation rate. Surface roughness also affects oxide electrical quality and channel mobility because in MOS transistors the current is confined to about the top 10 nm of the silicon layer parallel to the surface. The effects of various metals on CMOS device properties are described in Table 26.6.

Segregation of contaminants between Si and SiO_2 has a major impact on the effects of metallic contamination: during thermal oxidation Al, Ca, Cr, Mg and Zn are incorporated into the oxide and contribute to oxide quality problems, whereas Fe, Cu and Ni diffuse in bulk silicon and contribute to decreased minority carrier lifetime.

Deep-level impurities act as majority carrier traps. The recombination velocity is maximum when deep-level energy is in the middle of the forbidden gap, therefore Zn, Cu, Au and Fe are especially harmful impurities.

A number of methods and materials have been investigated as replacements for thermal oxide. Nitrided oxide (NO) and oxidation of nitrided oxide (ONO) are evolutionary developments based on thermal oxidation. New alternatives are CVD and ALD films, a major paradigm shift. Hafnium oxide (HfO_2) and hafnium silicates ($HfSi_xO_y$) are prime candidates. If, during deposition of the high-ε material, silicon dioxide is formed at the interface, the system that is formed is a SiO_2/high-ε two-layer structure, which must be analyzed as capacitors in series. Interfacial silicon dioxide formation is difficult to avoid because high-ε dielectrics are oxides, and oxygen is present in some form or another during their deposition.

Equivalent oxide thickness (EOT) is often used in describing high-ε materials which replace silicon dioxide. The equivalent oxide thickness is given by

$$\text{EOT} = \frac{\varepsilon_{SiO_2}}{\varepsilon_{\text{high } k}} \times t_{\text{high } k} + t_{SiO_2} \quad (26.6)$$

where t_{SiO_2} is the interfacial silicon dioxide thickness, if any.

Zirconium oxide (ZrO_2, $\varepsilon \approx 23$) film 6 nm thick has EOT $\approx$1 nm, under the assumption of zero interfacial SiO_2. Even a 1 nm SiO_2 layer will cause a drastic effect on EOT. Furthermore, dielectric constants of very thin films are different from bulk values, or from values measured for thicker films (recall Figure 5.2). It is also possible that very thin films are amorphous, but slightly thicker films are polycrystalline, as shown in Figure 26.13 for ZrO_2. The interfacial oxide is also visible in these TEM micrographs. Note that we have used classical capacitance formulas above: in the 3 nm thickness range a quantum mechanical description should be used for accurate results.

26.5.2 Self-aligned gate

Gate linewidth scaling is a combined lithography and etching problem: namely, feature size in resist vs. etched feature size. Etching is also related to gate oxide thickness: polygate etching has to stop on the thin gate oxide. The length of a gate-level conductor is only a few microns, or tens of microns, and low resistivity is not a major requirement. Instead, ease of patterning and thermal stability in contact with oxide are primary concerns. The gate pattern is, together with contact holes, the most demanding lithographic and etching challenge of modern ICs.

The self-aligned polygate was a major milestone in MOS evolution: S/D diffusions were automatically aligned to the gate. But as transistor scaling continued, more complex doping patterns were called for. One motivation was to reduce hot-electron effects: high electric fields in the channel accelerate electrons to high energies, and these electrons can degrade the gate oxide. In order to reduce these high electric fields, the lightly doped drain (LDD) structure was introduced. In LDD, S/D implantation is done in two steps.

Table 26.6 Metal contamination effects in MOS devices. Adapted from Hattori (1998)

Metallic species	Contamination effects in MOS
Heavy metals (Cu, Fe, Ni)	Junction leakage current increase
	Lifetime degradation
	Oxide dielectric strength failure
Alkali metals (Na, K, Ca, etc.)	Threshold voltage shift
Transition metals (Al)	Interface state increase
Noble metals (Au)	Lifetime degradation

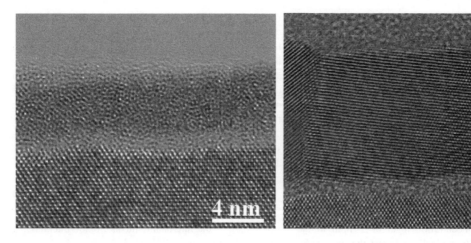

Figure 26.13 ALD ZrO$_2$: the 4 nm thick film is amorphous but the 12 nm thick film is polycrystalline. Reproduced from Kukli *et al.* (2007), copyright 2007, Elsevier

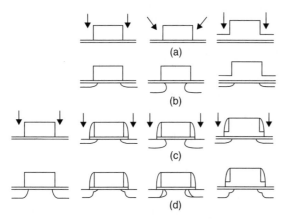

Figure 26.14 Gate-implant possibilities: (a) standard; (b) lightly doped drain LDD; (c) large-angle tilt device (LATID) and (d) inverse-T gate. Reproduced from Stinson, M. & Osburn, C.M. (1991), by permission of IEEE

After polygate etching, a self-aligned, low-energy, low-dose implant is done, followed by CVD oxide deposition and spacer etching (Figure 26.14). This spacer shifts the second high-dose S/D implant further away from the gate edge, where the highest electric field occurs. This minimizes hot-electron damage to thin gate oxide.

Process flow for LDD structure

- Polysilicon gate etching
- Implantation for S/D extension (20 keV, 10^{13} cm^{-2})
- CVD oxide conformal deposition (250 nm)
- Anisotropic oxide plasma etch for spacer formation
- Etch damage removal/cleaning
- Implantation for source/drain (50 keV, 10^{15} cm^{-2})

The spacer etching end point is difficult to see because the most abundant material under CVD oxide is the field oxide, and no selectivity is possible between two oxides. Some field oxide loss is therefore inevitable, and the spacer etch may etch some silicon in S/D areas.

In addition to junction depth, junction profile must be tailored more carefully in deep submicron CMOS. Large-angle tilted (halo) implants are extended beneath the gate. Various double implant scenarios are depicted in Figure 26.14.

26.5.3 Junction depth

Shallow junction formation is the interplay between implantation and annealing (recall Figure 15.7). Junction quality means controllable and reproducible junction depth, low leakage current and good (ideal) forward characteristics. Low sheet resistance requirements necessitate a high degree of electrical activation of dopants. Low leakage current requirements equal efficient damage removal and a low level of contamination. Solid solubility sets limits on activation and plays a role in damage dissolution. Clearly the demands are at odds with a typical damage annealing approach.

Point defects are critical for diffusion: vacancies created by the implantation process add to thermally generated vacancies and enhance diffusion. Boron diffusion is dependent on silicon self-interstitials which are created, for instance, during thermal oxidation. Boron diffusion under an oxidizing atmosphere is thus faster than in an inert atmosphere.

Points defects created during implantation offer fast diffusion paths. This is known as transient-enhanced diffusion (TED). If defects can be annealed away rapidly, TED is eliminated and thermal diffusion determines the doping profiles. Rapid thermal annealing (RTA) is a solution to this problem. A very short high-temperature step (seconds, or even subsecond, at 1000 – 1100 °C) eliminates TED and activates dopants, without too much diffusion (Figure 2.12). RTA will be further discussed in Chapter 32.

26.5.4 Self-aligned silicide

Titanium, cobalt and nickel silicides ($TiSi_2$, $CoSi_2$, $NiSi$) are used in advanced CMOS (Figure 26.11) to reduce resistance. Metal is deposited all over but metal-silicon reaction only takes place on the polygates and S/D areas; no reaction takes place on oxides (recall Figure 7.15). In scaled devices spacers are narrow and careful process optimization is needed to prevent silicide overgrowth and shorting of gate and S/D. When junction depths are scaled down, silicide thickness needs to be scaled accordingly. Therefore nickel, with the least consumption of silicon, is currently favored.

26.5.5 Contact to silicon

The selectivity requirement for contact hole etching is related to junction depth. If the selectivity between oxide and silicon is poor, oxide etching might reach through the shallow junction. With better selectivity, etching will stop with minimal silicon loss. Etching selectivity of oxide against silicide is much higher than selectivity against silicon, which also makes silicided contacts beneficial from a process integration point of view.

Contact resistance R_c is given by

$$R_c = \frac{\rho_c}{WL} \quad (26.7)$$

where ρ_c is contact resistivity, and W and L are contact dimensions.

Contact resistivity depends on barrier height (0.55 eV half band gap of silicon) and silicon doping concentration (2×10^{20} cm^{-3} maximum dopant solubility) which cannot be changed. Therefore metal-to-silicon contact resistivities cannot be much less than 10^{-7} hm-cm^2. This translates to about 0.1 ohms for 1×1 μm contacts. Metal-to-silicide and metal-to-metal contact resistivities are in

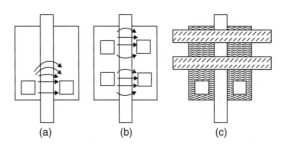

Figure 26.15 (a) MOS-transistor current paths in non-silicided contact; (b) current paths in multiple contact non-silicided contacts and (c) silicided contacts. In the case of silicided contacts, metal lines can run over the transistor, leaving greater freedom for signal routing. Adapted from Liu, R. Metallization, in C.Y. Chang & S.M. Sze (eds.) (1996), by permission of McGraw-Hill

the 10^{-8} ohm-cm^2 range, and this is one added benefit of silicides in submicron technologies. As shown in Figure 26.15, the silicide-to-silicon contact area is much larger than the contact hole area.

In Table 26.7 two processes, 130 nm and 65 nm CMOS, are compared. The 130 nm process is typically processed on wafers of 200 mm diameter, while the 65 nm process is done on 300 mm wafers. There were roughly 7 years between the processes: 130 nm dates from 2000, 65 nm from 2007.

26.6 SOI MOSFETs

SOI devices were commercially introduced in the late 1990s. SOI wafers command less than 10% of the silicon wafer market, in terms of money, and much less in terms of area. In spite of the high price of SOI, SOI devices are found in a wide variety of applications, especially in high-performance, low-power and high-operating-temperature devices.

Vertical isolation comes naturally in SOI, and lateral isolation is much easier than in bulk devices: it is sufficient to etch the SOI device layer and form isolated silicon islands for each device (Figure 26.16). No wells are needed, and many high-temperature process steps are eliminated. This benefit becomes more pronounced with each successive generation, while bulk devices require more and more process steps. Because pn junctions are eliminated, leakage currents and latch-up are also eliminated, enabling low-power operation. Elimination of parasitic capacitances leads to faster devices, and it is especially in low-power and high-speed applications that SOI excels.

Performance and power benefits over bulk devices can be 20–50%, which is considerable, but not necessarily

Table 26.7 Scaling trends from 130 to 65 nm. Adapted from International Technology Roadmap for Semiconductors (http://www.itrs.net)

Technology generation	130 nm	65 nm
Half-pitch (processor)	150 nm	65 nm
Physical gate length L_g	65–100 nm	25–37 (HP vs. LP)[a]
L_g variation (3σ)	6 nm	2.5 nm
Gate oxide thickness	1.3–2.4 nm	0.6–1.4 nm (HP vs. LP)[a]
Drain extension depth	27–45 nm	12–19 nm
Contact junction x_j	48–95 nm	18–37 nm
Spacer thickness	48–95 nm	18–37 nm
Drain extension junction abruptness	7.2 nm/dec	2.8 nm/dec
R_s drain extension, PMOS	400 ohm/sq	760 ohm/sq
R_s drain extension, NMOS	190 ohm/sq	360 ohm/sq
Silicide thickness	36 nm	14 nm
Silicide sheet resistance	4.2 ohm/sq	10.5 ohm/sq
Channel doping	4×10^{18} cm^{-3}	2.3×10^{19} cm^{-3}

[a]HP for high performance, LP for low power (portable applications).

enough to compensate for the high price of starting wafers. Extra benefits come from manufacturing economics: chip yield depends on the number of process steps (Equation 1.2), and because SOI devices require fewer steps, yields should be higher. Similarly, yield depends on chip area (Equation 1.3), and SOI circuits take up less area. Taken together with the higher prices commanded by more powerful chips, SOI has made inroads.

26.7 Thin-Film Transistors

Thin-film transistors (TFTs) are MOS devices made of deposited films. Great variety exists for substrates, semiconductors, conductors, insulators and passivation, as shown in Figure 26.17. While TFTs can be processed on silicon wafers (and this is often done in developing new devices), the real interest in TFTs comes from the fact that any substrate can be used. And for large-area electronics, like displays, glass and polymer sheets, or steel foils, are used instead. Then a number of limitations step in, for example temperature budget and metal contamination. Limitations that hold for glass plates hold for most parts also to TFTs made on steel foils, but there are some differences too. Higher processing temperatures can be used from a mechanical strength point of view, but iron contamination is a concern. Steel is a conducting material and an electrical insulator layer must be deposited before any electrical devices. Iron contamination from steel and sodium from glass both necessitate an ion barrier. If one film can act as electrical insulation, ion barrier and smoothing layer, the better. Steel surface smoothness is inferior to glass, and planarization may be needed. Spin-coated polyimide has been used as a barrier and planarization layer, reducing the roughness from 60 to 2 nm on a steel substrate. Spin-on glass is capable of a similar performance. In both cases multiple coatings can be done to improve planarization.

TFTs come in two basic geometries: bottom gate (Figures 26.18, 26.20, 26.22) and top gate (Figure 26.19). The channel material can be for example PECVD amorphous silicon or polysilicon, and many organic materials are used as channel materials. However, carrier mobilities vary a lot: electron mobility in SCS is about 500 cm^2/V-s, polysilicon about 100 cm^2/V-s, a-Si:H about 1 cm^2/V-s and organic films between 0.001 and 1 cm^2/V-s. Carbon nanotubes (CNTs) show promising

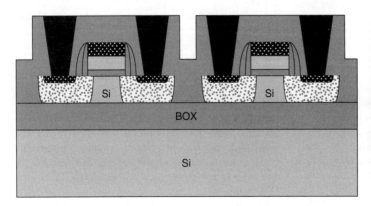
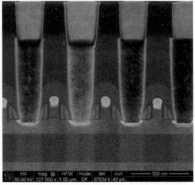

Figure 26.16 SOI MOSFET with first-level metal, schematic and TEM. Courtesy Brandon Van Leer, FEI Company4

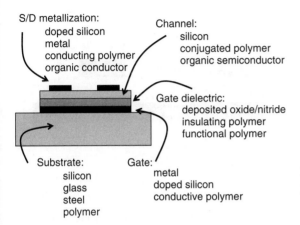

Figure 26.17 Choice of materials for TFTs. Adapted from Leising et al. (2006)

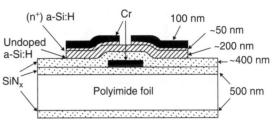

Figure 26.18 Bottom gate silicon TFT on polyimide. Reproduced from Gleskova et al. (2001) by permission of The Electrochemical Society

The TFT of Figure 26.18 is a prototypical bottom gate device. Its specialty is the use of polyimide foil as a substrate, otherwise it is identical to TFTs made on glass plates for displays. The maximum processing temperature has been limited to 150 °C due to the polymer substrate. Another bottom gate TFT for displays will be presented in Figure 37.4.

mobilities higher than silicon with the flexibility of polymers.

26.7.1 Silicon channel TFTs

The most common channel material in TFTs for flat-panel displays is polysilicon (LTPS, for low-temperature polysilicon) deposited on glass. Silicon deposition usually results in an amorphous state, and a separate crystallization step is performed. Silicon self-implantation and a long low-temperature (600 °C) anneal is one possibility, but one that results in fairly low mobility. MILC, or metal-induced lateral crystallization, is being actively studied. Nickel induces crystallization, but nickel has to be made inactive after the process, by for example phosphorus gettering. Another option is excimer laser anneal (ELA). The highest mobilities have been achieved by ELA, but results are sensitive to the direction of laser scanning; that is, whether current flows in the direction of grains or against them.

Dielectrics are (PE)CVD deposited. Maximum process temperatures are limited to about 500 °C because of glass softening. TFT performance can be improved by the same techniques used in silicon MOSFETs. Most often this involves increased mask counts. An eight-mask TFT process includes lightly doped drains (LDDs) for NMOS, and similar improved TFTs have been made with self-aligned silicides and with CMP being used in lab devices, but it is not really suitable because of cost considerations and large-area limitations. Plasma etching uniformity across meter-wide panels can also be problematic. But because linewidths are several micrometers, and thin films fairly thin, wet etching is suitable for most etching steps.

26.7.2 Polymer TFTs

Polymers can serve all the roles needed in MOSFETs. Additionally, polymer TFT fabrication involves a number of new technologies, including ink jetting (Chapter 23) and micromolding (Chapter 18), but also traditional methods like spin coating, PECVD, sputtering and evaporation.

We will describe four different polymer TFTs: the first one is made on a polymer substrate with non-standard materials and is transparent. The second one is made of standard materials except for a pentacene polymer active channel. The third one uses gold for metallization but is otherwise fully polymeric. The fourth one is an all-polymer TFT with a polymer substrate, polymer active channel, polymer conductors, polymer gate dielectric and polymer passivation.

Transparent and flexible transistors can be made if transparent conductors are used on transparent polymer foil. Figure 26.19 shows a top gate TFT on PET film. The source, drain and gate are all made of ITO (Indium-doped Tin Oxide) and the channel of a-IGZO (amorphous Indium Gallium Zinc Oxide). Yttrium oxide is used as a gate dielectric.

The organic semiconductor pentacene for the channel is shown in Figure 26.20. Evaporation of pentacene is performed under 5×10^{-7} mbar pressure and 0.1 nm/s deposition rate. Lift-off is used to form a pentacene pattern. Tantalum pentoxide, Ta_2O_5, has been used as the

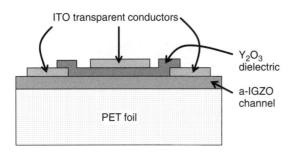

Figure 26.19 Transparent top gate TFT on PET foil. Adapted from Hosono (2007)

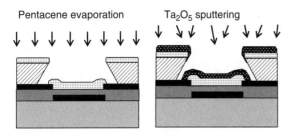

Figure 26.20 Bottom gate TFT with pentacene active channel and Ta_2O_5 passivation. Evaporated pentacene covers a narrower area than sputtered tantalum pentoxide. Redrawn after Goettling et al. (2008)

passivation, patterned using the same lift-off resist as the pentacene channel.

The polymer transistor of Figure 26.21 has polyimide in two different roles, as final substrate and as gate dielectric, with pentacene as the channel and parylene as the passivation. Silicon is used as a carrier during processing only.

It is important, as in all thin-film deposition methods, to consider the underlying layer surface structure because the crystallinity of the deposited film depends on the underlying material. Oxygen plasma treatment of polyimide improves the crystallinity of pentacene, and the crystalline channel material has higher charge carrier mobilities than amorphous ones.

The all-polymer TFT shown in Figure 26.22 differs not only in materials, but also in its fabrication: everything is done in wet solutions, and no vacuum processing is used. Three photomasks are used, with a minimum linewidth of 2 μm. Polyimide film is used as a substrate. Metals have been replaced by conductive polyaniline (PANI) polymer.

The bottom PANI (200 nm thick) is a conductor and serves as S/D. The remaining parts of PANI are converted to an insulator by DUV exposure, with a

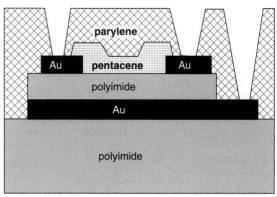

Figure 26.21 Bottom gate TFT on polyimide substrate. Spin-coated polyimide as gate dielectric, pentacene as active channel material and parylene as passivation. Metallization of Cr/Au. Adapted from Feili et al. (2006)

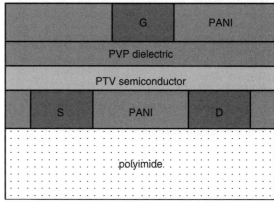

Figure 26.22 All-polymer top gate transistor. S, D, G are conductive PANI, the other PANI layers are insulating. Redrawn after Matters et al. (1999)

difference in resistivity of 11 orders of magnitude. The active channel is made of semiconducting polymer PTV (polythienylenevinylene) and its thickness is 50 nm. The gate dielectric is made of PVP (polyvinylphenol), and is 250 nm thick, and again applied by spin coating. A second layer of PANI serves as the gate, and part of it is UV processed to become an insulator, just like the bottom PANI.

In this practically room temperature process thermal issues are not critical but understanding polymer solubilities in solvents is of paramount importance. PTV solvent must be chosen not to dissolve the PANI layer. PTV itself is insoluble in most solvents, which makes it easy to choose solvent for PVP. PVP, again, is highly crosslinked and not soluble, so the second PANI layer spinning is easy. The top gate has an additional role: it blocks UV radiation and thus protects the PTV semiconductor from environmental degradation.

Contact hole technology is completely unorthodox: holes are pierced mechanically, and the top PANI makes contact with the bottom PANI. Circuits with 300 transistors have been fabricated, with a maximum operating frequency of 200 Hz. So polymer transistors have to find completely new markets because they will never compete with silicon ICs.

26.7.3 Carbon nanotube FETs

Carbon nanotube (CNT) transistors will not be limited by carrier mobility; in fact CNTs beat silicon, by a factor of 10. This is true for individual tubes that have been individually selected and contacted, but in a fashion not worthy of production. The alternative approach uses random CNT networks. These show mobilities similar to polysilicon, in the range of 50 cm^2/V-s. CNT growth is still an art, and results from different research groups differ widely. CNT transistors on polymer substrates offer flexibility and transparency and in this respect they compete with polymer TFTs. A schematic CNT-FET is shown in Figure 26.23.

Both silicon wafers and polymer sheets are used as substrates. Deposition methods for CNTs include wet methods like dielectrophoresis, spraying and spinning, as well as high-temperature CVD methods. Aerosol deposition enables deposition directly from a CVD reactor to polymer substrates, and many other combination methods are used: high-temperature deposition and transfer bonding or stamping to a polymer substrate. A variety of surface functionalization techniques have been tried to attach the tubes to the substrates. Gate dielectrics include traditional thermal and CVD oxides, anodized alumina, ALD oxides Al_2O_3 and TiO_2, SAMs and parylene among others. Gate oxide thicknesses have ranged from 2 to 900 nm. Contact metallizations are usually made by noble metals, namely gold, palladium and platinum. Passivation layers include PECVD nitride, ALD alumina and polymers among others. These devices are in the active research phase and no commercial launches have yet been made.

26.8 Integrated Circuits

Transistors are at the core of ICs, and MOS transistors are the most common ones. Enormous R&D effort over the last 50 years has resulted in magnificent strides in MOSFET performance, with operating frequencies rising from kilohertz to gigahertz and size shrinking from 30 μm to 30 nm. Consequently MOS transistors have conquered new application areas where they were initially thought unusable, for example high-frequency operation. Bipolar transistors, the topic of the next chapter, still offer some advantages over MOSFETs, especially in combining high-speed, low-noise and high-current-carrying capability.

Both MOSFETs and bipolars are just part of the story: transistors have to be wired together to create circuits. This used to be a minor finishing step in early ICs, but today the 10 levels of metallization that connect the billion transistors need in fact more process steps than the transistors themselves. A polymer TFT with 1000 transistor circuits is to be considered as advanced, and research is still centered on transistors, not on circuits.

26.9 Exercises

1. Where in CMOS could you find the following sheet resistances:
   0.05 ohm/sq
   0.5 ohm/sq
   5 ohm/sq
   50 ohm/sq
   500 ohm/sq
   5000 ohm/sq?
2. Silicon dioxide forms readily during Ta_2O_5 deposition because oxygen is present in all oxide deposition processes. What is the effective capacitance of SiO_2/Ta_2O_5 composite? ($\varepsilon = 25$ for Ta_2O_5, $\varepsilon = 4$ for SiO_2.)
3. Equivalent oxide thicknesses (EOTs) of 1.9, 2.3 and 3.1 nm have been measured for HfO_2 films 2, 4 and 8 nm thick, respectively. What is the interfacial SiO_2 thickness when the dielectric constant of HfO_2 is 20?

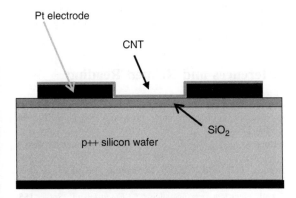

Figure 26.23 Bottom gate CNT transistor

4. Gate oxide thickness in 1 μm CMOS is 20 nm. On S/D areas it is thinned during gate poly plasma etching, but regrown during poly oxidation. Calculate the oxide thickness under the following assumptions:
   (a) poly etch rate 250 nm/min
   (b) poly thickness 250 nm
   (c) Si:SiO$_2$ etch selectivity 20:1
   (d) overetch time 20 s
   (e) reoxidation at 900 °C for 10 min (dry).
5. Design fabrication processes for the DMOSFET and UMOSFET shown below.

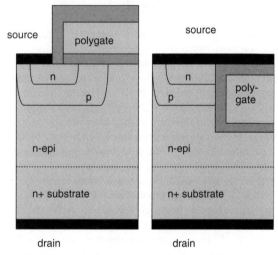

Adapted from Baliga (2001)

6. Compare the area of CMOS inverters made by two different lithography tools with a different mix of capabilities: (a) 4 μm resolution and 1 μm alignment; (b) 4 μm resolution and 2 μm alignment.

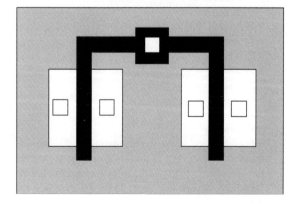

7. Compare minimum CMOS inverter area for:
   (a) a non-self-aligned Al gate
   (b) a self-aligned polysilicon gate
   keeping all other factors identical.

8. Design a fabrication process for the top gate TFT shown below. Maximum process temperature is 350 °C.

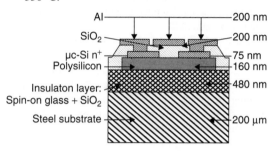

Reproduced from Wu et al. (1999) by permission of AIP

9. Design a fabrication process for the MOS-controlled thyristor of Figure 14.6.

Simulator exercises:

10. Ion implantation of boron at 40 keV with a dose of 10^{13} cm^{-2} is done for CMOS p-well formation. The wafers are 4 ohm-cm phosphorus doped. The well depth (= position of pn junction) is designed to be 5 μm. What diffusion times/temperatures should be used?
11. CMOS S/D implantation is made with arsenic (50 keV, 5×10^{15} cm^{-2}). The designed junction depth is 0.4 μm. Find the implant activation conditions when 40 nm of dry oxide forms during activation.
12. Shallow junctions are needed for advanced CMOS. Compare B-implanted p$^+$/n and As-implanted n$^+$/p shallow junctions (5×10^{15} cm^{-2} dose) when the substrate doping level is 5×10^{17} cm^{-3}.
13. Check your simulator for sheet resistances, junction depths and film thicknesses of the 5 μm CMOS process. Make sure to select a proper cross-section for your 1D simulation.

References and Related Reading

Appenzeller, J. (2008) Carbon nanotubes for high-performance electronics – progress and prospect, *Proc. IEEE*, **96**, 201–211.

Baliga, J. (2001) The future of power semiconductor device technology, *Proc. IEEE*, **89**, 822–832.

Brunco, D.P. *et al.* (2008) Germanium MOSFET devices: advances in materials understanding, process development and electrical performance, *J. Electrochem. Soc.*, **155**, H552–H561.

Burghard, M., H. Klauk and K. Kern (2009) Carbon-based field-effect transistors for nanoelectronics, *Adv. Mater.*, **21**, 2586–2600.

Chabinyc, M.L. *et al.* (2005) Printing methods and materials for large-area electronic devices, *Proc. IEEE*, **93**, 1491–1499.

Chesboro, D.G. *et al.* (1995) Overview of gate linewidth control in the manufacture of CMOS logic chips, *IBM J. Res. Dev.*, **39**, 189.

DiBenedetto, S.A. *et al.* (2009) Molecular self-assembled monolayers and multilayers for organic and unconventional inorganic thin-film transistor applications, *Adv. Mater.*, **21**, 1407–1433.

Feili, D. *et al.* (2006) Flexible organic field effect transistors for biomedical microimplants using polyimide and parylene C as substrate and insulator layers, *J. Micromech. Microeng.*, **16**, 1555–1561.

Flexible electronics (2005), *Proc. IEEE*, **93**, August, special issue.

Gleskova, H. *et al.* (2001) 150 °C amorphous silicon thin-film transistor technology for polyimide substrates, *J. Electrochem. Soc.*, **148**, G370.

Goettling, S., B. Diehm and N. Fruehauf (2008) Active matrix OTFT display with anodized gate dielectric, *J. Display Technol.*, **4**, 300–303.

Gottlob, H.D.B. *et al.* (2006) Scalable gate first process for silicon on insulator metal oxide semiconductor field effect transistors with epitaxial high-k dielectrics, *J. Vac. Sci. Technol.*, **B24**, 710–714.

Hattori, T. (ed.) (1998) **Ultraclean Surface Processing of Silicon Wafers**, Springer.

Hori, T. and T. Sugano (eds) (1997) **Gate Dielectrics and MOS ULSIs: Principles, Technologies and Applications**, Springer.

Hosono, H. (2007) Recent progress in transparent oxide semiconductors: materials and device application, *Thin Solid Films*, **515**, 6000–6014.

Huff, H.R. and D.C. Gilmer (eds) (2004) **High Dielectric Constant Materials: VLSI MOSFET Application**, Springer.

Kahng, D. (1976) A historical perspective on the development of MOS transistors and related devices, *IEEE Trans. Electron Devices*, **23**, 655.

Kukli, K. *et al.* (2007) Atomic layer deposition of ZrO_2 and HfO_2 on deep trenched and planar silicon, *Microelectron. Eng.*, **84**, 2010–2013.

Leising, G. *et al.* (2006) Nanoimprinted devices for integrated organic electronics, *Microelectron. Eng.*, **83**, 831–838.

Liu, R. (1996) Metallization, in C.Y. Chang and S.M. Sze (eds), **ULSI Technology**, McGraw-Hill.

Locquet, J.P. *et al.* (2006) High-K dielectrics for the gate stack, *J. Appl. Phys.*, **100**, 051610.

Matters, M. *et al.* (1999) Organic field effect transistors and all-polymer integrated circuits, *Opt. Mater.*, **12**, 189–197.

Nicollian, E.H. and J.R. Brews (2002) **MOS Physics and Technology**, Wiley.

Parent, D. *et al.* (2005) Improvements to a microelectronic design and fabrication course, *IEEE Trans. Educ.*, **48**, 497–502.

Plummer, J.D., M.D. Deal and P.B. Griffin (2000) **Silicon VLSI Technology**, Prentice Hall.

Polymer electronics (2009) *Proc. IEEE*, **97**, October, special issue.

Quirk, M. and J. Serda (2000) **Semiconductor Manufacturing Technology**, Prentice Hall.

Stinson, M. and C.M. Osburn (1991) Effects of ion implantation on deep-submicrometer, drain-engineered MOSFET technologies, *IEEE Trans. Electron Devices*, **38**, 487.

Wolf, S. (1990) **Silicon Processing for the VLSI Era**, Vol. 2, **Process Integration**, Lattice Press.

Wolf, S. (1995) **Silicon Processing for the VLSI Era**, Vol. 3, **The Submicron MOSFET**, Lattice Press.

Wu, M. *et al.* (1999) High electron mobility polycrystalline silicon thin-film transistors on steel-foil substrates, *Appl. Phys. Lett.*, **75**, 2244.

Zavodchikova, M.Y. *et al.* (2009) Carbon nanotube thin film transistors based on aerosol methods, *Nanotechnology*, **20**, 085201.

27

Bipolar Transistors

Both transistors and integrated circuits were initially made by bipolar technologies. The MOS transistor was conceived and patented in the 1920s, well before the bipolar transistor (1947), but the MOS transistor was not realized until 1960. Bipolar transistors gradually lost their competitiveness to MOS in the 1960s, but they are still used in many applications where special combinations of high speed, low noise and high current-carrying capability are needed.

Bipolar transistors are traditionally fabricated on <111>. Early epitaxial growth techniques made use of miscut <111> wafers (Figure 4.10) in order to nucleate silicon better. There is no fundamental reason why bipolars could not be fabricated on <100>, and, in fact, BiCMOS circuits which have both bipolar and MOS transistors are fabricated on <100> wafers. This is done to optimize the quality of MOS gate oxide: its quality is better on <100> -orientated silicon. This has to do with the arrangement of atoms on the silicon surface, and the resulting Si–O bonds and their spatial restrictions.

Bipolar transistors are vertical devices, that is currents are transported perpendicularly to the wafer surface, whereas MOS transistors are lateral devices with currents parallel to the wafer surface. The standard buried collector (SBC) bipolar transistor is shown in Figure 27.1. It exemplifies the importance of epitaxy and diffusions in bipolar fabrication.

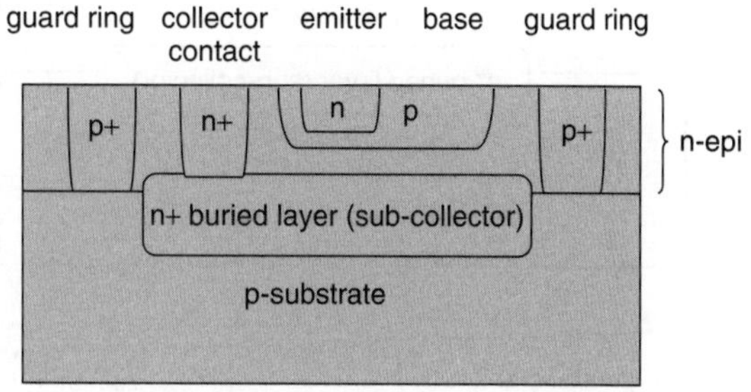

27.1 Fabrication Process of SBC Bipolar Transistor

Bipolar transistor fabrication was already touched upon in Chapter 14 where the UV photodiode process was described (Figure 14.3). A more detailed outline of the SBC process is given below. Before that a short excursion to discuss epitaxy on processed wafers is undertaken.

Figure 27.1 Standard buried collector (SBC) bipolar transistor: n-epitaxial layer on p-substrate (note that diffusions would really be much wider laterally

Buried layers are formed by either ion implantation or thermal diffusion. Oxide acts as a mask for thermal diffusion, but oxide is involved in the implanted process as well: during implant annealing a thin thermal oxide is grown to prevent dopant outdiffusion. Before epitaxy this oxide has to be removed. As a consequence, a step is formed on the wafer surface, and this can cause a pattern shift and distortion in the growing epitaxial layers (it can also cause growth defects if oxide removal is incomplete, or if implant damage is not fully annealed). When the epitaxial film growth from the edges of a pattern are in the same direction, patterns shift laterally. Structures can experience a shift in one direction and distortion in the direction orthogonal to the shift. In the extreme case the epitaxial layer "planarizes" patterns in what is known as washout. Alignment problems will be encountered in all cases.

Buried layers are sources of dopants, and autodoping from buried layers must be considered. An isolated heavily doped region can dope areas many millimeters away in the downstream direction of the epitaxial gas flow. When buried layers are tightly and uniformly spaced,

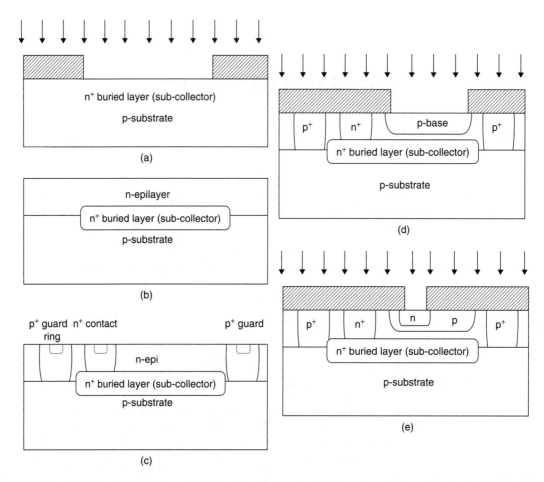

Figure 27.2 (a) Mask 1: buried layer formation by antimony ion implantation. (b) Growth of epitaxial phosphorus-doped n-type layer. (c) Masks 2 and 3: p^+ guard ring and n^+ subcollector contact diffusions: lateral spreading of diffusion is approximately equal to epilayer thickness. (d) Mask 4: ion implantation for base. (e) Mask 5: ion implantation for emitter

autodoping non-uniformity is reduced, but the change in doping level must be accounted for. Buried layers are heavily doped because their role is to minimize collector resistance. Heavy doping will change the lattice constant slightly, and there is a danger of misfit dislocations. Epitaxial growth procedures (to be discussed in Chapter 34), the gases used and the epitaxial equipment result in different degrees of shift, distortion and autodoping.

There are many bipolar technologies, but we will discuss a technology known as SBC bipolar which has been widely used for decades, and even though current bipolars do not immediately look like it, they share many of the SBC basic features.

The starting wafer is a lightly doped p-type wafer (Figure 27.2). Photomask 1 defines the area of the buried collector. This buried layer (subcollector) is doped to high concentration either by ion implantation or by furnace diffusion. If implantation is done, an annealing step must be done to remove damage and recover a perfect silicon surface for epitaxy. Antimony is often used as the buried layer dopant because of its low vapor pressure and consequently low evaporative losses during the subsequent epitaxial growth step.

Wafer cleaning after the buried collector is crucially important for the success of epitaxy. A lightly doped n-type layer is deposited on the wafer. Phosphine (PH_3) gas dopes the epilayer n-type during growth. It is additionally doped by outdiffusion from the buried layer (recall Figure 6.9).

Table 27.1 Bipolar transistors, three generations/technologies. Adapted from Muller and Kamins (1986)

Layers (dopants)	Amplifying junction isolated	Switching junction isolated	Switching oxide isolated
Substrate (B)			
Resistivity (ohm-cm)	10	10	5
Orientation	(111)	(111)	(111)
Buried layer (Sb/As)			
R_s (ohm/sq)	20	20	30
Updiffusion (µm)	2.5	1.4	0.3
Epitaxial film (P)			
Thickness (µm)	10	3	1.2
Resistivity (ohm-cm)	1	0.3–0.8	0.3–0.8
Base (B)			
R_s (ohm/sq)	100	200	600
Diffusion depth (µm)	3.25	1.3	0.5
Emitter (P/As)			
R_s (ohm/sq)	5	12	30
Diffusion depth (µm)	2.5	0.8	0.25

Photomask 2 defines the guard rings which isolate neighboring collectors by reverse-biased pn junctions. Guard rings are formed by boron ion implantation or diffusion. Photomask 3 defines n^+ contact diffusion (known as plug or sinker). Phosphorus is implanted. Implantation depths are about 200 nm only, whereas the epitaxial layer thickness can be anything up to 10 µm. Both p- and n-type dopants are driven to design depth by a thermal diffusion step, at very high temperatures, up to 1200 °C. Deep diffusions must be done early in the process because they require the highest thermal load. A lot of the silicon area is used for device isolation in SBC: the sideways diffusion distance of the p^+ guard ring is equal to the epitaxial layer thickness because diffusion is an isotropic process. The buried collector will experience updiffusion: a micrometer or two depending on the exact conditions during these diffusions.

Photomask 4 defines base areas. Ion implantation is used to introduce the dopants onto the wafer because it offers better control of doping concentration. It is crucial to anneal implant damage away quickly so that base width is controlled by thermal diffusion, not transient enhanced diffusion. It is customary to add to the process an extra step which will ensure a shallow, high doping area for good electrical contact to the p-base.

The emitter is defined by photomask 5. Emitter implantation and anneal are critical for device speed. Base transit time depends on base width, which is determined by both the base and emitter diffusions (transistor speed depends on capacitive charging as well, not just on base transit time). Oxides which have served as diffusion masks are etched away and new thermal oxide is grown. The remaining process steps for contacts holes and metallization are identical to those in MOS fabrication, Figures 26.8 and 26.9.

Contacts to diffusions are defined by using mask 6. Oxide etching is performed either by BHF or by plasma. After photoresist stripping and cleaning, aluminum is sputtered to provide electrical connections. Lithography step 7 defines aluminum wire patterns. After aluminum etching and photoresist stripping, a PECVD oxide and/or nitride passivation layer is deposited. The last mask (8) defines bonding pad openings in the passivation layer. Then, the wafer is ready for testing.

27.2 Advanced Bipolar Structures

Bipolar transistor scaling is not as straightforward as in the case of MOS and the number of transistors per chip is not the main driving force for bipolar technologies, but rather performance. Two different aspects of bipolar scaling will be discussed briefly: vertical scaling, which concentrates on base and emitter structures; and lateral scaling, which is related to isolation between transistors.

Vertical scaling is related to transistor speed via base transit time: a thin base equals faster operation. If the thermal budget can be reduced, less diffusion will take place, enabling base scaling to the 100 nm range, as opposed to micrometers in SBC transistors.

Lateral scaling is related to transistor speed too, because advanced isolation structures eliminate junction capacitances and allow faster switching. Trench isolation is used in advanced bipolars, as in MOS, but in bipolar technology the trenches need to be deeper, therefore the technology is dubbed DTI, for deep trench isolation.

27.2.1 Polyemitter bipolar transistor

Base width is the difference of two diffusion depths: both base and emitter diffusion must be considered. A general strategy is to eliminate high-temperature steps. Using polysilicon as an emitter, less silicon is consumed in making the emitter. Dopants diffuse out of the heavily doped polysilicon emitter and reach just the very top layer of single crystal silicon, ensuring electrical continuity between the polysilicon and single crystal silicon. This approach has a number of benefits: the single crystal silicon emitter will not be implanted, therefore defects from implantation, and transient enhanced diffusion, are eliminated. Elimination of implant annealing reduces the high-temperature steps and unwanted base diffusion. The polyemitter also eliminates the danger of aluminum spiking: if the emitter is very thin, aluminum might spike through it, destroying the device (recall Figure 7.9). Polysilicon 200 nm thick, for example, between the aluminum and emitter/base junction, eliminates the aluminum spiking problem.

27.2.2 Self-aligned polyemitter bipolar transistor

Bipolar transistor fabrication can utilize the same self-alignment principles as CMOS. One of the many self-aligned polysilicon emitter processes is presented in Figure 27.3. It employs self-alignment to the maximum, with three implants self-aligned to each other.

The thick (600 nm) recessed LOCOS isolation oxide is made first. A thin pad oxide (10 nm) is grown, followed by 75 nm LPCVD nitride. After nitride etching, a second LOCOS oxide is grown, this time 200 nm thick. The LOCOS nitride is not removed after field oxidation. Instead, polysilicon spacers are formed on nitride by conformal LPCVD polysilicon deposition and anisotropic etching in chlorine plasma. Boron implantation is carried out to form a heavily doped external base (p^{++}), with energy high enough to penetrate the 200 nm thick LOCOS oxide. Polysilicon spacers are etched away, with a high selectivity against oxide and nitride. Another boron implantation forms a link (p^+) between the external and intrinsic base. The p^+ and p^{++} areas are self-aligned to each other like the source/drain and source/drain extension in a LDD MOS. Nitride is etched away in CF_4 plasma, selectively against oxide. The oxide beneath the nitride protects single crystal silicon from being etched by fluorine. Oxide is then removed selectively against silicon in HF. The oxide has, of course, also a role as a stress-relief layer in the LOCOS structure. The third boron implantation forms the shallow active base. Because it is done last, it experiences the least thermal load and consequently least diffusion.

LPCVD polysilicon is deposited for the emitter. It is heavily phosphorus doped by ion implantation. The anneal drives phosphorus dopants from the n^+ polysilicon emitter into single crystalline silicon. The emitter reaches into the single crystal silicon only to a depth of a few tens of nanometers. Metallization of this transistor is left as exercise.

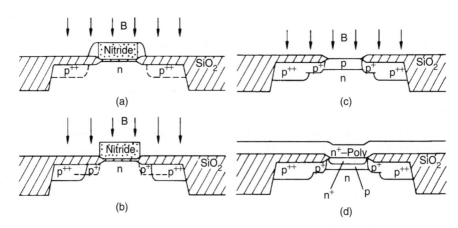

Figure 27.3 Self-aligned single poly bipolar transistor with n^+ polyemitter and base with $p^{++}/p^+/p$ areas. Reproduced from Chen *et al.* (1988) by permission of IEEE

27.2.3 Self-aligned double poly bipolar transistor

Phosphorus-doped polysilicon can act as a diffusion source for the emitter, and correspondingly boron-doped poly can act as a doping source for the p-base. This double poly process offers self-alignment as well, but differently from the previous example. The process flow is shown below and the resulting device in Figure 27.4.

Process flow for self-aligned double poly bipolar transistor

Process step	Comments
Base poly deposition	Boron-doped LPCVD poly, 200 nm
CVD oxide deposition	Thickness 200 nm
Lithography	Non-critical alignment
Etching of CVD oxide/poly stack	Need to change etch chemistry
Base link diffusion (p^+)	Doped poly acts as solid source
Boron implantation	Intrinsic base doping
Intrinsic base diffusion	Damage anneal and activation
CVD oxide 2 deposition	Conformal
Oxide spacer etching	Anisotropic, selective against silicon
Emitter poly deposition	In situ phosphorus doping
Emitter outdiffusion	Single crystal silicon doped

The link base doping level is independent of intrinsic base doping. The former can now be doped to minimize resistance. The link base has to be in electrical contact with the intrinsic base, and lateral diffusion is the desired effect. CVD oxide is needed on top of base poly because it will insulate the base link poly and emitter poly later on. Etching a double layer stack of oxide (with fluorine) and poly (with chlorine) adds complexity. Etching base poly leads to some loss of underlying single crystal silicon, because there is no etch selectivity between the polysilicon and single crystal silicon. But the intrinsic base has not yet been made, so no harm done. The intrinsic base implant dose, energy and annealing are optimized irrespective of link base properties. The thickness of the CVD oxide spacer determines the lateral distance between the link base and intrinsic base. This can be made very small because no lithography is involved. Spacer etching must be highly selective against silicon, because the intrinsic base has now been done, and if silicon is consumed in spacer etching, the intrinsic base will be thinned. Emitter poly is doped in situ in order to reduce the thermal budget: the poly LPCVD temperature is about 600 °C, compared to about 950 °C for poly doping by thermal diffusion or implantation annealing. The emitter will be automatically aligned to the base, too.

27.3 Lateral Isolation

In SBC bipolar devices are isolated from each other by guard ring diffusions (Figure 27.1). The diffusion depth has to be equal to the epilayer thickness, and guard rings take up a lot of area. The LOCOS isolation shown in Figure 27.3 becomes possible when epilayer thicknesses become similar to thermal oxide thicknesses (a micrometer). Oxide isolation improves not only area usage but also transistor speed because sidewall capacitances are minimized.

Trench isolation, which is even more area efficient than LOCOS, is used for high-performance bipolars. In bipolar technology deep trenches of 5 μm are typical, and the technology is called DTI, for deep trench isolation. CMOS trench isolation (trenches about 0.3 μm deep; Figure 26.12) are aptly called STI, for shallow trench isolation. Area usage for isolation becomes independent of epilayer thickness, limited only by lithography and trench etching. Trench filling (Figure 11.10) is usually done in two steps: a thin liner is grown/deposited first, followed by the filling material. For instance, thermal oxidation forms the liner, and TEOS or undoped polysilicon is used to fill the trench. One variant of

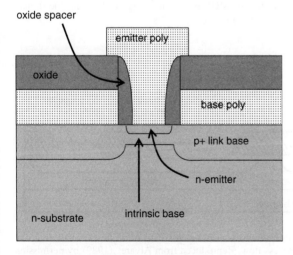

Figure 27.4 Double poly self-aligned bipolar transistor: link base and emitter in single crystal silicon are doped by the p^+ and n^+ polysilicon layers, respectively

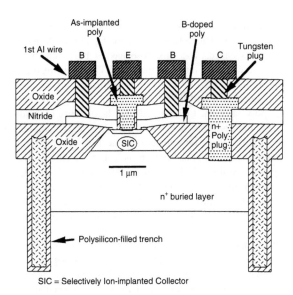

Figure 27.5 Deep trench isolation (DTI) bipolar. Reproduced from Ugajin (1995) by permission of IEEE

many trench isolated bipolar transistors is shown in Figure 27.5. It makes use of four polysilicon layers: for trench filling, link base doping, the emitter and buried layer contact plugs. Some of these poly layers can be used as resistors in analog devices.

27.4 BiCMOS Technology

BiCMOS tries to combine the best of both bipolar and CMOS: high speed, low noise and high current-carrying capacity of the former with the integration density and low power consumption of the latter. BiCMOS has been approached from both directions: taking a full-blooded bipolar process and adding CMOS to it, or taking CMOS as a starting point and adding process modules to create bipolar transistors. The latter approach is more prevalent but often fails to take advantage of the bipolars' best features. Unfortunately, the cost would rise too much if all features of both processes were combined; some performance tradeoff has to be accepted.

In the BiCMOS shown in Figure 27.6 an n^+ doping step is used to form both NMOS source/drain areas and bipolar emitters and collector contacts; similarly, a p^+ doping step creates both PMOS S/D and bipolar base contacts. Only the p-base diffusion step is needed in addition to standard CMOS steps. Elimination of the buried layer and epitaxy lead to an increase in collector resistance and lower operating frequency for bipolars, but the fabrication process is greatly simplified.

Heterojunction bipolar transistors (HBTs) involve SiGe epitaxy to create a heterojunction of Si–SiGe. BiCMOS can take advantage of HBT structure too, and in the process flow the HBT BiCMOS process is based on an analog CMOS process with resistors, capacitors and inductors. Four photomasks are added to the CMOS process to create bipolar elements: buried collector doping, emitter window opening, emitter poly patterning and base poly patterning. Changing the germanium content in SiGe affects the band gap, and thus offers possibilities for tailoring bipolar transistor performance. Epitaxy of SiGe is demanding because it must be carried out at low temperature, and in situ wafer cleaning is essential for its success. The resulting BiCMOS is shown in Figure 27.7 and the dopant depth profile in Figure 27.8.

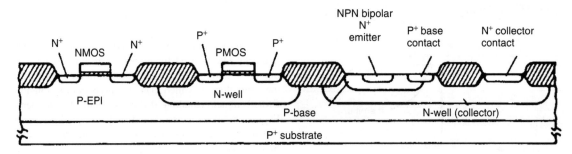

Figure 27.6 Simple BiCMOS technology: triple diffused-type bipolar transistor added to a CMOS process with minimal extra steps: only p-base diffusion mask is added to CMOS process flow. Reproduced from Alvarez (1989) by permission of Kluwer

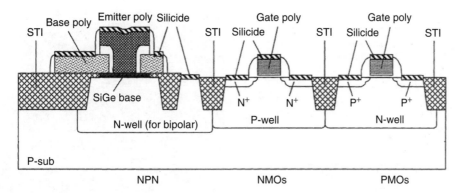

Figure 27.7 CMOS process with SiGe HBTs added. Reproduced from Sato *et al.* (2003), copyright 2003, by permission of IEEE

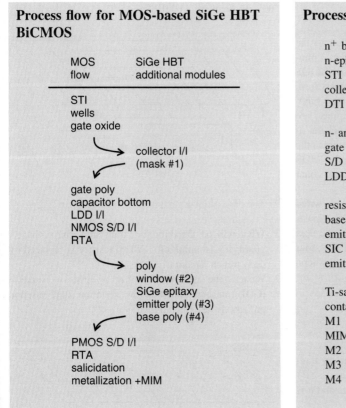

As an alternative, BiCMOS process flow that uses a bipolar process as a starting point and adds CMOS steps in between is also shown.

Bipolar, CMOS and metallization modules are kept unchanged as far as possible, and brought together to create BiCMOS.

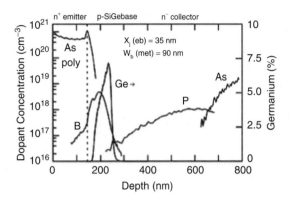

Figure 27.8 SiGe HBT doping profile under the emitter measured by SIMS. Reproduced from Cressler (2005b), copyright 2005, by permission of IEEE

27.5 Cost of Integration

As a rule of thumb, the cost is directly related to the number of photolithography steps. Evolution of a 13 photomask, 1 μm digital CMOS process into a 1 μm BiCMOS process can be done in several ways. In its simplest form only a base implant photomask is added. If true bipolar performance is needed, a buried layer and epitaxy are needed and the collector is made separately from the n-well. If analog elements like resistors are required, the mask count still increases, but this is true for both CMOS and bipolars alike. With these additions the mask count rises by 20–50%, and the cost similarly.

When transistors are ready, they have to be wired together to create circuits. This is true for CMOS, bipolar and other technologies. That will be the topic of the next chapter. In the BiCMOS process of Figure 27.9 four levels of metal are used, plus an additional polysilicon layer for resistors and an extra dielectric film for metal–insulator–metal capacitors.

27.6 Exercises

1. SBC is pictured below. Calculate the minimum transistor area under the following assumptions:
   - Minimum lithographic linewidth L is 3 μm, and it is the width of E, C and B.
   - Emitter is square; base length is 2× width, and collector length is 3× width.
   - Epilayer thickness is 5 μm.
   - Buried layer updiffusion is 1 μm.
   - Base diffusion depth is 1.5 μm.
   - Emitter diffusion depth is 0.5 μm.

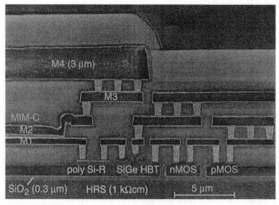

Figure 27.9 BiCMOS: high-resistivity SOI substrate, 0.25 μm NMOS and 0.3 μm PMOS, SiGe bipolars, polysilicon resistors, and MIM capacitors realized using metal 1 and metal 2. Reproduced from Washio (2003), copyright 2003, by permission of IEEE

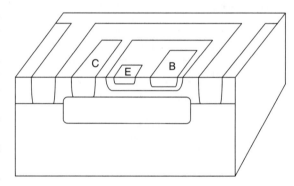

2. What will be the minimum transistor area if the p^+ guard ring isolation of a SBC transistor is replaced by deep trench isolation?
3. What is the area of the collector diffusion isolation (CDI) transistor shown below when the same baseline process as described above is used?

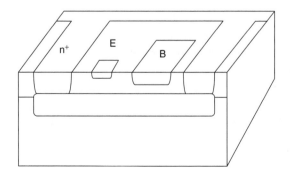

4. Perform front-end simulations to obtain the sheet resistances and diffusion depths of switching of the junction isolated transistor described in Table 26.1.
5. Design metallization process steps for the polyemitter transistor. This is the same device as shown in Figure 27.3.

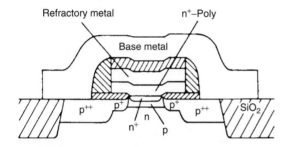

Reproduced from Chen *et al.* (1988) by permission of IEEE

6. Analyze the main fabrication steps of the bipolar transistor shown below.

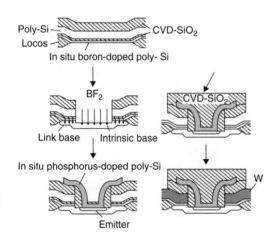

Reproduced from Onai *et al.* (1997) by permission of IEEE

References and Related Reading

Alvarez, A.R. (ed.) (1989) **BiCMOS Technology**, Kluwer.

Chen, T.-C. *et al.* (1988) An advanced bipolar transistor with self-aligned ion-implanted base and W/poly emitter, *IEEE Trans. Electron Devices*, **35**, 1322.

Cressler, J.D. (ed.) (2005a) **Silicon Heterostructure Handbook: Materials, Fabrication, Devices, Circuits and Applications of SiGe and Si Strained-Layer Epitaxy**, CRC Press

Cressler, J.D. (2005b) On the potential of SiGe HBTs for extreme environment electronics, *Proc. IEEE*, **93**, 1559–1582.

Lane, W.S. and G.T. Wrixon (1989) The design of thin-film polysilicon resistors for analog integrated circuits, *IEEE Trans. Electron Devices*, **36**, 738.

Muller, R.S. and T.I. Kamins (1986) **Device Electronics for Integrated Circuits**, John Wiley & Sons, Inc.

Onai, T. *et al.* (1997) 12ps ECL using low-base-resistance Si bipolar transistor by self-aligned metal/IDP technology, *IEEE Trans. Electron Devices*, **44**, 2207–2212, fig. 2.

Reisch, M. (2003) **High-Frequency Bipolar Transistors**, Springer.

Sato, F. *et al.* (2003) A 0.18 μm RF SiGe BiCMOS technology with collector-epi-free double-poly self-aligned HBTs, *IEEE Trans. Electron Devices*, **50**, 669.

Ugajin, M. (1995) Very-high f_t and f_{max} silicon bipolar transistors using ultra-high performance super self-aligned process technology for low energy and ultra-high-speed LSI's, *IEDM'95*, p. 735.

Washio, K. (2003) SiGe HBT and BiCMOS technologies for optical transmission and wireless communication systems, *IEEE Trans. Electron Devices*, **5**, 656.

Wolf, S. (1990) **Processing for the VLSI Era**, Vol. 2, **Process Integration**, Lattice Press.

28

Multilevel Metallization

Multiple levels of metallization offer possibilities for circuit designers to route signals over transistors, and thus to reduce the area needed for wiring. We will first discuss multilevel metallization for submicron technologies (0.8, 0.5, 0.35 and 0.25 μm) based on aluminum wiring with tungsten via plugs (Figure 28.1). The intermetal dielectric is oxide, and it is planarized by CMP. We will then delve into copper metallization which emerged in the late 1990s. There CMP is used too, but this time to polish copper. While transistors get speedier the smaller they are, metallization behaves differently: RC time delays increase with downscaling because thinner dielectrics increase capacitance and narrower and thinner wires have higher resistances.

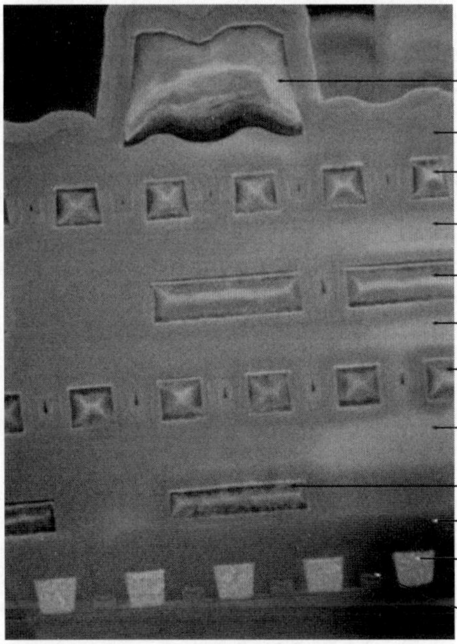

28.1 Two-Level Metallization

Two-level metallizations are extensions of one-level metallizations, with additional dielectric and metal films and only minor conceptual differences. The process continues after first metal as follows:

Figure 28.1 Cross-sectional view of six-level metal structure (M0 is metal zero). Reproduced from Koburger et al. (1995) by permission of IBM

Process flow for two-level metallization	
Intermetal dielectric deposition	PECVD oxide
Planarization	Spin-on-glass with etchback
Via hole lithography and etching	CHF$_3$ plasma oxide etch
Second metal deposition	TiW/Al sputtering
Second metal lithography and etching	Cl$_2$-based plasma etching
Passivation	PECVD nitride
Bonding pad patterning (litho and etch)	CF$_4$ plasma etch

Introduction to Microfabrication, Second Edition Sami Franssila
© 2010 John Wiley & Sons, Ltd

Contact hole etching of oxide against silicon demands a highly selective etch process because both oxide and silicon are etched by fluorine. Contacts between metal levels (known as via holes) are easier from an etching point of view: fluorine-based oxide etching will stop automatically once aluminum is reached. Because there is metal on the wafer, cleaning solutions after via etching are limited. The second-metal step coverage in the via hole is often critical. Fortunately, via holes are larger than contact holes, and aspect ratios are therefore smaller.

There are a number of practical aspects in two-level metal processes which demand attention. Each additional (PE)CVD step adds to thermal loads, film stresses and plasma damage. Aluminum lines experience thermal expansion and are under compressive stresses. These stresses are relaxed by hillocks: protrusions of aluminum sticking out. Hillocks can sometimes be micrometers high.

Two-level metallization cannot be extended to three levels because the topography of the wafer becomes more pronounced after each level, and the gap filling capability of (PE)CVD dielectric deposition as well as sputtering step coverage in via holes will reach their limits. Planarization helps, but it is no panacea: the surface may become flat, which eliminates optical lithography depth-of-focus problems, but, as shown below in Figure 28.2, it creates problems in via hole etching and sputtering because holes will be of different depth.

All devices need metallization, and logic circuits usually require the most complex routing, while memories suffice with three levels of metal. Even superconducting devices require multiple levels of metallization if they are complex logic circuits (Figure 28.3).

28.1.1 Spin-coated inorganic films

Spin-on-glasses (SOGs) are silicon-containing polymers which can be spun and then cured to produce a silicon dioxide-like glassy material (they are sometimes known as SODs, for spin-on dielectrics, which includes polymers, too). Numerous commercial formulations for SOGs exist, adjusted for molecular weight, viscosity and final film properties for specific applications. Two basic types of SOGs are the organic and inorganic. The inorganic SOGs are silicate based and the organics are siloxane based.

Upon curing the reaction at about 400 °C silicate SOGs turn into an oxide-like material which is thermally stable and does not absorb water accordingly. They are, however, subject to volume shrinkage during curing, leading to high stresses (~ 400 MPa). This limits silicate SOGs to thin layers, about 100–200 nm. Multiple coating/curing cycles can be used to build up thickness, at the cost of quite an increase in the number of process steps. Adding phosphorus to SOG introduces changes similar to the phosphorus alloying of CVD oxide films. The resulting films are softer and exhibit less shrinkage, and are better in filling gaps. However, water absorption increases, which results in less stable films. The gap-filling capability of SOGs is related to viscosity: low viscosity equals good gap fill, but, unfortunately, it is correlated with high shrinkage, too.

Organic SOGs based on siloxane (Figure 28.4) do not result in pure SiO_2-like material, but contain carbon even after curing. By tailoring the carbon content, the material properties can be modified for lower stress (~ 150 MPa) and consequently thicker films. Siloxane films are, however, polymer-like in their thermal stability, and 400 °C is a practical upper limit.

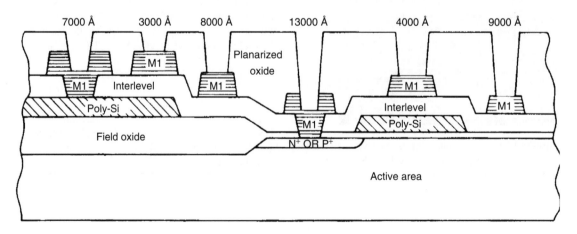

Figure 28.2 Variable via depth results from planarization. Reproduced from Brown (1986) by permission of IEEE

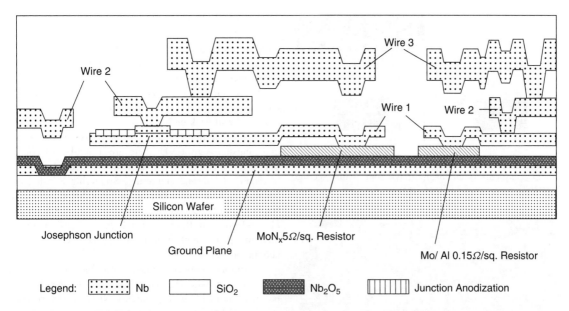

Figure 28.3 Multilevel metallization of a superconducting IC. Reproduced from Abelson and Kerber (2004), copyright 2004, by permission of IEEE

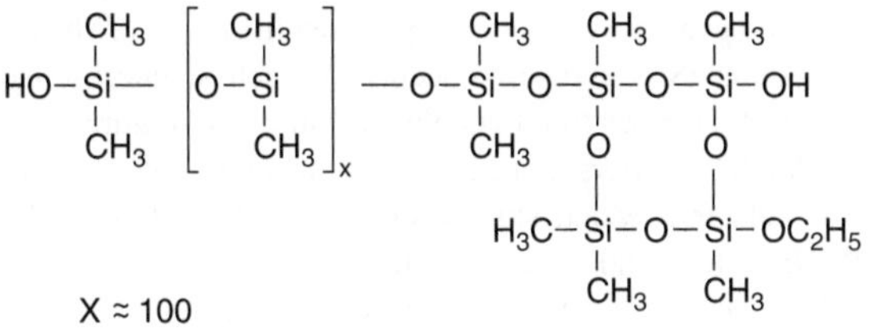

Figure 28.4 Structure of siloxane

28.2 Planarized Multilevel Metallization

True multilevel metallization starts at three levels of metal. Historically this occurred in the late 1980s when submicron CMOS technologies were introduced. In 0.25 μm technology up to six levels of metal are used in ASICs and logic chips, three levels in memory chips. In the 45 nm technology generation there can be 10 levels of metal.

A fully planar structure can be created when contact and via holes are filled by CVD tungsten, and excess tungsten is removed, by etchback or by CMP (Figure 16.1). The number of metal levels can be increased simply by repeating the process over and over again because all levels are planar, Figure 28.1.

Back-end process integration differs from that of front-end in the sense that the thermal budget concept has a very different meaning. Whereas the front-end thermal budget is about the temperature–diffusion relationship, the back-end thermal budget is about the temperature–stress relation. For n-level metallization there will be $2n$ steps at 300–400 °C (for each layer CVD tungsten and PECVD oxide steps), with room temperature steps (etching, spin coating, CMP) in between. Stress, strain, adhesion, hillocks, voids and cracks have to be understood.

28.2.1 Contact/via plug

In order to get planarized metallization, CVD W-plug fill has been adopted. Because CVD-W has excellent step coverage, the via hole will be completely filled. In order to improve adhesion, a Ti/TiN adhesion layer is deposited before tungsten. Excess metal is etched or polished away, leaving a planar surface. The second metal (Ti/TiN/Al) is then sputtered (Figure 28.5).

The SEM micrograph of Figure 28.6 shows the structure of a planarized multilevel metallization scheme. The top aluminum wiring levels are very planar. Tungsten has been used for local interconnects (in the length scale ~10 μm). All dielectric layers have been etched away to reveal the metallization for analysis (e.g., for failure analysis).

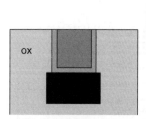 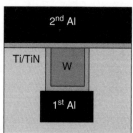

Figure 28.5 Aluminum bottom metal with Ti/TiN/W contact plug after etchback (left) and with second Ti/TiN/aluminum metal layer (right)

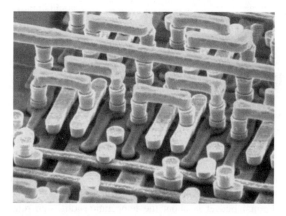

Figure 28.6 Multilevel metallization with all dielectric layers etched away. TiSi$_2$/poly gates, tungsten plugs and local wires, Al global wires. Reproduced from Mann *et al.* (1995) by permission of IBM

When vias can be stacked on top of each other in a multilevel metallization scheme, a lot of area can be saved, and freedom of wire routing increases. In Figure 28.6 tungsten plugs can be seen on top of each other. The top-level plugs are somewhat larger than the bottom plugs, ensuring overlap. Misalignment is still there, but because the surfaces are planar, it does not lead to topography build-up.

28.3 Copper Metallization

All ICs used aluminum for metallization till 1997, and most still do, but copper was introduced into high-performance applications from the 0.25 μm generation on. Copper resistivity is clearly smaller than that of aluminum, 1.8 vs. 3 mohm-cm, and like aluminum, it is an exceptional material that thin film resistivity can be very close to bulk value. However, copper has many drawbacks and limitations. It diffuses rapidly in both silicon and silicon dioxide, and new barrier materials have to be invented. Copper cannot be plasma etched, so it has to be patterned by polishing (CMP). Copper is an impurity that is harmful for silicon transistors, so the whole process line has to be designed to prevent copper from reaching silicon. This means that lithography, etching, CVD, etc., are duplicated for fabricating front-end and back-end.

Whereas aluminum deposition is always by sputtering and tungsten is by CVD, there are a number of copper deposition methods available: namely, electroless, electroplating, CVD and sputtering. Sputtering is ruled out because of poor step coverage and inability to fill holes, but it can still be used to deposit a thin seed layer for electrodeposition. Both CVD and electrodeposition methods can fill the high aspect ratios encountered in deep submicron devices.

To eliminate copper diffusion into oxide, one solution is to use non-oxide dielectrics, like nitride or polymers, but this is not without its problems. Nitride dielectric constant is fairly high ($\varepsilon_r \sim 7$) and polymers are not stable enough. As a compromise, oxide dielectric layers with nitride or carbide (SiC) barriers are used. These layers have an advantage in that they act as etch and polish stop layers. The general issues of copper metallization are shown in Figure 28.7, and cross-sectional electron microscope views of a copper-filled via plug are shown in Figure 28.8.

Metallic barriers can be used to separate copper from the dielectric. Much studied choices include TiN, W:N, W:N:C, TaN and TaSiN. Metallic barriers are thin: for 90 nm technology the barriers need to be below 10 nm. The resistivities of the barrier and plug are critical in the 100 nm range because the full benefit of low-resistivity copper cannot be realized if a high-resistivity barrier reduces the effective resistivity of the plug. Barrier deposition is by for example ALD, which has excellent conformality. Seed layer deposition requirements are not as strict: thickness uniformity is not mandatory, only film continuity. With more and more layers and materials, the number of materials interfaces is going up, and all these interfaces must be characterized for stability, reactions, diffusion, stresses, etc.

Polyimide is a very stable polymer and has been tried as the intermetal dielectric in copper metallization. Copper is clad in tantalum barriers and polyimide is protected by nitride etch stop layers as shown in Figure 28.9. Copper is completely clad by either tantalum or nitride and never in contact with the polyimide. Contact to silicon is still made by Ti/TiN/W plug, to prevent the danger of silicon contamination.

CMP selectivity between copper and tantalum is very high, which means that removal of tantalum leads to

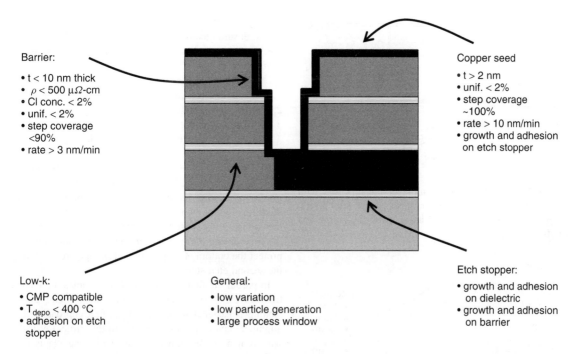

Figure 28.7 Copper–low-k metallization schematic for 90 nm technology. Adapted from Smith *et al.* (2002)

Figure 28.8 Copper via filling: left, with ALD TiN barrier; right, with ALD W:N:C barrier. Reproduced from Smith *et al.* (2002), copyright 2002, by permission of Elsevier

long overpolish times (cf. long overetch times). CMP non-idealities, dishing and erosion have to be analyzed. Dishing is strongly linewidth dependent but rather insensitive to pattern density, whereas oxide erosion is very strongly pattern density dependent and only mildly linewidth dependent, as shown in Figure 28.10. CMP dishing and erosion in the 20 nm range are targeted for 100 nm technologies.

28.4 Dual Damascene Metallization

Damascene metallization relies on etching via plugs in oxide, filling those plugs with copper, with CMP for removal of excess metal. In dual damascene this idea is developed further. First, very thick oxide is deposited. Then, two lithography and two etching steps define vias and wires. Copper is then deposited and fills both the via

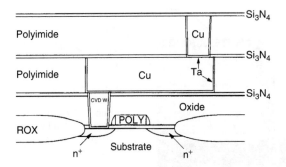

Figure 28.9 Cu/polyimide multilevel metallization with Ta barriers, W plugs and silicon nitride polish stop layers. Reproduced from Small and Pearson (1990) by permission of IBM

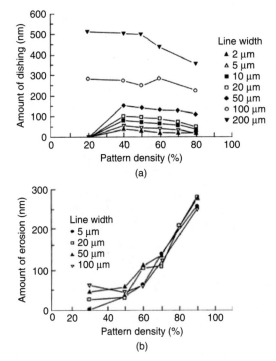

Figure 28.10 Dishing of copper and erosion of oxide. Source: Steigerwald J. M., et al, Chemical–Mechanical Planarization of Microelectronic Materials, © Wiley, 1997. This material is used by permission of John Wiley & Sons, Inc.

holes and the wires. CMP finishes the process as usual. This sequence is shown in Figure 28.11.

Dual damascene introduces novel process integration features. The thick dielectric consists of multiple layers: namely, barriers/etch stops and the actual thicker films which ensure electrical insulation. It is possible to make the vias and trenches in four different ways, as shown in Figure 28.12:

1. Full via first (etching through the thick dielectric).
2. Partial via first (etching halfway).
3. Wire first (etching halfway).
4. Partial wire first (etching a hard mask only).

Full via first (Figure 28.12a) is problematic because a very deep via hole of high aspect ratio is produced in the first step, making second photoresist spinning difficult. Additionally, the bottom hard mask needs to tolerate two etch steps: it is exposed at the end of the via etch and all the time during the trench (wire) etch. One solution is to protect the bottom of a via with undeveloped resist during the second etch step.

In partial via first (Figure 28.12b) via holes are etched till mid etch stop layer in the first step. Metal trench etching is easier than in the full via first approach. Misalignment can cause a grave error in this structure: if the wire trench is misaligned so much that the via is partially photoresist covered, the area of metal contact will be small and erratic.

The metal wire first (Figure 28.12c) approach does not need a top hard mask. Wires are etched down to the middle hard mask. The second lithography has to be done in a recess, and lithography depth of focus may pose problems.

The partial metal wire first (Figure 28.12d) approach needs a top hard mask. In the first step the top hard mask is etched and resist is then stripped. The next lithography step (for the via) can now be done on a practically planar surface (this approach is used in DRIE MEMS regularly, see Figure 21.17). After etching the top dielectric layer with a resist mask, the resist is stripped, and the wire trench and bottom half of the via are etched using hard mask only. Misalignment in the via lithography step can cause problems similar to "partial via first" described above.

28.5 Low-k Dielectrics

Dielectric constant (ε or k) can be reduced by modifying oxides or by switching to other materials. With SiO_2-based dielectrics (with $\varepsilon_r \approx 4$) there is an evolutionary development down to about $\varepsilon_r \approx 2.7$. The first approach is to deposit fluorine-doped oxide by CVD. This will lead to $\varepsilon_r \approx 3.6$. Carbon doping with CH_3 groups in silicon dioxide, designated as SiOC:H, can bring the dielectric constant down to about 2.7. The composition of SiOC:H films is typically 20–25 at. % Si, 30–40% O, 15% C and

Figure 28.11 Dual damascene metallization: left, two lithography and two etching steps define vias and wires in oxide; middle, vias and wire trenches filled by metal in one deposition step; right, metal polishing yields a planar surface. Courtesy Jorma Koskinen

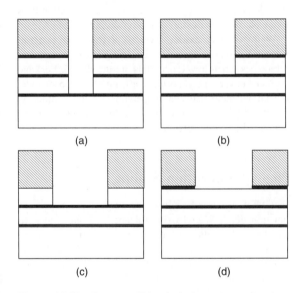

Figure 28.12 Four possible dual damascene processes with etch stop layers: (a) full via first; (b) partial via first; (c) wire first; (d) partial wire first

20–40% H. These films are well-known dense inorganic materials, compatible with existing CVD tools, processes and metrology.

Siloxanes and silsesquioxanes are common materials from spin-on planarization. Methyl silsesquioxane (MSQ) is attractive dielectric film because its very low dielectric constant of $\varepsilon_r \approx 2.6$. In SOD planarization the spin film is most often etched away, but it can used as a permanent part of the device. This leads to the whole new characterization of siloxanes. For instance, during subsequent sputtering steps outgassing from SODs can poison the metal, leading to contact problems.

Switching to polymers is a discontinuous shift: it requires a lot of work in materials science, process technology, metrology, process integration, equipment and reliability. For instance, adhesion and interface stability with metals need to be assessed and etching and polishing processes have to be developed. Sufficient mechanical strength of low-k films is essential for successful CMP. Fluoropolymers, aromatic hydrocarbons, poly(arylene ethers), parylene and PTFE offer dielectric constants down to $\varepsilon_r \approx 2$.

Porous inorganic materials, with $\varepsilon_r \approx 2$ are known as ULKs, for ultralow k. Pores can be made by controlled evaporation, nanophase separation or drying. Aerogels and xerogels are dried silica, with over 90% air in them. They promise further improvements in ε.

The ultimate dielectric is air (or vacuum) with $\varepsilon_r \approx 1$. There are some practical problems with air, however: the mechanical strength is not very good, thermal conductivity is poor and long-term stability is questionable. In spite of these drawbacks, gas-filled and vacuum dielectric structures have been demonstrated.

A wide repertoire of measurements are needed to characterize novel candidate materials (Table 28.1). PECVD boron nitride was measured for some 15 properties (Table 7.2). New polymeric low-k materials need to be evaluated

Table 28.1 Characterization needs for new dielectrics

Parameter	Comment
CMP rate	Young's modulus 1–10 GPa, high polish rates
T_g/T_d	Glass transition/decomposition temperatures (about 450 °C)
Plasma resistance	Organic materials etched in oxygen plasma
Cleaning resistance	Photoresist removers and solvents
Shrinkage	Volume changes upon heat treatment as solvents evaporate
Adhesion	Scotch tape test is the first hurdle
Outgassing	Even cured films may release gases into sputtering vacuum
Porosity	Tightly controlled for reproducible ε
Pore size	Pores that are too big behave like pinholes
Shelf life	Decomposition during storage not unlike photoresists
Viscosity	Film thickness depends on viscosity (and spin speed)
Impurities	The (alkali) metals have to be measured
CTE	Thermal expansion of polymers highly variable
Loss tangent	Electrical losses at high frequencies must be understood

for 15 more parameters before they can be accepted in manufacturing.

Modulated photoreflectance methods, already in use in implant dose monitoring, are useful for multilayer analysis when time-resolved mode is employed. A short laser pulse heats the sample, which then expands locally, giving off sound waves. Optical reflectivity changes due to propagating sound waves are measured on the wafer surface. Time-resolved measurements can distinguish between reflections from various interfaces in the sample, enabling multilayer measurement of both metals and dielectrics.

CMP of soft and porous materials with Young's moduli in the 1–10 GPa range is difficult because these materials are mechanically weak. They are also subject to peeling by shear forces, especially when multiple layers of materials are present (and there can be tens of layers in a multilevel structure). Polymeric abrasives have been tried as replacements for silica and alumina for soft material polishing. Cleaning remains a major problem for low-k materials, after CMP cleaning, after etch cleaning and photoresist strip. Many wet chemical cleaning solutions are out of the questions because they penetrate pores and cause swelling. Measurements of pore size and porosity are needed for the reproducibility of ultralow-k materials. Various methods are being developed; candidates include gas phase, optical, X-ray, positron and neutron methods.

When new materials are introduced, they are evaluated in several phases. Initial tests are carried out on planar wafers using blanket films. Basic physical and chemical characteristics are measured: namely, dielectric constant, shrinkage, moisture absorption, uniformity of deposition, blanket etching and polishing. Simple one-level test structures are then applied to check patterning issues (etch, strip) and interface stability under various process steps (metallization, CMP, etch). Multilevel test structures include electrical tests and more complex interaction tests like etch and polish stop, adhesion during CMP, etc.

While thermal oxide serves as a reference material when CVD oxides are evaluated, PECVD oxides serve as references when low-k materials are developed. Leakage current between neighboring lines, interline capacitance, breakdown field between copper lines, metal continuity, metal bridging, line resistance uniformity, etc., are all compared to oxide reference processes.

28.6 Metallization Scaling

In CMOS front-end scaling, the vertical parameters of junction depth x_j and oxide thickness t_{ox} are scaled to smaller and smaller values, leading to improved transistor performance. In the back-end, however, scaling is mostly detrimental. If metal lines are made thinner, the resistivity increases and linewidth scaling works in the same direction. If the dielectric thickness is scaled down, the capacitance between metal layers increases, leading to increased RC time delays. At 1 μm linewidths transistor delays are more significant than wiring delays, but the situation changes somewhere around 0.2 μm technology, and below 100 nm wiring delays clearly dominate transistor delays.

A simple model for back-end interconnect wire scaling is shown in Figure 28.13 and the RC time delays are described by

$$\tau = RCL^2 \quad C = \frac{\varepsilon W L}{T} \quad R = \frac{\rho L}{HW} \quad (28.1)$$

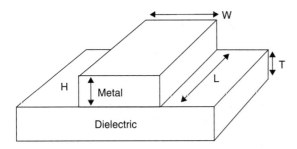

Figure 28.13 Wire geometry for simple RC time delay model

Table 28.2 Back-end scaling trends

CMOS generation	0.35 μm	0.25 μm	0.18 μm	0.13 μm
Min. metal linewidth (μm)	0.4	0.3	0.22	0.15
Min. space (μm)	0.6	0.45	0.33	0.25
Metal thickness (μm)	0.7	0.6	0.4	0.4
Dielectric thickness (μm)	1	0.84	0.70	0.6

where L is line length and resistance R and capacitance C are per unit length.

Scaled local connection lengths are given by L/n ($n > 1$) because smaller devices are closer to each other. Long-distance connections do not scale, however, because chips are not getting any smaller; quite the contrary, in fact, because more and more functions are being crammed on a chip. In our simple model we will assume constant line length, L. Scaled capacitance and resistance are given by

$$C' = \frac{\varepsilon(W/n)L}{T/n} = C \qquad (28.2)$$

$$R' = \frac{\rho L}{(H/N)(W/N)} = n^2 R \qquad (28.3)$$

RC time delay τ' is then given by

$$\tau' = n^2 RC \qquad (28.4)$$

Because scaling factor n is larger than unity, time delays are increasing. When linewidths are scaled down, film thicknesses are scaled down, in order to keep aspect ratios about the same, which is not an unreasonable assumption since very tall but narrow metal lines would be difficult to make. And because chip sizes (L) are increasing, time delays are bound to increase. Historical scaling trends are listed in Table 28.2.

In order to battle RC time delay, aluminum ($\rho \approx 3$ μohm-cm) has been replaced by copper ($\rho \approx 1.8$ μohm-cm), and silicon dioxide dielectrics ($\varepsilon_r \approx 4$) have been replaced by low-k dielectrics ($1 < \varepsilon_r < 4$).

In the era of 5 μm CMOS the front-end contributed most of the process steps and most of the cost of processing. The two levels of metal were a small finishing touch. Today the back-end dominates both the number of steps and costs. The number of metal levels (including passive devices) is up to 12, and it is expected to increase to 14 by 2020.

Metal wire width and thickness increase for the upper levels. While first metal level M1 is very narrow and thin, the successively higher levels of metal have more relaxed design rules. Aspect ratios of metal lines do not change appreciably: roughly 2:1 aspect ratios are common, and this will not change in the foreseeable future.

One limitation becoming more acute is resistivity. Ohm's law is no longer valid for small dimensions, below a few hundred nanometers. Resistivity increases rapidly below 100 nm (Figure 28.14). Copper is an exceptional metal because its thin-film resistivity is close to bulk copper resistivity, but at small dimensions grain boundary and sidewall scattering start to dominate, and 50 nm copper lines exhibit a resistivity of 3 μohm-cm, well above 1.8 μohm-cm bulk resistivity.

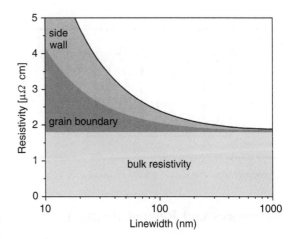

Figure 28.14 Copper resistivity as a function of linewidth. Courtesy of The Semiconductor Industry Association. The International Technology Roadmap for Semiconductors, 2007 Edition. International SEMATECH:Austin, TX, 2007

28.7 Exercises

1. If a via plug of 2:1 aspect ratio in 0.25 μm technology has a resistance of 0.4 ohms, is it made of tungsten or copper?
2. What is copper via resistance in 0.1 μm technology? Plot via resistance as a function of different tantalum nitride barrier thicknesses.
3. What is the breakdown field requirement for low-k dielectrics?
4. What is the effective dielectric constant of a nitride/BCB/nitride (20/500/20 nm) stack when $\varepsilon = 7$ (nitride) and $\varepsilon = 2.5$ (BCB).
5. What is the etch or polish selectivity needed in a low-k approach that uses nitride etch/polish stop layers 20 nm thick on 500 nm low-k material?
6. Dual damascene with etch stop layers can be done in four ways (Figure 28.11). How do these schemes differ with respect to alignment?
7. What etching processes were used to prepare the sample for the SEM micrograph in Figure 28.5? What selectivities and other criteria are required of those etching processes?
8. In order to reduce substrate coupling, inductor coils should have low dielectric constant material underneath. In the pictured device air gaps are left in the structure. Explain the fabrication process!

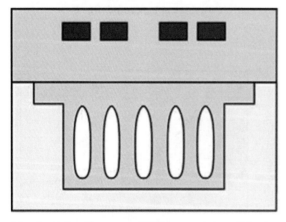

Reproduced from Farcy *et al.* (2008) by permission of Elsevier

References and Related Reading

Abelson, L.A. and G.L. Kerber (2004) Superconducting integrated circuit fabrication technology, *Proc. IEEE*, **92**, 1517.

Anand, M.B. *et al.* (1997) Use of gas as low-k interlayer dielectric in LSI's: demonstration of feasibility, *IEEE Trans. Electron Devices*, **44**, 1965.

Borst, C.L., W.N. Gill and R.J. Gutmann (2002) **Chemical-Mechanical Polishing of Low Dielectric Constant Polymers and Organosilicate Glasses: Fundamental Mechanisms and Application to IC Interconnect Technology**, Springer.

Brillouet, M. (2006) Challenges in advanced metallization schemes, *Microelectron. Eng.*, **83**, 2036–2041.

Brown, D. (1986) Trends in advanced process technology, *Proc. IEEE*, **74**, 1678 (special issue on integrated circuit technologies of the future).

Chen, W.-C. *et al.* (1999) Chemical mechanical polishing of low-dielectric constant polymers: hydrogen silsesquioxane and methyl silsesquioxane, *J. Electrochem. Soc.*, **146**, 3004.

Davis, J.A. *et al.* (2001) Interconnect limits on gigascale integration (GSI) in the 21st century, *Proc. IEEE*, **89**, March, 305 (special issue on limits of semiconductor technology).

Elers, K.-E. *et al.* (2002) Diffusion barrier deposition on a copper surface by atomic layer deposition, *Chem. Vapor Depos.*, **8**, 149–153.

Farcy, A. *et al.* (2008) Integration of high-performance RF passive modules (MIM capacitors and inductors) in advanced BEOL, *Microelectron. Eng.*, **85**, 1940–1946.

Furuya, A. *et al.* (2005) Ta penetration into template-type porous low-k material during atomic layer deposition of TaN, *J. Appl. Phys.*, **98**, 094902.

Helneder, H. *et al.* (2001) Comparison of copper damascene and aluminum RIE metallization in BICMOS technology, *Microelectron. Eng.*, **55**, 257–268.

Ho, P.S., W.W. Lee and J. Leu (2002) **Low Dielectric Constant Materials for IC applications**, Springer.

Hsu, H.-H. *et al.* (2001) Electroless copper deposition for ultralarge-scale integration, *J. Electrochem. Soc.*, **148**, C47.

Ishikawa, K. *et al.* (2008) Advanced method for monitoring copper interconnect process, *IEEE Trans. Semicond. Manuf.*, **21**, 578–584.

ITRS 2007: The International Technology Roadmap for Semiconductors, 2007 Edition. International SEMATECH: Austin, TX, 2007. http://www.itrs.net/

Ivanov, I.P., I. Sen and P. Keswick (2006) Electrical conductivity of high aspect ratio trenches in chemical-vapor deposition W technology, *J. Vac. Sci. Technol.*, **B24**, 523–533.

Koburger, C.W. *et al.* (1995) A half-micron CMOS logic generation, *IBM J. Res. Dev.*, **39**, 215.

Kriz, J. *et al.* (2008) Overview of dual damascene integration schemes in Cu BEOL integration, *Microelectron. Eng.*, **85**, 2128–2132.

Laurila, T. *et al.* (2000) Failure mechanism of Ta diffusion barrier between Cu and Si, *J. Appl. Phys.*, **88**, 3377.

Maex, K. *et al.* (2003) Low dielectric constant materials for microelectronics, *J. Appl. Phys.*, **93**, 8793–8841.

Mann, R.W. *et al.* (1995) Silicides and local interconnections for high performance VLSI applications, *IBM J. Res. Dev.*, **39**, 403.

Rao, G.K. (1993) **Multilevel Interconnect Technology**, McGraw-Hill.

Satta, A. *et al.* (2002) Enhancement of ALCVD TiN growth on Si–O–C and a-SiC:H films by O -based plasma treatments, *Microelectron. Eng.*, **60**, 59–69.

Small, M.B. and D.J. Pearson (1990) On-chip wiring for VLSI, *IBM J. Res. Dev.*, **34**, 858.

Smith, S. *et al.* (2002) Physical and electrical characterization of ALCVD TiN and WN_xC_y used as a copper diffusion barrier in dual damascene backend structures, *Microelectron. Eng.*, **64**, 247–253.

Steigerwald, J.M., S.P. Murarka, and R.J. Gutman (1997) **Chemical Mechanical Planarization of Microelectronic Materials**, John Wiley & Sons, Inc.

Wrschka, P. *et al.* (2000) Chemical mechanical planarization of copper damascene structures, *J. Electrochem. Soc.*, **147**, 706.

Zantye, P.B., A. Kumar and A.K. Sikder (2004) Chemical mechanical planarization for microelectronics applications, *Mater. Sci. Eng.*, **R45**, 89–220.

29

Surface Micromachining

Isotropic etching leads to undercutting of structures. This is generally considered a drawback of isotropic etching, but surface micromachining takes advantage of undercutting. Undercutting releases bridges, cantilevers and membranes that can be used as resonators, switches, movable mirrors, variable capacitors and in many other roles.

Instead of etching the substrate wafer to release structures, it is customary to use two thin films: an underlying sacrificial film and upper structural film. By etching the underlying film away, the structural film is released. The air gap formed by etching away the sacrificial film can serve as the dielectric in a capacitive pressure sensor, as an isolator in a RF switch and as a tunable optical path in an interferometer. The sacrificial film has to fulfill two major requirements: it has to tolerate the deposition process of the structural film; and selective etch process for the two materials must exist. The structural film has be mechanically stiff enough and reasonably stress free, otherwise it will bend, buckle or curl up when released. After released structures have been made, wafer processing becomes very difficult. Wet processes should be eliminated because liquid drying may lead to structure sticking during drying. Particles may also get stuck under released structures, preventing movements. Therefore wafer dicing, which uses water for cooling and generates silicon dust, is sometimes done before the release etch.

Note
The z-scale has been grossly exaggerated in the figures below to show the thin films more clearly; typical thin-film thicknesses in surface micromachining are in the micrometer range.

29.1 Single Structural Layer Devices

The simplest devices consist of two films only, one sacrificial, one structural. Such structures are ready-made in SOI wafers: the device silicon plays the role of structural layer and the buried oxide serves as the sacrificial layer. The extra benefit, in addition to simplicity, is that the excellent mechanical properties of single crystalline silicon, for example low stress, are available.

In the single axis piezoresistive accelerometer shown in Figure 29.1, doping of the SOI device layer is chosen to suit piezoresistor specifications. When the proof mass bends, the narrow piezoresistor between P1 and the proof mass will stretch (and its resistance increases) and the piezoresistor between P4 and the proof mass will be compressed (and its resistance decreases). In order to eliminate

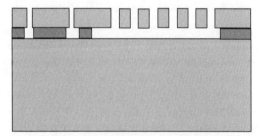
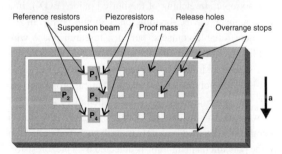

Figure 29.1 Piezoresistive accelerometer: cross sectional and top views of DRIE etched SOI device. Buried oxide etching for release. Reproduced from Eklund and Shkel (2007) by permission of IOP

Introduction to Microfabrication, Second Edition   Sami Franssila
© 2010 John Wiley & Sons, Ltd

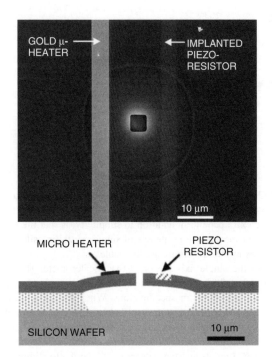

Figure 29.2 Thermally excited 10 MHz dome resonator, top view and cross-sectional view. Reproduced from Reichenbach et al. (2006), copyright 2006, by permission of IEEE

out-of-plane movement, the SOI device layer should be thicker than the piezoresistors are wide, but this condition is easily fulfilled by moderate linewidths (e.g., 3 μm) and a suitable SOI device layer (e.g., 15 μm). In its simplest implementation, this device can be fabricated in a single lithography step, followed by device silicon DRIE and isotropic HF etching of the buried oxide (BOX). In a more realistic device, bonding pads should be metallized for more reliable contacts, but in a research phase wire bonding can be done directly to silicon.

A structure similar to SOI can be made by CVD oxide and LPCVD polysilicon. Some mechanical properties are lost, but some degrees of freedom bought. Figure 29.2 shows a thermally excited dome resonator. Polysilicon membrane is excited by a gold microheater, and deflection of the polysilicon membrane is detected by a piezoresistor. The process flow is as follows.

Process flow for dome resonator

- Silicon wafer (no special requirements)
- CVD oxide deposition (phosphorus doped)
- LPCVD polysilicon deposition (undoped)
- Lithography for piezoresistor
- Ion implantation for piezoresistor
- Photoresist stripping and wafer cleaning
- Annealing for implant activation
- Lithography for gold heater
- Cr/Au deposition and lift-off
- HF isotropic oxide etching

Thermal oxidation could be done after piezoresistor implantation, to insulate the gold heater from the piezoresistor, but undoped polysilicon is practically an insulator. Both gold and silicon resistors tolerate the HF release etch well and no protection is needed. The mechanical properties of polysilicon are important because poly is an active mechanical material subject to 10 MHz vibrations.

After undercut etching, photoresist spinning is no longer possible, because it would spread and stick uncontrollably under the released layer. Therefore all lithographic steps must be performed before release etching. It is possible to use peeling masks (Figure 21.17) which are made ready but used only later on. In the optical grating shown in Figure 29.3 submicrometer features are made first: RIE of silicon is done to create a submicrometer grating, 500 nm deep, followed by aluminum metallization and aluminum patterning. SOI device layer (15 μm thick) lithography and etching then follow. It would be impossible to etch structures 15 μm deep first and then coat them with thin resist that would allow lithography of submicron features. In this device the SOI device layer etching is performed using the aluminum as an etch mask, and finally BOX isotropic release etching in HF vapor, because aluminum

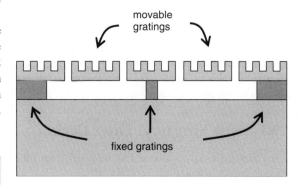

Figure 29.3 Optical spectrometer in SOI. Adapted from Sagberg et al. (2007)

is attacked by HF solutions. Underetching time is critical because it has to release the movable grating elements, but it must not release the fixed elements.

29.2 Materials for Surface Micromachining

The sacrificial layer has to tolerate the deposition process of the structural layer, and it has to be removable selectively with respect to the structural layer. Table 29.1 lists some commonly used pairs of structural and sacrificial layers. Silicon surface micromechanics utilizes LPCVD polysilicon as a structural layer, and oxides, especially CVD-PSG, as sacrificial layers. LPCVD nitride can be used as an additional structural or insulating layer.

Polysilicon, silicon dioxide and nitride, and sputtered metals are limited to a few micrometers in thickness. This is too small for many applications. Thick polysilicon can be used, but this material is not just thicker polysilicon: it is deposited differently compared to LPCVD "thin" polysilicon. Thick poly ("epi-poly," "poly-SOI") is deposited in an epitaxial reactor, with deposition rates of 1–5 μm/min vs. 10 nm/min for LPCVD poly. Therefore thicknesses of 10–50 μm are possible, similar to SOI device layer thicknesses. But because deposition is on an amorphous oxide, the resulting film will not be epitaxial (single crystalline). Electroplated metals and resists can also be applied in layers tens of micrometers thick.

Free-standing mechanical structures can be made of metals. Both sputtering (for Al) and electroplating (for Cu, Ni, Au) are used for structural layer deposition, and photoresist and many metals can be used as the sacrificial layers, for example all non-noble metals can be selectively etched underneath gold. Etch selectivity between resist and metal is practically infinite, but large structures are always difficult to release because the sacrificial etching relies on lateral diffusion in a restricted space. Free-standing spans of metals are smaller than those of poly and nitride structures.

If silicon dioxide is used as a sacrificial material, the removal etch has to be HF based. This limits the metals that can be used for device metallization. A first approach is to use metals that survive HF etching: gold, nickel and molybdenum are candidates. Second, different HF-based solutions are different in selectivity relative to aluminum, as shown in Table 29.2. Third, the metal can be protected by a thin film or photoresist. The fourth approach is to deposit metal after release etching, but a shadow mask or dry film resist lift-off has to be used because resist spinning cannot be done. Sacrificial etching is preferably the last process step because the released structures may bend, resonate, stick, break or otherwise be damaged in further processing steps.

There are many choices for sacrificial oxides. Thermal oxide, SiO_2, is seldom used, because it is the densest and has the lowest etch rate. Various deposited oxides etch much faster, even by a factor of 20, as shown in Table 29.2. Phosphorus-doped silicon glass (PSG) is a common choice. There is also the choice of HF solution: while 49% HF (highest standard concentration) has the highest oxide etch rate, it also attacks other materials, and therefore rate is not the only selection criterium.

Many combinations of materials have been used in making clamped–clamped beams (also known as microbridges). A prototypical RF switch is shown in Figure 29.4. It is made by depositing metal on top of photoresist, patterning the metal and washing the resist away in oxygen plasma. The capacitive RF switch consists of grounded lines and dielectric-coated RF signal lines. When a voltage is applied between the free-standing metal bridge and the RF signal lines, the bridge bends down. Capacitive coupling occurs because there is no metal-to-metal contact (RF switches with metal-to-metal contacts will be described in section 29.4).

Table 29.1 Materials for surface micromachining

Structural film	Sacrificial film(s)	Sacrificial etch(es)
Polysilicon	CVD oxide	HF, HF vapor
Silicon nitride	Oxide; Al	HF; NaOH, H_3PO_4
Nickel	Cu; resist	HCl; oxygen plasma
Aluminum	Resist	Oxygen plasma
Gold	Cu; resist	HCl; oxygen plasma
Copper	Resist	Oxygen plasma
Parylene	Resist	Acetone, other solvents
SU-8	Cu; Al	HCl; NaOH, H_3PO_4
Diamond	Cu	HCl

Table 29.2 HF-based wet etch rates (nm/min) for selected materials at room temperature

Etchant	Material					
	SiO_2	TEOS	PSG	Si_3N_4	Al	Mo
HF (49%)	1763	3969	4778	15	38	0.15
NH_4F:HF (7:1) (BHF)	133	107	1024	1	3	0.5
HF:H_2O 1:10	48	157	922	1.5	320	0.15

Source: Kim, B.-H. *et al.* (1999)

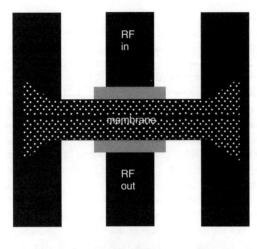

Figure 29.4 RF switch with nickel membrane, top and cross–sectional views: air gap h_a created by photoresist sacrificial etching

When the bridge is up, the gap h_a is large and capacitance very small:

$$C_{\text{off}} = \frac{1}{(h_D/\varepsilon_D A) + (h_a/\varepsilon_0 A)} \quad (29.1)$$

where h_D is the thickness of the dielectric layer on the signal electrode. The signal is shunted when the nickel bridge is pulled down. When the bridge makes contact with the dielectric, the gap h_a goes to zero, and on-capacitance is defined by dielectric thickness and permittivity. Use of materials with a high dielectric constant is desirable, but it is also important to maintain surface smoothness so that the contact distance can actually be made zero: a rough surface of the dielectric would mean that h_a is non-zero even in a contact situation.

29.3 Mechanics of Free-Standing Films

The structural layer needs to be of sufficient mechanical strength and proper stress state when released.

Free-standing beams and plates will bend depending on their stress state, as shown in Figure 5.17. A series of released beams with different lengths can act as a stress monitor.

Tensile stresses are preferred for free-standing structures. Depending on film mechanical properties, anything from 10 micrometer span lengths (for electroplated gold) to centimeters (for silicon nitride) are possible for structural layer dimensions. The length also depends on film thickness and stress, and on the details of release etching. The spring constant of a clamped–clamped beam (both ends firmly attached to an anchor, with distributed load) is calculated from

$$k = 32EW\left(\frac{t}{L}\right)^3 \quad (29.2)$$

where E is Young's modulus, t the beam thickness, L the beam length and W the beam width. Silicon and metals have high elastic moduli (Young's moduli) in the range of 100–200 GPa, but structures with very low spring constant can be made of polymers, which have values of E of a few gigapascals only. The drawback of low spring constants is the small restoring force, and therefore the danger of sticking. The resonant frequency of a simple clamped–clamped beam is given by

$$f = \frac{1}{2\pi}\sqrt{\frac{k}{m}} \approx \frac{W}{L^2}\sqrt{\frac{E}{\rho}} \quad (29.3)$$

Micromachined beams with 1–100 N/m spring constant and sizes in the range of tens to hundreds of micrometers result typically in kilohertz to megahertz resonant frequencies.

Residual stresses can easily dominate the resonant frequency. Frequency of a resonator with straight flexures (Figure 29.10) is given by Equation 29.4. Stress effect become significant especially for long, narrow beams. Stress control is one of the key issues in surface micromachining. It involves materials selection, deposition process optimization, annealing effects, geometrical design of released structures, and the release process.

$$f = \frac{1}{2\pi}\sqrt{\frac{4EtW^3}{ML^3} + \frac{24\sigma_r tW}{5ML}} \quad (29.4)$$

where M is shuttle mass and σ_r is the residual stress in the thin film. With thin film stresses easily in 100 MPa range, their effects can be drastic.

Compressively stressed clamped–clamped beams will remain straight for a while when stress is applied, but after a critical stress the beam will buckle. This critical stress, σ_{cr}, is given by

$$\sigma_{cr} = \frac{\pi^2 E t^2}{3L^2(1-\nu)} \quad (29.5)$$

where ν is the Poisson ratio of beam materials, and t the thickness and L the length of the beam. Stubbier beams tolerate more stresses, but then again, the actuation force has to stronger. If the beam is electrically conductive, and it is used as a thermal actuator, Equation 29.5 can be used to evaluate how high temperature (or how large thermal expansion) can be tolerated before critical stress is achieved.

29.4 Cantilever Structures

Cantilevers are very common structures. They are used in force sensors (including AFMs) and in many chemical and biological sensors where the added mass on the cantilever is detected either as a deflection or resonant frequency change. Two different techniques for surface micromachined cantilevers are depicted in Figure 29.5, showing the timed etching of sacrificial underlayer vs. a two lithography step process that is insensitive to under-etch time.

The single mask process of Figure 29.5 depends on timed etching: too much overetching would eliminate the anchor altogether and detach the cantilever from the substrate. Cantilever and anchor dimensions are closely related: the etch undercut must be long enough to release the cantilever, but short enough for the anchor to remain.

In the two-mask process the structural layer is attached to the substrate, and etch timing becomes irrelevant because the structure acts as its own anchor. Extended overetching does not destroy the structure, but poor etch selectivity between the layers may change the dimensions of the structural layer.

These free-standing structural layers serve many mechanical, electromechanical, thermal, fluidic and optical functions as mechanical resonators, mirrors, thermal isolation valves, cantilever sensors, RF switches, etc. Movement of the released structure can be either in the plane of the wafer, or out of the plane. Etching sacrificial layers is also useful for creating static structures, like channels and nozzles in microfluidics.

Additional lithography and etching steps can be used to make more complex cantilevers. For instance, a nitride cantilever with a rigid paddle at the end is shown in Figure 29.6. Even though the cantilever bends, the paddle will remain flat. This design is used in the biosensor of Figure 20.26 to improve the laser reflection signal.

The simple cantilever is not a device, but it can be transformed into a practical in-line RF switch (Figure 29.7). Its fabrication process is given below.

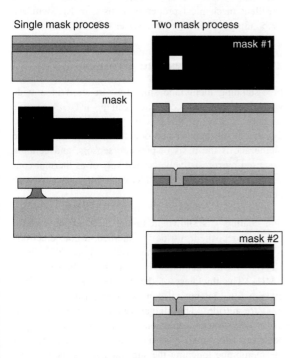

Process flow for gold switch

- Cr/Au anchors and pull-down electrode patterning (20/250 nm thick)
- Cu and Au sputtering (1000 nm/500 nm)
- Au lithography and wet etching
- Cu lithography and wet etching
- Thick-resist lithography for Au electroplating
- Au electroplating (8 μm thick)
- Resist removal
- Cu release etching

Figure 29.5 Polysilicon cantilevers with CVD oxide sacrificial layer: single mask process depends on timed etching while two-mask process is insensitive to oxide etch time

Because gold is sputtered right after copper in the same vacuum, no adhesion layer is needed. Gold wet etching in aqua regia is difficult, especially when it comes to resist adhesion, but 500 nm is difficult to

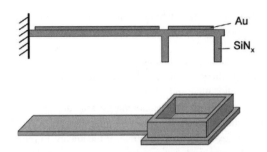

Figure 29.6 Cantilever sensor stiffening by a filled groove. Reproduced from Yue et al. (2004) by permission of IEEE

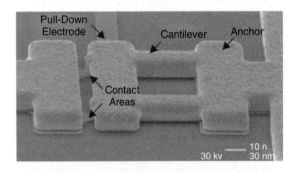

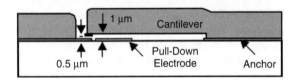

Figure 29.7 Gold cantilever switch (with copper sacrificial layer). Reproduced from Rebeiz and Muldavin (2001), copyright 2001, by permission of IEEE

pattern by lift-off too. There are many wet etch solutions for isotropic copper etching.

The spring constant of this design is larger than 100 N/m for a cantilever 75 μm long and 30 μm wide. The gap between the two gold contacts is 500 nm, and 30 V is needed to actuate the switch.

Thin-film stresses should be minimized, otherwise the gap between switch contacts, for instance, would become indeterminate. In an ingenious curl switch design stresses are in fact taken advantage of. The switch consists of an aluminum top electrode sandwiched between two PECVD oxide layers (Figure 29.8). The top oxide is tensile stressed, and upon release etching it will result in a sizable upward curl. Off-capacitance is very high because

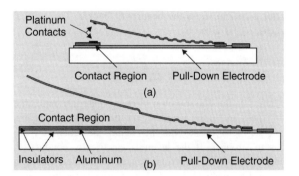

Figure 29.8 Curl switch: in-line (a) and capacitive (b) versions. Reproduced from Rebeiz and Muldavin (2001), copyright 2001, by permission of IEEE

the gap is now 10–15 μm above the bottom electrode. And the actuation voltage is reasonable at 80 V because the initial gap near the anchor end is still quite small.

In many cases the same functionality can be achieved by bulk or surface micromachining. The variable capacitor of Figure 17.2 is an aluminum membrane released by back-side etching, but a similar device could be made by front-side release etching. Single crystal silicon with its excellent mechanical properties is used in RF switches, by bonding silicon to a glass wafer (Figure 30.26).

29.5 Membranes and Bridges

Free-standing membrane structures are used in a wide variety of microdevices. They can be used as passive supports in thermal devices, but they can also serve as dynamical mechanical and optical elements. A movable membrane is the basis of pressure sensors, and bending is sensed piezoresistively, capacitively or as thermal conductance change due to surfaces moving closer to each other (Figure 20.20).

An optical modulator ("vertically moving antireflection coating") is shown in Figure 29.9. Oxide and nitride are deposited, nitride is patterned, a gold electrode is deposited and patterned, and oxide is etched away in HF. The nitride membrane is electrostatically actuated at 1 MHz. The thickness of the nitride is $\lambda/4n$ (which is 194 nm for 1550 nm telecom wavelength assuming $n = 2.00$ for nitride). The air gap (with $n = 1$) is designed to be $m\lambda/4$ for antireflection action. For $m = 1$ the gap becomes very small, 388 nm, and there is a danger of the membrane touching the substrate, so for example an $m = 4$ device is much easier to fabricate, corresponding to 1.55 μm sacrificial oxide thickness.

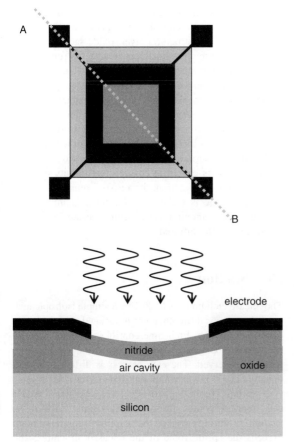

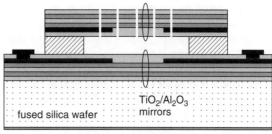

Figure 29.10 Visible light Fabry–Pérot interferometer: ALD dielectric mirrors made of TiO_2 and Al_2O_3. Polymer sacrificial layer defines the gap. Redrawn from Blomberg et al. (2009)

Figure 29.9 Optical modulator: nitride membrane is electrostatically actuated and air cavity optical length changes. Cross-section along line AB. Redrawn from Goossen et al. (1994)

Because silicon is transparent in the infrared wavelengths, it will travel through the wafer without any actual optical window.

In the visible wavelengths the substrate has to be transparent. In the Fabry–Pérot interferometer of Figure 29.10, a fused silica wafer has been used. Dielectric mirror λ/4 layers are made of TiO_2/Al_2O_3. Required layer thicknesses are smaller in the visible wavelengths. ALD, with atomic layer thickness control, is a method of choice for such films. ALD film quality is good also for very thin films, without any post treatment. The third benefit is low deposition temperature: ALD films can be deposited on polymers.

Movable membranes are also display devices: the gap defines the color (as in Fabry–Pérot!) and zero gap equals a black pixel (such a device is quite naturally processed on a glass wafer). Adaptive optics is based on movable, steerable membranes (Figure 17.7) in up–down mode, but many optical telecom switches, projection displays and scanners utilize torsion-mode membrane mirrors (Figure 21.12).

Electrostatic actuation is easy to implement with metallic thin films, or using the silicon substrate as the other electrode. The force and actuation voltage are related, however, by the nonlinear relationship

$$F = \frac{\varepsilon A V^2}{2(d-x)^2} \quad (29.6)$$

As the gap changes from its initial value d, the force rapidly increases because of the $(d-x)^2$ term. This eventually leads to the plates touching each other, a phenomenon known as pull-in. The pull-in voltage for a membrane depends on the spring constant (k), the air gap dimension (g_0) and the dielectric constant of air (ε), that is

$$V_{\text{pull-in}} = \sqrt{\frac{8kg_0^3}{27\varepsilon}} \quad (29.7)$$

Gaps range from tens of nanometers to hundreds of micrometers. Typical actuation voltages range from a few volts to hundreds of volts. As a way to improve tunability, double gap designs have been implemented. The actuation electrode gap is large but the signal electrode gap is small. This way the movable electrode can be moved very close to, and even to contact with, the signal electrode without pull-in. See Exercise 29.12.

A comb drive with interdigitated fixed and movable electrodes (Figure 21.2) is a versatile sensor and actuator.

The capacitance of a comb drive depends on capacitor plate area and distance, as usual, but compared to a bulk micromachined capacitor, we now have the freedom to increase the number of comb fingers to increase capacitance.

If large areas need to be released, perforations need to be employed (Figure 29.11). Sacrificial etching can then penetrate under a large plate from multiple points. Obviously, these extra holes have different consequences: in electrostatic actuation fringing fields will cover small perforations (hole diameter <3–4 × gap size), and there is no loss of electrostatic force. Capacitance is, however, reduced in the down state. Young's modulus is reduced and so is residual stress, even to half the original value. Perforations affect gas damping: air can flow through the perforations, while in the case of a continuous membrane the only escape route is over the edges. Perforated membranes can therefore have higher operating frequencies.

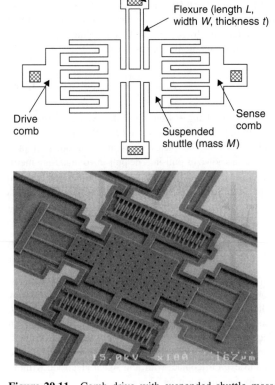

Figure 29.11 Comb drive with suspended shuttle mass. Plate release has been aided by using perforations in the plate. Reproduced from Bustillo et al. (1998) by permission of IEEE

Sideways movements can be realized also by thermal actuation. The microrelay shown in Figure 29.12 is made with a polysilicon structural layer, oxide sacrificial layer process, with a few modifications. Current is run through the polysilicon V-shaped actuator beam, which expands, moves laterally and makes contact with the signal electrodes. In order to eliminate electrical contact from the electrically heated actuator to the signal line, a silicon nitride insulator block is made before poly deposition. This also reduces thermal conduction from the hot actuator beam. The sacrificial oxide is etched in two steps: first a small recess is etched to ensure discontinuity of gold; and after gold evaporation, the resist is removed and all remaining oxide is etched by HF. Infrared microscopy can be used to monitor underetching because silicon is transparent in the infrared.

29.6 Stiction

The release etch process looks like a simple isotropic etch but it has many difficulties not associated with isotropic patterning etching. Etch time control is difficult because etch front propagation under the structural layer cannot often be observed. The etch process is diffusion limited in nature and slows down in long and narrow release gaps.

A serious limitation for the wet chemical release etch process comes from stiction (from "sticking + friction"): during drying the capillary force strength exceeds the spring force of the released structures, and the free-standing cantilever makes contact with the substrate and adheres. Equation 29.8 gives the critical length for a cantilever to stick:

$$L_{\text{crit}} = \sqrt[4]{\frac{3}{4} \frac{Et^3 g^2}{\gamma \times \cos\phi}} \qquad (29.8)$$

where t is cantilever thickness and g the gap, γ the liquid surface tension and ϕ the contact angle. For a silicon beam ($E = 160\,\text{GPa}$) of 1 μm thickness and 1 μm gap, using water ($\gamma = 72\,\text{mJ/m}^2$) for rinsing, this critical length comes to approximately 40 μm.

Stiction prevention has many alternative approaches, as hinted by Equation 29.8. One is to use thicker beams, but LPCVD polysilicon thicknesses are limited to a micrometer or two. Polysilicon and SOI beams can be made thicker. The gap could be increased, but the actuation voltage then becomes larger. It is also difficult to increase the gap when oxide is used as a sacrificial layer because its thickness is also limited to a few micrometers. Contact angle modification can be done by silane SAMs or fluoropolymers. This strategy

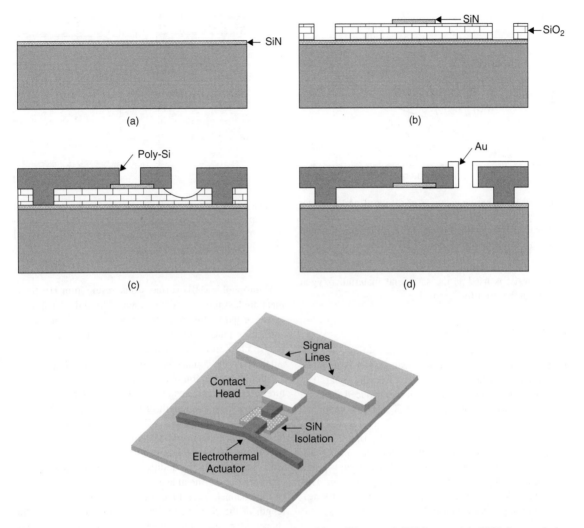

Figure 29.12 Sideways thermal actuator/microrelay. Reproduced from Wang *et al.* (2001), copyright 2001, by permission of IEEE

works best for in-use stiction prevention, once drying has been successful. Replacing water is one way to go. The elimination of capillary forces by going to dry release is yet another approach. And traditional drying can be replaced by other methods. As a fifth element, microstructures can be designed stiffer.

29.6.1 Stiffer structures

Using shorter beams, thicker beams or making non-flat beams, U-shaped (or T- or H-shaped), helps because they are stiffer. This requires an additional lithography step, but this has to be balanced against the efforts needed in other stiction prevention schemes.

29.6.2 Alternative liquids

Water has an exceptionally large surface tension, which leads to strong capillary forces. Changing water to isopropanol or some other low-surface-tension liquid will reduce stiction. The same strategy works for porous materials: in order not to collapse the thin pore walls, surface tension must be reduced.

29.6.3 Dry release

If silicon (single crystalline or thin film) is used as the sacrificial material, isotropic SF_6 plasma and XeF_2 are suitable. If oxide is used, anhydrous HF vapor can be used, but its etch rate is lower than with aqueous HF.

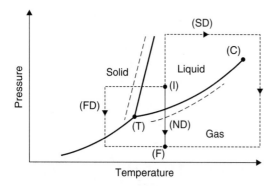

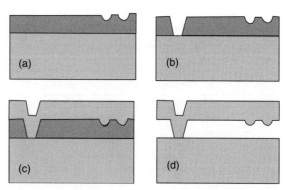

Figure 29.13 Thermodynamics of drying: I, initial stage; F, final stage; N, normal drying; FD, freeze drying; SD, supercritical drying. Reproduced from Bellet and Canham (1998) by permission of Wiley-VCH

Figure 29.14 Three-mask process for cantilever: (a) mask 1, dimples in oxide; (b) mask 2, anchor in oxide; (c) structural poly deposition and mask 3, poly pattern; (d) sacrificial oxide layer etching

If photoresist is used as the sacrificial material, oxygen plasma can be used for removal.

29.6.4 Alternative drying

Sublimation or critical point drying both sidestep the liquid drying step. In sublimation, rinsing water is replaced by tert-butanol, which has a subzero freezing point. The wafer is then frozen and heated under reduced pressure in a regime where solid tert-butanol turns to vapor directly (sublimation). This is known as freeze drying (FD) (Figure 29.13). In critical point drying CO_2 (b.p. 31 °C) is used. At the critical point the phase balance can be chosen to favor solid–vapor transformation over solid–liquid transformation by using pressure as the parameter. This is shown below as route SD (for supercritical drying). Normal drying is indicated as ND.

29.6.5 Surface microstructures

Because stiction depends on surface smoothness (on the microscale) and flatness (on the macroscale), just like wafer bonding, corrugated or otherwise patterned surfaces can prevent stiction, Figure 29.14. This approach requires extra process steps that need to be integrated into the process flow. Sometimes existing process steps can be utilized by making minor mask changes to create protrusions or other extensions.

Avoiding stiction during the fabrication process is one thing; avoiding stiction during device operation is another. RF switches operate by making contact between two surfaces. Both metal-to-dielectric contacts (Figure 29.4) and metal-to-metal contacts (Figure 29.6) are used. Some switches conduct sizable currents in contact, which may lead to the welding together of two metals.

Hydrophobic (HB) surfaces that result from HF treatment are less prone to stiction than hydrophilic (HL) ones, as suggested by Figure 22.9: the bond strength of the HL surface is much higher than that of HB surfaces. This is also clear from Equation 29.8: the $\cos \phi$ term is very small when the contact angle is high, making L_{crit} longer.

Stiction prevention is important also when static structures are made: if the capillary force is too large compared to the mechanical strength of the structure, the mechanical structure will collapse. Microfluidic channels must be designed with this in mind. Figure 29.15 presents two alternative ways for the fabrication of microfluidic channels with electrodes. In the traditional surface micromachining approach the sacrificial layer is etched away from the ends of the channel. Thus time increases as the square of the length of the channel. In the micromolding approach the shape of the channel is etched into silicon and coated with parylene. A carrier wafer with electrodes is bonded over the channel replica, and the carrier wafer is removed. No stiction can happen because no liquids are present.

29.7 Multiple Layer Structures

Many materials with different properties create functionalities, but every new material interface is a potential problem: adhesion, thermal expansion mismatch, chemical reactions, etc. Working with single material is therefore beneficial, but not many materials offer enough functionalities.

Figure 29.16 shows a microfabricated acoustic crystal. The active layer consists of tungsten scatterers embedded in a silicon dioxide matrix. The large difference in

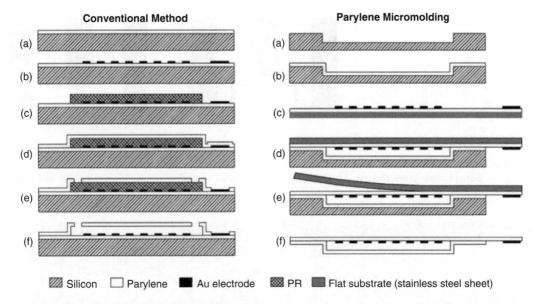

Figure 29.15 Surface micromachining with sacrificial layers vs. parylene molding for fabrication of a microchannel with electrodes. Reproduced from Noh *et al.* (2004), copyright 2004, by permission of Elsevier

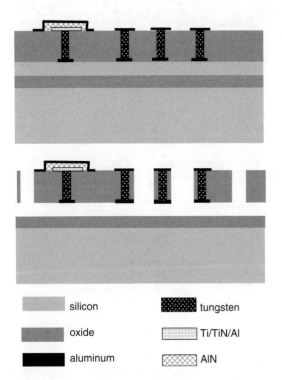

Figure 29.16 Acoustic band gap (ABG) crystal. Tungsten scatterers embedded in oxide, with polysilicon as sacrificial layer. Redrawn after Olsson *et al.* (2008)

sound velocity and density between the materials is essential. The piezoelectric aluminum nitride couplers on the sides are used for feeding acoustic energy into and out of the device. Polysilicon is chosen as the sacrificial layer and SF_6 isotropic plasma for dry release. This means that tungsten must be protected because it is readily etched by fluorine. Aluminum caps protect the tungsten at the top and bottom, and the sidewalls are covered by oxide. CMP is used after CVD-W in order to achieve a planar surface.

Two structural layer processes offer similar device and fabrication benefits also in metal micromechanics. Electroplating processes are basically room temperature processes and a wide variety of materials, including polymers, can be used. A simple process for free-standing metal structures is shown in Figure 29.17, and 3D copper coils and an air-isolated nickel-core transformer are shown in Figure 29.18.

29.8 Rotating Structures

Adding more layers enables rotating structures to be made. A center-pin process utilizes two structural and two sacrificial layers (Figure 29.19). Now, because of mechanical constraints, poly 1 becomes the movable element and poly 2 serves as the fixed element which bounds the rotating element made of poly 1. The first sacrificial layer defines the gap to the substrate,

380 Introduction to Microfabrication

Figure 29.17 Electroplated free-standing structure: left, first resist patterning and seed metal deposition, followed by second, thick-resist patterning; middle, electroplating; right, stripping of second resist, seed metal etching and stripping of the first resist

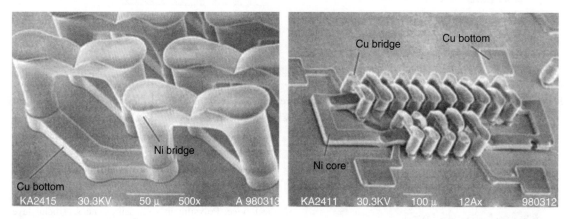

Figure 29.18 SEM micrographs of 3D metal electroplated structures. Reproduced from Yoon *et al.* (1998) by permission of Institute of Pure and Applied Physics

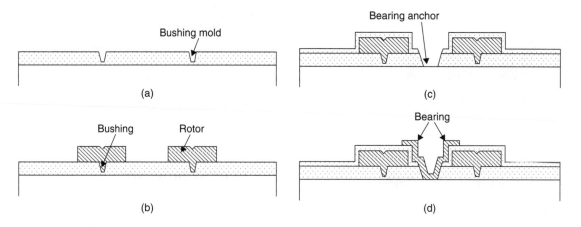

Figure 29.19 Cross-sectional schematics demonstrating the center-pin bearing process: (a) after patterning of the bushing mold in the first sacrificial layer; (b) after deposition and patterning of poly 1; (c) after deposition of the second sacrificial layer and anchor region definition; (d) deposition and patterning of poly 2, followed by oxide release etching. Courtesy Mehran Mehregany

Surface Micromachining 381

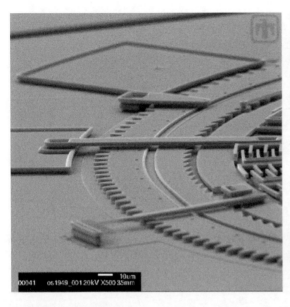
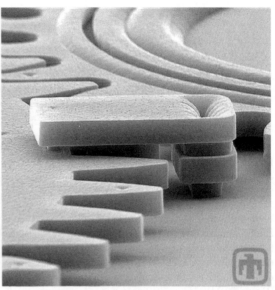

Figure 29.20 Mechanical gears made in multiple layer polysilicon process. Courtesy Sandia National Laboratories

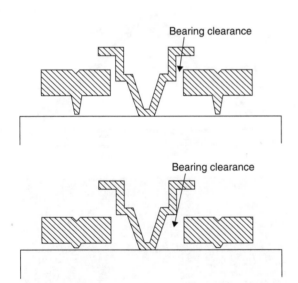

Figure 29.21 Cross-sectional schematics demonstrating two types of center-pin bearings that may result after release: (a) self-aligned; and (b) non-self-aligned. Courtesy Mehran Mehregany

and the second sacrificial layer determines the spacing between the poly layers.

Surface micromachining with polysilicon can be extended into multiple layers, as was already shown in Figure 16.9. Mechanical gears made in a multilevel polysilicon process are shown in Figure 29.20.

The concept of self-alignment is useful in released structures as well. The center-pin and rotor can be made in a self-aligned manner (Figure 29.21). It depends on the relative thickness of the structural and sacrificial layers. The poly 2 pin can be made to limit the movements of the poly 1 rotor in the lateral direction. In the opposite case the rotor can wobble because the center-pin is too high.

29.9 Hinged Structures

Structures that pop up from the plane of the wafer can be made by various different methods. Mechanical hinges can be made in a two structural layer process, or with polymeric hinges in a one-layer process. Such structures have applications for example as large stroke movable mirrors and as legs for microrobots.

In the polymeric hinge process a polyimide hinge connects a fixed plate and a movable plate. The movable plate can be actuated by for example thermal expansion of the polymer. In the poly staple process two layers of poly are used. The first poly forms the mirror plate, and the second poly forms a staple which allows the first poly to move but constrains its movement (Figure 29.22). Actuation can be accomplished by for example a comb drive which pushes up the movable mirror (Figure 29.23).

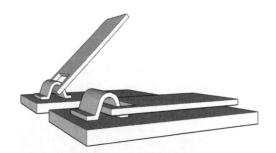

Figure 29.22 Two-poly staple hinge. Adapted from Pister *et al.* (1992) by Jorma Koskinen

In order to position the movable mirrors at proper angles, latches are designed into the mirrors. When the desired movement is performed, the polysilicon elements latch to each other, fixing the structure at the desired position, for example 45° or perfectly vertical (Figure 29.24).

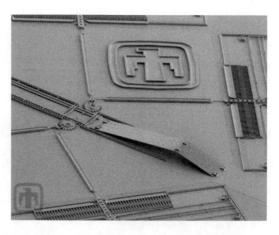

Figure 29.23 Comb-drive actuator–gear system lifts up a hinged mirror. Courtesy Sandia National Laboratories

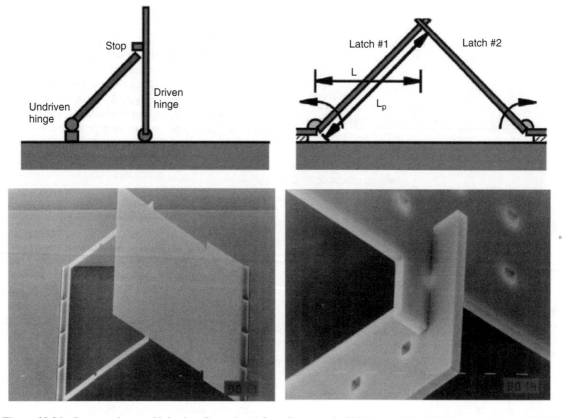

Figure 29.24 Pop-up mirrors with latches. Reproduced from Syms *et al.* (2001), copyright 2001, by permission of IEEE

29.10 CMOS Wafers as Substrates

Because all the functionality of surface micromachined structures is in the deposited thin films, any types of wafers can be used as substrates. CMOS wafers can therefore be used as substrates with some limitations, money and thermal budget being the foremost. But in devices where integration of drive or readout electronics is essential, like many array devices, the CMOS wafer is an obvious choice for a starting material. Figure 29.25 shows a micromirror array fabrication process. Each mirror has its own driving transistor underneath.

Process flow for micromirror array

- CMP planarization
- CVD oxide mirror stoppers
- Sacrificial photoresist
- Lithography for support posts
- Mirror metal deposition
- Mirror patterning
- Sacrificial resist removal

Integrated sensors benefit from the amplification of weak signals locally next to the sensor and only then are they transferred for further signal processing. Infrared pixel arrays are especially well suited for this as are many other imaging and display devices, like fingerprint sensors and Braille actuators. Monolithic integration of MEMS with electronics will be discussed further in Chapter 39.

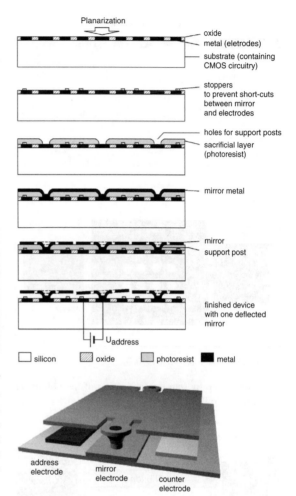

Figure 29.25 Micromirror array on CMOS wafer, and detail of a single mirror. Reproduced from Ljungblad *et al.* (2001) by permission of Elsevier

29.11 Exercises

1. What etch selectivity is needed to release a silicon nitride plate 1 μm thick and 50 μm wide by sacrificial oxide etching (49% HF, rate 2 μm/min), if the plate thickness variation due to etching has to be smaller than the nitride deposition non-uniformity of 3%?
2. Develop a fabrication process for the RF switch of Figure 29.4.
3. What is the ratio of on-capacitance and off-capacitance of the switch in Figure 29.4, assuming 100 nm oxide dielectric and 4 μm air gap.
4. Calculate the thicknesses and etch depths required to make a self-aligned rotor of Figure 29.21.
5. Develop a fabrication process for the polysilicon hinged mirror structure shown in Figure 29.22.
6. Design a fabrication process for the curl switch of Figure 29.8.
7. How many photomasks are needed to fabricate the copper–nickel coils of Figure 29.18?
8. What is the effect of patterning process tolerances to comb-drive resonant frequency when $W = t = 2$ μm?
9. Analyze the comb-drive resonant frequency shift due to residual stresses.
10. How can you fabricate the capacitive microphone shown below? The aluminum membrane mask is also shown.

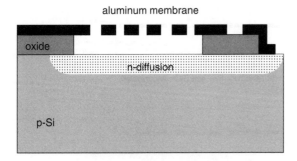

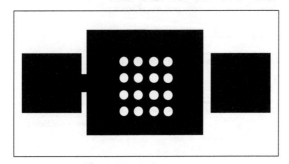

Adapted from Ganji and Majlis (2009)

11. Explain step by step the fabrication of the bolometer shown below.

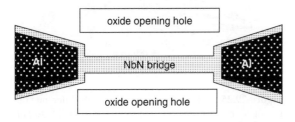

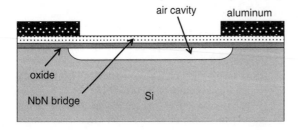

12. If the following condition holds, the suspended electrode shown below can be brought into contact with the signal electrode:

$$d_3 \leqslant \frac{d_2}{2} + \frac{d_1}{2\varepsilon_r}$$

where ε_r refers to the dielectric constant of the material covering the actuation electrodes and d_1, d_2 and d_3 are the three dielectric thicknesses (d_1 is solid material, d_2 and d_3 are air). Calculate some realistic dimensions for this condition to take place.

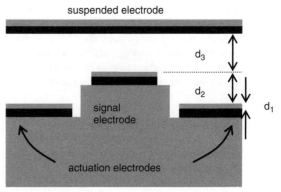

Redrawn after Nieminen et al. (2002)

13. The micromirror shown below is made in a six-mask process on a completed CMOS wafer. Detail the process steps.

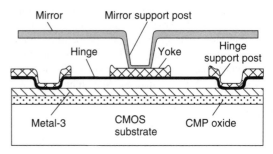

Reproduced from van Kessel et al. (1998) by permission of IEEE

References and Related Reading

Bellet, D. and L. Canham (1998) Controlled drying, Adv. Mater., **10**, 487.

Blomberg, M., H. Kattelus and A. Miranto (2009) Electrically tunable surface micromachined Fabry-Perot interferometer for visible light, Proccedings of Eurosensors 2009, p. 191.

Bühler, J., F.-P. Steiner and H. Baltes (1997) Silicon dioxide sacrificial layer etching in surface micromachining, J. Micromech. Microeng., **7**, R1–R13.

Bustillo, J. et al. (1998) Surface micromachining for microelectromechanical systems, Proc. IEEE, **86**, 1559.

Eklund, E.J. and A.M. Shkel (2007) Single-mask fabrication of high-G piezoresistive accelerometers with extended temperature range, J. Micromech. Microeng., **17**, 730–736.

Ganji, B.A. and B.Y. Majlis (2009) Design and fabrication of a new MEMS capacitive microphone using a perforated aluminum diaphragm, *Sens. Actuators*, **A149**, 29–37.

Goossen, K.W., J.A. Walker and S.C. Amey (1994) Silicon modulator based on mechanically-active anti-reflection layer with 1 Mbit/sec capability for fiber-in-the-loop applications, *IEEE Photonics Technol. Lett.*, **6**, 1119–1121.

Huang, I.-Y. et al. (2009) Development of a wide-tuning range and high Q variable capacitor using metal-based surface micromachining process, *Sens. Actuators*, **A149**, 193–200.

Kim, B.-H. et al. (1999) MEMS fabrication of high aspect ratio track-following microactuator for hard disk drive using silicon on insulator, Proceedings of MEMS 1999, p. 53.

Ljungblad, U. et al. (2001) New laser pattern generator for DUV using a spatial light modulator, *Microelectron. Eng.*, **57–58**, 23–29.

Löchel, B. et al. (1996) Ultraviolet depth lithography and galvanoforming for micromachining, *J. Electrochem. Soc.*, **143**, 237.

Malek, C.K. and V. Saile (2004) Applications of LIGA technology to precision manufacturing of high-aspect-ratio microcomponents and -systems: a review, *Microelectron. J.*, **35**, 131–143.

Nieminen, H. et al. (2002) Microelectromechanical capacitors for RF applications, *J. Micromech. Microeng.*, **12**, 177–186.

Noh, H.-S., Y. Huang and P.J. Hesketh (2004) Parylene micromolding, a rapid and low-cost fabrication method for parylene microchannel, *Sens. Actuators*, **B102**, 78–85.

Olsson, R.H. et al. (2008) Microfabricated VHF acoustic crystals and waveguides, *Sens. Actuators*, **A145–146**, 87–93.

Pister, K. et al. (1992) Microfabricated hinges, *Sens. Actuators*, **A33**, 249.

Rebeiz, G.M. and J.B. Muldavin (2001) RF MEMS switches and switch circuits, *IEEE Microw. Mag.*, December, 59.

Reichenbach, R.B. et al. (2006) RF MEMS oscillator with integrated resistive transduction, *IEEE Electron Device Lett.*, **27**, 805–807.

Roy, S. et al. (2002) Fabrication and characterization of polycrystalline SiC resonators, *IEEE Trans. Electron Devices*, **49**, 2323.

Sagberg, H. et al. (2007) Two-state optical filter based on micromechanical diffractive elements, Proceedings of IEEE Optical MEMS and Nanophotonics, 2007, pp. 167–168.

Soh, M.T.K. et al. (2005) Evaluation of plasma deposited silicon nitride thin films for microsystems technology, *J. Microelectromech. Syst.*, **14**, 971–977.

Syms, R.R.A. et al. (2001) Improving yield, accuracy and complexity in surface tension self-assembled MOEMS, *Sens. Actuators*, **A88**, 273.

van Kessel, P.F. et al. (1998) A MEMS-based projection display, *Proc. IEEE*, **86**, 1687.

Wang, Y. et al. (2001) A low-voltage lateral MEMS switch with high RF performance, *J. Microelectromech. Syst.*, **13**, 902–911.

Yoon, J.-B. et al. (1998) Monolithic fabrication of electroplated solenoid inductors using three-dimensional photolithography of a thick photoresist, *Jpn. J. Appl. Phys.*, **37**, 7081.

Yue, M. et al. (2004) A 2-D microcantilever array for multiplexed biomolecular analysis, *J. Microelectromech. Syst.*, **13**, 290–299.

30

MEMS Process Integration

MEMS devices come in a bewildering variety, in regard to structures, materials and functions. Whereas all CMOS technologies are close relatives, MEMS devices are made with a multitude of closely related, distantly related and unrelated technologies. Pressure sensor operation can be based on piezoresistive, capacitive, thermal conductance or resonance mechanisms, and while the first three share some structural features and fabrication steps, the fourth bears more resemblance to gyroscopes and RF oscillators. Accelerometers are made of single crystal silicon, of bulk, epitaxial and SOI type, of polycrystalline silicon and of electroplated metals. Electrospray emitters have been realized in silicon, glass, quartz, SU-8, PDMS, PMMA and many other polymers. Microneedles have been made in both out-of-plane and in-plane configurations, from single crystal and polysilicon, silicon dioxide, metals, polymers. Micromirrors are most often made of silicon, but also of metals or metallized nitride membranes. Mirrors come in in-plane (horizontal) and out-of-plane (vertical) versions too.

Bulk and SOI MEMS structures have high aspect ratios and highly complex 3D shapes resulting from etching and wafer bonding. These put new requirements on subsequent lithography, doping and thin-film steps, and introduce novel metrology requirements. MEMS devices with through-wafer holes pose process limitations: for instance, in spinning a vacuum chuck holds the wafer, and DRIE reactors use backside helium cooling. Through-wafer structures require double-sided processing and even without through-holes, there is often a need to align structures on the two sides of the wafer. Double-sided alignment is also mandatory for structured wafer bonding.

MEMS devices are not solid state devices: they have free-standing, moving, rotating and vibrating structures, like the grids of old-fashioned vacuum tubes. These moving structures create challenges for the subsequent processing and packaging steps. Capillary forces in drying, silicon dust and vibrations during dicing, or stresses in encapsulation may damage delicate mechanical structures. Closed cavities can sometimes be handled without problems, but high temperatures and changing pressures during fabrication can cause some design limitations, especially when the cavity roof is a thin diaphragm. Despite this seeming variety, there are many generic structures, design principles and widely used techniques for realizing them. These are the topics of this chapter.

30.1 Silicon Microbridges

A simple device like a silicon microbridge can be made in numerous different ways, depending on wafer choice, bulk <100>, <110>, <111> wafers, or from epitaxial or SOI material. Both DRIE and wet etching can be used and many process variations are possible with different combinations of front-side and back-side processing (two examples in Figure 30.1). Table 30.1 lists nine different processes that can be used to make single crystal silicon microbridges (clamped–clamped beams), but this list is

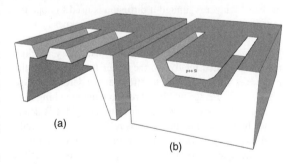

Figure 30.1 Silicon microbridges: (a) front-side KOH definition and timed back-side KOH release; (b) front-side bridge definition by p^{++} diffusion and front-side KOH release

Introduction to Microfabrication, Second Edition Sami Franssila
© 2010 John Wiley & Sons, Ltd

Table 30.1 Single crystal silicon bridges

Material	Bridge definition	Bridge release	Figure
Bulk <100>	Front-side KOH	Back-side KOH	30.1
p^{++} <100>	Front-side p^{++} doping	Front-side KOH	30.1
<111>	Front-side DRIE twice	Front-side KOH	20.34
Bulk <100>	Front-side DRIE	Front isotropic plasma	21.13
Bulk <100>	Front-side doping	Porous silicon etching	23.26
SOI	Front-side DRIE	Handle wafer isotropic	21.14
SOI	Front-side DRIE	BOX etching in HF	29.1
SOI	Front-side DRIE	Notching effect	21.28
Cavity SOI	Front-side DRIE	None required	30.18

by no means exhaustive! The final structure will be the same to a first approximation, but there are many details that differ, for example range of bridge thickness, degree of gap control underneath the bridge, silicon doping level, single-sided vs. double-sided lithography, etc. These will be briefly analyzed below.

The simple silicon bridge of Figure 30.1a suffers from many shortcomings: timed etching is used to define bridge thickness, and double-sided lithography is needed, even though the alignment is non-critical as the back-side opening can be made large. Both DRIE and KOH versions can be done with similar ease (but with different sidewall angle) and bridge doping can be varied over a large range. The <110> version is done similarly.

In the p^{++} etch stop version (Figure 30.1b) the bridge dimensions are determined mainly by the boron doping profile, but also by etch selectivity between lightly doped and highly doped silicon. This can be maximized by suitable selection of wet etchant, its concentration and temperature (Table 20.2), but there will be some loss of the p^{++} material and the sidewall will be rough. The thickness of the p^{++} layer is limited if diffusion is used to dope the top silicon. If epitaxy is used, then any thickness can be made, but a DRIE step is needed to etch through the p^{++} layer. In front-side release it is also mandatory to align the bridges tilted relative to the wafer flat, not along the main axes of the crystal, to effectuate undercutting (see Figure 20.9). Bridge width and required underetching time are correlated, and the gap underneath the bridge is determined by the etching time, so it is related to bridge width.

In the <111> bridge it is also mandatory to carefully align the bridges to crystal planes, and to use DRIE in the first step because <111> silicon cannot be etched by KOH. The process is described in Figure. Bridge thickness, gap and doping level can be freely chosen: the first DRIE step defines the bridge thickness and the second DRIE step defines the gap underneath the bridge. Sidewalls need oxide or nitride protection, and an additional RIE of oxide or nitride is required. Bridge width is not

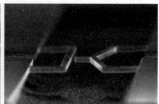

Figure 30.2 Microbridges by p^{++} etch stop and SOI. SEM micrographs courtesy Kestas Grigoras and Lauri Sainiemi, Aalto University

related to gap, but excessive undercut takes place in some crystal orientations, wasting area (Figure 20.34).

Combined anisotropic DRIE and isotropic RIE (Figure 21.13) can be used to make narrow and thick bridges only: because the isotropic release etch step eats underneath the bridge, wide bridges will have diminished thickness and rather poor thickness control because the undercut etch rate is not controllable to any great accuracy.

The gap of a SOI bridge is limited by the fact that BOX thicknesses are at most a few micrometers. Such small gaps are prone to stiction and dry HF-vapor release is often employed. Because BOX etching extends laterally, some area may become non-usable because the silicon is undercut. The gap can be increased without a limit if additional steps are taken to etch through BOX and the handle wafer is etched (Figure 30.2).

In yet another variant, the SOI handle wafer is wet etched to increase the gap. Isotropic 1:3:8 etchant (Section 11.12 on silicon etching) is selective between highly and lightly doped silicon, with the highly doped silicon etched faster. If the SOI device layer is lightly doped, and the handle highly doped, isotropic wet etch can be done without any protection of device layer bridges. This same etch is utilized in piezoresistor fabrication (see Figure 30.7).

SOI device layer etching against BOX leads to notching (see Figure 21.28). If narrow bridges are made, notchings will meet under the beam and released bridges result. The underside of this bridge is not flat, just like that of the bridge in Figure 21.13 but for a different reason: notching is different from isotropy. Generally, however, notching must be avoided in order to have proper control of beam shape and dimensions.

Simple devices like microbridges have many other applications in addition to resonators: they can function as mechanical actuators, switches, infrared light sources, microreactors for chemical synthesis, and as thermal detectors for explosive vapors, to name but a few. Cantilevers for AFM and chemical sensors are fabricated using exactly the same techniques as microbridges, but there are some additional ways of making cantilevers because the cantilever geometry offers more undercutting possibilities than clamped–clamped bridges.

30.2 Double-Sided Processing

In single side polished (SSP) wafers the backside is rough, with micrometer peak-to-valley heights. Only very coarse structures can be processed on such surface, and thick resist has to be used to avoid excessive resist thinning over topography. The same argument applies to thin films: for instance thick electroplated copper and thick poly are rough and not suitable for fine line lithography. Both sides of double side polished (DSP) wafers are mirror polished to sub-nanometer RMS roughness. However, the side which was polished last is of better quality than the other side, and DSP wafers are therefore not fully symmetric. This has implications especially to bonding which is critically dependent on surface quality.

Total thickness variation (TTV) is critical in MEMS through-wafer etched structures. If beams or diaphragms 10 μm thick need to be fabricated, 5 μm TTV results in totally unacceptable thickness control. MEMS wafer TTV values of 1 μm are typical, and 0.5 μm can be specified for demanding applications.

30.2.1 Double-sided lithography

Double-sided lithography comes with three degrees of difficulty:

- arrays without alignment
- non-critical alignment
- critical alignment.

Regular array structures on the wafer back side without alignment with the front can be seen in the solar cell of Figure 1.14. In non-critical alignment the major function of the device is determined by structures on one side only, and coarse auxiliary structures are made on the other side, as in the p^{++} etch stop defined nozzle of Figure 20.7.

Critical alignment involves device functions that are highly dependent on the accuracy of pattern position, for example symmetric resonating mass (see Figure 30.4) or positioning piezoresistors to the point of maximum deflection of a pressure sensor diaphragm, which will be discussed presently (Figure 30.7).

Double-sided lithography is done on one side at a time: the resist application on top, alignment and exposure on top, development, rinsing and drying. Then, depending on device structure, either etching of the front side is done, or back-side lithography is performed.

Back-side lithography involves the application of back-side resist, which means that the front side of the wafer is placed in vacuum contact with the spinner chuck. The front side must be protected. Photoresist is often used but it cannot be used for patterning after being vacuum chucked.

The alignment mechanism in double-sided lithography relies on image processing. An image of the mask alignment marks is stored, and the wafer is then inserted between the mask and the alignment microscope, so the alignment marks on the wafer are aligned with the stored mask alignment marks (Figure 30.3). Alignment accuracy is about 1 μm at best, and usually a few microns.

A seismic mass beam is made by double-sided lithography and anisotropic wet etching (Figure 30.4a). Due to misalignment, the resulting structure is asymmetric (Figure 30.4b), and beam length is ill-defined. Because deflection of a beam depends on length cubed, even small changes can result in large deviations. But even if alignment is perfect, the beam length is not necessarily correct (Figure 30.4c): mask undercut can make the beam longer. This undercut can come from switching from KOH to TMAH, their (100):(111) selectivities being about 200:1 and 30:1. Defects in silicon wafers can also affect crystal plane selectivity, as do impurities in etchants.

30.2.2 Etching

Wet etching (and wafer cleaning) in a tank takes place on both sides simultaneously. It may be useful to etch from both sides, either for symmetry reasons, or for doubling the apparent etch rate. If the etching depth is shallow, it is not necessary to protect the back side, but if deep etching is done on the front side, the back side needs to be protected. Single wafer plasma etching clears film on the front and leaves the back side film intact, but if wet etching is preferred (e.g., because of surface quality considerations) the back side must be protected.

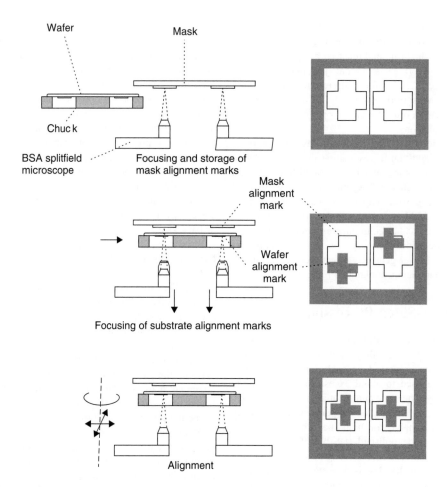

Figure 30.3 Double-sided alignment. Courtesy Suss Microtech GmbH

Protection by spin-coated polymers is a quick and easy method. Photoresist is suitable for many applications, like mask oxide etching in BHF, but aggressive etchants like KOH require either inorganic films (oxide, nitride) or more stable polymers. Blue tape common in wafer dicing can also be used as a protective layer, but removal of the tape can be difficult if fragile free-standing structures are present on the wafer.

A single wafer holder which exposes only one side of the wafer to the liquid is a universal solution. In electrochemical etching or deposition this holder also provides the necessary electrical contacts to the wafer. However, wafer front-side area is wasted by the clamp, and single wafer processing is more expensive than batch processing. With the holder the top side processing and materials can be selected from a device operation point of view, and no extra protective coatings are needed during processing.

30.2.3 Patterning of 3D structures

Because photoresist spinning on deep trenches and holes is difficult, or even impossible, depending on the through-hole topology, other patterning approaches must be used. Laminated dry resists can sometimes be used to overcome trenches (Figure 9.8). There is no problem with resist flowing, because the resist is dry. However, dry resists are limited in resolution: aspect ratios are roughly 1:1, and with 20–50 µm as the typical dry resist thickness, it is clear that it is best suited for non-critical lithography only. Spray coating of resist works for wet-etched deep structures with 54.7° angles, though exposure focus depth is another issue.

Sharp corners resulting from DRIE are difficult to coat, even by spray coating. Other solutions to patterning over severe topography have already been introduced: peeling masks (nested masks) are the standard approach in

MEMS Process Integration 391

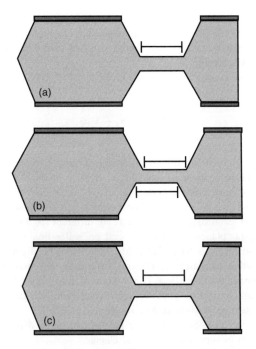

Figure 30.4 Alignment of seismic mass beam: (a) perfect alignment; (b) misalignment; (c) mask undercut

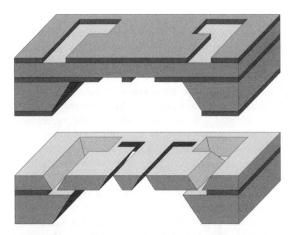

Figure 30.6 Double-sided etching of SOI. Reproduced from Sanz-Velasco *et al.* (2006), copyright 2006, by permission of Elsevier

both wet etching (Figure 20.4) and DRIE (Figure 21.17). Lithography is carried out twice on a planar surface, before any deep etching.

Shadow masks (Figure 23.17) enable metallization of wafers with severe topography or even wafers with through holes. However, pattern size control over severe topography may not be very good because of flux divergence. But if the shadow mask itself is a silicon wafer patterned to match the 3D geometry already fabricated, patterning accuracy is regained (Figure 30.5).

The catalyst layer at the bottom of the flow channel of the microreactor in Figure 17.1 was deposited by a 3D shadow mask. A shadow mask is also useful for patterning on membranes, especially if thin membranes have been made. The heater and temperature sensors of the microreactor in Figure 17.1 were also patterned by shadow masks. One limitation is that they are best suited when only a small percentage of area needs to be metallized. Making shadow masks for large areas would involve problems similar to making large-area membranes. In Figure 30.6 asymmetric single crystal silicon beams are made by double-sided lithography and etching of SOI wafers. After lithography on the bottom of the handle wafer is accomplished, anisotropic wet etching will proceed as usual.

30.3 Membrane Structures

How many different ways are there to make membranes? Many of the methods to make bridges have limitations regarding width, and membrane fabrication must be able to produce wide structures. Some of the bridge fabrication processes can be extended by perforating the membrane with release holes, but then some applications may become impossible, like pressure sensors. There are a number of different approaches to membrane fabrication:

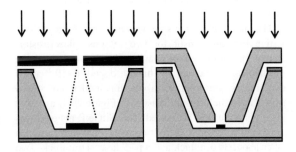

Figure 30.5 Conventional 2D and micromachined 3D silicon shadow masks compared. Redrawn from Brugger *et al.* (1999)

- back-side wet etching (as discussed in Chapter 20)
- back-side DRIE (especially with SOI)
- front-side release (perforated membranes)
- bonding over a cavity.

With back-side etching many membrane thicknesses can be made, and there are many choices of materials. With wet etching, typical membrane materials are silicon and nitride. With dry etching, other materials can be used: for example, aluminum oxide is extremely tolerant of fluorine plasma, and membranes of various kinds can be made. Metallic membranes like aluminum (Figure 17.2) or gold can also be made, but the mechanical strength of many metals is not sufficient for large-area free-standing membranes. Silicon supporting struts can be left to stiffen the structure, as shown in Figures 20.27 and 20.31.

With front-side release, single-sided lithography suffices, and etching times are much shorter than with back-side release. Again the underlying film has to be very selectively etched relative to the membrane. Protective capping layers can be used (Figure 29.16). The sacrificial layer can be either silicon wafer itself or some deposited film. Oxides are typically limited in thickness, dry release by HF vapor is a difficult step, and stiction is a problem in wet HF-based release. Polysilicon layers (and "epi-poly" especially) can be made much thicker and released easily in SF_6 plasma or XeF_2 dry release.

30.4 Piezoresistive Pressure Sensor

The piezoresistive pressure sensor is one of the oldest MEMS devices, dating back to the 1960s. In piezoresistance mechanical stress causes a change in resistance. Silicon has two major benefits over metals for piezoresistors: the piezoresistance coefficient is larger, and it is available with both positive and negative values, for p-type and n-type doping, respectively. Also, silicon is compatible with high process temperatures, unlike metals. One example of a piezoresistive pressure sensor is shown in Figure 30.7. Many important process integration issues can be seen in a simple piezoresistive pressure sensor.

For square membranes stress depends on pressure and membrane dimensions according to

$$\sigma = \frac{1.02 pa^2}{h^2} \quad (30.1)$$

where a is the membrane edge length and h its thickness. It is mandatory to be able to control membrane thickness accurately. While thicker layers (>50 μm) can be made by timed etching, especially if starting wafer thickness and TTV are tightly specified, thinner membranes necessitate some sort of etch stop structures. Diffused etch stop layers are limited in thickness because diffusion is a slow process and concentration falls off rapidly. Epitaxial layers enable practically any thickness to be used. And there is the choice between electrochemical etch stop and using p^{++} SiGeB etch stop, which is slightly more expensive but easy to integrate with standard TMAH or KOH etching.

The change in piezoresistor resistance is given by

$$\frac{\Delta R}{R} = \pi_1 \sigma_1 + \pi_2 \sigma_2 \quad (30.2)$$

where the π are the longitudinal and transverse piezoresistance coefficients, and the σ stresses. In the longitudinal case electric current is in the same direction as stress, while in the transverse case they are perpendicular. Piezoresistance coefficients are different for different crystal planes, and they depend on wafer doping type (and level) as well (Table 30.2). Piezoresistor alignment relative to wafer crystal axes determines the maximal sensitivity. Polysilicon is a mixture of many crystal orientations and its piezoresistance coefficient will depend on texture in the film. As a rough guide, 30×10^{-11} Pa^{-1} can be used. Stresses in the megapascal range will then result in very small resistance change.

Piezoresistive sensors are sensitive to temperature changes because the silicon temperature coefficient of resistivity is fairly high, 0.0015/°C. Resistance change due to temperature change is easily larger than change due to the piezoresistive effect. Integration of a temperature sensor or some temperature compensation scheme must be used. A diode can be used as a thermometer which does not depend on stress.

Piezoresistors need to be placed where the maximum deflection occurs. Due to misalignment, the resistor could be placed on a solid, non-bending part of the silicon, resulting in a null signal. But there are other ways in which things can go wrong: wafer thickness does not affect membrane thickness when etch stop is used, but the wafer can be too thick. The pressure-sensitive membrane is now too small, and the maximum deflection point is shifted.

The simple p^{++} etch stop does not work for a piezoresistive pressure sensor for two reasons: piezoresistors cannot be fabricated in heavily doped silicon, and the mechanical properties of highly doped (>10^{18} cm^{-3}) diaphragms are inferior to low or moderately doped material. In the pressure sensors of Figure 30.7, a double layer epi-wafer has been used. This Si:Ge:B material was described in Figure 22.6. It consists of a lightly doped layer about 30 μm thick and a p^{++} layer 4 μm thick. Two selective etches are used: KOH etching stopping on p^{++} and $HF:HNO_3:CH_3COOH$ (1:3:8) wet etching which etches highly doped silicon but not lightly doped silicon.

Table 30.2 Piezoresistance coefficients

	n-type			p-type		
	[100]	[110]	[111]	[100]	[110]	[111]
π_l (10^{-11} Pa^{-1})	−102	−31	−7.5	6.6	72	94
π_t (10^{-11} Pa^1)	53	−18	6	−1	−66	44

Process flow for piezoresistive pressure sensor

Step	Comment
Wafer selection: p-type silicon	n-type piezoresistors
Si:B:Ge epi and n-type lightly doped epi	3 µm + 30 µm
Lithography and implantation	Phosphorous doping for piezoresistors
Resistor diffusion in dry oxidation	Oxide grows on both sides
LPCVD nitride	Nitride deposited on both sides
Lithography for resistor contacts	Minimum size 3×3 µm
Etching for contacts	Nitride and oxide RIE
Metal sputtering	Al 500 nm thick
Lithography and Al etching	H_3PO_4 wet etching
PECVD nitride	Protects Al on front side
Photoresist spinning on front side	Mechanical protection
Photoresist spinning on back side	Preparing for through-wafer etching
Lithography on back side	Membrane-to-resistor alignment
Nitride and oxide etching on back side	RIE for nitride, HF for oxide
Photoresist stripping	Both sides simultaneously
KOH silicon etching	Si:B:Ge layer acts as an etch stop
HF:HNO$_3$:CH$_3$COOH isotropic etching	Etch p^{++} epi selectively against n-Si
Etch nitride and oxide on back	Reveal silicon for anodic bonding
Anodic bonding to a glass wafer	Glass wafer pre-drilled

Note: cleaning and resist stripping steps not listed.

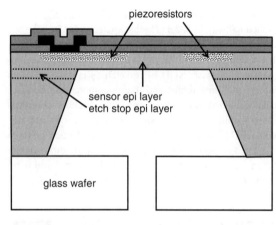

Figure 30.7 Piezoresistive pressure sensor

30.4.1 Microphones

Microphones are pressure sensors, too. They consist of a rigid backplate and a movable membrane. Again, many variants exist, both bulk micromachined and surface micromachined. Some involve wafer bonding for counter-electrode, others use sacrificial layer etching for air gap formation. We will present a two-wafer, wet-etched, single crystal silicon membrane version with wafer bonding (Figure 30.8), and a sacrificial layer version with polysilicon membrane and single crystal silicon backplate (Figure 30.20).

Process flow for bonded microphone

- Membrane (top) wafer:
  - Thermal oxidation and oxide patterning
  - Nitride deposition and top side nitride etch
  - Membrane metallization (Cr/Au)
  - KOH etch halfway
- Silicon backplate (bottom) wafer:
  - Oxidation
  - Peeling mask

- KOH etching
- Oxidation
- Metallization (Cr/Au)
• Final processing:
- Gold–gold thermocompression bonding
- KOH etching
- Aluminum metallization through shadow mask

The top membrane wafer is metallized by Cr/Au and etched halfway in KOH. The backplate wafer is patterned and etched to a depth of 20 μm to define the acoustic gap. It is a timed etch, and the gap is critical for microphone response. Otherwise, KOH silicon etching is standard. Both wafers are quite robust and thermocompression bonding can be carried out without problems. Etching continues after bonding. The only other materials present are silicon dioxide, silicon nitride and gold, and none of them is etched by KOH. KOH etching will stop automatically when the nitride membrane is reached, and it would seem to be non-critical. However, acoustic hole size depends critically on etch timing: too long an overetch will rapidly widen the holes (recall Figure 20.23). The aluminum metallization is done by shadow mask after completing the process (because no resist spinning is possible on holes, and because aluminum does not tolerate KOH etching).

Figure 20.19 presented a similar microphone process. There, however, both wafers are processed to completion before bonding. Delicate handling of the thin silicon nitride membrane is then mandatory in bonding.

30.5 Tilting and Bending Through-Wafer Etched Structures

Many MEMS structures require through-wafer etching. It is important to consider when this step should be done because wafer mechanical strength is compromised as a consequence, and some process options are no longer possible. Wafer thickness variation comes into play in many ways, as in simple nozzles, which were discussed in Section 20.7. Piezoresistive pressure sensors similarly showed many wafer thickness-dependent features. Sometimes using SOI wafers can eliminate thickness-related problems, sometimes epiwafers help, and sometimes p^{++} etch stops can be used. Complete redesign of the devices can of course also overcome these issues, but probably shifts problems elsewhere in process design.

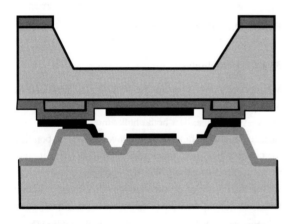

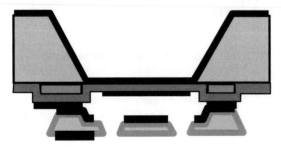

Figure 30.8 Microphone: bonding, final KOH etching, metallization. Compare to Figure 20.19. Adapted from Scheeper *et al.* (2003)

30.5.1 Mirrors

Mirrors can be thought of as membranes, in the plane of the wafer, but the modes of operation differ: some mirrors are for up–down parallel-plate movement without bending, and resemble pressure sensor membranes. The optical modulator shown in Figure 29.9 relies on this mode of operation. A torsionally moving mirror was shown in Figure 29.25. This mirror is a digital device: it has only two stable positions, while the angle of analog mirrors depends on the actuation force. The actuation force required depends on many factors, including the distance between the electrodes and the mirror, as well as the stiffness of the torsion bars.

A tilting mirror with a tilting frame was shown in Figure 1.2. This design enables free pointing directions. Mechanically strong, yet flexible springs are needed for such designs, and silicon is a good choice. Pop-up mirrors which result in 45° or 90° mirrors were described in Figure 29.24, and another type of pop-up mirror will be seen in Figure 39.2: a planar mirror with large pop-up stroke. In addition to membrane-type horizontal mirrors, vertical mirrors have been made. In Figure 21.3 a sliding mirror was made by DRIE, and in Figure 21.24 DRIE together with anisotropic wet etch smoothing were used to finish static laser mirrors.

Micromirrors have several optical design criteria: planarity, smoothness and reflectivity. Smoothness is an atomic-scale concept and planarity is a large-area concept. Reflectivity is maximized by using non-oxidizable metals like gold. For mirrors planarity can be quantified by radius of curvature (ROC) using Stoney's formula, Equation 5.20.

SOI device layers can provide a wide range of thicknesses, and 20 μm already results in large ROCs on the order of meters. Thin mirrors are typically made of silicon nitride or polysilicon. Thicknesses are limited to a few micrometers. Planarity is not good because the mirror bends. This can be combated by stiffening the structure by using 3D structures instead of planar films. An optically detected cantilever was made stiffer and more planar by plasma-etched U-grooves filled with LPCVD nitride (Figure 29.6).

The spring constant for flexures of width w and thickness t is given by Equation 29.2. The spring constant is very sensitive to thickness variations. The SOI device layer can be used to define torsion bar (flexure) thickness. But it would often be desirable to have a rather thick mirror, for example the full thickness of the SOI device layer, and much thinner flexures. One solution is to use polysilicon for flexures, so it is then poly deposition that defines flexure thickness, not etch thinning of bulk or SOI silicon. Because poly flexures can be very thin, spring constants can be made small.

A single crystal silicon mirror made in the device layer of a SOI wafer, Figure 30.9. The flexure geometry is similar to those in Figure 1.2. Three KOH etch steps sculpt the 100 mm thick device layer (compare to Figure 20.21). The mirror thickness is 14 mm and radius of curvature over 50 cm. In the third KOH etching step real-time monitoring is used to accurately control the suspension thickness (3–4 mm). Next, an auxiliary Al layer is deposited and patterned over the etched topography.

It keeps the structure together until the final release step. SOI wafer is then anodically bonded to a Pyrex wafer. The SOI handle wafer is etched away. Gold mirror is deposited and patterned (100 nm thick, with a 2 nm Ti adhesion layer). A silicon DRIE etch defines the gimbal, flexures and the mirror plate. Aluminum serves as an etch stop which eliminates notching. Aluminum wet etching completes the process. The Al wet etchant penetrates the cavity and careful drying is needed.

30.5.2 AFM needles and tips

AFM cantilevers need to bend and therefore they need to be released at some point in their fabrication process. Again, it is critical to decide when the through-wafer step takes place. As in mirrors, it is paramount to protect the active side of the device during back-side processing. Resist coating is one option, another is to prepare the back-side etching mask before finishing the front-side processing, so that no lithography will be needed after completing the front side. Another issue in front protection is KOH/TMAH compatibility: resist is not enough, but fluoropolymers may do, or PECVD nitride, or a special holder will be needed.

A AFM needle made on a SOI wafer is described in Figure 30.10. Cantilever and tip are fabricated in the SOI

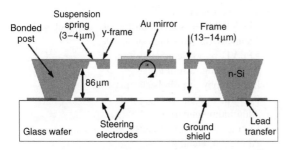

Figure 30.9 Two-axis single crystal silicon mirror made in SOI device layer, Reproduced from Dokmeci *et al.* (2004), by permission of IEEE

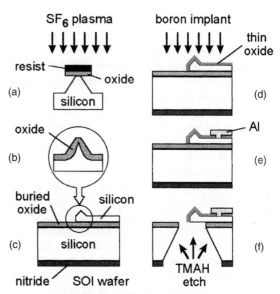

Figure 30.10 AFM cantilever–tip fabrication. Reproduced from Chui *et al.* (1998) by permission of IEEE

device layer, and BOX serves as an etch stop layer for the back-side release etch. A similar process was shown cursorily in Figure 20.2.

Process flow for AFM cantilever–tip

- SOI wafer with device layer 5 μm thick
- Thermal oxidation
- LPCVD nitride
- Etch nitride from front side
- Lithography for the tip
- Etch oxide
- Etch silicon isotropically (+ resist strip)
- Thermal oxidation for tip sharpening
- Lithography to define the cantilever
- DRIE of device silicon (+ resist strip)
- Thermal oxidation for passivation
- Lithography for piezoresistors
- Boron implantation for resistors (+ strip)
- Lithography for contact
- Boron implantation for contacts (+ strip)
- Implant activation in RTA
- Aluminum deposition and patterning
- Polyimide protective coating on front
- Back-side nitride patterning
- Back-side TMAH anisotropic etch
- Buried oxide etching
- Polyimide plasma removal

The SOI wafer enables precise and easy control of silicon cantilever thickness: this is essential for mechanical devices in order to control cantilever resonance frequency and stiffness. However, in this process the tip fabrication process (etching and oxidation) introduces some uncertainty into cantilever thickness. Sharp tips could of course be fabricated by simple anisotropic wet etching (Figure 20.15), but oxidation leads to sharper tips, and the process is better controlled by oxidation time than by etch time (Figure 13.14).

Boron implantation (40 keV, 5×10^{14} cm^{-2}) is used in piezoresistor formation. In order to improve electrical contact to the piezoresistor, an additional ion implantation (40 keV, 5×10^{15} cm^{-2}) is done to increase doping concentration in the contact, to ensure low-resistance contact to aluminum. The passivation oxide thickness is chosen to block the contact enhancement ion implantation. Rapid thermal annealing (RTA, 10 s at 1000 °C; 0.4 μm diffusion depth) is used to remove implantation damage.

Spin-coated polyimide is used to protect the silicon tip and aluminum metallization during the back-side silicon etch. Additionally, a single wafer chuck protects the front side. This eats up area on the wafer front side and leads to fewer chips.

The first lithography step could be back-side nitride etch mask opening, even though it will be used very late in the process. This would minimize the process steps after finishing the front-side processing.

30.6 Needles and Tips, Channels and Nozzles

30.6.1 Ink jet nozzle

Despite all the good features of anisotropic wet etching, through-wafer structures take up a lot of silicon area. Nozzles fabricated by anisotropic through-wafer wet etching cannot be packed close to each other, and for ink jet printers other nozzle geometries have been studied. One such geometry is the top shooter shown in Figure 21.30: one through-etched reservoir can supply ink to many nozzles fabricated by front-side processing. Another geometry is the side shooter, which is not limited by wafer thickness or etch geometries. All critical dimensions of the device are defined by front-side processes, and one non-critical through-wafer etching completes the process. One such design is described in Figure 30.11.

Process flow for ink jet

- Thermal oxidation, 1 μm thick
- Lithography 1: chip area definition
- Oxide etching
- Boron diffusion, 2 μm deep
- Lithography 2: chevron pattern, 1 μm width
- RIE of silicon, 4 μm deep
- Anisotropic silicon etching to undercut p^{++} chevrons
- Thermal oxidation
- LPCVD nitride deposition for chevron roof sealing
- Etchback (or polishing) of nitride
- LPCVD polysilicon deposition
- Poly doping, 20 ohm/sq
- Lithography 3: poly heater pattern
- Polysilicon etching
- Aluminum sputtering
- Lithography 4: metal pads
- Aluminum etching
- Passivation: CVD oxide 1 μm + PECVD nitride 0.3 μm
- Lithography 5: opening of bonding pads

- RIE of nitride and oxide
- Lithography 6: pattern for gold lift-off
- Evaporation of Cr/Au
- Lift of Cr/Au
- Lithography 7: fluidic inlet definition on the back side
- Anisotropic etching through the wafer from the back

Note: Resist stripping and cleaning steps omitted.

Figure 30.12 Cavity sealing by CVD: plasma-etched, chevron-shaped access holes are closed by LPCVD nitride deposition. Reproduced from Chen and Wise (1997) by permission of IEEE

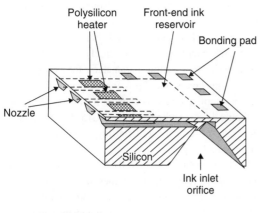

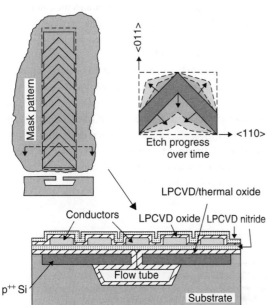

Figure 30.11 Side-shooting ink jet. The chevron structure enables both anisotropic underetch and roof sealing. Reproduced from Chen and Wise (1997) by permission of IEEE

Boron-doped silicon provides mechanical strength for the structure, compared to the nitride membrane, which can only be about 1 µm thick vs. micrometers for the silicon roof. The chevron patterns open fast etching silicon crystal planes which enable undercutting on the <100> wafer. Figure 30.12 shows what the chevrons look like before and after sealing.

Etchback thinning of nitride is done to improve thermal speed: the closer the heater resistor to the flow tube, the faster the heating. Polysilicon is heavily doped and will serve as a heater electrode. Less resistive aluminum cannot be used because of KOH etching. Gold on bonding pads makes wire bonding easy, and gold protects the front side during back-side anisotropic etching (areas that are not gold covered are either nitride or oxide, which are resistant to alkaline etchants). Through-wafer etching is non-critical because it will stop automatically on the bottom oxide of the flow tube.

30.6.2 Microneedles

Microneedles come in two major varieties: in plane and out of plane. In-plane needles can be very long, millimeters, while the out-of-plane variety is limited to a few hundred micrometers. Out-of-plane 2D arrays of needles can be made easily, and in-plane needles can be arranged in a linear array. Applications vary from drug delivery, blood sample extraction and electrospray to fluid mixing. A number of out-of-plane microneedles were presented in Chapter 21. An in-plane hollow needle was also introduced (Figure 21.16).

A solid in-plane microneedle made by bulk micromachining and intended for biomedical electrical stimulation studies is shown in Figure 30.13. It is fabricated in a modified p-well CMOS process, with the needle defined by the p^{++} etch stop. Fabrication of similar microneedles on SOI by DRIE is considerably simpler.

Hollow in-plane needles can be made by a number of techniques. For example, p^{++} silicon as mechanical support, PSG as the sacrificial layer and nitride as the roof of the hollow channel. Similar structures can be made by other combinations of materials. In Figure 30.14 parylene is used as the floor, AZ photoresist as the sacrificial material defining the channel and parylene as the roof. The needle is used in electrospray ionization (ESI) in mass spectrometry. The process flow for the hollow parylene ESI needle is given below.

Process flow for hollow parylene needle

- Thermal oxidation
- Front-side protection by resist
- Back-side lithography
- Back-side oxide etching
- Back-side silicon etching
- Oxide roughening etch
- First parylene deposition
- First parylene patterning
- AZ resist application
- Lithography defining channel
- Second parylene deposition
- Aluminum sputtering and patterning
- Second parylene plasma etching
- Aluminum mask removal
- Resist dissolution
- Oxide etching

Aluminum is needed as an etch mask for parylene RIE; photoresist cannot be used because the etch selectivity between two polymers is inadequate. In order to improve parylene adhesion, the oxide surface was roughened using gaseous BrF_3 etching, and then silane SAM was applied. Channel height is quite freely chosen by changing the resist spin coating parameters, for example 3 μm for parylene layers and 5 μm for the sacrificial AZ resist. For alternative narrow channel formation techniques, see Figure 17.8 (bonding), Figure 21.15 (buried channel), Figure 25.3c (vertical sacrificial layer) and Figure 29.15 (parylene molding).

Fluidic connection to a KOH-etched hole is possible with a capillary, but dead volumes will result from the mismatch of macroscopic capillary and microfabricated inlet. As a solution, a PDMS cast connector can be used (Figure 30.15). Using exactly the same mask as for backside hole etching, a mould is created for PDMS casting (see section 18.4 for PDMS casting). A dummy fiber is inserted and PDMS is cured. PDMS is peeled off silicon mould, and dummy fiber is removed, too. Capillary is inserted and the PDMS piece is inserted into fluidic chip. Because PDMS is self-adhesive, and because the shapes are exactly matching, a tight plug is created. But is it still reusable. If stronger bond/pressure tolerance is needed, oxygen plasma activation of PDMS can be used, but then bonding becomes irreversible.

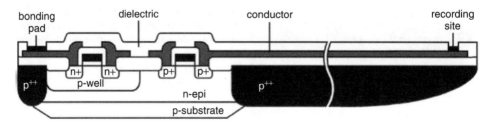

Figure 30.13 In-plane integrated CMOS microneedle for electrophysiological measurements by Ji and Wise (1992). Reproduced from Brand (2006), copyright 2006, by permission of IEEE

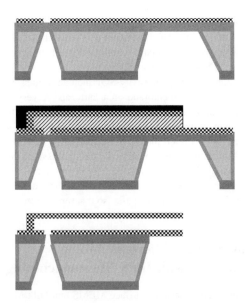

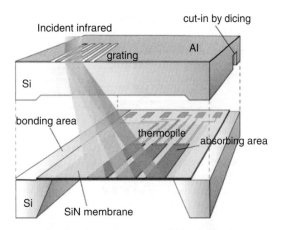

Figure 30.16 Infrared spectrometer. Reproduced from Kong *et al.* (2001), copyright 2001, by permission of Elsevier

Figure 30.14 Parylene hollow needle for ESI. Redrawn from Licklider *et al.* (2000)

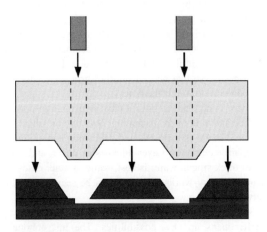

Figure 30.15 PDMS fluidic connector with exact fit into KOH etched silicon. Courtesy Ville Saarela, Aalto University, by permission of Elsevier

30.7 Bonded Structures

Bonding serves many functions, and this section covers those bonding applications that are performed in the middle of the process, while those that are done at the end of process for packaging will be discussed later on in this chapter.

A bonded microturbine was shown in Figure 1.17. The rotor is made of the middle wafer and released after it has been bonded to the top and bottom wafers. Narrow tethers of silicon dioxide are etched away to enable the rotor to move. A bonded microreactor was shown in Figure 17.1. Silicon and glass wafers were bonded to form the microreactor, and processing continued after bonding by deposition and patterning of heater and sensor resistors. No wafer thinning was used in these applications.

As another example, a spectrometer is discussed here (Figure 30.16). It relies on silicon optical properties in the infrared: silicon is transparent above 1.1 μm wavelengths, and a silicon wafer is suitable as a window for an IR spectrometer. A diffraction grating is fabricated on a window wafer, which is bonded to a thermopile detector wafer. In order to access bonding pads on the detector wafer, a dicing saw cut-in has been made on the window wafer, and as the last step of the process, the silicon piece above the pads is diced away. Without the cut-in, extreme precision would be needed in dicing, with the danger of cutting the thermopile detector metallization.

Bonding is used in many cases for the creation of critical structures, like capacitive sensor gaps (Figure 17.2). In capacitive micromachined ultrasonic transducers (cMUTs) bonding is essential for forming the active silicon membrane of the device (there are similar devices made by sacrificial layer techniques; as you may have guessed, it is again a question of polysilicon vs. single crystalline silicon properties and processing). After bonding a SOI wafer over etched cavities, the handle wafer and BOX are etched away, leaving the device silicon layer. Aluminum electrodes and passivation oxide are processed on this membrane (Figure 30.17).

The capacitive gap is defined by the KOH-etched depth. Fusion bonding was done under ambient conditions, meaning that air was trapped inside the cavity.

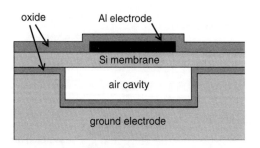

Figure 30.17 Ultrasonic transducer: SOI wafer bonded over cavity, handle wafer and BOX etched away. Aluminum electrode processing on SOI device layer. Adapted from Huang *et al.* (2003).

During bonding oxygen reacted with silicon, which means that the cavity pressure is below 1 atm and the membrane is deflected downward (recall Figure 17.14). This induces stress in the membrane, which has to be accounted for when an actuation voltage is applied.

A single crystal silicon bridge is the mechanical element in a piezoelectric resonator (Figure 30.18). It is fabricated by bonding a device wafer into a handle wafer with cavities (cavity SOI). No notching effect will take place when silicon etching stops at the air cavity. Another benefit is the elimination of stiction: no release etch, no stiction. The benefit of piezoelectric actuation is that gaps are not critical: piezoelectric actuation is independent of gap width and the DC bias is not needed at all. In capacitive actuation 100 nm gaps would be needed for actuation with a 10 V DC bias over the gap.

Design variables in cavity-SOI include silicon membrane thickness, membrane area and gap. A very wide range of dimensions can be made. In theory all SOI device layer dimensions are available, and membrane sizes range from 10 micrometers to millimeters. Practical process limitations (pressure in cavity etc.) necessitate design rules which limit the available cavity depths and sizes and silicon thicknesses. Note that alignment marks are etched on the back of the handle wafer, so that top side structures can be aligned with the cavity.

30.8 Surface Micromachining Combined with Bulk Micromachining

The distinction between surface MEMS and bulk MEMS is not clear cut. For example, the pneumatic bubble pump of Figure 30.19 exhibits elements of both surface and bulk machining: the structural polysilicon has been released by HF etching of oxide, but the fluidic manifold has been etched through the wafer by KOH. The sidewalls of the surface micromachined channel are hydrophilic, but the bulk silicon sidewalls are coated hydrophobic so that water from the surface channel does not penetrate through the hole.

There is no need to have a clear demarcation between surface and bulk micromachining: both rely on silicon and silicon dioxide as materials and deposition and etching as the main techniques. Layer thicknesses differ, and sometimes thick silicon wafer is best, while in other cases thin deposited poly works best. SOI can be seen as a hybrid between the two.

Similarly, combining anisotropic wet etching with DRIE opens up new possibilities. The microphone of Figure 30.20 combines acoustic holes made in a single crystal bulk silicon wafer by DRIE and the pressure-sensitive membrane made of LPCVD polysilicon. Phosphorus-doped silica glass PSG is used as a sacrificial layer. The process flow for the device is given below.

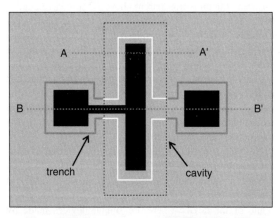

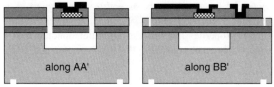

Figure 30.18 Piezoelectric length–extensional bar resonator on cavity SOI: AlN piezoelectric and molybdenum metallization. Adapted from Jaakkola *et al.* (2007)

Process flow for sacrificial layer microphone

- LPCVD nitride deposition
- Poly deposition, doping, patterning
- Acoustic hole lithography
- Poly/nitride/<Si> DRIE, 20 μm deep

- CVD oxide (PSG) deposition
- CVD oxide lithography and etching
- Poly membrane deposition and doping
- Lithography and etching of poly
- Metallization
- Back-side lithography and silicon DRIE
- PSG removal through the acoustic holes

The normal strengths and limitations of surface micromachining apply here: only one wafer is used, saving money compared to two-wafer bonding (see microphones of Figures 20.19 and 30.8). Polysilicon membrane stress has be to controlled, but because it is done early in the process, any kind of stress-relief anneal is applicable (Figure 25.9). Acoustic gap height is determined by CVD oxide (PSG) deposition, which can be accurately controlled and easily verified by measurement. However, the maximum gap height is limited to a few micrometers (2 μm in this case). And because the gap height is small, stiction problems must be considered. When PSG is deposited into the acoustic holes, the surface is not planarized (Figure 5.16). Poly deposited over PSG will then have small spikes, which make stiction unlikely. The spikes come "free of charge," without any lithography step.

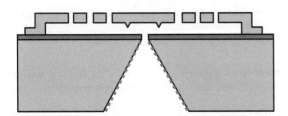

Figure 30.19 Bubble pump: air pressure from the KOH-etched, SAM-coated hydrophobic manifold pushes liquids in the hydrophilic surface channel. Adapted from Tas *et al.* (2002)

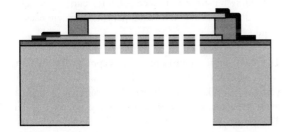

Figure 30.20 Polysilicon membrane microphone, backplane with acoustic gaps by DRIE in single crystal silicon. Adapted from Dehé (2007)

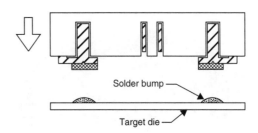

Figure 30.21 Fabrication process for polysilicon molded structures: <Si> DRIE, CVD oxide, poly deposition and patterning, metallization, bonding, sacrificial oxide etching. Reproduced from Horsley *et al.* (1998) by permission of IEEE

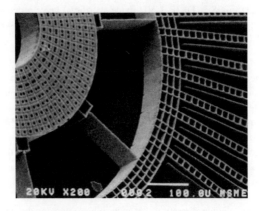

Figure 30.22 Polysilicon actuator fabricated by the process shown in Figure 30.20. Reproduced from Horsley *et al.* (1998) by permission of IEEE

DRIE of silicon, oxide deposition into silicon trenches, followed by polysilicon LPCVD, creates versatile microstructures. Metals can be deposited on poly. The basic scheme is shown in Figure 30.21. The CVD oxide acts as a sacrificial layer. Bonding such a structure to a carrier wafer (by eutectic or solder bonding) and etching the oxide away leaves polysilicon microstructures, like the hard disk drive read/write head actuator shown in Figure 30.22.

There is an alternative way to go as well: single crystal silicon underneath the polysilicon structures is etched away, in a fashion similar to Figure 21.14, leaving freestanding polysilicon structures. Polysilicon beams of high aspect ratio made this way are shown in Figure 30.23.

30.9 MEMS Packaging

Packaging provides protection to the chip and, at the same time, connectivity to the outside world. Information,

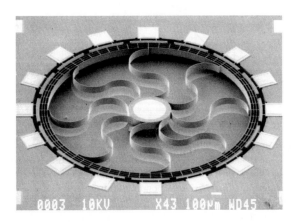

Figure 30.23 Polysilicon beams (4 μm wide, 80 μm high) made in <Si> DRIE, CVD oxide sacrificial layer process. Reproduced from Ayazi and Najafi (2000), by permission of IEEE

energy and matter must flow in and out of the chip in a controlled manner: for instance, the environment must interact with the sensor, energy has to be provided to the sensor (unless it is a so-called self-generating type like a photodiode) and a signal must be carried out of the package. In magnetic field sensors or accelerometers this is easy, because magnetic fields and gravity "penetrate" the package.

Often a cavity with known atmosphere is needed for proper device operation, for example a micromechanical RF switch needs to be operated in a protected atmosphere for stability (no water condensation) and at a known pressure for mechanical damping. RF MEMS devices are somewhat akin to ICs: once the package is finished, the devices are internal to the system and only need to be protected from the environment. In optical devices the package must be optically transparent, but additionally, in the case of mechanically moving mirrors, the pressure in the cavity affects mechanical damping. The pressure must be maintained over the lifetime of the device, for example 10 years. In IR devices there are intriguing possibilities because silicon is IR transparent, and the signal can enter through the wafer back side.

In chemical sensors and fluidics packaging is much more problematic: direct contact with liquid or gas is usually required, which means that the chip is exposed to the outside world. The lifetime of the device has two rather different elements: storage time and usage time. For example, a biochip or humidity sensor may lie on a shelf for six months and then has to function properly for an hour. Many chemical and biological chips are disposable devices, and for good reason, but many of them are miniaturized instruments for continuous use, like the flame ionization detector of Figure 1.10. Such devices must be compatible with the fluids in question, for example no unwanted adsorption on surfaces, no residues from analytes, no catalytic metals in contact with fluids, no leaching from bonding materials, etc.

ICs are solid state devices and their packaging is generic and simple: both plastic and hermetic packages are independent of chip design and technology. Wafer dicing relies on 20 000 rpm saw blades, which might make MEMS structures resonate; dicing saws are water cooled, which may lead to contamination and stiction; and silicon dust may block cavities and gaps.

The zero-level package is a structure that seals the MEMS part from the ambient. It is preferably applied on a wafer scale, and the packaged wafer then proceeds to dicing and assembly. But chip-scale packaging is still practiced in R&D. One reason is yield: if wafer-scale packaging goes wrong, the yield might be zero; with chip-scale packaging there are more trials and thus the chances of success increase.

There are a number of dichotomies that must be considered in selecting a zero-level package:

- wafer bonding vs. capping by thin-film deposition
- silicon cap wafer vs. glass cap wafer
- direct bonding vs. intermediate bonding
- electrical feedthroughs via cap wafer vs. via device wafer
- hermetic sealing vs. non-hermetic sealing.

Bonding for packaging is limited by the fact that the finished devices must not be affected. For example:

- Pre-bonding cleaning chemicals or plasmas must not attack devices.
- Bonding must not introduce incompatible materials.
- Bonding temperature must not change device properties.
- Discharges must be prevented during anodic bonding.
- Bonding must not introduce paths for current leakage.
- Bonding must not introduce parasitic capacitances or inductances.

The two basic approaches are capping by bonding and capping by deposition, but they come in many variants. One key issue is whether electrical contact is made through the device wafer or the capping wafer. The former is more demanding because through-wafer structure needs to be integrated with the device. Of course it may come naturally as a part of device processing. Figure 30.24 shows three ways of contacting: vias through cap and device wafer, plus a brute force

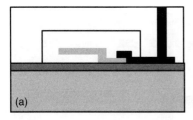 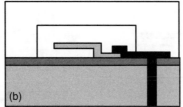 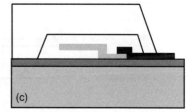

Figure 30.24 Zero-level packages by bonding: (a) electrical contact through via in cap wafer; (b) electrical contact through via device wafer; (c) electrical contact through large opening in cap wafer

method where a large cut is made in the cap wafer, to open contact to the device wafer. Note that Figure 30.24 is highly schematic and the cap wafer can be silicon or glass, and the vias vertical or otherwise. The electrical contacting can be metal, polysilicon, etc.

Electrical signal wires out of the package must be sealed in the capping process. Wiring by diffusions in silicon have two important benefits: they leave the surface flat, essential for many bonding processes; and they tolerate any imaginable bonding temperature. On the downside, there is the high resistance of diffused lines. Metal bonding requires an additional isolation layer, otherwise bonding metal would short the signal metal. In this respect adhesive bonding and glass frit bonding are the best: the soft polymer or glass covers the wire, and the height difference between wire and no-wire sites is only a fraction of the adhesive or glass-frit thickness, having no effect on the bonding process.

Hermeticity is one major criterion for packaging. The cavity should really be vacuum tight and hold its pressure for years. This kind of requirement is typical for devices which depend on resonance frequency stability for their operation. One such SOI resonator device is shown in Figure 30.25. It relies on anodic bonding and electrical contacts through the capping glass wafer.

The bonding process itself produces gases and these accumulate in the cavity. Additional gases come from outgassing (e.g., from CVD films or polymeric materials, including adhesives and glass-frit binders). One way to combat outgassing is to insert a getter in the cavity. Getters are porous and highly reactive materials which capture and hold gases. For example, titanium is effective in capturing oxygen, by reaction to form TiO_2.

The measurement of cavity pressure is no easy task because of leaks and gettering. In fact, resonant microstructures in the cavity are used as vacuum gauges: because frequency is very sensitive to pressure, it can be used for vacuum measurement. This, of course, depends critically on the stability of the resonator: any drift in the

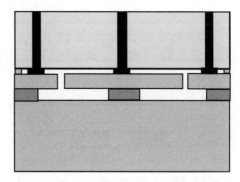

Figure 30.25 SOI resonator sealed with glass cap wafer by anodic bonding. Electrical feedthroughs in glass wafer. Adapted from Kaajakari *et al.* (2006)

mechanical quality factor, surface charging or film deposition on the resonator will change the resonant frequency.

Different bonding techniques suit different demands. In the RF switch shown in Figure 30.26, various bonding techniques have been used. The SOI starting wafer has been made by fusion bonding. The SOI wafer is anodically bonded to a glass wafer, and the gap (and hence the actuation voltage) between the two is defined by the SOI device layer etched depth. Glass-frit bonding is used in packaging. Glass-frit thickness is not well defined, but its function is to provide a hermetic package; its thickness does not affect device operating parameters. The resulting switch is quite expensive, because two silicon and two glass wafers are used. On the other hand, its reproducibility and reliability are good because the flexible element is made of single crystal silicon. Many of the surface micromachined switches discussed in Chapter 29 are very sensitive to stresses and have poor reproducibility.

30.9.1 Capping by deposition

Capping a MEMS structure by deposition faces a number challenges: it has to result in a mechanically robust, thick

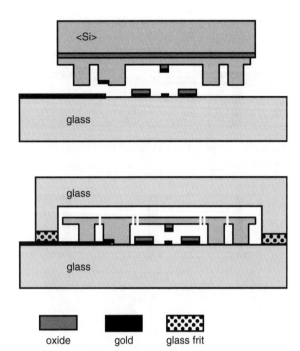

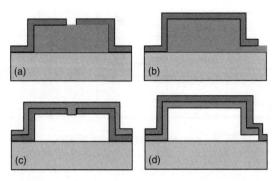

Figure 30.27 Cavity roof deposition and patterning, and sacrificial etching: (a) through vertical RIE access hole; (b) through horizontal access hole; (c, d) sealing by film deposition

ultimate vacuum is needed inside the cavity, evaporation is the deposition method of choice. Contrary to CVD, no (potentially) harmful gases will be incorporated into the

Figure 30.26 RF switch: flexible element in SOI device layer, anodically bonded to glass wafer which holds gold metallization. Glass cap wafer by glass-frit bonding. Adapted from Sakata *et al.* (1999)

and dense enough layer to protect the moving parts and to prevent gas and liquid diffusion, and the deposition process and material have to be compatible with the other process steps and structures.

In sealing flow channels (Figures 21.15, 30.12) good step coverage is mandatory to seal the channels. In encapsulating moving MEMS structures, care should be taken to minimize deposition on the movable structures, and in fact poor step coverage processes are preferred. In order to reduce the influence of the sealing film on the structural films, the sealing film should be as thin as possible. This is often best achieved with horizontal access holes, rather than with plasma-etched vertical holes. The two basic access hole types are compared in Figure 30.27. The horizontal access hole minimum dimension is determined by film thickness, which can easily be made small, compared to lithographically determined, plasma-etched access holes.

An absolute pressure sensor is a simple example of a sealed cavity: the cavity holds the reference pressure. If an

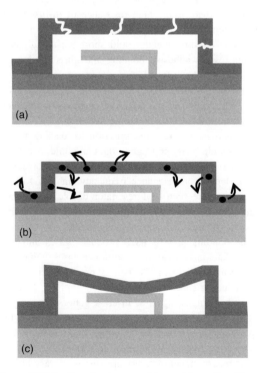

Figure 30.28 Capping layer problems: (a) cracks; (b) outgassing; (c) collapse. Adapted from Li *et al.* (2009)

cavity during evaporation. Due to the directional nature of evaporation, horizontal access holes have to be used. If polysilicon is used for sealing, hydrogen will remain in the cavity (Equation 5.1). Hydrogen, however, is usually not a concern, as it can escape during annealing.

Stresses in capping layers need careful optimization. If the capping film is under too high a tensile stress, cracks may develop (Figure 30.28a). Compressive stresses on the other hand lead to collapse (buckling) if critical stress is exceeded (Figure 30.28c). Because the temperature of capping layer deposition has to be fairly low in order not to affect the devices, the film density will be rather low. This may lead to leaks and outgassing (Figure 30.28b).

Much academic research in MEMS is about devices, their design, fabrication and characterization. However, in the marketplace the cost of a silicon chip is estimated to be only 30% of the price of MEMS. Zero-level packaging is one way to reduce costs; more generic packaging is possible for encapsulated chips.

30.10 Microsystems

Bulk micromachining utilizes anisotropic wet etching and DRIE for deep silicon structures. This is beneficial if large masses are needed, as in accelerometers or vibrating energy harvesters. Surface micromachining relies on thin films, and structure heights are 1% of those of bulk micromachined devices. Thin films combined with isotropic etching enable cavities and channels to be made without bonding, eliminating the cost of a capping wafer. However, the mechanical properties of thin films are no match for single crystal silicon. SOI MEMS solve this: the processing techniques are identical to surface MEMS but the resulting material is single crystal silicon.

The very name of MEMS contains the idea that the devices contain more than one function. Membrane bending is sensed as electrical resistance change (piezoresistive pressure sensor), electrical actuation voltage is pulling a beam down (RF switch), radiation heats up a resistor, which is sensed as current change (bolometer), thermal expansion of air deflects a membrane (Braille actuator), an electric field breaks up water flow into droplets (electrospray), thermal expansion bends a cantilever beam which deflects and writes a recess into a polymer film (AFM thermomechanical memory).

In microfluidics the concept of integration is embodied in the concept of lab-on-a-chip. It is the fluidic counterpart of an integrated circuit. The basic motivation for integration is the same: do as many operations as possible on a chip, because for working with minute amounts of liquids that is the way to go, as it is with weak electrical signals. The basic operations of fluidics and electronics are very much the same. Both systems have active and passive devices: transistors and pumps, diodes and valves, resistors and constrictions, guard rings and filters, reset systems (washing in fluidics), controlled metering (charge pumps vs. injectors), etc. A complete fluidic system would for example filter out large particles, concentrate the filtrate, separate the compounds, add fluorescent label, and detect them. Or it would run dozens of identical tests in parallel. Today very few such truly large-scale integrated fluidic systems exist, but then again ICs of the 1960s were not that great by today's standards.

30.11 Exercises

1. How would you fabricate the following single crystal silicon microbridges:
   (a) very thin, lightly doped and wide bridge
   (b) stubby, high aspect ratio bridge of n-type silicon
   (c) extremely narrow bridge of arbitrary aspect ratio?
2. Which devices in this chapter are made on DSP wafers?
3. If vertical-walled through-wafer structures are made, what is the minimum grating period (line + space) that can be realized by: (a) DRIE; (b) <110> wet etching; (c) <100> wet etching? What is the limiting the factor in these processes?
4. How could oxide membranes be made and what would they be good at? What are the limitations?
5. Explain quantitatively how thick films are needed to close the 1 μm wide chevrons of Figure 30.12.
6. Explain step by step the fabrication process for the RF switch of Figure 30.26.
7. Design a fabrication process for the accelerometer of Figure 17.2b. Include metallization, too.
8. Design a fabrication process for the 3D silicon shadow mask shown in Figure 30.5!
9. What is the number of AFM tips that could be fabricated on a 1 cm^2 chip by the process described in Figure 30.10? What if DRIE were used instead of wet etching?
10. Explain step by step the fabrication of the microphone of Figure 30.8.
11. Analyze the fabrication process for the nanoholes shown in Figure 13.13.
12. Design a fabrication process for the fountain pen of Figure 24.11.

13. Design a fabrication process for the fiber optic interferometer shown below!

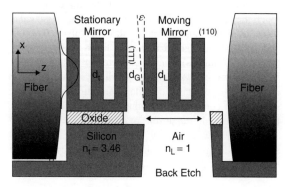

Reproduced from Lipson and Yeatman (2007), copyright 2007, IEEE

References and Related Reading

Allen, J.J. (2005) **Micro Electro Mechanical System Design**, CRC Press.
Ayazi, F. and K. Najafi (2000) High aspect-ratio combined poly and single-crystal silicon (HARPSS) MEMS technology, *J. Microelectromech. Syst.*, **9**, 288–294.
Brand, O. (2006) Microsensor integration into systems-on-chip, *Proc. IEEE*, **94**, 1160–1176.
Brugger, J. *et al.* (1999) Self-aligned 3D shadow mask technique for patterning deeply recessed surfaces of micro-electro-mechanical systems devices, *Sens. Actuators*, **76**, 329.
Chen, J. and K.D. Wise (1997) A high-resolution silicon monolithic nozzle array for inkjet printing, *IEEE Trans. Electron Devices*, **44**, 1401.
Chui, B.W. *et al.* (1998) Low-stiffness silicon cantilevers with integrated heaters and piezoresistive sensors for high density AFM thermomechanical data storage, *J. Microelectromech. Syst.*, **7**, 69.
Dehé, A. (2007) Silicon microphone development and application, *Sens. Actuators*, **A133**, 283–287.
Dokmeci, M.R. *et al.* (2004) Two-axis single-crystal silicon micromirror arrays, *J. Microelectromech. Syst.*, **13**, 1006–1017.
Esashi, M. (2008) Wafer level packaging of MEMS, *J. Micromech. Microeng.*, **18**, 073001.
Gad-el-Hak, M. (ed.) (2005) **MEMS Handbook**, 2nd edn, CRC Press.
Gerlach, G. and W. Dötzel (2008) **Microsystem Technology**, John Wiley & Sons, Ltd.
Graf, A. *et al.* (2007) Review of micromachined thermopiles for infrared detection, *Meas. Sci. Technol.*, **18**, R59–R75.
Horsley, D.A. *et al.* (1998) Design and fabrication of an angular microactuator for magnetic disk drives, *J. Microelectromech. Syst.*, **7**, 141.

Huang, Y. *et al.* (2003) Fabricating capacitive micromachined ultrasonic transducers with wafer-bonding technology, *J. Microelectromech. Syst.*, **12**, 128–137.
Jaakkola, A. *et al.* (2007) Piezotransduced single-crystal silicon BAW resonators, IEEE Ultrasonics Symposium 2007, p. 1653.
Ji, J. and K.D. Wise (1992) An implantable CMOS circuit interface for multiplexed microelectrode recording arrays, *IEEE J. Solid-State Circuits*, **27**, 433–443.
Kaajakari, V. (2009) **Practical MEMS**, Small Gear Publishing.
Kaajakari, V. *et al.* (2006) Stability of wafer level vacuum encapsulated single-crystal silicon resonators, *Sens. Actuators*, **A130–131**, 42–47.
Kong, S.H., D.D.L. Wijngaards and R.F. Wolffenbuttel (2001) Infrared micro-spectrometer based on a diffraction grating, *Sens. Actuators*, **A92**, 88–95.
Kovacs, G.T.A. (1998) **Micromachined Transducers Sourcebook**, McGraw-Hill.
Li, Q. *et al.* (2009) Assessment of testing methodologies for thin-film vacuum MEMS packages, *Microsyst. Technol.*, **15**, 161–168.
Liang, C. and Y.-C. Tai (1999) Sealing of micromachined cavities using chemical vapor deposition methods: characterization and optimization, *J. Microelectromech. Syst.*, **8**, 135–145.
Licklider, L. *et al.* (2000) A micromachined chip-based electrospray source for mass spectrometry, *Anal. Chem.*, **72**, 367–375.
Lipson, A. and E.M. Yeatman (2007) A 1-D photonic band gap tunable optical filter in (110) silicon, *J. Microelectromech. Syst.*, **16**, 521–527.
Modafe, A. *et al.* (2005) Embedded benzocyclobutene in silicon: an integrated fabrication process for electrical and thermal isolation in MEMS, *Microelectron. Eng.*, **82**, 154–167.
Oh, K.W. and C.H. Ahn (2006) A review of microvalves, *J. Micromech. Microeng.*, **16**, R13–R39.
Pham, N.G. *et al.* (2004) Photoresist coating methods for the integration of novel 3-D RF microstructures, *J. Microelectromech. Syst.*, **13**, 491–499.
Sakata, M. *et al.* (1999) Micromachined relay which utilizes single crystal silicon electrostatic actuator, Proceedings of IEEE MEMS 1999, pp. 21–24.
Sanz-Velasco, A. *et al.* (2006) Sensors and actuators based on SOI materials, *Solid-State Electron.*, **50**, 865–876.
Saarela, V. *et al.* (2005) Re-usable multi-inlet PDMS fluidic connector, *Sens. Actuators*, **B114**, 552–557.
Scheeper, P.R. *et al.* (2003) A new measurement microphone based on MEMS technology, *J. Microelectromech. Syst.*, **12**, 880–891.
Senturia, S. (2004) **Microsystem Design**, Springer.
Sparks, D., S. Massoud-Ansari and N. Najafi (2005) Long-term evaluation of hermetically glass frit sealed silicon to Pyrex wafers with feedthroughs, *J. Micromech. Microeng.*, **15**, 1560–1564.
Spearing, S.M. (2000) Materials issues in microelectromechanical systems (MEMS), *Acta Mater.*, **48**, 179–196.

Tas, N.R. et al. (2002) Nanofluidic bubble pump using surface tension directed gas injection, *Anal. Chem.*, **74**, 2224–2227.

Tiggelaar, R.M. et al. (2005) Fabrication of a high-temperature microreactor with integrated heater and sensor patterns on an ultrathin silicon membrane, *Sens. Actuators*, **A119**, 196–205.

Venstra, W.J. et al. (2009) Photolithography on bulk micromachined substrates, *J. Micromech. Microeng.*, **19**, 055005.

Witvrouw, A., H.A.C. Tilmans and I. De Wolf (2004) Materials issues in the processing, the operation and the reliability of MEMS, *Microelectron. Eng.*, **76**, 245–257.

Yun, S.-S. et al. (2009) Fabrication of morphological defect-free vertical electrodes using a (110) silicon-on-patterned-insulator process for micromachined capacitive inclinometers, *J. Micromech. Microeng.*, **19**, 035025.

31

Process Equipment

Microfabrication equipment component technologies differ completely: critical elements in lithography are optics and mechanics, plasma etchers depend on RF and vacuum technologies, and epitaxy is about temperature and flow control. The common aspects relate to process outcomes: uniformity across the wafer, run-to-run reproducibility, yield, rate and throughput.

The size of microfabrication equipment tends to be inversely proportional to the size of structures it makes: small tabletop instruments can pattern and etch 3 μm lines, but tools for 100 nm technology require garage-sized behemoths with multimillion-dollar price tags (Figure 31.1). Price tags for individual tools are up to $50 million today (lithography tools being the most expensive), even though $200 000 can still buy a system suitable for academic research purposes, be it a mask aligner, a furnace or a plasma etcher.

Microfabrication cost structures are not always obvious: the cost of diamond thin films is similar to silicon thin films, because both are deposited in similar CVD systems, and the gas cost difference of silane vs. methane is negligible. Cost difference may arise from gas purity: 99.9999% pure gas is pricier than 99.95% pure.

Cost of ownership is a combination of capital and operating costs, and yield. If the tool produces scrap, the cost of good chips rapidly goes up. If the tool produces excellent results but only a few wafers per hour, it is not much of a production tool. Production monitoring measurements are sometimes done inside the tool, enabling real-time control of the process, but most often monitoring is done after the fact: wafers are measured after the completion of processing.

31.1 Batch Processing vs. Single Wafer Processing

Microfabrication economies were earlier claimed to result from batch processing: tens of wafers with hundreds of chips are processed simultaneously in for example a furnace or wet etch bench. But scaling down linewidths has put increasing demands on process control, and single wafer tools have superseded batch equipment in many process steps. Besides, batch equipment for large wafers becomes prohibitively cumbersome.

Wet processing in a tank is a prototypical batch process: a full cassette of wafers is processed simultaneously (see Figure 11.8). Wafer cleaning and non-patterning etching (e.g., removal of sacrificial oxide by HF) are widely done in batch-mode wet processing even in the most advanced processes. Wet etching for patterning (e.g., H_3PO_4-based aluminum etching or BHF etching of oxide) is not an option when linewidths are below 3 μm because process control is difficult in batch wet processing: no in situ monitoring is possible, and wafer-to-wafer variations are often encountered. However, model-based control with ionic strength and temperature measurements can be used to improve wet etch rate control to some extent.

In batch processing both the uniformity over the batch and the uniformity across the wafer must be observed. Variation comes from wafer position in a batch system: flow patterns of gases and liquids over wafers depend on wafer position, and the thermal environment may also be position dependent. The first and last wafer have only one neighbor, but others are sandwiched between two wafers.

During the 3 inch era most wafer processing was batch, and a major shift started at 100 mm wafer size. Robotic loading/unloading is simple in single wafer systems, and they are more amenable to fabrication automation, including data gathering. Film thicknesses have been scaled down with linewidths, and thinner films require less process time in deposition and etching, which works in favor of single wafer processing. However, single wafer systems hardly ever approach batch system throughputs which can be up to 200 wafers per hour (WPH), and in PECVD and implant applications 500 WPH. It may also well be that at the back end of the process wafers are so expensive that manufacturers do not want to risk a lot by batch

---

Introduction to Microfabrication, Second Edition Sami Franssila
© 2010 John Wiley & Sons, Ltd

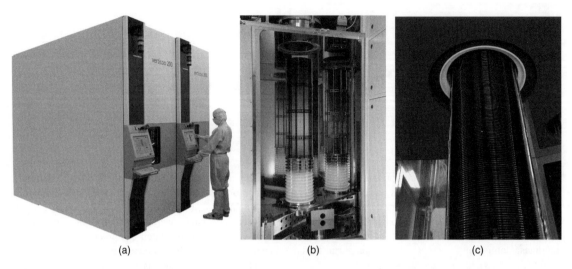

Figure 31.1 Vertical furnace: (a) overview; (b) closeup and (c) detail of wafer boat. Courtesy Centrotherm

processing: a 200 mm wafer with 300 chips is worth thousands of dollars and if a batch of 25 wafers is lost at the end of the process, the financial loss is considerable. It will also take time to fabricate a replacement lot, typically three to six weeks. This can be an even greater burden than the money lost if delivery time is used as the criterion for choosing a chip supplier.

In single wafer processing wafer-to-wafer repeatability is a major issue. The first-wafer effect means that the system has not stabilized, therefore the first wafer experiences for example a lower temperature or more concentrated chemicals. In addition to batch and single wafer processing, various combinations are being used, as given in Table 31.1.

Single feature processing is so slow that it is relegated to special applications only. Throughputs of a few wafers per hour are considered good for direct write processes. Single chip processing is done only in lithography, using reduction steppers and scanners.

Single wafer processing benefits from easy process development because fewer wafers are needed and batch effects are eliminated. Robotic handling from cassette to cassette and in situ monitoring without averaging over the batch enable a much higher degree of process control than batch systems. There are various combination systems, for instance high-current ion implanters load a batch of wafers on a rotating holder, but the beam scans one wafer at a time, and rotation of the holder takes care of batch processing. In epitaxy single wafer and batch tools coexist, but in plasma etching and sputtering single wafer tools are the norm in mainstream IC production.

Table 31.1 Granularity of processing

Single feature processing
   Direct writing for research and pilot production
   Mask making by e-beam or laser beam
   Mask repair, chip repair, chip customization

Single chip processing
   Steppers and scanners
   Better alignment and resolution

Single wafer processing
   Easy automation
   In situ monitoring
   Plasma etching, sputtering, (PE) CVD,
      medium-current implantation (MCI)

Batch processing
   Wet cleaning, oxidation, thermal CVD (oxide, poly, nitride)

Combinations
   Load multiple wafers but process one wafer at a time (HCI, CVD, sputter)

31.2 Process Regimes: Temperature and Pressure

Two major process parameters are pressure and temperature. Many microfabrication processes are vacuum/low-pressure processes (CVD, RIE, sputtering, implantation), some are room ambient processes (lithography, wet cleaning) and high-pressure oxidation is an exception. The temperature scale extends from 1200 °C diffusions to 850 −1100 °C oxidation, 300–900 °C CVD to room temperature processes (plasma etch, sputtering,

Process Equipment 411

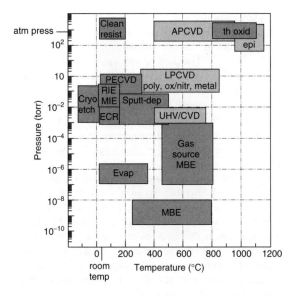

Figure 31.2 Equipment classified on temperature–pressure axes. Reproduced from Rubloff and Boronaro (1992) by permission of IBM

Table 31.2	Methods for heating
Method	Equipment
Resistance heating	Furnace
Induction heating	Epitaxial reactor
Photon heating	Rapid thermal processing (RTP)
Conduction	Hotplates; single wafer PECVD
Convection	Resist ovens; wafer back-side gas flow

implant, lithography, wet cleaning). Some etch processes use subzero temperatures: sublimation drying in MEMS works at about $-20\,°C$, and cryogenic DRIE at $-120\,°C$.

Many room temperature processes can be run at higher temperatures for special purposes: sputtering at $450\,°C$ for improved step coverage, implantation at $800\,°C$ for SIMOX wafers, or plasma etching at elevated temperature to reduce residues. Figure 31.2 shows major processes on a temperature–pressure chart. High-temperature/low pressure processes are difficult because of outgassing of adsorbed gases from vacuum components during high-temperature operation.

Evaporation and molecular beam epitaxy require the best vacuum (10^{-8}–10^{-11} torr). Etching and PECVD put rather modest demands on vacuum technology (1 mtorr to 1 torr). Sputtering systems require very low base vacuum (e.g., 10^{-8} torr), but when argon is introduced, operating pressures are 0.1–10 mtorr. The effects of a vacuum on thin-film quality will be discussed in Chapter 33.

A vacuum has many uses in microfabrication. It is a way to keep surfaces clean: if there are not many molecules around, very few will impinge on wafer surfaces. The same protective effect can be achieved by an inert gas atmosphere: a 100% nitrogen or argon atmosphere means that reactive gases like oxygen or water vapor are present only in residual amounts.

Vacuum quality is important for transport processes, as will be discussed in Chapter 33. In a high vacuum atoms and molecules will not experience collisions and will take direct routes. In a rough vacuum atoms and molecules will experience many collisions before arriving at their destination, and the arrival angles will be widely distributed. This can be detrimental or beneficial, depending on the application.

There are five main methods to heat the wafers currently in use, as listed in Table 31.2. Some prominent examples are also shown.

The first three methods are used in high-temperature processes and the latter two in low-temperature processes. Some degree of heating and/or temperature control is desirable in almost all tools. In all plasma equipment there is plasma heating (and latent heat release from condensation in deposition tools) and in ion implantation the beam flux can heat the wafer considerably.

In most tools wafers lie horizontally on a chuck (electrode, susceptor) and the chuck is heated. Old hotplates had no active control of wafer-to-chuck contact, and there was an inevitable air mattress between the wafer and the hotplate, but today the degree of thermal contact can be controlled as wished (with hotplate price tags of tens of thousands of dollars). Wafer clamping ensures intimate contact and efficient heat transfer. Both mechanical clamping and electrostatic clamping (ESC) are used. In the former, pins hold the top side of the wafer, which limits usable wafer area, and there is the danger of contamination from the clamp pins. Mechanical clamping is widely used because it is much simpler than ESC. If no clamping is done in RIE, the temperature can easily rise to about $120\,°C$, the photoresist glass transition temperature, in a few minutes. Steady state temperatures can be kept below $40\,°C$ indefinitely by back-side cooling. Clamping is also essential when wafers are processed in a vertical position (in for instance ion implanters where the long beam line can only be built horizontally) or when wafers are processed face down (as in CMP and in evaporation).

Heating (and cooling) can also be effectuated by fluid flow. For instance, hot argon gas is employed in sputtering systems to ramp up wafer temperatures to 400–500 °C, in a time scale of 10 seconds. In etchers the wafer back side is often cooled by helium flow. Some of these gases leak into the process chamber, and the type of heating/cooling gas has to be compatible with the process.

31.3 Cluster Tools and Integrated Processing

In cluster tools several process chambers are connected to each other, either serially or by means of a central transfer chamber. Figure 31.3 shows a PVD multichamber system. It incorporates a pre-clean module, multiple reactor modules and a cooldown module, all connected to a central handler chamber. Multiple identical reactor modules enable increased throughput, or alternatively two different processes can be run without risk of cross-contamination. Central handler reliability is crucial for cluster operation. A photograph of a three-chamber sputtering tool is shown in Figure 31.4. The motivation for separate chambers in this case has to do with contamination: reactive sputtering of AlN has to be kept separate from aluminum deposition, otherwise the aluminum target would be poisoned by nitrogen.

Cluster tools can be applied to any process sequence in principle, but in practice similar processes are integrated: similar temperature or similar vacuum or similar ambient in general. A titanium adhesion layer below platinum is another old example of integrated processing: the titanium surface is kept clean under vacuum, and platinum, which is deposited immediately after titanium, adheres to it well, whereas platinum would not adhere to an oxidized titanium surface, which would result immediately if a titanium wafer were taken out of a vacuum chamber and transferred to another deposition system.

Integration of thermal oxidation with sputtering or CMP with PECVD would be awkward, but PECVD and plasma etching, or RTO and RTCVD, can be combined fairly easily.

Integrated processing involves chaining process steps into longer sequences. Process integration is also about chaining process steps into sequences, but in a different sense: process integration is device related, whereas integrated processing is the tool's view of step chaining. In integrated processing steps follow each other under strictly controlled conditions in a vacuum, inert gas or some other well known ambient. This principle has long been used in silicon epitaxy: surface cleaning by HCl or H_2 gas is done in the same reactor chamber as the deposition itself, to guarantee an oxide-free surface. As shown in Figure 31.5, conventional processing involves a number of separate steps, with storage and cleaning steps, which can be eliminated by integrated processing.

Integrated processing has both scientific and manufacturing benefits. It enables a much higher degree of control over materials, interfaces and surfaces. This helps us to understand what is really going on in our processes. In manufacturing it makes savings in several ways: cleaning steps can be minimized because the wafer conditions are known all the time; wait and storage steps are eliminated; and cycle time is reduced.

31.4 Measuring Fabrication Processes

There are three different aspects that can be measured in a fabrication process: tool, process and wafer. Tool

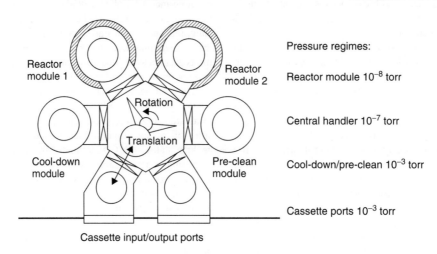

Figure 31.3 Multichamber vacuum cluster for PVD. Reproduced from Grannemann (1994) by permission of AIP

Figure 31.4 Cluster tool with central handler and three chambers (two with four sputter targets and one with a single target). Courtesy von Ardenne

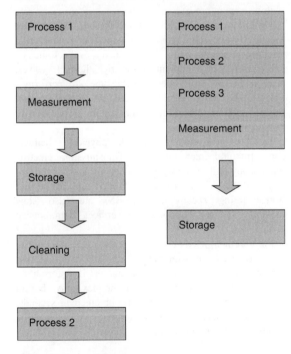

Figure 31.5 Conventional step-by-step process compared to an integrated sequence. In integrated processing wafers are under controlled ambient all the time

parameters like RF power, mass flow, process time or electrode temperature are easily measured. Process measurements deal with ionic strength in a cleaning solution, electron and ion energies in plasma, or ion dose. In lithography exposure time is usually fixed, but of course exposure depends on UV energy, which drifts with lamp lifetime. Indirect measurements are often much simpler than direct measurements: for example, vacuum chamber base pressure is a good indication of vacuum quality, but mass spectrometry (usually called RGA, for residual gas analysis) can actually identify the residual atoms and molecules, for instance if the residue is water vapor or pump oils. This is significant in understanding vacuum–film interactions as well as in troubleshooting.

Very few measurements are actually done on the wafers during processing. This is understandable because process chamber conditions are often harsh, for example RF fields, corrosive gases or high temperatures. Wafer temperature in RTA can be measured by pyrometry during processing. In ultrahigh-vacuum conditions surface spectroscopies can be used to monitor deposition processes in real time: reflection high-energy electron diffraction (RHEED) and low-energy electron diffraction (LEED) are routinely employed in MBE systems to check crystallinity of the growing film. Unfortunately, most deposition processes are operated under conditions where such systems cannot be used.

Measurements can be classified into four categories according to their immediacy:

1. In situ: during wafer processing in the process chamber.
2. In-line: after wafer processing inside the process tool (e.g., exit load lock).
3. On-line: in the wafer fab by wafer fab personnel.

4. Ex situ: in a dedicated analytical laboratory by expert users.

In situ resist development monitoring with an interferometric end point detector can improve linewidth control considerably. It can compensate for changes in exposure dose, resist (de)composition, developer concentration, temperature or resist bake drifts and shifts which could easily result in 10% develop time differences and in similar linewidth variation.

Plasma etching is almost always monitored in real time, in order to determine the end point and to prevent excessive etching of the substrate or the underlying film. Optical emission spectroscopy (OES) is commonly used: the intensity of some suitable excited species in the plasma is monitored with optical systems including a wavelength-selective detector. In fluorine plasmas a signal at $\lambda = 704$ nm (from excited fluorine atoms) can be used. During etching the signal is small because there is little free fluorine: most of it is bound as reaction products like SiF_4 or WF_6. At the etching end point the free fluorine intensity increases because it is no longer consumed by the reaction. A more selective method would be to monitor the reaction products themselves. This must be developed for every process individually. The nitrogen signal (396 nm) is suitable for monitoring nitride etching: there will be a sharp drop in the nitrogen signal when all the nitride has been etched away. OES does not, however, measure wafers but gives average values over the reactor. Film thickness during deposition or etching can be measured by for example ellipsometry or interferometry, but such systems are not commonplace.

One of the oldest applications of in situ monitoring is the quartz crystal microbalance (QCM) in film thickness control during evaporation and sputtering. The QCM is placed in the same atom flux as the wafers, and therefore it experiences the same film deposition. The mass change is detected as a frequency change and converted to film thickness. The resonant frequency of the QCM is determined by crystal thickness x and sound velocity v_{tr} given by

$$f = \frac{v_{tr}}{2x} \quad (31.1)$$

For a quartz wafer of 500 μm thickness, with transverse wave velocity of 3340 m/s, this translates to 3.3 MHz. The frequency change due to thickness increase Δx is given by the derivative of Equation 31.1,

$$\Delta f = \frac{-2f^2 \Delta x}{v_{tr}} \quad (31.2)$$

Taking into account the fact that the deposited film density differs from quartz (but neglecting its different elastic properties), we get thickness change from frequency change as

$$\Delta x = -\frac{v_{tr} \rho_{quartz} \Delta f}{2f^2 \rho_{film}} \quad (31.3)$$

With an easily detectable frequency shift of 1 ppm, the minimum thickness change that can be measured by the QCM is a fraction of a nanometer. The temperature sensitivity of the QCM is 0.5 ppm/°C, which has to be accounted for because deposition is usually accompanied with a temperature rise.

On-line measurements constitute the bulk of measurements in wafer fabrication. These include standard film thickness measurements (ellipsometry, reflectometry), sheet resistance, implant dose by thermal waves, step height by profilometry, etc. Some are performed in seconds, like sheet resistance or film thickness, others require a few minutes for sample preparation or pumpdown (SEM, AFM).

Ex situ measurements include physical, chemical and structural measurements. Transmission electron microscopy (with a TEM), secondary ion mass spectrometry (SIMS), X-ray diffraction (XRD) and Rutherford backscattering spectrometry (RBS) are also slow methods, and can be bought as services from outside contractors.

Measurement needs change over the process lifecycle: in R&D phase measurement needs are manifold, but requirements lax; it does not matter if information is obtained in an hour or even the next day, but in manufacturing the results must be obtained in seconds. Differences relate also to measurement spot size and the number of measurement spots, as shown in Table 31.3.

Table 31.3 Measurement needs

	R&D	Pilot production	Volume manufacturing
Samples	Anything	Full wafers (monitors)	Full wafers (scribe line measurement)
Analysis spot	Anything	Not a concern	Test site
Measurement time	Anything	Minutes/hours	Seconds/minutes

Surface analytical methods are problematic because sample transfer from process chamber to analytical chamber takes some time and gases and vapors adsorb on the sample surface and disguise the original surface signal. In-line tools do exist for integrated surface analysis, for example a RIE etch chamber connected to an X-ray photoelectron spectrometer (XPS), but such systems are for basic research only.

31.5 Equipment Figures of Merit

Equipment figures of merit include various aspects such as process, capital cost, labor, consumables, etc.. Some of the most important ones are discussed briefly below.

31.5.1 Uptime/downtime

Uptime is an overall measure of equipment availability. Uptime is reduced by both scheduled and non-scheduled maintenance. Recalibration/test wafers required to set the process running after a disruption can contribute significantly to downtime. Scheduled system cleaning is often mandatory for deposition equipment, to prevent film flaking from the chamber walls. This is sometimes done after every wafer (by plasma cleaning, not by mechanical cleaning, which would necessitate opening the chamber), with a drastic effect on uptime but with higher yield. Uptimes vary from almost 100% for wet benches to 90% for furnaces and plasma etchers, 80% for implanters and 40% for PECVD.

31.5.2 Utilization

Utilization is a measure of equipment use: actual productive hours of all available hours. General purpose tools like lithography have high utilizations, and the more dedicated tools lower ones. A $10 million lithography tool must not wait for a $1 million resist coater, but the resist coater can sit idle waiting for a stepper. The rapid thermal processor for silicide anneal is used twice during a CMOS process, and its utilization is the lowest of all tools, together with the dedicated wet bench for selective metal etching.

31.5.3 Throughput

How many wafers per hour (WPH) can the system handle? Single wafer tools have throughputs of for example 50–100 WPH, but batch tools can handle up to 200 WPH. This is very much dependent on the process: if a LPCVD polysilicon process is run at 635 °C the deposition rate is four times higher than at 570 °C. Similarly, if film thickness to be deposited is doubled, deposition time is doubled. Throughput, however, might not change much if overheads (loading, pump down, temperature ramp, etc.) are high relative to deposition time. In etching throughput can be severely reduced even if film thickness remains unchanged, but overetch requirements change due to topography (section 11.12.3 on spacers).

31.5.4 Footprint

How big is it? Cleanroom space is premium priced: $10 000 per square meter is the price range for a class 1 (Federal Standard) cleanroom. In most cases, just the front panel of the system is in the cleanroom, the rest of the tool being in the service area, which has more relaxed particle cleanliness requirements.

31.5.5 MTTF, MTBA, MTBC

How long will the equipment work before failure? Do operators need to interfere with its operation? How often does it have to be cleaned? These questions are operationalized by MTTF (Mean Time To Failure), MTBA (Mean Time Between Assists) and MTBC (Mean Time Between Cleans).

MTBC depends on the process: particle counts (on test wafers) are checked regularly, and increased counts indicate a cleaning need. But the acceptable particle count depends on chip size, sensitivity of the particular process step to particle contamination (subsequent steps may be a cleaning step that effectively removes particles) or just engineering judgment about the acceptable level of particles. Particle counts in individual process steps cannot easily be correlated with process yield, therefore short-loop test runs with specially designed test structures are used to check the effects of individual process steps.

31.6 Simulation of Process Equipment

Process simulation covers length scales of a few micrometers in both the lateral and vertical directions. In process equipment simulation the length scale is defined by tool size and can be up to a meter. In practice this scale difference means that tool simulation is carried out independently of process simulation. In tool simulation 3D geometry is the norm, but of course all symmetries in the tool geometry are utilized to reduce the computational load.

A typical tool simulation includes temperature distribution, flow patterns and plasma properties. Mass, momentum, energy and charge balances are calculated. Plasma modeling is difficult because it involves so many parameters: collision cross-sections, ionization, attachment, recombination, dissociation, etc. These plasma reactions must then be combined with surface reactions (deposition or etching). Taken together, these determine for instance PECVD film uniformity. For reactors operating in the mass transport-limited regime, flow patterns are of utmost importance. For reactors operating in the surface reaction-limited regime, thermal design is a high priority.

31.7 Tool Lifecycles

Tool development takes a long time: from the first proof-of-concept tool to multiple orders for volume manufacturing takes easily 10 years. The proof-of-concept tool is homemade or modified equipment that demonstrates the key features of a new process, usually for small wafer size. For e-beam lithography it might be a new electron column design; for a plasma etcher it might be a new RF coupling scheme. The alpha tool is a purpose-built system which has the new key elements designed in from the beginning. The alpha tool is missing productivity features like robotics and software, but is designed for the final wafer size. The reliability of the alpha tool is not comparable to production tools – it is a test bed for process research, not for production. Alpha tools are not shipped to outsiders. The beta tool is a fully equipped version, with essentially all the features that will make the final product distinct. Beta tools are shipped to select customers who are willing to bear part of the burden of testing new equipment in order to benefit from new technology. Beta customers provide productivity-related data that is difficult or even impossible to acquire at the tool manufacturer: What is uptime in production-like conditions running thousands of wafers? Is wafer yield comparable to existing or competing designs? What are the field servicing requirements?

Both academic and industrial labs buy equipment for R&D, but what will happen when a successful new process needs to be scaled up for production? The popular answer today is that the basic design of the process chamber (e.g., spinner bowl geometry, sputter cathode design, etcher gas manifold, RTA lamp configuration) is fixed. Research labs buy the very basic configuration, essentially the process chamber only (obviously this works better for some tools than others, and not at all for optical lithography). Later on, when the process is transferred to manufacturing, productivity features like cassette-to-cassette automation, multiple chambers and advanced software can be added. This reduces the risk of new equipment purchase for the industry, and it allows academic labs to do industrially relevant research without the need to invest in volume manufacturing tools.

31.8 Cost of Ownership

Difficulties in tool performance assessment have led to the introduction of a new figure of merit, the cost of ownership (CoO), which tries to put all tools on an equal footing, calculated over the lifetime of the tool. Equipment capital investment has very little meaning in cost calculations if other major factors like yield and throughput are neglected. CoO is an estimate of all costs associated with a certain piece of equipment, and it can be used to compare different mixes of fixed and running costs. Yield, or alternatively cost/good chip, is of paramount importance. CoO is defined as

$$\text{CoO} = \frac{\text{capital cost} + \text{operating cost} + \text{yield cost}}{\text{throughput} \times \text{uptime}}$$

(31.4)

It thus incorporates both the purchase price, which may sound huge, and the running costs, which include many items that may contribute considerably over the lifetime of the equipment. Running costs consist of a mixture of chemicals/gases/slurries (the CMP slurries market is bigger in dollar terms than the market for CMP equipment), consumables (e.g., O-rings), power consumption (high in epitaxy), personnel costs (very variable, depending on for example cleaning frequency), measurement costs (high if product wafers cannot be used and separate monitor wafer are needed), calibration/monitoring wafer costs (e.g., when the system is frequently cleaned and must be requalified).

Consider two hypothetical RIE etchers, A and B, with the features in Table 31.4 to see how seemingly minor differences add up and show that purchase price is only one consideration among many.

Table 31.4 Cost-of-ownership for RIE systems A and B

	A	B
Purchase price	1 000 000 €	1 500 000 €
Operating costs	150 000 €/yr	120 000 €/yr
5 years costs	1 750 000 €	2 100 000 €
Uptime	85%	90%
Throughput	45 WPH	55 WPH
Wafers/5 yrs	1.68 M	2.17 M
Yield	99%	99.8%
Good wafers	1.66 M	2.16 M
Cost/good wafer	1.06 €	0.97 €

Exercises

1. [ox]idation furnace is ramped up at 10 °C/min from [stan]d-by temperature of 800 °C and ramped down [to] a process temperature at 5 °C/min, what is the [proc]ess time (a) for 15 nm dry oxide at 900 °C; (b) for [10]0 nm wet oxide at 1000 °C?

2. [C]alculate the minimum deposition rate that can be [m]onitored by a QCM sensor if the wafers are heated [b]y the deposition process at 3 K/min.

3. What is the throughput of a sputtering system configured as shown in Figure 31.3 for TiN/Al metallization? (TiN in reactor 1, 100 nm thickness, 100 nm/min, Al in reactor 2, 1000 nm, at 500 nm/min, pre-clean etching and cooldown 30 s each, each transfer 15 s, and loadlock pump downtime 15 s.)

4. What is the throughput of a sputtering system configured as shown in Figure 31.3 for the deposition of 300 nm aluminum in both chambers, with pre-clean and cooldown modules used identically as preparation chambers for reactors 1 and 2?

5. How could metallization be monitored in the exit loadlock of a sputtering system?

6. Which methods could be used for the following measurement tasks:
   (a) oxide pinhole density
   (b) thickness of nominally 30 nm thick titanium
   (c) photoresist thickness uniformity
   (d) sputtered aluminum step coverage
   (e) implanted arsenic dose
   (f) particle removal efficiency in NH_4OH/H_2O_2 wet cleaning
   (g) Ta_2O_5 film deposition
   (h) ion implantation of boron into a phosphorus-doped wafer
   (i) silicon dioxide thinning in etching
   (j) mask oxide undercutting in KOH etching of <100> silicon
   (k) copper electroplating
   (l) photoresist sidewall angle?

7. If two PECVD systems are identical except for utilization, what will the price of a 70% utilization tool be relative to a 50% tool?

References and Related Reading

Barna, G.G. *et al.* (1994) MMST manufacturing technology – hardware, sensors and processes, *IEEE Trans. Semicond. Manuf.*, **7**, 149.

Grannemann, E. (1994) Film interface control, *J. Vac. Sci. Technol.*, **B12**, 2741.

Loewenstein, L. *et al.* (1994) First-wafer effect in remote plasma processing: the stripping of photoresist, silicon nitride and polysilicon, *J. Vac. Sci. Technol.*, **B12**, 2810.

May, G.S. and C.J. Spanos (2006) **Fundamentals of Semiconductor Manufacturing and Process Control**, John Wiley & Sons, Inc.

Moslehi, M.M. *et al.* (1992) Single-wafer integrated semiconductor device processing, *IEEE Trans. Electron Devices*, **39**, 4–32.

Rubloff, G.W. and D.T. Boronaro (1992) Integrated processing for microelectronics science and technology, *IBM J. Res. Dev.*, **36**, 233.

Schuegraf, K. (2003) Single-wafer process technology: enabling rapid SiGe BiCMOS development, *IEEE Trans. Semicond. Manuf.*, **16**, 121.

Wood, S.C. (1997) Cost and cycle time performance of fabs based on integrated single-wafer processing, *IEEE Trans. Semicond. Manuf.*, **10**, 98.

32

Equipment for Hot Processes

Thermal treatments constitute a major fraction of front end processes. Thermal oxidation, diffusion and implant annealing all call for temperatures around 1000 °C. Batch furnaces, horizontal and vertical, with loads of up to 200 wafers are traditional workhorses of thermal processing. More recently single wafer rapid thermal processors (RTP) have come on the scene, and laser annealing has also emerged. These new developments enable very high temperature ramp rates and combinations of very high process temperatures with very short process times, on the order of milliseconds and seconds instead of hours as in traditional furnaces.

32.1 High-Temperature Equipment: Hot Wall vs. Cold Wall

Two main varieties of high-temperature systems exist: hot wall and cold wall. Hot wall systems remain hot constantly. They are typically heated resistively, like horizontal furnaces. In cold wall systems only the wafers are heated, and the rest of the system stays cool, which enables faster temperature ramp rates and less deposition on the walls (because chemical reactions are exponentially temperature dependent). Heating can be achieved by inductive coils (as in epitaxy), by a susceptor/bottom electrode that is kept at a high temperature (as in PECVD) or by lamps (RTP). In analogy with kitchen equipment, an oven is a hot wall system, a microwave oven is a cold wall system. Warm wall systems do exist: system walls are heated unintentionally by the process but they remain at a much lower temperature than the wafers.

Large thermal masses in traditional furnaces provide excellent temperature uniformity, but very slow temperature ramp rates: 0.1 °C temperature uniformity and 5–10 °C/min ramp-up rates, and even slower cooling rates. New vertical furnaces (Figure 31.1) have higher ramp rates: tens of degrees per minute. In hot wall CVD systems deposition takes place on all hot surfaces, wafers and walls alike. Successive depositions build up thick films on the walls. Film cracking and flaking are especially probable when the system temperature is ramped up or down, or when a different film with a different stress state or CTE is deposited on the first one.

32.2 Furnace Processes

A horizontal oxidation furnace is shown schematically in Figure 32.1 and photographically in Figure 32.2. Quartz tubes sit inside resistive heater elements which ensure a uniform temperature. Typically four gas lines feed into a tube: oxygen (for dry oxidation), hydrogen (for wet oxidation), nitrogen (for inert protection during wafer loading and temperature ramping and cooling) and dichloroethane (DCE, for cleaning the tube). Because the mixture of oxygen and hydrogen is explosive, a burn box ensures that all hydrogen is burned, and no potentially dangerous mixture is formed.

Thermal oxidation is shown graphically in the time–temperature graph of Figure 32.3 and the process is detailed in Table 32.1. Wafer cleaning before all high-temperature processes is essential, but in order also to guarantee tube cleanliness, DCE cleaning is done before critically important oxidations. Chlorine cleaning could also be done but because chlorine is a corrosive gas, its handling is more complicated than that of DCE. The cleaning process reduces metallic contamination, much like RCA-2 clean, which uses HCl. Alternatively, chlorine-containing gases can be used during oxidation.

Actual oxidation time is only a fraction of total process time, for example 30%. An optional post-oxidation anneal (POA) has been included in the process flow. It densifies the film, but does not, to a first approximation, affect its thickness. POA can also be used to tailor fixed oxide charges (Q_f): while the oxidation temperature is by and

---

Introduction to Microfabrication, Second Edition   Sami Franssila
© 2010 John Wiley & Sons, Ltd

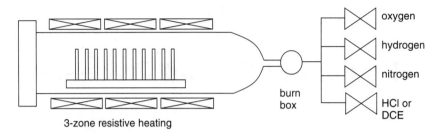

Figure 32.1 Horizontal oxidation furnace

Figure 32.2 Horizontal oxidation/diffusion furnace for 150 × 150 mm silicon solar cells. Courtesy Centrotherm

Table 32.1 Gate oxidation (25 nm thick dry oxidation)

Wafer cleaning RCA-1 ($NH_4OH:H_2O_2$) organic impurity removal
Wafer cleaning RCA-2 ($HCl:H_2O_2$) metallic impurity removal
Dip in dilute HF (1/100; 15 s) native oxide removal
Rinse and dry
Load at 800 °C
Boat insertion speed 25 cm/min; N_2 flow to flush water vapor
Ramp temperature from 800 to 950 °C in N_2/O_2 (15 min, or 10 °C/min)
Introduce oxygen flow through mass flow controller (4 slpm)
Oxidize for 35 min at 950 °C (target thickness 25 nm)
Shut off oxygen flow; introduce nitrogen
Post-oxidation anneal (POA) in nitrogen for 20 min at 950 °C (densification)
Cooldown to 800 °C, 40 min in nitrogen (4 °C/min)
Unload wafers at 800 °C; total process time 110 min
Ellipsometry/reflectometry measurement for thickness and uniformity

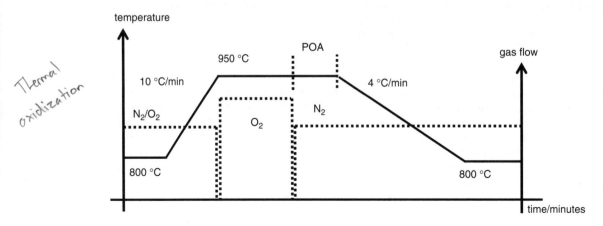

Figure 32.3 Temperatures and gas flows during oxidation in a horizontal furnace

Table 32.2 Comparison of furnace processes and RTP

	Furnace	RTP
Load	Batch	Single wafer
Throughput	High	Low
Cycle time	Long	Fast
Heating	Resistive	Lamp/radiation
Temperature uniformity	Excellent	Fair to good
Temperature gradients	Small	Large
Temperature measurement	Indirect	Direct

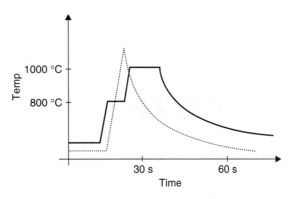

Figure 32.4 Temperature profile in rapid thermal annealing: solid curve, 1000 °C, 10 s anneal; dotted curve, 1100 °C spike anneal (zero-time anneal)

large determined by thickness requirements, POA temperature can be higher, which leads to reduced Q_f density.

32.3 Rapid Thermal Processing/Rapid Thermal Annealing

RTP (also known as RTA, for rapid thermal annealing) systems have emerged to address the issues of thermal budget, process time and gas phase impurity control.

Open tube furnaces are flushed with nitrogen during wafer loading, and this is usually effective in removing residual water vapor. It is useful to have a small, controlled oxygen flow during ramp-up to prevent thermal nitridation of the silicon surface and accept minor oxidation instead, but of course this is not applicable for very thin oxides.

However, even 100 ppm of residual water vapor will change the dry oxidation rate. Double tubing is used if better atmospheric control is required, but loadlocked systems must be used when exacting atmospheric control is mandatory. RTP systems are single wafer systems, and it is easy to implement loadlock and to control the gas phase in small volumes.

Furnace processes are slow, and the ion implantation monitoring cycle is very slow: after implantation photoresist is stripped, the wafer is cleaned and annealed in a furnace, and then sheet resistance is measured by a four-point probe. This easily takes 2 hours in a traditional furnace, but it can be accomplished in an hour in a RTP system: the time at high temperature is drastically reduced, but the stripping and cleaning constitute a sizable fraction of total time. The dominant implant monitoring method today is thermal wave because it can be done immediately after implantation, without any wafer preparation or processing.

But rapid in RTP really means thermal budget control: RTP can be very short but at high temperature, to minimize dopant diffusion. "Traditional" RTP annealing times are tens of seconds, called "soak" anneals, meaning anneal times of tens of seconds, while "spike" anneals are for example a second only (Figure 32.4). This is very fast compared to 30/60 min furnace anneals (FAs). In order to reduce unwanted diffusion during annealing, a high-temperature/short-time combination has been refined to a zero-time anneal (also known as a flash anneal), where times are in milliseconds.

In a silicide anneal (Section 7.10) oxygen must be eliminated and this is easier in a single wafer tool. $TiSi_2$ formation was the first RTA application. Titanium is very reactive with oxygen, and even minor residual oxygen of 5 ppm can result in titanium oxidation instead of silicidation.

Rapid heating is realized by three alternative methods: switching on powerful lamps, rapidly transferring the wafer(s) into a hot zone, or, for millisecond anneal, using lasers (either CO_2 or solid state lasers). Three designs for RTP are shown in Figure 32.5.

Tungsten halogen lamps deliver a kilowatt or two and a bank of lamps is needed, while a single xenon arc lamp can deliver tens of kilowatts. Ramp rates on the order of 50–300 °C/s are used in RTP, a factor of 1000 higher than in horizontal furnaces. Arc lamp output is in the visible and near infrared; the tungsten lamp spectrum extends to 4 µm. This leads to some differences in processes because high-energy photons can contribute to, for example, oxidation.

Lamp geometry is important for uniform processing. Large thermal non-uniformities, for example center-to-edge temperature difference, may reach 100 °C during ramping, which will result in detrimental crystal slips when the elastic deformation limit is exceeded, as discussed in connection with Equation 22.1. Cooling is usually by natural convection and 50 °C/s is typical. This cannot be much affected.

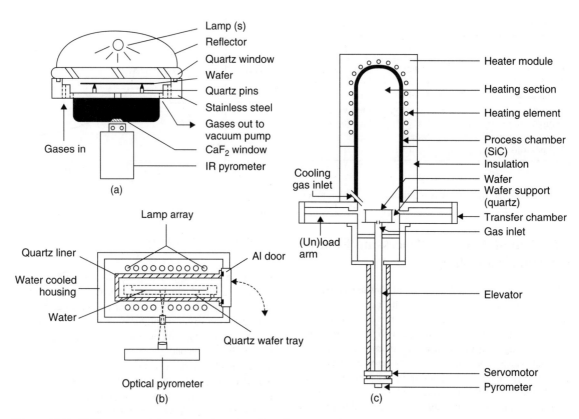

Figure 32.5 RTP systems: (**a**) arc-lamp heated, cold-wall system; (**b**) tungsten-lamp heated, warm-wall system and (**c**) resistively heated fast ramp, hot-wall system. Reproduced from Roozeboom, F. & Parekh, N. (1990), by permission of AIP

The lamp's spectrum has implications for temperature measurement: pyrometry is a non-contact method that can monitor wafer temperature in real time, but its operating wavelength must not overlap with the heating source. Pyrometry is based on the Stefan–Boltzmann law of emitted power:

$$P = \varepsilon \sigma T^4 \quad (32.1)$$

where $\sigma = 5.6697 \times 10^{-8}$ W/m^2-K^4 is the Stefan–Boltzmann constant.

Emitted power increases very rapidly as temperature increases because of the fourth-power dependence. Emissivity ε ranges from $\varepsilon = 1$ for an ideal black body to $\varepsilon = 0$ for a white body. Silicon emissivity is strongly dependent on doping level (charge carrier density), temperature and wafer thickness in the range up to about 600 °C. Above 600 °C silicon has a reasonably constant emissivity of about 0.7, but minor changes in emissivity result in large temperature errors. For example, oxide films on silicon act as interference filters and change the emissivity from 0.71 to 0.87 when oxide thickness increases from 0 to 400 nm. Below 600 °C thermocouples are employed. Thermocouples suffer from RTP thermal cycling and contact to silicon is not necessarily reproducible. Metallic contamination from a thermocouple is also an issue.

In addition to annealing, RTP can be used for oxidation (known as RTO) and for CVD (RTCVD). Rapid thermal oxidation is not significantly faster than furnace oxidation when it comes to oxidation rates, but from the equipment point of view it is: the loading–ramping–oxidation–cooling cycle can be a few minutes compared to hours in furnace processing. Figure 32.6 shows two problems with RTO: illumination is not uniform, but lamp geometry can be deduced from oxide thickness data. Gas flow patterns are also seen in the thickness data: incoming gas cools down the wafer, leading to thinner oxide near the gas inlet. Wafer edges are cooler than the center, but this is a natural consequence of cooling in general.

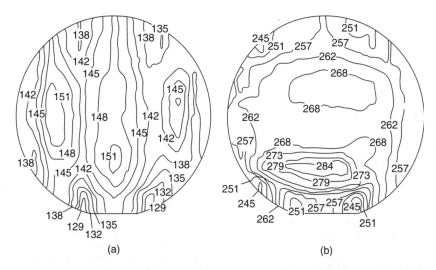

Figure 32.6 Rapid thermal oxidation uniformity: (**a**) vertical lamp bank geometry can be seen in oxide thickness (Å) chart and (**b**) gas-flow patterns are seen in oxide thickness: incoming gas cools the wafer near the flat, and wafer edges are cooler than the centre. Reproduced from Deaton, R. & Massoud, Z. (1992), by permission of IEEE

A hybrid technology between resistively heated furnaces and RTA is the fast ramp furnace. A heater, typically made of silicon carbide, is kept at very high temperature and the wafers are rapidly brought into its vicinity. The massive radiation source emits at much longer wavelengths than RTP lamps, and thermal equilibrium is possible. This arrangement can significantly reduce wafer emissivity variation and temperature non-uniformities. Ramp rates for fast-ramping systems are 10–100 °C/s, somewhat lower than in RTP.

32.4 Furnaces vs. RTP Systems

Furnaces are batch tools: even a small furnace loads 25 wafers, and large furnaces take hundreds of wafers. This is clearly very productive in a certain sense, but it necessitates large production batches. The alternatives are running furnaces with half-loads, or wafers have to wait until sufficient wafers are ready for the furnace.

RTP systems are single wafer tools, and cycle times are fast, though throughputs depend on batch size: RTP excels when small lots, for example five wafers, are processed, but the benefits compared to furnaces start to disappear when larger batches are processed.

Some modern devices could not be made without RTP: in ion implantation activation for shallow junctions (Chapter 26) the minimization of diffusion necessitates RTP. Similarly, silicide annealing is always done in RTP, for different reasons: gas atmosphere control is superior in RTP.

32.5 Exercises

1. What must oxygen flow be in a horizontal batch furnace to make sure that oxidation is not mass transfer limited? Write out and justify the assumptions you need in your solution.
2. If reproducibility and other uncertainties in a batch loading furnace limit the shortest practical oxidation time to 15 minutes, what is thinnest gate oxide that can be grown at 1000, 950, 900 and 850 °C? What are the corresponding CMOS linewidths?
3. How rapid is RTP? Calculate how short heat pulses will result in thermal equilibrium of the whole silicon wafer. Thermal diffusivity in silicon is 0.80 cm^2/s at room temperature and 0.1 cm^2/s at 1400 °C.
4. What temperature error does a change in emissivity from 0.71 to 0.87 cause in rapid thermal oxidation?
5. What power wattage does an RTP system for 300 mm wafers need if the maximum operating temperature is 1200 °C?
6. Anneal time and junction depth are connected by $x_j = k \times (Dt)^{1/3}$. If junction depth is about 100 nm in 0.25 μm technology, and the corresponding anneal time is 10 s, what is the anneal time for 0.1 μm technology? What is the junction depth?

Simulator exercises:

7. Rapid thermal oxidation (RTO) data is given below. How does RTO compare to furnace oxidation?

Constant time 30 s		Constant temperature 1050 °C	
Temp.	Thickness	Time	Thickness
950°C	44 Å	30 s	75 Å
1050°C	75 Å	150 s	158 Å
1150°C	145 Å	270 s	240 Å

Data from Deaton and Massoud (1992).

8. A typical implant activation anneal in a furnace is 950 °C/30 min, but in RTA a much higher temperature and much shorter time are used. Compare the junction depths that can be made by RTA and FA. Use implant conditions of 20 keV boron at 10^{15} cm^{-2} into a phosphorus-doped wafer of 10^{15} cm^{-3}.

References and Related Reading

Bensahel, D. *et al.* (2001) Front-end, single wafer diffusion processing for advanced 300mm fabrication line, *Microelectron. Eng.*, **56**, 49.

Bratschun, A. (1999) The application of rapid thermal processing technology to the manufacture of integrated circuits-an overview, *J. Electron. Mater.*, **28**, 1328 (special issue on RTP).

Deaton, R. and Z. Massoud (1992) Manufacturability of rapid-thermal oxidation of silicon: oxide thickness, oxide thickness variation and system dependency, *IEEE Trans. Semicond. Manuf.*, **5**, 347.

Endoh, T. *et al.* (2001) Influence of silicon wafer loading ambient on chemical composition and thickness uniformity of sub-5nm thick oxides, *Jpn. J. Appl. Phys.*, **40**, 7023.

Fair, R.B. (1996) Conventional and rapid thermal processes, in C.Y. Chang and S.M. Sze, **ULSI Technology**, McGraw-Hill.

Fischer, A. *et al.* (2000) Slip-free processing of 300 mm silicon batch wafers, *J. Appl. Phys.*, **87**, 1543.

Futase, T. *et al.* (2009) Spike annealing as second rapid thermal annealing to prevent pure nickel silicide from decomposing on a gate, *IEEE Trans. Semicond. Manuf.*, **22**, 475–481.

Gossmann, H.-J.L. (2008) Junction formation and its device impact through nodes: from single to coimplants, from beam line to plasma, from single ions to clusters, and from rapid thermal annealing to laser thermal processing, *J. Vac. Sci. Technol.*, **B26**, 267–272.

Malik, I.J. *et al.* (2007) Analysis of low temperature RTP needs for IC metallization, *Microelectron. Eng.*, **84**, 2729–2732.

Roozeboom, F. and N. Parekh (1990) Rapid thermal processing systems: a review with emphasis on temperature control, *J. Vac. Sci. Technol.*, **B8**, 1249.

Saga, K. *et al.* (1997) Influence of silicon-wafer loading ambients in an oxidation furnace on the gate oxide degradation due to organic contamination, *Appl. Phys. Lett.*, **71**, 3670.

33

Vacuum and Plasmas

When we talk about vacuum processes, pressures can be anything from slightly below atmospheric pressure down to 10^{-11} torr. Both the scientific motivation and the technical realization of these different vacuum regimes call for a multitude of concepts. Pumps of different designs, suitable for different vacuum ranges, must be employed. Residual gases will, however, always be present, but their effects must be understood. Plasmas in microfabrication are always low-pressure plasmas, and therefore sputtering, RIE and PECVD are discussed in this chapter. There are many units for pressure (and flow) and the reader is referred to the conversion tables in Appendix B.

33.1 Vacuum Physics and Kinetic Theory of Gases

The transport of ejected atoms or ions from the target to substrate requires a vacuum to prevent collisions and flux divergence. Mean free path (MFP, λ), Equation 33.1, is the distance traveled by atoms between collisions and is an important measure of molecular transport:

$$\frac{1}{\lambda} = \sqrt{2} \times \pi d^2 n \tag{33.1}$$

where n is atom density and d the molecular diameter.

This can approximated for diatomic molecules around 300 K as λ (m) $\approx 5 \times 10^{-5}/P$ (torr) which gives $\lambda \approx$ 65 nm for nitrogen ($d = 0.375$ nm) at room temperature and 1 atm (760 torr) pressure, and 5 cm at 1 mtorr pressure. The Knudsen number, Kn, relates mean free path and reactor chamber size:

$$Kn = \frac{\lambda}{L} \tag{33.2}$$

where L is the characteristic dimension of the chamber. Kn > 1 equals collisionless transport across the vacuum vessel. This regime is the molecular flow regime, and MBE obviously operates in this regime. In the regime Kn < 0.01 fluid dynamics has to be taken into account.

Contamination from the gas phase to the surface can be estimated from kinetic gas theory. The impingement rate of molecules on the surface is given by

$$z = \frac{P}{\sqrt{2\pi mkT}} \tag{33.3}$$

where P is pressure, m mass and T absolute temperature.

If the residual gas is assumed to be nitrogen ($m =$ 28 amu), then at 10^{-6} torr (1.33×10^{-4} Pa) $z = 3.8 \times 10^{18}$/m²-s. A monolayer of residual gas will be absorbed on the sample surface in a time scale given by

$$t_{\text{monolayer}} = \frac{N_{\text{surface}}}{\delta z} = \frac{\sqrt{N_{\text{volume}}^3}}{\delta z} \tag{33.4}$$

where δ is the sticking probability and N_{surface} the density of surface sites. For silicon, N_{volume} is 5×10^{28} m^{-3} and N_{surface} is about 10^{19} m^{-2}. Under the conditions described above, the monolayer formation time is about 1 s under the assumption of unity δ, which gives the shortest possible monolayer formation time. In Figure 33.1 background pressure and sticking coefficient are used to display monolayer formation time. For oxygen, the sticking coefficient is estimated to be about 0.1 (but it is strongly temperature dependent). Residual gases are not similar in their effects: oxygen, water vapor and hydrocarbons are much more problematic than nitrogen, carbon monoxide, carbon dioxide or argon. The sticking coefficient can be tailored by surface preparation: for instance, HF-last treated surfaces are much more resistant to water adsorption than RCA-1 treated surfaces.

Adsorbed species have a characteristic desorption time which is exponentially dependent on activation energy,

---

Introduction to Microfabrication, Second Edition Sami Franssila
© 2010 John Wiley & Sons, Ltd

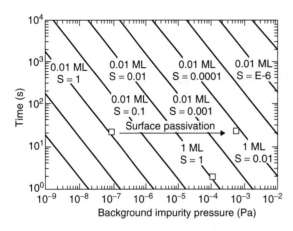

Figure 33.1 Monolayer (ML) and 0.01 ML formation times as a function of pressure and sticking coefficient (S). Surface can be passivated by for example HF treatment. Reproduced from Grannemann (1994) by permission of AIP

according to

$$\tau = \frac{1}{\nu}\exp\left(\frac{E_a}{kT}\right) \quad (33.5)$$

The order of magnitude for frequency factor ν is $10^{13}\,\text{s}^{-1}$, which describes a simple harmonic oscillator with frequency kT/h. Chemisorbed species have an E_a of about 1 eV and physisorbed species 0.4 eV, which translate roughly, at room temperature, to hours and microseconds, respectively.

33.2 Vacuum Production

Starting from the ideal gas law, Equation 33.6, we can get a feeling for vacuum production:

$$p = \frac{NkT}{V} \quad (33.6)$$

Vacuum production means a decrease in the number of atoms N over time, dN/dt. We use the following definitions:

  particle density $n \equiv N/V$ (in atoms/m³)
  flux $J \equiv dN/dt$ (in atoms/s)
  pumping speed $S \equiv -J/n$ (in m³/s)

Pumping speed is also known as volumetric flow rate.
The time evolution of pressure can be written as

$$\frac{dp}{dt} = \frac{dN}{dt}\frac{kT}{V} = -\frac{nSkT}{V} \quad (33.7)$$

which can be solved to yield

$$p = p_0 \exp\left(-\frac{St}{V}\right) \quad (33.8)$$

Pressure drops exponentially over time with characteristic time τ proportional to V/S.

Low–medium vacuum (10^5–0.1 Pa) can be produced by rotary vane pumps, rotary piston pumps, Roots blowers and sorption pumps, which are known collectively as roughing pumps. High vacuum (0.1–10^{-4} Pa) is produced by capture pumps (cryopumps, getter pumps) and momentum transfer pumps (turbomolecular pumps, diffusion pumps). Capture pumps capture and hold all the gas and therefore they need forepumps because of their limited holding capacity. And they have to be regenerated regularly. Momentum transfer pumps, on the other hand, require roughing pumps because they cannot start operation at ambient pressure.

An analogy between vacuum pumping and emptying a water bucket is the following. Initially each cup of removed water will decrease the water level in the bucket by a cupful until almost all the water is removed. After that the remaining water lies in small cusps and irregularities of the bucket, and each removed volume is less than a full cup. Therefore pressure drop (and water-level drop) are gradually diminished, and finally flatten altogether.

In an evaporator there is just the atmospheric gas that has to be pumped out, but in sputtering and UHV–CVD systems we feed in process gases intentionally and must be able to pump them out. Despite a similar base vacuum, the process vacuum in sputtering and UHV–CVD is 1–10 mtorr, three orders of magnitude higher than the base vacuum, and 10–100 Pa-l/s pumps can be used.

Crossover is the pressure where the high-vacuum pump is connected to the chamber. For capture pumps this is calculated from the torr–liter specification (Pa-l/s) by dividing by the chamber volume. Capture pumps hold the pumped material, therefore knowledge of chamber volume is essential. Capture pumps often bring the pressure down faster than roughing pumps, because the pumping speed of a mechanical roughing pump gets worse at lower pressures.

The ultimate pressure that can be reached by a pumping system is determined by pumping speed and vacuum chamber leak rate. We need the concept of conductance to estimate this: conductance is flow divided by gas density difference on the two sides of the vacuum system. Its unit is thus m³/s. Conductances add like capacitors in series, that is

$$\frac{1}{C_{total}} = \frac{1}{C_1} + \frac{1}{C_2} \quad (33.9)$$

Maximum conductance is limited by orifice opening, and further limited by tube conductance that leads from the orifice. The number of atoms leaking in from outside is given by

$$\frac{dN}{dt} = J = -C\Delta n \tag{33.10}$$

For high vacuum, $\Delta n \sim n$ (zero molecules on the high-vacuum side). For STP conditions $n = 2.4 \times 10^{25}/\text{m}^3$. Identifying flux J as the leak, and using the ideal gas law, we get Equation 33.11 which relates pumping speed and pressure:

$$pS = kTJ_{\text{leak}} = kTnC \tag{33.11}$$

The ultimate pressure that can be reached is then given by

$$p_{\text{ultimate}} = \frac{kTnC}{S} \tag{33.12}$$

If leak rate is $3.8 \times 10^{15}\,\text{s}^{-1}$ and a 1000 l/s pump is employed, the base pressure is about 1.6×10^{-5} Pa or 1.2×10^{-7} torr. Ultimate base pressures are produced by cryopumps or getter pumps, with values in the range of 10^{-11} torr. MBE systems operate at such base pressures.

The theoretical maximum pumping speed is derived from kinetic theory as

$$S = \frac{A}{4}v_{\text{ave}} \qquad v_{\text{ave}} = \sqrt{\frac{8kT}{\pi m}} \tag{33.13}$$

where A is inlet area and v_{ave} the molecular average speed. This represents the case where all atoms are impinging in one direction only, with no return flux. Real-life pumping speeds of diffusion pumps can be 50% of the theoretical maximum value, but rotary pumps fare much worse. Pumping speed is usually specified for nitrogen, and the light gases of hydrogen and helium are especially difficult to pump.

Gases will adsorb on surfaces when energetically favorable surface sites are available. Adsorbed gases are "surface gases" as opposed to "volume gases." The latter are related to chamber volume; the former to chamber wall area. This points to the importance of surface finish in vacuum chamber manufacturing: minimum surface area means a smooth surface with the least possible number of sites for adsorption. Water vapor is especially difficult to remove because of its high latent heat and low desorption rate. Heating is standard procedure to desorb adsorbed gases. Therefore high-vacuum chamber materials and surfaces, valves and all other components must be compatible with baking, which is done to outgas adsorbed species.

Bringing down the pressure is achieved by a multiple stage vacuum system. The sputtering system may have three levels of vacuum:

1. Vacuum cassette lock, pumped down to 0.1 torr range by a mechanical pump.
2. Transfer chamber, pumped down to 10^{-5} torr by a turbopump.
3. Process chamber, cryopumped to 10^{-9} torr.

If the transfer and process chambers take only one wafer at a time, the volume to be pumped can be made very small. In a batch deposition system, vacuum vessel volume is easily 100 liters; the corresponding pumpdown time is hours, and somewhat less with a loadlock.

Loadlocks come in two designs: single loadlocks, or separate entry and exit loadlocks. The former are used when process time is long compared to transfer time. Loadlocks serve many purposes: they protect the main chamber from atmospheric gases (especially water vapor) and they also protect cleanroom personnel from harmful or toxic gases that have been used in the process. They can also protect the wafers from the atmosphere: for instance, after aluminum plasma etching chlorine residues remain on the wafer (in the resist and on aluminum surfaces), and if the wafer is taken into cleanroom air with 45% humidity, the chlorine will react with water vapor to form HCl, according to

$$2\,\text{AlCl}_3 + 3\,\text{H}_2\text{O} \Rightarrow \text{Al}_2\text{O}_3 + 6\,\text{HCl} \tag{33.14}$$

Hydrogen chloride will etch aluminum locally. This is corrosion. Exit loadlock can be equipped with a plasma source, and photoresist can be stripped in oxygen plasma. This results simultaneously in aluminum surface oxidation to very passive Al_2O_3.

33.3 Plasma Etching

Plasma etching (RIE) has been discussed in Chapters 11 and 21 from the viewpoints of process performance and device applications. This section emphasizes equipment issues. Plasma generation has a major role in etching, sputtering, ion implantation, photoresist stripping and PECVD. The plasmas used in microfabrication are low-temperature, low-density plasmas (about $10^{10}\,\text{cm}^{-3}$ ion density), compared to welding or fusion plasmas, for example. In microfabrication a high-density plasma (HDP) means an ion density in excess of $10^{11}\,\text{cm}^{-3}$. The degree of ionization is still fairly low: at 1 mtorr pressure, 1 atom in 10 000 is ionized.

Plasma etching has a very high number of parameters that need to be controlled (recall Figure 21.10). This makes plasma etching difficult, both experimentally and simulation-wise. Furthermore, the machine parameters affect plasma parameters, which together with surface reactions determine the final outcome: rate, selectivity and other process responses of interest.

33.3.1 Direct plasmas

Plasma etch reactors can be classified in various ways, and the following is just one.

The parallel-plate diode reactor with two electrodes, one powered, one grounded, is a basic construction for an etcher (see Figure 11.9). Wafers are placed on electrodes that produce the plasma; plasma density, sheath voltage and the ion bombardment that hits the wafers are thus dependent on each other and cannot be controlled independently. Despite this seemingly inconvenient state of affairs, this arrangement is very widely used because of its simplicity. RF generators use the industrial standard 13.56 MHz frequency to create plasmas of typically 10^{10} cm^{-3} density.

33.3.2 Remote plasmas

In remote plasmas generation and biasing are separated. One power source is used to generate the plasma, far removed from the wafer, and another power supply is used to apply a small bias voltage to control ion bombardment. (It is possible to have radicals only, and no ions !) In this decoupled design very high RF power can be used to create plasma glow far away from the wafer. Ion ombardment of the wafer is not affected. Because high density of ions ($10^{11} - 10^{12}$ cm^{-3}) and radicals equals a high active species concentration, HDPs offer higher etch rates. DRIE reactors use 2–10 kW ICP (Inductively Coupled Plasma) for generation and a few watts of CCP (Capacitively Coupled Power) for biasing. A cryogenic ICP DRIE reactor is shown in Figure 33.2. Many of its features are common to all HDP etchers, for example helium back-side cooling and mechanical wafer clamping.

High etch rate and lower damage, easier photoresist removal and higher selectivities favor HDP reactors. Remote plasma reactors are often difficult to scale to large diameters because of the physical separation between plasma and wafer, whereas parallel-plate reactors naturally have the plasma "aligned" to the wafer.

33.4 Sputtering

Sputtering processes for thin-film deposition were discussed in Chapter 5, with scant regard to actual sputtering equipment. The oldest and simplest of sputter deposition

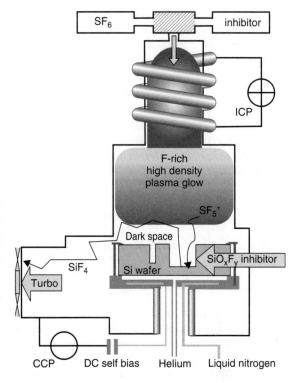

Figure 33.2 Cryogenic ICP high-density plasma etcher. Reproduced from Jansen et al. (2009) by permission of IOP

systems is the DC diode system which consists of a negatively biased plate (target cathode) which is bombarded by argon ions at about 100 mtorr pressure (Figure 33.3a). In order to get a high deposition rate, high sputtering power has to be used, which leads to high-voltage operation. This is undesirable because of damage to thin oxides.

In order to improve DC diodes, RF diode systems were introduced. RF sputtering systems usually work at 13.56 MHz. They can be used to deposit dielectrics, something that is not possible with DC systems due to charging. Electrons oscillating in the RF field couple energy more efficiently to the plasma, and higher deposition rates are possible in RF than in DC, at the same power levels. However, a very high voltage of 2000 V is used.

Magnetron sputtering has emerged as the main configuration. A magnet behind the target creates a field which confines electron movement, therefore ionization is much more efficient, leading to high deposition rates at low power (5–20 kW is used, depending on target size). Voltages in magnetron systems are, for example, 500 V (and argon ion energies 500 eV), clearly lower than in RF

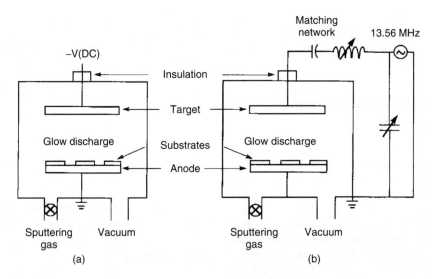

Figure 33.3 Schematic sputtering systems: (a) DC and (b) RF. Reproduced from Ohring, M. (1992), by permission of Academic Press

diodes. Magnetron sputtering systems work at about millitorr pressures (0.1–10 mtorr), with argon flows of 10–100 sccm. From the impurity viewpoint, however, sputtering systems are described by their base pressures, which are 10^{-7}–10^{-9} mtorr because high-purity argon sputtering gas (99.9999%) contributes less than background gases.

Sputtering systems have, in addition to plasma generation and vacuum subsystems, many other features: wafers can be heated and they can be shielded by from the plasma by shutters (especially during plasma ignition, when the discharge is unstable), as shown in Figure 33.4. In addition to argon, other gases can be introduced, to enable reactive sputtering. Sputtering from a titanium target in an Ar/N$_2$ atmosphere results in TiN, and from a Ta target and Ar/O$_2$ in Ta$_2$O$_5$. The latter is an insulator, and a RF sputtering system would be needed if a Ta$_2$O$_5$ target was used. In both cases the film is not stoichiometric tantalum pentoxide. An oxygen anneal after sputtering would be needed to fine-tune the stoichiometry.

33.4.1 Sputter etching and bias sputtering

If the voltages in a sputtering system are switched, and power is applied to the wafer electrode instead of the target, the wafers will experience argon ion bombardment. This is called sputter etching. (Sputtering systems can be turned into true plasma etch systems by introducing reactive gases instead of argon. The term RSE, for reactive sputter etching, was used for early plasma etching systems.)

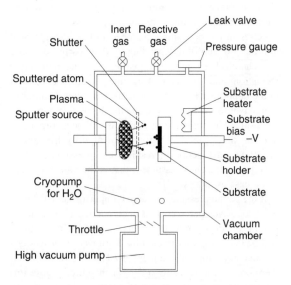

Figure 33.4 Sputtering system. Reproduced from Parsons (1991) by permission of Academic Press

If the wafer electrode is biased during sputtering (by a separate power supply), wafers will experience simultaneous deposition and etching. This will generally densify the film because ion bombardment knocks out loosely bound film (and impurity) atoms. The geometry of structures is important because argon ion etching depends on the angle of incidence: convex corners are etched faster, and faceting occurs. Smoothing of sharp corners is beneficial for step coverage in the next deposition step, but the total deposition rates in such deposition–etch processes are understandably slow.

33.5 Residual Gas Incorporation into Deposited Film

Standard evaporation is done in high vacuum, but it can be done in a gaseous atmosphere where, instead of 10^{-6} mbar, 10^{-4} mbar is used. Evaporation in a nitrogen atmosphere results in highly porous films, for example gold with a density of 3 g/cm^3 has been made (bulk gold density is 19 g/cm^3). Similar platinum films have also been made. These are used in infrared microsystems as absorbers, and are known as black gold and black platinum. Nitrogen prohibits adatom movements by forming strong bonds, and columnar porous films will result. Excellent IR absorbance is achieved, but further processing such films is difficult because high porosity results in poor mechanical strength. Therefore their preparation should be the final, or almost final, step in the process.

Gases in the deposition chamber will be incorporated into the growing film depending on their partial pressure and chemical activity. In Equation 33.4 the monolayer formation time was given. This time should be compared to the deposition rate for a monolayer of film: if impurity monolayer formation is rapid and the deposition rate slow, a considerable amount of residual gas will be incorporated into the film. Table 33.1 gives the impurity fraction in the film as a function of residual gas pressure. Unitary sticking coefficient is assumed, that is the values present worst case estimates.

If the partial pressure of reactive impurities is 10^{-6} torr, one monolayer per second is incorporated (~ 0.1 nm/s). Even if the deposition rate is very high at 100 nm/s, there will be 0.1% impurity in the film. Purities of typical starting materials for PVD are 99.999%. Poor vacuum can therefore contribute many orders of magnitude more impurities into film than the target materials. Of course not all impurities are equal: some manifest themselves much more strikingly than others. For example, oxygen in aluminum will rapidly lead to aluminum oxide formation and useless film, but a similar concentration of argon (always present in sputtering!) can be neglected in a first approximation. At base pressures of 10^{-9} torr target purity starts to become the limiting factor.

Deposition rates in batch systems are usually much slower than in single wafer systems: a difference by an order of magnitude is not unusual, therefore throughput rather than deposition rate is often mentioned for batch systems. But the calculations above show that film quality is related to deposition rate, not to throughput.

Table 33.1 Fraction of foreign atoms incorporated into growing film

Partial pressure (torr)	Deposition rate (nm/s)			
	0.1	1	10	100
10^{-9}	10^{-3}	10^{-4}	10^{-5}	10^{-6}
10^{-8}	10^{-2}	10^{-3}	10^{-4}	10^{-5}
10^{-7}	10^{-1}	10^{-2}	10^{-3}	10^{-4}
10^{-6}	1	10^{-1}	10^{-2}	10^{-3}
10^{5}	10	1	0.1	0.01

33.6 PECVD

PECVD reactors are very much like plasma etchers. From the hardware point of view the heated electrode is the main difference. Other aspects, like RF generators, reactive gases, pumping systems, etc., are similar (Figure 5.7). In etching HDPs offer enhanced etch rates; in PECVD HDP means a higher deposition rate and/or improved film quality. The typical PECVD temperature is 300 °C, but there is no fundamental lower limit to deposition temperature. Processes at 100 °C have been demonstrated, but film properties are strongly temperature dependent. Especially, the hydrogen content of the films increases rapidly as temperature is lowered, and the films become less dense.

The above discussion was about first-order effects only: the effects of pressure, power and reactant gas flows can be rather complex. An increase in RF power initially increases deposition rate, because more reactant gases are ionized, fragmented and available for reaction, as seen for PECVD nitride deposition in Figure 33.5a. A further increase in power leads to a leveling of the rate, however, as more and more ion bombardment causes sputtering of the growing film. Decreasing pressure leads to a smaller deposition rate (Figure 33.5b). Lower pressure leads to a higher electron energy and different decomposition reactions and a different degree of ionization of the reactants. The effect of NH$_3$ flow is quite subtle (Figure 33.5c): when more NH$_3$ is available, it reacts with silane to form Si(NH$_2$)$_4$, while the competing reaction product Si(NH$_2$)$_3$ is the important precursor for film growth. Increasing silane flow (Figure 33.5d) leads to an increased number of silicon radicals, therefore the deposition rate goes up. At the same time, the film becomes silicon rich, reducing stress. However, the film starts to lose some of its beneficial nitride properties, like etch resistance in KOH.

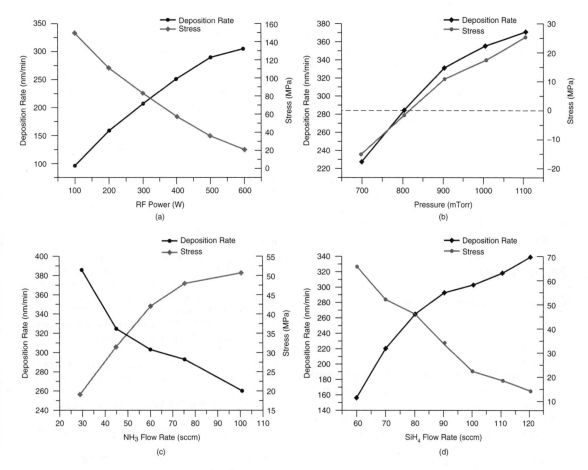

Figure 33.5 The effects of PECVD process parameters on deposition rate and film stress. Reproduced from Wei *et al.* (2008), copyright 2008, by permission of Elsevier

Deposition takes place not only on wafers but on reactor walls, and on the electrodes too. It is standard procedure to etch these deposited layers away at regular intervals, for example after every wafer, or after a certain thickness has been deposited, or when deposition temperature is changed, or when material to be deposited is changed. PECVD similarity to RIE is evidenced by the fact that the introduction of CF_4 or NF_3 gas into a PECVD reactor chamber turns it into an etch system. In situ cleaning of the PECVD chamber can thus be accomplished easily. NF_3 gas has a nice feature in that it decomposes into gaseous products only, whereas CF_4 and SF_6 are potential sources of carbon and sulfur residues. NF_3 is, however, toxic, and hard to handle. It is also a greenhouse gas, just like fluorinated hydrocarbons.

Utilization is a measure of reactant usage. It is the ratio of reactant atoms embodied in the thin film to all incoming source gas atoms. Utilization cannot even approach 100% because the main convective flow will always carry with it some reactants, and only a fraction will diffuse through the boundary layer and participate in film-forming reactions. Some metal–organic precursor molecules undergo disproportionation reactions, and only 50% of source gas atoms are available for deposition in the best case.

33.7 Residence Time

The effects of pressure and flow can be deduced from residence time τ,

$$\tau = \frac{p}{p_0} \frac{V}{F} \frac{273}{T} \quad (33.15)$$

where p_0 is a reference pressure of 1 atm.

Residence time is the characteristic time that a molecule spends in the reactor before being pumped away. The concept is useful in analyzing all sorts of reactors because residence time is very general. Increasing pressure leads to increased residence time, which translates to a higher deposition (or etching) rate: molecules have a higher probability of being incorporated into the film if they spend more time in the reactor. Increasing flow rate will sweep the molecules away faster, leading to a smaller τ and lower deposition rate. There are, however, many other aspects to consider: higher pressure means more collisions and less ion bombardment, which affects film structure and stresses, and when flow rates change, the relative proportions of ionized and excited species change, again affecting film growth, so optimization of plasma reactors involves a great many variables.

33.8 Exercises

1. What is the Knudsen number in
   (a) sputtering
   (b) evaporation
   (c) MBE
   (d) RIE?
2. What is the maximum theoretical pumping speed of a diffusion pump with a vacuum flange of 10 cm diameter?
3. If the water molecule sticking coefficient is 0.01 and water partial pressure is 10^{-4} Pa, how long will it take to form a monolayer?
4. What must the leak rate be in a MBE system be in order to achieve 10^{-11} torr base pressure?
5. In sputtering about 10–20 mW/cm^2 of energy is supplied to the surface (heat of condensation, kinetic energy of sputtered particles, ion and electron bombardment and ion neutralization each contribute about 2–5 mW/cm^2). By how much do wafers heat up during sputtering?
6. What would be the crossover pressure for film purity to become target purity dependent when a 99.9999% pure target (6N) is used?
7. How deep into an aluminum sputtering target will 500 eV argon ions penetrate?
8. In a pulsed (Bosch) process DRIE chamber the volume is 50 liters, flow rate is 200 sccm and operating pressure is 20 mtorr. What is the shortest possible pulsing period?
9. If 5 kW power is applied to an aluminum sputtering target of 200 mm diameter, what is the maximum possible deposition rate?
10. An XPS measurement takes 15 minutes. What is the pressure in a XPS chamber?

References and Related Reading

Cote, D.R. *et al.* (1995) Low-temperature CVD processes and dielectrics, *IBM J. Res. Dev.*, **39**, 437.

Grannemann, E. (1994) Film interface control in integrated processing systems, *J. Vac. Sci. Technol.*, **12**, 2741.

Hess, D.W. (1990) Plasma-material interactions, *J. Vac. Sci. Technol.*, **A8**, 1677.

Jansen, H.V. *et al.* (2009) Black silicon method X: a review on high speed and selective plasma etching of silicon with profile control: an in-depth comparison between Bosch and cryostat DRIE processes as a roadmap to next generation equipment, *J. Micromech. Microeng.*, **19**, 033001.

Lee, J.T.C. *et al.* (1996) Plasma etching process development using in situ optical emission and ellipsometry, *J. Vac. Sci. Technol.*, **B14**, 3283.

Loewenhardt, P. *et al.* (1999) Plasma diagnostics: use and justification in an industrial environment, *Jpn. J. Appl. Phys.*, **38**, 4362.

Mahan, J.E. (2000) **Physical Vapor Deposition of Thin Films**, John Wiley & Sons, Inc.

Nguyen, S.V. (1999) High-density plasma chemical vapor deposition of silicon-based dielectric films for integrated circuits, *IBM J. Res. Dev.*, **43** (1–2), 109 (special issue on plasma processing).

Ohring, M. (1992) **The Materials Science of Thin Films**, Academic Press.

Parsons, R. (1991) Sputter deposition processes, in J.L. Vossen and W. Kern (eds), **Thin Film Processes II**, Academic Press.

Rossnagel, S.M. (1999) Sputter deposition for semiconductor manufacturing, *IBM J. Res. Dev.*, **43** (1–2), 163.

Somorjai, G.A. (1998) From surface materials to surface technologies, *MRS Bull.*, May, 11.

Waits, R.K. (2001) Edison's vacuum coating patents, *J. Vac. Sci. Technol.*, **A19**, 1666.

Wei, J. *et al.* (2008) A new fabrication method of low stress PECVD SiN$_x$ layers for biomedical applications, *Thin Solid Films*, **516**, 5181–5188.

34

CVD and Epitaxy Equipment

Thermal CVD processes share many equipment features with oxidation and diffusion furnace processes, whereas PECVD is more akin to plasma etching. Epitaxial processes to be discussed here are limited to flow-type silicon CVD epitaxy processes which share many features with thermal CVD.

CVD reactors are classified by their operating pressure range:

- atmospheric pressure, APCVD
- sub-atmospheric, SACVD 10–100 torr
- low-pressure, LPCVD at ~1 torr
- ultrahigh vacuum, UHV–CVD, 10^{-6} torr base pressure

In UHV reactors the actual process pressures are 1–10 mtorr when gases are flowing, very much like magnetron sputtering systems. In both cases a good base vacuum (of 10^{-6}–10^{-9} torr level) is mandatory for removing residual gases from the chamber.

34.1 Deposition Rate

The two main differences between PVD and CVD reactions are in fluid dynamics and temperature dependence: in PVD fluid dynamics need not be considered, but CVD processes are flow processes with complex fluid dynamics. PVD processes are temperature insensitive as far as film deposition rate is concerned, but for example higher temperature can lead to impurity desorption (higher purity films) or annealing (lower stress). CVD processes are chemical processes, their rates obeying Arrhenius behavior. Activation energy E_a can be extracted from the Arrhenius formula when the deposition rate has been determined at several temperatures. The magnitude of the activation energy gives hints of possible reaction mechanisms.

Two temperature regimes can be found for most CVD reactions: when the temperature is low, the surface reaction rate is slow, and an overabundance of reactants is available. The reaction is then surface reaction limited. Surface reaction-limited processes usually result in uniform films with good step coverage, an obvious advantage.

When the temperature increases, the surface reaction rate increases exponentially and above a certain temperature all source gas molecules react at the surface. The reaction is then in the mass transport-limited regime (also known as the diffusion-limited regime) because the rate is dependent on the supply of new species to the surface. Fluid dynamics of the reactor then plays a major role in deposition uniformity and rate. These two cases are shown in Figure 34.1. The activation energy of a surface reaction-limited process is much higher than that of mass transport-limited process.

In PECVD low temperatures can be used: plasma activation ensures sufficient reactive species even at low temperatures, typically at about 300 °C, but even down to 100 °C (but temperature strongly affects film quality). Whereas typical activation energies for thermal

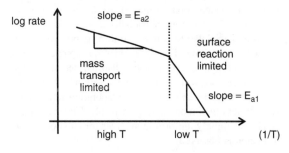

Figure 34.1 Surface reaction-limited vs. mass transfer-limited CVD reactions

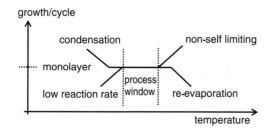

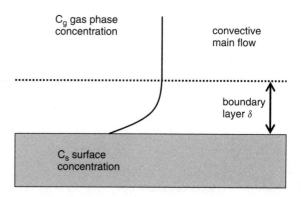

Figure 34.2 Process window for ALD is usually wide in temperature: it is limited by low reaction rate and condensation at low temperatures and by not self-limiting growth and re-evaporation of precursors at high temperatures

Figure 34.3 Model of gas phase deposition

CVD processes are 2 eV (200 kJ/mol), PECVD activation energies are a fraction of that, for example 0.3 eV for amorphous silicon deposition. The PECVD deposition rate is only weakly temperature dependent.

ALD reactions are self-limiting and film thickness is calculated from the number of pulses. But ALD reactors can be operated in a continuos CVD mode, so the equipment are related. In CVD deposition the rate is exponentially temperature dependent according to the Arrhenius formula, but in ALD there is a (wide) process window where the rate is independent of temperature. For example, the rate for $SrTiO_3$ deposition has been measured as 0.3 Å/cycle from 225 to 325 °C. In this regime exactly one layer of precursors is deposited. The uniformity of ALD is exceptionally good, with <1% uniformities reported for both within the wafer and wafer to wafer. This is typical of all surface reaction-controlled processes, for example LPCVD polysilicon and silicon nitride, too.

The ALD operating temperature is limited from below by two mechanisms (numbers refer to Figure 34.2): low temperature leads to a low reaction rate (1), and precursor condensation on the surface leads to excessive deposition (2). The former leads to less than monolayer deposition, the latter to non-self-limiting deposition of unwanted composition. The upper operating temperature is also limited by two mechanisms: thermal decomposition of the precursors, which results in deposition in the normal CVD fashion (3); and a high re-evaporation rate, which leads to less than monolayer growth (4). Under the right conditions, highly uniform monolayer (or submonolayer) formation is observed.

34.2 CVD Rate Modeling

CVD can be modeled with a simple model that resembles the Deal–Grove model of thermal oxidation (compare Figures 13.2 and 34.3). Flux J of reactants from the gas flow to the surface is controlled by diffusion through the boundary layer, and film deposition takes place at the wafer surface. Flux from the gas phase to the surface is given by

$$J_{\text{gas-to-surface}} = h_g(C_g - C_s) \qquad (34.1)$$

where h_g is the gas phase transport coefficient, C_g the gas phase concentration and C_s the surface concentration of reactants. The surface reaction rate is assumed to be directly proportional to reactant concentration, as shown by

$$J_{\text{surface-reaction}} = k_s C_s \qquad (34.2)$$

Under steady state conditions the fluxes are equal, $J_g = J_s$, and concentrations are given by

$$C_s = \frac{C_g}{1 + (k_s/h_g)} \qquad (34.3)$$

Conversion from fluxes to rate is given by $R = J_s/n$ where n is atom density in the film.

From Equation 34.3 we can recognize the two familiar regimes of mass transport-limited deposition (Equation 34.4) and surface reaction-limited deposition (Equation 34.5):

$$C_s = \frac{h_g}{k_s} C_g \quad \text{mass transport limited, } k_s \gg h_g \qquad (34.4)$$

$$C_s = C_g \quad \text{surface reaction limited, } k_s \ll k_g \qquad (34.5)$$

In the mass transport-limited case the reaction rate at the surface is very high and leads to local depletion of reactants in the gas phase. The inadequate supply can be due to a gas flow rate that is too small, but it can also be caused by too slow a diffusion through the boundary layer. Surface reaction-limited cases are characterized by an oversupply of reactants in the vicinity of the surface, but the slow surface reaction cannot consume all of them.

The gas phase transport coefficient, h_g, can be gauged as follows: Fick's law (Equation 14.2) states that flux J is proportional to diffusivity D and to concentration gradient dC/dx. If we identify dx with the boundary layer thickness δ, and combine Equations 14.2 and 34.1, we get the flux from

$$J_{\text{gas-to-surface}} = -\frac{D}{\delta} C_g \qquad (34.6)$$

The boundary layer is the region of stagnant fluid near the wall. Boundary layer thickness δ is given by

$$\delta = \sqrt{\frac{\eta L}{v\rho}} \qquad (34.7)$$

where η is the viscosity, v the fluid velocity, ρ the density and L the characteristic dimension of the system (e.g., chamber diameter). The boundary layer thickness increases along the flow, and it is thinner at the inlet and thicker at the exhaust end of the reactor.

For an atmospheric system at about $1000\,°\text{C}$ the values are $D \approx 10\,\text{cm}^2/\text{s}$, $L \approx 100\,\text{cm}$, $\eta \approx 10^4$ Poise (g/cm-s) and $\rho \approx 10^{-4}\,\text{g/cm}^3$ ($\rho \propto 1/T$). We get an approximate boundary layer thickness of 3 cm which is close to values found in real systems, for example epitaxy reactors. The gas phase transfer coefficient is then $h \approx 3\,\text{cm/s}$.

If we lower the operating pressure by a factor of 1000, diffusivity increases a 1000-fold, as seen from

$$D \propto T^{3/2}/P \qquad (34.8)$$

However, the boundary layer thickness increases because density decreases (and velocity increases), but because of the square root dependence (Equation 34.7, 34.8), this opposing trend is about one order of magnitude only. The diffusivity increase clearly dominates and the flux of reactants to the surface is greatly enhanced. A reaction which was transport limited at atmospheric pressure can be turned into a surface reaction-controlled one by operating at reduced pressure. This is the reason for using LPCVD for polysilicon deposition: excellent uniformity and step coverage are enabled by surface reaction-controlled deposition.

In order to get a feeling for the temperature dependence we have to compare k_s and h_g as a function of temperature. Chemical reactions obey Arrhenius behavior with exponential dependence, and thus surface reaction-limited deposition is strongly temperature dependent (high E_a).

The gas phase transport coefficient h_g is proportional to D which has a $T^{3/2}$ temperature dependence. This explains the shallower slope in the transport-limited regime of Figure 34.1.

34.3 CVD Reactors

APCVD reactors operate in transport-limited mode and flow geometries are important for film uniformity. LPCVD reactors operate in the surface reaction-controlled regime and wafers can be packed closely, which increases system throughput. LPCVD reactors are similar to oxidation tubes and both LPCVD and oxidation tubes can be fitted into the same furnace stack (compare Figures 32.1 and 34.4).

Flow, temperature and pressure are the important CVD reactor design criteria. Practically all CVD processes use toxic, corrosive and flammable fluids like ammonia, silane, dichlorosilane, hydrides and metal organics. Reactor designs include double piping, inert gas flushing and venting and other safety features. Some of the reaction byproducts are harmful to pumps and mechanical constructions, which translates to special care in materials selection. Environmental, safety and health issues will be discussed in Chapter 35.

As an example of a CVD process, silicon nitride deposition is shown in Table 34.1. If wafers come directly from another furnace operation (e.g., LOCOS pad oxide growth) no cleaning is required. The time limit for a new clean can be set for example at 2 hours. Otherwise standard cleaning is required, for example RCA-1 and RCA-2.

CVD furnace systems are hot wall systems, meaning that deposition also takes place on the walls. This leads to film build-up and flaking problems. Tubes thus have a limited lifetime and need to be replaced regularly.

Deposition leads to reactant depletion, and the rate in the entrance zone is higher than near the exit. Increasing boundary layer thickness towards the tube end also reduces deposition rate. This is compensated by increased temperature (= increased rate of chemical reaction). Heating elements are arranged in three zones, for example T1: $747\,°\text{C}$, T2: $750\,°\text{C}$, T3: $753\,°\text{C}$ for LPCVD silicon nitride. This temperature ramp compensates for reactant depletion along the tube.

In polysilicon LPCVD this three-zone system results in a grain size gradient along the length of the tube. In so-called flat–poly systems the temperature is kept constant and gas introduction is made more uniform by an elaborate gas distribution system. Alternatively "poly" can be deposited amorphously at $570\,°\text{C}$ to eliminate grain size gradients.

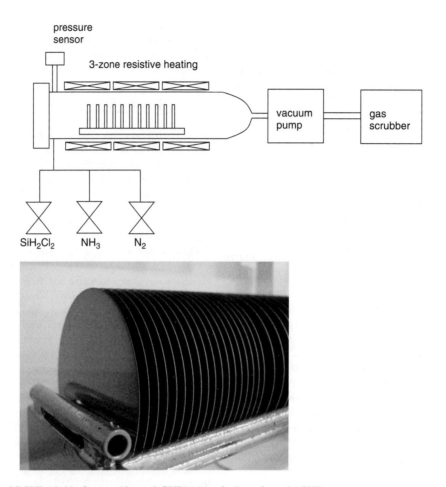

Figure 34.4 LPCVD nitride furnace (thermal CVD). A wafer boat for poly CVD

Table 34.1 LPCVD of silicon nitride (Si_3N_4)

Load the boat, fill with dummy wafers to equalize load and flow patterns

Ramp temperature from 500 to 750 °C under nitrogen flow, 50 min (5 °C/min)

Pump to vacuum and perform leak check, 2 min

Introduce ammonia NH_3, stabilize flow at 30 sccm, for 1 min

Introduce dichlorosilane, SiH_2Cl_2, flow 120 sccm, deposition starts

Deposit at 300 mtorr for 25 min (4 nm/min deposition rate)

Cool down to 700 °C (10 min)

Take boat out

Monitoring: film thickness and refractive index by ellipsometer

34.4 CVD with Liquid Sources

Most CVD processes use simple source gases like silane and hydrides, but there is the possibility of using liquid precursors. A widely used liquid source for CVD is TEOS (tetraethoxysilane) for oxide deposition. Liquid is heated in a container to increase its vapor pressure, and then a carrier gas, nitrogen, helium or hydrogen, is bubbled through the liquid and the precursor vapors are carried along by the carrier gas stream. This same method is applied also in gas phase diffusion: dopants like $POCl_3$ are introduced with bubbling, and wet oxidation can be done by bubbling nitrogen carrier gas through water.

When the precursors are metal–organic compounds (MOs), the technique is termed MOCVD, also known as MOVPE, for metal–organic vapor phase epitaxy. It is widely used in III–V compound semiconductor epitaxy, with group III elements supplied as MOs, like trimethyl gallium Ga(CH$_3$)$_3$ or triethyl aluminum Al(C$_2$H$_5$)$_3$, while group III precursors are usually hydrides, AsH$_3$ and PH$_3$.

MOCVD has also been studied for metal deposition. Copper has been deposited from precursors like vinyltrimethylsilane hexafluoroacetylacetonate, VTM-SCu(hfac), or Cu(I)-β-diketonate. Conformal deposition is possible and filling holes of high aspect ratio has been demonstrated. Trimethyl aluminum source gas has been used for MOCVD of aluminum. It would be beneficial to deposit aluminum films with copper alloying (0.5–4%), but this complicates MOCVD even further.

The problems with MOCVD are both practical and fundamental. The vapor pressure has to be right, the precursor must not react with other gases or materials present in the system, and its decomposition reactions must be reproducible. There is always the danger of carbon incorporation into the film when MOs are used as source materials. On the practical side, purity must be high, and this is difficult for complex compounds like MOs. Many MOs are extremely reactive with oxygen, and premature contact with oxygen will destroy the reagents.

34.5 Silicon CVD Epitaxy

Silane gases (SiH$_x$Cl$_{4-x}$, $x = 0$–4) can all be used for epitaxy, but the temperature regimes are different. Growth temperature is a compromise between rate (thickness) and thermal budget (dopant diffusion during growth). Temperature is closely related to substrate/epi interface steepness: a higher deposition temperature offers a higher growth rate but at the expense of more thermal diffusion. Autodoping from the substrate and from the buried layers has also to be considered.

Because silicon homoepitaxy is a CVD reaction, the same laws about mass transport and surface reaction-limited deposition apply to it. In Figure 34.5 these two regimes are clearly visible. Different source gases have different useful temperature ranges but practically identical activation energies in the surface reaction-limited regime. Most epitaxy reactors, however, operate in the mass transport-limited regime, and gas flow design in the reactor in crucial.

Epitaxy is not necessarily a high-temperature process. It has traditionally been so, but epitaxy as such can be carried out at any temperature. In situ cleaning of the wafer has been a factor for high temperatures: HCl or H$_2$ gas phase cleaning processes work better at elevated temperatures. Surface composition, however, is also dependent on the preceding cleaning step, and if that can be modified to reduce native oxide growth, the in situ cleaning temperature can be lowered.

34.6 Epitaxial Reactors

Reactors can be classified according to gas flow patterns relative to wafers (Figure 34.6). The gas flow can be parallel to the wafer surface, as in barrel (or hexode)

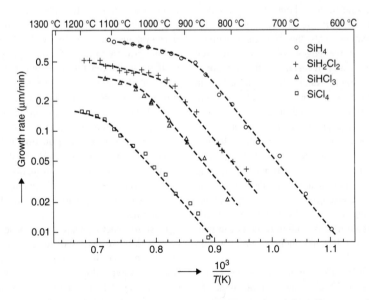

Figure 34.5 Epitaxial growth for different SiH$_x$Cl$_{4-x}$ source gases. Reproduced from Everstyen (1967) by permission of Philips

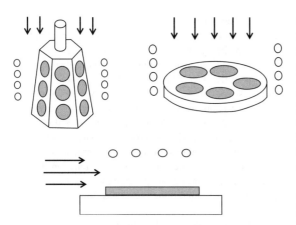

Figure 34.6 Barrel reactor, vertical wafers, parallel flow; pancake reactor, horizontal wafers, perpendicular flow; single wafer reactor, horizontal wafer, parallel flow. Lamps or RF coils for heating are shown, but the reactor chamber is not

reactors where the wafers are placed vertically. The wafers can be placed on a horizontal susceptor, with perpendicular gas flows, as in a pancake reactor (or disk reactor). In a single wafer reactor the wafer is placed horizontally, and a lateral gas flow is parallel to the wafer surface.

Two wafer heating methods, namely induction (RF coils) and lamp heating, are used. Lamp heating can be used in all major reactor types. The wafer surface is hotter than the back side because lamps heat the wafers from the top, and the wafers are bowed upward at the centre. Induction heating heats the graphite susceptor, and wafers bow up at the edges, which is countered by designing curved wafer recesses in the susceptor. Induction heating is more suited for sustained high temperatures (90 min is usual for power devices.), and lamp heating for short depositions/thin layers.

As for much other equipment, there are both batch and single wafer reactors on the market. Both designs coexist because they have different strengths regarding film thickness, growth rate, interface abruptness or doping uniformity. Batch reactors typically have about 1 µm/min growth rates and are preferred for thick-layer applications (up to 200 µm in some power devices) where interface sharpness is not an issue. Batch loading reactors can take for instance 30 wafers of 100 mm diameter, or 12 wafers of 200 mm diameter.

Single wafer reactors offer high growth rates, for example 5 µm/min at 1120 °C using trichlorosilane. In addition to the steep interface due to short deposition time, single wafer reactors are superior with respect to film uniformity, 1% across the wafer for thickness, 4% for resistivity. A rotating susceptor is easy to implement in a single wafer tool. It has several consequences. Rotation ensures good uniformity, and rotation can be used to reduce boundary layer thickness: the velocity in Equation 34.7 is now the relative velocity of the gas and the rotating wafer. A thinner boundary layer results in a higher deposition rate, and the evaporated dopants from buried layers will rapidly diffuse to the main gas flow and be swept away.

Epi reactors operate at atmospheric pressure but reduced pressure, typically 50–100 torr, can also be used. Reduced pressure operation adds to equipment complexity, and it is used for demanding applications only, including SiGe epitaxy (which differs from silicon epitaxy in regard to process temperature which is only about 700 °C).

Gases used in epitaxy are extremely pure: carrier hydrogen must be free of oxygen and water below the 100 ppb level. Silane purity is measured by resistivity: above 3000 ohm-cm. Dopant gases are very dilute: 100 ppm phoshine or diborane in hydrogen is typical. All piping for process gases must be made of stainless steel because chlorosilanes and HCl are aggressive gases. Electropolishing down to nanometer surface roughness is used in piping to eliminate particle contamination.

Epi reactors are power hungry: keeping wafers at about 1100 °C consumes hundreds of kilowatts, which turns into waste heat, 80–90% of it into cooling water and the rest to hot exhaust gases. These gases are unused silanes (typical silane utilization is 10–30%) and hydrogen, which can account for 99% of the flow. Gas treatment is done by burn systems, wet scrubbers or by thermal decomposition.

The growth process for an epilayer 13 µm thick in a single wafer reactor is shown in Figure 34.7. As can be seen, the actual deposition is just a fraction of total process time: the remainder is spend on heating, cooling and cleaning. These steps are essential for epitaxial film quality. Prebake has many effects: native oxide is removed (according to Equation 6.1) and the surface layer will become depleted in dopants and oxygen (similar to denuded zone formation, Figure 22.3). Any damage from preceding ion implantation steps is annealed away. All this improves crystal quality and minimizes autodoping.

In some reactors wafers are loaded upright and their back sides are exposed to gas flows, and substrate autodoping can be significant (Figure 6.9). The back sides of heavily doped wafers are usually protected by CVD oxide (LTO) film to prevent the evaporation of dopant into the reactor. In addition to intentional doping

CVD and Epitaxy Equipment 439

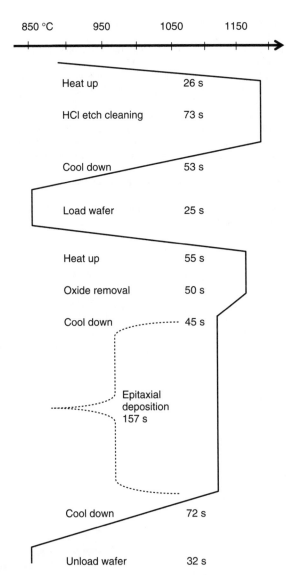

Figure 34.7 Single wafer epitaxy reactor running SiHCl$_3$ process. Actual deposition time is 30% of total time. Deposition rate is about 5 µm/min, or film thickness is 13 µm.

and autodoping, films on reactor walls release some dopant. This is known as the reactor memory effect.

Even though silicon growth in epi reactors is typically in the transport-limited regime, dopant incorporation can be in the surface reaction-limited regime, which necessitates accurate temperature control. Temperature uniformity is also very important because even minor temperature differences lead to crystal slips when silicon yield strength is exceeded (Equation 22.1).

34.7 Control of CVD Reactions

Pressure has profound effects on the mechanism of film deposition. While temperature affects rate in a predictable manner (Arrhenius behavior), pressure has more subtle effects: the rate limiting step can change from surface reaction limited to transport limited by a pressure change. Many factors contribute to this: in single wafer reactors the flow rate can be made high, and the boundary layer becomes thin, increasing the deposition rate. In batch reactors low pressure is used, and the diffusivity increase switches the reaction from transport limited to surface limited. The benefits are increased uniformity over batch. The drawbacks include reduced deposition rate and the extra cost brought about by vacuum requirements.

Depending on the application and reactor design, it may be advantageous to operate in the transport-limited regime where temperature dependence is small but flow control must be accurate. On the other hand, in the surface reaction-limited regime the uniformity of deposition becomes independent of fluid dynamics, but critically temperature dependent.

34.8 Exercises

1. What is the Knudsen number in
   (a) APCVD
   (b) LPCVD
   (c) UHV–CVD?
2. Polysilicon LPCVD activation energy E_a is 1.7 eV. What happens to the deposition rate if, instead of standard 630 °C deposition, 570 °C is used?
3. If the gas phase transfer coefficient h is 3 cm/s, and the surface reaction coefficient $k = 5 \times 10^7 \exp(-1.7\,eV/kT)$ (in cm/s), at what temperature does the reaction turn from transport controlled to surface controlled?
4. What is the cost of an epiwafer of 150 mm diameter if the single wafer process described in Figure 34.7 costs $2 million, running costs are $800 000 a year (gas and graphite costs are significant, much higher than labor costs) and starting wafer cost is $20?
5. Single wafer reactor, 150 mm wafer size, deposits oxide with a rate 200 nm/min. What is the utilization of silane if the flow of silane is 150 sccm silane and overerabundance of N$_2$O is present?

6. Nitride LPCVD is midzone is at 750 °C. What thickness difference does a 6 °C temperature difference between front and rear zones indicate if $E_a = 1.9\,\text{eV}$?
7. What is the thinnest layer that could reasonably be deposited using the PECVD parameters of Table 7.2, assuming a single wafer reactor volume of 5 liters?
8. What is total gas consumption in the process shown in Figure 34.7?

References and Related Reading

Cote, D. R. *et al.* (1995) Low-temperature chemical vapor deposition processes and dielectrics for microelectronic circuit manufacturing at IBM, *IBM J. Res. Dev.*, **39**, 437.

Crippa, D., D. R. Rode and M. Masi (2001) **Silicon Epitaxy**, Academic Press.

Elers, K.-E. *et al.* (2006) Film uniformity in atomic layer deposition, *Chem. Vapor Depos.*, **12**, 13–24.

Everstyen, F. C. (1967) Chemical-reaction engineering in the semiconductor industry, *Philips Tech. Rep.*, **29**, 45.

Low, C. W. *et al.* (2007) Characterization of polycrystalline silicon-germanium film deposition for modularly integrated MEMS applications, *J. Microelectromech. Syst.*, **16**, 68–77.

Middleman, S. and A. K. Hochberg (1993) **Process Engineering Analysis in Semiconductor Device Fabrication**, McGraw-Hill.

Ohring, M. (1992) **The Materials Science of Thin Films**, Academic Press.

Vossen, J. and W. Kern (1991) **Thin Film Processes, II**, Academic Press.

35

Cleanrooms

Cleanrooms were initially a solution to particle contamination reduction and were not invented for microelectronics but for fine mechanical assembly. Later on temperature and humidity control for improved reproducibility in lithography were recognized. Other features have been added over the years, so a modern cleanroom is a system of facilities which ensure contamination-free processing under very stable environmental conditions.

Particle size distributions in cleanroom air, in process gases, in DI water and in wet chemicals all have the same basic characteristics: 4–8 times more particles are detected if the detection threshold is halved. Therefore, if the minimum linewidth is halved, the number of particles that are potential killers increases 4–8 times.

In microfabrication cleanrooms (wafer fabs) particle cleanliness can be achieved by applying overpressure to blow dirt outside, but virus labs need to contain everything. In both pharmaceutical and microfabrication cleanrooms a great many harmful and even toxic chemicals are used, but fortunately the silicon wafers themselves are not dangerous to humans. But just as with pharmaceuticals, worker contact with the product should be minimized to reduce contamination.

35.1 Cleanroom Construction

Cleanrooms are build in an onion-like fashion: there is the outer shell of the building, with the cleanroom inside, and the cleanroom is partitioned into areas of different levels of cleanliness and environmental control. A cleanroom facility consists of the actual cleanroom plus all the supporting facilities for air-conditioning, water, chemical delivery, etc. An overall view of a cleanroom is shown in Figure 35.1. Cleanroom takes up three floors: ground floor and 2nd floor for supporting facilities, with the actual cleanroom on the first floor. Other main features of cleanrooms are:

- overpressure (50 Pa) for keeping particles outside
- filtered air (99.9995% at 0.15 µm particle size)
- heating/cooling/humidification/drying of incoming air
- laminar (unidirectional) airflow in the working areas
- materials compatibility
- mechanical and electrical interference minimization
- working procedures.

Cleanroom envelopes of walls, floor, ceiling, etc., need to be made of materials compatible with the overall objective of environmental control. The walls must not outgas, must be easy to clean and have to be easily removed for installing equipment. They must also be tight because cleanliness is partly ensured by a slight overpressure which prevents the outside air from entering. The ceiling consists of blank elements and filter elements. The higher the proportion of filter elements, the better the cleanliness.

A raised perforated floor is essential for unidirectional (laminar) flow conditions: air from ceiling filters can travel unidirectionally. If particles are generated in the cleanroom, they will be transported away directly through the floor, hopefully not interfering with the wafers. Return air will travel laterally under the raised floor and be returned either in the service aisles or in separate return air ducts.

Vibration isolation is important for lithography and microscopy. Massive air handling units generate vibrations, and therefore mechanical separation of air circulation fans from other parts of the building is needed. Sensitive process areas for lithography are established on isolated concrete slabs extending down to bedrock. Elimination of mechanical disturbances is just part of the story: static electricity can destroy delicate chips and stray electric fields disturb sensitive measurements, which translates to fiber optic lighting.

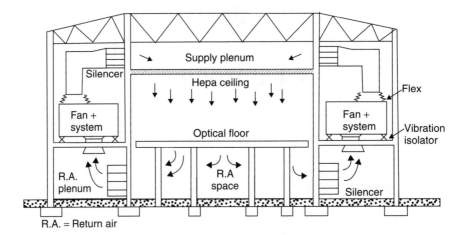

Figure 35.1 Cleanroom: laminar airflow from HEPA filter ceiling, optical floor and return air (RA) space Reproduced from Whyte (1999) by permission of John Wiley & Sons, Ltd

Much process equipment produces excessive heat loads, for example furnaces in the range of 100 kW, and this heat has to be removed in order to maintain a constant temperature in the cleanroom. Most of the waste heat is taken away by cooling water. The design of a cleanroom must therefore always include knowledge about the processes that will running there. This applies to other functions as well: for instance, epitaxial reactors are very high on electric power consumption. Electric supplies should have a backup system: at least the toxic gas and fire alarm systems should be connected to an uninterruptible power supply (UPS), but sometimes critical processes are also protected by UPS: it is worthwhile not switching off oxidation of 200 wafers unexpectedly.

Static electricity elimination, acid neutralization, acid regeneration, waste chemical storage, particle counters, air quality monitors and various other systems are required to operate a cleanroom. The cleanroom can be regarded as a single big instrument because proper cleanroom conditions can only be fulfilled when all subsystems are running.

35.2 Cleanroom Standards

Cleanrooms are classified mainly on the basis of particle counts per volume. Older specifications like Fed. Std 209 specify particles/cubic foot, and designation "Class 100" refers to 100 particles, 0.5 µm, per cubic foot, as shown in Table 35.1.

Newer ISO standards employ units of particles/m³ (conversion factor: $1\,m^3 = 35.3\,ft^3$). ISO standard

Table 35.1 Simplified Fed. Std 209D airborne particle cleanliness classes (particles/ft³)

Class	1	10	100	1000	10 000
No. of particles 0.5 µm	1	10	100	1000	10 000
No. of particles 0.1 µm	35	350	3500	35 000	350 000

cleanliness Class N with particle concentration C_n (particles/m³) is calculated from

$$C_n = 10^N \times \left(\frac{0.1\,\mu m}{D}\right)^{2.08} \quad (35.1)$$

where D is particle size in µm. The proper way to specify cleanroom cleanliness is therefore: Class X (at Y µm particle size). In the ISO cleanroom classification in Figure 35.2 it can be seen that smaller particles are allowed in greater numbers than larger ones.

Federal Standard and ISO correspondence is given in Table 35.2. It also gives average air changes that are required in the cleanroom.

Data in Table 35.3 lists cleanroom features, and, as can be seen, there is really a lot more to a controlled environment that particle cleanliness.

Cleanliness is defined for three different stages of cleanroom construction:

1. As-built: cleanroom construction is finished, but no tools installed.
2. Static: with process tools installed and running, but no personnel.
3. Operational: with people working in the cleanroom (Figure 35.3).

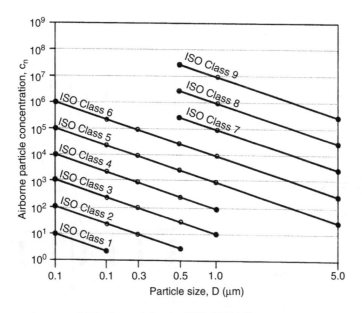

Figure 35.2 Cleanroom class vs. particle size and density (ISO 14644-1)

Table 35.2 Cleanroom classes and air change frequency

ISO 6	Class 1000	150–240/h
ISO 5	Class 100	240–480/h
ISO 4	Class 10	300–540/h
ISO 3	Class 1	360–540/h

Table 35.3 Fed. Std Class 1 cleanroom. Adapted from Cheng and Jansen (1996)

Feature	Values
Cleanliness, process area	<35 particles/m^3, >0.10 µm
Temperature, lithography	$22 \pm 0.5\,°C$
Temperature, other areas	$22 \pm 1.0\,°C$
Humidity, lithography	$43 \pm 2\%$
Humidity, other	$45 \pm 5\%$
Air quality:	
Total hydrocarbons	<100 ppb
NO$_x$	<0.5 ppb
SO$_2$	<0.5 ppb
Envelope outgassing	6.3×10^8 torr-l/cm^2/s
Pressure	Typical 30 Pa relative to outside
Acoustic noise	<60 dB
Vibration	3 µm/s (8–100 Hz)

Figure 35.3 ISO 4 (Class 10) cleanroom. Courtesy VTT

As-built tests should indicate about one class better cleanliness than the designed operational class. Laser scattering of sampled air is used to measure particle counts. There are some methodological problems in the best cleanrooms: there are simply too few particles to get good statistics.

Cleanrooms need not be large halls or rooms; in fact cleanroom area should be minimized because it is expensive to maintain laminar, constant temperature airflow. One way to reduce the need for a high-cleanliness area is shown in Figure 35.4: cleanliness is locally higher where

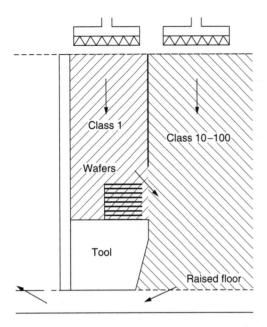

Figure 35.4 Air flows over a tool: wafers are kept in Fed. Std Class 1 area while the rest is Fed. Std Class 10–100 Reproduced from Rubloff and Boronaro (1992) by permission of IBM

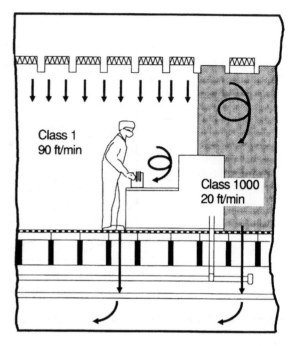

Figure 35.5 Cleanroom and gray area: ISO 3 (Class 1) area for wafer processing, ISO 6 (Class 1000) turbulent flow in service aisle Reproduced from Whyte (2001) by permission of John Wiley & Sons, Ltd

wafers are handled. In Figure 35.5 another solution is shown: the actual wafer handling takes place in a high-class cleanroom but the equipment is situated behind a partition; in service area (also known as gray area) cleanliness is much less (e.g., ISO 3 vs. ISO 6), and in the ISO 6 area airflow is turbulent, not laminar.

35.3 Cleanroom Subsystems

35.3.1 Air

Air handling consists of four major blocks:

- extraction unit
- make-up air unit
- recirculation unit
- filter fan units.

In the first phase air is filtered of coarse objects and humidification or dehumidification is performed. Airborne pollutants like SO_x, NO_x and ammonia are removed by activated carbon filters. Cooling coils and heaters are used to stabilize air temperature. Successive stages of filtration remove finer and finer particles. The final filter is called HEPA (High Efficiency Particle) or ULPA (Ultralow Penetration Air): it is installed in the cleanroom ceiling. ULPA filters have 99.9995% filtration efficiency at particles greater than 0.12 µm. Filter efficiencies can also be classified according to most penetrating particle size (MPPS). Filter defectivity (pinholes) is also a major concern. Filter fan units have another function in addition to particle filtering: they make the airflow laminar.

Airflow velocity and air change requirements increase with cleanliness. As indicated in Table 35.2, 500 air changes per hour is typical of modern cleanrooms. Once the air has been processed, it is recirculated, with only 10% of replacement air introduced at each cycle.

The settling velocity of particles on wafers depends on many factors. For larger particles (in the micrometer range), the gravitational settling velocity (Equation 35.2) is an important parameter:

$$u_g = \frac{\rho g d_p^2}{18\mu} \quad (35.2)$$

where ρ is the density (actually the density difference between the particle and the fluid), g is 9.8 m/s², d_p is the particle size and μ the air viscosity, 1.8×10^{-5} Pa·s.

Smaller particles do not necessarily settle at all: they remain airborne because of Brownian motion. A word of caution: not all particles that deposit on the wafer remain there, and not all small particles are fatal to devices.

The particle deposition rate J on a wafer which is parallel to the airflow is given by

$$J = nu \qquad (35.3)$$

where n is the particle density ($1/m^3$) and u is the sum of gravitational and diffusive settling velocities. For submicron particles the settling velocity is on the order of 10^{-3} cm/s.

35.3.2 DI water

De-ionized water (DI water, DIW), also known as ultrapure water (UPW), is a major subsystem because of the enormous water consumption in modern IC fabrication. A big fab uses a million cubic meters of ultrapure water a year. Water is purified in a multistep process that involves many different techniques, to get rid of many different impurities (Table 35.4). Coarse sand and carbon filtering remove larger particles, while reverse osmosis and ion exchangers remove salts, and UV treatment kills bacteria. DI water quality is monitored by resistivity measurements: 18 Mohm-cm is required. Regular bacteria checks are also performed, as well as particle tests.

35.3.3 Gas systems

Gas system requirements include particle specifications (which set limits on the choices of materials for piping, valves, regulators, mass flow controllers, etc.), leak rates (static leak test, helium leak test) and gas impurity tests.

Bulk gases (also known as line gases or house gases) are gases shared by many tools. These include nitrogen, oxygen, hydrogen, argon and compressed air. Nitrogen is especially widely used, both in processes and as an inert protective gas. Four purity classes of nitrogen can be offered, for different applications:

1. Process nitrogen: furnace annealing or reactive sputtering, 99.99999% (7N) purity.
2. Dry nitrogen: venting and flushing of process chambers, 99.999% (5N) purity.
3. Pistol nitrogen: for drying.
4. Pump nitrogen: as ballast for pumps.

Specialty gases are used by dedicated equipment, and they are supplied from gas bottles in a one-to-one distribution topology. These include for example SF_6 and Cl_2 for etchers, SiH_2Cl_2 and NH_3 for nitride LPCVD, SiH_4 and N_2O for PECVD oxide, PH_3 for doped polysilicon LPCVD and WF_6 for tungsten CVD. Ion implanter gas consumption is very small, and AsH_3, PH_3 and BF_3 minibottles are usually located inside the implanter cabinet. Implanter gases can also be supplied from SDS (Safe Delivery System) sources: the dopant gases are absorbed in the solid absorber material in the bottle and released by application of temperature or under pressure.

35.4 Environment, Safety and Health (ESH)

Various gases, chemicals and tools are sources of potential health hazards to cleanroom personnel. Ion implanters operate at 200 kV and are sources of X-rays (gamma rays may be emitted in hydrogen implantation); plasma systems may leak microwave energy and UV radiation and wet etch and plating baths may contain cyanides. These hazards are dealt with in many different ways.

Strong mineral acids like H_2SO_4, HNO_3, H_3PO_4 and HCl are routinely used. Normal burn hazards are associated with them, and they must be neutralized after use. HF is different because its effect is not immediate but delayed; it does not attack skin but bone. Special care is needed for all HF-containing liquids and separate disposal of HF is required.

Solvents and organics come from various sources: HMDS, which is used as a priming agent before photoresist coating is released into cleanroom air (HMDS is the main airborne pollutant in many cleanrooms); solvents which are released from resists upon baking; IPA and acetone used for drying and cleaning. Solvents are major reasons for fires in cleanrooms.

Process exhausts remove unwanted thermal and mass flows from the cleanroom. Acid vapors from wet benches

Table 35.4 Production of DI water

- Sand filter
- Active carbon filter
- Particle filtering at 3 μm
- Softening of water
- RO: reverse osmosis
- CEDI: continuous electrical deionization
- UV treatment
- Ion exchangers
- Particle filtering at 0.2 μm
- Storage tank
- Continuous DI water circulation in the cleanroom loop

are removed and safely disposed of in plastic ducts, while solvent exhausts are removed in stainless steel ducts. Separate piping is required to prevent explosive mixing. In most cases cleanroom systems protect wafers from humans, but in wet benches protection of humans from chemicals is required. Acid vapors will be cleaned by gas abatement systems (solid absorber, combustion system and/or gas effluent washing machine, or wet scrubber) before release to the outside air.

In many processes the utilization of source gases is very low and the outpumped flow consists mostly of unused source gas. These gases, for example SiH_4 from a LPCVD system, may be incinerated or diluted. Silane is spontaneously flammable. It is used at 100% concentration in LPCVD polysilicon, but in PECVD systems it is usually dilute, 1–5% SiH_4 in nitrogen, argon or helium. Wet oxidation is usually done by in situ-generated water from H_2 and O_2 gases (Figure 32.1). Hydrogen/oxygen mixtures are flammable between 4% and 75% hydrogen, and hydrogen content in exhaust gases needs to be controlled, by combustors or by other gas abatement systems.

A toxic gas alarm system is required because many of the gases used in semiconductor processing are extremely toxic: hydrides, PH_3, AsH_3, B_2H_6 are lethal in the low ppm concentrations and chlorine was used in World War I as a battlefield gas. Many chlorine-containing gases react with humid air to form hydrogen chloride (HCl) which is similarly toxic and corrosive. Table 35.5 lists some of the common toxic and dangerous gases used in microfabrication.

Plasma etcher and CVD pumps and pumps oils can accumulate considerable amounts of unknown compounds: for example, products from reactions between etch gases and photoresist. Pumping oxygen is a safety concern: oxygen can explode if it reacts with pump oil. Therefore, either most plasma and CVD equipment uses inert perfluorinated pump oils (Fomblin, Krytox), or else dry pumps are employed. Dry pumps are beneficial also because they tolerate more corrosive and abrasive chemicals than standard mechanical pumps.

Fire detection in a cleanroom cannot be done in the same way as in normal office rooms because the high cleanliness prevents particle-based detection and ionization detectors in the ceiling from seeing anything because of the unidirectional downflow. Local sampling and thermal detection are used. Fire extinguishing must be accomplished without generating particles because damage from extinguishing might be intolerable to the cleanroom as a whole. Carbon dioxide or water mist systems are used.

Alarm strategies in a microfabrication cleanroom need to be carefully planned. In the case of a toxic-gas alarm

Table 35.5 Toxic gases in semiconductor manufacturing

	TLV (ppm)	IDLH (ppm)	Other properties
NH_3	25	300	DO: 0.04–50 ppm
Cl_2	0.5	10	DO: 0.03–0.4 ppm
HCl	5	50	
HF	3	30	
BF_3	1	25	[a]
SiH_4	5	N/A	ER: 1.37–96%
GeH_4	0.2	N/A	
$SiCl_2H_2$	[b]	N/A	ER: 4.1–99%
AsH_3	0.05	3	DO: 0.5–4 ppm, garlic
PH_3	0.3	50	DO: 0.01–5 ppm, fishy
B_2H_6	0.1	15	DO: 1.8–3.5 ppm, sweet

[a] Reacts to form HF upon contact with moisture.
[b] Reacts to form HCl.
TLV = Threshold Limit Value: no adverse effects for prolonged exposure.
IDLH = Immediately Dangerous to Life and Health: 30 min escape time to ensure no permanent health effects.
ER = Explosive Range (% by volume in air).
DO: Detectable Odor.
N/A: Not Applicable.
Source: Baldwin, D.G., M. Williams and P.L. Murphy (2002).

personnel need to be evacuated, but it does not necessarily mean that oxidation furnaces have to be shut down. If a lot of 200 wafers is lost in the case of an unplanned shutdown, massive damage will be incurred. In the case of a fire alarm, air circulation needs to be closed down because otherwise it would spread the fire efficiently, but it is important to keep exhausts operational. If the fire originated from a wet bench (which is the usual case), then the wet bench exhaust will at least remove hot acid and/or solvent vapors, but there is a danger that the fire will spread along the exhaust ducts.

35.5 Cleanroom Operating Procedures

A cleanroom must include not only the structures and airflows, but also procedures for transferring people and materials. People enter cleanrooms in zones of increasing cleanliness:

- pre-change zone
- change zone
- entrance zone.

In the pre-change zone extra clothing is removed, personal belongings like mobile phones are set aside, a glass of water is drunk (to reduce particle counts in exhaling).

Shoes are cleaned, covered or changed to special cleanroom shoes. Hair is covered by a hairnet (disposable hat). A sticky mat is often used to remove dust from the soles of shoes. In the change zone a cleanroom overall and face mask are put on. Temporary gloves may be used in this zone. Upon entry into the entrance zone shoes are covered by cleanroom boots, and final gloves are put on. Additionally, goggles may be applied. Another sticky mat and possibly an air shower are passed on the way to the cleanroom proper. The details of these procedures vary from cleanroom to cleanroom, but the aforementioned chain of events is typical of ISO 3 (Class 1) cleanrooms. In ISO 5–6 (Classes 100–1000) cleanrooms more relaxed gowning procedures are applicable.

A similar, but somewhat reverse, procedure of increasing cleanliness applies when new tools, wafers, sputtering targets or any other material are transported into the cleanroom. The transportation packaging is removed in a non-clean area. Wood chips or wood fibers from cardboard boxes should be minimized. Cleanroom tools and materials are packed in a multilayer fashion. The outermost plastic packaging is removed in a semi-clean area, and the plastic packaging underneath is possibly cleaned before being taken into the cleanroom. The innermost packaging layer is removed under cleanroom conditions. Obviously, tools and materials for cleanrooms have to be packed in a cleanroom!

Contamination can come from air, water, chemicals or people, but also from wafers. Major contamination problems arise if wafers are processed improperly. For example, if wafers with metal on them are cleaned in a SC-1/SC-2 cleaning bath, metals will be etched and the cleaning bath contaminated. Similarly, if a gold-containing wafer is processed in a high-temperature tool, the tool will be contaminated. If contact holes in oxide are etched in a HF bath intended for native oxide removal before oxidation, organic contamination from the resist will be carried over into the oxidation furnace. Working in a cleanroom must be very disciplined. It is not enough for you to understand the desired processes – you must also understand the collateral damage from mistaken processes. These may ruin your wafers, and everything that is being processed after you.

35.6 Mini-Environments

In the mini-environment approach a small cleanroom is built locally around the tools or the wafers. It is easier to maintain high purity locally over a small area than in a big room. At one extreme the wafer box is the cleanroom, filled with high-purity nitrogen. Compared to the cleanroom, it has two benefits: nitrogen is inert, so reactive impurities from the atmosphere are eliminated; and the gas is stagnant in the box, so particles do not move, as they would do in the laminar airflow of the cleanroom.

Elimination of the cleanroom itself has been toyed with: if all tools were to use a standard interface, wafers could be carried in mini-environment boxes from tool to tool, and they would never see the cleanroom air, in which case the cleanroom would become redundant. Wafer fabs with such standard mechanical interfaces (SMIFs) have been built, but cleanrooms have not been made redundant because conversion of all process and measurement tools has been elusive.

35.7 Exercises

1. What is the linear air velocity if cleanroom air in exchanged 360 times per hour?
2. What class of cleanroom would be suitable for (a) 1 μm and (b) 0.1 μm CMOS production?
3. What would be the gravitational settling velocities of 1 μm and 0.1 μm particles?
4. Explain how the factors listed in Table 35.3 affect wafer processing!
5. If a 0.5 liter bottle (under 50 bar pressure) of boron trifluoride (BF_3) leaks into a 1000 m^2 cleanroom, will it be immediately dangerous to health?
6. How many particles will be deposited on a 200 mm wafer in an ISO Class 2 cleanroom in a hour?

References and Related Reading

Baldwin, D.G., M. Williams and P.L. Murphy (2002) **Chemical Safety Handbook for the Semiconductor/electronics Industry**, 3rd edn, OEM Press.

Cheng, H.P. and R. Jansen (1996) Cleanroom technology, in C.Y. Chang and S.M. Sze (eds), **ULSI Technology**, McGraw-Hill.

Middleman, S. and A.K. Hochberg (1993) **Process Engineering Analysis in Semiconductor Device Fabrication**, McGraw-Hill.

Rubloff, G.W. and D.T. Boronaro (1992) Integrated processing for microelectronics science and technology, *IBM J. Res. Dev.*, **36**, 233.

Whyte, W. (ed.) (1999) **Cleanroom Design**, John Wiley & Sons, Ltd.

Whyte, W. (2001) **Cleanroom Technology**, John Wiley & Sons, Ltd.

Zhou, Z. (2004) From classrooms to the real engineering world: the training program in the microelectronics research center at Georgia Tech, *IEEE Trans. Educ.*, **47**, 114–120.

36

Yield and Reliability

Microfabrication is a statistical business: some devices always fail due to random errors and fluctuations in process performance. Understanding these losses is a life and death issue in wafer fabs. Yield loss can mean for example non-functional chips or wafer breakage. Catastrophic process problems can destroy a batch of 100 wafers, for example if gas runs out during film deposition, resulting in no film at all, or a film with unknown properties.

Device testing after manufacturing classifies chips into good and bad, but there are failure mechanisms that only manifest themselves over longer periods of time. Some of these can be found in accelerated testing: running devices under extreme conditions (higher than rated voltage, humidity and temperature), which will reveal some systematic problems. However, if the accelerated conditions change the failure mechanism, the tests are of little use. Random failures are even more difficult to screen out.

Reliability can be taken to mean performance as designed, resistance to failure or avoiding unexpected failure. In contrast to yield, it is about long-term performance in the field. Metallic wires subject to high current density will be damaged, so electromigration lifetimes must be assessed. Locally thin regions of oxides will be subject to higher electric fields than areas of specified thickness, and this may limit the reprogramming cycles that the device can tolerate. Metallic movable mirrors will have memory effects if they are switched to one side more often than the other, leading to reduced contrast or color shift. Vacuum packages for MEMS resonators may leak and resonance frequency drift to unusable values. Solar cells are subject to "photobleaching," a conversion efficiency drop due to high-energy UV photons. In all these cases the goal is to understand the physical basis of ageing mechanisms, and to tailor the processes to minimize them.

Yield and reliability are connected: poor process control leads to low yields and more failing devices, and devices with poorly controlled properties are subject to larger variation and more failures. For instance, in a well-controlled process gaps are always closed, but in a poorly controlled process they are sometimes closed and sometimes open. Process gases, cleaning chemicals, rinsing water and other fluids can be trapped in the keyholes (Figure 5.16d), and these fluids can be released or react in an unpredictable manner when environmental conditions change, for example the temperature rises.

36.1 Yield Definitions and Formulas

Yield is success rate. It can be calculated at different points of the process, and different yield numbers obtained. In all cases yield is the quotient "good outcomes/total." Fab yield takes into account the number of wafers out of the process, divided by wafer starts. Note, however, that for example 10% of wafers in fab to be used for monitoring and testing and not contribute to saleable chips, even in theory, but fab yield for prime wafers approaches 99%.

Die yield, or chip yield, is the fraction of functional chips on a wafer. In one survey die yields ranged from 46% to 92% for 0.5 cm^2 devices. Again, not all chips on the wafer are product chips: some chips are dedicated to process monitoring test structures (identical in all products, to gather statistical data on the process) and some are product-specific test structures.

Yield (Y) is a product of yields of individual process steps (Y_i), according to

$$Y = \prod_i Y_i \qquad (36.1)$$

Total yield can never be better than the yield of the lowest yielding step. If CVD yield is 99%, CMP yield 95% and bonding yield 90%, the total yield of this process sequence will be 85% only.

A six mask student lab PMOS process with 50 steps and 95% step yield results in 8% total yield and hundreds of transistors for measurements. But an industrial 50 step MEMS process would probably have 99.5% step yield and 78% total yield. A CMOS process with 500 steps requires at least 99.9% step yields. However, one single badly yielding step, with say 70% yield, will limit the total yield to less than 70%; therefore, process development effort must be carried out on all process steps.

Yield can also be viewed as a product of systematic and random components:

$$Y_{\text{total}} = Y_{\text{systematic}} \times Y_{\text{random}} \quad (36.2)$$

Random yield loss comes from process errors and equipment malfunctioning, and the systematic yield loss from process capability limitations. All processes have variation (across the wafer, wafer to wafer, lot to lot) and devices cannot be designed to tolerate tails of statistical distributions.

SRAM is the prototypical test vehicle for process development: in a regular memory array of transistors it is easy to locate the electrical fault, and to investigate it by optical, physical and chemical means and to correlate it with a physical defect, a particle, a residue, corrosion or linewidth change.

36.2 Yield Models

Random yield loss has been described by many models. The Poisson distribution (Equation 1.3) is the simplest model: exponential dependence on area and defect density. This holds fairly well for small chips and/or low defect densities, as can be seen in Figure 36.1, but clearly the data shows that discrepancies rapidly develop if the model is applied to chips that are a little bit larger.

A more general model takes defect clustering into account and models yield as

$$Y_{\text{random}} = \left(1 + \frac{AD_0}{\alpha}\right)^{-\alpha} \quad (36.3)$$

where α is the cluster factor.

Cluster factor α presents the tendency of defects to cluster; that is, they are not randomly distributed but tend to cluster. The values of α are usually considered trade secrets, and companies are reluctant to reveal their yield statistics. Figure 36.2 compares yield models to cluster factors. A cluster factor $\alpha = \infty$ corresponds to the Poisson distribution, and $\alpha = 1$ results in Seeds' model, that is

$$Y = \frac{1}{1 + AD} \quad (36.4)$$

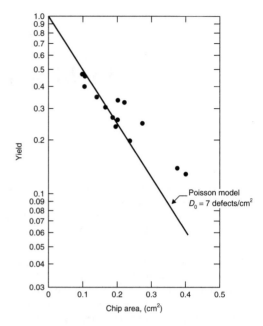

Figure 36.1 Poisson distribution of chip yield: good fit for small chips. Reproduced from Cunningham (1990) by permission of IEEE

Another yield model is known as Murphy's model, that is

$$Y = \left(\frac{1 - e^{(-AD)}}{AD}\right)^2 \quad (36.5)$$

Chip size A is a result of two opposing trends: as linewidths are scaled down, chip area should decrease; but because more and more functionality and memory are desired, the number of transistors on a chip increases so fast that chip area in fact is constantly increasing. Defect density must be scaled down with decreasing linewidths because the product DA must be made smaller. Not all particles will destroy a chip: some are too small and some land on non-critical sites. The model shown in Figure 36.3 estimates yield DRAM yield when particle fatality is a parameter that is set from 10% to 20%.

36.3 Yield Ramping

Process research for a new generation of chips starts about 10 years before commercial introduction. It involves the exploration of new technologies and materials, and novel device structures. About 5 years before introduction, equipment should be available in single units, and 2–3 years before introduction pilot production

Yield and Reliability

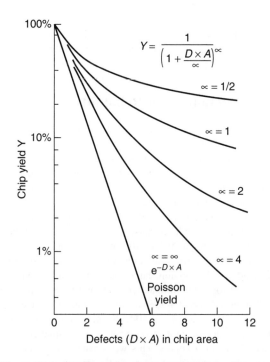

Figure 36.2 Yield models compared: cluster factor α ranges from 0.5 to infinity. Reproduced from Carlson and Neugebauer (1986) by permission of IEEE

quantities of equipment will be purchased, say five units in a major company.

Working devices are ready a year or two before introduction. This implies device and equipment readiness, but does not give any indication of systematic or random yield. Depending on device type and company culture, 10–20 lots, each taking 1–3 months (running partly in parallel), are fabricated and analyzed. Production start is the date when every lot produces functioning devices. Figure 36.4 shows yield data over time. The yield was ramped rapidly from an initial 20% to nearly 100% in less than six months.

Yield is related to a particular process characterized by its linewidth or process technology generation. It is not constant over the device lifecycle: at product introduction yield is low and it rises with production volumes. Some schematic values for processes at different stages of maturity are given in Table 36.1.

The yield ramp phase often determines commercial success or failure. Commodity devices like DRAMs have a market price, and because fab investments are similar for same-generation technology, the difference in revenue comes mostly from yield in the early phase. The IC industry has been able to prosper in spite of dire predictions about yield-limited economics. In fact, statistics show that yield ramp rates have been steeper for new, small linewidth processes (Figure 36.5). This is partly due to multiple fabs, where everything is copied from

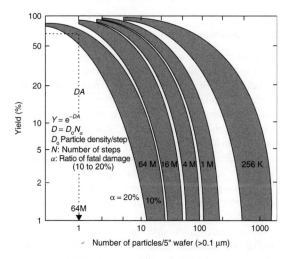

Figure 36.3 Particle-induced yield loss in DRAMs according to the Poisson model. Note that only 10–20% of particles are assumed to cause fatal damage to chips. Reproduced from Hattori (1998)

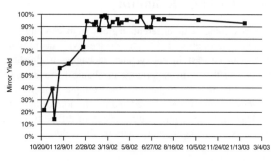

Figure 36.4 Yield over time for an optical beam steering device, with 100 000 polysilicon micromirrors. Reproduced from Romig *et al.* (2003), copyright 2003, Pergamon

Table 36.1 Yields (%) at different stages of maturity

	Y_{random}	$Y_{systematic}$	Y_{total}
Introduction	20	80	16
Ramp-up phase	80	90	72
Mature	95	99	94

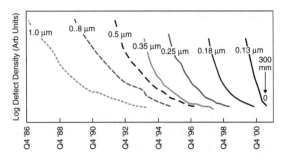

Figure 36.5 Defect density reduction is getting faster and faster for each successive technology generation. Reproduced from Natarajan *et al.* (2002) by permission of Intel

an existing fab, and experience accumulates faster than in one-of-a-kind fabs.

Yield stability during ramp-up and production is mandatory because otherwise there is no yardstick for process development efforts. Gross variations in yield would mean that even major process improvements might be rejected if yield variation and process improvement have opposite signs. Similarly, cosmetic improvements might get approval even though the effect came from normal yield variation. Yield decreases at the end of the product lifecycle: it is caused by phasing out the process and decreased engineering effort.

36.4 Package Reliability

Devices come with different degrees of packaging and reliability concerns: solid state devices with no moving parts can be cast in epoxy, and packaging is finished. Of course they will experience mechanical movements, induced by thermal expansion, for example. But MEMS devices are much more varied and much more difficult to package. Devices come in five classes of difficulty when it comes to packaging and reliability difficulty:

1. Solid state devices, no moving parts.
2. Devices with channels/cavities but no moving parts.
3. Devices with moving but non-contacting parts.
4. Devices with contacting surfaces.
5. Devices with contacting and rubbing surfaces.

Many microfluidic devices belong to the second class, for example nozzle devices, or microreactors. Pressure sensors, microphones, accelerometers, IR detectors, cMUTs and RF resonators belong to the third class. Digital micromirrors, microvalves and pumps and RF switches belong to the fourth class. Gears, turbines and pop-up mirrors are members of the fifth class. Devices in this last class remain to be commercialized, and very few devices in the fourth class are commercially available, apart from digital micromirrors and a few RF switches.

36.5 Metallization Reliability

Metal-to-metal contacts in in-line RF switches are subject to deformation and welding. Deformation is not detrimental as such, because it leads to good contact and low contact resistance, but the more intimately the surfaces make contact, the likelier it is that they will stick together. This is especially important for "hot contacts" that carry significant current (e.g., 100 mA) during switching. While cold contacts (small currents) can be switched 10^9 times, hot contacts might survive 10^7 switchings only.

There are materials solutions to metal–metal contacts: instead of gold, rhodium can be used; it is harder, requires a higher force for intimate contact, and is less prone to adherence. Contact force of course changes contact resistance: for example, a 100 mN force might lead to a 0.2 ohm Au–Au contact resistance, but a smaller force may result in 0.4 ohms. This will change over time: the contacting surfaces are not perfectly smooth, and initially only the highest points (asperities) contribute to the contact; later on the repeated contacts of surfaces lead to increased contact area (and smaller resistance). This situation is analogous to CMP (Figure 16.3).

36.5.1 Electromigration

Electromigration is atom movement due to electron momentum transfer. Electrons dislodge metal atoms from the lattice, and these atoms consequently move and accumulate at the positive end of the conductor and leave voids at the negative end. This is shown both schematically and in the SEM micrograph in Figure 36.6. This effect is encountered in aluminum conductors when current densities approach the level of mega-amps per square centimeter, but copper and tungsten tolerate higher current densities.

Electromigration depends on a number of factors: macroscopic factors include geometry of the lines, their width, shape, area as well as passivation. Microscopic factors include grain size, texture, alloy solutes and their precipitation at grain boundaries and interfaces. Solutes like copper in aluminum increase resistance to electromigration because $CuAl_2$ formed at grain boundaries blocks diffusion at grain boundaries. What is more, grain size and linewidth are not independent: when grain size and linewidth become equal (typically when the thickness-to-width ratio is about unity), the number of grain boundaries is strongly reduced,

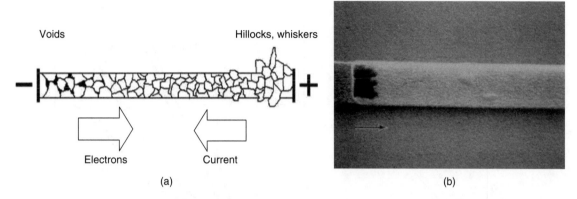

Figure 36.6 Electromigration: atoms are transported from the anode end of a wire towards the cathode with electron wind. Voids are left at the anode end, and hillocks form towards the cathode end: (a) schematic. Figure courtesy Antti Lipsanen, VTT; (b) SEM micrograph of Al lines (4 µm wide). Reproduced from Hu, C.-K. *et al.* (1993), by permission of American Inst of Physics Reproduced from Hu *et al.* (1993) by permission of American Institute of Physics

leading to a so-called bamboo structure with one grain extending across the line. In polycrystalline material grain boundary diffusion is important, and elimination of grain boundaries will affect electromigration.

Mean time to failure (MTTF) due to electromigration is given by

$$\mathrm{MTTF} = AJ^{-n}e^{(E_a/kT)} \quad (36.6)$$

where A is a constant depending on the wire geometry and metal microstructure, J the current density and E_a the activation energy. The factor n is not known very accurately, but $n = 1.7$ is used for aluminum. For aluminum thin films E_a is on the order of 0.5–0.8 eV, whereas for bulk aluminum it is 1.4–1.5 eV. As a general trend, the higher the activation energy, the better the electromigration resistance. It can be roughly estimated on the basis of metal melting point T_m: the higher the melting point, the higher the electromigration resistance. To put it another way: high melting point equals high bond strength. At room temperature, which is $T_m/3$ for aluminum, aluminum atoms have a reasonable probability for diffusion. For tungsten, room temperature corresponds to $T_m/10$, and electromigration is orders of magnitudes less. But for copper the situation is different: instead of bulk diffusion it is surface diffusion that matters, therefore control of interfaces and barriers is of paramount importance.

In Figure 36.7 the incubation time of resistance increase is plotted for Al–2% Cu lines and tested at 225 °C to see how fast the resistance increase starts. This is again critically dependent on current density.

36.6 Dielectric Defects and Quality

Even though the interface between silicon and thermally grown silicon dioxide can be reproducibly fabricated, it is far from ideal. The interface trapped charges are caused by broken bonds (from structural defects, oxidation-induced defects and contamination). Because they are at the interface, the potential in silicon will charge or discharge them. Interface trapped charges can be reduced by forming gas anneal.

There is always some positive fixed charge in the vicinity of the interface, and it is related to silicon ionization during the oxidation process. There are also trapped charges, which can be positive or negative, caused by energetic electrons or ionizing radiation, and there can be mobile charges from contamination, most notably from Na^+ ions.

The electric field that oxide can sustain is usually reported by breakdown voltage: >10 MV/cm is considered to be the intrinsic breakdown field. This is also termed C-mode failure; B-mode failure happens at 2–8 MV/cm and A-mode below 2 MV/cm. An example of oxide breakdown statistics is shown in Figure 36.8.

A-mode failures are gross defects: pinholes and voids. B-mode failures are more benign and more subtle, like oxide thinning, trapped charges or metal contamination-induced defects. C-mode failures are intrinsic to oxide structure but can be affected by nanoscopic defects like increased surface and interface roughness. These different oxide defects are shown in Figure 36.9. A-mode failures are seen as yield loss; B-mode failures as reliability problems in accelerated testing or in the field; C-mode failures become important at the end of product lifetime.

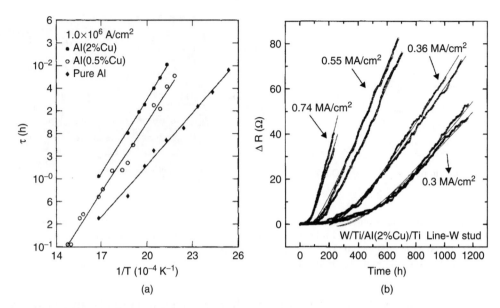

Figure 36.7 (a) Mean time to failure of 2.5 μm wide Al, Al (0.5 wt% Cu) and Al (2 wt% Cu) lines at different temperatures with 1 MA/cm² current density. Reproduced from Hu, C.-K. *et al.* (1993), by permission of AIP. (b) Incubation time before resistance increase sets in at 255 °C. From Hu, C.-K. *et al.* (1995a), by permission of Elsevier

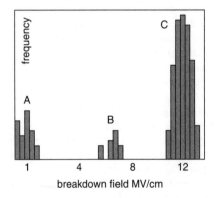

Figure 36.8 Oxide breakdown distribution: A-mode at low field; B-mode at medium field; C-mode at high field

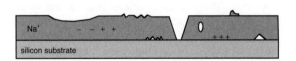

Figure 36.9 Oxide defects (left to right): Na⁺ mobile charge, thinning, fixed charge, surface and interface microroughness, pinhole, void, interface charge, particle, stacking fault. Adapted from Schröder (1998)

Metals are responsible for many of the defects described above. If the surface is contaminated, silicates like $MgSiO_4$ or silicides like Cu_3Si and $NiSi$ can be formed, rather than silicon dioxide. Their formation consumes silicon and therefore oxide will be locally thinner. Unreactive metals dissolve in the growing oxide, which leads to decreased intrinsic breakdown strength. Sodium (Na) contamination leads to an increased oxidation rate, whereas iron (Fe) and aluminum (Al) lead either to an increase or decrease depending on the level of contamination and time. Metals can also catalyze the reaction SiO_2 (s) + Si (s) $\Rightarrow$ 2 SiO (g) (which takes place under low oxygen partial pressure, e.g., during ramp-up in a furnace), leading to oxide evaporation and pinhole-like defects.

Oxide dielectric strength is tested by a number different experimental set-ups:

- Ramped voltage: the voltage between the MOS gate and substrate is linearly increased (0.1 or 1 V/s) until the oxide breaks down. Breakdown voltage V_{BD} is defined as the voltage where a sudden voltage drop occurs.
- Time-to-breakdown under constant current (TTBD, t_{BD}): constant, preset current is fed into the insulator, and voltage is recorded as a function of time. TTBD is the time when a sudden voltage drop occurs.
- Charge-to-breakdown (Q_{BD}): in a constant current test $Q_{BD} = J_{injected} \times t_{BD}$. Good oxides exhibit values of 10 C/cm², but this is dependent on injected current.

36.7 Stress Migration

Electromigration is studied by accelerated tests under current densities that are higher than normal at elevated temperatures. But voids appear in metal lines at elevated temperatures even when no current runs through them. This is known as stress-induced voiding, or stress migration. The driving force is the gradient in the strain field: some atoms find it energetically favorable to move to voids.

The source of stress is thermal expansion mismatch between the metal and the encapsulating (PE)CVD dielectric. Strain (elongation) is proportional to the CTE and temperature difference, which for aluminum translates to 1% linear elongation, or about 3% volume change when 300 °C PECVD is done. This elongation corresponds to stresses of about 1 GPa (as can be estimated from Equation 4.1). Aluminum lines expand during PECVD, and they are fixed at their elongated state because of the mechanical stiffness of deposited oxide/nitride layers. This high tensile stress can be relaxed by cracks, and once a crack is formed, it tends to grow.

Compressive stresses in aluminum can be relaxed via hillock formation. Hillocks are small protrusions. Their size can be up to micrometers, which is equal to the insulator thickness between two levels of metallization. If some mechanically stiffer film prevents relaxation in the vertical direction, hillocks can grow laterally, and, again, a micrometer is a very sizable distance relative to line spacing. In both cases hillocks can short two metal lines. Low-temperature processing helps to reduce hillocks (and stress and electromigration). Alloying aluminum with copper is also helpful to minimize hillock formation because it blocks grain boundary diffusion.

36.8 Die Yield Loss

Because microfabricated devices are small and complex, usually no repair is possible or feasible. There are a few exceptions: big memory arrays with redundant cell blocks can be repaired by disconnecting malfunctioning blocks and connecting redundant blocks. Another repair operation in regular use is photomask repair: because mask writing is slow and expensive, it makes sense to repair a few defective sites rather than rewrite the whole mask plate. Masks are also inspected 100% and therefore defects are caught early on.

The fishbone diagram in Figure 36.10 depicts contributors to die yield loss. As can be seen, the yield loss mechanism can be difficult to pinpoint because both design and manufacturing contribute to it, in addition to

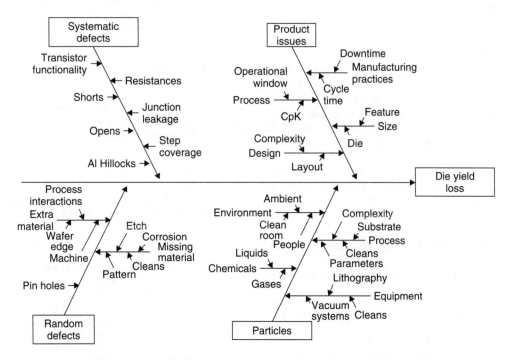

Figure 36.10 Factors influencing die yield loss. Reproduced from Rao (1993) by permission of McGraw-Hill

random processes. Despite the enormous complexity of modern microdevices, they are routinely manufactured at yields of over 90%.

36.9 Exercises

1. Compare the number of 0.5 cm² chips on 100 mm and 150 mm wafers to the 6 mm edge exclusion rule. Repeat for 2 cm² chips on 200 mm and 300 mm wafers with 3 mm edge exclusion.
2. If linewidth is halved but the same old cleanroom is used, what will happen to yield?
3. Use Minesweeper (for Windows, or XMine for Unix) as a tool to simulate fabrication yield: chips are 1×1, 2×2, 3×3, 4×4, 5×5 or 6×6 areas on the grid. Vary defect density (= the number of mines) and check how defect density and chip size are related.
4. What is the extrapolated yield of a new 2 cm² chip if $D = 2 \text{ cm}^{-2}$ using the Poisson model, measured from a large sample of small chips ($<0.6 \text{ cm}^{-2}$). What is the yield if Murphy's model is used instead? How about Seeds' model?
5. If 64 Mbit DRAM chips are 2 cm², what must the fab defect density be?
6. Accelerated tests for chips are run at elevated temperatures in order to find failures faster. Acceleration factor temperature (AFT) is given by

$$\text{AFT} = e^{(E/kT_{\text{operation}})}/e^{(E/kT_{\text{test}})}$$

Using an activation energy of 0.7 eV, what acceleration factor does 175 °C present? (Temperatures are junction temperatures, and typical values are 55 °C for consumer and 85 °C for industrial electronics.)
7. If Al (2% wt Cu) lines have MTTF of 400 h at 255 °C, what is their expected lifetime under standard operating conditions?
8. If a TiW/Al (50 nm/400 nm) line experiences a void in aluminum, by how much will line resistance increase? (TiW resistivity is 150 μohm-cm).

References and Related Reading

Carlson, R.O. and C.A. Neugebauer (1986) Future trends in wafer scale integration, *Proc. IEEE*, **74**, 1741.
Cunningham, J.A. (1990) The use and evaluation of yield models in integrated circuit manufacturing, *IEEE Trans. Semicond. Manuf.*, **3**, 60.
Diebold, A.C. (1994) Materials and failure analysis methods and systems used in the development and manufacture of silicon integrated circuits, *J. Vac. Sci. Technol.*, **B12**, 2768.
Gardner, D.S. and P.A. Flinn (1988) Mechanical stress as a function of temperature in aluminum films, *IEEE Trans. Electron Devices*, **35**, 2160.
Hattori, T. (ed) (1998) **Ultraclean Surface Processing of Silicon Wafers**, Springer.
Hayashi, M., S. Nakano and T. Wada (2003) Dependence of copper interconnect electromigration phenomenon on barrier metal materials, *Microelectron. Reliab.*, **43**, 1545–1550.
Hu, C.-K. (1995) Electromigration failure mechanism in bamboo-grained Al(Cu) interconnections, *Thin Solid Films*, **260**, 124.
Hu, C.-K. *et al.* (1993) Electromigration of Al(Cu) two-level structures: effect of Cu kinetics of damage formation, *J. Appl. Phys.*, **74**, 969.
Hu, C.-K. *et al.* (1995) Electromigration and stress-induced voiding in fine Al- and Al-alloy thin- film lines, *IBM J. Res. Dev.*, **39**, 465.
Hu, S.M. (1991) Stress-related problems in silicon technology, *J. Appl. Phys.*, **70**, R53–R80.
Kim, K.O., M.J. Zuo and W. Kuo (2005) On the relationship of semiconductor yield and reliability, *IEEE Trans. Semicond. Manuf.*, **18**, 422–429.
Li, Q. *et al.* (2009) Assessment of testing methodologies for thin-film vacuum MEMS packages, *Microsyst. Technol.*, **15**, 161–168.
Melzner, H. and A. Olbrich (2007) Maximization of good chips per wafer by optimization of memory redundancy, *IEEE Trans. Semicond. Manuf.*, **20**, 68–76.
Natarajan, S. *et al.* (2002) Process development and manufacturing of high-performance microprocessors on 300mm wafers, *Intel Technol. J.*, **6**, 14–22.
Rao, G.P. (1993) **Multilevel Interconnect Technology**, McGraw-Hill.
Romig, A.D., M.T. Dugger, and P.J. McWhorter (2003) Materials issues in microelectromechanical devices: science, engineering, manufacturability and reliability, *Acta Mater.*, **51**, 5837–5866.
Schroder, D.K. (1998) **Semiconductor material and device characterization**, 2nd edn, John Wiley & Sons, Inc.
Spanos, C.J. (1992) Statistical process control in semiconductor manufacturing, *Proc. IEEE*, **80**, 819.
Stapper, C.H. and R.J. Rosner (1995) Integrated circuit yield management and yield analysis: development and implementation, *IEEE Trans. Semicond. Manuf.*, **8**, 95.
Tabata, O. and T. Tsuchiya (eds) (2008) **Reliability of MEMS: Testing of Materials and Devices**, Wiley-VCH Verlag GmbH.
Tan, C.M. and A. Roy (2007) Electromigration in ULSI interconnects, *Mater.Sci.Eng.*, **R58**, 1–75.
Warwick, C. and A. Ourmazd (1993) Trends and limits in monolithic integration by increasing the die area, *IEEE Trans. Semicond. Manuf.*, **6**, 284.
Witvrouw, A., H.A.C. Tilmans, and I. De Wolf (2004) Materials issues in the processing, the operation and the reliability of MEMS, *Microelectron. Eng.*, **76**, 245–257.
Yue, J.T. (1996) Reliability, in C.Y. Cheng and S.M. Sze (eds), **ULSI Technology**, McGraw-Hill.

37

Economics of Microfabrication

This chapter deals with the economics of microfabrication, the cost structures and trends of IC, MEMS, solar cell, display and magnetic data storage manufacturing. These devices are driven by very different motives: large-scale integration for CMOS, large area for displays, novel functionality for MEMS, low cost for solar cells, extreme throughputs for hard disk drives, etc. Starting material costs, equipment cost and operating costs differ. Silicon wafers are just a few percent of the cost of an IC but roughly 30% of the cost of a crystalline silicon solar cell. Lithography is a central cost and performance issue in ICs and hard disk read heads, while large-area substrate processing drives thin-film solar cells and flat-panel displays.

All dollar values in this and the following chapters are bound to be crude approximations for many reasons. Some numbers are proprietary and can only be gauged by outsiders, for example costs are trade secrets but prices are public knowledge. Currency fluctuations make worldwide comparisons time dependent. Business cycle ups and downs can result in 30% increases annually as well as decreases in sales volumes. This has been especially true for DRAM memories, but solar cells have experienced similar-sized fluctuations.

37.1 Silicon

There are only half a dozen companies making high-purity polysilicon, and this industry has very high entry barriers: the facilities are big and expensive, and it takes three years to get a new facility up and running. Companies making CZ silicon wafers or MC solar wafers buy their electronic-grade polysilicon on long-term contracts for something like $50 per kg. As in the oil industry, there is a spot market where prices can gyrate wildly, and in 2007–2008 solar boom poly spot prices shot to $300 per kg.

Some 20 companies make silicon wafers and 1000 fabs (and hundreds of university labs) process devices on those silicon wafers. Silicon wafers are tailored to customer needs, and a wafer manufacturer has thousands of wafer specifications. Sometimes customer-specific features are crucial for device performance, like SOI device layer thickness in scaled-down SOI MOSFETs or interferometer optical devices; sometimes it is customized gettering, or novel back-side particle specification or edge exclusion minimization, and sometimes it is just convenience, like specialized markings for wafer tracking or non-standard flats.

In order to ensure wafer supply, device manufacturers have a second source of wafers. This second-source supplier may supply for instance 30% of wafers, and in case the primary supplier has problems, the second source is expected to boost deliveries, so this 30% may not present more than 20% of capacity of the second source. Obviously the more specialized the wafers used, the more difficult it is to find a second source, and in order to be the first on the market, it may pay to rely on a single innovative wafer supplier.

The annual usage of $5\,\text{km}^2$ of silicon wafers translates to roughly 200 million wafers. The 300 mm wafers account for about 30 million wafers, 200 mm for 70 million, 150 mm for 60 million, 125 mm for 25 million and 100 mm for 10 million. Figure 37.1 shows the general trends of wafer size usage over decades.

Wafer size transitions have been important events in IC history: all tooling has to be upgraded, and every process has to be available and qualified for the new wafer size, including all metrology and testing as well. Wafer sizes used to last for 5–7 years, but more recently it seems that lifecycles are getting longer. The transition from 300 mm to 450 mm has been envisioned to start in 2012, but nothing is certain yet. In 2009 there were 100 fabs running

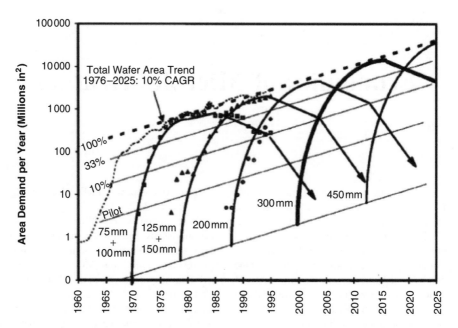

Figure 37.1 Wafer usage evolution over the years. Reproduced from Doering and Nishi (2001), copyright 2001, by permission of IEEE

300 mm wafers. The first ones were built in 2000, and it seems that 300 mm wafers will continue to be used for quite some time to come.

37.2 IC Costs and Prices

Based on a few numbers, like IC sales ($250 billion), silicon area (5 km^2) used and the number of photomasks (700 000 annually), many interesting statistics can be calculated. The average price of a starting silicon wafer turns out to be $40 (about $0.1 per cm^2), and that of processed silicon is $5 per cm^2. The latest generation technologies (e.g., 45 nm CMOS) command higher prices, for example $10 per cm^2, while older technologies are much cheaper, though price depends on volumes, of course.

The photomask industry produces some 700 000 photomasks a year, which translates to 30 000 mask sets with an average price per set of over $100 000 according to data from VLSI Research Corp. Average price is misleading, though, because of the escalation in mask prices: for 0.25 μm CMOS technology a mask set costs $50 000, for 65 nm, $1 million, and for 45 nm, $2 million. A mask set is used to expose 5000 wafers on average, but because R&D uses masks too, production series are in fact longer. The number of IC designs is even larger because some mask sets contain many designs. Additionally, field programmable gate arrays (FPGAs) are customized by users.

If a processed 90 nm CMOS wafer sells for $5000, production runs smaller than 100 wafers (which might comprise 100 000 chips) do not cover even the nonrecurring mask costs. Semicustom chips solve the mask price problem, at least partially: front-end processing, and therefore the transistors, are identical in all products, and chips are customized by a few customer-specific metallization masks. In the best case only one mask is product specific, while all the other masks are shared between many products. Of course, semicustom chips cannot use silicon area optimally, but the cost reduction relative to full custom design is significant, especially for small series products. For very special chips direct write may be an option: with a speed of 1 WPH for electron beam lithography, not many wafers can be processed, but if a handful of chips suffice, direct write makes sense because mask cost is eliminated. There is an analogy in the CD business: if you need 1, 10 or 100 copies, they are burned in a CD player; if you need 1000 or more, it makes sense to fabricate a master and use injection molding.

Many companies have captive mask shops for a variety of reasons. It makes sense to collaborate with a nearby unit in developing new technology, and it may be advantageous to keep these developments secret from outsiders. It is also useful to be able to understand the cost structure

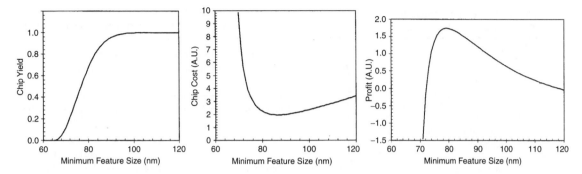

Figure 37.2 Chip yield, chip cost and profit as a function of linewidth. Reproduced from Mack (2007) by permission of John Wiley & Sons, Ltd

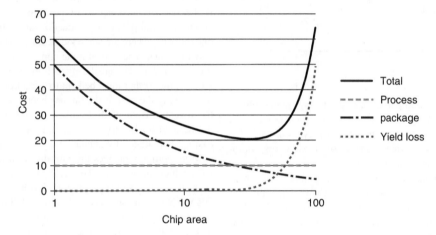

Figure 37.3 Chip size vs. costs: constant process cost; huge packaging cost for small chips, and huge yield loss cost (waste) for large chips

of a supplier: a captive mask shop is seldom the main mask supplier, rather a reference.

Yield and profit are closely related, but somewhat counter-intuitively. As shown in Figure 37.2, chip yield goes down if feature size is scaled down too aggressively. Cost per good chip goes up dramatically when the number of good chips declines. Maximum profit, however, does not coincide with minimum chip cost, because smaller linewidth equals higher performance, and these chips command higher prices. Therefore, in this case, maximum profits are generated at 80 nm linewidth, even though 85 nm minimizes production costs.

Looking further at the cost structure, the cost of silicon chips can be seen to consist of three elements:

- cost of wafer processing (both capital and running costs)
- cost of scrap (yield loss)
- cost of package.

If chip size increases, package cost is reduced because fewer chips need to be packaged, but scrap cost (yield loss) increases with chip size (Figure 37.3). The cost of processed silicon per square centimeter has remained more or less constant for over 30 years, which is remarkable considering the growth in complexity of fabrication processes. This cost always refers to the most advanced, yet established, process technology of its day – older technologies are cheaper.

37.3 IC Industry

There are many levels of actors in any industry. The electronics industry as a whole is worth about $1200 billion annually, and IC production about $250 billion, which means that ICs comprise about 20% of the value of electronic products.

Table 37.1 Equipment industry (2007 and 2008 averages in billions of dollars)

Process equipment:	27
Lithography	6.5
Deposition	5.7
Etch	4.3
Inspection and measurement	3.1
Resist processing	1.8
Surface preparation/clean	1.8
Implantation	1.1
CMP	1.1
Thermal processing	1.1
Other wafer processing	0.3
Assembly and packaging	2.4
Testing	4.3
Fab facilities	1.6
Other	0.9
Total	36

Source: SEMI.

The big IC companies are producing large-volume products: microprocessors, signal processors and memories. A much larger number of players produce RF devices, analog devices, power semiconductors, optoelectronics, sensor interface circuits, etc.

There are many types of IC manufacturers: merchant manufacturers design and produce their own chips for sale. Of the top 20 manufacturers, 16 are merchants. Captive manufacturers are divisions of large companies, and they cater for their internal customers. In the 1980s every major electronics and aerospace company had a captive CMOS division, but today captives usually offer special technologies, not mainstream CMOS. Foundries are companies without their own chip designs: their product is fabrication service. Fabless companies are companies that design chips, but do not own fabrication facilities; foundries make the chips. Only one foundry and three fabless companies are in the top 20 IC companies.

Behind the semiconductor industry is the $30–40 billion equipment industry (Table 37.1; because of the cyclical nature of the semiconductor business, equipment sales experience large fluctuations, especially in 2007 and 2008, for which the latest statistics are available). Materials like resists, gases, cleaning chemicals, sputtering targets, CMP slurries, etc., add another $10 billion. CMP slurries represent a $1.7 billion business, larger than the CMP equipment business.

The IC industry has been growing 17% annually for over 30 years, whereas the electronics industry as a whole grows only 7% annually. For the IC industry to keep growing at its historical rate, the IC content of electronics has to rise at the expense of discrete devices, circuit boards, connectors, displays, switches, keyboards, or else IC growth will slow down. Is it reasonable to expect the proportion of ICs to rise to 30% or to 50% of all electronics, as it is now in handheld devices like MP3 players?

Mainframe computers	1980s	10% of value consists of ICs
Personal computers	1990s	20–30% of value consists of ICs
Handheld devices	2000s	40–50% of value consists of ICs

Measures from IC manufacturing can be used to check if the rate of introduction of novel devices is slowing down. The ramp rate of production at high volumes is one measure. There are some indications that this might be slowing down. The cost of a fab compared to the revenue it is assumed to generate during its lifetime is another. Obviously, the former must be kept to a fraction of the latter, but recently the cost of the fab has been rising faster than the revenue. Both of these measures are tricky because the IC industry is very cyclical, and long-term trends are easily camouflaged by annual or quarterly fluctuations.

More complex devices are introduced at regular intervals, which means that the R&D effort must grow for each successive device generation: development of 1 Mbit DRAM has been estimated to have cost $200 million dollars, and $1.5 billion dollars for 1 Gbit DRAM. So far the market size has grown steadily, which means that there have always been customers for more memory and more processing power, therefore the interval between introductions of new generations has been steady.

37.4 IC Wafer Fabs

IC wafer fabs can be massive: high-volume fabs have monthly production of 50 000–100 000 wafers (wafer starts per month, WPM). In a high-volume fab there are separate furnace tubes for gate oxidation, other dry oxides, wet oxides and polysilicon oxides; in a research lab the division might be gate oxide vs. other oxides, or dry oxides vs. wet oxides. Fabs have plasma etchers dedicated to oxide, poly, aluminum and tungsten. In a university lab with two plasma etchers the division is based on fluorine vs. chlorine processes (or between clean and not-so-clean processes). LPCVD processes have dedicated tubes for poly, nitride and oxides, and this holds for small fabs and labs alike because thin-film interactions would ruin reproducibility. In a research lab one sputtering system can take care of all metal

Table 37.2 Fab investment (millions of dollars) for volume manufacturing (top fab of its day)

1957	0.2
1967	2.5
1977	10
1987	100
1997	1000
2007	3000
2017	?

depositions, but production sputters are dedicated to certain films or film stacks exclusively.

Wafer fab cost has increased exponentially with decreasing linewidth, Table 37.2. Cleanrooms have become more expensive as the size of a killer particle has gone down, but equipment is the most expensive part of a fab. One recent estimate gave capital investment in tools as equivalent to 80% of revenue that the fab is going to generate in its lifetime.

The IC industry is faced with a number of challenging issues, in fab economics, device structures and packaging. Fab cost is not only high, but the amortization times are very short, 5–7 years only. Lithography cost especially is rising very fast, with $50 million price tags for lithography tools. One solution to rising costs and risks is called "copy exactly": the second and successive fabs are exact copies of the first. It is, however, only really applicable to the biggest companies. This approach will mean faster learning curves and higher yields (see Figure 36.4). Previously the thinking was that the first fab is an experiment and improvements are made in each successive fab. But rapid yield ramp is a crucial element in getting to the market early, and constant process and tool changes can be detrimental. Of course, there is no escaping the fact that sometimes fabs with new technology and new equipment have to be built.

37.4.1 Fab operation

Regular wafers are run in lots of 25 or 50 wafers. Cycle time (CT) is the time in days it takes to complete a lot. Process time (PT) is the time that the wafer is actually being processed. Cycle times vary based on process complexity, degree of automation and capacity utilization, but 3–7 weeks is typical. The ratio of cycle time to process time, CT/PT, is a measure of fab efficiency. For standard processing CT/PT is about 2, which means that wafers spend half of the time in a queue and storage.

For batch processes like oxidation many batches are combined, which leads to higher CT/PT. Sometimes a lot is made up of 24 wafers together with a monitor wafer. The monitor wafer is not physically one and the same wafer but an allocation only: in gate oxidation it is a prime wafer which then continues to polysilicon deposition, poly doping and polysilicon etching, and exits after that. A new monitor wafer starts at first interlevel dielectric deposition, and it is then used as a contact hole etch monitor and as the first-metal resistance and step coverage monitor. This monitor is not a prime wafer, but a monitor quality wafer.

In addition to the device and process-specific monitor wafers which are run with the product wafers, a lot of other monitor wafers are running in a wafer fab. These are used in:

- equipment qualification (e.g., after maintenance)
- regular monitoring (e.g., particle tests, film thickness/uniformity)
- process development (e.g., modifying an existing process step)
- short-loop test wafers (e.g. via-chain test).

In the start-up phase of a new fab product wafers may in fact represent less than half of all wafers. Test/monitor wafers are often reclaimed wafers. Reclaimed wafers have been recycled after processing: thin films have been etched away, and the wafers may have been repolished and inspected. Reclaimed wafers have been through various process steps, especially thermal processes, which affect the properties of the wafer bulk, for example oxygen precipitation and wafer curvature. Reclaimed wafers are cheaper choices for non-critical processes: as thin-film thickness monitors, as equipment qualification wafers, or as particle test wafers.

CT and PT are intimately coupled to batch vs. single wafer tool combinations in a fab. Most front-end processes are batch and most back-end processes single wafer. For batch processes PT is "overhead + batch time," which is fairly constant, but for single wafer processes PT is "overhead + lot size × single wafer time," and lot size has a major effect. All single wafer fabs have been experimented with and record CTs of three days have been demonstrated for 0.25 μm CMOS. There are no single wafer fabs running in volume production, but in order to reduce the risks associated with billion-dollar fabs, the minifab concept has been created. Minifabs are low-volume fabs with mostly single wafer and some small batch equipment (batch size of 25 wafers in thermal processes vs. 200 wafer furnace batches in high-volume fabs). Such minifabs are expected to be more agile because CTs will be shorter, and production scheduling is going to be more flexible.

Other ways to reduce CT include lot status and priority classification schemes. Hot lots (or rush lots) are priority lots that receive preferential treatment in the fab. When a hot lot arrives at a process tool, it is processed in front of the queue. Hot lot CT may be 30% less than that of a regular lot. Super hot lots (or bullet lots) are even more prioritized: process equipment is reserved for the super hot lot so that it can be processed as soon as it arrives. For a super hot lot, CT/PT is thus 1. The catch is that not too many such lots can run at any one time, otherwise production scheduling goes berserk.

Yields of hot lots tend to be consistently better than those of standard lots. This can be explained by a simple particle deposition model: hot lots spend less time in the wafer fab, and there is less time available for particles to be deposited on the wafers.

37.5 MEMS Industry

The MEMS industry is worth roughly $8 billion a year and products employing MEMS devices sell for about $100 billion. MEMS are a much more poorly defined concept than ICs, and different sources give different values. Table 37.3 lists major MEMS products.

MEMS fabs use some 4 million 150 mm wafers a year and smaller amounts of 100, 125 and 200 mm wafers. For bulk micromechanics scaling to 200 mm is not an attractive option, because through-wafer etching of thick wafers wastes area. In surface micromechanics wafer size increases productivity just as in ICs, and waveguide optical microsystems are fabricated on 200 mm wafers because the chips are large due to large radii of curvature. Besides, IC companies are abandoning old 200 mm wafer fabs, and those are converted to MEMS fabs.

Accelerometers and microphones are typically square millimeters in size, but a 96-channel capillary electrophoresis "chip" takes up a whole wafer. Calculation of MEMS chip costs takes into account the following: wafer usage, industry turnover and silicon share of MEMS device cost. A typical cost breakdown could be 30/30/30/10 for silicon, ASIC, package and testing. If silicon cost is taken as 30%, then MEMS silicon price is about \$3 per cm^2, roughly the same as that of ICs. While fewer mask levels are used in MEMS, use of epi, SOI and capping wafers, and low-throughput DRIE, push up MEMS prices.

Annual production volumes of many MEMS chips are in the 1000 to 100 000 range, corresponding to 0.01–1% of monthly production of a big IC fab. These chips include various aerospace, biomedical and instrumentation system applications where MEMS are providing a small, but often essential, element which makes the product distinct. For instance, pacemakers have silicon accelerometers, for detecting physical exercise, but the market is small both in the number of devices and for the money involved.

The automotive industry has long been the driving force for new MEMS devices. Pressure sensors are used for power train management, tire pressures and fuel economy applications. Accelerometers and gyroscopes are used for passive safety: crash sensing, rollover detection for launching airbags and tensioning seatbelts; as well as for active safety: electronic stability control (ESP), antilock braking (ABS) and traction control (TCS). The same devices also contribute to ride comfort via electrically controlled suspension (ECS). Single axis accelerometers sell for a few dollars and three-axis versions for a little bit more. Some 100 million units are supplied annually to the automotive industry.

Consumer electronics requirements are quite different from automotive ones, and cost pressures much more severe. About a billion cell/mobile phones are made annually, and many of them contain silicon microphones and FBAR/BAW duplexers. The total cost of hardware for the cheapest cell phones is less than $10, and that includes the display, PCB and keyboard in addition to ICs and MEMS. Therefore it is no wonder that MEMS microphones for cell phones sell for 50 cents (and 300 million are sold annually). Accelerometers for display orientation sensing are also sold by the hundreds of millions. Their prices are similar to microphones. More elaborate orientation sensing is needed in gaming and virtual reality, and digital camera image stabilization systems rely on MEMS gyros.

If RF switches, varactors, local oscillators, delay lines and resonators could be produced with high enough quality factors and good temperature stability, there would be a huge market. For instance, a five-band cell phone has some 20 RF components, and even if only a few of

Table 37.3 MEMS market (millions of dollars) by products, 2010 estimate

Ink jet heads	1610
Accelerometers	1122
Microfluidics	1051
Pressure sensors	1041
Gyroscopes	945
Microdisplays	746
RF MEMS	499
MOEMS	270
Microbolometers	254
Microphones	193
Microtips and probes	134
Others	103
Total	7968

Data from Yole Development.

those were replaced by RF MEMS devices, the market would be a billion devices a year. Many new devices are being developed for the cell phone market: namely, microzoom optical devices for cameras, and compasses and barometric pressure sensors for sports models.

Microfluidic/BioMEMS devices have potentially large markets, if they can be made cheap enough for disposable applications, like point-of-care measurements in health monitoring where $10 might be a reasonable price for a test, which translates to a dollar or two for the bare chip.

37.6 Flat-Panel Display Industry

Worldwide production of flat-panel TVs is about 100 million units, but they are outnumbered by computer displays, and small displays for MP3 players and cell phones are made by the billions.

Flat-panel display fabrication deals with large substrates. Glass panes of 2160 mm × 2460 mm, or 5.2 square meters, are being introduced. If TVs are made, six or eight fit on a pane, but if displays for cell phones or digital cameras are produced, hundreds can fit on a single "mother pane." Photomasks are large too, though smaller than mother panes, and step-and-repeat (Figure 10.2) is usually used (and just as in the IC industry, scanners are also available). In contrast to IC step-and-repeat, however, flat-panel steppers are 1× machines, not reduction optical tools. 42 inch masks cost tens of thousands of dollars (and a mask set for a five- or seven-mask process hundreds of thousands). Small displays can be made with substantially cheaper 6 or 7 inch masks (which can hold for example four or six displays). They are getting more expensive, however, because for instance AMOLED displays require 1–2 μm linewidths and more mask levels: 10 today and more in the future.

The number of lithography steps for LCD TFT is typically five. Some companies have even reduced this, sometimes even without losing performance, because polysilicon mobility has been improved simultaneously. Fabrication of a five-mask TFT process is shown in Figure 37.4. The process flow details the process steps.

Process flow for five-mask bottom gate PMOS TFT

- Buffer layer: 100 nm CVD oxide/nitride
- Metal gate sputtering
- Gate patterning, mask 1
- Gate oxide CVD
- a-Si:H 50 nm by PECVD
- a-Si patterning, mask 2
- Anneal hydrogen out
- Crystallization (by laser)
- S/D doping
- Activation anneal
- S/D metallization
- S/D patterning, mask 3
- Passivation film
- Contact patterning, mask 4
- Hydrogenation
- Transparent electrode deposition
- ITO pattering, mask 5

Gate oxide thickness, which is a critical MOS transistor parameter, is not aggressively scaled in TFTs. CVD oxide quality is inferior to thermal oxides, and for example breakthrough voltages are 3–5 MV/cm vs. 10 MV/cm for CMOS gate oxides. Roughnesses of polysilicon (in top gate) or sputtered chromium (in bottom gate) are not amenable to very thin oxides.

Linewidth scaling and mobility improvement both contribute to greater integration possibilities: it is now possible to fabricate driver transistors on the TFT panel itself, instead of using separate driver ICs, which must be mounted on the edges of the glass panel. With an integrated SRAM pixel memory the power consumption of a LCD can be reduced by two orders of magnitude. LCD backlight efficiency can be improved by microlenses that are fabricated and aligned to pixels, to focus the light to the pixels, instead of uniform illumination. It is also possible to integrate polysilicon ambient light sensors which control backlight intensity. Both techniques reduce power consumption, which is a critical issue in portable devices.

Direct writing is being considered for both displays and solar cells. Patterning processes using lithography and laser direct write are contrasted in Figure 37.5. Laser tool cost is in the $10 million range, which may sound expensive, but the lithography tool price for 42 inch displays is also very expensive. The resist dispensing tool, and development, etch and stripping wet benches for large panes, are costly, too. Additionally, chemical consumption in the wet processing of 5 m^2 panes is formidable. Throughputs of laser systems are highly variable, depending on film thickness and etch/ablation rates. There is always the possibility to use parallel laser beams, and in fact eight-beam systems exist. Writing times of 10–100 seconds per 42 inch display, or 100–1000 seconds per mother pane, make laser systems competitive in certain applications, like plasma display ITO electrode patterning.

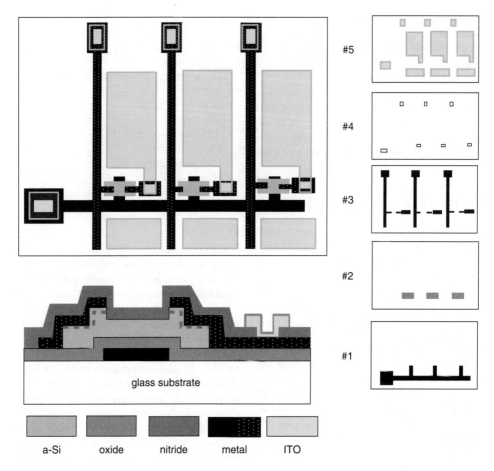

Figure 37.4 Bottom gate TFT process: top view, cross-section of a single transistor and the five photomasks. Adapted from Ukai (2007)

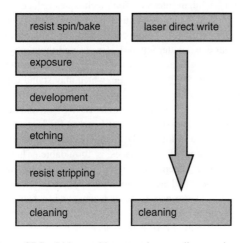

Figure 37.5 Lithographic patterning vs. direct write

37.7 Solar Cells

The solar cell industry (also known as PV, for photovoltaics) is still dominated by crystalline silicon cells (Figures 1.14, 24.9, 25.2) as of this writing, but thin-film solar cells (Figure 24.10) are rapidly catching up. In 2008 multicrystalline and single crystalline silicon accounted for over 80% of all solar cells.

Solar cells are the ultimate low-cost devices. CZ silicon can be used for solar cells, but multicrystalline (MC) silicon is dominant. It is made by casting 200 kg of electronic-grade silicon into a crucible and solidifying from the bottom up. The crucible is then broken and the silicon ingot sawn first into 25 rectangular 5 or 6 inch blocks, which are wire sawn to square wafers. This is good for filling space: round wafers leave empty spaces between them. Solar wafers are thin (e.g., 200 μm) in order to make more wafers out of the same silicon ingot. MC wafers cost one-tenth of similar-sized CZ wafers.

There is intensive research ongoing to develop cheaper silicon purification methods. The chlorosilane gas phase purification with distillation and CVD conversion described by Eqs. 4.2–4.4 is extremely efficient but also energy consuming. Starting from upgraded metallurgical-grade silicon (UMG) which has typically metallic impurities at 10 ppm levels, a simple plasma torch method can purify that to sub-ppm levels.

Microfabrication technologies are used for performance, like thermal oxidation for passivation and PECVD nitride antireflective coatings, but cheaper techniques like phosphoric acid diffusion sources and screen printing of conductive pastes are used for cost minimization. Another reason for screen-printed metallization is current density: it is easy to screen-print thick conductors (10 μm), while PDV metals are limited to a micrometer or so.

Light capture can be improved by nitride antireflection coatings (Figure 25.2). Texturing can also be used to reduce reflections: in Figure 1.14 lithography and KOH etching were used to create inverted pyramid structures, but random pyramids formed by non-lithographic KOH etching serve almost as well, and eliminate photomask, resist processing and lithography equipment cost. Figure 37.6 shows a cross-section of an interdigitated back contact cell. It has 100% collection efficiency on the front side because all metallization is on the wafer back side. The limitation with this design is the need for high-quality silicon, because the charge carriers have to diffuse all through the wafer before they are collected at the electrodes.

Solar cell linewidths are relaxed, in the 50 μm range. This enables cheap mask technologies and screen printing. Some solar cells are made completely without lithography (Buried collector cell, Figure 24.9, is laser patterned). Another lithography reduction technique is to fire metal through a dielectric: no contact holes are defined, but the metal is patterned. A firing (annealing) step drives the metal through the dielectric where metal lines are patterned. Yet another way to eliminate lithography is to make point contacts by laser doping and to contact those by the laser firing of metallization.

Laser processing has found various other applications in solar cell production: laser grooves (Figure 24.9) enable narrow (and deep) metallization which minimizes front surface area lost to metallization. Diffusion is an immersion process and the edges of the wafer are also doped. This creates a current path that has to be eliminated, and laser cutting is used routinely in this edge isolation.

Light absorption can also be increased by fabricating tandem and multiple cells: using different materials in combination can increase the use of the spectrum: one layer is more sensitive to IR wavelengths, the other to visible light. $Ga_{0.35}In_{0.65}P/Ga_{0.83}In_{0.17}As/Ge$ triple junction solar cells (similar to Figure 6.2) have shown 40% efficiency when run under 400× concentrated sunlight. Single junction crystalline silicon cells can reach 22% efficiency at best (the theoretical limit for a single junction cell is 29%), multicrystalline silicon cells have record efficiencies around 17% and thin-film cells around 10%. Note that cells may show 22% efficiency for 1 cm^2 area when fabricated in a research lab, but when the same device is mass produced on large wafers, using cost-optimized tools and processes, the efficiency is probably only 17%.

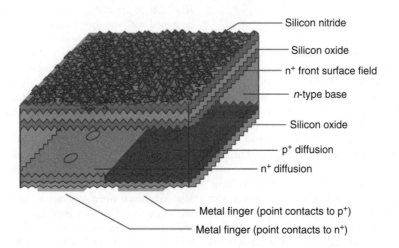

Figure 37.6 Interdigitated back contact cell. Reproduced from Neuhaus and Münzer (2007) (Creative Commons)

While IC makers can differentiate their products by adding new features, and in memories speed can be used to command a higher price, all solar cells produce exactly the same final outcome, power. There is very little in competition except cost per watt. This can improved by increasing efficiency, through better optical capture of photons, higher conversion to charge carriers, and reduced losses of those carriers before reaching the electrodes. These translate to light confinement structures and low-loss cover glass.

The main metric for solar cells is the cost of a cell for peak wattage (W_p). This approached 1\$ per W_p in 2009, down from \$2 in 2002 and \$8 in 1992. The solar electricity system cost is roughly double the cell cost: it includes mechanical structures, inverters, cabling, possibly batteries, etc., and these are highly system size dependent. Solar irradiation on the Earth's surface is about $1\,kW/m^2$ under the best conditions and 20% conversion efficiency then translates to $200\,W/m^2$. But the sun does not shine 24/7, and 3000 hours per year is a reasonable estimate for sunny climates, while much less is available further north. Assuming a lifetime of 20 years (different sources assume 15–25 years), the cost of solar electricity is 5–10 cents per kWh. The cost of capital must also be added and annual maintenance costs are estimated to be 1% of investment cost. This is a very rough calculation and it is valid for dollar cents and euro cents alike.

Coal-fired power plants can produce electricity at about 5–10 cents per kWh but local tariffs for consumers include grid costs and taxes, bringing the price up to 30 cents per kWh, making solar power already economically viable in some regions. If the trends of solar cell performance are any guide, solar electricity should become competitive with coal-fired electricity around 2013–2017.

37.8 Magnetic Data Storage

Magnetic data storage uses microfabrication techniques for many of its devices. The write head is an inductive microfabricated coil, and the read head is a complex magnetic multilayer stack (Figure 7.19). A MEMS positioner (Figure 30.22) keeps the read/write head on track. Scaling of storage density in the last 50 years has been impressive: in 1960 it was $10\,kbit/in^2$, in 1980 it was $10\,Mbit/in^2$, in 2000 it had reached $1\,Gbit/in^2$ and in 2010 it is $100\,Gbit/in^2$.

Thin-film read/write head (TFH) fabrication for magnetic data storage shares surprisingly many aspects with IC fabrication, especially the steady growth in the number of process steps, the number of thin films (up to 20) and the steady (and very steep) decrease in linewidths: from 1990 to 2000 the minimum linewidth in TFH fabrication

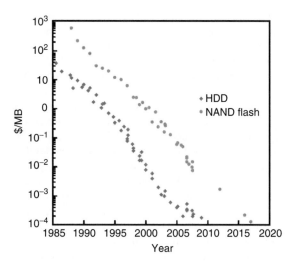

Figure 37.7 Memory bit cost development, hard disk drive (HDD) vs. NAND flash memory. Reproduced from Burr (2008) by permission of IBM Technical Journals

came down from 5 to $0.5\,\mu m$, and in 2010 it is equal to IC linewidths at around $50\,nm$. This means that hard disk drive memory density increases faster than semiconductor memory density. This linewidth reduction goes hand in hand with cost/bit price development, which is pictured in Figure 37.7. The cost of a stored bit on a hard disk drive is about 1–10% of that of flash memory storage.

The thin-film head consists of two parts: a giant magnetoresistance (GMR) sensor read head and an inductive coil write head (Figure 37.8). Film thicknesses in GMR heads are very demanding: the thinnest films are about $1\,nm$ thick. The whole stack consisting of 10 layers totals less than $50\,nm$ thick. Sputtering systems for TFHs are equipped with multiple targets. What is more, each one must have approximately the same deposition time, otherwise the in-line system would be choked by one thick layer. Instead of increasing deposition rate, three identical modules can be used, to deposit a third of the thickness, for example $5\,nm$ each. All the films must be done in one sputtering system because atmospheric oxygen would destroy subnanometer films.

Thickness control is, however, only part of the problem with GMR heads. Interface sharpness needs to be maintained during the 10 year lifespan, so no mixing is allowed during either processing or use. Smoothness is also paramount when nanometer and subnanometer films are made. As discussed previously, thin-film properties are thickness dependent, and for example the PtMn antiferromagnetic layer has to be at least $15\,nm$ thick in order to have the desired pinning effect. New materials

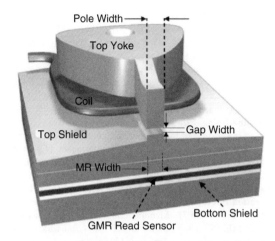

Figure 37.8 The read/write head of a HDD, with GMR read head and inductive coil write head. Reproduced from Childress and Fontana (2005), copyright 2005, by permission of Elsevier

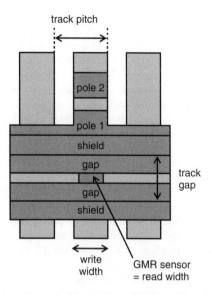

Figure 37.9 Three lanes of data with the head overlaid. Redrawn from Childress and Fontana (2005)

will probably be needed when thinner layers are required in the future.

The density of magnetic recording depends on two factors: track pitch, or how closely neighboring tracks can be written, and GMR track gap, or how thin the magnetic read head can be made (Figure 37.9). The actual magnetic structure is about 30 nm thick, but shields on both sides take up space. The sensor height is taken as half-pitch.

Production volumes for magnetic memory disks are huge: about a billion per year. This is reflected in throughput requirements for equipment: 100 WPH would be good for a IC sputtering system, but in magnetic media disk production 1000 WPH is more typical. Fortunately, the magnetic layers are rather thin, with a total thickness below 200 nm for perpendicular recording, as was shown in Figure 7.18.

One future development in magnetic media is expected to be patterned media: each bit is stored in a magnetic area of its own, eliminating interference from neighboring bits which are present in continuous film. Currently there are some 60 grains for a bit, which equals a grain size smaller than 10 nm. A leading candidate for patterning the disks is NIL (see Chapter 18). Resolution of NIL is extremely good: as long as a master can be obtained, NIL can stamp even 5 nm patterns. In patterned media applications no alignment is required, which reduces the cost of a NIL system considerably. But the master fabrication remains problematic: writing a terabit (10^{12}) master may take a month with an electron beam.

37.9 Short Term and Long Term

Semiconductor demand is inelastic in the short run, and this explains some counter-intuitive aspects of price swings: often prices go up hand in hand with volumes. When the PC or cell phone market is strong, demand surges and chip prices go up because ramping up chip production takes time. Downswings show inelasticity as well: when for example memory chip prices go down, there is no sizable upsurge in demand. This price reduction can be due to the fact that the chip is nearing the end of its lifecycle, and nobody wants to increase usage of a soon-to-be-extinct chip.

But on a more fundamental level it has more to do with long design lifecycles: it is not possible to rapidly switch chips on a whim, therefore a certain inertia is built into the system. These are short-term effects, but in the long run costs and prices do follow predictable trends, like memory cost per bit and solar cell peak watt cost falling 25% annually.

37.10 Exercises

1. The investment for a large-volume CMOS wafer fab was $1 billion (year 2000, 0.25 μm technology, 200 mm wafer size). Fab running costs are $1 million a day. Assuming 30 000 wafer starts per month (WPM), what is the cost of the finished silicon?

2. Calculate the mask cost contribution to silicon area price if 0.25 μm CMOS with 25 photomasks at $3000 per mask plate is used and each mask set is used to fabricate 50/500/5000/50 000 wafers?
3. Maskless lithography by direct writing is expensive because it is very slow, but there is no photomask cost. Assuming identical capital investment ($10 million) and running costs ($1 million a year) for both optical and direct write lithography systems (crude approximations), and 100 WPH for optical and 1 WPH for DW on 300 mm wafers, what would be the number of wafers where DW becomes competitive with optical lithography for 90 nm CMOS if the mask set cost is assumed to be $500 000?
4. If a multilevel metallization process consists of 10 layers, how many oxide plasma etchers are needed to etch contact holes if the oxide etch rate is 1 μm/min, in a 100 000 WPM fab?
5. How many LPCVD tubes are needed in a 30 000 WPM fab if the process has five polysilicon layers?
6. If we assume an average big fab to have 50 000 WPM capacity, how many such fabs are there in the world?
7. How many silicon crystal pulling systems are there in the world?
8. What is the price per kilogram (or carat price) of thin-film diamond if PECVD capital cost is $500 000 and running costs are $100 000 a year? Take 10 nm/min for the deposition rate on 150 mm wafer size in a single wafer system.
9. What is the bit density of the system shown in Figure 37.9.
10. TFT itself takes up very little area compared to pixels, and the transistor packing density increase offered by self-alignment is not important. What are the benefits of self-alignment in TFT fabrication for displays?

References and Related Reading

Beaucarne, G. *et al.* (2006) Epitaxial thin-film Si solar cells, *Thin Solid Films*, **92**, 511–512 533–542.

Berglund, C.N., C.M. Weber and P. Gabella (2009) Benchmarking the productivity of photomask manufacturers, *IEEE Trans. Semicond. Manuf.*, **22**, 499–506.

Burr, G.W. (2008) Overview of candidate device technologies for storage-class memory, *IBM J. Res. Dev.*, **52**, 449–464.

Childress, J.R. and R.E. Fontana (2005) Magnetic recording read head sensor technology, *C.R. Physique*, **6**, 997–1012.

Degoulange, J. *et al.* (2008) Multicrystalline silicon wafers prepared from upgraded metallurgical feedstock, *Sol. Energy Mater. Sol. Cells*, **92**, 1269–1273.

Doering, R. and Y. Nishi (2001) Limits of integrated circuit manufacturing, *Proc. IEEE*, **89**, 375.

Garg, D. *et al.* (2005) An economic analysis of the deposition of electrochromic WO_3 via sputtering or plasma enhanced chemical vapor deposition, *Mater. Sci. Eng.*, **B119**, 224–231.

Hamakawa, Y. (ed.) (2004) **Thin-Film Solar Cells: Next Generation Photovoltaics and Its Applications**, Springer.

Lawes, R.A. (2007) Manufacturing costs for microsystems/MEMS using high aspect ratio microfabrication techniques, *Microsyst.Technol.*, **13**, 85–95.

Leonovich, G.A. *et al.* (1995) Integrated cost and productivity learning in CMOS semiconductor manufacturing, *IBM J. Res. Dev.*, **39**, 201.

Liehr, M. and G.W. Rubloff (1994) Concepts in competitive microelectronics manufacturing, *J. Vac. Sci. Technol.*, **B12**, 2727.

Mack, C. (2007) **Fundamental Principles of Optical Lithography**, John Wiley & Sons, Ltd.

Neuhaus, D.-H. and A. Münzer (2007) Industrial silicon wafer solar cells, *Adv. OptoElectron.*, 24521.

SEMI, Semiconductor Equipment and Materials International, http://www.semi.org.

Saito, M. (2009) Global semiconductor industry trend VIDM versus foundry approaches, *Proc. IEEE*, **97**, 1658–1660.

Suzuki, T. (2006) Flat panel displays for ubiquitous product applications and related impurity doping technologies, *J. Appl. Phys.*, **99**, 111101.

Terris, B.D. and T. Thomson (2005) Nanofabricated and self-assembled magnetic structures as data storage media, *J. Phys. D: Appl. Phys.*, **38**, R199–R222.

Ukai, Y. (2007) TFT-LCD manufacturing technology - current status and future prospects, International Workshop on the Physics of Semiconductor Devices (IWPSD), pp. 29–34. VLSI Research: https://www.vlsiresearch.com/more/CompanyBackground.html?rdr=about_vlsi

38

Moore's Law and Scaling Trends

This chapter deals with the past, present and future of silicon integrated circuits, concentrating on CMOS logic and memory, which are the driving force of scaling to smaller linewidths and higher device densities. Device physics, materials, fabrication processes, reliability issues and manufacturability are discussed, along with trends, limits, opportunities and threats to continued scaling.

Early transistors could be made with just five elements, Si, B, P, O, Al; fabrication of 0.18 μm CMOS uses 14 elements. In addition to the aforementioned elements, N, As, Ti, W, Co, Ta, Cu, C and F are used. Some of these materials have been adopted easily, like the addition of a TiW barrier underneath aluminum, or n-type doping by arsenic instead of phosphorus, which were implemented with the same equipment as the older versions. Introducing tungsten plugs necessitated a novel CVD equipment, which was a major departure from previous metallization schemes which were all based on PVD techniques. CMP was another paradigm shift, a completely new way of doing things, not an evolution of SOG and etchback planarization. Copper metallization was again a major transition, but because of the experience with oxide and tungsten CMP, its adoption was made easier. New barrier metals had to be invented, though, to eliminate contamination risks.

Device structures have been intimately involved with new materials and tools. The CMOS self-aligned polysilicon gate was a major change, because both gate material and doping technology changed. Silicides were also profound changes because new material and new process technology, namely rapid thermal annealing, were introduced hand in hand. Strained silicon wafers, deposited gate oxides like HfO_2 and metal gates are the current major changes being implemented. Taken together, these developments, both revolutionary and evolutionary, have contributed to the realization of microchips with a billion devices.

38.1 From Transistor to Integrated Circuit

Transistor fabrication in the 1950s was crystallography and metallurgy, not microfabrication. Junction formation was an alloying process that does not share many features with modern transistor fabrication. Pallets of indium, a p-type dopant, were attached to both sides of an n-type semiconductor piece, the diffusion step was performed, and metal wires attached to the two p-type and one n-type regions; *voilà*, the pnp transistor was ready.

The modern key concepts of microfabrication, namely diffusion masking by an oxide layer, photolithographic patterning, wet etching of the oxide and the use of evaporated aluminum as a conductor, emerged in the mid 1950s mostly at Bell Laboratories and at Fairchild Semiconductor. These techniques were put together in what is known as the planar process for transistor fabrication, by Jean Hoerni.

The integrated circuit was invented twice, simultaneously and independently. Jack Kilby of Texas Instruments demonstrated integrated circuits in 1958 and filed for a patent in early 1959. However, Kilby used germanium transistors and gold wire bonds for connecting the devices. Six months later Robert Noyce at Fairchild based his invention on the planar process, using evaporated aluminum for metallization and silicon dioxide as an insulator, and realized the first device that became the forerunner of current ICs.

There were many objections to ICs at the beginning of the 1960s, as Jack Kilby (1976) reminiscences:

1. Electronics designs would become hard to change once the circuits had been etched onto silicon.
2. Electronics engineers would be out of work because all design would shift to IC manufacturers.
3. Transistors are low-power devices which are suitable only for some special applications.

Introduction to Microfabrication, Second Edition Sami Franssila
© 2010 John Wiley & Sons, Ltd

4. ICs do not use optimum materials: NiCr resistors are better than silicon resistors, and Mylar capacitors are superior to oxide capacitors.
5. The yield of transistors is low (e.g., 80%) and if, say, 20 of them are made on a single chip, the combined yield will be miniscule.

Argument number 1 still holds today: custom circuits especially take a long time to design: one year is typical, and changes are hard to make. This is, however, a small price to be paid for the enormous gains in speed and functionality. We now know that argument number 2 was groundless as ICs propelled the electronics industry into supergrowth.

Argument 3 was wrong; some people had seen it already in the 1950s: Bob Wallace of Bell Labs stressed:

> Gentlemen, you've got it all wrong! The advantage of the transistor is that it is inherently a small-size and low-power device. This means that you can pack a large number of them in a small space without excessive heat generation and achieve low propagation delays. And that's what we need for logic applications. The significance of the transistor is not that it can replace vacuum tube but that it can do things that the vacuum tube could never do! (Ross 1997)

Many MEMS and nanodevices today are miniaturized versions of existing devices. Sometimes smaller size is useful because it results in, for example, smaller power consumption or higher speed. But it is equally important to look for new applications where new physical phenomena, new combinations of speed and power, can be utilized, or where macroscopic counterparts do not exist, or where the scale economies of microfabrication have not yet been utilized.

The whole can be more than the sum of its parts. The very concept of integration seems to have escaped the attention of supporters of argument number 4. ICs paved the way for more powerful electronic systems. And the savings in assembly costs quickly more than compensated the higher cost of ICs.

Argument 5 was mathematically valid, but it was based on the technology of its day, and it did not anticipate the tremendous strides in microfabrication technologies. The success of ICs has been dependent on the fact that, in spite of continuous miniaturization and complexification of the manufacturing process, the yield of individual transistors on ICs has improved dramatically. In 1960 the yield of 50% for individual devices resulted in a 3% yield for a five-transistor IC. Today, in 2010, chips with a billion devices are manufactured with about 90% yields, which translates to practically 100% yield for individual devices, and this is for 32 nm devices, compared to the 30 μm devices of 1960. Integrated microfluidic systems face a similar situation: while pumps, valves and mixers may have reasonable yields, systems consisting of many such devices have rather low yields.

The early proponents of the IC had to balance between two options:

1. Suitable only for price-insensitive applications like military or space technology.
2. Will be cheap in the future once the technology matures.

Early growth was of course along the first argument because somebody had to pay for the chips, but at the end of the 1960s the second argument was finally realized, and the IC became a household term.

38.2 Historical Development of IC Manufacturing

In addition to scaling the lateral and vertical dimensions, a multitude of other refinements have taken place in IC manufacturing during the past 50 years. These involve new materials for both metallization and dielectrics, new equipment designs, new control measurements and inspections tools, new contamination control strategies as well as new devices, Table 38.1.

Lithography has evolved from $1\times$ contact/proximity printers to $1\times$ projection tools to $5\times$ step-and-repeat systems to $4\times$ step-and-scan machines. Batch wet etching has been replaced by single wafer plasma etching. Thermal diffusion has been replaced by ion implantation. Some processes have remained fairly unchanged, like wet cleaning and thermal oxidation. The industry has been quite conservative, with very few radical changes in any one technology generation.

In the 1960s practically everything in IC production was proprietary: the product, process flow, process recipe and even process equipment were developed in-house and kept as trade secrets. Then equipment manufacturers emerged and started offering everybody the same tools. The recipe remained proprietary. Then equipment manufacturers started selling not only the equipment but also the recipe. Some equipment manufacturers are even selling uptime, not hardware but availability. Little by little as the technology matured, even process flow became a commodity, and only the product differentiated manufacturers. This opened up the market for both

Table 38.1 Historical development of IC processes

1960s–1970s

30 µm to 3 µm linewidths
Proximity and projection 1× lithography at
 $\lambda = 436$ nm
Fewer than 10 lithography steps
Wet etching
Doping by furnace diffusion
Batch processing
(Pure) aluminum metallization; one level of metal
Si, O, N, P, B, Al needed
Wafer size increase from 1 inch to 3 inches

1980s

3 µm to 1 µm linewidths
Step-and-repeat lithography at $\lambda = 365$ nm
 introduced at 1.2 µm
10–15 lithography steps
Plasma etching replaces wet etching for critical
 steps
Ion implantation for doping
Single wafer equipment emerging, first in plasma
 etching
Two levels of metallization
SOG and resist etchback planarization
Silicides introduced
New elements: As (n-doping), Cu (in Al:Cu alloy),
 Ti, W (in TiW barrier)
100/125/150 mm wafer size

1990s

Linewidths 1 µm to 0.25 µm
20–25 lithography steps for advanced CMOS
High-density plasma (HDP) equipment for etching
 and deposition
W plugs by CVD with TiN barrier
CMP oxide planarization
Cu metallization introduced in damascene structure
Number of metal levels increasing up to seven in
 logic circuits
150/200 mm wafer size

2000s

Linewidths 0.25 µm to 45 nm
30 lithography steps for advanced CMOS
Step-and-scan lithography with $\lambda = 248$ nm
 introduced at 0.25 µm
Phase shift masks (PSMs) adopted at 0.18 µm
New elements: Co (in $CoSi_2$), F (in SiOF), Ta (in
 TaNSi barrier for Cu)
Copper becoming standard for high-performance
 circuits
Low-k dielectrics introduced in multilevel
 metallization
200/300 mm wafer size

Table 38.1 (continued)

2010s

Linewidths 32, 22 and 15 nm
Lithography DUV immersion 193 nm, maybe EUV
High-k deposited gate dielectrics with EOTs <1 nm
Metal gates, different for NMOS and PMOS
Copper metallization in commodity chips
Vertical and multigate devices
Strain engineering
Over 10 levels of metallization in logic circuits
Wafer size 300 mm, maybe transition to 450 mm

foundries and fabless companies. Now even the product is beginning to be commoditized: companies are selling IP blocks (for intellectual property); designs can be traded.

The commoditization of process flow has been carried furthest in CMOS, and foundries are CMOS houses. The cost of process development increases with every technology generation, and companies form joint ventures to share the development risk. This reduces the number of independent CMOS process flows. There are differences, of course: bulk and SOI processes, aluminum and copper metallization, cobalt and nickel silicides. The same linewidth CMOS can often be found both as a low-power version for battery-powered operation and as a high-performance version for the desktop. The main difference is in gate oxide thickness: slightly thicker gate oxide in the low-power version (2.4 vs. 1.3 nm in 130 nm CMOS) reduces leakage currents 100–1000-fold, while maximum frequency goes down by only 50%.

Adding features like analog capability (resistors, capacitors, inductors), bipolars (for RF operation at very high frequencies) and embedded memory for programmability is done in a modular fashion: the basic CMOS process flow is kept as it is, and the additions are implemented in modules. Systems with embedded memory cannot be optimized for both logic and memory, so, depending on whether the system is memory or logic dominated, sacrifices are made in the minority part, and performance is optimized according to the requirements of the majority device type.

38.3 MOS Scaling

Linewidth scaling has been very predictable, and it captures most of the public's imagination of scaling, together with memory capacity increase. But practically every aspect of CMOS is being scaled and modified constantly. The schematic of Figure 38.1 shows the main elements of

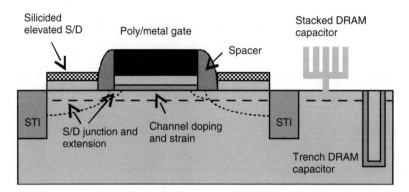

Figure 38.1 Schematic of a modern transistor

MOS and DRAM. Aspect ratios are increasing in DRAM: both stacked designs and trench designs call for extreme aspect ratios. Shallow junctions necessitate small thermal budgets, and selective epitaxial elevated source/drain areas need ultimate surface preparation control.

Gate oxide thickness used to relate to gate length as about $L/50$ for a long time. Oxide and later nitrided oxides were used, but the end is in sight for two reasons: first, tunneling currents through thin oxides increase leakage to unacceptable levels; second, boron from the doped polysilicon gate diffuses through the oxide and dopes the channel. Deposited oxides like ALD HfO_2 are being implemented, but there is most often silicon dioxide between the high-k oxide and silicon, for example 0.4 nm thick. This is partly inevitable because oxide forms when oxygen is present, and partly it has to do with the interface quality: the scattering of channel electrons at the interface can be minimized by SiO_2, leading to higher channel mobility. Then again, it has to be 0.4 nm in a controllable fashion.

Source/drain junction depths (x_j) were scaled with linewidths as $L/5$ for decades, but more recently it has been difficult to scale down x_j as aggressively as linewidth. Junction formation involves low-energy ion implantation together with annealing. Damage anneal times have become shorter and shorter, with millisecond flash anneals now in use and microsecond laser anneals being studied. In order to minimize diffusion, some non-activated dopants and some remaining damage may have to be tolerated, but this leads to increased leakage currents. In addition to junction depth, junction abruptness is an optimization parameter.

Gate linewidth is considerably less than the technology node might indicate. Lithographic half-pitch, which describes the lithography capability, and gate width are therefore two separate concepts. Narrow gates are made by resist trimming (Figure 10.17). The polysilicon gate is being replaced by metal gates, or, to be more precise, by multilayer polysilicon/metal stacks. This introduces many etching problems: some of the metal halides may not be volatile enough, leading to micromasking and contamination. Gate etch selectivity against the nanometer-thick gate oxide must be extremely high, and multistep etching is essential. In situ interferometry can be used to monitor the remaining gate thickness in real time, so that high-rate etching can be switched to a high-selectivity end point step. The allowed variation for gate linewidth is 12%, but this includes contributions from lithography, resist trimming and gate etching.

With shallower junctions, silicide formation has become limited, with 10 nm silicon consumption reserved for silicide. NiSi is replacing $CoSi_2$, because less silicon is consumed in nickel silicidation. NiSi can also be formed at lower temperatures, with a 13 ohm/sq design value for silicide sheet resistance. Contact resistance at the silicon/silicide interface is a key issue, with a target value for contact resistivity of 7×10^{-8} ohm-cm^2 for a 45 nm technology node.

One solution to the shallow junction contacting problem is to make elevated junctions. This involves selective epitaxial growth on S/D areas. Silicide thickness can then be chosen irrespective of junction depth. Selective epitaxy is a difficult step, however. Another solution to silicide formation is to use SiGe on S/D areas: it is more easier silicided.

In earlier generations pre-gate cleaning was the ultimate cleaning challenge, but now post-gate cleaning is becoming more critical. This cleaning must remove metallic residues from metal gate etching, while not attacking the nanometer-thick oxide. Spacer etching ends against the thin gate oxide again, and preferably it should not consume it. If it does, cleaning before epitaxy for elevated S/D becomes critical because the etching has compromised surface quality.

Oxide and silicon loss in cleaning are limited to 0.3 nm per step in 45 nm node. The old cleaning methods relied on etching away some silicon to undercut contaminants (Figure 12.4) but clearly this can no longer be done. To reduce etching, more dilute solutions are being used. Instead of ammonia/peroxide (SC-1), ammonia/water might be used, and instead of HCl/peroxide (SC-2), dilute HCl clean can be used. Organics removal is based more and more on ozonated water and not sulfuric acid. Photoresist stripping would preferably be done in wet solutions to eliminate plasma damage, but resists are often hardened in plasmas and implantation and must be removed by plasma. However, in addition to oxygen, new plasma chemistries are being developed, and completely new approaches are being explored, like cryogenic aerosols.

In order to implement materials which cannot withstand front-end high-temperature steps, a replacement gate has been formulated (Figure 38.2). Oxide or nitride serves in place of the metal gate during the high-temperature steps. After completion of S/D implant activation anneals, the first dielectric layer is deposited and planarized. The dummy gate is etched away, gate dielectric is deposited, and the final metal gate is deposited (followed by CMP). The replacement gate makes return of the aluminum gate possible, but refractory metals are more likely candidates. The added process complexity is quite considerable, and oxidation/oxide deposition into the groove left by dummy gate etching is by no means easy or straightforward.

With 3 billion transistors in graphics processors, the design complexity is enormous, and the same applies to device testing. CMOS was originally a solution to power consumption: CMOS logic consumes energy only during switching, but the sheer number of devices means that excessive amounts of waste heat are generated in advanced chips. Chip cooling has two elements: hotspot cooling and overall cooling. Waste heat powers are approaching 200 W in high-performance processors, whereas processors for battery-powered devices consume 10 W or even less than a watt.

38.4 Departure from Planar Bulk Technology

Historically, transistor intrinsic speed has increased 17% annually ($\tau = CV/I$, where C is capacitance per micrometer gate length, V the operating voltage and I the drain current). This trend is now coming to an end, as far as planar bulk silicon devices are concerned. For SOI and 3D transistors the trend is expected to continue, maybe until 2020. After that transistor development is expected to continue, but probably the historical pace cannot be continued unless dramatic departures from current MOS offer room for future speed increases.

When MOS transistors are made extremely small, the ability of the gate to control the current in the channel is diminished. This can be overcome if two (or more) gates are used instead of one, as shown in Figure 38.3. Fabrication of these devices is not obvious, and the two-gate version can exist in various configurations, with the gates parallel to the silicon surface or vertical. One approach is FINFET: the channel is formed in an etched piece of silicon, and gate oxide is grown on three sides of the fin. This is a formidable etching problem: the quality of the sidewall has to be good enough so that nanometer-thick oxide can be deposited on it.

Another technology for making multiple gate devices is tunnel epitaxy. The SiGe layer is selectively etched away to create a tunnel, which can be refilled by epitaxial silicon. Selective epitaxial growth (SEG) and epitaxial lateral overgrowth (ELO) processes are used, as shown in Figure 38.4. In addition to the sacrificial dummy channel, dummy gates made of nitride were used originally,

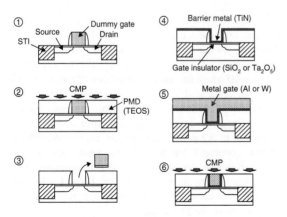

Figure 38.2 Replacement gate process. Reproduced from Yagishita *et al.* (2001) by permission of IEEE

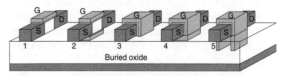

Figure 38.3 SOI MOSFETs with: 1, one gate; 2, two gates; 3, three gates; 4, four gates; 5, extended three gates. Reproduced from Park and Colinge (2002) by permission of IEEE

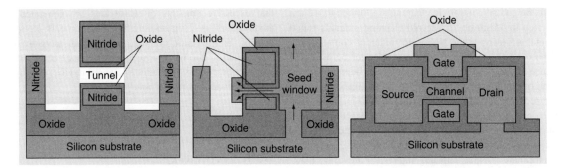

Figure 38.4 Selective epitaxial growth (SEG) to fabricate double gate MOS: epitaxy proceeds laterally in the tunnel once a suitable seeding surface is available. Adapted from Wong *et al.* (1997) by permission of IBM Technical Journals

later replaced by the gate proper. CMP was also needed to planarize after SEG.

The MOS saturation current is given by

$$I_{\text{dsat}} = \frac{WC_{\text{inv}}\mu(V_g - V_{\text{th}})^2}{2L} \quad (38.1)$$

The first term is determined by lithography; oxide thickness determines inversion capacitance; V_t is a doping-level issue; and V_g is related to operating voltage and power consumption. The mobility (μ) is the key to channel engineering. Strain affects charge carrier mobility and there are many ways to exert strain. Global strain can be implemented by SiGe epitaxy: as discussed in connection with epitaxial growth (Section 6.1), $\text{Si}_{1-x}\text{Ge}_x$ alloys have lattice constants larger than silicon and they are under compressive stress, and consequently silicon on $\text{Si}_{1-x}\text{Ge}_x$ will be under tensile stress. This tensile stress introduces an energy split in the conduction band of silicon, which leads to mobility enhancement, for electrons by a factor of two, for holes by a factor or four (depending on germanium content, doping level and field strength). Higher operating frequency can thus be obtained from MOSFETs without lithographic scaling.

Local strain can also be produced by nitride spacers or some other thin films. Using two different nitride deposition steps enables the even tailoring of stress independently for NMOS and PMOS. Wafers of different crystal orientations behave differently with regard to strain, and <110> is considered a candidate material because of its high hole mobility. However, electron mobility in <110> is detrimentally reduced due to strain.

Implementing new features like strain necessitates new measurements. The general outline given in Chapter 2 will apply: the methods must be considered based on many criteria. Spatial resolution for localized strain measurement, strain sensitivity, sample preparation needs (if any) and strain gradient measurement capability must be considered. Many methods are available, with different features: wafer curvature measurement is very sensitive and requires no sample preparation, but is applicable to global strain only. X-ray diffraction can be carried out on a 100 μm scale, is non-destructive, and does not require sample preparation, while nanobeam diffraction can resolve 100 nm areas; it is applicable to local strain measurements, but is sample destructive.

The same performance-enhancing techniques of silicon MOSFETs can mostly be applied to TFTs as well. For instance, LDD structures for better short channel behavior can be made, with a few extra masking steps. Vertical TFTs are also possible, as shown in Figure 38.5. Channel length is now determined by a-SiN$_x$:H deposited thickness, not by lithography.

38.5 Memories

DRAMs are capacitors and capacitance is proportional to area and dielectric constant and inversely proportional to dielectric thickness. The first designs were polysilcon–insulator–semiconductor (PIS) capacitors (Figure 38.6). The same polysilicon served as the transistor gate and capacitor top plate. The bottom plate was a diffused silicon area. Initially DRAM scaling involved area scaling with insulator thickness reduction to keep the capacitance unchanged. Planar design continued for another 10 years but with a polysilicon–insulator–polysilicon (PIP) structure. This gave more freedom to design the DRAM capacitor because the capacitor dielectric could be optimized irrespective of the MOS gate.

At the end of the 1980s, 3D structures came on the scene: in order to keep capacitance reasonably high, area could not be reduced any further, but by going to 3D

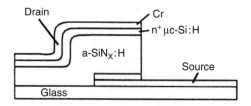

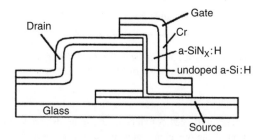

Figure 38.5 Vertical TFT. Reproduced from Chan and Nathan (2005), copyright 2005, American Institute of Physics

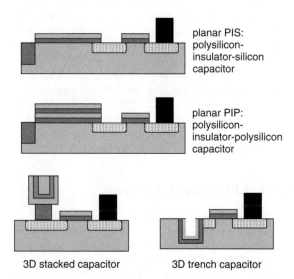

Figure 38.6 Evolution of DRAM from 1980 to 2005: from planar single poly to planar double poly to 3D stacked and trench designs (which can be PIP, PIS or MIM). Adapted from Gerritsen et al. (2005)

structures silicon chip area could be still reduced, while capacitor area was not affected. Two competing solutions to 3D storage node emerged: the trench capacitor and the stacked capacitor. In the trench capacitor a deep trench of high aspect ratio is etched into silicon, forming a PIS capacitor. In the stacked design a pillar or crown is deposited on the wafer, resulting in a PIP capacitor. A DRAM cell for 1 bit, including the capacitor and transistor, occupies an area of $6F^2$ for the stacked design and $8F^2$ for the trench version, where F is the minimum lithographic feature size. Capacitor area can be $2\,\mu m^2$ yet occupy only 100 nm × 50 nm silicon area. Over the years both trench depth and crown height have constantly increased to keep capacitance constant, while cell area has been reduced. Aspect ratios of trench capacitors are approaching 100:1 and stacked capacitors exhibit 40:1 pillars.

Recently silicon dioxide and nitrided oxide have been replaced by materials with a high dielectric constant, like Al_2O_3, Ta_2O_5, HfO_2 and ZrO_2. These have dielectric constants of 10–30, while that of silicon dioxide is 4. The equivalent oxide thickness of DRAM capacitors is roughly 1 today, with an extrapolated trend to 0.3 by 2013. More exotic future materials include BST (barium strontium titanate) and STO (strontium titanate) which have dielectric constants of over 100. However, as shown in Figure 5.2, the dielectric constant is a function of film thickness, and the full potential of these new materials may not be realized. Another development is toward MIM capacitors: metals as both the top and bottom electrodes. Extreme step coverage is required for electrode deposition in both the trench and stacked designs and ALD is a prime candidate. Titanium nitride is the standard choice: it has long been used in contact plugs as a barrier and adhesion promotion layer. Research is being conducted on platinum, iridium oxide, ruthenium and others.

Flash memory has recently taken over as the device with the smallest half-pitch, replacing DRAM. In 2009, flash half-pitch was 40 nm while DRAM half-pitch was 50 nm. The market for non-volatile memory (NVM) has been good in recent years, with MP3 players and digital cameras demanding more NVM, pushing flash NAND developments.

Flash scaling is getting difficult: bits are stored as charge in the floating gate and injecting those holes into the floating gate requires an electric field of roughly 1 V/nm, but tunnel oxide cannot be made much thinner (presently 6–7 nm in NAND and 8–9 nm in NOR flash) because injection gradually degrades the oxide, limiting programming cycles (erasure is by Fowler–Nordheim tunneling of electrons out of the floating gate). Interpoly oxide is 10–13 nm thick because a rougher poly surface results in a lower quality (Figure 13.6), but the erase voltage is only 0.3 V/nm, making life easier. However, the slow erase is one of the reasons why other non-volatile memory technologies are being investigated as a replacement for flash memory.

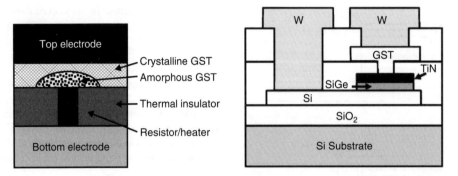

Figure 38.7 Phase change memory (PCM): left, rapid heating and quenching amorphizes the GST layer, changing resistance; right, device implementation. Reproduced from Lee et al. (2008) by permission of ECS – The Electrochemical Society

As an example of a radical departure from previous memory technologies, phase change memory (PCM, also known as ovonic memory) has been proposed. PCM operation is based on resistivity changes between the amorphous (high resistance) and crystalline (low resistance) phases of GST, germanium antimony telluride, $Ge_2Sb_2Te_5$. While the material sounds exotic, it is well known in rewritable DVDs (where reflectivity difference between the two phases is used), which makes its adoption more likely. The operation is based on localized heating of GST, as shown in Figure 38.7. A fast high-current pulse melts the GST layer, and rapid cooling results in an amorphous structure. The reverse operation is achieved by a smaller amplitude, slow (10–100 ns) current pulse.

PCM offers many intriguing advantages: first of all, it is at least an order of magnitude faster than flash, and it scales nicely; the heating power needed to change a bit becomes smaller and smaller when the cell size is scaled smaller. At 180 nm node a 450 µA programming current is needed, but at 50 nm only 250 µA is required. And PCM has been reported to endure 10^{13} erase/write cycles; even though the consensus estimate is 10^8, it is much more than 10^5 cycles for flash memories. There are problems, of course: the programming temperatures are 600 °C, and thermal stresses will be considerable, so reliability needs to be assessed.

Speed and memory capacity can be traded for each other: SRAMs are the fastest non-volatile memories, but each cell is about $50F^2$ in size, while the PCM cell is $6F^2$ and flash is $1-2F^2$. But if PCM can provide the speed advantage it has indicated over flash, it may be a winner.

Totally new materials are not accepted easily, and it takes a long time to gather reliability data, say, on copper metallization or deposited high-k oxides, before customers will be convinced. This holds back the adoption of many technologies that use difficult or exotic materials, for example ferroelectric RAM memories (FeRAMs), which use PZT, $Pb(Zr,Ti)O_3$, for the storage layer and IrO_2 or $SrRuO_3$ as electrodes. FeRAM is a non-volatile memory which stores information as changes in the polarization state of a ferroelectric capacitor. FeRAM has many attractive properties like high speed, low voltage and low power. A drawback of FeRAM is cell size: $10-20F^2$ area per cell. Magnetic tunnel junction memories (MRAMs) sense resistance changes when the magnetic moment is switched in a ferromagnet–insulator–ferromagnet triple layer structure. The materials challenges are formidable but MRAM promises to be indefinitely rewritable like DRAM, as fast as SRAM and non-volatile like flash.

38.6 Lithography Future

Several lithography technologies have been advertised over the decades as the future lithography. Optical lithography was predicted in the mid 1980s to be able to print 0.5 µm lines, no smaller, and the hunt was on to find alternatives. We will shortly discuss what it takes to overthrow optical lithography, and why optical lithography has been so pervasive.

First of all, massive infrastructure exists for optical lithography. This includes investments made in lens fabrication, focusing and alignment mechanics, metrology tools, simulation software, photoresists, photomasks and other related technologies over the decades. Scanner optics and mechanics have been made by the thousands, bringing in experience in volume manufacturing that no emerging technology can hope to match. Immersion lithography with water enables numerical apertures in excess of 1, and oils would raise this to 1.35, extending the life of optical lithography by another generation or two, enabling 22 nm generation devices (2012 onward) to be printed optically. Even though narrow linewidths have been demonstrated by many techniques, there are additional requirements, like exposure area, which is currently 8.58 cm^2 in optical lithography, so that big memories and processors fit within a single exposure field. It is possible that current 4× demagnification will be abandoned in favor of a larger demagnification factor (when wafer steppers first appeared, both 10× and 5× were used). Advantages include less polarization effects and cheaper masks and lenses, but, unfortunately, smaller chip area.

All RET technologies, PSM, OAI, SRAF and OPC (Section 10.6), are used in advanced optical lithography. In order to push optical lithography even further, they are optimized for each and every mask level independently. Each mask will have customized off-axis illumination, optimized for its pattern shape and direction, and carefully analyzed for polarization. Different types of phase shift structures, embedded PSM, alternating PSM and complementary PSM will be used. Exposure dose can be adjusted across the slit and along the scan. And in order to make RET implementation easier, designs may have to be limited to certain simple basic shapes which are amenable to analysis and adjustment.

Further into the future, double exposure and double patterning are being investigated. Double exposure means that two exposures are done, with two different masks, and one etch step accomplishes the final structure. Double patterning refers to a method where lithography and etch are done twice, to pattern one layer. These approaches are very critical to overlay. Spacer lithography (Figure 11.14) is another way to scale linewidth beyond optical limits.

A new light source needs to be developed and qualified for next generation optical lithography: at 157 nm the F$_2$ laser is a candidate, but at 126 nm the choices are wide open. Quartz has served well as an optical material, both for stepper and scanner lenses as well as for masks, but at 157 nm new materials are needed, for example CaF$_2$ lenses. The high thermal expansion coefficient of CaF$_2$, 19 ppm/°C, presents major problems with thermal control. As far as resists are concerned, it is not clear if evolutionary approaches are feasible or whether completely new resists have to be developed. These major changes pose a radical departure from previous generation optical lithography, and perhaps a window of opportunity for competing technologies may open.

The throughput of optical lithography is impressive, approaching 200 wafers per hour (WPH). This was not always the case: when step-and-repeat reduction optical systems were introduced around 1980, they could print 25 WPH, while the then mainstream 1× projection tools printed 100 wafers. New candidates are in very much the same situation: they are selling extendibility into future generations, even though their productivity today is less than that of current tools. But in a few years' time those old tools will no longer be able to do the job, so it might be worthwhile to start gaining experience with the new technology.

38.6.1 Extreme ultraviolet lithography (EUVL)

Extending optical lithography from DUV to EUV involves many more changes than in previous wavelength reductions. A completely new source of irradiation is needed, and even though pulsed plasma laser sources exist, their power output is insufficient and their long-term reliability and running costs uncertain. At 40 W power today, throughput is only 10 WPH (partly also due to the low reliability of new technology), but increasing that to 200 W will mean 50 WPH, and improving resist sensitivity from 15 to 7 mJ/cm^2 will double that to 100 WPH, and so on.

The whole optical path has to be in a vacuum to eliminate atmospheric absorption. This presents additional resist requirements: it has to be vacuum compatible. A shift from refractive optics (which would absorb too much EUV) to reflective optics is a major paradigm shift. In an EUVL system light of 13.5 nm wavelength is reflected from the mask and projected onto a wafer using reflective optics. EUVL may win because its process robustness is good, while the optical lithography process window is very narrow, as shown in Figure 10.4, and the situation is getting worse with every successive generation.

EUV mask technology will be radically different, with nanolaminate multilayer reflectors (shown in Figure 7.20). Manufacturability of large-area masks with stacks of 80 layers consisting of 4 nm layers with 0.4 nm thickness control is an open question. After the masks have been made, inspection and repair techniques must be available. If they are not, the masks cannot be qualified and defect analysis cannot be done, and if improvements in mask technology are not forthcoming, nobody will adopt them in production.

The infrastructure for EUVL must come into existence before it will be a serious competitor for optical lithography and its extensions. These additional technologies include software for proximity correction, overlay metrology and new photoresists. The short wavelength means that a lot of energy is hitting the mask, and because the nanolaminate layers are very thin, their defect generation and its effects on reflectivity need to be understood. One additional requirement for EUVL is backward compatibility with optical technology: mix-and-match lithography must be available because there are non-critical mask levels which must be made on cheaper, older, technology. Estimates put the EUVL system price near $100 million, as opposed to $50 million for the optical one, for systems designed for 22 nm generation.

38.6.2 X-ray lithography (XRL)

The decisive difference between 13.5 nm EUVL and 1 nm XRL is not really wavelength: many people would accept 13.5 nm as X-rays, but XRL as understood in microfabrication means transmissive exposure through the mask. X-ray reduction optics do not exist, which necessitates 1× photomasks. This is a major drawback for XRL. Add to this the blocking layers that need to be thick to effectively block X-rays: heavy elements like tungsten or gold are used. Aspect ratios of chrome lines on an optical reticle for 50 nm linewidths on a wafer are about 0.5, whereas in XRL they are 20:1. XRL has many advantages over optical lithography: the exposure field is larger and XRL is relatively insensitive to small particles because, for example, 0.5 μm silicon particles are relatively transparent to X-rays. Traditional X-ray sources are not bright enough to produce reasonable throughputs, so new sources have been developed: namely, synchrotron radiation storage rings and laser plasmas. This leads to enormous starting costs for XRL systems.

38.6.3 Electron and ion projection lithographies

Because direct writing with electron or ion beams is slow, masked versions have been sought. In electron and ion projection lithographies (EPL, IPL) a broad beam illuminates the mask, and the main problem is again the mask: electrons and ions need to be admitted through the mask at selected sites and blocked elsewhere. This leads to masks with thick (blocking) areas and thin or open (transparent) areas. Thin areas need to be made of materials of low atomic weight for good transmission, with a thickness on the order of 1 μm. And they must preferably be several square centimeters across for large chips to fit in a single exposure field. Thick blocking layers on these thin membranes cause stresses and pattern distortions. Shadow mask-like structures with open areas are excluded because making donut-shaped objects would require two masks and exposures. The mask will be heated by the incoming beam, just like the photomask in optical lithography, but, additionally, ions or electrons lead to mask charging and damage. Electron scattering masks, instead of absorbing masks, have been developed for EPL. This eliminates many of the thickness, stress and heating problems. The throughput of ion and electron projection systems remains an open question because the current systems are research tools and not close to production.

38.6.4 Nanoimprint lithography (NIL)

NIL is the latest contender to replace optical lithography. As discussed in Chapter 18, NIL has the potential to print even 5 nm structures, so basically it is capable of future IC demands. This is important because any technology that only offers solutions to the next generation and is not scalable to future generations has precious little chance of being successful. NIL resists exist, but imprinting throughput remains low compared to optical. The main attractions are simplicity and price: for a few million dollars the same linewidths can be produced as with a $50 million optical tool. But while NIL linewidths are impressive, alignment in NIL has lagged behind, and when alignment systems are added, the price goes up steeply.

All new lithographies face serious mask technology challenges, but NIL is in a class of its own because the master actually makes contact with the resist, and this generates serious concerns about master lifetime. One solution is to deposit a thin fluoropolymer layer to ensure master–resist demolding, and to rework the coating, and continue using the same master, which is hopefully intact. Competition between full wafer thermal NIL and step-and-repeat UV NIL is open. The step-and-repeat master has to tolerate 100 times more contacts for the same number of chips. On the other hand, making chip-sized masters is much cheaper and faster than making 300 mm masters. Inspection technology for 3D objects is always very demanding, and the fact that NIL masters are 1× means that they contain lines that are one-quarter the size of optical reticle lines, and inspection of these is already very demanding.

38.7 Moore's Law

The development of ICs seemed to follow a regular pattern: double the number of devices on the chip every year. In 1965 Gordon Moore remarked on this pattern. His observation was based on rather few data points since the birth of the IC, but the conclusion became famous. Later the prediction was revised to doubling every 18 months, and this version has been especially long lasting. It has been dubbed Moore's law, even though it is only an empirical pattern without fundamental justification. Table 38.2 lists the evolution of the technology over the years, giving linewidth decrease and memory capacity increase.

Moore's 1965 prediction extended till 1975 and his extrapolation was quite accurate. The trend has continued

Table 38.2 Moore's law

Year	Transistors/chip	Memory	Linewidth	Wafer size
1959	1		30 µm	0.5 in
1960	2			
1961	4			
1962	8			1 in
1963	16			
1964	32		20 µm	1.5 in
1965	64			
1968	256		12 µm	2 in
1970	1024	1 k	8 µm	
1973	4096	4 k	5 µm	
1975	16 384	16 k	3 µm	3
1979	65 536	64 k	2 µm	
1983	262 144	256 k	1.5 µm	100 mm
1986	1 048 576	1 M	1.2 µm	125 mm
1989	4 194 304	4 M	0.8 µm	150 mm
1992	16 777 216	16 M	0.5 µm	
1995	67 108 864	64 M	0.35 µm	
1998	268 435 456	256 M	0.25 µm	
2000	536 870 912	512 M	0.18 µm	200 mm
2002	1 073 741 824	1 G	0.13 µm	300 mm
2004	2 147 483 648	2 G	90 nm	
2006	4 294 967 296	4 G	65 nm	
2008	8 589 934 592	8 G	45 nm	
2010	17 179 869 184	16 G	32 nm	

Note: DRAM used to be the memory device with the highest bit density but since the turn of the century NAND flash has been the leader: the 16 Gbit refers to NAND, while the largest DRAMs are 2 Gbit only. Flash memory cell area is $2-4F^2$ but each cell stores 2 bits, for $1-2F^2$ per bit, vs. $6-8F^2$ for DRAM.

Table 38.3 Scaling trends

Year	2010	2012	2014	2016
Memory half-pitch (nm)	45	36	28	22
Processor gate in resist (nm)	30	24	19	15
Processor gate after trimming (nm)	18	14	11	9
Gate CD control 3σ (nm)	1.9	1.5	1.2	0.9
Resist thickness (nm)	70–130	55–100	45–80	35–65
Metal 1 half-pitch (nm)	45	32	28	23
Metal 1 aspect ratio	1.8	1.8	1.9	2
RC time delay for 1 mm line (ns)	2.1	3.5	6.4	10.6
Copper barrier thickness (nm)	3.3	2.6	2.1	1.7
Oxide thickness EOT (nm)	0.9	0.9	0.8	0.55
EOT thickness control, 3σ (%)	$<\pm 4$	$<\pm 4$	$<\pm 4$	$<\pm 4$
Spacer thickness (nm)	9.9	7.7	6.1	5
Silicon and oxide loss per cleaning step (nm)	0.09	0.09	0.06	0.06
Surface roughness (nm)	0.2	0.2	0.2	0.2
Critical surface metals (10^{10} cm^{-2})	0.5	0.5	0.5	0.5
Critical particle diameter (nm), starting wafers	22.5	17.9	14.2	11.3
Particles/cm^2	<0.15	<0.32	<0.16	<0.16

Source: ITRS 2007.

at approximately the predicted speed, give or take some fluctuations. Since the turn of the millennium the pace has been even faster than predicted by Moore's law. Memory chips are best suited for Moore's law studies because the law is about production economics: chip size and cost minimization. Processors are governed by quite different laws: they are design heavy, rather than manufacturing driven, and proprietary architectures are not subject to ultimate price reductions. It should be borne in mind that sometimes product demonstration date is used (when the first fully functional chips are fabricated), sometimes production start date is used, and sometimes peak production year is stated.

Shrink versions make the situation more complex: the first functional 1 Gbit DRAMs were demonstrated using 0.18 µm technology, but production versions have been made at smaller linewidths: 0.13 µm to 0.10 µm. Add to this minor differences between companies reporting linewidth (or half-pitch) and it is fair to accept a few years of discrepancies in Moore's law data.

Table 38.3 considers future forecasts. It assumes that the technology develops more or less according to predictable paths. If the technology goals of linewidths, film thicknesses, particle counts and others can be met, Moore's law will be valid for another decade.

Moore's law was originally proposed in the era of bipolar transistors and it has held up well in the era of PMOS, NMOS and CMOS, and seems set to hold for the next decade of strained-silicon, SOI–CMOS and other evolutionary MOS technologies. Moore's law is about device packing density and cost, not about any particular technology. There have been a number of dubious extensions of Moore's law: it has been said to apply to computing power, which is not true because computer architecture is not part of it. Despite its non-fundamental nature, it is one of the few predictions about the future of technology that have held for almost 50 years.

38.8 Materials Challenges

Nominal or design width is just an idealization of a microstructure. The physical structure in silicon or in thin-film material adds its own features. These effects are the more pronounced, the narrower the linewidth or the thinner the film. The smaller the details we study, the more the effects come into play.

Line edge roughness (LER) is becoming significant compared to linewidth. In the extreme it is partly a materials limitation: chrome, photoresist and the thin film on the wafer are granular to some extent, and for instance polycrystalline materials may be etched at slightly different etch rates for different crystal orientations, so this preferential etching contributes to LER. It is not known exactly which factors contribute to LER, and how to quantify it, and it is not even certain how LER contributes to device performance.

When linewidth scaling is continued, the relative importance of physical effects changes. Current conduction in a $1 \times 1\,\mu m$ cross-section of a conductor line is fully characterized by the classical ohmic description. Narrower lines, and thinner films, reach a limit where the surface scattering contribution to resistance becomes important, and in the 10 nm size range quantum effects come into play, when single electron conduction can be seen. The characteristic scale for non-classical effects is given by the mean free path, which is 40 nm for copper and 15 nm for aluminum. However, some deviation from classical behavior has been seen even at 500 nm, probably due to grain boundary reflections, and at 100 nm linewidths copper resistivity has been reported to increase by 100%, to 4 µohm-cm, Figure 28.14.

Film thickness downscaling in the back end is driven by the need to keep aspect ratios reasonable, even though RC time delays inevitably increase as resistance increases in thinner wires, and capacitance increases when dielectric thickness is scaled down. The ultimate limits are fairly close in back-end scaling: copper is as close to the minimum resistivity as any metal can practically be, and with dielectrics, $\varepsilon_r = 1$ (vacuum) is not so far away, with $\varepsilon_r = 2$ materials being introduced. Superconducting wiring was touted in the early 1990s as a solution to resistance problems, but enthusiasm waned rapidly when

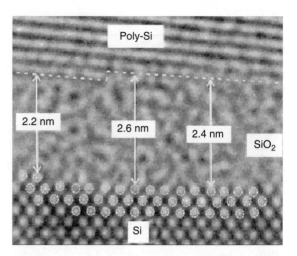

Figure 38.8 Quantized gate oxide thickness: 2.2 nm, 2.4 nm and 2.6 nm represent possible thicknesses. Reproduced from Buchanan (1999) by permission of IBM

the difficulties of high-T_c superconductor deposition and structural control became apparent. Optical interconnects are now being explored, though at a very experimental stage.

Scaling to atomic dimensions leads to inevitable limitations. The gate oxide thickness is approaching such limits: because atoms are discrete, the gate oxide thickness is "quantized": we cannot have any gate oxide thickness, only integral multiples of atomic dimensions (Figure 38.8). Putting it another way, each transistor will have its own microscopic oxide thickness pattern and consequently idiosyncratic microroughness which affects channel mobility and tunneling currents.

38.9 Statistics and Yield

Yield is tied to the number of process steps, which has been increasing constantly. With 25 lithography steps, and about 500 steps altogether, individual step yield has to be very high indeed. This is putting more and more demands on metrology: process monitoring precision and speed have to be increased so that more wafers can be checked. With polymeric thin films, film thickness and density are not enough, so pore size and pore size distribution must be known. Film behavior in CMP, plasma ashing and wet cleaning need to be understood, as do thin-film deposition and wet chemical penetration into nanometer pores.

Despite aggressive linewidth scaling, chip area keeps increasing. The number of defects per chip has to remain constant or decrease, which means that defect density has to be scaled down more aggressively than linewidth. Chip area increases because of the economic incentive to integrate as many functions as possible on the chip, in order to reduce packaging and assembly costs (Figure 37.3). At the moment is seems that lithographic lenses are limiting the increase in chip size: it has not been possible to simultaneously improve resolution and increase lens field size at the same pace. This, of course, applies mostly to evolutionary scaling of refractive optical systems; reflective optics, XRL and EPL have their own scaling trends.

Chemicals, DI water, gases and sputtering targets have been "scaled" to higher and higher purity levels. Metal impurity levels have been reduced by a factor of 100 in four technology generations. Measurement of minute impurities must be available for gases, liquids and solids. Cleanrooms have been "scaled" to higher and higher standards of purity. Cleanliness today is so high that particle measurements have hit a barrier: there are simply not enough particles to statistically assess particle purity. With cleanroom costs increasing, there has been an incentive to find alternative operation modes. Integrated processing is one such approach, keeping the wafers under controlled ambients at all times.

Statistics with extremely large or extremely small quantities can produce some surprises, even before the ultimate limits. In a circuit with 1 billion devices the tails of statistical distributions can easily cause circuits to fail: there are 20 devices which have a variation larger than six standard deviations. In very small volumes the distribution of atoms becomes a source of variation: in 100 nm linewidth MOS transistors, the volume under the gate is about 100 nm × 500 nm × 10 nm (L_{eff} × W_{eff} × inversion layer thickness), and the channel doping level is $N_A \approx 10^{18}$ cm^{-3}, which translates to about 500 dopant atoms only. The small number of dopants in itself leads to detectable fluctuations in threshold voltage, but also the random positions of dopant atoms must be considered. The standard deviation of threshold voltage V_{th} is given by

$$\sigma(V_{th}) = \frac{3.19 \times 10^{-8} t_{ox} N_A^{0.4}}{\sqrt{L_{eff} W_{eff}}} \quad (38.2)$$

Continued scaling to smaller dimensions, together with an increase in the number of devices per chip, rapidly leads to situations where not all devices switch.

Linewidth and gate oxide scaling are the most visible parts of scaling, but there are many other parameters that are continuously been pushed forward. The energy consumption of a logic operation was 10 nJ in 1960, 1 pJ in 1980 and only 1 fJ in 2000. Operating voltage, which was 5 V for many generations (5 μm to 0.8 μm), is now being reduced rather regularly, and 1 V operation will soon be usual for non-battery-powered devices, too. The number of metallization levels for logic is rapidly going up. Since the 0.5 μm generation, when three levels of metals were standard, one level of metallization has been added in almost every generation, leading to eight levels in 0.1 μm technology. The corollary trend is that of output pin-count increase, to thousands, which has led to various ball-grid-like packaging solutions.

38.10 Limits of Scaling

The death of Moore's law has been much discussed but newer predictions of IC scaling have often proven inaccurate, even in the quite short term: in 1994 it was predicted that 0.1 μm technology would become available in 2007, microprocessor chips would have 350 million transistors and operate at 1 GHz with 1.2 V, which was very pessimistic about the linewidth and speed. In 1986 it was predicted that 16 Mbit DRAMs would be available at the turn of the millennium, but actually it was 256 Mbit. Around 1980 the prediction was that optical lithography could not print lines smaller than 1 μm and in 1989 the end of optical lithography was predicted for 1997. Quite regularly, the end of optical lithography has been predicted to be 10 years in the future, and this same prediction still holds. Also in 1989 it was assumed that silicon dioxide as the gate oxide would be replaced by high-k dielectrics starting from 1993, but nobody dared to do it before 2008. Long-term predictions have been off by a far wider margin: the 1984 linewidth predictions for 2007 were 0.1 μm (optimistic) and 0.5 μm (pessimistic), yet in 2007, 65 nm was in production.

How long can this scaling continue? If all goes according to Moore's law, in 2059, the 100th anniversary of the integrated circuit, we will have:

- 0.25 nm minimum linewidth
- 0.004 nm gate oxide thickness
- 2 mV operating voltage
- 64 exabit memories (exa = 10^{18}).

Obviously a scaled version of the current MOS transistor cannot be the device described above, for instance the atomic size is about 0.1 nm. But remember: Moore's law is independent of device technology. The first working 1 μm MOSFET was reported in 1974, and about 15 years later 1 μm devices entered mass production. The first 100 nm device was unveiled in 1987, and about 15 years later 100 nm devices were being fabricated. At the beginning of the third millennium, 10 nm devices exist in laboratories and are extrapolated to enter production before 2020. Extrapolation, however, is a tricky business.

38.11 Exercises

1. Price per bit has been scaled down at a rate of about 30% a year. If 1 Gbyte of DRAM memory cost about $50 in 2010, how much will it cost in 10 years' time?
2. Given the scaling trend predicted by Moore's law, when will CMOS gate oxide be one atomic diameter thick?
3. The speed of light sets the ultimate limit on signal speed. How close is this limit?
4. The price of the refractive lens used in a wafer stepper has increased rapidly over the years: $25 000 in 1986, $102 000 in 1989, $294 000 in 1992, $670 000 in 1995, $1.5 million in 1998. What is the price of a stepper lens today?
5. A DRAM trench memory cell for 1 bit takes up an area of $8F^2$, where F is the lithographic linewidth. What is the chip size of 1 Gbit DRAM?
6. DRAM trench capacitors are cylindrical holes with high aspect ratios. What is the aspect ratio in a 0.15 μm linewidth process if the capacitor oxide thickness is 5 nm and the capacitance is 40 fF?
7. How does resist sidewall error contribute to linewidth control?
8. The chip area of a 2 billion transistor is $3.5\,cm^2$. What is the area of a single transistor and what is the linewidth?
9. How does PCM scale down? Which factors improve with scaling and which become more problematic?
10. If NMOS and PMOS gates were fabricated from different metals (optimized for their respective devices), how many process steps would be added compared to a n^+/p^+ dual gate (Figure 38.2)?

References and Related Reading

Asenov, A. *et al.* (2003) Simulation of intrinsic parameter fluctuations in decananometer and nanometer-scale MOSFETs, *IEEE Trans. Electron Devices*, **50**, 1837 (special issue on nanoelectronics).
Bohr, M.T. *et al.* (2007) The high-k solution, *IEEE Spectr.*, October, 29–35.
Brueck, S.R.J. (2005) Optical and interferometric lithography – nanotechnology enablers, *Proc. IEEE*, **93**, 1704–1721.
Buchanan, M. (1999) Scaling the gate dielectric: materials, integration and reliability, *IBM J. Res. Dev.*, **43**, 245.
Burr, G.W. *et al.* (2010) Phase change memory technology, *J. Vac. Sci. Technol.*, **B28**, 223.
Chan, I. and A. Nathan (2005) Amorphous silicon thin-film transistors with 90° vertical nanoscale channel, *Appl. Phys. Lett.*, **86**, 253501.
Chang, L. *et al.* (2003) Moore's law lives on, *IEEE Circuits Devices Mag.*, **1**, 35.
Chiang, C. and K. Jamil (2007) **Design for Manufacturability and Yield for Nano-Scale CMOS**, Springer
Delhougne, R. *et al.* (2004) Selective epitaxial deposition of strained silicon: a simple and effective method for fabricating high performance MOSFET devices, *Solid-State Electron.*, **48**, 1307–1316.
Doering, R. and Y. Nishi (2001) Limits of integrated circuit manufacturing, *Proc. IEEE*, **89**, 375.
Galatsis, K., R. Potok and K.L. Wang (2007) A review of metrology for nanoelectronics, *IEEE Trans. Semicond. Manuf.*, **20**, 542–548.
Gerritsen, E. *et al.* (2005) Evolution of materials technology for stacked-capacitors in 65nm embedded-DRAM, *Solid-State Electron.*, **49**, 1767–1775.
Gottlob, D.B. *et al.* (2006) Scalable gate first process for silicon on insulator metal oxide semiconductor field effect transistors with epitaxial high-k dielectrics, *J. Vac. Sci. Technol.*, **B24**, 710–714.
Henderson, R. (1995) Of life cycles real and imaginary: the unexpectedly long old age of optical lithography, *Res. Policy*, **24**, 631.
Hisamoto, D. *et al.* (2000) FinFET - a self-aligned double-gate MOSFET scalable to 20nm, *IEEE Trans. Electron Devices*, **47**, 2320.
Huff, H.R. (2002) An Electronics division retrospective 1952-2002 and future opportunities in the twenty-first century, *J. Electrochem. Soc.*, **149**, S35–S58.
Integrated circuit technologies of the future (1986) *Proc. IEEE*, **74** (12), special issue.
ITRS, (2007) The International Technology Roadmap for Semiconductors, 2007 Edition. International SEMATECH: Austin, TX, http://www.itrs.net/
Kemp, K. and S. Wurm (2006) EUV lithography, *C.R. Physique*, **7**, 875–886.
Keyes, R.W. (2001) Fundamental limits of silicon technology, *Proc. IEEE*, **89**, 305 (special issue on limits of semiconductor technology).
Kilby, J. (1976) The invention of the integrated circuit, *IEEE Trans. Electron Devices*, **23**, 648.
Lacaita, A.L. (2006) Phase change memories: state-of-the-art, challenges and perspectives, *Solid-State Electron.*, **50**, 24–31.
Lee. S.-Y. *et al.* (2008) Bilayer heater electrode for improving reliability of phase-change memory devices, *J. Electrochem. Soc.*, **155**, H314–H318.
Moore, G. (1965) Cramming more components onto integrated circuits, *Electronics*, **38** (available at http://www.intel.com/technology/mooreslaw/).

Park, J.-T. and J.-P. Colinge (2002) Multiple-gate SOI MOSFETs: device design guidelines, *IEEE Trans. Electron Devices*, **49**, 2222.

Ross, I. (1997) The foundations of the silicon age, *Bell Labs Tech. J.*, **2**, 3 (50th anniversary issue on the invention of the transistor).

Skotnicki, T. *et al.* (2005) The end of CMOS scaling, *IEEE Circuits Devices Mag.*, January, 16–26.

Sotomayor Torres, C.M. (2003) **Alternative Lithography**, Springer.

Theuwissen, A.J.P. (2008) CMOS image sensors: state-of-the-art, *Solid-State Electron.*, **52**, 1401–1406.

Wolf, S. (2002) **Silicon Processing for the VLSI Era**, Vol. **4**, *Deep-Submicron Process Technology*, Lattice Press.

Wong, H.-S., K. Chan and Y. Taur (1997) Self-aligned (top and bottom) double-gate MOSFET with a 25nm thick silicon channel, *IEDM Tech. Dig.*, 427.

Wong, H.-S.P. (2002) Beyond the conventional transistor, *IBM J. Res. Dev.*, **46**, 133 (special issue on scaling CMOS to the limit).

Yagishita, A. *et al.* (2001) Improvement of threshold voltage deviation in damascene metal gate transistors, *IEEE Trans. Electron Devices*, **48**, 1604.

39

Microfabrication at Large

Integration of different technologies is a major trend all over microfabrication. Analog–digital ICs (or mixed signal circuits) integrate resistors and capacitors with MOS or bipolar transistors; BiCMOS integrates bipolars and CMOS; and microprocessors integrate more and more SRAM (which in fact takes up most of the silicon area, up to 90%, in microprocessors). MEMS integrate mechanical and electrical functions by definition. Microsensors for mechanical, optical, chemical and magnetic quantities most often produce an electrical output signal which opens up possibilities to process, store and transmit those signals with microelectronics, which may be integrated on the same chip. With high-performance thin-film transistors now available, the display drivers and other electronics can be integrated with the flat-panel display, enabling much thinner packages for displays.

Integration of two technologies enhances performance but adds to process complexity: roughly speaking, a 20% mask count increase leads to a 20% cost increase. A surface micromachined airbag accelerometer integrated with BiCMOS readout electronics has been commercialized and is manufactured in significant volumes. Many MEMS devices are produced in small numbers, while millions of IC chips are made in a month in a big fab. This discrepancy often leads to hybrid integration: IC and MEMS are fabricated separately, and integration takes place at package level. The reason for this is economical: adding advanced CMOS capability to a MEMS production line would be too expensive, and adding special MEMS process modules to a smoothly running CMOS process flow would be disruptive to the work flow.

39.1 New Devices

Microfabricated devices have a number of benefits compared to classic or macroscopic devices: small size, low cost, high speed (of electron transit time across bipolar base, or of microreactor thermal ramp time), low power consumption (and low reagent consumption in chemical microsystems) and high device packing density (of DRAM cells or attached DNA strands) all relate to the exceptional possibilities offered by microfabrication. One of the special benefits of microfabrication is the completely different cost structure compared to real-world manufacturing. Material usage is minuscule and almost any material can be used if it can be micromachined, because its price is not a limiting factor.

New classes of devices are being introduced in microfabricated versions, as are novel devices with no macroscopic counterparts. New names for devices and categories are popping up: nanoelectromechanical systems (NEMS) (Figure 1.2), adaptive optics (Figure 17.8), biosensors (Figure 20.28), microacoustics (Figures 7.11, 20.19, 30.8, 30.19), micro power systems (Figure 1.10), microrockets (Figure 11.7) or DNA–CMOS hybrids (Figure 39.1). DNA sensors have been made by locally attaching DNA probe strands onto transistors and covering the chip by PDMS microfluidic channels. The change in threshold voltage is monitored as DNA strands pair with their counter-strands.

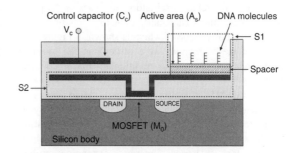

Figure 39.1 DNA pairing reaction changes transistor threshold voltage. From Barbaro *et al.* (2006), by permission of IEEE

Introduction to Microfabrication, Second Edition Sami Franssila
© 2010 John Wiley & Sons, Ltd

Microfabrication possibilities are everywhere: polyester fiber spinnerets in the textile industry are nowadays microfabricated pieces; a micromachined interferometer measures carbon dioxide concentration for heating, ventilation and air-conditioning applications. Superconducting quantum interference devices are measuring weak magnetic fields generated in the human brain, enabling new views of human decision making, pain, pleasure and cognition. Acoustic microsensors monitor mechanical machinery for sounds of cracks and imbalances. MEMS microphones have become standard in cell/mobile phones but only partly for their small size and not for their sound quality, rather because of their ease of mounting, which is similar to ICs, while traditional microphones require special assembly.

Non-CMOS technologies ride on CMOS processes to some extent: linewidths are lagging behind CMOS, but steadily getting narrower, too. RF and analog circuits are some two generations (5–6 years) behind CMOS in linewidth, high-voltage and high-power circuits maybe by three or four generations (10 years), mechanical microsensors are further away, and many other MEMS devices are still using 1980s' linewidths. However, the high aspect ratios make MEMS processes very demanding and very different from 1980s' CMOS.

39.2 Proliferation of MEMS

MEMS technologies are expanding in many directions, and subfields are emerging all the time: RF MEMS, powerMEMS, optical MEMS, BioMEMS, etc. Mirrors pop up in many applications both figuratively and literally: in Figures 29.22-24 and 39.2 mirrors do pop up from the plane of the silicon wafer.

RF MEMS cover a wide range of devices, from low-noise reference oscillators to passive coplanar waveguides, filters, antennas, inductors, phase shifters and switch arrays, one of which is shown in Figure 39.3. Many of these functions were earlier handled by either semiconductor devices, or traditionally machined metal parts.

PowerMEMS similarly include a wide range of completely different devices, although fabricated by the same technologies. Energy harvesting from vibrations employs mechanical microdevices, utilizing for instance piezoelectrics. Thermal energy scavenging utilizes temperature differences and thermoelectric materials like bismuth telluride. Various thrusters, turbines and nozzles are used in microrockets, but also in more down-to-earth applications like fuel injection. Lithium ion batteries have been made by thin-film deposition technologies. Fuel cells have been made by various technologies,

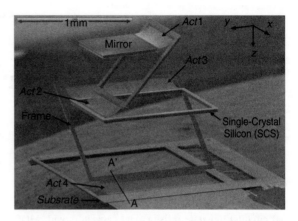

Figure 39.2 Planar pop-up micromirror with four actuator stages. From Jain and Xie (2006), copyright 2006, by permission of Elsevier

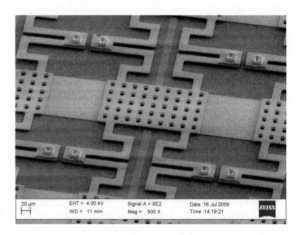

Figure 39.3 High-impedance tunable metamaterial: an array of silicon bridges. Reproduced from Sterner *et al.* (2010), copyright 2010, by permission of IEEE

including wet-etched, DRIE and electrochemically etched versions (Figure 39.4). The benefits of microfabrication in fuel cells come from channel dimension control: the microcapillaries (fuel channels) are made by etching silicon (electrochemical or DRIE) so small that liquids penetrate the channels by capillary action, irrespective of cell direction, and no pumping is needed to supply fuel.

39.3 Microfluidics

While most microfluidic devices are fairly simple by mask counts (17.1, 18.11, 18.24, 19.12, 25.7), more complex

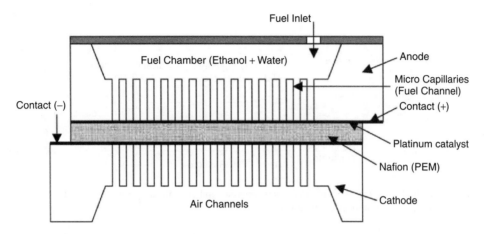

Figure 39.4 Micro fuel cell: a proton conducting membrane sandwiched between two wafers with electrochemically etched pores. Reproduced from Aravamudhan *et al.* (2005), copyright 2005, by permission of Elsevier

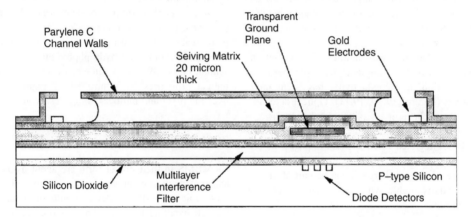

Figure 39.5 Microfluidic separation chip with integrated optical detection. Reproduced from Webster *et al.* (2001), by permission of the American Chemical Society

devices have been made, for example the 13-mask separation device with integrated fluidic filters and optical detection with optical filters (Figure 39.5).

Microreactors come in various forms: sometimes fast temperature ramp rates are needed, sometimes working with small sample amounts is beneficial because of dangerous substances like fluorine or radioactive elements, and sometimes multiple parallel reactions need to be followed. PCR (Polymerase Chain Reaction) is a DNA amplification reaction that requires 95 °C → 58 °C → 72 °C ramping. Two basic reactor configurations are used: a temperature ramp reactor (Figure 19.13 glass reactor; Figure 25.7 SU-8 reactor) and a continuous flow reactor, Figure 39.6. The former are scaled-down versions of traditional PCR instruments but the continuous flow reactor is completely different,

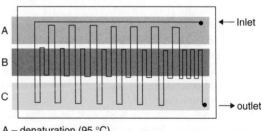

A – denaturation (95 °C)
B – primer extension (72 °C)
C – primer annealing (e.g. 58 °C)

Figure 39.6 Continuous flow PCR: DNA melting in zone A (95 °C), extension in B (72 °C) and annealing in C (58 °C). Reproduced from Obeid *et al.* (2003) by permission of the American Chemical Society

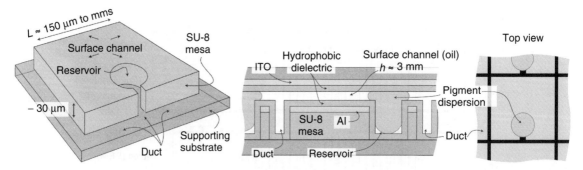

Figure 39.7 Electrofluidic display based on mechanical pressure and electowetting forces. Reproduced from Heikenfeld *et al.* (2009) by permission of Nature Publishing Group

enabled by microtechnology. Three temperature blocks are maintained at constant temperatures of 95, 58 and 72 °C, and the liquid travels in a fluidic channel between the zones. After 30 cycles the output liquid contains enough DNA for analysis.

Integration of detection with fluidics is important and many solutions have been implemented over the years. Electrical detection, like conductivity, has been implemented for CE (Figures 19.10, 19.11). Optical detection, for example absorbance or laser-induced fluorescence, is widely used. Various degrees of integration are possible. In Figure 18.9 integrated PDMS lenses improve the signal by focusing, but detection is external. Because lens integration comes free of charge, it can be part of the fluidic system, and disposable, while the expensive detection system is reusable.

Microrockets have chemical action but the desired end result, thrust, is physical and chemistry is just a way to achieve that. Fluidic logic circuits have been made, and fluidics is used to cool high-power ICs and lasers. While ink jets are mostly IT output devices, they have chemical and biological applications as well: printing nucleotides for DNA arrays, and dispensing very small amounts of precious biological reagents. Ink jets are in fact very rapid evaporators: picoliter drops dry in milliseconds (depending on liquid properties) and rapid evaporation is a convenient concentration method.

Most microfluidic devices work with liquids but there are many gas phase fluidic systems as well. One of the earliest microfluidic systems was a gas chromatograph, in the late 1970s. Mass spectrometric analysis has given rise to a number of microdevices. These devices either spray a droplet cloud or vaporize a sample from a micronozzle to a mass spectrometer. Ionization is based on many different principles, for example high voltage (in electrospray and corona ionization), high flow velocity (in supersonic ionization), temperature (in thermospray), or photoionization.

Fluidics is also strongly present in novel displays and electronic paper for e-book readers. These devices share many features with flat-panel displays, obviously, but surface modification with hydrophobic thin films is also essential. Figure 39.7 shows an electrowetting display. Pixel optical properties are controlled by pigment spreading over the pixel and pullback into a reservoir.

39.4 BioMEMS

BioMEMS in their purest form are about using mechanical transduction or movement in a biological context. A prototypical bioMEMS device could be a microgripper for cell handling (Figure 39.8). Thermal actuation using resistors enables the gripper to close its claws.

The patch clamp chip (Figure 21.20) holds a single cell in place while electric cell membrane measurements are

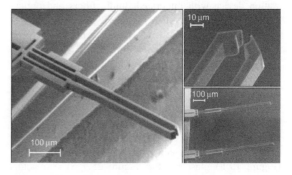

Figure 39.8 Thermally actuated cell gripper (20 μm thick) fabricated in SU−8. Reproduced from Chronis and Lee (2005), copyright 2005, by permission of IEEE

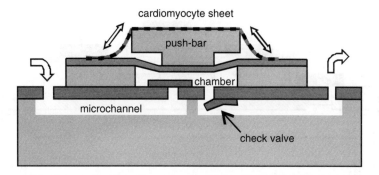

Figure 39.9 Cardiomyocyte micropump: integration of live cells with MEMS moving parts. Adapted from Tanaka *et al.* (2007)

made. Microneedles are available for drug delivery and sample extraction (Figures 1.13, 21.16). The line between microfluidics and BioMEMS chips is very vague, because most biologically interesting phenomena take place in the liquid phase. A microneedle for neural measurements with CMOS electronics was shown in Figure 30.13. It is not a fluidic device, but it must be compatible with body fluids.

Cell culturing is another big application area of BioMEMS: cell compartments, air and nutrient channels, stimulus channels for supplying drugs, toxins or genetic material are microfabricated on a chip. Additionally, sensors for monitoring changes induced by the stimulus are included: for instance pH, temperature and oxygen. The cardiomyocyte micropump (Figure 39.9) is an exemplification of another level of integration: live cells are integrated with silicon microfabricated parts for a pumping action. The contraction of cardiomyocytes cells induces movement of the push-bar, changing the pump chamber volume.

39.5 Bonding and 3D Integration

Silicon wafers used to be made of silicon but today wafers are much more complex objects. Layer transfer techniques enable thin layers of expensive or hard to make materials to be transferred on common substrates, like SiC on Si, silicon on quartz and germanium on oxidized silicon, which results in GeOI, germanium on insulator. Bonded wafers with silicide interlayer have been demonstrated for RF circuits, Figure 39.10. Layer transfer often necessitates temporary bonding: the thin layers need a support wafer for transfer or for processing, and it must be debonded easily. This is obviously quite a departure from traditional bonding which aims at permanent (and often hermetic) bonding.

An alternative way to increase transistor packing density without resorting to ever smaller linewidths is to stack wafers on top of each other. 3D integration has been around for decades because it is such an attractive idea. It is possible to thin CMOS wafers down after processing, and align those thinned wafers on top of other CMOS wafers, to realize 3D integration. In addition to mechanical joining of the wafers (bonding), the wafers have to be joined electrically, too.

Vias (known as TSV, for through-silicon vias) can be made from top or bottom side. Electrical contacts can then be made on wafer backside. This saves area on wafer front side, improves device packing density and for example photosensor chips have more effective area for light gathering. TSV are also essential in MEMS packaging: electrical contacts to cavities can be made through device or capping wafers (Figure 30.24).

After finishing device processing, two major possibilities are open for TSV: etching narrow vias (for example 3 μm wide, 30 μm deep) on the front side, followed by wafer thinning, or, alternatively, wafer thinning to 100 μm followed by etching large vias (e.g. 10 μm) on the backside, Figure 39.11 a and b. In both cases aspect ratios are approximately 10:1. Alignment accuracy of the front side vias is much better, but thinning is more demanding. After etching, an insulating layer is deposited, followed by conductor deposition (often in two steps: a seed layer plus the thick current carrying material). Front side via filling is easier because less material needs to be deposited, for example 1.5 μm thick CVD tungsten can be used to close 3 μm via (see Figure 39.12). Backside and combined front & back vias (Figure 39.11c) are filled by electroplated copper. Closing will be easy because of the thin part, but two lithography and two etching steps are required. If TSV is done on a capping (or carrier) wafer, there is more freedom in filling: for example highly doped CVD polysilicon can be used as the conductor. The TSV wafer

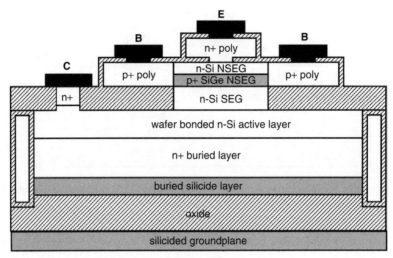

Figure 39.10 Bipolar transistor on silicon on insulator with buried silicide conductive layer and silicide ground plane. Reproduced from Bain *et al.* (2005), by permission of IEEE

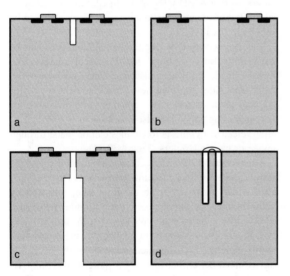

Figure 39.11 Thru-silicon vias: a) top side narrow via; b) backside wide via; c) combined front/back via; d) annular via. Wafer thinning after TSV etch in a and d; thinning before etching in b and c

can then tolerate practically all possible further processing steps. In Figure 39.11d an annular TSV is shown. There is no need for depositing any conductor because the highly doped central silicon pillar itself is used as the conducting element. However, thick insulator is needed. Note that this approach is not amenable to most device wafers because silicon has to be highly conductive in order to have low enough via resistance. In Figure 39.12 integration of three CMOS wafers is shown. The first bond is face-to-face, but the second one is face-to-back. CVD-W deposition temperature of 400 °C is low enough and it can be done on completed wafers.

Flexible electronics is a rapidly growing topic. Various polymer devices (antennas, logic, displays) have been processed on thin polymer sheets to enable bending and even stretching. Silicon is not out of the race for flexible electronics. One approach is shown in Figure 39.13. After completion of the CMOS (or MEMS) process, the wafer is coated with a polymer layer, and the silicon wafer is etched away almost completely, leaving only small islands where the active devices are located. Another polymer layer is coated on the back side, sandwiching the silicon.

39.6 IC–MEMS Integration

Silicon is just one possible substrate for MEMS, but it is the one that promises integration with electronic (e.g., CMOS circuitry) and optical (e.g., photodiodes) functions. This section discusses some general integration issues encountered with IC–MEMS integration.

MEMS and CMOS can be made side by side, with real estate reserved for both. Integration benefits come from closeness: for example, preamplifiers can be made close to sensors. The other way is to build MEMS on top of

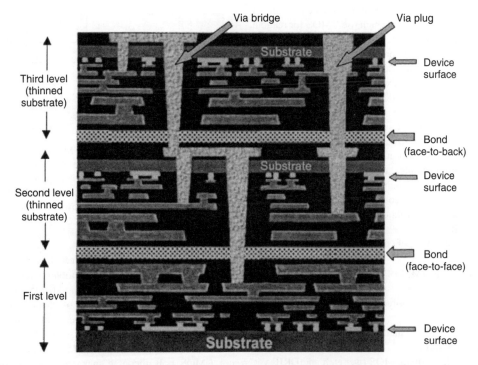

Figure 39.12 Chip stacking by wafer thinning and adhesive bonding. Via plugs filled by CVD tungsten. Reproduced from Lu *et al.* (2000) by permission of Materials Research Society

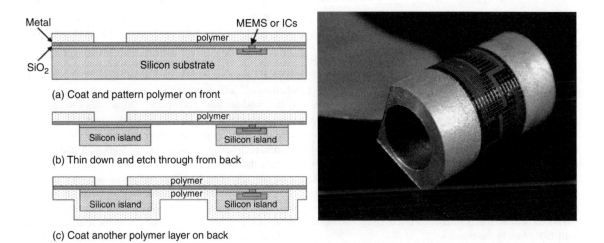

Figure 39.13 Flexible silicon circuits by thinning and silicon island polymer capping. Reproduced from Katragadda and Xu (2008), copyright 2008, by permission of Elsevier

CMOS. Classes of devices making most benefit of this include various array devices, which use CMOS for readout, for example photodetectors, infrared imagers, fingerprint sensors and steerable micromirror arrays (Figure 29.25).

CMOS wafers can be treated like any other substrates, even though they are very expensive ones: a CMOS wafer might cost for example $500 (0.8 μm CMOS on 150 mm wafer) vs. $20 for a bulk wafer, $40 for an epiwafer, $100 for a SOI wafer. Usually the topmost metallization layer is not planarized, but CMP is needed when CMOS is used as a substrate. CMOS transistors have to be protected from chemical contamination. This has been successfully done by combined oxide/nitride passivation and polymeric protective coating, and even KOH etching can be accomplished without any deleterious effects on CMOS.

There are many ways of combining CMOS and MEMS:

- MEMS before IC (MEMS release after CMOS completion).
- MEMS using CMOS polygate as the mechanical structure.
- MEMS and CMOS interleaved (optimized for both CMOS and MEMS).
- MEMS in silicon after CMOS (wet etch and DRIE versions).
- MEMS in thin films using CMOS multilevel metallization.
- MEMS postprocessing on top of CMOS (poly, polySiGe, metal, polymers).

All of these have their strengths and weaknesses, but in all cases process complexity increases and cases of successful commercialization of integrated systems remain limited.

In the MEMS-before-IC approach, MEMS devices are processed and covered (by for example TEOS oxide), so hopefully they will not be adversely affected by the hundreds of process steps it takes to complete the IC. The IC process temperatures limit severely the selection of materials for MEMS-first integration: silicon, polysilicon, oxide and nitride are really the only candidates. Contacting the MEMS part to the IC part is preferably done by diffusions because metal/silicon interfaces cannot be made until fairly late in the process.

The plug-up process shown in Figure 39.14 is a SOI MEMS–IC process which consists of the following main modules:

1. MEMS structure processing and encapsulation.
2. CMOS process.
3. MEMS structure release.

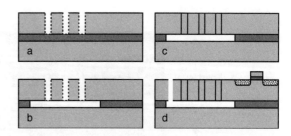

Figure 39.14 Integration of MEMS and electronics on SOI: a) DRIE and semipermeable polysilicon deposition; b) buried oxide etching through semi-permeable poly; c) deposition of standard polysilicon and CMP; d) IC processing and release hole etching by DRIE. Adapted from ref. Kiihamäki

There is no topography increase in SOI MEMS steps, and the sealed cavities do not pose problems for subsequent CMOS processing if the CMOS and MEMS parts are side by side on a wafer. The behavior of trapped gases inside cavities requires attention, however, and planarity of the prefabricated membrane is not guaranteed.

The simple approach to MEMS–IC integration is to use CMOS polysilicon to make mechanical structures; that is, to leave the IC process as it is and add suboptimal MEMS modules. CMOS gate polysilicon is typically 0.25 μm thick, whereas micromechanical poly is 1–2 μm thick. CMOS gate poly is optimized for poly/SiO$_2$ interface properties and it is highly doped. Micromechanical poly is designed for minimal stresses and stress gradients. But the integrated electronics can be used to correct for some of the performance losses in sensing.

True interleaved fabrication offers the greatest challenges and greatest benefits because the best of both worlds are combined. Often this integration means a significant increase in mask count and process complexity. Only products with long production runs can be made this way, because of the work and expense of process development.

In MEMS using CMOS layers there are two basic approaches: using the thin films of CMOS as in surface micromechanics; and using lower films or a silicon substrate as sacrificial layers. In both cases the release etching is done from the front side. If back-side DRIE is done, it is possible to use single crystal silicon for mechanical elements and thin films for actuation (e.g., heater resistors and sense electrodes). These approaches are shown in Figure 39.15. The pop-up mirror of Figure 39.2 is fabricated by the method shown in Figure 39.15b.

Wet etching of silicon after CMOS completion is shown in Figure 39.16. The pn junction etch stop is used to make

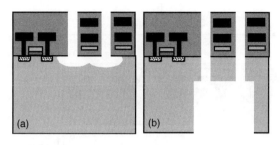

Figure 39.15 Two ways to do post-CMOS MEMS: (a) thin-film MEMS by front release; (b) single crystal silicon MEMS by back-side DRIE. Adapted from Jain and Xie (2006)

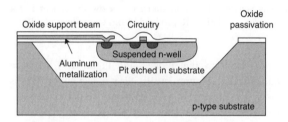

Figure 39.16 Post-CMOS wet etching with electrochemical etch stop to protect n-well of CMOS part. Reproduced from Kovacs et al. (1998) by permission of IEEE

sure that the circuitry in the n-well is not affected in KOH etching. This approach sacrifices some silicon area, but its beauty is in the fact that only KOH etching is needed after completion of CMOS.

The largest repertoire of choices is available in MEMS postprocessing after completion of CMOS. One choice is poly-SiGe with CVD oxides as sacrificial layers. These processes are limited by metal–silicon contact interface stability, and about 450 °C is the limit of usable temperature. The advantage of poly-SiGe over polysilicon lies especially in the lower processing temperatures and larger process window for stress-relief anneal.

Sputtered aluminum with photoresist sacrificial layers is an inherently low-temperature process. The problems come from the mechanical stability of polycrystalline aluminum in dynamic applications: in micromirror applications 10 years of operation correspond to 10^{10} mirror flips. Even when no mechanical fatigue sets in, aluminum mirrors have memory effects and alloys with better mechanical properties are used, for example TiAl.

Electroplated metals, typically copper and nickel, have been used to make gyroscopes and other mechanical elements on top of CMOS. Unlike sputtered aluminum, electroplated film thickness can be considerable, and seismic masses and stiff vertical structures can be made.

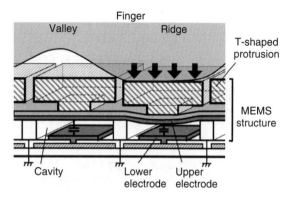

Figure 39.17 Fingerprint sensor on CMOS. Reproduced from Sato et al. (2005), copyright 2005, by permission of IEEE

PECVD nitride membranes can be deposited at 200–300 °C, which is low enough for CMOS. In IR imagers a sacrificial Al-layer and nitride structural layer are used. Each IR pixel has an amplifier transistor right beneath it. The nitride membrane is released to provide thermal isolation. The stresses in the nitride have to be optimized, with all the other layers on top of it, including the absorbers.

Fingerprint feature sizes are fairly large. A pixel size of 50 μm is adequate, and even very old-fashioned, cheap and high-yielding CMOS can be used as a substrate. In the sensor shown in Figure 39.17, electroplated metals and thick photosensitive polyimide have been used to fabricate movable membranes for capacitive sensing of localized forces.

More than two technologies can be combined, but at the expense of increased mask count, of course. The integrated microhotplate gas sensor pictured in Figure 39.18 combines bulk silicon micromachining, chemically sensitive resistors and SOI CMOS transistors. However, simple SOI wafers were not usable in this application because device silicon thickness needs to be about 1.5 μm, therefore epitaxial deposition of silicon on top of a SIMOX SOI wafer was used. Anisotropic wet etching of silicon was used for vertical thermal isolation, with SOI buried oxide as an etch stop layer. But the sensor, operating at 350 °C, has to be also laterally thermally isolated from the readout electronics. This is achieved by trench isolation, a technique borrowed from advanced IC technologies. CMOS circuits on a SOI device layer take care of signal processing, but MOSFETs are also used as heaters. This was done in order to simplify the process: platinum heaters would have added a new material and new cleaning and contamination concerns. Contacting, however, introduces exotic materials: the sensor material, porous palladium-doped SnO_2, makes contact with gold

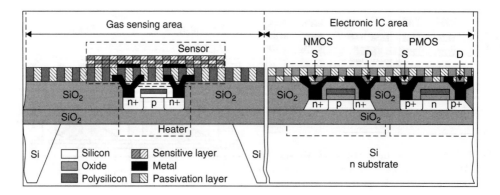

Figure 39.18 Integrated SOI CMOS microhotplate resistive gas sensor. The MOS transistor below the sensor is for heating; the readout electronics is situated beside the sensor for thermal isolation. Reproduced from Gardner *et al.* (2001) by permission of John Wiley & Sons, Ltd

electrodes, which make contact with the electronics. In order not to contaminate the SOI CMOS part, Au, Pd and SnO_2 depositions have to be made as postprocessing steps, and they put extra demands on barriers.

39.7 Microfabricated Devices for Microfabrication

As the microfabrication industries have expanded and applications widened, microfabrication has been fed into its own infrastructure. Microfabricated devices are being used in the fabrication of future microfabricated devices. The atomic force microscope (Figure 2.4) is already the established standard for many measurements within microfabrication. Similar to the AFM, the four-point probe (4PP; Figure 2.9) has been miniaturized (Figure 39.19). Size reduction in 4PP is not just miniaturization for its own sake but extends the measurement range into sizes not previously possible. As an application, 4PP analysis of laser annealing of ion implantation damage is shown in Figure 39.19b: resistivity is non-uniform, which indicates problems with laser scanning. Incomplete activation is therefore the reason, because charge carrier mobility is constant across the scanned area.

Residual gas analysis (RGA) in vacuum chambers is one application where microsystems have already been commercialized. Instead of bulky traditional mass spectrometers, vacuum residual gases are analyzed by microfabricated mass spectrometers (Figure 39.20). Their

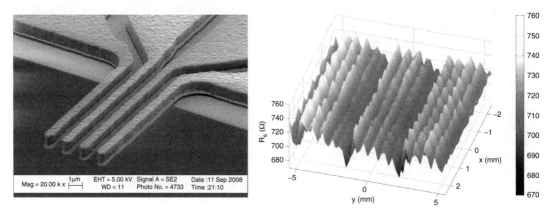

Figure 39.19 Micro 4PP: left, SEM micrograph; right, 4PP data. Reproduced from Petersen *et al.* (2010) by permission of the American Institute of Physics

Figure 39.20 Microfabricated mass spectrometer. Reproduced from Wapelhorst *et al.* (2007), copyright 2007, by permission of Elsevier

performance does not match that of traditional instruments (mass resolution especially is poor, and mass range limited) but the lower price makes it possible to install RGAs in all vacuum equipment, for routine monitoring. In the past, RGA was a special tool that was used in troubleshooting and system check-ups by professionals. While a pressure sensor is a much simpler and cheaper device for vacuum monitoring, RGAs can tell which gases are present (hydrocarbons from pump oils vs. water vapor), which can be valuable information in troubleshooting.

Micromirror-based patterning systems (23.3, 29.25, 39.21) offer many benefits over existing systems. No physical mask plates are needed, reducing cost and time needed to make microstructures. Both expensive high resolution systems and simple systems have their own applications, in ICs and microfluidics, respectively.

DNA microarrays consist of sample spots containing DNA strands, made by for example microfabricated silicon pin spotters or by ink jetting. In those methods the DNA strands have been synthesized before they are applied on a chip. The microfabrication alternative is to synthesize the DNA strands on the chip one nucleotide at a time (i.e. adding one of the four bases A,C,G,T). A set of 25 masks is used to create DNA strands with 25 nucleotides, by adding one nucleotide in each lithographic step. DNA arrays have been made commercially by optical lithography since mid-1990's. These chips can have over one million spots with different DNA strands. Increasing the number of spots and decreasing their size is easy but increasing the mask count, and hence the length of DNA strands, is very expensive. Virtual masks provided by cheap micromirror devices enable pixels in the 10 µm size range to be printed very economically. DNA strands with up to 100 nucleotides are fabricated this way. Combining virtual masks with microfluidic reactors for DNA synthesis makes it possible to integrate DNA chip fabrication into a single system.

Micromirror array (also known as spatial light modulator, SLM), excimer laser, optics and mechanics make up an advanced pattern generator system capable of exposing 80 nm pixels (Figure 39.21). An array of 2048×512 consists of individual mirrors 16×16 µm in size, resulting in a chip 33×8 mm. Actuation speed is 2 kHz. The feature size and patterning speed are enough to write advanced photomasks, and of course direct writing is also an option, but throughput is the limiting factor. Microsystems and nanotechnology are still in a nascent state, and there are many contenders for the main devices and device classes. Some of them may reach CMOS-like volumes and markets, some will remain niche applications, but most will never enter the manufacturing stage. This is how evolution in technology imitates natural evolution: the more variation there is and the more experiments are conducted, the more likely it is that some viable applications and technologies will emerge and will reproduce into many future generations.

39.8 Exercises

1. Find out about permeable polysilicon and explain how the plug-up process of Figure 39.14 works.
2. If 5 µm diameter, 50 µm deep TSV is filled by copper, what is the via resistance? What if the filling is by poly? By tungsten?

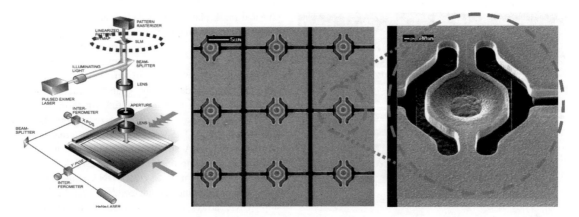

Figure 39.21 Micromirror array (SLM, for Spatial Light Modulator) for deep submicron pattern generation: virtual masks created in a 2048 × 512 mirror array are projected onto a wafer or a mask blank. Courtesy Micronic Mydata and Fraunhofer Institute, by permission of the American Institute of Physics

3. Draw the mask layout views and develop the fabrication process for the microgripper of Figure 39.8.
4. What will be the critical steps in fabricating the wafer stack of Figure 39.12?
5. How would you fabricate the PCR chip of Figure 39.6?
6. How is the starting wafer of Figure 39.10 fabricated? What limitations does it impose on the bipolar fabrication process, if any?
7. Break down the fabrication of the hot plate sensor of Figure 39.18 into the main process modules, and explain in which order they are undertaken!
8. Select polymers and design the fabrication process for flexible silicon electronics of Figure 39.13!

References and Related Reading

Aravamudhan, S., A.R.A. Rahman and S. Bhansali (2005) Porous silicon based orientation independent, self-priming micro direct ethanol fuel cell, *Sens. Actuators*, **A123–124**, 497–504.

Bain, M. *et al.* (2005) SiGe HBTs on bonded SOI Incorporating buried silicide layers, *IEEE Trans. Electron Devices*, **52**, 317–324.

Barbaro, M. *et al.* (2006) A CMOS fully integrated sensor for electronic detection of DNA hybridization, *IEEE Electron Device Lett.*, **27**, 595–597.

Brand, O. (2006) Microsensor integration into systems-on-chip, *Proc. IEEE*, **94**, 1160–1176.

Chronis, N. and L.P. Lee (2005) Electrothermally activated SU-8 microgripper for single cell manipulation in solution, *J. Microelectromech. Syst.*, **14**, 857–863.

Dunn-Rankin, D., E.M. Leal and D.C. Walther (2005) Personal power systems, *Prog. Energy Combust. Sci.*, **31**, 422–465.

Fedder, G.K. *et al.* (2008) Technologies for cofabricating MEMS and electronics, *Proc. IEEE*, **96**, 306–322.

Gardner, J.W., V.V. Varadan and O.O. Awadelkarim (2001) **Microsensors, MEMS and Smart Devices**, John Wiley & Sons, Ltd.

Garrou, P., C. Bower, and P. Ramm (eds) (2008) **Handbook of 3D integration**, Wiley-VCH Verlag GmbH.

Heikenfeld, J. *et al.* (2009) Electrofluidic displays using Young–Laplace transposition of brilliant pigment dispersions, *Nat. Photonics*, **3**, 292–296.

Hierlemann, A. and H. Baltes (2003) CMOS-based chemical microsensors, *Analyst*, **128**, 15–28.

Jain, A. and H. Xie (2006) A single-crystal silicon micromirror for large bi-directional 2D scanning applications, *Sens. Actuators*, **A130–131**, 454–460.

Katragadda, R.B. and Y. Xu (2008) A novel intelligent textile technology based on silicon flexible skins, *Sens. Actuators*, **A143**, 169–174.

Kiihamäki, J. *et al.* (2004) "Plug-up" – a new concept for fabricating SOI MEMS devices, *Microsyst. Technol.*, 10, 346–350

Ko, W.H. (2007) Trends and frontiers of MEMS, *Sens. Actuators*, **A136**, 62–67.

Kovacs, G.T.A. *et al.* (1998) Bulk micromachining of silicon, *Proc. IEEE* **86**, 1543.

Koyanagi, M., T. Fukushima and T. Tanaka (2009) High-density through silicon vias for 3-D LSIs, *Proc. IEEE*, **97**, 49–59.

Kwon, Y. *et al.* (2008) Evaluation of BCB bonded and thinned wafer stacks for three-dimensional integration, *J. Electrochem. Soc.*, **155**, H280–H286.

Lu, J.-Q. *et al.* (2000) 3D integration using wafer bonding, Advanced Metallization Conference 2000, San Diego, paper V3.

Machida, K. *et al.* (2001) A novel semiconductor capacitive sensor for a single-chip fingerprint sensor/identifier LSI, *IEEE Trans. Electron Devices*, **48**, 2273.

Obeid, P.J. *et al.* (2003) Microfabricated device for DNA and RNA amplification by continuous-flow polymerase chain reaction and reverse transcription-polymerase chain reaction with cycle number selection, *Anal. Chem.*, **75**, 288–295.

Petersen, D.H. *et al.* (2010) Review of electrical characterization of ultra-shallow junctions with micro four-point probes, *J. Vac. Sci. Technol.*, **B28**, C1C27–C1C33.

Poupon, G. *et al.* (2009) System on wafer: a new silicon concept in SiP, *Proc. IEEE*, **97**, 60–69.

Reuss, R.H. *et al.* (2005) Macroelectronics: perspectives on technology and applications, *Proc. IEEE*, **93**, 1239–1256.

Sato, N. *et al.* (2005) Novel surface structure and its fabrication process for MEMS fingerprint sensor, *IEEE Trans. Electron Devices*, **52**, 1026–1032.

Sedky, S. (2007) SiGe: an attractive material for post-CMOS processing of MEMS, *Microelectron. Eng.*, **84**, 2491–2500.

Song, J. *et al.* (2009) Solid-state microscale lithium batteries prepared with microfabrication processes, *J. Micromech. Microeng.*, **19**, 045004.

Sterner, M., G. Stemme and J. Oberhammer (2010) Nanometer-scale flatness and reliability investigation of stress-compensated symmetrically-metallized monocrystalline-silicon multi-layer membranes, IEEE International Conference on Nano/Micro Engineered and Molecular Systems, 2010.

Takeuchi, H. *et al.* (2005) Thermal budget limits of quarter-micrometer foundry CMOS for post-processing MEMS devices, *IEEE Trans. Electron Devices*, **52**, 2081–2086.

Tanaka, Y. *et al.* (2007) Biological cells on microchips: new technologies and applications, *Biosens. Bioelectron.*, **23**, 449–458.

Topol, A.W. *et al.* (2006) Three-dimensional integrated circuits, *IBM J. Res. Dev.*, **50**, 491–506.

Wapelhorst, E., J.-P. Hauschild and J. Müller (2007) Complex MEMS: a fully integrated TOF micro mass spectrometer, *Sens. Actuators*, **A138**, 22–27.

Webster, J.R. *et al.* (2001) Monolithic capillary electrophoresis device with integrated fluorescence detector, *Anal. Chem.*, **73**, 1622–1626.

Wise, K.W. (2007) Integrated sensors, MEMS, and microsystems: reflections on a fantastic voyage, *Sens. Actuators*, **A136**, 39–50.

Appendix A

Properties of Silicon

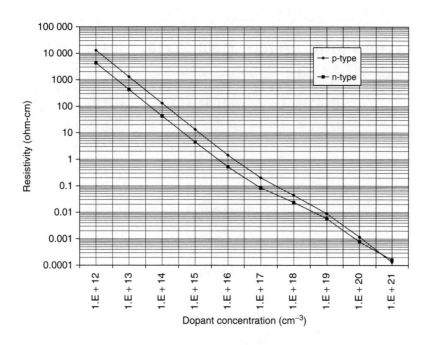

Structural and mechanical

Atomic weight	28.09
Atoms, total (cm^{-3})	4.995×10^{22}
Crystal structure	Diamond (fcc)
Lattice constant (Å)	5.43
Density (g/cm^3)	2.33
Density of surface atoms (cm^{-2})	(100) 6.78×10^{14}
	(110) 9.59×10^{14}
	(111) 7.83×10^{14}
Young's modulus (GPa)	190 (111) crystal orientation
Yield strength (GPa)	7
Fracture strain	4%
Poisson ratio, ν	0.27
Knoop hardness (kg/mm^2)	850

Electrical

Energy gap (eV)	1.12
Intrinsic carrier concentration (cm^{-3})	1.38×10^{10}
Intrinsic resistivity (Ω-cm)	2.3×10^5
Dielectric constant	11.8
Intrinsic Debye length (nm)	24
Mobility (drift) (cm^2/V-s)	1500 (electrons)
	475 (holes)
Temperature coeff. of resistivity (K^{-1})	0.0017

Introduction to Microfabrication, Second Edition Sami Franssila
© 2010 John Wiley & Sons, Ltd

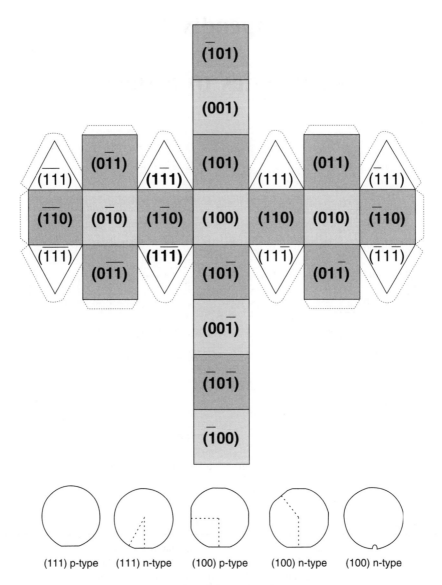

Wafer flats and notches for identifying wafer orientation and doping type

Thermal		**Optical**		
Coefficient of thermal expansion (°C^{-1})	2.6×10^{-6}	Index of refraction	3.42	$\lambda = 632$ nm
Melting point (°C)	1414		3.48	$\lambda = 1550$ nm
Specific heat (J/kg-K)	700	Energy gap wavelength	1.1 µm	(transparent at larger wavelengths)
Thermal conductivity (W/m-K)	150	Absorption	$> 10^6$ cm^{-1}	$\lambda = 200$–360 nm
Thermal diffusivity	0.8 cm^2/s		10^5 cm^{-1}	$\lambda = 420$ nm
			10^4 cm^{-1}	$\lambda = 550$ nm
			10^3 cm^{-1}	$\lambda = 800$ nm
			< 0.01 cm^{-1}	$\lambda = 1550$ nm

Appendix B

Constants and Conversion Factors

Atomic mass unit amu	1.66×10^{-27} kg
Electron charge e	1.602×10^{-19} C
Avogadro's constant N_A	6.022×10^{23}/mol
Boltzmann constant k	1.38066×10^{-23} J/K = 8.6544×10^{-5} eV/K
Faraday constant F	96 500 As/mol ($F = e \times N_A$)
Gas constant R	8.3144 J/Kmol ($R = k \times N_A$)
Gas molar standard volume	22.4 l/mol ($V_m = RT_0/p_0$)
Permittivity of vacuum ε_0	8.854×10^{-12} F/m
Speed of light c	2.9979×10^8 m/s
Stefan–Boltzmann constant σ	5.67×10^{-8} W/m^{2}K^4

Pressure conversion

From \ To	Pa	Torr	atm	mbar
		multiply by		
Pascal (Pa)	1	7.5×10^{-3}	9.87×10^{-6}	10^{-2}
Torr (mmHg)	133	1	1.316×10^{-3}	1.33
atm	1.013×10^5	760	1	1013
mbar	100	0.75	9.87×10^{-4}	1

Flow conversion

	Pa m^3/s	Torr l/s	sccm
Pa m^3/s	1	7.5	592
Torr l/s	0.133	1	78.9
sccm	1.69×10^{-3}	1.27×10^{-2}	1

Conversion factors

T/K = 273.15 + t/°C
1 eV = 1.6×10^{-19} J
1 eV $\times N_A$ = 96.5 J/mol = 23.06 kcal/mol
1 cal = 4.184 J
1 N = 10^5 dyne
1 Pa = 1 N/m^2 = 10 dyne/cm^2
1 μm = 10^{-6} m = 1000 nm = 0.001 mm
1 Å = 0.1 nm = 1×10^{-10} m
1 mil = (1/1000) inch = 25.4 μm

Appendix C

Oxide and Nitride Thickness by Color

Color chart for thermal SiO$_2$ films under daylight fluorescent lighting

Thickness (μm)	Color	Order
0.05	Tan	
0.07	Brown	
0.10	Dark violet to red-violet	
0.12	Royal blue	
0.15	Light blue to metallic blue	
0.17	Metallic to yellow-green	i
0.20	Light gold or yellow	
0.22	Gold	
0.25	Orange to melon	
0.27	Red-violet	
0.30	Blue to violet-blue	
0.31	Blue	
0.32	Blue to blue-green	
0.34	Light green	
0.35	Green to yellow-green	
0.36	Yellow-green	ii
0.37	Green-yellow	
0.39	Yellow	
0.41	Light orange	
0.42	Carnation pink	
0.44	Violet-red	
0.46	Red-violet	
0.47	Violet	
0.48	Violet-blue	
0.49	Blue	
0.50	Blue-green	
0.52	Green (broad)	
0.54	Yellow-green	
0.56	Green-yellow	iii
0.57	Yellowish	
0.58	Light orange	
0.60	Carnation pink	

Color chart for thermal SiO$_2$ films under daylight fluorescent lighting

Thickness (μm)	Color	Order
0.63	Violet-red	
0.68	Bluish	
0.72	Blue-green to green	iv
0.77	Yellowish	
0.80	Orange	
0.82	Salmon	
0.85	Dull light red-violet	
0.86	Violet	
0.87	Blue-voilet	
0.89	Blue	
0.92	Blue-green	v
0.95	Dull yellow-green	
0.97	Yellow to yellowish	
0.99	Orange	
1.00	Carnation pink	

Source: Pliskin, W. and E. Conrad (1964) Non-destructive determination of thickness and refractive index of transparent films, *IBM J. Res. Dev.*, **1**, 43.

Color chart for Si$_3$N$_4$ tungsten filament microscope illumination

0–20 nm	Silicon
20–40 nm	Brown
40–55 nm	Golden brown
55–73 nm	Red
73–77 nm	Deep blue
77–93 nm	Blue
93–100 nm	Pale blue
100–110 nm	Very pale blue
110–120 nm	Silicon
120–130 nm	Light yellow

Color chart for Si_3N_4 tungsten filament microscope illumination

130–150 nm	Yellow
150–180 nm	Orange red
180–190 nm	Red
190–210 nm	Dark red
210–230 nm	Blue
230–250 nm	Blue-green
250–280 nm	Light green
280–300 nm	Orange yellow
300–330 nm	Red

Source: Reizman, F. and W. van Gelder (1967) Optical thickness measurement of SiO_2-Si_3N_4 films on silicon, *Solid-State Electron.*, **10**, 625.

Index

Bold font indicates main entry
T indicates table

100 silicon, 40, 251
110 silicon, 40, 249
111 silicon, 40, 250
113 silicon, 317
130 nm, CMOS, 340T
1:3:8 etch, 136, 393
1D, one-dimensional simulation, 30
2D, two-dimensional simulation, 31
2D, two-dimensional growth, 78
3D, three-dimensional growth, 78
3D, three-dimensional simulation, 32
4PP, four-point probe, 19, 174, 494
5N (99.999 % purity), 143, 445
65 nm, CMOS, 340T
7N, (99.99999% purity), 143, 445

Aalto, Alvar, 301
ablation, 302
absorption, 37T
abrasive, 182
accelerometer, 192, 248, 369
activation energy, **5**, 131, 167, 168, 433
adatoms, 78
adhesion, 80, **84**, 216, 230, 322, 412
adhesion promotion, 103
adhesive bonding, 196, 219
AES, Auger electron spectroscopy, **23**, 80, 85
AFM, atomic force microscope, **17**, 187, 238, 305, 395
ALD, atomic layer deposition, 4, **53**, 217, 361, 375, **434**, 472, 475
alignment, **104**, 319, 389
   bonding 198
   critical, 319
   design rules, 319
   double side, 390
   marks, 11, 104, 108
alpha-tool, 416

aluminum,
   etching, 137, 427
   gate, 329
   MEMS, 196, 370, 371T
   metallization, 393, 394, 396, 452
   polishing, 184
   properties, 61T, 83
aluminum nitride, 4, 25, 49, 53, 91, 197, 412, 475
aluminum oxide, 4, 49, 53, 375, 475
ammonia-peroxide clean, 145T
amorphization, 177, 322
amorphous state, 4, 69, 338, 476
   silicon, 62, 168, 325, 340
   polymers, 205
anisotropic plasma etching, 130
anisotropic wet etching, 130, 237
annealing, **325**
   contact improvement, 83, 327
   CVD films, 199, 325
   equipment, 421
   forming gas, 332
   implant damage, 177
   laser, 303, 421
   millisecond, 421
   post-deposition, 77
   post-oxidation, 157, 420
   RTA, 88, 421
   silicide, 88
   stress tailoring, 325
   thermal budget, 326
anodic bonding, **194**, 225, 231, 393
APCVD, atmospheric pressure CVD, 433
antireflection coating, 120, 316, 374
APM, ammonia-peroxide mixture, 145T
ARC, antireflection coating, 120, 315
ARDE, aspect ratio dependent etching, 266
Arrhenius equation, **5**, 50, 130, 435

arsine, AsH$_3$, 179, 446T
ashing, same as resist stripping, 111
aspect ratio, 7, 208, 255, 358, 365
aspect ratio dependent etching, ARDE, 266
atomic clock, 234
atomic force microscope, AFM, **17**, 187, 238, 305, 395
atomic layer deposition, ALD, 4, **53**, 217, 361, 375, 434, 472, 475
Auger electron spectroscopy, AES, 23
autodoping, 73, 80, 85
back-end of the line, BEOL, 6, 315, 330
bake, **104**, 119, 143, 208
bamboo structure, 453

BARC, bottom antireflection coating, 120
barrel reactor, 438
barrier, 84, 361
base (of a bipolar transistor), 347
batch processing, 409
BCB, benzocyclobtadiene, 64T, 205
BCP, block co-polymer, 284
BEOL, back-end of the line, 315, 330
BESOI, Bond-etchback SOI, 275
beta-tool, 416
BHF, buffered HF, 131T
BiCMOS, 352
BioMEMS, 488
bipolar transistors, 30, 172, **347**, 349T, 490
binary mask, 121
bird's beak, 158
blanket wafer, 26
block co-polymer, 284
BMD, bulk microdefects, 272
BOE, buffered oxide etchant, 131T
bolometer, 140, 244, 384
bond alignment, 198
bonding, **191**, 200T, 399, 402
  adhesive, 196, 219
  anodic, 194, 231, 393
  eutectic, 196
  fusion, 193, 276, 399
  glass frit, 200
  glass, 230
  localized, 218
  metallic, 195
  polymer, 217
  solder, 246, 401
  solvent, 218
  thermal, 217
  thermocompression, 195, 394
bond strength, 133, 199
bonding pad, 61, 326
boron etch stop, 240
boron nitride, 82T
borosilicate glass, 226T

Bosch process, 258, 266
bottom gate TFT, 340
bottom coverage, 57, 90
boundary layer, 51, 128, **435**
bow, 274
BOX, buried oxide in SOI, 276
BPSG boron-phosphorous doped silica glass, 52
Braille, 220
breakdown field, 154, 453
brush scrubbing, 145
buffered HF, 131
bulk microdefects, BMD, 272
bulk micromachining, 45, 237, 400
buried layer, 347
buried oxide, BOX, 347

CA, contact angle, 149
cantilever, 238, 373, 378, 396
capacitance, 63, 255, 334, 337, 364, 372, 474
capacitor, 192, 323, 328, 354, 384
capillary electrophoresis, CE, 232
capillary forces, 376
capping, 402, 404
CAR, chemically amplified resist, 118
carbon nanotube, 53, 343
cavity, 279, 400
CD, compact disc, 211, 303
CD, critical dimension, **17**, 125, 138, 479
CDI, collector diffusion isolation, 354
CE, capillary electrophoresis, 232
channeling, 176
chemically amplified resist, CAR, 118
chemical mechanical polishing, **183**, 360, 364, 383
chemical vapor deposition, CVD, **50**, 64T, 87, 434
chemisorption, 77
chip, 11, 450, 459
chip yield, 450, 459
chrome, 98
chromium, 61T, 84, 159, 227
cleaning, **143**, 185, 324, 419, 439, 473
cleanroom, 10, **441**
cluster tool, 412
CMOS, 10, **330**
  as substrate, 383, 492
  fabrication, 329, 470
  MEMS integration, 398, 490
  scaling, 471
  wafer selection, 336
CMP, chemical mechanical polishing, 183, 360
cMUT, capacitive micromachined ultrasonic transducer, 399
CNT, carbon nanotube, 53, 343
cobalt silicide, 83, 88, 472
COC, cyclic olefin copolymer, 205
coefficient of thermal expansion, CTE,

anodic bonding, 195
   polymers, 206T
   silicon, 37T
   stresses, 59
   thin films, 61T, 64T, 65
cold wall reactor, 419
collar, 319
collector, 347
collimated sputtering, 80
comb-drive, 256, 376, 382
combustor, 235
contact angle, 149
contact, **60**, 326, 332, 452
contact hole, 57, 66, 320, 332, 359
contact resistance, 339, 452
contact lithography, 103, 109
contamination, **143**, 324, 441
contrast, 118
CoO, cost of ownership, 416
COP, crystal originated particle, 272
COP, cycloolefic polymer, same as COC, 205
copper,
   deposition, 49, 54, 56
   etching, 135
   interfaces, 83
   MEMS, 371, 373, 380, 493
   metallization, 360, 471, 489
   oxidation, 325
   polishing, 184
   resistivity, 61T, 365
corner effects, 159, 243, 320
corrosion, 427
$CoSi_2$, 83, 88, 471
cost of ownership, CoO, 416
critical alignment, 389
critical dimension, CD, 17, 125, 138, 479
critical length, 376
critical lithography, 318
critical point drying, 378
crucible, 49, 271
cryocooler, 233
cryogenic etching, 258, 428
crystal originated particle, COP, 272
crystal pulling, 36, 272
crystal structure, 39, 237
c-SOI, cavity SOI, 400
CTE, coefficient of thermal expansion, 37T, 59, 61T, 64T, 65, 195, 206T
curl switch, 9
CVD, chemical vapor deposition, **50**
   equipment, 433
   mechanism, 53
   MEMS, 370
   nitride, 62, 64T
   oxide, 62, 64T
   polysilicon, 62, 373
   plasma enhanced, 52
   rate, 434
   reactors, 435
   tungsten, 52, 87, 469
cycle time, 461
CYTOP, amorphous fluoropolymer, 205
Czochralski silicon, CZ, 36, 317

damage, 177
damascene, 361
dangling bonds, 157
dark field, 98, 110
dark field microscopy, 15
DCS, dichlorosilane SiH_2Cl_2, 72, 436
Deal-Grove oxidation model, 154
deembossing, 215
deep trench isolation, DTI, 351
deep UV, $\lambda < 300$ nm, 117
defect
   crystalline, 44
   density, 450
   etching, 131
   oxide, 454
deflection, equation, 198, 392
demolding, 215
denuded zone, DZ, 273
depth of focus, DOF, 117
design rules, 318
desorption, 425
development (of resist), 106
DF, dark field (mask), 98, 110
DHF, dilute HF, 131T
diamond, 52, 280, 295, 409
diamond-like carbon, DLC **53**, 80, 280
diaphragm (membrane), 245, 248, **391**
diazonapthoquinine, DNQ, 106, 208
diborane, 62, 86, 438, 446T
die, 11, 455
dichlorosilane SiH_2Cl_2, 72, 436
dicing, 11, 303, 307
dielectrics, 48, **63T**, 64T, 364
die yield, 455
diffusion barrier, 84
diffusion, 165, 314
diffusivity, thermal, **302**, 424
Dill parameters, 118
dip pen, 306
direct bonding, 191
direct writing, 93, 299, 301
dishing, 185
dislocation, 44
display, 375, 463, 488
disposable mold, 294

DIW, de-ionized water, 10, **445**
DLC, diamond-like carbon, 53, 80
DNA chip, 233, 234, 322, 485, 487
DNQ, diazonapthoquinine, 107
DOF, depth of focus, 117
dogbone, 319
dopant, 38, 165
doping profile, 22, 168, 171
double poly (bipolar), 351
double side processing, 314, 389
double side polished wafers, DSP, 45, 313, 389
down force, 182
down-time, 415
drain, 10, 176, **332**, 337
DRAM, dynamic random access memory, 451, **474**, 479
DRIE, deep reactive ion etching, 230, **255**, 268T, 365, 388T, 493
drilling, 308
drive-in, 166
dry cleaning, 146
dry etching, 128, 317
dry oxidation, 153
drying, 146, 378
DSP, double side polished wafers, 45, 313, 389
DTI, deep trench isolation, 351
dual damascene, 363
DUV, deep ultra violet, λ < 300 nm, 117
DVD, 211, 476
DZ, denuded zone, 273

EBL, electron beam lithography, **95**, 478
EBR, edge bead removal, 107
ECD, electrochemical deposition, 54
ECM, electrochemical machining, 309
edge bead removal, EBR, 107
edge exclusion, 11
edge rounding, 43, 276
EDM, electro discharge machining, 308
EDP, ethylene diamine pyrocathecol, 238, 240T
EDX, energy dispersive X-ray analysis, 23
EG, extrinsic gettering, 273
EGS, electronic grade polysilicon, 35
EPW, ethylene diamine pyrocathecol water, 238, 240T
ELA, excimer laser annealing, 303
electrochemical deposition, ECD, 54
electrochemical etching, 241, 291
electrochemical etch stop, 240, 493
electroless deposition, 55
electromigration, 452
electron beam lithography, EBL, 95
electron projection lithography, EPL, 478
electroplating, 54, 380
electropolishing, 292
electronic stopping, 174

electrospray, 219, 399
ellipsometry, 17, 64
ELO, epitaxial lateral overgrowth, 76, 473
EM, electromigration, 452
embossing, 212, 228
emissivity, 422
emitter, bipolar transistor, 347
emitter push, 170
EMPA, electron microprobe analysis, 24
end point, 135
energy dispersive X-ray analysis, EDX, 23
energy loss, 174T
EOR, end of range damage, 177
EOT, equivalent oxide thickness, 337
epitaxial lateral overgrowth, ELO, 76, 473
epitaxial wafers, 275
epitaxy, **69**, 275, 347, 393, 437, 473
epoxy, 205
EPW, ethylene diamine pyrocathecol water, 238, 240T
equipment, 409
equipment industry, 460
equivalent oxide thickness, EOT, 337
erosion, 185
ERR, etch rate ratio, same, 127
ESCA, electron spectroscopy for chemical analysis, 24
ESH, Environment, safety & health, 179, **445**
ESI, electrospray ionization, 219, 399
etchback, 135
etch
    gases, 133T
    mask, 134, 239, 259, 261
    mechanism, 133
    profile, 128
    rate, 138, 255
    rate ratio, 127
    reactors, 428
    residues, 134, 267
    selectivity, 127
    stop, 240, 275, 392
etching
    anisotropic plasma, 130
    anisotropic wet, 130
    DRIE, 255
    dry, 128
    isotropic, 130
    plasma, 127
    RIE, 128
    wet, 127, 227
eutectic bonding, 196
EUVL, extreme ultra violet lithography, 477
evaporation, **49**, 289, 291
exposure, 96, 105, 108
exposure field, 116, 477

F, feature size, 476, 479
FA, furnace annealing, 421
fab (IC fabrication facility), 460
fabless company, 460, 471
Fabry-Perot interferometer, 375
failure analysis, 26
fatal defects, 451
FBAR, film bulk acoustic resonator, 87
F/E, focus/exposure matrix, 117
FEB, focused electron beam, 301
Fed. standard, 442
FEOL, front-end of the line, 6, 315, 329
FET, field effect transistor, 10, 31, 32, **334**, 329, 473, 478, 490
FGA, forming gas anneal, 332
FIB, focussed ion beam, 299
Fick's law, 168
field oxide, FOX, 331
fingerprint sensor, 493
flame ionization, 7
flash anneal, 421
flash memory, 156, 466, 475
flat (wafer flat), 11, 41, 191, Appendix B
flat panel display, FPD, 463
flatness, 191, 274
flexible electronics, 341, 491
float zone silicon, FZ, 39, 317
flow/reflow, 326
fluidic, 486
 channels, 197, 210, 211, 219, 220, 316, 379
 connector, 399
 diode, 221
 sieve, 204, 316
 valve, 209
fluoropolymers, 56, 150, 205, 215
focal plane deviation, FPD, 274
focus depth, 116
focussed ion beam, FIB, 299
footprint, 415
forming gas (N_2/H_2), 332
Foturan, 234
foundry, 460, 471
four-point probe, 4PP, 19, 174, 494
FOX, field oxide, 331
FPD, focal plane deviation, 274
FPD, flat panel display, 463
front-end, 6, 315, 329
front-side micromachining, 492
FTIR, Fourier-transform infrared spectroscopy, 74, 82
fuel cell, 487
furnace, 154, 410, 419
fused silica, 97, 226
fusion bonding, 193, 276
FZ, float zone silicon, 39, 317

galvanic deposition, 54
gap fill, 186
GaAs, gallium arsenide, 69, 134
gas phase transport coefficient, 434
gas sensor, 494
gate, 329, 337, 473
gate oxide, 325, 331, **337**, 420, 480
Gaussian beam profile, 302
generation (CMOS node), 12, 479
germanium, 70
getter, 403
gettering, 272
giant magnetoresistance GMR, 90, 466
GLAD, glancing angle deposition, 291
glass, 51, 98, **225**, 340
glass frit bonding, 200, 404
glass transition temperature, T_g, 106, **206**, 210, 218, 226
glass wafers, 227T
g-line, $\lambda = 436$ nm, 117
global planarization, 186
GMR, giant magnetoresistance, 90, 466
gold, 61T, 195, 197, 246, 370, 374, 394, 494
grain boundary, 4, 167, 453
grain size, 79
grinding, 181, 308
GST, germanium antimony telluride, 303, 476
guard ring, 347
gyroscope, 258

h-line, $\lambda = 405$ nm, 108
handle wafer, 276
hard mask, 134
haze, 44
HBT, heterojunction bipolar transistor, 72, 352
HCI, high current implantation, 177
HDD, hard disk drive, 466
HDP, high density plasma, 428
HEI, high energy implantation, 177
HEPA, High Effciency Particulate Air filter, 444
hermeticity, 201, 327, 403
heteroepitaxy, 69
HfO_2, 53, 91, 337, 472, 475
Hg-lamp, 105
high-current implanter, HCI, 177
high density plasma, HDP, 428
high index planes, 40, 238
high-k dielectric materials, 337
high vacuum, 49, 426
hillock, 455
hinged structures, 381
HIPOX, high pressure oxidation, 155
HKMG, high k, metal gate CMOS, 473
HMDS, hexamethyl disilazane, 103
Hoerni, Jean, 469
homoepitaxy, 72

horizontal furnace, 154, 420
hot embossing, 212
hot lot, 461
hot plate, 104, 208
hot wall reactor, 419
HPM, hydrochloric acid-peroxide mixture, 145T
HRTEM, high resolution transmission electron microscope, 15
HTO, high temperature oxide, 51
HV, high vacuum, 49, 426
hydrochloric acid, 131, 145T
hydrofluoric acid, 131T, 145T, 371
hydrogen implantation, 278, 281
hydrophilic, 143, 149, 193, 400
hydrophobic, 143, 149, 277, 400

IBE, ion beam etching, 138, 289
IC, integrated circuits, 329, 343, 458, **470**
ICECREM simulator, 30, 75, 161, 179
ICP, inductively coupled plasma, 428
IDHL, immediately dangerous to health and life, 446
IG, internal gettering, 273
I/I, ion implantation, 173, 277
i-line, λ = 365 nm, 117
imprinting, 201, 213
impingement rate, 425
in situ monitoring, 413
infinite source diffusion, 168
infrared, 82, 208, 399
ingot, 38
injection molding, 211
ink jet, 268, 396
ink jetting, 306
indium tin oxide, ITO, 302, 341, 463
InP, indium phosphide, 280
integrated circuits, 329, 343, 458, 470
integrated passives, 323
integrated processing, 412
interconnect, 326, 332, 364
interfaces, 3, 83
interfacial oxide, 83
interference, 118, 284
interferometer, 19, 375
intermetal dielectrics, 85, 357
International Technology Roadmap for Semiconductors, ITRS, 340, 482
interstitial diffusion, 167
interstitialcy diffusion, 167
ion beam etching, IBE, 138, 289
ion cut, 277
ion implantation, 173, 277
ion milling, 138, 289
ion projection lithography, IPL, 478
IPA, isopropyl alcohol, -propanol, 146, 238, 243
IR, infra red, 82, 208, 399

island growth, 78
ISO standard, 443
isopropyl alcohol, IPA, 146, 238, 243
isotropy, 130
ITO, indium tin oxide, 302, 341, 463
ITOX, internal oxidation, 278
ITRS, International Technology Roadmap for Semiconductors, 340, 482

junction, 169, 338, 472

Kapton, polyimide, 204
keV, kiloelectron volt, 173
Kilby, Jack, 469
killer defect (fatal defect), 451
Knudsen number, 425
KOH, potassium hydroxide, **237**, 240T, 268T, 293, 388

laminar flow, 441
lapping, 41
laser pattern generation, 97
laser processing, 218, **301**, 464
latex sphere equivalent, LSE, 147
lattice constant, 37, 69
layer transfer, 196
layout rules, 319
LDD, lightly doped drain, 337
LER, line edge roughness, 125, 480
lift-off, 232, 288
LIGA, 55, 216
light field, 98, 110
lightly doped drain, LDD, 337
limited source diffusion, 169
line edge roughness, LER, 125, 480
linewidth, 12T, **18**, 21, 106, 109, 116, 117T, 138, 479
liner oxide, 336
Linhard solution, 173
lithography, **103, 115**, 318, 476
  block copolymer, 284
  colloidal, 286
  contact, 109
  direct write, 93
  double sided, 389
  electron beam, 95
  electron projection, 478
  EUV, 477
  holographic, 284
  interferometric, 284
  ion beam, 299, 478
  microcontact printing, 287
  nanoimprint, 213, 478
  optical, 103
  projection, 115
  proximity, 109

    stereo, 284
    UV, 103
    X-ray, 478
load lock, 427
loading effect, 265
LOCOS, local oxidation of silicon, 157, 330
LOR, lift-off resist, 289
lot, 461
low-k dielectric materials, 363, 364T
low-temperature bonding, 194
LPCVD, low-pressure CVD, 52, 62, 371, **433**, 436
LSE, latex sphere equivalent particle size, 147
LTO, low temperature oxide, 51

magnetic recording, 80, 466
magnetron sputtering, 428
mask, photomask, **97**, 123, 318
    alignment, 104, 390
    applications, 104, 321
    cost, 97, 458
    count, 354, 485
    defects, 99
    repair, 98, 301
    virtual, 101, 495
mask, etch mask, 134, 239, 262, 314
mass spectrometry, 495
mass transport limited, 131, 433
master, 203, 215
MBE, molecular beam epitaxy, 71, 411
MC, Monte Carlo simulation, 90, 179
MCI, medium current implanter, 177
MCZ, magnetic Czochralski silicon, 39
mean free path, 49, **425**
medium-current implanters MCI, 177
megasonic cleaning, 145
membrane, same as diaphragm, 245, 391
memory, 466, 474, 479
MEMS, microelectromechanical systems, 1, 387, 462, 486
metal contamination, 149, 337, 454
metal gate, 473
metal micromechanics, 371
metallic bonding, 195
metallic thin films, 60
metallization, 326, 332, 364
metal-semiconductor contacts, 326, 339
MFP, mean free path, 425
MGS, metallurgical grade silicon, 36, 465
microbridges, 387
microchannels, 210, 211, 219, 220, 379
microcontact printing, μCP, 287
microcrystalline, 5
microelectromechanical systems, MEMS, 1, 387, 462, 486
microfluidics, 486

microhotplate, 247, 494
microlens, 210, 228
microloading effect, 266
micromirror, 196, 256, 257, 285, 382, 383, 384, **394**, 486
micron, same as micrometer
microneedle, 8, 261, 264, **397**
microphone, 246, 384, 394, 401
micropump, 401, 489
microreactor, 191, 233, 250
microrocket, 131
microsystems, 1, 387, 462, 486
microturbine, 10
microvalve, 209
microvoid, same as COP, 272
Miller index, 40
MIM, metal-insulator-metal capacitor, 354
MIMIC, micromolding in capillaries, 211
mini-environment, 447
minifab, 461
misalignment, 320, 391
miscut, 41, 72, 317
mix-and-match lithography, 318
ML, monolayer, 53, 426
MLM, multilevel metallization, 359
mobility, 19, 37T, 168, 340
MOCVD, Metal Organic CVD, 436
modulated photoreflectance, 175
MOEMS, microoptoelectromechanical systems, 1, 256, 257, 261, 370, 374, 375
molding, 203, 209
    glass, 228
    injection molding, 211
    lost mold, 294
    micromolding, 211
    replica molding, 209
MOSFET, Metal Oxide Semiconductor Field Effect Transistor
    devices, 10, 31, 32, **334**
    fabrication, 329, 473
    MEMS integration, 490
    scaling, 333, 478
MOVPE, Metal Organic Vapor Phase Epitaxy, 436
molecular flow, 425
molybdenum, 48T, 51T, 61T, 91
monocrystalline, 4, 35
monolayer, ML, 53, 425
Monte Carlo simulation, 179
Moore, Gordon, 478
Moore's law, 12, 469
MRAM, magnetic RAM memory, 476
MTBA, mean time between assists, 415
MTBC, mean time between cleans, 415
MTTF, mean time to failure, 415, 453

multicrystalline, 5, 465
Murphy's yield model, 450
MW, molecular weight, 203

NA, numerical aperture, 115, 117
nanocrystalline, 5
nanoimprint (lithography) NIL, 213, 478
nanolaminate, 91
nanowire, 159
native oxide, 4, 138, 143, 438
NEMS, nanoelectromechanical systems, 485
negative resist, 93, 106, 109
nested mask, 239, 262
neutron transmutation doping, NTD, 165
nickel, 55, 61T, 371, 380
nickel silicide, 88, 472
NiCr, 320, 470
NiFe, 54
NIL, nanoimprint lithography, 213, 478
NiSi, 88, 472
NIST, National Institute of Standards and Technology, 26
nitride thin films, 62, 64T, 239, 371, 387
NO, nitrided oxide, 157, 337
node, CMOS technology generation, 12, 479
non-conformal step coverage, 57
non-critical lithography, 318
non-uniformity, 25
non-volatile memory, NVM, 475
novolak resist, **106**, 208
Noyce, Robert, 469
nozzle, 261, 263
NSOM, near-field scanning optical microscope, 15
nuclear stopping, 173
nucleation, 78
numerical aperture, NA, 115, 117
NVM, non-volatile memory, 156, 475

OAI, off-axis illumination, 122
O2P, oxygen plasma, 112T, 161, 371, 378
O2P, oxygen precipitate, 272
oblique angle evaporation, 291
OED, oxidation enhanced diffusion, 170
OES, optical emission spectroscopy, 414
ohmic contact, 326
ONO, oxidized nitrided oxide, 157
OPC, optical proximity correction, 123, 477
optical emission spectroscopy, OES, 414
optical MEMS, 256, 257, 261, 370, 374, 375
optical microscopy, 15
optical proximity correction, OPC, 123, 477
optofluidics, 488
organic contamination, 148
Ormocer, 196, 204
oven, 104, 208

overetch, 138
overlay, 108
overplating, 55
overpolishing, 183
OTS, octadecane trichlorosilane, 56, 307
ovonic memory, 476
oxidation, **153**, 160
oxidation enhanced diffusion, OED, 170
oxidation sharpening, 160
oxide, 160
   breakdown, 154, 453
   defects, 157, 454
   stress, 158
   thin films, 51, 62, 64T
oxidized nitrided oxide, ONO, 157, 337
oxygen precipitate, 272
oxynitride, 52, 86, 325, 332
ozone, 112, 146, 148

PAB, post apply bake, prebake, 104
PAC, photoactive compound, 106
packaging, 200, 327, 401, 452
PACVD, plasma assisted CVD, same as PECVD, 52
pad, bonding pad, 61, 326
pad, polishing pad, 182
PAG, photoacid generator, 118
palladium, 250, 493
PANI, polyaniline, 206, 342
parabolic growth, 155
particle contamination, **146**, 147T, 196, 279, 442
parylene, 65, 205, 206T, 371T, 379, 398
passivation, 63, 66, **327**, 332
pattern density effects, 183, 265
pattern generation, 93, 284
PC, polycarbonate, 204
PCB, printed circuit board, 283
PCM, phase change memory, 476
PCR, polymerase chain reaction, 233, 322, 487
PDA, post deposition anneal, 77
PDMS, poly(dimethyl)siloxane
   bonding, 192, 218
   devices, 210, 211, 217, 220, 322, 398
   material, 150, 204, 206T
   molding, 209, 215, 295
PEB, post exposure bake, 105
PECVD, Plasma Enhanced CVD **52**, 63
   amorphous silicon, 62, 81
   boron nitride, 82T
   oxide, 64T
   nitride, 64T
   step coverage, 57
peeling mask, 239, 262
pellicle, 117
pentacene, 341
Permalloy, 54

permeability, 201, 220
PET, polyethylene terephthalate, 205, 220, 342
phase diagram, 83
phase shift mask, PSM, 121, 477
phosphine, 86, 178, 438, 446T
phosphoric acid, 131T, 137
phosphorus doped silica glass, PSG, 51, 165, 327, 371T
photoacid generator, PAG, 118
photodiode, 166
photolithography, see also lithography, **103,** 115, 318, 476
photomask, **97,** 123, 318
photonic crystal, 187
photoresist, **106,** 117, 318, 389
   dry film, 108
   e-beam, 96
   negative, 93, 106
   novolak, 106, 208
   positive, 93, 106
   profile, 109
   removal, 111
   requirements, 112T
   spin coating, 107
   spray coating, 108
   stripping, 111, 112T
   submicron, 117
   SU-8, 192, 196, 205, 208, 219, 371, 488
   thick, 207
   trimming, 123
photostructurable glass, 234
physical cleaning, 145, 185
physical vapor deposition, PVD, **48,** 90
piezoelectric, 400
piezoresistance, 369, 392
PIII, plasma immersion ion implantation, 178
pinhole, 65, 99
PIP, polysilicon-insulator-polysilicon capacitor, 475
PIS, polysilicon-insulator-silicon capacitor, 475
Piranha, sulphuric acid peroxide mixture, 145T
pitch, 109
pitting, 83
planarization, 186, 359
plasma,
   cleaning, 150
   CVD (PECVD), 52, 81
   equipment, 427, 430
   etching, 132, 255
   oxidation, 161
   stripping, 112
plating, 54
platinum, 25, 61T, 85, **321,** 374, 487
plug, 60, 66, 359
PMMA, polymethyl methacrylate, 96T, 205
POA, post oxidation anneal, 157, 420

$POCl_3$, 52, 169
point defect, 44, 170
Poisson ratio, 37T, 373
Poisson yield model, 450
polarity (of photomask), 318
polishing, 48, 181
polycrystalline, 4, 69, 73, 78, 338
polysilicon, 17, 53, **62**
   crystal structure, 79
   emitter, 350
   epipoly, 371
   gate, 329
   LPCVD, 62
   oxidation, 156
   MEMS, 370, 371, 373, 396, 400
   resistivity, 62, **168,** 320, 322
   trench filling, 351
polycarbonate, PC, 205
polycide, 136
polydimethyl siloxane, PDMS, 192, 205, 206T 209, 210, 211, 215, 217, 218, 220, 398
polyimide, 205, 342, 360, 363, 396, 493
polymer, 64, **203** (see also separate entries for parylene, polyimide, PDMS, SU-8)
   bonding, 217
   chemical structure, 205
   devices, 217
   properties, 206T
   sacrificial layer, 371, 375
   structural layer, 371, 378, 398
   transistors, 307, 341
   viscosity, 207
porous silicon, 291
post-oxidation anneal, POA, 157, 420
positive resist, 106
post exposure bake, PEB, 105
powder blasting, 288
PowerMEMS, 486
power devices, 9, 169, 344
ppb, parts per billion, 21
ppm, parts per million, 21
ppma, parts per million atoms, 21, 38, 271
ppt, parts per trillion, 21
precipitate, 44, 170, 272
precursor, 53
predeposition, 166
pressure sensor, 198, 246, 392
Preston model, 183
prime wafers, 449
priming, 103
process equipment, 409
process integration, 313
process latitude, 109, 117
process simulation, 30

profile,
　depth, 22
　diffusion, 168
　etch, 128, 138
　resist, 109
profilometer, 17
projected range, 173
projection lithography, 115
proximity correction, 121
proximity effect, 96
proximity lithography, 109
PSG, phosphorous doped silica glass, 51, 165, 327, 371T
PSi, porous silicon, 291
PSL, polystyrene latex sphere, 147
PSM, phase shift mask, 121, 477
PTFE, polytetrafluoroethylene, 205, 215
pull-in voltage, 375
pumping speed, 426
PV, photovoltaics, 464
PVD, physical vapor deposition, 48, 78, 90
Pyrex glass, 195, **225**, 231, 393, 395, 403

QCM, quartz crystal microbalance, 414
quartz, (fused silica), 97, 226

radius of curvature, 17, 60, 395
range, 174
raster scan, 94
rapid thermal annealing, RTA, 88, 339, 421
rapid thermal processing, RTP, 421
rate limiting step, 54, 131, 433
RBS, Rutherford backscattering spectrometry, 22, 80
RCA clean, 145T, 331, 473
RC-delay, 365
RCL-chip, 323
reactive ion etching, RIE, 128, 132, 255
reclaim wafers, 461
reflective notching, 121
reflectometry, 18
reflow, 326, 332
refractive index, 18, 37T, 47, 64T, 82, 87, 206T
refractory metals, 137
relay, 377
release etch, 369, 376
release (from mold), 215
reliability, 449
remote plasma, 428
replacement gate, 473
replication, 203
residence time, 431
residual gas analyzer, RGA, 494
resist, see also photoresist, 93, 95, **106**, 117, 207, 318, 389
resist removal, 112T

resistivity, **19**
　diffused layers, 168
　DI-water, 445
　metals, 61T, 320
　polysilicon, 62, 168, 320
　silicon, 36, 168
　silane, 438
resistors, **320**, 322, 323, 333, 354, 369, 397, 493
resolution, 97, 109, 117
resonance frequency, 372
resonator, 19, 87, 200, 370, 400
RET, resolution enhancement techniques, 121, 477
reticle, 116
retrograde profile, 109, 215, 257
reverse engineering, 26
rework, 127
RF-MEMS, 402, 486
RF-switch, 9, 372, 373, 404
RGA, residual gas analyzer, 494
RIE, reactive ion, **128**, 132, 230, 255, 427
RIE lag, 266
rinsing, 146
RMS, root mean square (roughness), 43, 187, 191
rotating structures, 10, 379
roughness, 43, 80, 187, 191, 227, 265, 389
RT, room temperature
RTA, rapid thermal annealing, 88, 421
RTO, rapid thermal oxidation, 423
RTP, rapid thermal processing, 421

SACE, spark-assisted chemical engraving, 309
sacrificial etching, 293, 369, **371**, 376
sacrificial layer, 369, 371
salicide, self-aligned silicide, 228, 339
SAM, self-assembled monolayer, **56**, 151, 307, 401
SAM, scanning acoustic microscopy, 199
SAMPLE simulator, 124
SBC, standard buried collector (bipolar transistor), 347
SC, standard clean, 145T
scaling, **479**
　CMOS, 334, 340
　metallization, 364
Scanning Electron Microscope, SEM, 15
scatterometry, **18**, 43, 147
sccm, standard cubic centimeters per minute
scCO2, supercritical carbon dioxide, 378
Schottky contact, 326
screen printing, 290
scribeline, 11
scrubber, 436
SCS, single crystal silicon, 3, 35, 271, 329
S/D, source/drain, 10, 176, 332, 337
sealing, 397, 402
secondary ion mass spectroscopy, SIMS, 22, 275, 354
seed layer, 55, 360, 380

Seeds yield model, 450
SEG, selective epitaxial growth, 75, 474, 490
segregation, in crystal growth, 39T
segregation, in oxidation, 160
selective deposition, 86
selective epitaxial growth, SEG, 75, 474
selectivity, 138
self-alignment, 228
 bipolar, 350
 MOS gate, 176, 337
 phase shift mask, 121
 rotor, 381
 silicide (salicide), 339
 TFT, 228
self-assembled monolayer, SAM, 56, 151, 307, 401
self-interstitial, 167
self-limiting depth, 237, 249
self-limiting growth, 53, 56
SEM, scanning electron microscope, 15
shadow mask, 290, 391
shallow trench isolation, STI, 336
sheet resistance, 19, 168
shelf-life, 111, 364
shrink version, 479
SiC, 52, 135, 360
SiCr, 323
sidewall spacer, 139, 338, 351
SiGe, single crystalline, 69, 438, 472
SiGe, polycrystalline, 86, 493
Si:Ge:B, 275, 393
silane, 36, 51, **72**, 438, 446T
silicide, 83, 87, 136, 227, 339, 472
silicon, **35**
 bulk wafers, 3
 crystal growth, 36
 economics, 457
 epitaxy, 71, 275
 plasma etching, 133, 136, **255**
 properties, 37T
 wafers, 41, **43T**, 273, 274T, 281, 313, **317**, 317T, 336T
 wet etching, 130, 136, **237**
silicon carbide, SiC, 52, 135, 360
silicon dioxide, SiO_2, **51, 153**
 buried, 276, 369, 396
 CVD, 51, 62, 337
 etching, 131T, 136, 371
 properties, 64T, 154
 reliability, 156, 453
 thermal, 153, 331, 337
silicon nitride, Si_3N_4, SiN_x, **51**
 device applications, 246, 247, 249, 323, 371, 374, 394, 396, 399, 493
 etch mask, 239
 LPCVD, 52, 436

PECVD, 52
 properties, 64T
silicon on insulator, SOI
 CMOS applications, 339, 490, 492, 494
 fabrication, 193, 275
 MEMS applications, 245, 261, 369, 395, 400, 492, 494
silicon on sapphire, SOS, 71
siloxane, 204, 358
siloxane bond, 194
silsesquioxane, 363
SIMOX, 278
SIMS, secondary ion mass spectrometer, 22, 179, 275, 354
simulation, **29**
 deposition, 33, 89
 diffusion, 170
 epitaxy, 74
 equipment, 415
 front end, 334
 ion implantation, 179
 lithography, 124
 Monte Carlo, 90, 179
 oxidation, 160
single-wafer processing, 409
SiOC:H, 362
slip, 271, 317
SLM, spatial light modulator, 101, 496
slpm, standard liters per minute
 slurry, 182
SM, stress migration, 455
Smart-cut, 277
SMIF, Standard Mechanical InterFace, 447
SOD, silicon-on-diamond, 280
SOD, spin-on dielectric, 56
soda lime glass, 97, 225
soft bake, 104
soft lithography, 287
SOG, spin-on-glass, 56, 358
SOI, silicon on insulator, 193, 245, 261, 275, 339, 369, 395, 400, 490, 492, 494
solar cells, 9, 304, 305, 315, 464
solubility, 166
solvent bonding, 218
SOS, silicon on sapphire, 71
source/drain, 10, 176, 332, 337
spacer, 139, 338, 351
spark assisted machining, 309
SPC, statistical process control, 318
spectrometer, 399
spin coating, 56, 103, 107
spin-on glass, SOG, 56, 358
spin processor, 146
SPM, scanning probe microscope, 16

SPM, sulphuric acid peroxide mixture, 145T
spray coating, 108
spray tool, 132
spreading resistance profiling, SRP, 20
spring constant, 372
sputtering, **50, 428**
   bias sputtering, 429
   collimated sputtering, 80
   equipment, 413, 428
   etching, 429
   reactive, 50
   step coverage, 57
   yield, 50, 86
SRAF, subresolution assist feature, 123
SRAM, static random access memory, 450, 476
SRP, spreading resistance profiling, 20, 74
SSP, single side polished wafer, 45, 281, 313, 389
stacking fault, 44
stamp, 204, 215
standard buried collector bipolar transistor, SBC, 347
standing waves, 119
steam oxidation, 153
steel, 36, 283, 295, 340, 344, 379
Stefan equation, 213
Stefan-Boltzman law, 422
stencil mask, 290, 391
step-and-scan, 116
step-and-repeat, 116
step-and-stamp, 215
step coverage, 33, 57
stepper, 116
stereolithography, 284
STI, shallow trench isolation, 336
sticking probability, 425
stiction, 376
stoichiometry, 5, 52
Stoney formula, 60
STO, strontium titanate, 49, 475
STP, standard temperature and pressure, 427
straggle, 137
Stranski-Krastanov growth mode, 78
stress, **58**, 158, 230, 325, 392, 431, 474
stress migration, 455
Stribeck diagram, 183
stripping, 111
structural layer, 369
SU-8 epoxy resist, 192, 196, 205, 208, 219, 371, 488
submicron, <1 μm
substitutional diffusion, 167
substrates, 2, 283, 317, 340
sulphuric acid peroxide clean, 145T
superlattice, 70
SUPREM simulator, 30
surface analysis, 22
surface energy, 193

surface micromachining, 369
surface preparation, 5, **150**, 325
surface processes, 51, 78, 128
surface reaction limited, 51, 131, 433
surface roughness, 43, 187, 191, 193, 227, 265, 389
Sylgard 184, 205

tantalum, 23, 48, 50, 61T, 362
tantalum nitride, 48, 360
Ta_2O_5, 86, 135, 160, 325, 342, 475
TAR, top antireflection (coating), 120
target, 50, 430
TCAD, technology CAD, 29
TCE, temperature coefficient of expansion, 37T, 61T, 64T, 195, 206T
TCO, transparent conducting oxide, 302, 305, 463
TCR, temperature coefficient of resistivity, 79, 321
TDS, thermal desorption spectroscopy, 25
TED, transient enhanced diffusion, 339
Teflon, 47, 65, 132, 205, 215
TEM, transmission electron microscope, 7, 15, 42, 71, 338, 480
temperature coefficient of resistivity, TCR, 79, 321
temperature programmed desorption, 25
TEOS, tetraethoxysilane $Si(OC_2H_5)_4$, 51, 371
test structures, 21, 26, 106, 318
texture, 79
TFH, thin film head, 90, 466
TFT, thin film transistor, 228, 307, **340**, 463
T_g, glass transition temperature, 106, 207, 226
T_f, flow transition temperature, 212
thermal actuator, 377, 488
thermal bonding, 191, **218**
thermal budget, 326
thermal conductivity, 37T, 61T, 226T, 280
thermal desorption spectroscopy, TDS, 25
thermal isolation, 493
thermal oxidation, 153
thermal waves, 175
thermocompression bonding, TCB, 195
thermopile, 246, 399
thermoplast, 204
thermoset, 204
thick resist, 100, **207**
thin films, **3, 47, 77**
   characterization, 80T, 82T
   deposition, 48, 78
   dielectrics, 63T
   metallic, 61T
   polymeric, 64
   stresses, 58
   structure, 77
thin film head, TFH, 467
thin film optics in resist, 118
thin film solar cell, 305

thin film transistor, TFT, 340
thinning, 193, 399
thiol, 197, 287
threshold limit value, TLV, 446T
threshold voltage, 331, 334, 481
throughput, 415, 416T
through silicon via, TSV, 257, **490**
TiN, titanium nitride, 60, 65, 80T, 91, 360
tip, 160, 238, 395
TIR, total indicator reading, 274
titanium, 61T, 65, 359, 403
titanium silicide, $TiSi_2$, 88
TiW, 85, 87, 357
TLV, threshold limit value, 446T
TMAH, tetramethyl ammonium hydroxide, 237, 240T, 389
top antireflection (coating) TAR, 120
top gate (TFT), 340
top surface imaging, TSI, 120
total indicator reading, TIR, 274
total thickness variation, TTV, 44, 227, 389
transfer bonding, 196, 489
transient enhanced diffusion, TED, 339
transition width, 74
transmission electron microscope, TEM, 7, 15, 42, 71, 338, 480
transparent conducting oxides, TCO, 302, 463
transport limited reaction, 131, 433
trench isolation, 336, 351
trichlorosilane, 72, 438
TSI, top surface imaging, 120
TSV, through silicon via, 257, **490**
TTV, total thickness variation, 44, 227, 389
tub, same as CMOS well, 330, 335
tungsten, 47, 50, 52, 61T, 87, 301, 359, 378
tungsten lamp, 421
turning, 308
twin-well, 335
TXRF, total reflection X-ray fluorescence, 25, 149

UHV, ultrahigh vacuum, 49, 426
ULK, ultra-low k dielectric, 363
ULPA, Ultra Low Penetration Air filter, 444
ultrahigh vacuum, UHV, 49, 426
ultrahydrophobic, 149
ultrasonic cleaning, 145
ultrasonic transducer, 400
UMG, upgraded metallurgical grade silicon, 465
undercutting, 129, 244, 369
uniformity, 25
unidirectional flow, 441
unintentional processes, 314
unlimited source diffusion, 168

up-time, 415, 416T
UPW, ultrapure water, 10, **445**
USG, undoped silica glass, 51
utilization, 415, 431, 438
UV-NIL, ultraviolet nanoimprint lithography, 214, 478
UV-lithography, 103, 115
UV-photodiode, 167

vacancy, 44, 167
vacancy cluster, 272
vacuum, 411, 425
vacuum pumps, 427
vector scan, 94
vertical furnace, 410
vertical transistor, 473, 475
via hole, 358
viscosity, 107, 207, 226, 435
void, 78
volatility, 133
volume change, 88, 153, 170

wafer, 3, 11, 12, **43**
 bonded, 193, 275
 bulk, 3
 DSP, 45, 281, 313
 cost, 45, 457
 edge rounding, 43
 epitaxial, 275T
 glass, 227T
 selection, 313, 317
 size, 12, 35, 336, 457
 SOI, 193, 275
 SSP, 45
 specifications, 43T, 336T
wafer fab, 12, 460
wafering, 41, 226
wafer starts per month, WPM, 460
Wallace, Bob, 470
warp, 274
waveguide, 4, 87, 257
Weir, 316
well, 330, 335
wet cleaning, 144, 473
wet etching, 130, 131T, 229
wet oxidation, 153
WIWNU, within-wafer non-uniformity, 25
WPH, wafers per hour, 409, 416T
WPM, wafer starts per month, 460
WSi_2, 136, 160
WTWNU, wafer-to-wafer non-uniformity, 25

x_j, junction depth, 169, 338, 472
XeF_2, 128, 133, 134
XPS, X-ray photoelectron spectroscopy, 24

XRD, x-ray diffraction, 24, 48, 474
XRF, x-ray fluorescence, 25
XRL, x-ray lithography, 109, 478
XRR, X-ray reflectivity, 18
X-Si, crystalline silicon, 35

yield, 11, **449**, 470, 480
yield loss, 455
yield models, 450
yield cost, 416, 459
yield ramping, 450
yield (sputtering), 50, 86

yield strength, 36
yield stress, 199
Young's modulus, **35**, 58, 62, 61T, 64T, 184, 199, 206T, 226, 279, 372
yttrium oxide, 71, 341

zero anneal, 421
zero level alignment mark, 318
zero level package, 403
zeta potential, 147
zone melting, 39
zone model, 79
ZrO_2, 337, 475